Smart Cities

Smart Cities

Foundations, Principles, and Applications

Edited by
Houbing Song, Ravi Srinivasan, Tamim Sookoor, and Sabina Jeschke

Registered Offices
John Wiley & Sons, Inc., 111 River Street, Hoboken, NJ 07030, USA

Editorial Office
111 River Street, Hoboken, NJ 07030, USA

For details of our global editorial offices, customer services, and more information about Wiley products visit us at www.wiley.com.

Wiley also publishes its books in a variety of electronic formats and by print-on-demand. Some content that appears in standard print versions of this book may not be available in other formats.

Library of Congress Cataloging-in-Publication Data

Names: Song, Houbing, editor. | Srinivasan, Ravi, editor. | Sookoor, Tamim, 1984- editor. | Jeschke, Sabina, editor.
Title: Smart cities : foundations, principles, and applications / edited by Houbing Song, Ravi Srinivasan, Tamim Sookoor, Sabina Jeschke.
Description: Hoboken, NJ : John Wiley & Sons, 2017. | Includes bibliographical references and index.
Identifiers: LCCN 2017000480 (print) | LCCN 2017008972 (ebook) | ISBN 9781119226390 (cloth) | ISBN 9781119226413 (pdf) | ISBN 9781119226437 (epub)
Subjects: LCSH: Cities and towns–Technological innovations. | City planning–Technological innovations. | Sustainable development. | Urban ecology (Sociology) | Telematics.
Classification: LCC HT166 .S5877757 2017 (print) | LCC HT166 (ebook) | DDC 307.76–dc23
LC record available at https://lccn.loc.gov/2017000480

Cover image: © johnason/Gettyimages
Cover design by Wiley

Set in 10/12pt WarnockPro by SPi Global, Chennai, India

10 9 8 7 6 5 4 3 2 1

Contents

Editors Biographies   *xxiii*
List of Contributors   *xxvii*
Foreword   *xxxiii*
Preface   *xxxv*
Acknowledgments   *xxxvii*

1   Cyber–Physical Systems in Smart Cities – Mastering
        Technological, Economic, and Social Challenges   *1*
        Martina Fromhold-Eisebith
1.1   Introduction   *1*
1.2   Setting the Scene: Demarcating the Smart City and Cyber–Physical
        Systems   *3*
1.3   Process Fields of CPS-Driven Smart City Development   *4*
1.4   Economic and Social Challenges of Implementing the
        CPS-Enhanced Smart City   *10*
1.5   Conclusions: Suggestions for Planning the CPS-Driven Smart
        City   *15*
        Final Thoughts   *17*
        Questions   *18*
        References   *18*

2   Big Data Analytics Processes and Platforms Facilitating Smart
        Cities   *23*
        Pethuru Raj and Sathish A. P. Kumar
2.1   Introduction   *24*
2.2   Why Big Data Analytics (BDA) Is Significant for Smarter Cities   *24*
2.3   Describing the Big Data Paradigm   *26*
2.4   The Prominent Sources of Big Data   *27*
2.4.1   The Salient Implications of Big Data   *27*
2.4.2   Information and Communication Infrastructures for Big Data and Its
        Platforms   *28*

2.4.3   Transitioning from Big Data to Big Insights   *29*
2.5   Describing Big Data Analytics (BDA)   *29*
2.6   The Big Trends and Use Cases of Big Data Analytics   *31*
2.6.1   Customer Satisfaction Analysis   *32*
2.6.2   Market Sentiment Analysis   *32*
2.6.3   Epidemic Analysis   *32*
2.6.4   Using Big Data Analytics in Healthcare   *33*
2.6.5   Analytics of Machine Data by Splunk   *34*
2.7   The Open Data for Next-Generation Cities   *38*
2.8   The Big Data Analytics (BDA) Platforms   *39*
2.8.1   Civitas: The Smart City Middleware   *42*
2.8.2   Hitachi Smart City Platform   *43*
2.8.3   Data Collection   *44*
2.8.4   Data Analysis   *44*
2.8.5   Application Coordination   *45*
2.9   Big Data Analytics Frameworks and Infrastructure   *45*
2.9.1   Apache Hadoop Software Framework   *46*
2.9.2   NoSQL Databases   *48*
2.10   Summary   *50*
    Final Thoughts   *51*
    Questions   *51*
    References   *52*

3   Multi-Scale Computing for a Sustainable Built
    Environment   *53*
    Massimiliano Manfren
3.1   Introduction   *53*
3.2   Modeling and Computing for Sustainability Transitions   *55*
3.2.1   Multilevel Perspective Modeling   *56*
3.2.2   Technological and Social Learning   *57*
3.2.3   Multidisciplinary System Thinking   *58*
3.2.4   Long-Term Thinking for the Built Environment   *59*
3.2.5   Data and Modeling Techniques   *63*
3.3   Multi-Scale Modeling and Computing for the Built Environment   *66*
3.3.1   Virtual Prototyping for Design Optimization   *67*
3.3.2   Performance Optimization Across Building Life Cycle Phases   *68*
3.4   Research in Modeling and Computing for the Built Environment   *70*
3.4.1   Building Energy Balance Analysis   *74*
3.4.2   Forward/Inverse Modeling and Visualization Techniques   *76*
3.4.3   Workflows and Integration of Modeling Strategies   *76*
3.4.4   Research Advances in Modeling and Computing   *81*
    Final Thoughts   *82*
    Questions   *84*
    References   *84*

4   Autonomous Radios and Open Spectrum in Smart Cities   *99*
         Corey D. Cooke and Adam L. Anderson
4.1   Introduction   *99*
4.2   Candidate Wireless Technologies   *101*
4.2.1   Open Spectrum   *101*
4.2.2   5G Wireless Technologies   *103*
4.2.3   Internet of Things (IoT)   *104*
4.3   PHY and MAC Layer Issues in Cognitive Radio
         Networks   *105*
4.3.1   Spectrum Sensing   *106*
4.3.1.1 Detection Methods   *106*
4.3.1.2 Cooperative Spectrum Sensing   *107*
4.3.1.3 Other Sensing Issues   *107*
4.3.2   Spectrum Management and Handoff   *108*
4.3.3   Rendezvous Problem   *109*
4.3.4   Coexistence   *109*
4.4   Frequency Envelope Modulation (FEM)   *110*
4.4.1   Network Self-Configuration   *112*
4.4.2   Physical Layer Performance   *113*
4.4.3   Experimental Results   *115*
4.5   Conclusion   *116*
         Final Thoughts   *117*
         Questions   *118*
         References   *118*

5   Mobile Crowd-Sensing for Smart Cities   *125*
         Chandreyee Chowdhury and Sarbani Roy
5.1   Introduction   *125*
5.2   Overview of Mobile Crowd-Sensing   *127*
5.2.1   Categories of Crowd-sensing   *127*
5.2.2   Architecture of Mobile Crowd-sensing   *127*
5.2.3   Applications of Mobile Crowd-sensing in Smart City   *131*
5.2.3.1 Applications in Infrastructure   *131*
5.2.3.2 Environmental Applications   *134*
5.2.3.3 Social Applications   *134*
5.3   Issues and Challenges of Crowd-sensing in Smart Cities   *135*
5.3.1   Task Assignment Problem   *135*
5.3.2   User Profiling and Trustworthiness   *139*
5.3.3   Incentive Mechanisms   *140*
5.3.4   Localized Analytics   *141*
5.3.5   Security and Privacy   *142*
5.4   Crowd-sensing Frameworks for Smart City   *144*
5.4.1   Here-*n*-Now Framework   *144*
5.4.2   Crowd-sensing Framework based on XMPP   *146*

5.4.3   McSense   *146*
5.4.4   Supporting Framework for Crowd-sensing Apps   *148*
5.5   Conclusion   *149*
           Final Thoughts   *149*
           Questions   *150*
           References   *150*

6   Wide-Area Monitoring and Control of Smart Energy
           Cyber-Physical Systems (CPS)   *155*
           Nilanjan R. Chaudhuri
6.1   Introduction   *155*
6.2   Challenges and Opportunities   *156*
6.2.1   Wide-Area Monitoring: Damping, Frequency, and Mode-shape
           Estimation   *156*
6.2.2   Wide-Area Damping Control: Latency Compensation   *157*
6.2.3   Wide-Area Damping Control with Wind Farms   *159*
6.3   Solutions   *159*
6.3.1   . Phasor Approach   *159*
6.3.2   Wide-Area Monitoring: Damping, Frequency, and Mode-Shape
           Estimation   *161*
6.3.3   Test System: 16-Machine, 5-Area System   *162*
6.3.3.1   Simulation Results   *163*
6.3.4   Wide-Area Damping Control: Latency Compensation   *166*
6.3.4.1   Simulation Results   *168*
6.3.5   Wide-Area Damping Control with Wind Farms   *169*
6.3.5.1   DFIG-based Wind Farm Modeling   *169*
6.3.5.2   Rotor Side Converter (RSC) Control   *171*
6.3.5.3   Grid Side Converter (GSC) Control   *171*
6.3.5.4   Control Input   *172*
6.3.5.5   Coordinated Control Design   *172*
6.3.5.6   Simulation Results   *172*
6.4   Conclusions and Future Direction   *173*
           Final Thoughts   *175*
           Questions   *175*
           References   *175*

7   Smart Technologies and Vehicle-to-X (V2X) Infrastructures
           for Smart Mobility Cities   *181*
           Bernard Fong, Lixin Situ, and Alvis C. M. Fong
7.1   Introduction   *181*
7.2   Data Communications in Smart City Infrastructure   *182*
7.2.1   Data Acquisition   *183*
7.2.2   Traffic Surveillance   *185*

7.3   Deployment: An Economic Point of View   *186*
7.3.1   Detecting Abnormal Events   *187*
7.3.2   Network Failure   *188*
7.3.3   Micromobility Data Communications   *190*
7.3.4   V2X Network Integration and Interoperability   *194*
7.4   Connected Cars   *195*
7.4.1   Multi-Hop Communication in V2X   *195*
7.4.2   Green V2X Communications in Smart Cities   *198*
7.4.3   Vehicular Communications Infrastructure Reliability   *200*
7.4.4   Business Intelligence in Connected Cars   *201*
7.5   Concluding Remarks   *202*
          Final Thoughts   *203*
          Questions   *203*
          References   *204*

8   Smart Ecology of Cities: Integrating Development Impacts on
          Ecosystem Services for Land Parcels   *209*
          Marc Morrison, Ravi S. Srinivasan, and Cynnamon Dobbs
8.1   Introduction   *209*
8.2   Need for Smart Ecology of Cities   *212*
8.3   Ecosystem Service Modeling (CO_2 Sequestration, PM10 Filtration,
          Drainage)   *214*
8.3.1   Overview of Ecosystem Services in Urban Contexts   *214*
8.3.2 CO_2 Sequestration   *215*
8.3.3   PM10 Filtration   *217*
8.3.4   Drainage   *218*
8.4   Methodology   *219*
8.4.1   Carbon Sequestration   *219*
8.4.2   Drainage   *224*
8.4.3   PM10 Filtration   *228*
8.5   Implementation of Development Impacts in Dynamic-SIM
          Platform   *231*
8.6   Discussion (Assumptions, Limitations, and Future Work)   *234*
8.7   Conclusion   *235*
          Final Thoughts   *236*
          Questions   *236*
          References   *236*

9   Data-Driven Modeling, Control, and Tools for Smart
          Cities   *243*
          Madhur Behl and Rahul Mangharam
9.1   Introduction   *245*
9.1.1   Contributions   *247*

9.1.2   Experimental Validation and Evaluation   *248*
9.2   Related Work   *248*
9.3   Problem Definition   *250*
9.3.1   DR Baseline Prediction   *250*
9.3.2   DR Strategy Evaluation   *251*
9.3.3   DR Strategy Synthesis   *251*
9.4   Data-Driven Demand Response   *252*
9.4.1   Data Description   *252*
9.4.2   Data-Driven DR Baseline   *253*
9.4.3   Data-Driven DR Evaluation   *253*
9.5   DR Synthesis with Regression Trees   *254*
9.5.1   Model-Based Control with Regression Trees   *254*
9.5.2   DR Synthesis Optimization   *256*
9.6   The Case for Using Regression Trees for Demand
         Response   *259*
9.7   DR-Advisor: Toolbox Design   *261*
9.8   Case Study   *263*
9.8.1   Building Description   *263*
9.8.2   Model Validation   *265*
9.8.3   Energy Prediction Benchmarking   *265*
9.8.4   DR Evaluation   *266*
9.8.5   DR Synthesis   *268*
9.8.5.1 Revenue from Demand Response   *271*
9.9   Final Thoughts   *271*
         Questions   *272*
         References   *272*

10   Bringing Named Data Networks into Smart Cities   *275*
         Syed Hassan Ahmed, Safdar Hussain Bouk, Dongkyun Kim, and Mahasweta
         Sarkar
10.1   Introduction   *275*
10.2   Future Internet Architectures   *278*
10.2.1   Data-Oriented Network Architecture (DONA)   *278*
10.2.2   Network of Information (NetInf)   *279*
10.2.3   Publish Subscribe Internet Technology (PURSUIT)   *281*
10.3   Named Data Networking (NDN)   *282*
10.4   NDN-based Application Scenarios for Smart Cities   *285*
10.4.1   NDN in IoT for Smart Cities   *285*
10.4.2   NDN in Smart Grid for Smart Cities   *287*
10.4.3   NDN in WSN for Smart Cities   *288*
10.4.4   NDN in MANETs for Smart Cities   *290*
10.4.5   NDN in VANETs for Smart Cities   *293*
10.4.6   NDN in Climate Data Communications   *296*

10.5   Future Aspects of NDN in Smart Cities   *297*
10.5.1   NDN Content/Data   *297*
10.5.2   Naming Content/Data in NDN   *298*
10.5.3   NDN Data Structures   *299*
10.5.4   NDN Message Forwarding   *299*
10.5.5   Content Discovery in NDN   *300*
10.5.6   NDN in Dynamic Network Topology   *301*
10.5.7   Content Caching in NDN   *301*
10.5.8   Security and Privacy   *302*
10.5.9   Evaluation Methods   *303*
10.6   Conclusion   *303*
             Final Thoughts   *304*
             Questions   *304*
             References   *304*

11   Human Context Sensing in Smart Cities   *311*
             Juhi Ranjan, Erin Griffiths, and Kamin Whitehouse
11.1   Introduction   *311*
11.2   Human Context Types   *312*
11.2.1   Physiological Context   *313*
11.2.2   Emotive Context   *314*
11.2.3   Functional Context   *315*
11.2.4   Location Context   *316*
11.3   Sensing Technologies   *317*
11.3.1   Video and Audio   *317*
11.3.2   Wearables   *320*
11.3.3   Smartphones   *324*
11.3.4   Environment   *328*
11.4   Conclusion   *331*
             Final Thoughts   *332*
             Questions   *332*
             References   *333*

12   Smart Cities and the Symbiotic Relationship between Smart
             Governance and Citizen Engagement   *343*
             Tori Okner and Russell Preston
12.1   Smart Governance   *344*
12.1.1   Smart Governance and Smart Cities   *347*
12.1.2   The Role of Planning & Design   *347*
12.2   Case Study – Somerville, Massachusetts   *348*
12.2.1   Slumerville to Somerville   *348*
12.2.1.1 Professionalizing City Hall   *349*
12.2.2   Planning Somerville   *352*

12.2.2.1 SomerVision   *352*
12.2.2.2 Somerville by Design   *352*
12.2.2.3 Smart Cities and Planning in Somerville   *363*
12.3   Looking Ahead   *365*
12.3.1   Lessons from Somerville   *365*
12.3.2   Recommendations   *367*
           Final Thoughts   *368*
           Questions   *370*
           References   *370*

13   Smart Economic Development   *373*
           Madhavi Venkatesan
13.1   Introduction   *373*
13.1.1   Significance of Educating for Sustainability   *374*
13.1.2   Economics in Cultural Context   *375*
13.1.3   Reconciling Economic Theory and Historical Context   *376*
13.1.4   Significance of Context   *377*
13.2   Perception of Resource Value, Market Outcomes, and Price   *378*
13.2.1   Market Distortions   *379*
13.2.2   Externalities   *380*
13.2.3   Common Goods   *382*
13.2.4   Market Prices   *383*
13.3   Conscious Consumption and the Sustainability Foundation of Smart
           Cities   *384*
13.3.1   Smart Economic Development   *386*
13.3.2   Next Steps Theory and Practice   *387*
           Final Thoughts   *388*
           Questions   *388*
           References   *388*

14   Managing the Cyber Security Life-Cycle of Smart Cities   *391*
           Mridul S. Barik, Anirban Sengupta, and Chandan Mazumdar
14.1   Introduction   *391*
14.2   Smart City Services   *393*
14.3   Smart Services Technologies   *394*
14.4   Smart Services Security Issues   *396*
14.5   Management of Cyber Security of Smart Cities   *397*
14.5.1   Scope and Cyber Security Policy Formulation   *398*
14.5.2   Cyber Security Requirements Identification   *400*
14.5.3   Risk Management   *400*
14.5.4   Detailed Security Policy Formulation   *401*
14.5.5   Security Measures Implementation   *401*
14.5.6   Cyber Security Incident Management   *401*

14.5.7   Service Continuity Management and Disaster Recovery   *402*
14.5.8   Cyber Security Metrics Generation   *403*
14.5.9   Audit and Compliance Checking   *403*
14.6   Discussion   *403*
14.7   Conclusion   *404*
          Questions   *404*
          References   *405*

15   Mobility as a Service   *409*
          Christopher Expósito-Izquierdo, Airam Expósito-Márquez, and Julio
          Brito-Santana
15.1   Introduction   *409*
15.2   Mobility as a Service   *413*
15.2.1   Millennials   *413*
15.2.2   Concept of Mobility as a Service   *415*
15.2.3   Transportation Infrastructures   *418*
15.2.4   Information and Communications Technologies   *420*
15.2.5   Interoperability   *422*
15.2.6   Autonomous Car   *423*
15.2.7   Connected Vehicle   *424*
15.2.8   Sharing Mobility   *425*
15.3   Case Studies on Mobility as a Service   *427*
15.3.1   UbiGo   *427*
15.3.2   car2Go   *428*
15.3.3   Uber   *430*
15.3.4   RideScout   *431*
15.4   Conclusions and Further Research   *432*
          Acknowledgments   *433*
          Final Thoughts   *433*
          Questions   *433*
          References   *434*

16   Clustering and Fuzzy Reasoning as Data Mining Methods for
          the Development of Retrofit Strategies for Building
          Stocks   *437*
          Philipp Geyer and Arno Schlueter
16.1   Introduction   *438*
16.1.1   Problem Description   *438*
16.1.2   Smart Cities and Data Mining   *438*
16.1.3   Approach   *438*
16.1.3.1 Clustering   *438*
16.1.3.2 Conditions of Data Mining in Building Stock   *439*
16.1.3.3 Contents of the Chapter   *439*

16.2   Method   *440*
16.3   Application Case   *442*
16.4   Data Sources and Preprocessing   *443*
16.4.1   Data Sources and Modeling   *443*
16.4.1.1 Public Databases and Datasets   *444*
16.4.1.2 Protocols and Surveys   *444*
16.4.1.3 Sensor Measurements and Bottom-Up Modeling   *445*
16.4.2   Construction of the Feature Space   *446*
16.4.2.1 Metamodeling   *446*
16.4.2.2 The Feature Space   *447*
16.5   Clustering   *448*
16.5.1   Nonspatial Clustering   *448*
16.5.2   Spatial Clustering   *451*
16.5.2.1 Managing Nonuniform Effect Dimensions   *452*
16.5.2.2 Results of One-Step Spatial Clustering   *452*
16.5.2.3 Two-Step Spatial Clustering Results   *454*
16.6   Fuzzy Reasoning   *456*
16.6.1   Nonspatial Fuzzy Reasoning   *456*
16.6.1.1 Ramp Membership Functions   *456*
16.6.2   Spatial Fuzzy Reasoning   *458*
16.7   Mixed Fuzzy Reasoning and Clustering   *459*
16.8   Postprocessing: Interpretation and Strategy
          Identification   *459*
16.8.1   Data Visualization   *459*
16.8.2   Data Mining Postprocessing   *461*
16.8.3   Transformation Strategy Development   *463*
16.8.4   Strategy Development   *463*
16.8.5   Policy Development and Retrofit Programs   *463*
16.9   Comparison and Discussion of Methods   *464*
16.9.1   Information   *465*
16.9.2   Scaling   *465*
16.9.3   Robustness   *466*
16.10   Conclusion   *467*
          Final Thoughts   *468*
          Questions   *468*
          Acknowledgments   *469*
          References   *469*

17   A Framework to Achieve Large Scale Energy Savings for
      Building Stocks through Targeted Occupancy
      Interventions   *473*
      Aslihan Karatas, Allisandra Stoiko, and Carol C. Menassa
17.1   Introduction   *474*
17.2   Objectives   *475*
17.3   Review of Occupancy-Focused Energy Efficiency Interventions   *476*
17.3.1 Knowledge-Based Interventions   *477*
17.3.2 Persuasion Interventions   *478*
17.3.3 Penalty Interventions   *479*
17.3.4 Technology Interventions   *480*
17.3.5 Building Energy Use Interventions in Energy Policy Design   *480*
17.4   Role of Occupants' Characteristics in Building Energy Use   *481*
17.5   A Conceptual Framework for Delivering Targeted
        Occupancy-Focused Interventions   *483*
17.5.1 Measuring the Impact of Occupancy Characteristics on Building
        Energy Use   *483*
17.5.2 Clustering Occupants' MOA Levels and Energy Use Profiles   *486*
17.5.3 Identifying Multilevel Building Energy Use Intervention
        Strategies   *487*
17.6   Case Study Example   *490*
17.7   Discussion   *493*
17.8   Conclusions and Policy Implications   *494*
        Questions   *496*
        Acknowledgment   *496*
        References   *496*

18   Sustainability in Smart Cities: Balancing Social, Economic,
        Environmental, and Institutional Aspects of Urban Life   *503*
        Ali Komeily and Ravi Srinivasan
18.1   Introduction   *503*
18.2   Sustainability Assessment in Our Cities   *506*
18.3   Sustainability in Smart Cities   *508*
18.4   Achieving Balanced Sustainability   *512*
18.4.1 Improving Procedural Balance   *512*
18.4.2 Improving Contextual and Temporal Balance   *512*
18.4.2.1 City Blocks as a Contextual Variable   *513*
18.4.3 Improving Integrational Balance   *518*

18.4.3.1 Institutional and Governing Aspect   *520*
18.4.4   Current Developments: Sustainability Information Modeling
         Platforms   *520*
         Final Thoughts   *521*
         Questions   *522*
         Appendix 1   *522*
         Appendix 2   *523*
         References   *531*

19   Toward Resilience of the Electric Grid   *535*
         Jiankang Wang
19.1   Electric Grids in Smart Cities   *536*
19.1.1   Structure of Power Systems   *537*
19.1.1.1 Vertical Structure   *537*
19.1.1.2 Transmission   *538*
19.1.1.3 Station and Substation   *539*
19.1.1.4 Distribution   *539*
19.1.2   Operation of Power Systems   *540*
19.1.2.1 Control   *541*
19.1.2.2 Scheduling   *542*
19.1.2.3 Protection   *542*
19.1.2.4 Distribution Automation   *543*
19.2   Threats to Electric Grids   *544*
19.2.1   Threats to the Physical Grid   *544*
19.2.1.1 Threats from Weather Hazards   *544*
19.2.1.2 Threats from Malicious Attacks   *545*
19.2.1.3 Models for Threats on the Physical Grid   *546*
19.2.2   Threats to the Cyber Layers of Power Systems   *547*
19.2.2.1 How Secure Is the Cyber Layer of Power Systems?   *547*
19.2.2.2 Classification of Cyber Attacks   *547*
19.2.2.3 Attacks on the Control Layer   *548*
19.2.2.4 Attacks on the Monitoring Layer   *551*
19.3   Electric Grid Response under Threats   *553*
19.3.1   (Unwanted) Physical Phenomena on Power Grids   *553*
19.3.1.1 Circuit Faults   *553*
19.3.1.2 Frequency Instability   *554*
19.3.1.3 Voltage Instability   *555*
19.3.2   Failing Mechanisms on Electric Grids   *556*
19.3.3   Modeling and Vulnerability Assessment Methods of Grid
         Response   *557*
19.4   Defense against Threats to Electric Grids   *558*
19.4.1   Recommendations for Grid Resilience Enhancement   *559*
19.4.2   Core Technologies for Power System Resilience   *560*

19.4.2.1 Emergency Control and Protection *560*

19.4.2.2 Restoration from the Distribution Level *561*

19.4.3   Development of Defense Methods against Threats *562*

19.4.3.1 Defense on the Physical Grid *562*

19.4.3.2 Defense on the Control Layer *564*

19.4.3.3 Defense on the Monitoring Layer *565*

Final Thoughts *567*

Questions *568*

References *568*

20   Smart Energy and Grid: Novel Approaches for the Efficient
         Generation, Storage, and Usage of Energy in the Smart Home
         and the Smart Grid Linkup *575*

*Julian Praß, Johannes Weber, Sebastian Staub, Johannes Bürner, Ralf Böhm,
Thomas Braun, Moritz Hein, Markus Michl, Michael Beck, and Jörg Franke*

20.1   Introduction *576*

20.2   Generation of Energy *576*

20.2.1   Aerodynamically and Aeroacoustically Optimized Small Wind
         Turbines *576*

20.2.2   Combined Heat and Power Micro Plants Using Organic Rankine
         Cycles *578*

20.3   Storage of Energy *581*

20.3.1   Thermal Storage Heating Systems *581*

20.3.2   Connecting Smart Homes to the Smart Grid *584*

20.4   Smart Usage of Energy *587*

20.4.1   Energy-Efficient Radiation Heating *587*

20.4.2   Comfort-Orientated Heating in Smart Homes–Overview and Field
         Study *593*

20.4.3   Smart Waste Heat Usage and Recovery from Refrigerators and
         Freezers *595*

20.4.4   Ventilation with Heat Recovery *597*

20.5   Summary *600*

Final Thoughts *600*

Questions *601*

References *601*

21   Building Cyber-Physical Systems – A Smart Building Use
         Case *605*

*Jupiter Bakakeu, Franziska Schäfer, Jochen Bauer, Markus Michl, and Jörg
Franke*

21.1   Foundations—From Automation to Smart Homes *606*

21.2   From Today's Technologically Augmented Houses to Tomorrow's
         Smart Homes *608*

21.2.1   Smart Home: Past, Present, and Future   *608*
21.2.2   CPS-Based Smart Home Automation: Design Challenges   *611*
21.3   Smart Home: A Cyber-Physical Ecosystem   *612*
21.3.1   Use Cases and Smart Home Scenarios   *613*
21.3.2   Interoperability for CPS-based Smart Home Environments   *615*
21.3.3   Decentralized Coordination and Cooperation: Applying
          Agent-Based Theory to Adaptive Architectural Environments   *618*
21.3.3.1 Multi-Agent Systems (MAS)   *619*
21.3.3.2 Potentials of MAS for Smart Building Applications   *620*
21.3.3.3 Applying MAS to Smart Building Environment   *621*
21.3.4   Application: A Decentralized Control in Private Homes Based on the
          Paradigms of the Industry 4.0   *624*
21.3.5   Application: Ambient Assisted Living (AAL)   *627*
21.4   Connecting Smart Homes and Smart Cities   *629*
21.5   Conclusion and Future Research Focus   *631*
          Final Thoughts   *632*
          Questions   *632*
          References   *633*

22   Climate Resilience and the Design of Smart Buildings   *641*
          Saranya Gunasingh, Nora Wang, Doug Ahl, and Scott Schuetter
22.1   Climate Change and Future Buildings and Cities   *642*
22.2   Carbon Inventory and Current Goals   *644*
22.3   Incorporating Predicted Climate Variability in Building Design   *646*
22.4   Case Studies   *648*
22.4.1   Modeling Methodology   *648*
22.4.2   Analysis of Climate Scenarios and Impacts   *651*
22.4.3   Results   *652*
22.4.3.1 Case 1: Southern Mississippi Space Center   *652*
22.4.3.2 Case 2: Chicago Multifamily Building   *658*
22.4.3.3 Case 3: Fort Collins Multifamily Building   *660*
22.4.4   Limitations of the Study and Future Work   *662*
22.5   Implications for Future Cities and Net-Zero Buildings   *662*
          Final Thoughts   *664*
          Questions   *664*
          References   *665*

23   Smart Audio Sensing-Based HVAC Monitoring   *669*
          Shahriar Nirjon, Ravi Srinivasan, and Tamim Sookoor
23.1   Introduction   *669*
23.2   Background   *671*
23.2.1   HVAC Failure Detection   *671*
23.2.2   HVAC Failures and Acoustics   *672*
23.2.3   Strategy for Acoustic-Based HVAC Fault Detection   *674*

23.3 The Design of SASEM *675*
23.3.1 Building a Low-Cost Sensing Platform *675*
23.3.2 Acoustic Modeling of HVAC Systems *678*
23.3.3 Decision Support System *682*
23.4 Experimental Results *685*
23.4.1 Spectral Analysis of HVAC Sounds *685*
23.4.2 Longer-Term Deployment *687*
Final Thoughts *689*
Questions *689*
Acknowledgement *690*
References *690*

24 Smart Lighting *697*
Jie Lian
24.1 Introduction *697*
24.2 Background *698*
24.3 Smart Lighting Applications *699*
24.4 Visible Light Communication (Smart Lighting Communication) System *701*
24.4.1 System Description *702*
24.4.1.1 Transmitter and Receiver Model *702*
24.4.1.2 Indoor VLC Channel Model *702*
24.4.2 VLC MIMO Technology *706*
24.4.2.1 Modulation Schemes in VLC Systems *708*
24.4.3 On–Off Keying (OOK) *708*
24.4.3.1 M-ary Pulse Amplitude Modulation (M-PAM) *709*
24.4.3.2 Pulse Position Modulation (PPM) *710*
24.4.4 Multiuser VLC Systems *710*
24.4.4.1 Multiple Access Schemes *710*
24.4.4.2 Cellular Structure for VLC Systems *713*
24.4.5 Practical Considerations *716*
24.4.5.1 Intersymbol Interference (ISI) *716*
24.4.5.2 Nonlinearity of LEDs *717*
24.4.5.3 Illumination Requirements and Dimming Control *717*
24.5 Conclusion and Outlook *718*
Final Thoughts *719*
Questions *719*
References *719*

25 Large Scale Air-Quality Monitoring in Smart and Sustainable Cities *725*
Xiaotan Jiang
25.1 Introduction *726*
25.2 Current Approaches to Air Quality Monitoring and Their Limitations *729*

25.3   Overview of a Cloud-based Air Quality Monitoring System   *731*
25.3.1   Data Sources   *731*
25.3.2   Data Representation and Storage   *731*
25.3.3   Air Quality Analytics Engine   *732*
25.3.4   APIs and Applications   *732*
25.4   Cloud-Connected Air Quality Monitors   *733*
25.4.1   Sensor Selection   *733*
25.4.2   Mechanical Design   *734*
25.4.3   Data Communication   *734*
25.4.4   Hardware Calibration   *734*
25.5   Cloud-Side System Design and Considerations   *736*
25.5.1   sMAP   *736*
25.5.2   Data Format, Authentication, Storage and Web Services   *737*
25.6   Data Analytics in the Cloud   *739*
25.6.1   Filtering Using Signal Reconstruction   *739*
25.6.2   Calibration Using Artificial Neural Networks   *740*
25.6.2.1 Neural Network Model   *741*
25.6.3   Online Inference Model   *742*
25.6.3.1 Gaussian Process   *742*
25.6.3.2 Gaussian Process Regression   *743*
25.6.4   Evaluation of Effectiveness   *744*
25.7   Applications and APIs   *748*
        Final Thoughts   *748*
        Questions   *751*
        References   *751*

26   The Smart City Production System   *755*
        Gary Graham, Jag Srai, Patrick Hennelly, and Roy Meriton
26.1   Introduction   *755*
26.2   Types of Production System: Historical Evolution   *757*
26.2.1   Pure Fordism 1920s Onward   *757*
26.2.2   Toyota Production System (TPS)   *758*
26.2.3   Post-Fordism   *759*
26.3   The Integrated Smart City Production System Framework   *761*
26.3.1   Smart City Infrastructure   *761*
26.3.2   Big Data   *762*
26.3.3   Industrial Internet of Things   *762*
26.4   Production System Design   *763*
26.4.1   Network Design   *763*
26.4.2   Redistributed Manufacturing (RDM)   *764*
26.4.3   Manufacturing Scale and Inventory   *766*
26.4.4   Distribution and Service   *766*
26.5   Chapter Summary   *767*

Final Thoughts *768*
Questions *768*
References *768*

27 Smart Health Monitoring Using Smart Systems *773*
Carl Chalmers
27.1 Introduction *773*
27.2 Background *775*
27.2.1 Advanced Metering Infrastructure *775*
27.2.2 Smart Meters *777*
27.2.3 AMI Implementation Challenges *781*
27.2.4 Patient Behavior and Uses *782*
27.2.4.1 Active Monitoring for Behavioral Changes with Dementia *783*
27.2.4.2 Active Monitoring for Behavioral Changes in Depression and Other Mental Illness *784*
27.2.4.3 Prediction for EIP *785*
27.2.5 Current Assistive Technologies *785*
27.3 Integration for Monitoring Applications *786*
27.3.1 Case Study *787*
27.4 Conclusion *788*
Final Thoughts *789*
Questions *789*
References *789*

28 Significance of Automated Driving in Japan *793*
Sadayuki Tsugawa
28.1 Introduction *793*
28.2 Definitions of Automated Driving Systems *794*
28.3 A History of Research and Development of Automated Driving Systems *795*
28.3.1 Classification of Automated Driving Systems *795*
28.3.2 The First Period of Automated Driving Systems *796*
28.3.3 The Second Period of Automated Driving Systems *798*
28.3.4 The Third Period of Automated Driving Systems *798*
28.3.5 The Fourth Period of Automated Driving Systems *801*
28.4 Expected Benefits of Automated Driving *804*
28.4.1 Safety *804*
28.4.2 Efficiency *804*
28.5 Issues of Automated Driving for Market Introduction *805*
28.5.1 Performance of Human Drivers *805*
28.5.2 What we Learned from Experiments *805*
28.5.3 Reliability and MTBF Requirements of the Systems and the Devices *806*

28.5.4   Issues on Human Factors   *807*

28.6   Possible Market Introduction of Automated Driving Systems in Japan   *808*

28.6.1   Population Issues   *808*

28.6.2   Small, Low-Speed Automated Vehicles   *809*

28.6.3   Automated Truck Platoons   *813*

28.6.4   Cooperative Adaptive Cruise Control   *814*

28.7   Conclusion   *815*

    Questions   *816*

    References   *816*

29   Environmental-Assisted Vehicular Data in Smart Cities   *819*

    Wei Chang, Huanyang Zheng, Jie Wu, Chiu C. Tan, and Haibin Ling

29.1   Location-Related Security and Privacy Issues in Smart Cities   *820*

29.2   Opportunities of Using Environmental Evidences   *822*

29.3   Challenges of Creating Location Proofs   *823*

29.4   Environmental Evidence-Assisted Vehicular Data Framework   *825*

29.4.1   System Model and Attack Model   *825*

29.4.2   Roadside Unit-Based Environmental Evidence Construction   *826*

29.4.3   Environmental Evidence-Assisted Application Models   *827*

29.4.3.1   Location Claim Verification   *827*

29.4.3.2   Privacy-Preserved Data Collecting   *827*

29.4.3.3   Environmental Index-Based Data Retrieval   *829*

29.4.4   Optimal Placement of Roadside Units   *830*

29.4.4.1   Problem Formulation   *831*

29.4.4.2   Properties   *832*

29.4.4.3   Approximation for the Optimal RSU Placement   *834*

29.4.4.4   Extension: Optimal RSU Placement with Package Loss   *836*

29.4.4.5   Performance Analysis   *837*

29.4.5   Time Synchronization among Roadside Units   *839*

29.5   Conclusion   *841*

    Final Thoughts   *841*

    Questions   *842*

    References   *842*

Index   *845*

Editors Biographies

Houbing Song received his M.S. degree in Civil Engineering from the University of Texas, El Paso, TX, in December 2006 and Ph.D. degree in Electrical Engineering from the University of Virginia, Charlottesville, VA, in August 2012.

In 2017, he joined the Department of Electrical, Computer, Software, and Systems Engineering, Embry-Riddle Aeronautical University, Daytona Beach, FL, where he is currently an Assistant Professor and the Director of the Security and Optimization for Networked Globe Laboratory (SONG Lab, www.SONGLab.us). He served as an assistant professor in the Department of Electrical and Computer Engineering, West Virginia University, Montgomery, WV, and the Founding Director of West Virginia Center of Excellence for Cyber-Physical Systems sponsored by West Virginia Higher Education Policy Commission, from 2012 to 2017. In 2007 he was an Engineering Research Associate with the Texas A&M Transportation Institute. He is the editor of four books, including *Smart Cities: Foundations, Principles, and Applications*, Hoboken, NJ: Wiley, 2017; *Security and Privacy in Cyber-Physical Systems: Foundations, Principles and Applications*, Chichester, UK: Wiley, 2017; *Cyber-Physical Systems: Foundations, Principles and Applications*, Waltham, MA: Elsevier, 2016; and *Industrial Internet of Things: Cybermanufacturing Systems*, Cham, Switzerland: Springer, 2016. He is the author of more than 100 articles. His research interests include cyber–physical systems, Internet of Things, cloud computing, big data analytics, connected vehicle, wireless communications and networking, and optical communications and networking.

Dr Song is a senior member of IEEE and a member of ACM. Dr Song was the very first recipient of the Golden Bear Scholar Award, the highest faculty research award at West Virginia University Institute of Technology (WVU Tech), in 2016.

Ravi S. Srinivasan received his Ph.D. and M.S. in Architecture (Building Technology) from the University of Pennsylvania (2010 and 2004, respectively); M.S. in Civil Engineering from the University of Florida (2002); and B. Architecture from National Institute of Technology, India (1996). He is a LEED Accredited Professional (since 2004), Certified Energy Manager (since 2008), and Green Globes Professional (since 2012).

He is an Assistant Professor of Low/Net Zero Energy Buildings at the M.E. Rinker, Sr School of Construction Management, at University of Florida (since 2011). He was an invited reviewer of MDes theses at the Graduate School of Design, Harvard University (2014), and a frequent guest lecturer to the Department of Architecture, School of Design, University of Pennsylvania (since 2011). He is the author of *The Hierarchy of Energy in Architecture: Emergy Analysis*, PocketArchitecture Technical Series (Routledge, Taylor & Francis, 2015) with Kiel Moe, Harvard University. He has contributed to over 60 scientific articles published in peer-reviewed international journals (impact factor > 2), chapters, and conference proceedings.

Dr Srinivasan's current research focuses on the development of Dynamic Sustainability Information Modeling (Dynamic-SIM) Workbench with integrative three-dimensional building modeling, simulation, and visualization platform using extensible virtual environments for net zero campuses and cities. He has won several awards: University of Florida's prestigious Excellence Award for Assistant Professors (2016), the Nancy Perry Teaching Excellence Award (2014), Innovation in Green Design Award (2010), Chartered Legend in Energy Award (2007), and Young CAADRIA Award (2006).

Tamim I. Sookoor was born in Colombo, Sri Lanka, in 1984. He received a B.E. degree in Computer Engineering from Vanderbilt University, Nashville, TN, in 2006 and M.S. and Ph.D. degrees in Computer Science from the University of Virginia, Charlottesville, VA, in 2009 and 2012, respectively.

He was a Computer Scientist at the US Army Research Laboratory from 2012 to 2014 and is currently a Senior Professional Staff at The Johns Hopkins University Applied Physics Laboratory, Laurel, MD. He has published in a number of peer reviewed conferences and journals. His research interests are cyber–physical security, energy-efficient smart buildings, embedded systems, networking, and distributed systems.

Dr Sookoor is a member of the Association for Computing Machinery (ACM). He was a recipient of the DoD SMART Fellowship in 2010.

Sabina Jeschke was born in Kungälv, Sweden, in 1968. She received her diploma in physics from Berlin University of Technology, Germany, in 1997. After her stay at the NASA Ames Research Center/California and the Georgia Institute of Technology/ Atlanta, she gained a doctoral degree on "Mathematics in Virtual Knowledge Environments" from said university in 2004.

She worked at Berlin University of Technology, Germany, as a Junior Professor from 2005 to 2007. Until 2009, she has been a Full Professor at the University of Stuttgart, at the Department of Electrical Engineering and Information Technology, and simultaneously Director of the Central Information Technology Services (RUS) and the Institute for IT Service Technologies (IITS). In 2009 she was appointed Professor at the Faculty of Mechanical Engineering, RWTH Aachen University, Aachen, Germany. Her research areas are inter alia distributed artificial intelligence, robotics and automation, traffic and mobility, virtual worlds, and innovation and future research. Sabina Jeschke is Vice Dean of the Faculty of Mechanical Engineering of the RWTH Aachen University, Chairwoman of the board of management of the VDI Aachen, and Member of the supervisory board of the Körber AG.

Prof. Dr Sabina Jeschke is a senior member of the Institute of Electrical and Electronics Engineers (IEEE) and a member and consultant of numerous committees and commissions, including the American Society of Mechanical Engineers (ASME), the Association for Computing Machinery (ACM), the American Mathematical Society (AMS), and the American Society for Engineering Education (ASEE). She is an alumnus of the German National Academic Foundation (Studienstiftung des deutschen Volkes) and Fellow of the RWTH Aachen University. In July 2014, the Gesellschaft für Informatik (GI) honored her with their award Deutschlands digitale Köpfe (Germany's digital heads). In September 2015 she was awarded the Nikola-Tesla Chain by the International Society of Engineering Pedagogy (IGIP) for her outstanding achievements in the field of engineering pedagogy.

List of Contributors

Doug Ahl
Seventhwave
Madison, WI 53705, USA

Syed Hassan Ahmed
School of Computer Science and
Engineering
Kyungpook National University
Daegu, Korea

Adam L. Anderson
Department of Electrical and
Computer Engineering
Tennessee Technological University
Cookeville, TN 38505, USA

Jupiter Bakakeu
Institute for Factory Automation and
Production Systems (FAPS)
Friedrich-Alexander University
Erlangen-Nürnberg
Erlangen, Germany

Mridul S. Barik
Department of Computer Science &
Enginering
Jadavpur University
Kolkata, India

Jochen Bauer
Institute for Factory Automation and
Production Systems (FAPS)
Friedrich-Alexander University
Erlangen-Nürnberg
Erlangen, Germany

Michael Beck
Institute of Separation Science and
Technology
Friedrich-Alexander University of
Erlangen-Nürnberg
Erlangen, Germany

Madhur Behl
Department of Computer Science
University of Virginia
Charlottesville, VA, USA

Ralf Böhm
Institute for Factory Automation and
Production Systems (FAPS)
Friedrich-Alexander University of
Erlangen-Nürnberg
Erlangen, Germany

Safdar Hussain Bouk
School of Computer Science and
Engineering
Kyungpook National University
Daegu, Korea

Thomas Braun
Institute for Factory Automation and
Production Systems (FAPS)
Friedrich-Alexander University of
Erlangen-Nürnberg
Erlangen, Germany

Julio Brito-Santana
University of La Laguna
San Cristóbal de La Laguna, Spain

Johannes Bürner
Institute for Factory Automation and
Production Systems (FAPS)
Friedrich-Alexander University of
Erlangen-Nürnberg
Erlangen, Germany

Carl Chalmers
Department of Computer Science
Liverpool John Moores University
Liverpool, UK

Wei Chang
Department of Computer Science
Saint Joseph's University
Philadelphia, PA 19131, USA

Nilanjan Ray Chaudhuri
School of Electrical Engineering and
Computer Science
Pennsylvania State University
State College, PA 16801, USA

Chandreyee Chowdhury
Department of Computer Science
and Engineering
Jadavpur University
Kolkata, India

Corey D. Cooke
Department of Electrical and
Computer Engineering
Tennessee Technological University
Cookeville, TN 38505, USA

Cynnamon Dobbs
Departamento de Ecosistemas y
Medio Ambiente
Pontifical Catholic University of
Chile
Santiago, Chile

Christopher Expósito-Izquierdo
University of La Laguna
San Cristóbal de La Laguna, Spain

Airam Expósito-Márquez
University of La Laguna
San Cristóbal de La Laguna, Spain

Alvis C. M. Fong
Department of Computer Science
Western Michigan University
Kalamazoo, MI 49008, USA

Bernard Fong
Faculty of Health and Environmental
Sciences
Auckland University of Technology
Auckland

Jörg Franke
Institute for Factory Automation and
Production Systems (FAPS)
Friedrich-Alexander University of
Erlangen-Nürnberg
Erlangen, Germany

Martina Fromhold-Eisebith
Department of Geography
RWTH Aachen University
Aachen, Germany

Philipp Geyer
Department of Architecture
Faculty of Engineering Science
Katholieke Universiteit Leuven
Belgium

Gary Graham
Dept – Business School
Leeds University Business School
Leeds, UK

Erin Griffiths
University of Virginia
Charlottesville, USA

Saranya Gunasingh
Seventhwave
Chicago IL, USA

Moritz Hein
Chair of Measurement and Control
Systems
University of Bayreuth
Bayreuth, Germany

Patrick Hennelly
Institute for Manufacturing
Cambridge University IfM
Cambridge, UK

Xiaofan Jiang
Department of Electrical Engineering
Columbia University
New York, NY 10027, USA

Aslihan Karatas
Department of Civil and
Environmental Engineering
University of Michigan
Ann Arbor, MI 48109, USA

Dongkyun Kim
School of Computer Science and
Engineering
Kyungpook National University
Daegu, Korea

Ali Komeily
School of Construction Management
University of Florida
Gainesville, FL 32611, USA

Sathish A.P Kumar
Department of Computing Sciences
Coastal Carolina University
Conway, SC 29528, USA

Jie Lian
Charles L. Brown, Department of
Electrical and Computer Engineering
University of Virginia
Charlottesville, VA 22904
USA

Haibin Ling
Department of Computer and
Information Sciences
Temple University
Philadelphia, PA 19122
USA

Massimiliano Manfren
Department of Industrial
Engineering (DIN)
University of Bologna
40136 Bologna, Italy

Rahul Mangharam
Electrical and Systems Engineering
University of Pennsylvania
Philadelphia, PA, USA

Chandan Mazumdar
Department of Computer Science &
Enginering
Jadavpur University
Kolkata, India

Carol C. Menassa
Department of Civil and
Environmental Engineering
University of Michigan
Ann Arbor, MI 48109
USA

Roy Meriton
Dept – Business School
Leeds University Business School
Leeds, UK

Markus Michl
Institute for Factory Automation and
Production Systems (FAPS)
Friedrich-Alexander University of
Erlangen-Nürnberg
Erlangen, Germany

Marc Morrison
M.E. Rinker, Sr. School of Building
Construction Management
University of Florida
Gainesville, FL 32611, USA

Shahriar Nirjon
Department of Computer Science
University of North Carolina at
Chapel Hill
Chapel Hill, NC 27599, USA

Tori Okner
Senior Associate
Principle Group
Boston, MA, USA

Julian Praß
Institute for Factory Automation and
Production Systems (FAPS)
Friedrich-Alexander University of
Erlangen-Nürnberg
Erlangen, Germany

Russell Preston
Design Director
Principle Group
Boston, MA, USA

Pethuru Raj
Infrastructure Architect, IBM Global
Cloud Center of Excellence
IBM India
Bangalore, India

Juhi Ranjan
University of Virginia
Charlottesville, USA

Sarbani Roy
Department of Computer Science
and Engineering
Jadavpur University
Kolkata, India

Mahasweta Sarkar
Electrical & Computer Engineering
Department
San Diego State University
San Diego, CA 92182, USA

Franziska Schäfer
Institute for Factory Automation and
Production Systems (FAPS)
Friedrich-Alexander University
Erlangen-Nürnberg
Erlangen, Germany

Arno Schlueter
Institute of Technology in
Architecture
ETH Zürich, Switzerland

Scott Schuetter
Seventhwave
Madison, WI 53705, USA

Anirban Sengupta
Department of Computer Science &
Enginering
Jadavpur University
Kolkata, India

Lixin Situ
Faculty of Health and Environmental
Sciences
Auckland University of Technology
Auckland

Tamim Sookoor
Senior Professional Staff, The Johns
Hopkins University Applied
Physics Laboratory
Laurel, MD, USA

Jag Srai
Institute for Manufacturing
Cambridge University IfM
Cambridge, UK

Ravi Srinivasan
M.E. Rinker, Sr. School of Building
Construction Management
University of Florida
Gainesville, FL 32611, USA

Sebastian Staub
Chair of Energy Process Engineering
Friedrich-Alexander University of
Erlangen-Nürnberg
Erlangen, Germany

Allisandra Stoiko
Department of Chemical Engineering
University of Michigan
Ann Arbor, MI 48109, USA

Chiu C. Tan
Department of Computer and
Information Sciences
Temple University
Philadelphia, PA 19122, USA

Sadayuki Tsugawa
Intelligent Systems Research Institute
National Institute of Advanced
Industrial Science and Technology
Tsukuba, Japan

Madhavi Venkatesan
Department of Economics
Bridgewater State University
Bridgewater, MA, 02325, USA

Jiankang Wang
Department of Electrical and
Computer Engineering
Ohio State University
Columbus, OH 43210, USA

Nora Wang
Pacific Northwest National
Laboratory
Washington, DC, USA

Johannes Weber
Institute for Factory Automation and
Production Systems (FAPS)
Friedrich-Alexander University of
Erlangen-Nürnberg
Erlangen, Germany

Kamin Whitehouse
University of Virginia
Charlottesville, USA

Jie Wu
Department of Computer and
Information Sciences
Temple University
Philadelphia, PA 19122, USA

Huanyang Zheng
Department of Computer and
Information Sciences
Temple University
Philadelphia, PA 19122, USA

Foreword

Currently, there is significant research on the topic of making the world smart. Whether it is smart homes, phones, watches, other wearables, or smart cities, progress in attaining new and increased levels of "smart" is accelerating. Many communities are addressing various issues in attempting to achieve the goal of a smart world. The main communities include the Internet of Things, mobile computing, pervasive computing, wireless sensor networks, and cyber–physical systems. Each of these communities can contribute significantly to this goal, but increased progress will occur if, and when, the communities interact and leverage results from their respective communities. In fact, as greater sophistication of smart technologies is achieved, more and more technical challenges sit at the center of all these areas.

Let me briefly mention some of these central and fundamental challenges. How can we deal with trillions of devices? What will fundamentally change when there is an average of thousands of devices per person on the planet (vs only two or three today)? What IoT architectures will be necessary to manage and control each administrative domain (i.e., domains that own and control devices, apps, and services) and their interactions? With hundreds of millions of apps running, largely independently developed, how can we ensure safety and efficient performance? For example, one app from traffic control in a smart city may want to turn a traffic light red, but an emergency service wants to turn it green. Enormous amounts of data are being collected. How can we create knowledge from this big data and use it effectively in optimizations? Who decides the optimization criteria and how do these criteria get balanced between the benefits for the public versus individuals? Smart worlds will not seem smart if they are not robust and dependable. How can long-lived services be continuously checked for correctness, robustness, and safety? Security challenges will abound due to both the openness of these smart systems and the fact that many devices have very limited capacity to house security technology. How can privacy policies be stated, checked, and coordinated across systems of systems? Many smart systems will intimately connect with humans in the loop.

How can we develop human behavior models and use them to improve smart services?

All these issues also relate to smart cities, as an exemplar of a smart system. In particular, the so-called smart cities have existed for some time. However, in the past, most smart city services have been limited and often only address one domain (e.g., transportation), or even just one aspect of a domain. For example, one smart city may have a transportation service focused on highways only or another smart city may have a "smarter" service that simultaneously considers traffic congestion and air pollution. In the future, with the increased deployment of sensors and actuators and the use of machine learning and data analytics, it is expected that a dramatic increase in the number and sophistication of services will occur. There will also be a greater integration across services. To meet the challenges of these future smart cities requires many research questions to be solved, including the ones listed here. This book covers smart city services in the domains of transportation, energy, buildings, manufacturing, climate and air quality, and healthcare. The book also addresses the human and economic aspects of smart cities. Reading this book will convey a broad knowledge of the current technology as well as the challenges faced for building smart cities of the future.

August 2016

John A. Stankovic
BP America Professor
Department of Computer Science
University of Virginia

Preface

As world urbanization continues to grow – 66% of the world's population is projected to be urban by 2050 – there is a worldwide trend toward the establishment of smart cities. This trend not only opens up significant opportunities but also creates many physical, social, behavioral, economic, and infrastructural challenges. To address these challenges in implementing smart cities, increased understanding of how to design, adapt, and manage smart cities in an intelligent and effective way is essential.

This edited book, *Smart Cities: Foundations, Principles, and Applications*, aims to present the scientific foundations and engineering principles needed to achieve the vision of smart and connected communities (S&CC), including smart cities, and various innovative applications and services to enable S&CC. Accordingly, this book is organized into three parts: foundations, principles, and applications.

Foundations: After presenting the opportunities and challenges of S&CC (Chapter 1), this part presents various scientific foundations of S&CC, including big data analytics (Chapter 2), multiscale computing (Chapter 3), communications and networking (Chapter 4), citizen science and crowdsourcing (Chapter 5), real-time control and adaptation (Chapter 6), mobile services (Chapter 7), and ecosystem services (Chapter 8).

Principles: This part presents various engineering principles of S&CC, including data-driven modeling (Chapter 9), future Internet architecture (Chapter 10), context-aware computing (Chapter 11), smart governance (Chapter 12), social impact (Chapter 13), smart economy (Chapter 14), cyber security and privacy (Chapter 15), smart mobility (Chapter 16), decision-making (Chapter 17), human factors (Chapter 18), and sustainability (Chapter 19).

Applications: This part presents various innovative applications and services of S&CC, including spanning energy (Chapters 20 and 21), building design and automation (Chapters 22 and 23), smart lighting (Chapter 24), environmental protection (Chapter 25), manufacturing (Chapter 26), healthcare (Chapter 27), and transportation (Chapters 28 and 29).

July 2016

Houbing Song, Ravi Srinivasan,
Tamim Sookoor, Sabina Jeschke

Acknowledgments

This book would not have been possible without the help of many people. First, we would like to thank all the contributors and reviewers of the book from all over the world for their valuable input. In addition, we would like to thank our editorial assistants, Ruth Hausmann and Alicia Dröge, both at RWTH Aachen University, who provided essential support at all stages of the editorial process of the book. Also we would like to thank Divya Narayanan and Brett Kurzman at Wiley, who helped shepherd us through the book-editing process. At last, we would like to acknowledge the support of the Cluster of Excellence Integrative Production Technology for High-Wage Countries at RWTH Aachen University, German Research Foundation, German Federation of Industrial Research Associations – AiF.

Special thanks go out to the following reviewers:
Syed Hassan Ahmed (Kyungpook National University)
Mahabir Bhandari (Oak Ridge National Laboratory)
Safdar H. Bouk (Kyungpook National University)
William Braham (University of Pennsylvania)
Christos G. Cassandras (Boston University)
Zhi Chen (Arkansas Tech University)
Xiuzhen Cheng (George Washington University)
Robert Dickerson (IBM)
David Doria (HERE)
Qinghe Du (Xi'an Jiaotong University)
Melike Erol-Kantarci (University of Ottawa)
Yaser P. Fallah (West Virginia University)
Bin Guo (Northwestern Polytechnical University)
Yu Jiang (Tsinghua University)
Burak Kantarci (University of Ottawa)
Krasimira Kapitanova (Palantir Technologies)
Ali Komeily (University of Florida)
Kristina Lahl (RWTH Aachen University)
Jaime Lloret (Universitat Politecnica de Valencia)

Hengchang Liu (University of Science and Technology of China)
Javad Mohammadpour (University of Georgia)
Mohammad Mozumdar (California State University, Long Beach)
Sirajum Munir (Bosch Research and Technology Center)
Shahriar Nirjon (University of North Carolina at Chapel Hill)
Kaoru Ota (Muroran Institute of Technology)
Alexander Paulus (RWTH Aachen University)
Manisa Pipattanasomporn (Virginia Tech)
Mohammad Shehadeh (RWTH Aachen University)
Mohammad Shojafar (Sapienza University of Rome)
Alexia Fenollar Solvay (RWTH Aachen University)
Vijay Srinivasan (Samsung R&D)
Lo'ai A. Tawalbeh (Umm Al-Qura University)
Huihui Wang (Jacksonville University)
Teresa Wu (Arizona State University)
Llewellyn Van Wyk (Council for Scientific and Industrial Research)
Yun Kyu Yi (University of Illinois at Urbana-Champaign)
Wei Yu (Towson University)
Guohui Zhang (University of New Mexico)
Yanxiao Zhao (South Dakota School of Mines and Technology)

1

Cyber–Physical Systems in Smart Cities – Mastering Technological, Economic, and Social Challenges

Martina Fromhold-Eisebith

Department of Geography, RWTH Aachen University, Aachen, Germany

CHAPTER MENU
Introduction, 1 Setting the Scene: Demarcating the Smart City and Cyber–Physical Systems, 3 Process Fields of CPS-Driven Smart City Development, 4 Economic and Social Challenges of Implementing the CPS-Enhanced Smart City, 10 Conclusions: Suggestions for Planning the CPS-Driven Smart City, 15

Objectives

- To broaden readers' understanding of the process fields of smart city development that can profit from enhanced information and communication technologies
- To explore in which respects cyber–physical systems (CPSs) may help in improving the coordinated control, regulation, and monitoring of different smart city processes, supporting sustainability objectives
- To introduce geographical perspectives into the debate on technology-driven smart city development, highlighting place-specific dynamics as well as upper-scale national and global influences on localized economic processes
- To point out risks and caveats that influence the societal acceptance of technology-driven smart city development and need to be regarded in related policies.

1.1 Introduction

The "smart city" notion has become synonymous with visions of future urban development, which is marked by the widespread digitization of services [1–4].

Smart Cities: Foundations, Principles and Applications, First Edition.
Edited by Houbing Song, Ravi Srinivasan, Tamim Sookoor, and Sabina Jeschke.
© 2017 John Wiley & Sons, Inc. Published 2017 by John Wiley & Sons, Inc.

A major objective of smart cities is to achieve triple sustainability in social, economic, and environmental issues [5]. Modern information and communication technologies (ICTs), the Internet, and the continuous expansion of data supply broaden the options for improving urban citizens' working and living conditions [6–8]. This chapter explores how cyber–physical systems (CPSs) in particular may enhance smart cities. In the field of manufacturing, CPS now represents innovative options for integrating ICT and networking systems into infrastructure, so as to more efficiently control and coordinate complex physical production processes and machine interactions [9, 10]. ICT devices that are embedded into products or components are able to monitor and direct physical processes in a self-regulating manner – directly communicating with each other via the Internet. CPS also has a vast potential beyond the industrial field. Within a smart city context, these systems could significantly augment urban services and supply infrastructure. Moreover, new CPS-driven manufacturing may emerge according to the trends of "urban production" [11].

Together with the requirements and potential implications of smart cities [1, 5], policy demands associated with this visionary goal are also eagerly explored [2, 12–15]. This chapter intends to broaden the conceptual and analytical views, as is required for effective policy making, by employing the perspectives of economic geography to CPS-supported smart city development. In principle, economic geographers focus on place-specific dynamics and systemic interactions between technology trends, economic challenges, social needs, and planning requirements [16]. Localized processes are seen as being embedded in wider, often global, economic, political, and societal settings [17, 18]. Accordingly, viable CPS technologies, along with socioeconomic and cultural settings at the regional level and other spatial scales (as discussed in [2]), are required if a smart city is to reach its full potential. In this context, the chapter explores the following key issues:

- In which ways can CPS help create a technologically enhanced smart city within the context of working and living in a manner that responds to environmental, economic, and societal challenges?
- Looking at the various fields in which CPS may support smart city development, how is the effective implementation of technologies influenced by economic conditions on a regional, national, and global scale and also by social attitudes?
- How can city–regional planning approaches meet these factors in order to arrive at economically and socially acceptable, hence viable, smart city policies?

The chapter's structure reflects this agenda. After defining the system setting (Section 1.2), the author identifies various process fields of working and living in a smart city that can profit from CPS-related applications and impulses (Section 1.3). Then it is discussed how economic conditions influence effective

CPS implementation in place-specific ways (Section 1.4). Additionally, social challenges of CPS-based smart city development that relate to aspects of acceptability, qualification, and adaptation are highlighted. The conclusions (Section 1.5) derive insights for policies, suggesting how economic and social well-being can be best promoted when implementing the CPS-enhanced smart city of tomorrow.

1.2 Setting the Scene: Demarcating the Smart City and Cyber–Physical Systems

The range of human activities that are affected by digitization increases every day, which amplifies the scope of ICT applications in urban development. Certain system limits need to be determined before we can concisely assess potential CPS applications in smart cities. This refers both to a proper understanding of the smart city notion and the CPS technology field. Only after the arena and its boundaries have been decided upon, it is possible to succinctly identify the relevant technology options.

A smart city can be defined in various ways (Albino *et al.* [1]: 6ff, compiled a list of over 20 different definitions). Despite differing perspectives held by public or private actors, there are common denominators [3–5, 8]. A smart city is first and foremost characterized by the strategic, systematic, and coordinated implementation of modern ICT applications in a range of urban functional fields (as elaborated below). This idea of a "digital city" involves both software and hardware components, such as sensors, meters, and other technical devices. Furthermore, the generation targeted and the use of inhabitants' knowledge, learning capacities, creativity, and human capital for innovations, in line with analytical and modeling skills, qualify smart cities as "intelligent cities" [19, 20]. The notion thus transcends a technological focus and explicitly bears an anthropocentric note: smart cities aim to comprehensively fulfill people's needs in terms of economic and social sustainability, happiness, and well-being [1, 5]. The connotation "green city" highlights the ecological objectives of reduced resource consumption, constrained pollution, and high process efficiencies [21]. Shared ideology and community governance strategies are important to stakeholders, especially regarding how the outcomes manifest themselves in public and politically promoted attitudes regarding improvements to the urban quality of life. We need to keep these manifold smart city qualities in mind when assessing the prospects of CPS implementation later on.

CPS technologies build on antecedent embedded systems, that is, computer-driven devices like medical, military, or scientific instruments, cars, and toys. Accordingly, CPSs use software and ICT networks in order to control, monitor, and coordinate complex physical processes, though this is mainly applicable to modern manufacturing [9, 10]. CPS can efficiently organize production within companies and also, through communication between components and machines, coordinate the value chain between different firms. Advancing from conventional computer-integrated manufacturing, CPS incorporates elements of self-awareness and self-regulation, wireless inter-machine adjustment, and complex data processing, thus integrating information from various stages and organizations within a production system. The Internet of Things forms a crucial CPS component, as objects can now carry digital information themselves and directly communicate with other objects, such as processing machines, via the Internet [22]. The expected results are a substantially modified division of labor in manufacturing and related services, consequently a transformed production landscape [23].

Existing literature discussing smart city qualities selectively refers to CPS components [7, 24, 25]. Big data sets, that is, large amounts of data on citizens' purchasing habits, mobility, and other behaviors, call for adequate information processing technologies and the expedient use of said technology [6, 8, 26]. Intelligent products or services like thinking machines, sensor-monitored smart homes, and self-regulating infrastructure are expected to significantly shape the face of smart cities [1–3, 27] and reach far beyond simple ICT applications. It seems that due to the more systematic exploration approaches available, urban CPS applications have more potential merits and also risks. As elaborated below, CPS based on advanced ICT can promote all smart city objectives in some way and ICT is essential for a truly integrated development. Only the flexible, self-regulating CPS qualities meet the requirements for cities that want to be able to resiliently react to future challenges [28].

1.3 Process Fields of CPS-Driven Smart City Development

Several scholars have already categorized the process fields that constitute a smart city (overview in [1]: 11ff; [5]). Drawing on these works, and also adding further urban arenas that could benefit from "smart" enhancement, this section proposes 11 fields that have the potential for CPS-driven processes: smart ICT infrastructure, smart energy, smart mobility, smart construction, smart security, smart metabolism, smart industries, smart economy, smart education, smart living, and smart governance. Each of these fields touches upon the aforementioned smart city objectives, and the ways in which CPS applications can substantially support efficient functionality and regional sustainability will be discussed in this chapter (see Table 1.1).

The first aspect, ICT infrastructure, forms the backbone of any smart city, and no CPS can work without it. A scaffold of high-capacity computers, communication nodes, and fast connections is crucial for linking various system elements together and driving their efficient cooperation as well as

Table 1.1 Smart City Process Fields, CPS Applications, and Sustainability Effects.

Process fields	Functions of CPS applications	Expected effects
Smart ICT infrastructure	Establish broadband or wireless connectivity and communication compatibility between all urban citizens and institutions based on hardware and software components	Overarching urban network of enhanced information flows and coordination that supports system efficiency
Smart energy	Implementation of smart grid technologies that control energy consumption and optimize coordination between decentralized power generation and utilization of renewables	Less consumption of fossil fuels and better integration of renewable energy production, leading to higher efficiency
Smart mobility	Coordinated goods logistics and e-mobility modes (cars, bikes, public transport), using geographical information, enable customized transport for citizens and companies	Reduction of gas and particulate matter (PM) emissions and of traffic congestions, also diminishing noise
Smart construction	Setting up buildings and settlements that, through system coordination, optimize the supply of amenities, consumption of resources, and living conditions	Reducing the individual use and waste of resources; increasing living quality for individuals and communities
Smart security	Coordination of advanced lighting and surveillance equipment that covers urban settlement areas, public spaces, and traffic lines and is controlled according to frequencies of use	Improving the overall quality of residence and living by preventing various types of (environmental) crime
Smart metabolism	Organizing a circular economy ("closing the loop") that optimizes the (re)use and recycling of resources, including water and waste management, by matching supply and demand	Less consumption of various resources, reduced emissions, and waste through functional eco-industrial networks

Table 1.1 (Continued)

Process fields	Functions of CPS applications	Expected effects
Smart industries	Selective integration of tailor-made goods manufacture and related services into urban development ("urban production"), colocating production, and consumption	Increased diversity of localized industrial activities and job opportunities, reducing home-to-work travel distances
Smart economy	Targeted orientation of entrepreneurship and business development toward new fields of ICT and CPS application in order to build up modern sectors that shape future trends	Laying the foundations for future economic competitiveness and for capacities to cater to local ICT/CPS needs
Smart education	Dedicated inclusion of smart city–related ICT/CPS technologies in education and R&D activities that support adequate qualification and skills of people, integrating all social classes	Better preparation of people for future economic tasks and job market demands, also supporting social inclusion
Smart consumption	Broadening options for a sharing economy as well as public monitoring and display of environment data, activating a reflected and resource-conscious consumption behavior	Reducing the individual waste of resources; increasing social cohesion and public awareness for environmental issues
Smart governance	Advance and coordinate ICT-based public services to all residents ("e-government") and offer new ways to influence and take part in political processes ("e-democracy")	Better quality and efficiency of public services, broadening the participation of citizens in political processes

Source: Author's depiction, partly drawing on information from Allwinkle and Cruickshank [12]; Komninos [20]; Lombardi *et al.* [29]; Matt *et al.* [11]; Neirotti *et al.* [14]; Albino *et al.* [1].

interoperability, like energy grids, mobility and logistics, and resource and waste management [14]. This field has been associated with smart city ideals right from the start, enabling all inhabitants to benefit from digital infrastructure and participate in knowledge acquisition, learning, and innovation [19, 20, 30]. The system-enhancing powers of this process field are at the heart of the notion:

> smart cities must integrate technologies, systems, services, and capabilities into an organic network that is sufficiently multi-sectorial and flexible for future developments, and moreover, open-access.
>
> (Albino *et al.* [1: 11]).

CPSs can substantially support the required coordination, openness, and flexible learning qualities of a modern ICT system [31]. None of the other smart city arenas operate in isolation, as energy and transport issues, construction, industrial development and the metabolism of resource inputs, and waste or emission outputs are intrinsically intertwined. CPSs are capable of coordinating infrastructure operation and information flows within and between these subsystems. The same holds for sensors and observational devices, like those mounted onto unmanned aerial vehicles (UAVs, or drones), which should serve various monitoring purposes simultaneously. In this context, the Internet occupies a key role as it has the potential to reach every citizen [32]. Debates concerning this process field focus on strong technology, thus involving various CPS elements: hardware and software, network solutions, extended options of advanced sensing [33], Internet of Things [7], cloud computing, ubiquitous Wi-Fi access, and real-time big data processing [8, 15, 26, 34, 35].

The interrelated dimensions of smart energy and mobility (Table 1.1) traditionally play a major role in smart city development regarding environmental concerns and sustainability objectives [5, 36, 37]. In the energy field, the installation of flexible, coordinated systems that link decentralized sources of energy production (based on photovoltaics, biogas, etc.) to individual dwellings with variable levels of energy consumption requires veritable "smart grid" solutions [11, 14]. CPSs offer important means for measuring, adjusting, and balancing energy supply in relation to demand. Likewise, visions of a "smart transport city" refer to ICT and geographical information systems that support flow management of goods and people, delivering "a self-operative and corrective system that requires little or no human intervention" [38: 48]. In addition to transport logistics coordination and intelligent structures of distribution, CPS applications enhance the monitoring and regulation of traffic flows using mobility sensors and smart parking management [1, 13]. E-mobility offers further scope for CPS utilization, matching the supply of e-transport modes, like e-bikes or Internet-connected "cyber cars," with individual "mobility-on-demand" needs [11].

The process field of smart construction is associated with customized control requirements of "smart homes" or "sustainable buildings" [39]. Objectives also include, but reach beyond, reduced energy consumption and emissions. Modern ICT applications form important elements of smart houses; the combination of various monitoring and control devices can optimize living conditions and environmental balance sheets [40]. The particular assistance demands of an aging population call for the implementation of ICT-controlled ambient assisted living technologies [41]. CPS applications were explicitly discussed as a means to improve the energy efficiency of buildings [24, 27], but they can also be useful beyond that. If entire settlements or housing clusters were equipped with CPS controls and sensors, it would be possible to create further reaching social sustainability effects by involving different households and generations of individuals. Here CPS integration may allow for energy and material flows within and between buildings. This moves city quarters toward achieving self-sufficiency and potentially turns them into hybrid energy storage systems for other parts of the city [11].

The smart mobility, smart construction, and smart ICT infrastructure fields are closely connected with the smart security process field, which is often mentioned in seminal smart city literature [1, 14, 29, 42]. Supposedly, modern observation and sensor-based surveillance technologies, in line with smart lighting installations or UAVs, significantly support the security and feeling of safety in urban citizens. CPSs aim to better combine, coordinate, and assess the information delivered by all the cameras, sensors, and environmental monitors placed in various parts of a city. The big data collection resources provided, through public Wi-Fi use, for example, may also feed into the smart security system, which in turn should be able to detect and foresee abnormal situations and potential risks [8].

Smart metabolism and smart industries (Table 1.1) refer to dimensions that have not yet received dedicated attention in smart city debates. This includes options to more efficiently organize industrial resource consumption in line with the "reduce, reuse, recycle" strategies and by establishing eco-industrial networks or parks where the waste, used water, or by-products of one industry serve as valued input for others [43]. Efficient regional material flow management requires sophisticated production planning, process monitoring, and logistics [44], all of which CPSs offers viable solutions for. Concerning the process field of smart industries, CPS technologies may also allow various manufacturing activities to return to urban locations [11]. Ongoing digitization will substantially change the methods of industrial production, in doing so establishing new combinations of advanced manufacturing and city–regional development in different parts of the world [45]. CPS applications can create extended options for "clean" production (for instance, using additive fabrication or 3D printing techniques) and bring tailor-made consumer goods production closer to the individualized demand of the open-minded, affluent urban citizen. Additionally, people profit from broadened job opportunities near their home. In general, CPS-enhanced coordination and control of industry processes may even improve the environmental impact and resource efficiency of urban production systems.

The suggested categorization of smart city process fields intentionally distinguishes between smart industries and a smart economy (Table 1.1). While the first notion looks at new ways of goods production in an urban context, the second one emphasizes new entrepreneurial opportunities that arise through the development of modern ICT and CPS technologies and importantly their adaptation to local applications [4, 29]. Companies face a diversified set of expanding business options in sectors such as engineering consulting, ICT system installation, implementation of CPS tools, big data processing, and development of mobile services and applications [46]. Growing clusters of firms primarily cater to local needs, supporting smart metabolism objectives of localized cycles of product and service delivery. Furthermore, globally competitive competences are created, which more broadly foster a city's economic sustainability.

The smart education field aims to technologically sustain and upgrade the smart regional economy. Several works emphasize the pivotal role of higher education institutions in delivering the qualifications and expertise needed to create a sustainable urban future [1, 47]. Human capital is regarded as one of the most important resources in smart city development [48, 49]. Given that major cities attract talented people from all over the world, the smart city vision promotes ideas of social inclusion across nationalities and cultures [50]. In line with education requirements for operating CPS applications, technological innovation is needed. The proclaimed openness of smart cities to communication and interaction calls for integrated skills regarding open innovation [51] and the competence to activate the triple helix university–industry–government collaboration [52]. The education system must also foster regional skills in order to purposefully interpret the big data stocks collected in a smart city. For implementing effective CPSs, qualified people who are able to derive meaningful insights from mobility, input–output, environmental, and other data sets are needed.

Finally, the process fields of smart consumption and smart governance address important sociopolitical qualities (Table 1.1). Smart city visions can only be put into practice when the majority of citizens are convinced of this idea and actively engaged in the relevant arenas of daily life [50]. Following Vanolo's logic [53], people need to develop a certain attitude toward "smart" solutions, dubbed "smart mentality." Sustainability-oriented smart consumption attitudes can grow over time and are driven by smart education. Modern ICT can also be employed in order to accelerate the achievement of sociopolitical goals. It helps make people better aware of major smart city issues in terms of environmental factors like the continuous monitoring and public display of air quality data using CPS. Furthermore, digital services enable people to actively shape a resource-efficient urban economy, for instance, through online platforms that support the local sharing of goods and services. Additionally, municipal administration can foster citizens' participative attitudes via ICT-based smart governance tools. Any smart city is expected to provide interoperable, Internet-based government services that offer ubiquitous access to public assistance. Citizens and businesses shall be enabled to closely follow, and even actively influence, government processes without barriers like language, education, or disabilities hindering them [1].

Summing up, CPSs that build on and expediently combine modern ICT appliances seem virtually indispensable in achieving the various goals of smart city development (right column in Table 1.1). Interrelatedly they can promote the economic, social, and ecological aspects of urban sustainability. The systemic, integrated, and adaptive nature of the ideal smart city and its spatially condensed options for ICT uses turn it into a perfect test site for exploring the potential benefits incorporated in CPSs. However, critical perspectives must also be taken into account.

1.4 Economic and Social Challenges of Implementing the CPS-Enhanced Smart City

In the large body of literature available regarding the opportunities of ICT-enhanced smart city development, hardly any critical approaches point out obstacles, caveats, and conflicting relationships between smart city process fields that could hamper smart city success [2, 50, 51]. It appears as if mankind must focus on tackling technological challenges associated with the creation of smart cities. The authors discuss how CPS "can extract the awareness information from the physical world and process this information in the cyber-world" [27, p. 1149], how to achieve data security [54], how to reduce urban CPS infrastructure risks [25], and even how devices, the Internet of things, and big data technology must be designed in order to be fit for service [6, 7, 55]. Concerning CPS technologies more specifically, issues of network functionality, efficiency, and failure prevention still need to be solved [9, 31, 56]. While these topics are crucial from an engineering perspective, we also need to highlight economic and social factors, barriers, and challenges that influence how smart cities can be realized. Academic scholars already recognized that technologies must adapt to human qualities:

> future CPSs will need to bolster a closer tie with the human element, through Human-in-the-Loop controls that take into consideration human intents, psychological states, emotions and actions inferred through sensory data.
>
> (Sousa Nunes *et al.* [57: 944])

In line with considerations regarding human adaptability to new ICT, the development of smart cities should be positioned within the context of different spatial scales of agency and decision power [2], in doing so employing institutional perspectives of economic geography [16]. Technology-oriented conceptualizations lead us to believe that the smart city just needs to emerge from its own roots, free to choose its trajectory based on endogenous assets and forces, as well as CPS preferences. The crucial role of exogenous national and global frameworks that specifically shape place-specific smart city options and patterns requires dedicated attention. Geographers regard (city) regions as spatial units marked by dynamics that reflect the interplay of upper-level influences and localized processes that are shaped by spatial proximity [17]. Accordingly, this section tries to assess how the CPS-enhanced smart city process fields are shaped by economic and social conditions in place- and scale-related ways (summarized in Table 1.2). Besides broadening conceptual views on smart cities, reasoning also offers insights for planning and policies.

By and large, the economic determinants of CPS-related smart city opportunities depend firstly on regional assets and location factors, secondly on the national framework, and thirdly on global competition [18]. Viewing the 11 smart city process fields through this lens tells us that localized assets are particularly important for implementing the basic ICT infrastructure and smart building construction and installing smart metabolism, smart industries and economy, and smart consumption. While CPS technologies may also be developed elsewhere, their customized installation and implementation in the city requires local, professional agency, and well-developed knowledge about a city's topography and functional structure. Likewise, the construction of smart homes and building clusters significantly profits from localized expertise. Regionally rooted companies know best which specialized partners nearby should be included in fulfilling the sophisticated requirements of housing settlements controlled and interconnected by modern ICT facilities [40].

In the field of smart metabolism, spatial proximity, even colocation in the same industry park, is essential for firms that want to exchange by-products and collectively use resources more efficiently [43]. Companies integrated in a CPS-coordinated material flow system can potentially reap strong economic benefits from "cash for trash" revenues and save on costs for resources, water, and waste treatment. Examples show that remarkable local marketing and reputation effects emerge from successfully establishing eco-industrial networks [58]. Smart industry strategies for utilizing CPS aim to bring production back to urban areas, and in order to better coordinate processes within and between firms, they rely on suitable location factors, such as a qualified labor market, supportive service partners, and chances to embed firms in local production systems [45]. In the cases of smart industry and smart economy, which are both driven by entrepreneurs that locally design, produce, and install CPS components, certain agglomeration and proximity advantages support

Table 1.2 Importance of Economic Context at Different Spatial Scales and Social Acceptance for Smart City Process Fields.

Process fields	Economic context at spatial scales	Social acceptance (++ = strong support to −− = strong opposition)
Smart ICT infrastructure	Regional knowledge for adopted implementation; global search for technologies	−−/+
Smart energy	National energy provision systems and incentive framework for renewable energies; global search for technologies, international energy provision	++/−
Smart mobility	National incentive framework and lead producers; global search for technologies; global players of production	++/−
Smart construction	Regional knowledge and partner networks; global search for technologies; global players of production	−−/+
Smart security	National regulation for devices, surveillance options, and data use; global search for technologies	−−
Smart metabolism	Regional material flows of colocating firms save transport and purchasing cost; alternative of higher value global sourcing	+
Smart industries	Regional factors of location, collaboration opportunities, and cluster advantages; global–local impact of CPS unclear, strong global competition in manufacturing industries	+/−−
Smart economy	Regional sources of entrepreneurship and business networks, cluster advantages; strong global competition in CPS technology field	++
Smart education	National systems of education and innovation; global knowledge flows and education/innovation exchanges	++
Smart consumption	Regional initiatives and communities of sharing; global cultures of sustainable living and virtual communities	++
Smart governance	National administrative requirements, incentive structures, and model initiatives	−−/+

Source: Author's depiction.

business development. Clusters of colocating firms that belong to the same value chain, for example, could potentially profit from regional interaction and knowledge collaboration as they support productivity, innovation, and competitiveness within the region [59]. A city that strategically promotes CPS-related industrial opportunities of this kind could reap high economic benefits. In the field of smart consumption, where the economic feasibility of ICT tools and services depends on large numbers of urban users, the region of residence forms the arena for joint initiatives and communities based on shared visions of sustainable urban life. Overall, the regional powers of smart city development are determined by urban population size, wealth, and qualification levels and by an industry structure that allows firms to profitably engage in CPS creation and implementation.

Quite a number of CPS-enhanced process fields, although locally installed, appear to depend predominantly on national economic conditions. The dominance of the national over the regional sphere of decision-making applies to the smart energy, mobility, security, and education fields and to smart governance in particular. Energy production and distribution by nature requires pipe and line systems that cover wider areas; therefore, it would be beneficial to combine sources at different locations and find a balance between them. While some renewable energies permit more decentralized, locally confined production (e.g., photovoltaics), their economic viability still relies on nationally determined and implemented incentives [60]. The availability of smart mobility systems, be it advanced transport logistics or e-vehicles, for urban use is also influenced by national contexts. The high prices set on these innovative products and services obstruct market success, that is, without political intervention. Similarly, the producers of mobility items add technical requirements of larger markets to national and markets of a wider scale. The same can be said for smart security, where hardware (intelligent sensors or UAVs) and software come from players operating at least at the national level. In this field, national regulations concerning permitted devices, limited surveillance rights, and data security play a crucial role. Looking at smart education, the knowledge base for human capital and innovation capacity is often formed according to national strategies and economic interests, as outlined in the national innovation system notion [61]. Finally, smart governance approaches must obey national requirements, as well as fulfill common administrative routines and functions. National governance organizations may also be influential in launching e-government incentives and initiatives.

With regard to the global scale of influence, a smart city is in principle more integrated into international goods and information flows than other cities [4, 50]. In almost all process fields, producers and users profit from effective, global research regarding suitable smart city solutions and also from embedding themselves in international supply networks (Table 1.2). In the cases of mobility and construction in particular, important global players tend to dominate the provision of high standard CPS modules. Moreover, smart industries and economies are particularly sensitive to global influences because of strong competition [17, 18]. Concerning metabolism, regionally closing the loop of material flows may not always be the most profitable option in comparison to efficient global sourcing, as local supplies may lack quality or the product might require further processing steps before use. The question of whether CPS-enhanced systems can sustain urban production in the face of global competitive pressure is debatable. Furthermore, emerging clusters of enterprises producing CPS components operate in fields where numerous regions and companies compete worldwide. Logically, smart education draws on international impulses embedded in global networks of scientific collaboration and human capital exchange [61]. Smart consumption also thrives on global trends that internationally promote sustainable lifestyles and the sharing of resources, which is only spurred on by the formation of Internet communities.

Fundamental caveats need to be raised regarding social issues associated with CPS-enhanced smart city transformation. It may be questioned as to whether large parts of an urban population are really ready to follow the "smart turn" in every aspect of their daily life – cultivating the commonly shared "smart mentality" [53]. While the practical implementation of smart city models proliferates worldwide [1, 37], a citizen's propensity to accept or actively promote a city's digitization still significantly varies between regions and nations. In some parts of the world, societies are generally more open toward smart city technologies than in others, as indicated by the differences between continents (with Southeast Asia taking a lead [62]) and countries, as is the case in Europe [36, 63]. The interregional differences detected through smart city benchmarking exercises clearly point at place-specific phenomena, particularly in technologically friendly and open-minded populations [64, 65].

At a city–regional level, societal attitudes are also assumed to vary according to different process fields (right column in Table 1.2). In the fields of smart security and smart governance in particular, large parts of the urban population may mistrust ICT means of control. Many people will not like an all-embracing, centrally controlled CPS infrastructure of sensors, monitors, and hovering UAVs, which in fact form the backbone of many smart city dimensions. Equally smartly constructed houses and intelligent settlements could heighten inhabitants' suspicions about increased monitoring and automatic self-regulation, or smart buildings could simply expect too much of the technical skills of their inhabitants. In the smart industry field, fears that CPS implementation leads to rationalization and abolition of jobs could cause opposition in large sectors of society, institutionalized and amplified through trade unions. The new options created by urban production, CPS manufacture, and maintenance selectively benefit a small group of people with specific skills.

The field of smart education in turn plays a key role for promoting CPS-enhanced smart city transformations. It is unequivocally associated with affirmative societal attitudes and allows for the inclusion of all social groups. It also generally conveys broader acceptance for essential urban sustainability objectives, like those pursued in the smart consumption field and generally profits from being driven from the bottom up by the people themselves. Smart education also provides opportunities for achieving crucial ICT- and CPS-related qualifications that promote other process fields, most notably the technology-driven entrepreneurship of a smart economy, which substantially enriches regional industry structure. The fields of smart energy and mobility, which already offer socially acknowledged sustainability benefits, can also profit from a highly educated population in well-paid jobs. Smart cities need people who cherish the high societal value of a good environment in order to introduce CPS-coordinated energy and mobility solutions, which may be more expensive and make life less comfortable than previous non-smart components. The positive influences of education may over time also affect all other process fields that initially saw social opposition, such as smart infrastructure, construction, and governance. Finally, slightly more positive social attitudes can be expected in the smart metabolism field, which, except for regional cycles of smart consumption, will not visibly affect people's daily life.

1.5 Conclusions: Suggestions for Planning the CPS-Driven Smart City

This chapter explores how visions of smart city developments could potentially profit from using CPS applications in order to enhance system efficiency. Eleven process fields are identified, in which intelligent hardware- and software-based monitoring, control, and sensing tools can substantially improve system coordination and self-regulation, thus supporting the creation of the sustainable, environmentally friendly city of tomorrow. CPS draws on modern ICT to facilitate the alignment and communication of different devices, vehicles, buildings, companies, and people. Geographical information systems, the Internet, and wireless smartphone technology offer ample potential for the development of smart urban solutions. Besides taking up dimensions previously associated with the smart city notion, such as enhanced ICT infrastructure, energy, mobility, security, and governance [1, 3], this chapter highlights further industry-related process fields that have previously been overlooked. This includes the fact that CPS can improve resource-efficient material flow coordination within and between value chains, revitalize sites of urban production, and expand entrepreneurial opportunities.

Furthermore, the chapter explains how the process fields of CPS-enhanced smart city formation are embedded in wider economic contexts on a regional, national, and global scale, and it also raises issues of social acceptability. It is crucial to bring up more skeptical considerations in order to counterweigh the often overly optimistic attitudes expressed in most smart city literature [11, 12, 15]. Works mainly depict diverse technological possibilities and close their eyes to limited human capabilities and the level of compliance necessary for the utilization of these new technologies. Employing a critical perspective to socioeconomic factors shows that some smart city process fields may be much easier to implement regionally than others, whose introduction, for example, may face societal barriers. Policies that intend to support the CPS-driven smart city should take these insights into account and adjust activities to the differentiated set of conditions in order to better achieve their goals. Several cases have already shown how important flexible policies and institutional constellations are for installing smart city elements in urban spaces [2, 42, 66].

It seems advisable to start by promoting initiatives through which modern ICT applications directly and openly improve people's working and living conditions without enacting too much regulation or control. Smart education performs a key role as it conveys qualifications, job opportunities, and attitudes that support CPS integration in all other smart city fields. A major objective must be to diminish digital divides between the various groups that compose urban society. In line with impulses through ICT- and CPS-oriented (spin-off) entrepreneurship and intelligent urban production, smart education can set into motion a functional, self-reinforcing smart city innovation ecosystem [67]. Citizen community activities and grassroots movements aiming to bring about sustainable living through smart consumption deserve dedicated technological support right from the start. Together, this setting helps overcome societal opposition and skepticism against the installation of comprehensive ICT infrastructure required for all process fields and importantly forms the backbone of a smart city. By promoting highly visible aspects likely to affect the citizens' quality of life, CPS directs monitoring and coordination tools to factors that are significant for improving the sustainability and resource efficiency of mobility and energy systems. This applies also to the provision of user-friendly e-government services and public information on urban environmental quality, which could be used to further motivate smart consumption. These foundations can then facilitate the growth and buildup of CPS-enhanced regulations and control in the system arenas of smart construction and material flow metabolism in eco-industrial parks. As citizens may for the most part oppose the widespread implementation of smart security surveillance and ubiquitous governance, related smart technologies should only be installed with the highest degree of sensitivity and through engagement with public communication channels.

Planning and implementing the various CPS-supported processes requires participants to bear in mind that decisive conditions of smart city development are set on the national, and sometimes global scale, rather than on the directives of city–regional authorities and actor groups. Recently, Angelidou [2] highlighted the related policy issues by discussing, among other things, the "national versus local strategy" choice. Moving on from that, smart city policies should fruitfully combine factors determined at different spatial levels, just as it seems viable to combine regional, national, and global assets in order to adequately promote innovation systems [61]. Initiatives clearly rooted in the locality and favored by proximity or cluster advantages, like those relating to smart consumption, economic entrepreneurship, and material flow, can expediently be connected with national initiatives. Often only national funding, incentives, and innovation efforts make sophisticated CPS solutions for renewable energy, mobility, or security systems possible, as is the case with related programs in higher education. Urban mobility and energy systems form part of a larger setting and need effective external connectivity. Yet, regional authorities should be ready to continuously learn about and locally adapt to these options, taking advantage of regional knowledge and networks that help customize ICT facilities and smart housing settlements. The technologies and skills that sustain the evolution of CPS-enhanced smart cities are definitely of a global nature [4, 20], which demands local authorities to closely follow international trends in all related fields. When promoting the fields of smart industries and economy locally, global competition should not be ignored.

In all these endeavors, actors must be patient and let processes emerge over time without forcing the urban population because

> the path to becoming a smart city is not a sprint. It is a marathon. Or, even an ultra-marathon. [...] Smart city officials realise that they need to begin by focusing on solving the most pressing issues facing their cities, but also need to have a longer term view of the end goal.
>
> (Taylor [68])

Only integrated planning that takes care of economic as well as social contexts and purposefully uses assets beyond a regional scale will create truly sustainable smart cities that make the best use of CPS applications.

Final Thoughts

Exploring technological and socioeconomic underpinnings of smart cities, this chapter provided a list of 11 process fields in which CPS, which enable intelligent hardware- and software-based monitoring, control, and regulation of urban activities, can improve system coordination and self-regulation. It was shown that CPS better connects devices, vehicles, buildings, companies, and people as well as improves resource-efficient material flow coordination, revitalizes urban production, and expands entrepreneurial opportunities. Employing perspectives of economic geography, the chapter pointed out the place-specific nature as well as interdependencies of regional, national, and global economic contexts for smart city development. It discussed issues of societal acceptance for CPS-enhanced, digitized urban activities and, finally, suggested how smart city planning and policies should deal with the socioeconomic challenges.

Questions

1   How can CPS be defined and what are the origins of this notion?

2   In which respects can CPS improve living and working in smart cities?

3   What qualifies a geographical perspective on technology-driven smart city development?

4   Which smart city process fields are most strongly influenced by national and global economic dynamics?

5   Why and in which ways should smart city policies account for societal barriers of CPS-enhanced smart city development?

References

1 Albino, V., Berardi, U., and Dangelico, R.M. (2015) Smart cities: Definitions, dimensions, performance, and initiatives. *Journal of Urban Technology*, **22** (1), 3–21.
2 Angelidou, M. (2014) Smart city policies: A spatial approach. *Cities*, **41** (suppl. 1), 3–11.
3 Angelidou, M. (2015) Smart cities: A conjuncture of four forces. *Cities*, **47**, 95–106.
4 Kourtit, K. and Nijkamp, P. (2012) Smart cities in the innovation age. *Innovation: The European Journal of Social Science Research*, **25** (2), 93–95.
5 Ferrero, F. and Vesco, A. (eds) (2015) *Handbook of research on social, economic, and environmental sustainability in the development of smart cities*, IGI Global, Hershey, PA.

6 Jara, A. (2015) Big data for smart cities with KNIME. *Software Practice & Experience*, **45**, 1145–1160.

7 Jin, J. (2014) Information framework for creating a smart city through Internet of Things. *IEEE Internet of Things Journal*, **1**, 112–121.

8 Townsend, A.M. (2013) *Smart Cities: Big Data, Civic Hackers, and the Quest for a New Utopia*, W.W. Norton & Company, New York.

9 Gunes, V. (2014) A survey on concepts, applications, and challenges in cyber-physical systems. *KSII Transactions on Internet and Information Systems*, **2014** (8), 4242–4268.

10 Monostori, L. (2014) Cyber-physical production systems: Roots, expectations and R&D challenges. *Procedia CIRP*, **17**, 9–13.

11 Matt, D.T., Spath, D., Braun, S. *et al.* (2014) Morgenstadt – urban production in the city of the future, in *Enabling Manufacturing Competitiveness and Economic Sustainability. Proceedings of the 5th International Conference on Changeable, Agile, Reconfigurable and Virtual Production, Munich 2013* (ed. M.F. Zaeh), Springer International Publishing, Berlin, pp. 13–16.

12 Allwinkle, S. and Cruickshank, P. (2011) Creating smarter cities: An overview. *Journal of Urban Technology*, **18** (2), 1–16.

13 Letaifa, S. (2015) How to strategize smart cities: Revealing the SMART model. *Journal of Business Research*, **68**, 1414–1419.

14 Neirotti, P., De Marco, A., Cagliano, A.C. *et al.* (2014) Current trends in smart city initiatives: Some stylised facts. *Cities*, **38**, 25–36.

15 Stratigea, A. (2015) Tools and technologies for planning the development of smart cities. *Journal of Urban Technology*, **22** (2), 43–62.

16 Hayter, R. and Patchell, J. (2011) *Economic Geography. An Institutional Approach*, Oxford Univ. Press, Oxford.

17 Coe, N.M., Kelly, P.F., and Yeung, H.W.C. (2013) *Economic Geography. A Contemporary Introduction*, 2nd edn, Wiley & Sons, Chichester.

18 MacKinnon, D. and Cumbers, A. (2011) *Introduction to Economic Geography. Globalization, Uneven Development and Place*, 2nd edn, Pearson and Prentice Hall, Harlow.

19 Komninos, N. (2002) *IntelligentCities: Innovation, Knowledge Systems and Digital Spaces*, Spon Press, London.

20 Komninos, N. (2011) Intelligent cities: Variable geometries of spatial intelligence. *Intelligent Buildings International*, **3** (3), 172–188.

21 Hammer, S., Kamal-Chaoui, L., Robert, A., and Plouin, M. (2011) Cities and green growth: a conceptual framework. OECD Regional Development Working Papers 08. Paris: OECD Publishing.

22 Sprenger, F. and Engemann, C. (eds) (2015) *Internet der Dinge. Über smarte Objekte, intelligente Umgebungen und die technische Durchdringung der Welt*, transcript, Bielefeld.

23 Herterich, M. (2015) The impact of cyber-physical systems on industrial services in manufacturing. *Procedia CIRP*, **30**, 323–328.

24 Kleissl, J. and Agarwal, Y. (2010) Cyber-physical energy systems: Focus on smart buildings, in *Proceedings of the 47th Design Automation Conference*, ACM, New York, pp. 749–754.

25 OCIA (2015) *The Future of Smart Cities – Cyber-Physical Infrastructure Risk*, Office of Cyber and Infrastructure Analysis, Washington, DC.

26 Kitchin, R. (2014) The real-time city? Big data and smart urbanism. *GeoJournal*, **79** (1), 1–14.

27 Gurgen, L., Gunalp, O., Benazzouz, Y., and Gallissot, M. (2013) Self-aware cyber-physical systems and applications in smart buildings and cities, in *Proceedings of the Conference on Design, Automation and Test in Europe*, San Jose, CA, EDA Consortium, pp. 1149–1154.

28 Baron, M. (2012) Do we need smart cities for resilience? *Journal of Economics & Management*, **10**, 32–46.

29 Lombardi, P., Giordano, S., Farouh, H., and Yousef, W. (2012) Modelling the smart city performance. *Innovation: The European Journal of Social Science Research*, **25** (2), 137–149.

30 Yovanof, G.S. and Hazapis, G.N. (2009) An architectural framework and enabling wireless technologies for digital cities & intelligent urban environments. *Wireless Personal Communications*, **49** (3), 445–463.

31 Broy, M., Cengarle, M.V., and Geisberger, E. (2012) Cyber-physical systems: Imminent challenges, in *Large-Scale Complex IT Systems. Development, Operation and Management* (eds R. Calinescu and D. Garlan), Springer, Berlin, pp. 1–28.

32 Komninos, N., Pallot, M., and Schaffers, H. (2013) Smart cities and the future Internet in Europe. *Journal of the Knowledge Economy*, **4** (2), 119–134.

33 Hancke, G.P. and Hancke, G.P. Jr. (2012) The role of advanced sensing in smart cities. *Sensors*, **13** (1), 393–425.

34 Mitton, N., Papavassiliou, S., Puliafito, A., and Trivedi, K.S. (2012) Combining cloud and sensors in a smart city environment. *EURASIP Journal on Wireless Communications and Networking*, **247**, 1–10.

35 Piro, G., Cianci, I., Grieco, L.A. *et al.* (2014) Information centric services in smart cities. *The Journal of Systems and Software*, **88**, 169–188.

36 Caragliu, A., Del Bo, C., and Nijkamp, P. (2011) Smart cities in Europe. *Journal of Urban Technology*, **18** (2), 65–82.

37 Shelton, T. (2015) The "actually existing smart city". *Cambridge Journal of Regions, Economy and Society*, **8** (1), 13–25.

38 Debnath, A.K., Chin, H.C., Haque, M.M., and Yuen, B. (2014) A methodological framework for benchmarking smart transport cities. *Cities*, **37**, 47–56.

39 Berardi, U. (2013) Clarifying the new interpretations of the concept of sustainable building. *Sustainable Cities and Society*, **8**, 72–78.

40 Ghaffarian Hoseini, A., Berardi, U., Dahlan, N. *et al.* (2013) The essence of future smart houses: From embedding ICT to adapting to sustainability principles. *Renewable & Sustainable Energy Reviews*, **24**, 593–607.

41 Rashidi, P. and Mihailidis, A. (2013) A survey on ambient-assisted living tools for older adults. *IEEE Journal of Biomedical and Health Informatics*, **17** (3), 579–590.

42 Mone, G. (2015) The new smart cities. *Communications of the ACM*, **58** (7), 19–21.

43 Boons, F.A. and Lambert, A.J.D. (2002) Eco-industrial parks: Stimulating sustainable development in mixed industrial parks. *Technovation*, **22**, 471–484.

44 Karl, U. (2004) *Regionales Stoffstrommanagement – Instrumente und Analysen zur Planung und Steuerung von Stoffströmen auf regionaler Ebene*, VDI Verlag, Düsseldorf.

45 Müller, B. and Schiappacasse, P. (2015) Advanced manufacturing – why the city matters, perspectives for international development cooperation, in *Industry 4.0 and Urban Development* (eds B. Müller and O. Herzog), acatech – National Academy of Science and Engineering, Munich, pp. 139–169.

46 Walravens, N. (2015) Qualitative indicators for smart city business models: The case of mobile services and applications. *Telecommunications Policy*, **39** (3–4), 218–240.

47 Tewdwr-Jones, M., Goddard, J., and Cowie, P. (2015) *Newcastle city futures 2065: Anchoring universities in cities through urban foresight*, Newcastle Institute for Social Renewal, Newcastle University, Newcastle.

48 Shapiro, J.M. (2006) Smart cities: Quality of life, productivity, and the growth effects of human capital. *Review of Economics & Statistics*, **88** (2), 324–335.

49 Thite, M. (2011) Smart cities: Implications of urban planning for human resource development. *Human Resource Development International*, **14** (5), 623–631.

50 Hatzelhoffer, L., Humboldt, K., Lobeck, M., and Wiegandt, C. (2012) *Smart City in Practice: Converting Innovative Ideas into Reality*, Jovis, Berlin.

51 Paskaleva, K.A. (2011) The smart city: A nexus for open innovation? *Intelligent Buildings International*, **3** (3), 153–171.

52 Leydesdorff, L. and Deakin, M. (2011) The triple-helix model of smart cities: A neo-evolutionary perspective. *Journal of Urban Technology*, **18** (2), 53–63.

53 Vanolo, A. (2014) Smartmentality: The smart city as disciplinary strategy. *Urban Studies*, **51** (5), 883–898.

54 Reddy, Y.B. (2015) Security and design challenges in cyber-physical systems, in *12th International Conference on Information Technology – New Generations*, IEEE, New York, pp. 200–205.

55 Lee, J.H., Phaal, R., and Lee, S. (2013) An integrated service-device-technology roadmap for smart city development. *Technological Forecasting and Social Change*, **80** (2), 286–306.

56 Huang, Z. (2015) Small cluster in CPS: Network topology, interdependence and cascading failures. *IEEE Transactions on Parallel and Distributed Systems*, **26** (8), 2340–2350.

57 Sousa Nunes, D., Zhang, P., and Silva, J.A. (2015) A survey on human-in-the-loop applications towards an Internet of all. *IEEE Communications Surveys & Tutorials*, **17** (2), 944–964.

58 Kalundborg Symbiosis (2015) http://www.symbiosis.dk/en (accessed 20 May 2016).

59 Porter, M. (2000) Locations, clusters and company strategy, in *The Oxford Handbook of Economic Geography* (eds G.L. Clark, M.P. Feldman, and M.S. Gertler), Oxford Univ. Press, Oxford, pp. 253–274.

60 Dewald, U. and Truffer, B. (2011) Market formation in technological innovation systems – diffusion of photovoltaic applications in Germany. *Industry and Innovation*, **18** (3), 285–300.

61 Fromhold-Eisebith, M. (2007) Bridging scales in innovation policies: How to link regional, national and international innovation systems. *European Planning Studies*, **15** (2), 217–233.

62 Thuzar, M. (2011) Urbanization in Southeast Asia: Developing smart cities for the future? *Regional Outlook*, **2011/2012**, 96–100.

63 Kourtit, K., Nijkamp, P., and Arribas, D. (2012) Smart cities in perspective – a comparative European study by means of self-organizing maps. *Innovation: The European Journal of Social Science Research*, **25** (2), 229–246.

64 O'Grady, M. and O'Hare, G. (2012) How smart is your city? *Science*, **335** (3), 1581–1582.

65 Winters, J.V. (2011) Why are smart cities growing? Who moves and who stays. *Journal of Regional Science*, **51** (2), 253–270.

66 Crivello, S. (2015) Urban policy mobilities: The case of Turin as a smart city. *European Planning Studies*, **23** (5), 909–921.

67 Zygiaris, S. (2013) Smart city reference model: Assisting planners to conceptualize the building of smart city innovation ecosystems. *Journal of the Knowledge Economy*, **4** (2), 217–231.

68 Taylor, S. (2015) *The Future of Smart Cities*. http://blogs.cisco.com/cle/the-future-of-smart-cities-2 (accessed 20 May 2016).

2

Big Data Analytics Processes and Platforms Facilitating Smart Cities

Pethuru Raj[1] and Sathish A. P. Kumar[2]

[1] Infrastructure Architect, IBM Global Cloud Center of Excellence, IBM India, Bangalore, India
[2] Department of Computing Sciences, Coastal Carolina University, Conway, SC, USA

CHAPTER MENU

Introduction, 24
Why Big Data Analytics (BDA) Is Significant for Smarter Cities, 24
Describing the Big Data Paradigm, 26
The Prominent Sources of Big Data, 27
Describing Big Data Analytics (BDA), 29
The Big Trends and Use Cases of Big Data Analytics, 31
The Open Data for Next-Generation Cities, 38
The Big Data Analytics (BDA) Platforms, 39
Big Data Analytics Frameworks and Infrastructure, 45
Summary, 50

Objectives

- Articulating how the distinct advancements in the field of big data analytics (BDA) is turning out to be highly beneficial for the establishment and sustenance of smarter cities
- There are big data platforms (open source and commercial grade) in order to speed up the process of bringing forth applications for searching, filtering, processing, mining, and analysis of big data. How these platforms come handy in conceptualizing and concretizing next-generation smarter cities applications
- How the actionable insights extracted out of city data heaps through the BDA platforms can be supplied to city administrators, consultants, service providers, city residents, and so on to take correct decisions
- Finally, how to bring in the right automation for various software applications and various devices by feeding the intelligence squeezed out of big data.

Smart Cities: Foundations, Principles and Applications, First Edition.
Edited by Houbing Song, Ravi Srinivasan, Tamim Sookoor, and Sabina Jeschke.
© 2017 John Wiley & Sons, Inc. Published 2017 by John Wiley & Sons, Inc.

2.1 Introduction

According to market analysts and researchers, the amount of different kinds of data getting captured and crunched is almost doubling every 2 years. Several technological advancements in the information technology (IT) space are being given as the principal reason for such a monumental growth of data across the globe. Data sources, scopes, structures, speeds, and sizes are invariably on the rise. The device ecosystem (personal as well as professional) is rapidly growing. In addition, all kinds of physical, mechanical, electrical, and electronic systems are being fundamentally and functionally enabled to do bigger and better things through a seamless and spontaneous integration with local as well as remotely held cyber–virtual–cloud applications, services, and data stores. With cloud environments emerging as the core and central IT infrastructure for hosting, delivering, and managing large and varied datasets, the data volume is going to be steadily and significantly growing.

Furthermore, every noticeable event, transaction, operation, interaction, collaboration, request and reply, and so on are being captured and persisted in storage appliances and arrays for real-time and posterior investigations. The digitized objects (alternatively smart and sentient materials) are capable of getting interconnected with one another in the vicinity and with remote ones. The *ad hoc* networks being formed out of these instrumented, interconnected, and intelligent entities and elements spit out a large amount of multi-structured data. In a nutshell, the realization of extremely connected physical objects is in our midst of grand foundation for big data, which is typically multi-structured, massive in volume, and mesmerizing in variety, velocity, and value. Big data storage solutions are prevalent these days due to technology breakthroughs. The most important activity on big data is to do synchronized and systematic analytics to correctly and readily emit big insights. BDA frameworks primarily comprise data ingestion, processing, mining, investigating and storage modules, toolsets, connectors, drivers, and adaptors that are made available by open-source and commercial-grade solution vendors. Due to extreme complexity induced by multiplicity and heterogeneity of big data derived from smart cities, BDA products, platforms, patterns, practices, and processes need to be derived and realized to do big data analysis easily and quickly.

2.2 Why Big Data Analytics (BDA) Is Significant for Smarter Cities

As widely experienced, data is the core of information and knowledge that can be wisely used on bigger and better things. For the knowledge era, data is being carefully collected, cleansed, classified, clustered, and conformed to simplifying and streamlining process toward its final destination. Big data storage solutions are feverishly prevalent these days. The most respectable activity on big data is to do synchronized and systematic analytics to correctly and readily emit big insights. BDA frameworks primarily comprising data processing and storage modules, toolsets, connectors, drivers, and adaptors are made available by open-source and commercial-grade solution vendors. Due to the extreme complexity induced by multiplicity and heterogeneity of big data, enabling BDA products, platforms, patterns, practices, and processes are being derived and released by IT professionals to do big data analysis easily and quickly.

Large cities and megacities are beset with scores of problems in housing, infrastructures, safety and security, transport, energy, communication, water, quality of life, and so on. The receding and recessionary economy put more stress and strictures on our declining and deteriorating cities. At the same time, cities also open up fresh possibilities and opportunities for thinkers and practitioners to contemplate and activate different things differently. Our cities need to change their structure and behavior significantly in order to be cogently fit with the distinctly identified ideals of the smarter world, a next-generation idea or vision being proclaimed and pursued vigorously and rigorously by leading IT infrastructure and product companies these days. This incredible notion of the smarter world is being presented as the next logical move by worldwide technology creators and service providers to be relevant in their long and arduous journey. There are several key drivers and decisive trends for the surging popularity of this game-changing concept. There is a series of enabling developments and advancements in realization technologies being unfolded in order to simplify and streamline the hitherto unknown path toward the desired and delectable transformation. The BDA discipline is in fast track and its contributions are leveraged in order to design and develop smarter cities. In this chapter, we discuss how BDA serves immeasurably and immaculately for the faster realization and sustenance of next-generation cities.

Our cities are becoming more complex these days, and hence integrated, insightful, and intelligent IT systems need to be in place to anticipatively monitor and manage the intricacies and intimacies of world cities. Today IT is penetrating into all kinds of industry verticals. For example, modern airlines are activated and automated through a host of IT systems. Similarly, as the complications and convulsions of today's cities are on the rise, the smart leverage of all the delectable advancements of IT has to be ensured elegantly [1]:

- Realizing intelligent operations is to originate from the boundaryless and ubiquitous access to and flow of data across multiple sources. There are tools to collect, correlate, and corroborate data accurately, analyze it rapidly, and see the resulting information visually anytime and anyplace to enable taking informed decisions and make agencies more nimble, transparent, and adaptive.
- All the poly-structured data getting gleaned are not value adding. There are repetitive and redundant data. City-specific data systems must understand

the difference between significant and non-significant data in their specific contexts, and hence data management platforms, practices, processes, and patterns are mandatory to attain the desired success.

- Master data management (MDM), city performance management (CPM), and BDA platforms, data virtualization and visualization tools, and predictive and prescriptive analytics capabilities can layer on top to deliver intelligent operations.

Thus, it is clear that IT, especially data analytics, along with a flexible and futuristic strategy is to play a very critical role in shaping up our sliding and sagging cities.

2.3 Describing the Big Data Paradigm

When the data size is becoming massive, there arise several business and technical challenges. Hence there is a formalism formulated and a discipline created in the IT circle to ponder about the ways and means of capturing, stocking, and subjecting the data for various purposes. Typically there are four important characteristics that are to define and defend the ensuing era of big paradigm:

- Volume – As indicated above, machine-generated data is growing exponentially in size compared with man-generated data volume. For instance, digital cameras produce a high-volume image and video files to be shipped, succinctly stored, and subjected to a wider variety of tasks for different reasons including video-based security surveillance. Research labs such as CERN generate massive data, avionics and automotive electronics too generate a lot of data, and smart energy meters and heavy industrial equipment like oil refineries and drilling rigs generate huge data volumes.
- Velocity – These days social networking and microblogging sites create a large amount of information. Though the size of information created and shared is comparatively small here, the number of users is huge, and hence the frequency is on the higher side, resulting in a massive data collection. Even at 140 characters per tweet, the high velocity of Twitter data ensures large volumes (over 8 TB per day).
- Variety – Newer data formats are arriving compounding the problem further. As enterprise IT is continuously strengthened with the incorporation of nimbler embedded systems and versatile cloud services to produce and provide premium and people-centric applications to the expanding user community, new data types and formats are evolving.
- Value – Data is an asset, and it has to be purposefully and passionately processed, prioritized, protected, mined, and analyzed, utilizing advanced technologies and tools in order to bring out the hidden knowledge that enables individuals and institutions to contemplate and carry forward the future course of actions correctly.

There are other characteristics such as data veracity, variability, and viscosity. Big data computing is the IT part of extracting and emitting actionable insights out of big data. This fast-emerging computing model is all about the technological solutions and their contributions in tackling the positive and the negative factors of big data.

2.4 The Prominent Sources of Big Data

As we discussed above, big data represents huge volumes of data in petabytes, exabytes, and zettabytes in the near future. As we move around the globe, we leave a trail of data behind us. B2C and C2C e-commerce systems and B2B e-business transactions, online ticketing and payments, web 1.0 (simple web), web 2.0 (social web), web 3.0 (semantic web), web 4.0 (smart web), still and dynamic images, and so on are the prominent and dominant sources for data. Sensors and actuators are deployed in plenty in specific environments for security and for enabling the occupants and owners of the environments to be smart. In short, every kind of integration, interaction, orchestration, collaboration, automation, and operation produces streams of decision-enabling data to be plucked and put into transactional and then into analytical data stores. As the world and every tangible in it are connected purposefully, the data generation sources and resources are bound to grow ceaselessly, resulting in heaps and hordes of data.

We have discussed the fundamental and fulsome changes happening in the IT and business domains. The growing aspect of service enablement of applications, platforms, infrastructures (servers, storages, and network solutions), and even everyday devices besides the varying yet versatile connectivity methods has laid down strong and simulating foundations for big interactions, transactions, automation, and insights. The tremendous rise in the data collection along with all the complications has instinctively captivated both business and IT leaders and luminaries to act accordingly and adeptly to take care of this huge impending and data-driven opportunity for governments, corporates, and organizations. This is the beginning of the much-discussed and discoursed big data computing discipline.

2.4.1   The Salient Implications of Big Data

This paradigm is getting formalized with the deeper and decisive collaboration among product vendors, service organizations, independent software vendors, system integrators, and research organizations. Having understood the strategic significance, all the different stakeholders have come together in complete unison in creating and sustaining simplifying and streamlining techniques, platforms and infrastructures, integrated processes, best practices, design patterns, and key metrics to make this new discipline pervasive and persuasive. Today the acceptance and activation levels of big data computing are consistently on the climb. However it is bound to raise a number of critical challenges, but at the same time, it is to be highly impactful and insightful for business organizations to confidently traverse in the right route if it is taken seriously. The continuous unearthing of integrated platforms is a good indication for the bright days ahead for the shining and strategic big data phenomenon.

The implications of big data are vast and varied. The principal activity is to do a variety of tool-based and mathematically sound analyses of big data for instantaneously gaining bigger insights. It is a well-known fact that any organization having the innate ability to swiftly and succinctly leverage the accumulating data assets is bound to be successful in what they are operating, providing, and aspiring. That is, besides instinctive decisions, informed decisions go a long way in shaping up and confidently steering organizations. Thus, just gathering data is no more useful, but IT-enabled extraction and squeezing of actionable insights in time out of those data assets serves well for the betterment of world businesses. Analytics is the formal discipline in IT for methodically doing data collection, filtering, cleaning, translation, storage, representation, processing, mining, and analysis with the aim of extracting useful and usable intelligence. BDA are the newly coined word for accomplishing various sorts of analytical operations on big data. With this renewed focus, BDA is getting more market and mind shares across the world. With a string of new capabilities and competencies being accrued out of this recent and riveting innovation, worldwide corporates are keenly jumping into the BDA bandwagon with all the optimism. This chapter is designed for demystifying the hidden niceties and ingenuities of the raging BDA.

2.4.2 Information and Communication Infrastructures for Big Data and Its Platforms

Big data is the general term used to represent massive amounts of data that are not stored in the relational form in traditional enterprise-scale databases. New-generation database systems based on **symmetric** multiprocessing (SMP) and **massive** parallel **processing (MPP) techniques**. These are being framed in order to store, aggregate, filter, mine, and analyze big data efficiently. The following are the general characteristics of big data:

- Data storage is defined in the order of petabytes, exabytes, and so on in volume to the current storage limits (gigabytes and terabytes).
- There can be multiple structures (structured, semi-structured, and less structured) for big data.
- Multiple types of data sources (sensors, machines, mobiles, social sites, etc.) and resources for big data.

- Data is time sensitive (near real time as well as real time). That means big data consists of data collected with relevance to the time zones so that time-sensitive insights can be extracted.

Thus, big data has created a number of rightful repercussions for businesses to give a prominent place for big data in their evolving IT strategy in order to be competitive in their dealings and decisions.

2.4.3 Transitioning from Big Data to Big Insights

The main mandate of IT is to capture, store, and process a large amount of data to output useful information in a preferred and pleasing format. With continued advancements in IT, lately, there arose a stream of competent technologies to derive usable and reusable knowledge from the expanding information base. The much-wanted transition from data to information and to knowledge has been simplified through the meticulous leverage of those IT solutions. Thus, data have been the main source of value creation for the past five decades. Now with the eruption of big data and the enabling platforms, corporates and consumers are eyeing and yearning for better and bigger value derivation. Indeed, the deeper research in the big data domain breeds a litany of innovations to realize robust and resilient productivity-enhancing methods and models for sustaining business value. The hidden treasures in big data troves are being technologically exploited to the fullest extent in order to zoom ahead of the competitions. The big data-inspired technology clusters facilitate the newer business acceleration and augmentation mechanisms. In a nutshell, the scale and scope of big data are to bring in big shifts. The proliferation of social networks and multifaceted devices and the unprecedented advancements in connectivity technologies have laid a strong and stimulating foundation for big data. There are several market analysts and research reports coming out with positive indications that bright days are ahead for BDA.

2.5 Describing Big Data Analytics (BDA)

Without an iota of doubt, the world's iconic cities have outlived empires, man-made annihilations, natural disasters, and enigmatic civilizations successfully and are continuously evolving to absorb all kinds of changes wrought in by a bevy of factors such as environmental shifts, population growth, business enterprising and industry clusters, technological advances, infrastructure impacts, and people's needs [1]. As a result, cities have become the most congested and complicated systems. Multiple distributed systems are intertwined together within any growing city environments, and hence, cities are being aptly touted as systems of systems. As diverse systems including millions of smart sensors and actuators interact with one another, the first and foremost implication is well known. That is, data in different sizes, structures, speeds, and scopes are being emitted in massive volumes. For drawing viable and value-adding insights for designing, developing, and deploying next-generation city systems and services for people, the captured and stocked data has to go through a series of specific processing, mining, and analyzing. Considering the hugeness of data getting gleaned, the traditional information management systems are often found inadequate and obsolete.

The timely arrival and the overwhelming acceptance of the Hadoop frameworks [2] are seen as a boon for deriving actionable insights that can empower transformational leaders by reducing complexity and enable the concerned to make informed decisions. Data is turning out to be an asset for city management. Data to information and to knowledge has become a challenge. Today every country head, county governor, metro mayor, city manager, or agency director is harnessing existing information to transform the city management. For example, in most cities, ambulances are stationed at a single and central location even though data often suggests that ambulances parked at specific locations around the city based on predicted events and historical needs would be able to respond to emergencies more quickly. To realize their vast economic, social, and cultural potential, cities clearly need to become substantially IT enabled. Especially BDA through its prime ability to ingest and crunch big data and to emit pragmatic insights in time is the need of the hour for cities to zoom ahead toward deftly and decisively fulfilling the evolving aspirations of city dwellers.

This recent entrant of BDA into the continuously expanding technology landscape has generated a lot of interest among industry professionals as well as academicians. Big data has become an unavoidable trend and it has to be solidly and succinctly handled in order to derive time-sensitive and actionable insights. There is a dazzling array of tools, techniques, and tips evolving in order to quickly capture data from diverse distributed resources and process, analyze, and mine the data to extract actionable business insights to bring in technology-sponsored business transformation and sustenance. In short, analytics is the thriving phenomenon in every sphere and segment today. Especially with the automated capture, persistence, and processing of the tremendous amount of multi-structured data getting generated by men as well as machines, the analytical value, scope, and power of data are bound to blossom further in the days to unfold.

Precisely speaking, data is a strategic asset for organizations to insightfully plan to sharply enhance their capabilities and competencies and to embark on the appropriate activities that decisively and drastically power up their short- and long-term offerings, outputs, and outlooks. Business innovations can happen in plenty and be sustained too when there is a seamless and spontaneous connectivity between data-driven and analytics-enabled business insights and business processes.

In the recent past, real-time analytics has gained much prominence, and several product vendors have been flooding the market with a number of elastic and state-of-the-art solutions (software as well as hardware) for facilitating on-demand, *ad hoc*, real-time, and runtime analysis of batch, online transaction, social, machine, operational, and streaming data. There are a number of advancements in this field due to its huge potentials for worldwide companies in considerably reducing operational expenditures while gaining operational insights. Hadoop-based analytical products are capable of processing and analyzing any data type and quantity across hundreds of commodity server clusters. Stream computing drives continuous and cognitive analysis of massive volumes of streaming data with sub-millisecond response times.

There are enterprise data warehouses, analytical platforms, in-memory appliances, and so on. Data warehousing (DW) delivers deep operational insights with advanced in-database analytics. The EMC Greenplum Data Computing Appliance (DCA) is an integrated analytics platform that accelerates the analysis of big data assets within a single integrated appliance. IBM PureData System [3] for Analytics architecturally integrates database, server, and storage into a single, purpose-built, easy-to-manage system. Then SAP HANA is an exemplary platform for efficient BDA. Platform vendors are conveniently tied up with infrastructure vendors especially cloud service providers (CSPs) to take analytics to the cloud so that the goal of analytics as a service (AaaS) sees a neat and nice reality sooner than later. There are multiple start-ups with innovative product offerings to speed up and simplify the complex part of big data analysis.

2.6 The Big Trends and Use Cases of Big Data Analytics

The future of business definitely belongs to those enterprises that swiftly embrace the BDA movement and use it strategically to their own advantages. It is pointed out that business leaders and other decision-makers, who are smart enough to adopt a flexible and futuristic big data strategy, can take their businesses toward greater heights. Successful companies are already extending the value of classic and conventional analytics by integrating cutting-edge big data technologies and outsmarting their competitors. There are several forecasts, exhortations, expositions, and trends on the discipline of BDA. Market research and analyst groups have come out with positive reports and briefings, detailing its key drivers and differentiators, the future of this brewing idea, its market value, the revenue potentials and application domains, the fresh avenues and areas for renewed focus, the needs for its sustainability, and so on. Here come the top trends emanating from this field.

Enterprises can understand and gain the value of BDA based on the number of value-added use cases and how some of the hitherto hard-to-solve problems can be easily tackled with the help of BDA technologies and tools. Every enterprise is mandated to grow with the help of analytics. As elucidated before, with big data, big analytics is the norm for businesses to take informed decisions. Several domains are eagerly enhancing their IT capability to have embedded analytics, and there are several reports eulogizing the elegance of BDA. The following are some of the prominent use cases.

2.6.1 Customer Satisfaction Analysis

This is the prime problem for most of the product organizations across the globe. There is no foolproof mechanism in place to understand the customers' feelings and feedbacks about their products. Gauging the feeling of people correctly and quickly goes a long way for enterprises to ring in proper rectifications and recommendations in product design, development, servicing, and support, and this has been a vital task for any product manufacturer to be relevant for their customers and product consumers. Thus, customers' reviews regarding the product quality need to be carefully collected through various internal as well as external sources such as channel partners, distributors, sales and service professionals, retailers, and in the recent past, through social sites, microblogs, surveys, and so on. However, the issue is that the data being gleaned is extremely unstructured, repetitive, unfiltered, and unprocessed. Extraction of actionable insights becomes a difficult affair here and hence leveraging BDA for a single view of customers (SVoC) will help enterprises gain sufficient insights into the much-needed customer mindset and to solve their problems effectively and to avoid them in their new product lines.

2.6.2 Market Sentiment Analysis

In today's competitive and knowledge-driven market economy, business executives and decision-makers need to gauge the market environment deeply to be successful in their dreams, decisions, and deeds. What are the products shining in the market, where is the market heading, who are the real competitors, what are their top-selling products, how are they doing in the market, what are the bright spots and prospects, and what are customers' preferences in the short- and long-term perspective through a deeper analysis legally and ethically? This information is available in a variety of websites, social media sites, and other public domains. BDA on this data can provide an organization with the much-needed information about strengths, weaknesses, opportunities, and threats (SWOT) for their product lines.

2.6.3 Epidemic Analysis

Epidemics and seasonal diseases like flu start and spread with certain noticeable patterns among people, and so it is pertinent to extract the hidden information to put a timely arrest on the outbreak of the infection. It is all about capturing all types of data originating from different sources, subjecting them to a series of investigations to extract actionable insights quickly and contemplating the appropriate countermeasures. There is a news item that says how spying on people data can actually help medical professionals to save lives. Data can be gathered from many different sources, but few are as superior as Twitter, and tools such as twitterhose facilitate this data collection, allowing anyone to download 1% of tweets made during a specified hour at random, giving researchers a nice cross section of the twitterverse. Researchers at Johns Hopkins University have been taking advantage of this tool, downloading tweets at random and sifting through this data to flag any and all mentions of flu or cold-like symptoms. Because the tweets are geo-tagged, the researchers can then figure out where the sickness reports are coming from, cross-referencing this with flu data from the Centers for Disease Control and Prevention to build up a picture of how the virus spreads and, more importantly, predict where it might spread to next.

In a similar line, with the leverage of the innumerable advancements being accomplished and articulated in the multifaceted discipline of BDA, myriad industry segments are jumping into the big data bandwagon in order to make themselves ready to acquire superior competencies and capabilities especially in anticipation, ideation, implementation, and improvisation of premium and path-breaking services and solutions for the world market. BDA brings forth fresh ways for businesses and governments to analyze a vast amount of unstructured data (streaming as well as stored) to be highly relevant to their customers and constituencies.

2.6.4 Using Big Data Analytics in Healthcare

The healthcare industry has been a late adopter of technology when compared with other industries such as banking, retail, and insurance. As per the trendsetting McKinsey report [4] on big data from June 2011, if US healthcare organizations could use big data creatively and effectively to drive efficiency and quality, the potential saving could be more than $300 billion every year.

- Patient Monitoring: Inpatient, Outpatient, Emergency Visits, and ICU – Everything is becoming digitized. With rapid progress in technology, sensors are embedded in weighing scales, blood glucose devices, wheelchairs, patient beds, X-ray machines, and so on. Digitized devices generate large streams of data in real time that can provide insights into patient's health and behavior. If this data is captured, it can be put to use to improve the accuracy of information and enable practitioners to better utilize limited provider resources. It will also significantly enhance patient experience at a healthcare facility by providing proactive risk monitoring, improved quality of care, and personalized attention. Big data can enable complex event processing (CEP) by providing real-time insights to doctors and nurses in the control room.

- Preventive Care for Accountable Care (ACO) – One of the key ACO goals is to provide preventive care. Disease identification and risk stratification will be very crucial to business function. Managing real-time feeds coming in from HIE, pharmacists, providers, and payers will deliver key information to apply risk stratification and predictive modeling techniques. In the past, companies were limited to historical claims and HRA/survey data but with HIE, the whole dynamic to data availability for health analytics has changed. Big data tools can significantly enhance the speed of processing and data mining.
- Epidemiology – Through HIE, most of the providers, payers, and pharmacists will be connected through networks in the near future. These networks will facilitate the sharing of data to better enable hospitals and health agencies to track disease outbreaks, patterns, and trends in health issues across a geographic region or across the world, allowing determination of source and containment plans.
- Patient Care Quality and Program Analysis – With the exponential growth of data and the need to gain insight from information comes the challenge to process the voluminous variety of information to produce metrics and key performance indicators (KPIs) that can improve patient care quality and Medicaid programs. Big data provides the architecture, tools, and techniques that will allow processing terabytes and petabytes of data to provide deep analytic capabilities to its stakeholders.

2.6.5 Analytics of Machine Data by Splunk

All your IT applications, platforms, and infrastructures generate data every millisecond of every day. The machine data is one of the fastest growing and most complex areas of big data. It is also one of the most valuable insights containing a definitive record of users' transactions, customer behavior, sensor activity, machine behavior, security threats, fraudulent activity, and more. Machine data hold critical insights useful across the enterprise:

- Monitor end-to-end transactions for online businesses providing 24×7 operations.
- Understand customer experience, behavior, and usage of services in real time.
- Fulfill internal SLAs and monitor service provider agreements.
- Identify spot trends and sentiment analysis on social platforms.
- Map and visualize threat scenario behavior patterns to improve security posture.

Making use of machine data is challenging. It is difficult to process and analyze by traditional data management methods or in a timely manner. Machine data is generated by a multitude of disparate sources, and hence, correlating meaningful events across these is complex. The data is unstructured and difficult to fit into a predefined schema. Machine data is high-volume and time-series based, requiring new approaches for management and analysis. The most valuable insights from this data are often needed in real time. Traditional business intelligence (BI), data warehouse, or IT analytics solutions are simply not engineered for this class of high-volume, dynamic, and unstructured data.

As indicated in the beginning, machine-generated data is more voluminous than man-generated data. Thus without an iota of doubt, machine data analytics is occupying a more significant portion in BDA. Machine data is being produced 24 × 7 × 365 by nearly every kind of software application and an electronic device. The applications, servers, network devices, storage and security appliances, sensors, browsers, compute machines, cameras, and various other systems deployed to support business operations are continuously generating information relating to their status and activities. Machine data can be found in a variety of formats such as application log files, call detail records, user profiles, KPIs, and clickstream data associated with user–web interactions, data files, system configuration files, alerts, and tickets. Machine data is generated by both machine-to-machine (M2M) and human-to-machine (H2M) interactions.

Outside of the traditional IT infrastructure, every processor-based system including HVAC controllers, smart meters, GPS devices, actuators and robots, manufacturing systems, and RFID tags and consumer-oriented systems such as medical instruments, personal gadgets and gizmos, aircraft, scientific experiments, and automobiles that contain embedded devices are continuously generating machine data. The list is constantly growing. Machine data can be structured or unstructured. The growth of machine data has accelerated in recent times with the trends in IT consumerization and industrialization. That is, the IT infrastructure complexity has gone up remarkably driven by the adoption of portable devices, virtual machines (VMs), bring your own devices (BYODs), and cloud-based services.

The goal here is to aggregate, parse, and visualize these data to spot trends and act accordingly. By monitoring and analyzing data emitted by a deluge of diverse, distributed, and decentralized data, there are opportunities galore. Someone wrote that sensors are the eyes and ears of future applications. Environmental monitoring sensors in remote and rough places bring forth the right and relevant knowledge about their operating environments in real time. Sensor data fusion leads to develop context and situation-aware applications. With machine data analytics in place, any kind of performance degradation of machines can be identified in real time, and corrective actions can be initiated with full knowledge and confidence. Security and surveillance cameras pump in still images and video data that in turn help analysts and security experts to preemptively stop any kind of undesirable intrusions. Firefighting can become smart with the utilization of machine data analytics.

The much-needed end-to-end visibility, analytics, and real-time intelligence across all of their applications, platforms, and IT infrastructures enables business enterprises to achieve required service levels, manage costs, mitigate security risks, demonstrate and maintain compliance, and gain new insights to drive better business decisions and actions. Machine data provides a definitive, time-stamped record of current and historical activity and events within and outside an organization, including application and system performance, user activity, system configuration changes, electronic transaction records, security alerts, error messages, and device locations. Machine data in a typical enterprise is generated in a multitude of formats and structures, as each software application or hardware device records and creates machine data associated with their specific use. Machine data also varies among vendors and even within the same vendor across product types, families, and models.

There are a number of newer use cases being formulated with the pioneering improvements in smart sensors, their *ad hoc* and purpose-specific network formation capability, data collection, consolidation, correlation, corroboration and dissemination, knowledge discovery, information visualization, and so on. Splunk is a low-profile big data company specializing in extracting actionable insights out of diverse, distributed, and decentralized data. Some real-world customer examples include:

- E-Commerce – A typical e-commerce site serving thousands of users a day will generate gigabytes of machine data, which can be used to provide significant insights into IT infrastructure and business operations. Expedia uses Splunk to avoid website outages by monitoring server and application health and performance. Today, around 3000 users at Expedia use Splunk to gain real-time visibility on tens of terabytes of unstructured, time-sensitive machine data (not only from their IT infrastructure but also from online bookings, deal analysis, and coupon use).
- Software as a Service (SaaS) – Salesforce.com uses Splunk to mine the large quantities of data generated from its entire technology stack. It has >500 users of Splunk dashboards from IT users monitoring customer experience to product managers performing analytics on services like "Chatter." With Splunk, SFDC claims to have taken application troubleshooting for 100,000 customers to the next level.
- Digital Publishing – NPR uses Splunk to gain insights of their digital asset infrastructure, to monitor and troubleshoot their end-to-end asset delivery infrastructure, to measure program popularity and views by the device, to reconcile royalty payments for digital rights, and to measure abandonment rates and more.

Figure 2.1 vividly illustrates how Splunk captures data from numerous sources and does the processing, filtering, mining, and analysis to generate actionable insights out of multi-structured machine data.

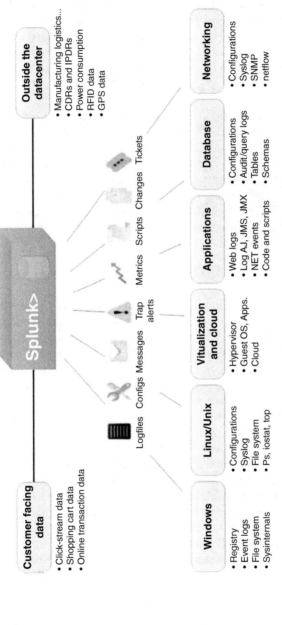

Figure 2.1 The Splunk reference architecture for machine data analytics.

Customer facing data
- Click-stream data
- Shopping cart data
- Online transaction data

Outside the datacenter
- Manufacturing logistics...
- CDRs and IPDRs
- Power consumption
- RFID data
- GPS data

Logfiles Configs Messages Metrics Scripts Changes Tickets
Trap alerts

Windows
- Registry
- Event logs
- File system
- Sysinternals

Linux/Unix
- Configurations
- Syslog
- File system
- Ps, iostat, top

Vitualization and cloud
- Hypervisor
- Guest OS, Apps.
- Cloud

Applications
- Web logs
- Log AJ, JMS, JMX
- NET events
- Code and scripts

Database
- Configurations
- Audit/query logs
- Tables
- Schemas

Networking
- Configurations
- Syslog
- SNMP
- netflow

Splunk Enterprise is the leading platform for collecting, analyzing, and visualizing machine data. It provides a unified way to organize and extract real-time insights from massive amounts of machine data from virtually any source. This includes data from websites, business applications, social media platforms, application servers, hypervisors, sensors, and traditional databases. Once your data is in Splunk, you can search, monitor, report, and analyze it, no matter how unstructured, large, or diverse it may be. Splunk software gives you a real-time understanding of what is happening and a deep analysis of what has happened, driving new levels of visibility and insight. This is called operational intelligence.

Most organizations maintain a diverse set of data stores – machine data, relational data, and other unstructured data. Splunk DB Connect delivers real-time connectivity to one or many relational databases and Splunk. Hadoop Connect delivers bidirectional connectivity to Hadoop. Both Splunk apps enable you to drive more meaningful insights from all of your data. The Splunk app for HadoopOps provides real-time monitoring and analysis of the health and performance of the end-to-end Hadoop environment, encompassing all layers of the supporting infrastructure.

2.7 The Open Data for Next-Generation Cities

Peter Hinssen [5] writes about the power of open data in establishing and sustaining smart cities. Cities are producing a lot of data. However, the data have to be open to be accessed and subjected to a series of deeper investigations to squeeze pragmatic knowledge. Cities of today are the true magnets, invigoratingly attracting people from everywhere to consciously explore and enjoy a variety of exciting opportunities. But cities are severely groaning under the weight of their incessant expansions and extensions. The city roads and streets are clogging up with congestion and traffic snarls, the city inhabitants consume electricity faster than utilities could produce, crime is increasingly difficult to anticipate and control, people are dissatisfied with civil services due to the fast life patterns, and so on. But the answer is nigh. City users with all their smartphones, cars, social interactions, houses, offices, energy consumption, online transactions, professional assignments, engagements, and so on leave behind a sumptuous amount of data. Not only people but also all the enabling machines, handhelds and digital assistants, street security and surveillance cameras, RFID readers, in-car devices, smart meters, and so on produce a lot of data every second. Now any responsible city has to make sense out of this fast-growing data treasure to comprehensively meet up the varying expectations. The insights extracted help administrators to chalk out viable and value-adding strategy and associated plans toward the betterment of cities and their inhabitants. The data-driven insights bring a lot of confidence and clarity toward effective and futuristic city planning. For example, instead of building more power plants, it is more judicious and analytical to have smart grids and smart meters in place to tackle rising energy requirements of cities. Enterprise resource planning (ERP)-kind of software solutions for cities (city resource planning (CRP)) can be a better option for future cities.

The author insists that all the aggregated data have to be open to being leveraged by different entities and stakeholders (government officials, non-government organizations (NGOs), individuals, etc.). This openness facilitates the realization of right intelligence in time, and the derived insights empower different service providers (communication, healthcare, energy, government, e-business, IT, etc.) to be proactive, preemptive, and prompt in all their obligations to their constituents. Smart cities are inherently capable of optimized usage of all kinds of city resources through the smart leverage of highly proven IT tools, platforms, and infrastructures well-connected (wired as well as wireless), purposefully collaborative, transparently sharing, elegantly sensitive, and responsive to city dwellers and residents.

2.8   The Big Data Analytics (BDA) Platforms

Integrated platforms are essential in order to automate several tasks enshrined in the data capture, analysis, and knowledge discovery processes. A converged platform comes out with a reliable workbench to empower developers to facilitate application development and other related tasks such as data security, virtualization, integration, visualization, and dissemination. Special consoles are being attached to new-generation platforms for performing other important activities such as management, governance, enhancement, and so on. Hadoop is a disruptive technology for data distribution among hundreds of commodity compute machines for parallel data crunching and any typical big data platform is blessed with Hadoop software suite.

Furthermore, the big data platform enables entrepreneurs; investors; chief executives; information, operation, knowledge, and technology officers (CXOs); and marketing and sales people to explore and perform experiments on big data, at the scale at a fraction of the time and cost, required previously. That is, platforms are to bestow all kinds of stakeholders and end users with actionable insights that in turn lead to consider and take informed decisions in time. Knowledge workers such as business analysts and data scientists could be the other main beneficiaries through these empowered platforms. Knowledge discovery is an important portion here, and the platform has to be chipped in with real-time and real-world tips, associations, patterns, trends, risks, alerts, and opportunities. In-memory and in-database analytics are gaining momentum for high-performance and real-time analytics. New advancements in the form of predictive and prescriptive analytics are emerging fast with the maturity and stability of big data technologies, platforms, infrastructures, tools, and finally a cornucopia of sophisticated data mining and analysis algorithms. Thus, platforms need to be fitted with new features, functionalities, and facilities in order to provide next-generation insights.

There is no doubt that consolidated and compact platforms accomplish a number of essential actions toward simplified big data analysis and knowledge discovery. However, they need to run in optimal, dynamic, and converged infrastructures to be effective in their operations. In the recent past, IT infrastructures went through a host of transformations such as optimization, rationalization, and simplification. The cloud idea has captured the attention of infrastructure specialists these days as the cloud paradigm is being proclaimed as the most pragmatic approach to achieving the ideals of infrastructure optimization. Hence with the surging popularity of cloud computing, every kind of IT infrastructure (servers, storages, and network solutions) is being consciously subjected to a series of modernization tasks to empower them to be policy based, software defined, cloud compliant, service oriented, networkable, programmable, and so on. That is, BDA is to be performed in centralized/federated, virtualized, automated, shared, and optimized cloud infrastructures (private, public, or hybrid). Application-specific IT environments are being readied for the big data era. Application-aware networks are the most sought-after communication infrastructures for big data transmission and processing. Figure 2.2 vividly illustrates all the relevant and resourceful components for simplifying and streamlining BDA.

As with DW, data marts, and online stores, an infrastructure for big data too has some unique requirements. The ultimate goal here is to easily integrate big data with enterprise data to conduct deeper and influential analytics on the combined dataset. As per the White paper titled "*Oracle: Big Data for the Enterprise*" [6], there are three prominent requirements (data acquisition,

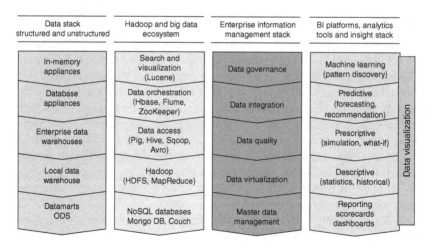

Figure 2.2 Big data analytics platforms, appliances, products, and tools.

organization, and analysis) for a typical big data infrastructure. NoSQL [7] has all these three intrinsically. You can find more on NoSQL databases (DBs) in the subsequent sections.

- Acquire Big Data – The infrastructure required to support the acquisition of big data must deliver low and predictable latency in both capturing data and in executing short and simple queries. It should be able to handle very high transaction volumes often in a distributed environment and also support flexible and dynamic data structures. NoSQL DBs are the leading infrastructure to acquire and store big data. NoSQL DBs are well suited for dynamic data structures and are highly scalable. The data stored in a NoSQL DB are typically of a high variety because the systems are intended to simply capture all kinds of data without categorizing and parsing the data. For example, NoSQL DBs are often used to collect and store social media data. While customer-facing applications frequently change, underlying storage structures are kept simple. Instead of designing a schema with relationships between entities, these simple structures often just contain a major key to identify the data point and then a content container holding the relevant data. This extremely simple and nimble structure allows changes to take place without any costly reorganization at the storage layer.
- Organize Big Data – In classical DW terms, organizing data is called data integration. Because there is such a huge volume of data, there is a tendency and trend gathering momentum to organize data at its original storage location. This saves a lot of time and money as there is no data movement. The brewing need is to have a robust infrastructure that is innately able to organize big data, process, and manipulate data in the original storage location. It has to support very high throughput (often in batch) to deal with large data processing steps and handle a large variety of data formats (unstructured, less structured, and fully structured).
- Analyze Big Data – The data analysis can also happen in a distributed environment. That is, data stored in diverse locations can be accessed from a data warehouse to accomplish the intended analysis. The appropriate infrastructure required for analyzing big data must be able to support deeper analytics such as statistical analysis and data mining on a wider variety of data types stored in diverse systems, to scale to extreme data volumes, to deliver faster response times driven by changes in behavior, and to automate decisions based on analytical models. Most importantly, the infrastructure must be able to integrate analysis of the combination of big data and traditional enterprise data to produce exemplary insights for fresh opportunities and possibilities. For example, analyzing inventory data from a smart vending machine in combination with the events calendar for the venue in which the vending machine is located will dictate the optimal product mix and replenishment schedule for the vending machine.

2.8.1 Civitas: The Smart City Middleware

The concept of the smart city has been drawing a lot of attention these days across the globe. Governments, research labs, product vendors, and service organizations are showing exemplary interest in collaborating with one another to make current cities more citizen friendly, efficient, and sustainable. As written elsewhere, several departments or divisions in a city environment need to team up together in order to bring in the perceptible changes. Therefore, the smart city paradigm can be summarized as a greatly complicated and distributed system. As we all experience, standard-compliant middleware solutions are being pressed into service for the seamless integration of diverse software applications and data sources in an IT environment. Here too, the role and relevance of highly competent middleware are on the climb for attaining the expected success in integrated cities.

City-specific applications are being developed, deployed in plenty, and delivered to users. Service middleware (service bus, hub, and fabric) in synchronization with data middleware (a growing collection of data adaptors) is prominent in the middleware space. Service delivery platform (SDP) is another important constituent in the service IT ecosystem inappropriately choosing and composing distributed services from multiple sources and delivering the people-centric, situation-aware, knowledge-filled, and cost-effective composite services dynamically to requesters. Thus, service discovery, integration, orchestration, management, and delivery aspects are classified as the IT portion of smart city establishment and sustenance strategy.

However, the traditional IT middleware might be found wanting for smart city projects because of the multiplicity- and heterogeneity-induced complexity. The authors [8] have designed a special middleware for integrating smart city services. This middleware provides services that range from environmental sensor deployment to the necessary hardware for high-performance algorithms devoted to extracting information from raw data. In order to cope with the multifaceted nature of the smart city paradigm, Civitas has been enhanced with reasoning capabilities. Leveraging reasoning capabilities enable the middleware to have few hard-coded features that are rather deduced from the available data. In this sense, Civitas is able to adapt to the deployed city, without requiring important modifications or adoption works. The main intention of this work is to promote tightly integrated systems and managed smart cities to simplify the IT environment for service developers. Figure 2.3 vividly illustrates the macro-level architecture of the middleware.

National governments across the globe, urban planners, and metro and city officials are keen on embarking on transitioning their cities into livable, lovable, green, and knowledge hubs for facilitating entrepreneurship. All along, IT has been a business enabler and there is a twist now. That is, IT is being prescribed and presented as the prospective enabler for our city systems also. Due to the inherent and growing heterogeneity, city-specific IT middleware solutions are

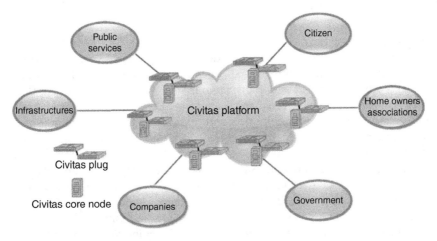

Figure 2.3 The reference architecture of Civitas platform.

to play a telling and transforming role in the days to unfurl. In this section, we discuss one such middleware and there will be many more to arrive and flourish considering the fact that there is a big market waiting for smart cities.

2.8.2 Hitachi Smart City Platform

The infrastructural requirements for our cities are constantly on the rise [9]. We have social infrastructures for services such as energy, water, and road. Now for the sustainability goals such as the arrest of greenhouse gas emissions into our fragile environment, existing social infrastructures need to be considerably extended and optimized. This involves the understanding patterns of usage or consumption and operationalizing appropriate decisions to ensure the different types of social infrastructure functioning efficiently. It is paramount therefore to collect and process a wider variety of operational and consumption data to extract the decision-enabling patterns. Because of the nature of this data, its frequency of collection, and the size of activity records, its quantity is expected to be large enough to justify the term "big data."

This smart city platform plays an important role in concise and clear understanding of the changing patterns of use and leveraging them precisely to ensure and enhance the efficiency of different types of social infrastructure. The dominant roles of the smart city platform are (i) data collection, (ii) data analysis, and (iii) coordination of the systems (applications) that activate, automate, and augment social infrastructures. There are incredible interrelationships between different social infrastructure systems to be captured and used. All the extracted knowledge needs to be at the center of attraction for implementing and sustaining next-generation operational applications for smart cities.

2.8.3 Data Collection

The smart city platform includes a database function. The details such as equipment performance and configuration data ought to be given the prime importance. Then due to the extra capability of M2M integration, the network topology data also plays a very indispensable role. Furthermore, the city platform also handles the collection and management of large quantities of other useful data such as the consumption and operational records, malfunctions, etc. For example, data on the supply of electric power is collected from sensors fitted on various instruments and equipment in power plants, transmission and distribution lines, and so on. Similarly, details of consumption are collected from sources such as smart meters installed in buildings, home energy management systems (HEMSs), building and energy management systems (BEMSs), and EV charging equipment. As the service area expands and with a number of users, the data getting generated and collected is going to be enormous.

The smart city platform also has a bus function that is used to collect control information. The bus buffers equipment control information sent from control applications and then forwards it to the destination device. The buffered data is also saved in the database for posterity. The in-memory data storage capability provides a zero-latency guarantee, meaning that the control information reaches the target device within the allocated time.

The smart city platform also collects journal data at intervals of between several seconds and several minutes, including data on equipment operation or alarms (notification of malfunction). This provides timely updates on whether equipment is operating normally. If malfunctioning, then the identification of the root cause is being facilitated through the journal data.

2.8.4 Data Analysis

The perpetual procedure is that data is systematically collected, cleaned, and cataloged for enabling appropriate analysis with the aim of producing actionable insights in time. There are several matured algorithms for data mining, machine learning, data interpolation, prediction, knowledge generation, and dissemination [10]. The data interpolation is all about interpolating the overall situation from collected sampling data. Electric power data, for example, may not be collected from all buildings. Instead, statistical analysis or other viable techniques could be leveraged to estimate the power consumption for the entire district.

Similarly, the collected data can be time-stamped and stored in a database. All kinds of changes in historical data can be readily analyzed to identify any usable trends. With more data, prediction accuracy is bound to go up significantly. The prediction capability helps to determine emerging and evolving trends in electric power or water use or for assessing conditions such as traffic congestion. The ultimate derivative of analytics is to extract knowledge to work on with all clarity and confidence. There are data visualization tools too in plenty these days to disseminate all the acquired knowledge in preferred formats.

2.8.5 Application Coordination

Data collected from social infrastructure and data obtained from the analysis are made available to applications to take account of the tactic as well as the strategic interrelationships between different aspects of the infrastructure including its control and usage pattern. This city-specific platform has the ability to empower application by sharing data among them.

One prominent example is the way in which greater use of EVs increases demands of electric power for charging vehicle batteries. Therefore by collecting all the information on EV use, it is made easier to determine factors such as where charging equipment has to be placed, at what times there is a higher demand for power, and so on. Similarly, interrelationships also exist between existing grid power and renewable energy. If power-usage information such as times of peak demand can be obtained, it is possible to determine the times when renewable energy will be used. Also, predictions about the deterioration of facilities can be made by coordinating information about the provision and use of electric power or EVs with enterprise asset management (EAM) systems. Thus, data capture and crush for information are going hand in hand. In the days ahead, there will be several smart city platforms from IT product vendors and software solution providers for transforming our clogged cities into smart and sustainable ones.

2.9 Big Data Analytics Frameworks and Infrastructure

There are majorly two types of big data processing: real-time and batch processing. The data is flowing endlessly from countless sources these days. Data sources are on the climb. Innumerable sensors, varying in size, scope, structure, smartness, and so on, are pouring data continuously. Stock markets are emitting a lot of data every second, and system logs are being received, stored, processed, analyzed, and acted upon ceaselessly. Monitoring agents are working tirelessly producing a lot of usable and useful data, business events are captured, knowledge discovery is initiated, information visualization is realized, and so on to empower enterprise operations. Stream computing is the latest paradigm being aptly prescribed as the best course of action for real-time receipt, processing, and analysis of online, live, and continuous data. Real-time data analysis through in-memory and in-database computing models is gaining a lot of ground these days with the sharp reduction in computer memory costs. For the second category of batch processing, the Hadoop technology is being recommended with confidence. It is clear that there is a need for competent products, platforms, and methods for efficiently and expectantly working with both real-time and batch data. There is a separate chapter for in-memory computing toward real-time data analysis and for producing timely and actionable insights.

In this section, you can read more about the Hadoop technology. As elucidated before, big data analysis is not a simple affair, and there are Hadoop-based software programming frameworks, platforms, and appliances emerging to tackle the innate complications. The Hadoop programming model has turned out to be the central and core method to propel the field of big data analysis. The Hadoop ecosystem is continuously spreading its wings wider, and enabling modules are being incorporated freshly to make Hadoop-based big data analysis simpler, more succinct, and quicker.

2.9.1 Apache Hadoop Software Framework

Apache Hadoop is an open-source framework that allows for the distributed processing of large datasets across clusters of computers using a simple programming model. Hadoop was originally designed to scale up from a single server to thousands of machines, each offering local computation and storage. Rather than rely on hardware to deliver high availability (HA), the Hadoop software library itself is designed to detect and handle failures at the application layer. Therefore, it delivers a highly available service on top of a cluster of cheap computers, each of which may be prone to failures. Hadoop is based out of the modular architecture and thereby any of its components can be swapped with competent alternatives if such a replacement brings noteworthy advantages.

Despite all the hubbub and hype around Hadoop, few IT professionals know its key drivers, differentiators, and killer applications. Because of the newness and complexity of Hadoop, there are several areas wherein confusion reigns and restrains its full-fledged assimilation and adoption. The Apache Hadoop product family includes the Hadoop Distributed File System (HDFS), MapReduce, Hive, HBase, Pig, ZooKeeper, Flume, Sqoop, Oozie, Hue, and so on. HDFS and MapReduce together constitute core Hadoop, which is the foundation for all Hadoop-based applications. For applications in BI, DW [11], and BDA, core Hadoop is usually augmented with Hive and HBase, and sometimes Pig. The Hadoop file system excels with big data that is file based, including files that contain nonstructured data. Hadoop is excellent for storing and searching multi-structured big data, but advanced analytics are possible only with certain combinations of Hadoop products, third-party products, or extensions of Hadoop technologies. The Hadoop family has its own query and database technologies, and these are similar to standard SQL and relational databases. That means BI/DW professionals can learn them quickly.

The HDFS is a distributed file system designed to run on clusters of commodity hardware. HDFS is highly fault tolerant because it automatically replicates file blocks across multiple machine nodes and is designed to be deployed on low-cost hardware. HDFS provides high throughput access to application data and is suitable for applications that have large datasets. As a file system, HDFS manages files that contain data. Because it is file based, HDFS itself does not offer random access to data and has limited metadata capabilities when compared with a DBMS. Likewise, HDFS is strongly batch oriented and hence has limited real-time data access functions. To overcome these challenges, you can layer HBase over HDFS to gain some of the mainstream DBMS capabilities. HBase is one of the many products from the Apache Hadoop product family. HBase is modeled after Google's Bigtable and hence HBase, like Bigtable excels with random and real-time access to very large tables containing billions of rows and millions of columns. Today HBase is limited to straightforward tables and records with little support for more complex data structures. The Hive meta-store gives Hadoop some DBMS-like metadata capabilities.

When HDFS and MapReduce are combined, Hadoop easily parses and indexes the full range of data types. Furthermore, as a distributed system, HDFS scales well and has a certain amount of fault tolerance based on data replication even when deployed on commodity hardware. For these reasons, HDFS and MapReduce can complement existing BI/DW systems that focus on structured and relational data. MapReduce is a general-purpose execution engine that works with a variety of storage technologies including HDFS, other file systems, and some DBMSs.

As an execution engine, MapReduce and its underlying data platform handle the complexities of network communication, parallel programming, and fault tolerance. In addition, MapReduce controls hand-coded programs and automatically provides multi-threading processes so they can execute in parallel for massive scalability. The controlled parallelization of MapReduce can apply to multiple types of distributed applications, not just analytic ones. In a nutshell, Hadoop MapReduce is a software programming framework for easily writing massively parallel applications that process massive amounts of data in parallel on large clusters (thousands of nodes) of commodity hardware in a reliable and fault-tolerant manner. A MapReduce job usually splits the input dataset into independent chunks, which are processed by the map tasks in a completely parallel manner. The framework sorts the outputs of the maps, which are then input to the reduce tasks, which in turn assemble one or more result sets.

Hadoop is not just for new analytic applications, it can revamp old ones too. For example, analytics for risk and fraud that is based on statistical analysis or data mining benefit from the much larger data samples that HDFS and MapReduce can wring from diverse big data. Furthermore, most 360° customer views include hundreds of customer attributes. Hadoop can provide insight and data to bump up to thousands of attributes, which in turn provides greater detail and precision for customer-based segmentation and other customer analytics. Hadoop is a promising and potential technology that allows large data volumes to be organized and processed while keeping the data in the original data storage cluster. For example, weblogs can be turned into browsing behavior (sessions) by running MapReduce programs (Hadoop) on the cluster and generating aggregated results on the same cluster. These aggregated results are then loaded into a relational DBMS system for analytical solutions to take care of.

HBase is the mainstream Apache Hadoop database. It is an open-source, non-relational (column-oriented), scalable, and distributed database management system that supports structured data storage. Apache HBase, which is modeled after Google Bigtable, is the right approach when you need random and real-time read–write access to your big data. This is for hosting of very large tables (billions of rows × millions of columns) on top of clusters of commodity hardware. Just as Google Bigtable leverages the distributed data storage provided by the Google File System, Apache HBase provides Bigtable-like capabilities on top of Hadoop and HDFS. HBase does support writing applications in Avro, REST, and Thrift.

2.9.2 NoSQL Databases

Next-generation databases are mandated to be non-relational, distributed, open source, and horizontally scalable. The original inspiration is the modern web-scale databases. Additional characteristics such as schema-free, easy replication support, simple API, and eventually consistent/BASE (not ACID) are also being demanded. The traditional relational database management systems (RDBMSs) use Structured Query Language (SQL) for accessing and manipulating data that reside in structured columns of relational tables. However, unstructured data is typically stored in key–value pairs in a data store and, therefore, cannot be accessed using SQL. Such data that are stored are called NoSQL data stores and are accessed via getting and putting commands. There are some big advantages of NoSQL DBs compared with the relational databases as illustrated in the page http://www.couchbase.com/why-nosql/nosql-database.

- Flexible Data Model – Relational and NoSQL data models are very different. The relational model takes data and separates it into many interrelated tables that contain rows and columns. Tables reference each other through foreign keys that are stored in columns as well. When looking up data, the desired information needs to be collected from many tables and combined before it can be provided to the application. Similarly, when writing data, the write needs to be coordinated and performed on many tables.

  NoSQL DBs follow a very different model. For example, a document-oriented NoSQL DB takes the data you want to store and aggregates it into documents using the JSON format. Each JSON document can be thought of as an object to be used by your application. A JSON document might, for example, take all the data stored in a row that spans 20 tables of a relational database and aggregate it into a single document/object. The resulting data model is flexible and easy to distribute the resulting documents. Another major difference is that relational technologies have rigid schemas, while NoSQL models are schemaless. Changing the schema once data is inserted is a big deal, extremely disruptive, and frequently avoided. However, the exact opposite of the behavior is desired in the big data era. Application developers need to constantly and rapidly incorporate new types of data to enrich their applications.

- High Performance and Scalability – To deal with the increase in concurrent users (big users) and the amount of data (big data), applications and their underlying databases need to scale using one of two choices: scale-up or scale-out. Scaling up implies a centralized approach that relies on bigger and bigger servers. Scaling out implies a distributed approach that leverages many commodities' physical or virtual servers. Prior to NoSQL DBs, the default scaling approach at the database tier was to scale up. This was dictated by the fundamentally centralized, shared-everything architecture of relational database technology. To support more concurrent users and/or store more data, you need a bigger server with more CPUs, memory, and disk storage to keep all the tables. Big servers tend to be highly complex, proprietary, and disproportionately expensive.

NoSQL DBs were developed from the ground up to be distributed and scale-out databases. They use a cluster of standard, physical, or virtual servers to store data and support database operations. To scale, additional servers are joined to the cluster and the data and database operations are spread across the larger cluster. Since commodity servers are expected to fail from time to time, NoSQL DBs are built to tolerate and recover from such failures making them highly resilient. NoSQL DBs provide a much easier and linear approach to database scaling. If 10,000 new users start using your application, simply add another database server to your cluster. To add 10,000 more users, just add another server. There is no need to modify the application as you scale since the application always sees a single (distributed) database. NoSQL DBs share some characteristics with respect to scaling and performance:

- Auto-Sharding – A NoSQL DB automatically spreads data across servers without requiring applications to participate. Servers can be added or removed from the data layer without application downtime, with data (and I/O) automatically spread across the servers. Most NoSQL DBs also support data replication, storing multiple copies of data across the cluster and even across data centers to ensure HA and to support disaster recovery (DR). A properly managed NoSQL DB system should never need to be taken offline, for any reason, supporting HA.
- Distributed Query Support – Sharing a relational database can reduce or eliminate in certain cases the ability to perform complex data queries. NoSQL DB systems retain their full query expressive power even when distributed across hundreds of servers.

- Integrated Caching – To reduce latency and increase sustained data throughput, advanced NoSQL DB technologies transparently cache data in system memory. This behavior is transparent to the application developer and the operations team, compared with relational technology where a caching tier is usually a separate infrastructure tier that must be developed to and deployed on separate servers and explicitly managed by the operations team.

There are some serious flaws on the part of relational databases that come in the way of meeting up the unique requirements of modern-day social web applications, which gradually move to reside in cloud infrastructures. Another noteworthy factor is that doing data analysis for BI is increasingly happening in clouds. That is, cloud analytics is emerging as a hot topic for diligent and deeper study and investigation. There are some groups in academic and industrial circles striving hard for bringing in the necessary advancements in order to prop up the traditional databases to cope up with the evolving and enigmatic requirements of social networking applications. However, NoSQL and NewSQL DBs are the new breeds of versatile, vivacious, and venerable solutions capturing the imagination and attention of many.

The business that needs to leverage complex and connected data is driving the adoption of scalable and high-performance NoSQL DBs. This new entrant is to sharply enhance the data management capabilities of various businesses. Several variants of NoSQL DBs have emerged over the past decade in order to handsomely handle the terabytes, petabytes, and even exabytes of data generated by enterprises and consumers. They are specifically capable of processing multiple data types. That is, NoSQL DBs could contain different data types such as text, audio, video, social network feeds, weblogs, and many more that are not being handled by traditional databases. These data are highly complex and deeply interrelated. Therefore, the demand is to unravel the truth hidden behind these huge yet diverse data assets besides understanding the insights and acting on them to enable businesses to plan and surge ahead

Having understood the changing scenario, web-based businesses have been crafting their own custom NoSQL DBs to elegantly manage the increasing data volume and diversity. Amazon's Dynamo and Google's Bigtable are the shining examples of home-grown databases that can store lots of data. These NoSQL DBs were designed for handling highly complex and heterogeneous data. The key differentiation here is that they are not built for high-end transactions but for analytic purposes.

2.10 Summary

Enterprises squarely and solely depend on a variety of data for their day-to-day functioning. Both historical and operational data have to be religiously gleaned from different and disparate sources and then cleaned, synchronized, and analyzed in totality to derive actionable insights that in turn empower enterprises to be ahead of their competitors. In the recent past, social computing applications throw a cornucopia of peoples' data. The brewing need is to seamlessly and spontaneously link enterprise data with social data in order to enable organizations to be more proactive, preemptive, and people centric in their decisions, discretions, and dealings. Data stores, bases, warehouses, marts, cubes, and so on are flourishing in order to congregate and compactly store different data. There are several standardized and simplified tools and platforms for accomplishing data analysis needs. Then there are dashboards, visual report generators, business activity monitoring (BAM), and performance management modules to deliver the requested information and knowledge of the authorized persons.

Data integration is an indispensable cog in that long and complex process of transitioning data into information and knowledge. However, as has been observed, data integration is not easy and rosy. There are patterns, products, processes, platforms, and practices galore for smoothening data integration goal. In this chapter, we had described the necessity of information architecture for next-generation BDA processes and platforms for smart cities.

Final Thoughts

The digitization process has resulted in massive volumes of polystructured data. The information and communication technology (ICT) space has grown up significantly in order to have the inherent strength to collect, ingest, process, mine, and analyze all kinds of city-generated data to squeeze out actionable insights that can be given to people, administrators, and other decision makers in order to tactically and strategically ponder and produce premium offerings for city dwellers, visitors, entrepreneurs, bureaucrats, and investors. Thus, next-generation data analytics platforms, processes, infrastructures, practices, and patterns are being insisted to speed up the setting up and sustaining smarter cities in the days ahead.

Questions

1 Define Big Data?

2 Why big data analytics?

3 Write about the big data processing, storage and analytics technologies and tools

4 Articulate the nuances of the Apache Hadoop framework

5 Why Cloud is being prescribed as the most appropriate IT infrastructure for big data storage and analytics?

6 Explain how smarter cities are bound to generate big data

7 How big data analytics leads to big insights?

8 How big insights contribute for the production and sustenance of smarter city applications?

9 List out the needs and the mechanisms for real-time processing of big data

10 How NoSQL and NewSQL databases contribute for big and real-time data analytics?

References

1 Dignan, J., Shah, N., and Tunvall, F. (2013) *Deriving Insight from Data for Smarter Urban Operations*, Ovum Publication.
2 Russom, P. (2011) *Hadoop: Revealing Its True Value for Business Intelligence* www.tdwi.org, TDWI Research (accessed 18 January 2017).
3 Harness the Value of Big Data to Build Smarter Infrastructures, a thought leadership white paper from IBM Software, 2013.
4 Big data: The Next Frontier for Innovation, Competition, and Productivity, McKinsey Global Institute, June 2011.
5 Hinssen, P., Open Data Power Smart Cities, published in www. datascienceseries.com (accessed 18 January 2017).
6 Oracle: Big Data for the Enterprise, a White paper by Oracle, 2011
7 NoSQL for the Enterprise, a White paper by Neo Technology, 2011.
8 Villanueva, F.J., *et al.*, Civitas: The Smart City Middleware, from Sensors to Big Data, Innovative Mobile and Internet Services in Ubiquitous Computing (IMIS-2013), Asia University, Taichung, Taiwan.
9 Iwamura, K., Tonooka, H., Mizuno, Y., and Mashita, Y. (2014) Big Data collection and utilization for operational support of smarter social infrastructure. *Hitachi Review*, **63** (1), 18–24.
10 Munirathinam, S. and Ramadoss, B., Big Data Predictive Analytics for Proactive Semiconductor Equipment Maintenance: A Review ASE BIG DATA/SOCIALCOM/CYBERSECURITY Conference, Stanford University, May 27–31, 2014.
11 Bloor, R. (2011) *Enabling the Agile Business with an Information Oriented Architecture*, The Bloor Group.

3

Multi-Scale Computing for a Sustainable Built Environment

Massimiliano Manfren

Department of Industrial Engineering (DIN), University of Bologna, Bologna, Italy

---

CHAPTER MENU

Introduction, 53
Modeling and Computing for Sustainability Transitions, 55
Multi-Scale Modeling and Computing for the Built Environment, 66
Research in Modeling and Computing for the Built Environment, 70

Objectives

- Introduce readers to the concept of multi-level perspective modeling in sustainability transitions planning.
- Introduce readers to technological and social learning dimensions in sustainability transitions.
- Introduce readers to the issue of long-term sustainability thinking for the built environment.
- Describe the most relevant characteristics and attributes of modeling techniques and data schemes for built environment performance modeling.
- Describe ongoing research on multi-scale computing for the built environment.
- Indicate possible paths of development for empirical data analysis methods to support sustainability transitions.

3.1 Introduction

The need to promote sustainable human settlements and to mitigate the spatial, demographic, social, economic, and environmental impacts, determined by the

Smart Cities: Foundations, Principles and Applications, First Edition.
Edited by Houbing Song, Ravi Srinivasan, Tamim Sookoor, and Sabina Jeschke.
© 2017 John Wiley & Sons, Inc. Published 2017 by John Wiley & Sons, Inc.

rapid global urbanization trend, is creating a concentration of research and development efforts in the built environment area. At the global scale, we can trace a path of awareness increase on sustainability matters, starting from the Brundtland Report "Our Common Future" [1] in 1987, passing through Kyoto Protocol stipulation in 1997, and arriving today at the UN Climate Conference of Parties (COP) 21 held in Paris at the end of 2015.

Sustainability is a broadly defined concept; the focus of initiatives was initially on greenhouse gas (GHG) emissions and ozone depletion, while today the attention is focused on the problem of climate change, following the ongoing research performed by the Intergovernmental Panel on Climate Change (IPCC). In its fifth report, IPCC claims "Ignorance can no longer be used as an excuse for no action" [2]. In other words, although we may not a have a complete understanding of the problem of climate change, we have to start prioritizing actions against it because of its potentially irreversible effects. Prioritize means being able to define policies that will have to cope effectively with the climate change problem, considering realistically social and economy constraints. Of course, single governments have to define their own policies, but we have to act together for the shared goal of limiting the increase of the average temperature of the atmosphere. What appears immediately evident from energy and emission statistics [3–5] is that our energy systems have to switch toward low-carbon sources (decarbonization), in particular renewable energy sources. In other words, we have to phase out fossil fuels by means of an increasing share of renewables. Further, we have to consider the issue of efficiency, roughly described by energy consumption with respect to GDP (energy intensity) and per capita [6]. An additional problem is represented by energy poverty, initially considered in Millennium Development Goals [7] and today in the Sustainable Development Goals [8, 9].

The relation among energy use, economic growth, and urbanization is complex, and several researches highlight how the urban dimension cannot be evaluated in pure economic terms [10].

From a sustainability perspective, cities constitute a priority for action, because the absence of effective planning and governance can cause dramatic consequences such as pollution and health issues, poverty, unemployment, and crime. At the same time, cities can become an incredibly successful driver of social development and economic growth. Of course, no single regulation or rigid planning scheme can govern the complexity of the evolution of a city. At the global level we are assisting to a rapid increase of the share of renewable energy production [11], but, on the efficiency side, the question is more debatable. While in some countries energy efficiency is increasing, we may have a negative effect on energy use and emissions in other countries (e.g., fastest growing countries). This phenomenon has been addressed with the term "carbon leakage" [12–14]. For these reasons, it is difficult to set up at the global level coherent policies aimed at sustainability transitions, which are necessarily path dependent.

However, the present situation is different from the past because of the concrete possibility of creating a learning society [15] and a circular economy [16, 17], where the essential principles of sustainability are set at the basis of society itself. We have to "think globally" and "act locally": "Sustainable Energy Action Plans" (SEAP) [18–22] are an example of value-driven decision-making, where appropriate indicators, monitored periodically, are used to assess the effect of local policies. Another interesting example is the "Common European Sustainable Building Assessment" (CESBA) [23] that gives a holistic view on the built environment sustainability problem, at the building and district scale. What seems fundamental today is creating a convergence among these good practices and the ongoing evolution in information and communication technology (ICT) and energy infrastructures [24], respectively, toward the emerging "Internet of things" (IoT) [25–28] and "Smart Grid" paradigms [29, 30]. Cities are important test fields for both IoT development [31–33] and for decentralized energy planning [20, 34]. Design, construction, and operation practices in the built environment can profoundly benefit from these evolutions, which does not involve a single sector but the whole economy and society.

The research presented articulates on two levels of analysis: general features of modeling and computing tools for sustainability transitions and specific features for built environment applications. The methodological approach adopted for the research is the following: first, general and specific elements are selected and described, and then relevant issues are identified, discussed, and synthesized. Finally, possible paths for future developments are given, starting from the insights learned in the analysis process.

3.2 Modeling and Computing for Sustainability Transitions

The definition of possible strategies for sustainability transitions requires the contribution of experts in several domains. Among the most relevant ones, we find energy planning, energy conversion and management, built environment, industrial systems, transportation systems, ICT, infrastructures, economy, and social sciences. Further, experts in modeling and technology foresight play a cross-disciplinary role by proposing new syntheses (from domain-specific knowledge) that are aimed at strategic decision-making (from short- to long-term planning). There are relevant examples of strategic projects and initiatives for the long-term planning of a low-carbon and sustainable society [35–38]. The transition from the present economic and societal paradigm to a sustainable one is a great challenge, which will encompass clearly the implementation of cleaner energy systems, but which will impact, more in general, the way we live, move, and work in a profound way, determining a structural change. New concepts such as circular economy [17] and learning society [15] are important elements to mobilize stakeholders, because this structural change affects environment, economy, and society (the three pillars of sustainability) simultaneously. Analyzing and modeling this change, at multiple levels, requires the evolution of present tools and methodologies, including more adequate techno-economic and socioeconomic evaluation methodologies [39]. For example, we can already see how in sustainability protocols and planning tools at the building, neighborhood, and urban scale [20, 23] decentralized energy planning [34, 40, 41] is increasingly becoming a crucial question. Energy modeling, as introduced before, is multidisciplinary and cross-sectoral; built environment applications can share, at least, a similar methodological approach with other sectors, such as industrial systems [42–45] with respect to accounting, simulation, and optimization models and tools. Further, energy demand, conversion, and storage processes are deeply related to sustainability. In some cases, the use of aggregate indicators for sustainability evaluation can be questionable [46–49] and can be improved further with respect to current practices [23]. Nonetheless, from a practical standpoint, it is necessary to unveil the connections among multiple aspects of sustainability (environment, economy, and society), multiple levels of analysis (e.g., technologies, infrastructures, policies), and performance indicators, which can affect specific planning, design, and operation choices for the built environment. The presence of a portfolio of possible choices is necessary to engage effectively stakeholders in value-driven decision-making [50, 51]. More in general, for the reasons outlined before, sustainability transitions clearly require a multilevel planning strategy to help redirecting the existing dynamics in economy, society, and technology, considering environmental constraints [39].

3.2.1 Multilevel Perspective Modeling

The evolution in complex technological systems is intrinsically path dependent. In particular, the future development of sustainable practices can be inhibited, or even prevented, by inappropriate choices made today. Therefore, transition modeling in a path-dependent process has to deal with the coevolution of the social, technological, industrial, and policy changes, that is, the so-called multilevel perspective (MLP). Some important examples of MLP modeling can be found in literature [52, 53]. The MLP approach considers three levels:

1) Energy infrastructures (i.e., energy systems and technologies)
2) Behavior (i.e., consumer and investor's choices)
3) Institutional factor (i.e., policy, regulation, and markets)

These three levels form a regime level that reflects on the governance of the system. A circular approach should be used to improve system performance with respect to key performance indicators (KPIs), resembling a Deming or plan–do–check–act (PDCA) for continuous improvement. In fact, PDCA cycle

Table 3.1 Description of the PDCA (Deming) cycle.

Action	Description
Plan	Establish the processes needed to achieve the policy defined (e.g., energy, environmental). Plan for change, identifying specific objectives (target or goals), defining steps, and expected results
Do	Implement the processes following the steps outlined in the plan, collecting data for the following steps: "check" and "act"
Check	Monitor and measure the processes and report the results, verify if the goals are met, and examine the results by comparing them with the expected ones. Investigate eventual deviations in the implementation of the plan and the appropriateness and completeness of the plan
Act	Act to maintain and improve the performance with respect to objectives, updating continually by restarting the cycle. If the objectives are met, they become the new baseline; otherwise the baseline objectives remain the same and the cycle restarts

is currently at the basis of international energy and environmental system management standards [1, 54] and is summarized in Table 3.1.

An important project aimed at integrating these theoretical principles and identifying the gaps in existing tools, models, and methodologies is Analysing Transition Planning and Systemic Energy Planning Tools for the implementation of the Energy Technology Information System (ATEst) [39]. In this project it has been clearly highlighted that most of the existing tools and methodologies are focused on the quantitative analysis of the development of energy infrastructures on different levels of analysis. While, on the one hand, there exist very good bottom-up energy system models and top-down macroeconomic models [55, 56] to support decision-making, on the other hand, tools and methods focused on the analysis of the behavior of consumers and investors are moderately or relatively poorly covered. Further, the area that presents the most serious deficiencies is the analysis of institutional factors driving decisions. Additionally, with respect to technological development (crucial for a path-dependent transition), modeling and analysis of the impacts of R&D policies is a major gap to support sustainable, low-carbon transitions, in particular for the built environment.

3.2.2 Technological and Social Learning

Technological learning curves [55, 57] may be used in models of transition. However, while two-factor learning curves account for some impact of R&D policies, they still do not consider the institutional factors around the development of new technologies. Further, learning curves only account for the cost of a technology that is already deployed in the market, not for the effectiveness of

R&D policies in supporting the development of breakthrough technologies and innovative practices. In fact, the implementation of two-factor learning curves requires technology-specific data, which might not be publicly available or even exist because of the very early stage of development of the technology itself. Further, technological solutions for the built environment (affecting design, construction, commissioning, operation, and dismantling phases), although constituted by single technologies, have to be necessarily optimized from the system point of view [58–61], considering the whole building life cycle [62, 63]. Cost-optimal design [64, 65] and efficient operation of buildings [22, 66–68] can be identified as good practices, but it is difficult to link them to ordinary tools for energy policy or planning, although specific tools are being developed for this purpose [69–73] and further research, in this direction, is possible.

The "traditional" way of modeling technological learning can add limited value to sustainability transition planning for the built environment, neglecting the complexity in design [74, 75] and operational optimization (including behavioral issues) [76–84], which can determine a huge variability of energy consumption, emission, and cost. Technological innovation should be information-centric [24, 26, 32, 33], and energy systems can benefit from the "Internet of things" principles in terms of infrastructures' operation [24] and direct feedback data to evaluate the behavioral and social impact of technologies [85, 86], progressively overcoming the limitations of current models. At the local level, protocols and legislation, acting on various scales, from buildings to district/neighborhoods and cities, can become complementary with models and tools [85], by sharing essential information and scopes, thereby filling the existing gaps, in particular with respect to behavioral and social learning dimensions.

3.2.3 Multidisciplinary System Thinking

Multidisciplinary system thinking is a key ingredient in the challenge of sustainability transitions. On the one hand, we have to develop models and tools that are able to deal with complexity, but, on the other hand, we have to keep our analysis as simple as possible, increasing the transparency with respect to modeling assumptions and uncertainty of results. The development of the research on multi-scale modeling for a sustainable built environment can embody a large part of the concepts previously described. First, the techno-economic side of the problem cannot be considered separately from the socioeconomic side of the problem, in particular with respect to policy questions regarding stakeholders' behavior and social acceptability of technical solutions. Further, existing sustainability protocols (which include multi-criteria evaluation) can represent a starting point for the evolution in sustainable design and operation practices. Protocols should be not thought as mere labeling schemes but should be increasingly integrated in local decision-making processes [19, 22, 73, 87], directly considering the multiple interactions among socioeconomic and techno-economic issues, whose effects can be analyzed periodically by means of data collected at multiple spatial and temporal resolutions. This is going to be the focus of international research initiatives [88] aimed at developing robust and multidisciplinary approaches to such analyses.

3.2.4 Long-Term Thinking for the Built Environment

From a general standpoint, considering the problem of the development of models and tools for the built environment, we should put emphasis on the following issues:

1) Model verification, validation, and calibration standards, including uncertainties
2) Model benchmarking using common standard for data analysis
3) Clear identification of quantities in models that are assumed (e.g., bound/unbounded, physically constrained), measured, or estimated (derived indirectly by measures)
4) Ability to perform efficiently *ex ante*/*ex post* analysis aimed at verifying effectiveness
5) Analysis of end users' behavior incidence and promotion of stakeholders' involvement, directly addressing relevant elements that affect energy consumption, investment, and so on
6) Integration of models and databases, including appropriate data related to
   a) Geography and land use
   b) Infrastructures (transportation, energy, water, and so on)
   c) Property, end use, business/activity, and operation profiles
   d) Environmental, energy, and economic impacts
   e) Social and behavioral aspects of energy and water use

Beyond models and tools, what appears to be missing today for the built environment is a comprehensive strategy, resembling the 4P model of Lean Production [89–91]:

1) Philosophy: long-term value-driven reflections, system thinking, and decision-making
2) People: stakeholders involved in the sustainability of the built environment
3) Processes: all the processes that have influence on the sustainability of the built environment
4) Problem solving: actions on processes and decision-making

With respect to long-term thinking, the planning of sustainability transitions clearly involves design, construction, and operation choices for built environment, as introduced before. Prioritization with respect to multiple objectives and initial investment cost remains critical dimensions, because buildings are generally designed, constructed, and operated by different entities (often with conflicting needs and different responsibilities), and financing schemes are not generally suited for medium- to long-term investment. In fact, costs across the building life cycle are distributed among different actors and processes as buildings are long-term assets. These elements determine a barrier toward innovative technologies and practices. Therefore, an additional effort should be placed on understanding innovation with respect to people and process. People involved in value-driven decision-making for sustainability, that is, stakeholders involved in sustainability assessment for the built environment [23, 50] are reported in Table 3.2. Stakeholders have to become aware of their role in decision-making. The strong or weak impact of the stakeholders in the key decisions is strongly dependent on the specific characteristics of the building, district/neighborhood, and city-scale processes, in particular with respect to end uses, business/activity, and property (social and economic dimension). Effective ways to mobilize stakeholders cannot be inscribed in a rigid scheme but have to be rather investigated with respect to the specific problems to be solved. The problems can be subdivided into two main categories:

1) Introducing effective strategies for design, construction, and operation [89], considering the gap, frequently encountered among measured and simulated building performance [76] (building-scale performance gap minimization, understanding and reducing inefficiencies)
2) Planning decentralized energy system and optimizing their operation (from building to district and city scale) [82, 84, 92–94]

These two general problems directly translate into computing tasks in the different processes and scales summarized in Table 3.3. The models that are used for these computing tasks should be aimed at maximizing the value of information (unveiling synergies across processes and scales) with respect to the multiple potential feedbacks that can be exploited to improve performance (i.e., multiple PDCA cycles).

A state of the art of building energy modeling can be found in [51, 74–76, 95–104], and a synthesis is presented in Figure 3.1. The first relevant distinction to be made is among top-down (econometric, technological) and bottom-up (engineering) models [105]. After that, an important subdivision is the one related to modeling strategies (white-box, gray-box, and black-box models), as explained in Table 3.4. Given the current modeling and computing capabilities, the essential elements that have to be shared across processes and scales are as follows:

1) Standardized methodologies, tools, and models
2) Effective data visualization techniques to interact with stakeholders and practitioners in decision-making processes
3) Appropriate feedback data, at different temporal and spatial resolutions, to be evaluated in PDCA cycles

Table 3.2 Role of stakeholders with respect to sustainability assessment.

Stakeholder group	Order assessment	Provide information for assessment	Elaboration of assessment	Use assessment results
Architects and designers	Weak	Strong	Weak	Strong
Banking sector				Medium
Certification entities			Strong	Strong
Community representatives				Strong
Consultants		Weak	Strong	
Contractors		Strong		Strong
Estate agents				
Facility managers		Strong		Strong
Funding providers	Strong			Strong
Insurers				Weak
Neighbors of the site	Strong			
National and regional authorities	Strong			Strong
Product manufacturers		Strong		
Property investors	Strong			Strong
Property valuers	Strong			Strong
Real estate developers	Strong	Strong	Strong	Strong
Researchers and academics				
Users of the building				Strong

Strong
Medium
Weak

Table 3.3 Use of information in modeling and computing for the built environment in different processes at different scales.

Process/scales	Building	District/neighborhood	City
Design	Feasibility study	Feasibility study	Urban planning
	Preliminary design	Preliminary design	Energy planning
	Detail design	Detail design	Energy policy
	Protocols/ratings	Protocols/ratings	
Construction	Functional testing	Functional testing	–
	Commissioning	Commissioning	
Operation	Continuous commissioning	Load forecasting	Load forecasting
	Control (HVAC, distributed generation, etc.)	Control (distributed generation)	Control (distributed generation)
	Load forecasting	Energy management	Energy management
	Energy management		

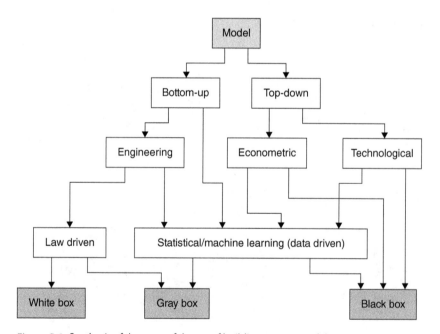

Figure 3.1 Synthesis of the state of the art of building energy models.

Table 3.4 Features of different types of models.

Model type	Description	Advantages	Disadvantages
White box	Detailed physics-based models (law driven) employing algebraic and differential equations (ODE, PDE)	Accuracy and precision, detailed physical description of phenomena	Computational effort, difficult and error-prone modeling and implementation process, difficult to calibrate, not suitable for control and real-time operation
Gray box	Simplified physics-based models (law driven) with algebraic equations and first-order ODE, whose parameters can be determined from measured data	Easier implementation with respect to white-box models, simplified physical description of phenomena, computational efficiency, easy to calibrate, suitable for control and real-time diagnostics	Not as accurate and precise as white-box models, error-prone implementation process, simplifications depend on the specific computing task, difficulty of generalization
Black box	Empirical models (data driven), based on little or no physical knowledge of the system, which rely on the available data to identify the model structure. They are suitable for predicting future behavior under a similar set of conditions	Computational efficiency and flexibility, simple implementation with respect to achievable accuracy for control and real-time diagnostics	Absence of a physical representation, opaque to the user, application specific

The last point is recalled also in Lean Production with the terms "Hansei" (process reflection) and "Kaizen" (continuous improvement) and is directly related to the elements outlined in the previous sections. Lean production was originally conceived for industrial organizations, but it has been successfully applied in many other fields [89, 91, 106, 107]. PDCA cycle (Table 3.1) is the method for putting "kaizen" principle into practice.

3.2.5 Data and Modeling Techniques

Modeling techniques should be seamlessly integrated with methodologies and tools for sustainability transitions. In other words, we can consider models to be appropriate if they can be successfully included in this general framework, which is emerging as a response to the pressing environmental, economic, and social issues. Focusing on the information exchange and decision-making processes, one of the key aspects of standardization is the development of appropriate KPIs to monitor performance quantitatively. KPIs and visualization techniques are essential as they provide a concise way to understand critical insights (for field practitioners and, more in general, stakeholders) on problems and to set quantitative baselines for improvement (PDCA). In sustainability protocols, for example, performance indicators are organized according to macro-areas (e.g., location, territory and site, process quality, environmental quality, social quality, economic quality) [23]. With respect to performance optimization in buildings, we can consider the following as the most important categories of indicators:

1) Energy (EnPI [1])
2) Emissions [108]
3) Cost [109]
4) Comfort (thermal/visual/acoustic comfort) [110–112] and indoor air quality (IAQ)
5) Efficiencies of energy conversion processes (heating, cooling, domestic hot water, lighting, appliances, etc.)

KPIs are generally calculated by means of computing processes executed by machines. In this sense, the interaction among humans and machines is also an important topic to be considered, as already recognized in industrial system with the term "autonomation" [90, 113]. The crucial problem is increasing the capability of self-diagnosis when a problem arises in an automated process; this self-diagnosis is a quality control measure. Autonomation concept is transposed for built environment applications in Table 3.5.

Table 3.5 Automation versus autonomation in the built environment.

Category	Automation	Autonomation
People	Manual processes are reduced and simplified but human supervision is still needed	Humans can perform multiple tasks and improve productivity and efficiency
Processes	Technologies (devices, machines, etc.) operate following fixed settings	Detection of errors and correction is autonomous, interacting in a smarter way with human
Efficiency	Waste due to malfunctioning is not diagnosed promptly. Errors are discovered later	Appropriate diagnosis and detection algorithms communicate to humans and prevent wastes. Errors are discovered and corrected earlier

Of course, in order to go more in depth in problem resolution, we need specific data schemes, and models (computational tools) have to be built based on their effective ability to solve the computational problems in the relevant processes outlined in Table 3.3. What appears to be immediately evident from the state of the art of building energy models [114–116] is that there are several possible improvements. The most relevant ones are:

1) Encapsulation of knowledge in computing "objects"
2) Topological interconnection capability among "objects"
3) Hierarchical modeling to establish a meaningful relation among "objects"
4) Generalized networking capability to increase scalability, from small- to large-scale computing tasks

These improvements are explained in detail in Table 3.6. There exist potentially multiple circular approaches (PDCA cycles) that can be implemented for

Table 3.6 Features of advanced modeling and computing tools for the built environment.

Feature	Description	Advantages
Encapsulation of knowledge	Knowledge is contained in a particular object (node), establishing connections (links) with other objects	Higher level of abstraction and compact representation
Topological interconnection capability	Models are built from interconnected objects, using rules that are more similar to human's way of structuring systems and knowledge (e.g., conservation equation, diffusion). The connection of objects defines the topological structure of the system	Effectively combine the encapsulated knowledge in objects Modeling separated from the solution process Easy assembling (at the interface level for the modeler, at the equation level for the software) Enhanced possibility of integrating domain-specific knowledge, transparency, and usability
Hierarchical modeling	The possibility of describing the model in a hierarchical view (starting from its topology), ideally like a graph with nodes and links	Manage complex multi-scale, multi-domain problems in a more transparent way
Generalized networking capability	The possibility of modeling a problem as a network goes beyond the simple object-oriented modeling	Possibility of using general-purpose algorithms conceived for problems that can be represented as networks

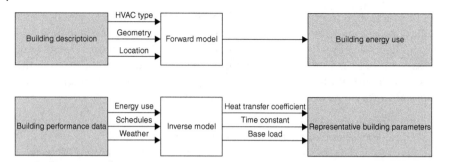

Figure 3.2 Forward and inverse building energy models.

the continuous improvement of built environment performance, depending on the specific application. It is important to underline the fact that, in order to implement them effectively, a much tighter integration and comparability among models has to be present (benchmarking, design comparison). In fact, we should be able to pass from model to simulated data (model output, forward approach) and from measured data back to models (model input, inverse approach) [116]. A synthetic scheme is reported as an example in Figure 3.2.

Today, this is time consuming especially when bottom-up detailed models (engineering models) are used [117, 118]. However, the ongoing research on model calibration techniques, to match simulated and measured data [51, 101, 119], can help derive fundamental insights on the performance of subsystems and building components [120–122], that is, detailed information on the real performance of technologies, including the interaction with the end users [123–126].

The essential features of methodological and computational tools for the built environment are described in detail in the following section.

3.3 Multi-Scale Modeling and Computing for the Built Environment

Modeling and computing tasks, reported in Table 3.3, can be grouped across different scales, with respect to processes:

1) Design (feasibility/preliminary/detail, protocols/ratings, energy planning, and policy)
2) Construction (functional testing and commissioning)
3) Operation (continuous commissioning, energy management, load forecasting, control)

The evaluation of performance (through KPIs) can be performed at the individual level (single building) or for clusters of buildings to find, respectively, specific or aggregated properties and to investigate variability across temporal and spatial scales. Spatial clustering can be:

1) Geographical (district/neighborhood aggregation)
2) Non-geographical (similar building in terms of shape, age, end use, business activity, etc.)

In this sense, relevant information can be collected at different levels – detailed and aggregated – for different purposes [127–130], which clearly require different temporal resolution of data (seasonal, monthly, daily, hourly). Starting from the design process, we encounter two fundamental issues. First, most of the decisions that have great impact on energy performance (and the related costs and sustainability) are made early in the design process, without clearly understanding how they will affect the whole life cycle. The practical problems more frequently encountered are as follows:

1) Misalignments between designer, contractor, and owner
2) Critical information provided after decisions are made
3) Building performance not thoroughly evaluated during initial design phase

After that, in the design stage it is difficult to quantify the impact of the behavior of end users, in terms of occupancy, energy usage patterns, comfort preferences, and, more in general, the potential evolution over time of end uses themselves. A more complete evaluation of these problems can be found in literature [76, 89]. These issues have to be tackled in order to reduce the gap between simulated and measured performance [18, 131].

3.3.1 Virtual Prototyping for Design Optimization

We have to be able to optimize building performance [77, 132] by exploring the potential outcomes of design solutions and operation strategies with respect to multiple KPIs already in the design phase. The optimization problem itself can be formulated as an optimal learning problem [133], if we think about the exploration of possibilities with respect to the immediate exploitation of knowledge and information. One of the biggest limitations in the past was the lack of user-friendly software to simulate quickly and accurately building performance in the early design phases [134] and to optimize it with respect to multiple criteria and scenarios, including uncertainty [135, 136]. However, recent software developments are pointing in this direction [137–141]. This topic is indeed complex, as it contains both the problem of optimizing the design and of simulating the optimal behavior of the building system during its life cycle, under uncertainty. These are not trivial tasks and are common problems encountered in energy systems design; nonetheless, we can briefly summarize the necessary evolution from current design practices to more advanced ones (i.e., rapid virtual prototyping (RVP)). In a traditional design process, activities are organized as follows:

1) Gather information (design requirements, product data, etc.).
2) Make intuitive decisions.
3) Develop design documentation.
4) Analyze results and discuss with stakeholders.
5) Revise decisions based on analysis and discussion.
6) Repeat steps 3–5, until final design is defined.

In RVP activities are designed as follows:

1) Gather information.
2) Simulate many different hypotheses (scenario, parametric, probabilistic, etc.).
3) Analyze results and discuss with stakeholders (sensitivity and uncertainty on KPIs, useful insights on decisions, etc.).
4) Make informed decisions (based on analysis of results, experience, and discussion).
5) Develop final design documentation.

The differences between the two processes are highlighted graphically in Figure 3.3. What appears immediately evident is that there is a waste connected to project revisions and iterations in traditional design processes. This is clearly determined by initial design choices based on intuition, limited number of alternatives considered, and difficult interaction with stakeholders, which are called to express their decision on a single hypothesis rather than on a more comprehensive spectrum of alternatives already considered before the development of the final design documentation.

In synthesis, the benefits of RVP are the possibility of having more informed decision-making processes and avoiding useless project iterations. Clearly, as outlined before, appropriate software is necessary but not sufficient. The presence of standards and protocols for model verification, validation, calibration, and data analysis [34, 142–146], together with performance benchmarking [147, 148] and appropriate training of practitioners, is essential to increase simulation reliability.

3.3.2 Performance Optimization Across Building Life Cycle Phases

Despite the potential of advanced modeling software and even in the presence of adequate standards and methodologies, some critical issues related to data and models for performance optimization across building life-cycle phases will remain, namely:

1) Interoperability of information from design to construction and operation phases [149]
2) Reusability of information across multiple domain-specific applications [150]
3) Simplification and computational tractability for the implementation in automated systems for commissioning, management, and control [66, 77]

The recent advances in the field of model calibration [51, 101] show how techniques involving the use of reduced-order inverse models (or surrogate models, meta-models, etc.) [18, 117, 119, 131] can represent partially a solution for the problems listed before.

In fact, the use of simple but robust surrogate models that are correctly constructed (e.g., trained on a few, carefully selected, simulations or on sufficient real measured data) has been already experimented in performance simulation and design of technical systems [151–153].

Another important contribution of the research on model calibration is the fact that, by investigating the mismatch between simulated and real performance, we can learn fundamental insights for design and operational optimization. In fact, in model calibration, we start with a spectrum of possible parameters, and we progressively reduce the difference between simulated and real data by acting simultaneously on multiple variables (even automatically, by means of an optimization process). This process is inherently similar (from a computational perspective) to the RVP described in Figure 3.3, where we start from multiple hypotheses and we refine our choices, using multi-criteria

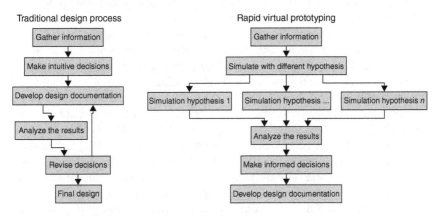

Figure 3.3 Traditional design and rapid virtual prototyping.

evaluation or optimization routines. In synthesis, in model calibration our goal is minimizing the difference among simulated and measured data, while in virtual prototyping our goal is optimizing choices with respect to selected criteria (e.g., either minimizing or maximizing KPIs under multiple constraints).

Therefore, there exist potential synergies to be exploited among advances in design [74, 75] (up to ZEB, NZEB, nZEB paradigms [61, 131, 154–156]) and energy management practices [18, 22, 66, 119]. Further, these synergies can be extended up to the control level [102, 149], because computation tools are already available for deriving control algorithms from detailed building energy models [157–159]. As a conclusion, the evolution of computational tools (techniques and software) is certainly contributing to improvement in design, commissioning, and operation practices (i.e., control, energy management), although there remain a lack of integration among tools and data, across process and scales, and a lack of comprehension by practitioners of its potential. In order to render more evident the potential of the evolution of computational tools and data schemes, research efforts have to be concentrated on specific cross-disciplinary aspects from design to operation [149], considering the fact that these methods have to support realistic transition pathways [160]. The relevant aspects to be considered are synthesized and analyzed in the following section.

3.4   Research in Modeling and Computing for the Built Environment

As introduced in Section 3.2.4, several modeling strategies (from top-down to bottom-up approaches) can be employed to describe the performance of buildings. For the specific applications considered in the research and summarized in Table 3.3, we have first to parameterize building data, in order to be able to deal with them across multiple scales (spatial and temporal) in a computationally efficient way (e.g., clustering information in a geographic or non-geographic way).

A fundamental concept for parameterization is that of "reference building" [58, 59, 69, 161] or "prototype building" [162, 163] suitable for multiple purposes. A reference building can be algorithmically defined [164] to simulate multiple configurations up to urban [83, 165–167] and building stock scales [168]. A list of the most relevant features to be considered in the construction of a reference (or prototype) building is reported in Table 3.7.

We have to draw a distinction among data/metadata; there are some data that are directly employed in modeling and computational processes and other data (metadata) that provide useful information about other data (e.g., property type, age, etc.). Metadata can be particularly helpful, especially in urban-scale

Table 3.7 Reference building dataset.

Category	Subcategory	Information
Location		Geographic position
		Climate condition
Property		Business/activity
		End use of zones
Geometry		Floor area
		Floor height
		Number of floors
		Orientation
		Aspect ratio
		Window-to-wall ratio
		Shading
Construction components	Opaque	Thermal transmittance
		Thermal capacity
	Transparent	Thermal transmittance
		Gain factor
		Thermal bridges
Air change	Infiltration	Standard conditions/operation
	Ventilation	Natural
		Mechanical
Technical systems	Heating/cooling/DHW	Emission
		Distribution
		Storage
		Control

Table 3.7 (Continued)

Category	Subcategory	Information
	Lighting	Appliances
		Control
	Air handling	Heating/cooling
		Humidification/dehumidification
	Production	Heating/cooling
		Electricity
Operation	Internal gains	Occupancy
		Lighting
		Appliances (electric, thermal)
	Set points	Temperature
		Relative humidity
	Schedules	Occupancy
		Appliances
		Lighting
		Heating/cooling/ventilation
		DHW
Energy		Delivered/imported
		Exported
		Production (on-site)
		Demand

Table 3.8 Issues for multi-scale spatial data aggregation.

Data	Issue	Proposed solution
Spatial data (GIS)	The parcel for property tax assessment does not identify exactly the building	Use of a building ID together with parcel and owner IDs, to enable the assignments of multiple data to the building
Property type	Property tax assessments represent an information generally available, but there is a lack of further details	Other data sources (e.g., construction permits) should be integrated with property data to enable further investigation, including a survey on the number of people
End use/ business activity	Property types are the main identifier of end use. In the case of mixed-use buildings, it is difficult to assign the correct parcel within the building geometry (e.g., which floors/zone correspond to which use)	In mixed-use buildings a description of the combination of property types, expressed in terms of percentage of surface (reducing to one prevailing end use when possible) can be useful to simplify building models from multi-zone to single-zone or nodal models. A single-zone or nodal model is much easier to handle from the numeric point of view
General data	The general data to describe a building include geometry (e.g., shape, area, volume), age, renovation, building components, technical systems. Many of them are not generally available	Collect the relevant data for the buildings, using multiple data sources. Archetypes and categorization can be used to simplify data collection and analysis, based on similarities

applications [84, 87, 169] where we do not have generally complete information, as briefly outlined in Table 3.8.

The essential KPI types for buildings are reported in Section 3.2.5. Energy, carbon dioxide emission, cost, and comfort are the essential categories of performance indicators to be studied for multi-scale applications (i.e., at different levels of aggregation). Essentially, these indicators can be employed in two fundamental categories of computational problems:

1) Techno-economic optimization at the building [58–60, 170] and district level [92–94, 171], following a multi-objective logic
2) Optimal interaction with energy infrastructure (in particular electric grid) [172, 173], considering load matching [174–178], energy storage [78, 104], dispatch of distributed energy resources [34, 179], and potential of flexibility in building operation [180, 181]

These problems are clearly the practical expression of the more general ones outlined in the previous section. While the formulation of problems in terms of optimization (stochastic optimization, if we consider uncertainties) has to be necessarily based on system-level KPIs (representing objective functions), which provide an aggregate result, it is also important to consider appropriate constraints represented in the form of multilevel metrics [120, 121]. These additional metrics can be also lumped parameters (aggregated quantities, macro-parameters) [77, 182, 183], commonly used in model identification. In any case, the fundamental goal is the accurate description of building energy dynamics, with different focuses, dependent on the specific processes and scales [127–129].

3.4.1 Building Energy Balance Analysis

If we analyze building energy balance [108], from the system level down to the subsystem level, we can draw first a distinction between two fundamental components:

1) Demand (heating, cooling, DHW, lighting, appliances, etc.)
2) Production (conversion technology, on-site distributed generation, etc.)

From the outside (ideally at the meter level), we can have two fundamental components that have to balance (algebraically) the previous two:

1) Energy delivered/imported (from energy infrastructures or directly by energy carriers as fuels)
2) Energy exported (to energy infrastructures)

The latter two can be directly accounted in terms of carbon dioxide emissions and cost, which can be used as KPIs in building design optimization problem (objective functions in optimization models) [75], as well as operational optimization (e.g., Smart Grid interaction) [77, 172].

Energy production (i.e., conversion processes) can be performed by means of technologies fed by electricity (e.g., heat pumps, chillers), fossil and renewable fuels (e.g., boilers, furnaces, CHP), and directly from on-site renewables (e.g., photovoltaics, wind turbines, solar thermal). These technologies, combined among themselves in an optimal way, can determine multi-generation or poly-generation systems [184–188], producing electricity, heating, and cooling at different temperature levels. Energy demand can be analyzed not only according to energy carriers in technical systems (electricity, heating, cooling) [189] but also with respect to:

1) Energy needs (e.g., space heating, space cooling, lighting)
2) System losses (e.g., emission, distribution)
3) Auxiliary energy (e.g., electricity from pumps, fans)

This subdivision enables a clearer identification of the efficiency issues from the demand side point of view, providing a meaningful structure for analysis. System losses and auxiliary energy demand are clearly determined by the choice of technical systems (types and sizes, depending on energy loads) as well as by operation modes (i.e., control strategies and settings) [189]. Energy needs can be grouped in the following categories:

1) Heating (sensible heat)
2) Cooling (sensible heat)
3) Ventilation (infiltration, natural, mechanical)
4) Humidification (latent heat transfer)
5) Dehumidification (latent heat transfer)
6) Domestic hot water (DHW)
7) Lighting (electric)
8) Appliances (electric, thermal)

The first five elements are directly related to the thermo-hygrometric balance of a building zone [190–193] considering sensible/latent heat and mass transfer. We can study heat and mass balance dynamics by means of algebraic and differential equations in the time and space domains in several ways, as described in Table 3.9.

Table 3.9 Heat and mass balance modeling in buildings.

Model type	Physical element	Application	Solution techniques	Input type	Computational speed
White box	Control volume/ surface	Heat transfer, computational fluid dynamic, diffusion	2D/3D finite elements/finite volume	Detailed	Low
	Building zone (zonal model)	Building energy simulation	1D finite differences, transfer functions, state space	Detailed	Low
Gray box	Building zone (zonal model)	Building energy simulation, predictive modeling and diagnostics	1D finite differences, transfer functions, state space	Simplified	Medium
	Building zone (nodal model)	Building energy simulation, predictive modeling and diagnostics, control	1D finite differences, transfer functions, state space	Simplified	High

3.4.2 Forward/Inverse Modeling and Visualization Techniques

As illustrated before, building modeling techniques can be applied in forward (law driven) [114], calibrated (law driven, model calibrated on real data) [51, 101], and inverse (data driven) modeling. Inverse modeling can be used to determine relevant components [131, 194] of the energy balance and lumped parameters [182, 183], following the general definitions given in Table 3.4.

Inverse models are built upon measured data, and it is fundamental to couple them with visualization techniques, in order not only to facilitate the derivation of insights (reflection on processes) directly from data and to improve the supervision by humans (Table 3.5). The necessity of the use of models and computing tools to derive insights is evident if we look at the articulation of data/metadata presented in Table 3.7 together with the data collected by automation systems and sensors (measured data), summarized in Table 3.10. A useful classification of parameters according to their meaning in modeling is given in [120].

With respect to visualization techniques, the most common ones are the following:

1) Scatterplots (e.g., hourly/daily monthly data) to visualize correlation and identify trends [18, 66, 131, 195–198]
2) Load curves (e.g., load matching, production and demand, load duration) [174, 175, 177]
3) Histograms and bin methods (e.g., distribution of quantities) [196, 197, 199]
4) Carpet plots or 2D/3D surface plots (e.g., variability of patterns and anomalies in time series) [200, 201]

These techniques can be used to perform tasks such as anomaly detection and association discovery in a graphical way (visual analytics).

3.4.3 Workflows and Integration of Modeling Strategies

Considering the state of the art of building performance modeling, the potential workflows are depicted synthetically in Figure 3.4 as a graph, employing the definitions previously reported. This graph excludes top-down models (econometric, technological) and focuses on bottom-up (engineering oriented) models, which are generally used in building design and operation practices. The fundamental distinction introduced is among:

1) Forward (design process, law driven)
2) Calibrated (model calibrated in the commissioning and operation phases on measured data)
3) Inverse (data driven model, trained on simulated or measured data)

If we exclude from the initial graph in Figure 3.4 the workflows exploiting the integration among white-, gray-, and black-box model types, we obtain three

Table 3.10 Data acquired by automation systems and sensors.

Element	Quantity	Unit	Description
Energy demand	Total fuel consumptions	W h, kW h, MW h, m³, l, kg	Cumulative value from meter reading
	Total energy consumption (heating and cooling)	W h, kW h, MW h	
	Total electricity consumption	W h, kW h, MW h	
	Total water consumption	m³, l	
Weather	Outdoor temperature	°C	Value from on-site weather station or weather data provider
	Outdoor relative humidity	%	
	Global irradiation on horizontal surface	W/m²	
Indoor conditions	Indoor temperature	°C	Value from selected reference zone
	Indoor relative humidity	%	
	CO₂ concentration	ppm	
	Illuminance	lux	
	Other detailed internal data (air velocity, VOC concentration, etc.)	ppm, m/s, and so on	
Systems	Supply/return air temperature – water circuits	°C	Value from system of the selected reference zone
	Supply/return air temperature – AHU	°C	
	Supply/return air relative humidity – AHU	%	
	Settings and signal for thermostats and humidostats	°C, %, on/off, 0/1	
	Flow meters	l/s, m³/h	
	Pressure meters	kPa, Pa	

Table 3.10 (Continued)

Element	Quantity	Unit	Description
	Control signals for pumps and fans	0–100%, 0–1, ON/OFF, 0/1	
	Control signals for presence sensors	0–100%, 0–1, ON/OFF, 0/1	
	Control signals for CO_2 sensors	0–100%, 0–1, ON/OFF, 0/1	
	Control signals for other detailed data metered	0–100%, 0–1, ON/OFF, 0/1	
Sampling and analysis interval	Data polling interval	s, min, h, d	Value depending on the type of data analysis
	Data acquisition length period	d, mo, yr	

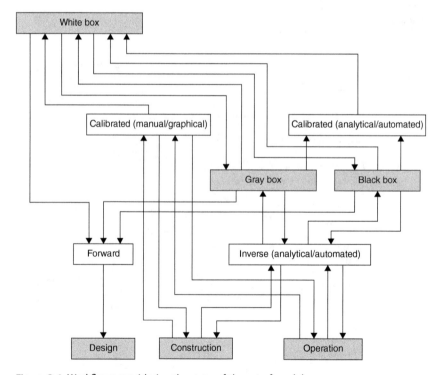

Figure 3.4 Workflows considering the state-of-the-art of models.

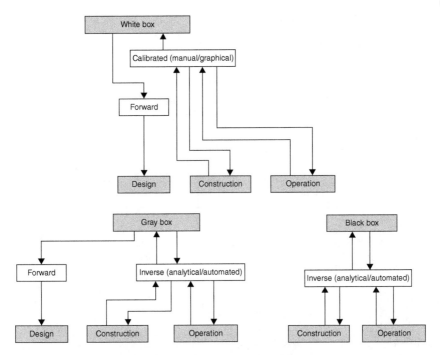

Figure 3.5 Separated workflows for different types of models.

subgraphs, which are shown in Figure 3.5. In this case, we can immediately see how white-box models have limited applicability with respect to construction/commissioning and operation processes, relying on time-consuming and error-prone procedures that depend on the experience of the modeler [51, 117]. Gray-box models can be used across the three processes, although within the limits outlined in Table 3.9 (simplified representation of physical phenomena). Finally, black-box models, which are efficient from the computational point of view and relatively simple to implement, are mainly used in the construction/commissioning and operation processes and have to rely necessarily on measured data. After having analyzed the workflows without considering model integration, we outline in Figure 3.6 the workflows that become possible with model integration. The additional workflows are the following:

1) Gray-/black-box model-based optimization, uncertainty, and sensitivity analysis (reduced-order models derived from white-box simulation data)
2) Gray-/black-box model-based commissioning and functional testing (reduced-order models derived from white-box simulation data and compared with measured data)

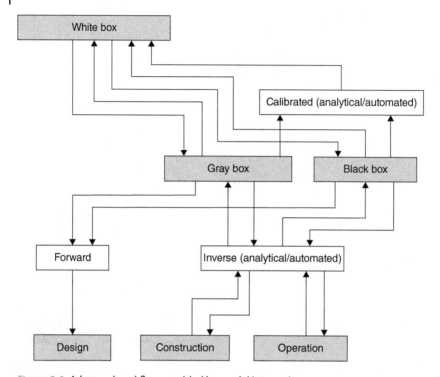

Figure 3.6 Advanced workflows enabled by model integration.

3) Gray-/black-box model-based calibration (reduced-order model-based automated calibration on measured data)
4) Gray-/black-box model-based control (reduced-order models trained on white-box simulation data, calibrated on measured data, and used for control)

The fact that model integration is crucial to bridge the gap among processes is evident [76, 149]. Given the evidences of the analysis, we can sketch two possible paths of development, starting from current practices:

1) Improvement of the existing tools by means of new theory aimed at enabling efficient and integrated workflows [74, 202, 203]
2) Improvement of gray-box modeling strategies (acknowledging their limits but keeping a physics-based representation) and synthesis of different approaches for integrated simulation, optimization, calibration, and control [77, 82, 84, 204–207]

The first path appears to be more ambitious as it requires new development in theory, while the second, which is not as pervasive as the first one, seems to be more suitable for short-/medium-term implementation. In the first case in fact, results can be potentially disruptive, but the effects probably would be gained in the medium/long term, while in the second case the interactions among multiple domains and applications can be easily exploited by using similar solution strategies.

3.4.4 Research Advances in Modeling and Computing

Considering the potential inherent to the integration of the modeling strategies, described in the previous section, we can envisage some research advances in modeling and computing for the built environment. First, we have to focus on continuous improvement with respect to current practices and on practitioners' training. In fact, it is necessary to start from current practices and to introduce new elements in them, which have to be coherent with the general objective of sustainability. In this sense, we can judge the effectiveness of modeling and computing strategies based on their ability to provide an appropriate representation of the different dynamics and domains at the building (occupancy, comfort conditions, natural/mechanical/hybrid ventilation, heating, cooling, lighting use, appliances use, etc.) and district/urban level (energy demand, energy conversion, energy storage, etc.). Models are appropriate if they are able to capture the fundamental elements governing phenomena, simply not only at the physical level but also at the behavioral level (physiological and psychological), and to define strategies to control them. Control of physical phenomena should be coupled with actions oriented toward the mobilization of stakeholders (Table 3.2) whose behavior and decisions can practically affect the sustainability of the built environment. For this reason, a specific effort should be put on semantic schemas, not only to remove ambiguities in the definition of entities in modeling but also to define quantities suitable for performance benchmarking and comparison. The advantages of an object-oriented modeling approach are summarized in Table 3.6. Knowledge encapsulated within objects should be complemented by appropriate data/metadata to clarify the context of applicability of modeling strategies and computing techniques. We can think about "smart" computing objects, which are using correctly all the available information. The current limitations are particularly evident if we consider problems where "two-way" domain coupling is necessary (i.e., bidirectional flow of information). With "smart" object-oriented modeling, we can overcome the deadlock of the development of single programs or multiple cooperating programs, in favor of a more open perspective. Further, considering the different model types described in Table 3.4, we have the possibility to employ specific model reduction strategies (from white box to gray box and black box), using coherent methodologies for verification, validation, and uncertainty analysis [208, 209]. Of course, there is the need to investigate formally the interaction among different domains and to define appropriate levels of abstraction and

Table 3.11 Examples of software to simulate, optimize, and control multiple dynamics from building to city scale.

Scale	Capabilities	Software
Building	Simulation/parametric simulation	E+, Trnsys, ESP-R, DOE 2, iEPlus
	Optimization	GenOpt
	Control	RBCM Toolbox, MLE+ Toolbox
District/ neighborhood	Simulation	UMI (Urban Modeling Interface)
	Simulation/optimization	Homer Energy
	Optimization	DER CAM (WebOpt)
City/region	Simulation/optimization	EnergyPlan

approximation in problem solving, dependenting of the specific task. Finally, we report in Table 3.11 a list of software to simulate, optimize, and control different domains and dynamics at the building, district, and city scales [34, 41, 75, 77, 165], together with examples of numerical techniques with applications in the built environment [115], summarized in Table 3.12. Numerical techniques can be used to manage the complexity of multi-scale problems (e.g., exploiting analogies in solution strategies), and optimization techniques, in particular, are interesting as they enable the use of uniform "low-level" modeling rules (linear/nonlinear programming, mixed-integer programming, convex programming, dynamic programming, stochastic programming, etc.) [133, 210–212] while developing specific interfaces at the "high level" (model deployment). There exist already evidences of the feasibility of the use of these techniques for large-scale applications [204–207].

Final Thoughts

The promotion of sustainable human settlements that are able to drive social development and economic growth is a multifaceted problem. Considering the urbanization trend at the global level, cities constitute a priority for research and development in sustainability transitions, which should necessarily face techno- and socio economic problems. Energy use and technology affect sustainability in its fundamental components, society, environment, and economy. For the built environment, in particular, conventional energy planning and technological learning models are not sufficient because of their inability to deal with issues such as the behavior of consumers (end users, occupants) and investors, as well as the institutional factors driving decisions, in particular,

Table 3.12 Numerical techniques for multiple applications in the built environment.

Category	Numerical techniques	Applications
Clustering	k-Means	Aggregated spatial and temporal properties of systems, identification of recurrent patterns (occupancy and load profiles), and so on
Regression	Multivariate linear and nonlinear regression, k-nearest neighborhood	Energy management and monitoring, model calibration, fault diagnosis and detection, technical systems operating curves, and so on
Surrogate modeling (reduced-order modeling or meta-modeling)	Neural network, support vector machines, Gaussian processes	Energy management and monitoring, model calibration, fault diagnosis and detection, technical systems operating curves, and so on
Time series	ARX, ARIMA, ARMAX, state space	Simulation of dynamic phenomena such as behavior of occupants, heat transfer, energy demand, energy conversion and storage, and so on
Optimization	Linear/nonlinear programming, convex programming, dynamic programming, heuristics	Optimization of stationary and dynamic problems, model predictive control of buildings, optimal dispatch of distributed energy resources (match demand and production), and so on

at the district and urban scale. These issues inherently highlight the need for a research in multi-scale computing for the built environment, which should contribute to the optimal design and operation of decentralized energy systems, phasing out fossil fuels by means of an increasing share of renewables in energy production and increasing efficiency in end-use energy demand. It is necessary to establish a methodological continuity among design and operation practices aimed, on the one hand, at reducing the gap among simulated and real performance of buildings and, on the other hand, at transforming buildings into intelligent nodes of our evolving infrastructures. Built environment evolution requires a long-term strategy because of its long-term assets but can exploit the short and medium term evolution of ICT and energy infrastructures, respectively, toward IoT and Smart Grid paradigms. These evolutions will involve the whole economy and society. Design, construction, and operation practices in the built environment can profoundly change with respect to present state, but there should be a clear commitment to sustainability from design phase up to the complete building life cycle. The improvement of practices should be based on incremental learning from field experience, considering both the KPIs employed and the modeling and computing processes performed, fostering a fruitful collaboration among stakeholders and practitioners.

Questions

1　Why are sustainability transitions and decarbonization fundamental?

2　Which ones are the most critical elements in sustainability transition pathways planning?

3　Why is it important to focus on cities and built environment?

4　What are the specific challenges of research and development in multi-scale computing for the built environment?

5　Which strategies can be pursued to overcome the present limitations in research and development?

References

1 Brundtland, G.H. (1987) Our common future—Call for action. *Environmental Conservation.*, **14**, 291–294.

2 Allen, M.R., Barros, V.R., Broome, J., Cramer, W., Christ, R., Church, J.A. et al. (2014) *IPCC Fifth Assessment Synthesis Report-Climate Change 2014 Synthesis Report.*

3 International Energy Agency (2015) *World Energy Outlook 2015*, International Energy Agency.

4 Energy Outlook 2015. Exxon Mobile. 2015.

5 Olivier, J.G., Janssens-Maenhout, G., Peters J. A. and Marilena, M. (2015) *Trends in global CO_2 emissions: 2015 Report: PBL Netherlands Environmental Assessment Agency Hague.*

6 de Wilde, P. and Coley, D. (2012) The implications of a changing climate for buildings. *Building and Environment.*, **55**, 1–7.

7 Sachs, J.D. (2012) From millennium development goals to sustainable development goals. *The Lancet*, **379**, 2206–2211.

8 Robert, K.W., Parris, T.M., and Leiserowitz, A.A. (2005) What is sustainable development? Goals, indicators, values, and practice. *Environment: science and policy for sustainable development*, **47**, 8–21.

9 Griggs, D., Stafford-Smith, M., Gaffney, O. *et al.* (2013) Policy: Sustainable development goals for people and planet. *Nature,* **495,** 305–307.

10 Dobbs, R., Smit, S., Remes, J. *et al.* (2011) *Urban world: Mapping the economic power of cities,* McKinsey Global Institute.

11 Sawin, J.L., Sverrisson, F., Chawla, K., Lins, C., Adib, R., Hullin, M., *et al.* (2014) Renewables 2014. Global status report 2014.

12 Babiker, M.H. (2005) Climate change policy, market structure, and carbon leakage. *Journal of International Economics.,* **65,** 421–445.

13 Bernard, A. and Vielle, M. (2009) Assessment of European Union transition scenarios with a special focus on the issue of carbon leakage. *Energy Economics.,* **31,** S274–S284.

14 Böhringer, C., Carbone, J.C., and Rutherford, T.F. (2012) Unilateral climate policy design: Efficiency and equity implications of alternative instruments to reduce carbon leakage. *Energy Economics.,* **34,** S208–S217.

15 Stiglitz, J.E. and Greenwald, B.C. (2014) *Creating a learning society: a new approach to growth, development, and social progress,* Columbia University Press.

16 Preston, F. (2012) *A Global Redesign? Shaping the Circular Economy* Energy, Environment and Resource Governance, Chatham House, London.

17 MacArthur, E. (2013) *Towards the Circular Economy,* Ellen Mac Arthur Foundation.

18 Paulus, M.T., Claridge, D.E., and Culp, C. (2015) Algorithm for automating the selection of a temperature dependent change point model. *Energy and Buildings,* **87,** 95–104.

19 Jalori, S. and Agami Reddy, T. (2015) A unified inverse modeling framework for whole-building energy interval data: Daily and hourly baseline modeling and short-term load forecasting. *ASHRAE Transactions,* **121,** 156.

20 Bertoldi, P., Cayuela, D.B., Monni, S., and de Raveschoot, R.P. (2010) *Existing Methodologies and Tools for the Development and Implementation of Sustainable Energy Action Plans (SEAP),* Publications Office.

21 Torres, P.B. and Doubrava, R. (2010) The Covenant of Mayors: Cities leading the fight against the climate change, in *Local Governments and Climate Change: Sustainable Energy Planning and Implementation in Small and Medium Sized Communities* (eds M. Staden and F. Musco), Springer, p. 91.

22 Lin, G. and Claridge, D.E. (2015) A temperature-based approach to detect abnormal building energy consumption. *Energy and Buildings,* **93,** 110–118.

23 Castillo, L., Enríquez, R., Jiménez, M.J., and Heras, M.R. (2014) Dynamic integrated method based on regression and averages, applied to estimate the thermal parameters of a room in an occupied office building in Madrid. *Energy and Buildings*, **81**, 337–362.

24 Katz, R.H., Culler, D.E., Sanders, S. *et al.* (2011) An information-centric energy infrastructure: The Berkeley view. *Sustainable Computing: Informatics and Systems*, **1**, 7–22.

25 Gubbi, J., Buyya, R., Marusic, S., and Palaniswami, M. (2013) Internet of Things (IoT): A vision, architectural elements, and future directions. *Future Generation Computer Systems*, **29**, 1645–1660.

26 Vermesan, O. and Friess, P. (2014) *Internet of Things-From Research and Innovation to Market Deployment*, River Publishers.

27 Glikson, A. (2011) *FI-WARE: Core Platform for Future Internet Applications*. Proceedings of the 4th Annual International Conference on Systems and Storage.

28 Vázquez, A.G., Soria-Rodriguez, P., Bisson, P. *et al.* (2011) *FI-WARE security: future internet security core* Towards a Service-Based Internet, Springer, pp. 144–152.

29 Lund, H. (2009) *Renewable energy systems: The choice and modeling of 100% renewable solutions*, Academic Press.

30 Fan, Z., Kulkarni, P., Gormus, S. *et al.* (2013) Smart grid communications: overview of research challenges, solutions, and standardization activities. *Communications Surveys & Tutorials, IEEE.*, **15**, 21–38.

31 Ramparany, F., Galan Marquez, F., Soriano, J., and Elsaleh, T. (2014) *Handling Smart Environment Devices, Data and Services at the Semantic Level with the FI-WARE Core Platform*. Big Data (Big Data), 2014 IEEE International Conference on: IEEE; 2014. pp. 14–20.

32 Schaffers, H., Komninos, N., Pallot, M. *et al.* (2011) Smart cities and the future internet: Towards cooperation frameworks for open innovation. *Future Internet Assembly*, **6656**, 431–446.

33 Zanella, A., Bui, N., Castellani, A. *et al.* (2014) Internet of things for smart cities. *Internet of Things Journal, IEEE*, **1**, 22–32.

34 Manfren, M., Caputo, P., and Costa, G. (2011) Paradigm shift in urban energy systems through distributed generation: Methods and models. *Applied Energy*, **88**, 1032–1048.

35 Knopf, B., Chen, Y.-H.H., De Cian, E. *et al.* (2013) Beyond 2020—Strategies and costs for transforming the European energy system. *Climate Change Economics.*, **4**, 1, 1340001, 1–38.

36 Creutzig, F., Goldschmidt, J.C., Lehmann, P. *et al.* (2014) Catching two European birds with one renewable stone: Mitigating climate change and Eurozone crisis by an energy transition. *Renewable and Sustainable Energy Reviews*, **38**, 1015–1028.

37 Energy Roadmap 2050. European Commission. 2011.

38 Hübler, M. and Löschel, A. (2013) The EU Decarbonisation Roadmap 2050—What way to walk? *Energy Policy*, **55**, 190–207.

39 Robertson, J.J., Polly, B.J., and Collis, J.M. (2015) Reduced-order modeling and simulated annealing optimization for efficient residential building utility bill calibration. *Applied Energy*, **148**, 169–177.

40 Mathiesen, B.V., Lund, H., and Karlsson, K. (2011) 100% renewable energy systems, climate mitigation and economic growth. *Applied Energy*, **88**, 488–501.

41 Connolly, D., Lund, H., Mathiesen, B.V., and Leahy, M. (2010) A review of computer tools for analysing the integration of renewable energy into various energy systems. *Applied Energy*, **87**, 1059–1082.

42 Fleiter, T., Worrell, E., and Eichhammer, W. (2011) Barriers to energy efficiency in industrial bottom-up energy demand models—a review. *Renewable and Sustainable Energy Reviews*, **15**, 3099–3111.

43 Olanrewaju, O.A. and Jimoh, A.A. (2014) Review of energy models to the development of an efficient industrial energy model. *Renewable and Sustainable Energy Reviews*, **30**, 661–671.

44 Kissock, F.S.K., Brown, P.D., and Mulqueen, P.S. (2011) Estimating industrial building energy savings using inverse simulation. *ASHRAE Transactions*, **117**, 348–356.

45 Masuda, H. and Claridge, D.E. (2014) Statistical modeling of the building energy balance variable for screening of metered energy use in large commercial buildings. *Energy and Buildings*, **77**, 292–303.

46 Scofield, J.H. (2009) Do LEED-certified buildings save energy? Not really…. *Energy and Buildings*, **41**, 1386–1390.

47 Newsham, G.R., Mancini, S., and Birt, B.J. (2009) Do LEED-certified buildings save energy? Yes, but…. *Energy and Buildings*, **41**, 897–905.

48 Alyami, S.H. and Rezgui, Y. (2012) Sustainable building assessment tool development approach. *Sustainable Cities and Society*, **5**, 52–62.

49 Ng, S.T., Chen, Y., and Wong, J.M. (2013) Variability of building environmental assessment tools on evaluating carbon emissions. *Environmental impact assessment review*, **38**, 131–141.

50 *Sustainability and Performance assessment and benchmarking of buildings.* http://cic.vtt.fi/superbuildings/ (accessed 24 January 2017).

51 Fabrizio, E. and Monetti, V. (2015) Methodologies and advancements in the calibration of building energy models. *Energies*, **8**, 2548.

52 Geels, F.W. (2005) Processes and patterns in transitions and system innovations: refining the co-evolutionary multi-level perspective. *Technological forecasting and social change*, **72**, 681–696.

53 Foxon, T.J., Hammond, G.P., and Pearson, P.J. (2010) Developing transition pathways for a low carbon electricity system in the UK. *Technological Forecasting and Social Change*, **77**, 1203–1213.

54 ISO 14001:2015 (2015) Environmental management systems – Requirements with guidance for use.

55 Bhattacharyya, S.C. and Timilsina, G.R. (2010) A review of energy system models. *International Journal of Energy Sector Management.*, **4**, 494–518.

56 Energy models https://www.ucl.ac.uk/energy-models/models (accessed 24 January 2017).

57 Salas, P. (2013) *Literature Review of Energy-Economics Models, Regarding Technological Change and Uncertainty*, University of Cambridge, Department of Land Economy, Cambridge Centre for Climate Change Mitigation Research.

58 Corgnati, S.P., Fabrizio, E., Filippi, M., and Monetti, V. (2013) Reference buildings for cost optimal analysis: Method of definition and application. *Applied Energy.*, **102**, 983–993.

59 Aste, N., Adhikari, R.S., and Manfren, M. (2013) Cost optimal analysis of heat pump technology adoption in residential reference buildings. *Renewable Energy.*, **60**, 615–624.

60 Hamdy, M., Hasan, A., and Siren, K. (2013) A multi-stage optimization method for cost-optimal and nearly-zero-energy building solutions in line with the EPBD-recast 2010. *Energy and Buildings*, **56**, 189–203.

61 Ferrara, M., Fabrizio, E., Virgone, J., and Filippi, M. (2014) A simulation-based optimization method for cost-optimal analysis of nearly zero energy buildings. *Energy and Buildings*, **84**, 442–457.

62 Desideri, U., Arcioni, L., Leonardi, D. *et al.* (2014) Design of a multi-purpose "zero energy consumption" building according to European Directive 2010/31/EU: Life cycle assessment. *Energy and Buildings*, **80**, 585–597.

63 Han, G., Srebric, J., and Enache-Pommer, E. (2014) Variability of optimal solutions for building components based on comprehensive life cycle cost analysis. *Energy and Buildings*, **79**, 223–231.

64 Commission Delegated Regulation (EU) No 244/2012 (2012) http://eur-lex .europa.eu/LexUriServ/LexUriServ.do?uri=OJ:L:2012:081:0018:0036:en:PDF (accessed 24 January 2017).

65 Guidelines accompanying Commission Delegated Regulation (EU) No 244/2012 (2012) http://eur-lex.europa.eu/legal-content/EN/TXT/PDF/? uri=CELEX:52012XC0419%2802%29&from=EN (accessed 24 January 2017).

66 Bynum, J.D., Claridge, D.E., and Curtin, J.M. (2012) Development and testing of an Automated Building Commissioning Analysis Tool (ABCAT). *Energy and Buildings*, **55**, 607–617.

67 Mills, E. (2011) Building commissioning: A golden opportunity for reducing energy costs and greenhouse gas emissions in the United States. *Energy Efficiency*, **4**, 145–173.

68 Neumann, C. and Jacob, D. (2010) *Results of the project building EQ. Tools and methods for linking EPBD and continuous commissioning*, Fraunhofer ISE.

69 Invert/EE-Lab – Modelling the energy demand for space heating and cooling in building stocks, http://www.invert.at/ (accessed 24 January 2017).

70 Kranzl, L., Müller, A., Toleikyte, A. *et al.* (2014) *Policy Pathways for Reducing Energy Demand and Carbon Emissions of the EU Building Stock until 2030*, EU Entranze Project http://www.entranze.eu/ (accessed 24 January 2017).

71 Kranzl, L., Müller, A., Toleikyte, A. *et al.* (2015) *What Drives the Impact of Future Support Policies for Energy Efficiency in Buildings?* ECEEE Summer Study Proceeding, 1391–1402.

72 The shift project, http://theshiftproject.org/this-page/rogeaulito (accessed 24 January 2017).

73 Dynemo project, https://www.ucl.ac.uk/energy-models/models/dynemo (accessed 24 January 2017).

74 Evins, R. (2013) A review of computational optimisation methods applied to sustainable building design. *Renewable and Sustainable Energy Reviews*, **22**, 230–245.

75 Nguyen, A.-T., Reiter, S., and Rigo, P. (2014) A review on simulation-based optimization methods applied to building performance analysis. *Applied Energy*, **113**, 1043–1058.

76 de Wilde, P. (2014) The gap between predicted and measured energy performance of buildings: A framework for investigation. *Automation in Construction*, **41**, 40–49.

77 Lehmann, B., Gyalistras, D., Gwerder, M. *et al.* (2013) Intermediate complexity model for Model Predictive Control of Integrated Room Automation. *Energy and Buildings*, **58**, 250–262.

78 Kossak, B. and Stadler, M. (2015) Adaptive thermal zone modeling including the storage mass of the building zone. *Energy and Buildings*, **109**, 407–417.

79 Aste, N., Leonforte, F., Manfren, M., and Mazzon, M. (2015) Thermal inertia and energy efficiency – Parametric simulation assessment on a calibrated case study. *Applied Energy*, **145**, 111–123.

80 Verhelst, C., Axehill, D., Jones, C., and Helsen, L. (2010) *Impact of the Cost Function in the Optimal Control Formulation for an Air-to-Water Heat Pump System*. 8th International Conference on System Simulation in Buildings (SSB), Liege, Belgium.

81 Verhelst, C., Degrauwe, D., Logist, F. *et al.* (2012) Multi-objective optimal control of an air-to-water heat pump for residential heating. *Building Simulation*, **5**, 281–291.

82 De Coninck, R., Magnusson, F., Åkesson, J., and Helsen, L. (2015) Toolbox for development and validation of gray-box building models for forecasting and control. *Journal of Building Performance Simulation*, **9**, 1–16.

83 Lauster, M., Teichmann, J., Fuchs, M. *et al.* (2014) Low order thermal network models for dynamic simulations of buildings on city district scale. *Building and Environment*, **73**, 223–231.

84 Fonseca, J.A., Nguyen, T.-A., Schlueter, A., and Marechal, F. (2016) City Energy Analyst (CEA): Integrated framework for analysis and optimization of building energy systems in neighborhoods and city districts. *Energy and Buildings*, **113**, 202–226.

85 Vesco, A. (2015) *Handbook of Research on Social, Economic, and Environmental Sustainability in the Development of Smart Cities*, IGI Global.

86 Dockins, T.M. and Huber, M. (2012) Social Influence Modeling for Utility Functions in Model Predictive Control. FLAIRS Conference 2012.

87 Keirstead, J., Jennings, M., and Sivakumar, A. (2012) A review of urban energy system models: Approaches, challenges and opportunities. *Renewable and Sustainable Energy Reviews*, **16**, 3847–3866.

88 IEA - EBC Annex 70 - Building Energy Epidemiology. https:// energyepidemiology.org/ (accessed 24 January 2010).

89 Miller, C. and Schlueter, A. (2013) *Applicability of Lean Production Principles to Performance Analysis Across the Life Cycle Phases of Buildings*. Proceedings of the 8th International Conference on Indoor Air Quality, Ventilation and Energy Conservation in Buildings (CLIMA 2013), Prague, Czech Republic 2013. pp. 1–10.

90 Krijnen, A. (2014) *The Toyota way: 14 Management Principles from the World's Greatest Manufacturer*, McGraw Hill.

91 Womack, J.P. and Jones, D.T. (2010) *Lean thinking: banish waste and create wealth in your corporation*, Simon and Schuster.

92 Fazlollahi, S., Bungener, S.L., Mandel, P. *et al.* (2014) Multi-objectives, multi-period optimization of district energy systems: I. Selection of typical operating periods. *Computers & Chemical Engineering*, **65**, 54–66.

93 Fazlollahi, S., Becker, G., and Maréchal, F. (2014) Multi-objectives, multi-period optimization of district energy systems: II—Daily thermal storage. *Computers & Chemical Engineering*, **71**, 648–662.

94 Fazlollahi, S., Becker, G., and Maréchal, F. (2014) Multi-objectives, multi-period optimization of district energy systems: III. Distribution networks. *Computers & Chemical Engineering*, **66**, 82–97.

95 Kavgic, M., Mavrogianni, A., Mumovic, D. *et al.* (2010) A review of bottom-up building stock models for energy consumption in the residential sector. *Building and Environment*, **45**, 1683–1697.

96 H-x, Z. and Magoulès, F. (2012) A review on the prediction of building energy consumption. *Renewable and Sustainable Energy Reviews*, **16**, 3586–3592.

97 Swan, L.G. and Ugursal, V.I. (2009) Modeling of end-use energy consumption in the residential sector: A review of modeling techniques. *Renewable and Sustainable Energy Reviews*, **13**, 1819–1835.

98 Harish, V. and Kumar, A. (2016) A review on modeling and simulation of building energy systems. *Renewable and Sustainable Energy Reviews*, **56**, 1272–1292.

99 Fumo, N. (2014) A review on the basics of building energy estimation. *Renewable and Sustainable Energy Reviews*, **31**, 53–60.

100 Foucquier, A., Robert, S., Suard, F. *et al.* (2013) State of the art in building modelling and energy performances prediction: A review. *Renewable and Sustainable Energy Reviews*, **23**, 272–288.

101 Coakley, D., Raftery, P., and Keane, M. (2014) A review of methods to match building energy simulation models to measured data. *Renewable and Sustainable Energy Reviews*, **37**, 123–141.

102 Henze, G.P. (2013) Model predictive control for buildings: a quantum leap? *Journal of Building Performance Simulation.*, **6**, 157–158.

103 Shaikh, P.H., Nor, N.B.M., Nallagownden, P. *et al.* (2014) A review on optimized control systems for building energy and comfort management of smart sustainable buildings. *Renewable and Sustainable Energy Reviews*, **34**, 409–429.

104 Yu, Z., Huang, G., Haghighat, F. *et al.* (2015) Control strategies for integration of thermal energy storage into buildings: State-of-the-art review. *Energy and Buildings*, **106**, 203–215.

105 EN 16212:2012 (2012) *Energy Efficiency and Savings Calculation – Top-down and Bottom-up Methods.*

106 Ballard, G. and Howell, G. (2003) Lean project management. *Building Research & Information*, **31**, 119–133.

107 Sacks, R., Koskela, L., Dave, B.A., and Owen, R. (2010) Interaction of lean and building information modeling in construction. *Journal of construction engineering and management*, **136**, 968–980.

108 ISO 16343:2013 (2013) *Energy Performance of Buildings – Methods for Expressing Energy Performance and for Energy Certification of Buildings.*

109 EN 15459:2008 (2008) *Energy Efficiency for Buildings – Standard Economic Evaluation Procedure for Energy Systems in Buildings.*

110 EN 15251:2008 (2008) *Indoor Environmental Input Parameters for Design and Assessment of Energy Performance of Buildings Addressing Indoor Air Quality, Thermal Environment, Lighting and Acoustics.*

111 ISO 7730:2008 (2008) *Ergonomics of the Thermal Environment – Analytical Determination and Interpretation of Thermal Comfort Using Calculation of the PMV and PPD Indices and Local Thermal Comfort Criteria.*

112 ANSI/ASHRAE 55:2013 (2013) Thermal Environmental Conditions for Human Occupancy.

113 Boakye-Adjei, K., Thamma, R., and Kirby, E.D. *Autonomation: The Future of Manufacturing.* Proceedings of The 2014 IAJC-ISAM International Conference.

114 Hensen, J.L. and Lamberts, R. (2012) *Building performance simulation for design and operation*, Routledge.

115 Reddy, T.A. (2011) *Applied Data Analysis and Modeling for Energy Engineers and Scientists*, Springer, US.

116 Krarti, M. (2012) *Weatherization and Energy Efficiency Improvement for Existing Homes: An Engineering Approach*, CRC Press.

117 *ASHRAE Research Project Report 1404-RP*, Measurement, Modeling, Analysis and Reporting Protocols for Short-term M&V of Whole Building Energy Performance.

118 Reddy, T.A., Maor, I., and Panjapornpon, C. (2007) Calibrating detailed building energy simulation programs with measured data—Part I: General methodology (RP-1051). *HVAC&R Research*, **13**, 221–241.

119 Manfren, M., Aste, N., and Moshksar, R. (2013) Calibration and uncertainty analysis for computer models – A meta-model based approach for integrated building energy simulation. *Applied Energy*, **103**, 627–641.

120 Yang, Z. and Becerik-Gerber, B. (2015) A model calibration framework for simultaneous multi-level building energy simulation. *Applied Energy*, **149**, 415–431.

121 Calleja Rodríguez, G., Carrillo Andrés, A., Domínguez Muñoz, F. *et al.* (2013) Uncertainties and sensitivity analysis in building energy simulation using macroparameters. *Energy and Buildings*, **67**, 79–87.

122 Yan, C., Wang, S., Xiao, F., and Gao, D.-c. (2015) A multi-level energy performance diagnosis method for energy information poor buildings. *Energy*, **83**, 189–203.

123 Sandels, C., Brodén, D., Widén, J. *et al.* (2016) Modeling office building consumer load with a combined physical and behavioral approach: Simulation and validation. *Applied Energy*, **162**, 472–485.

124 Yan, D., O'Brien, W., Hong, T. *et al.* (2015) Occupant behavior modeling for building performance simulation: current state and future challenges. *Energy and Buildings*, **107**, 264–278.

125 Gaetani, I., Hoes, P.-J., and Hensen, J.L. (2016) Occupant behavior in building energy simulation: Towards a fit-for-purpose modeling strategy. *Energy and Buildings*, **121**, 188–204.

126 Tagliabue, L.C., Manfren, M., Ciribini, A.L.C., and De Angelis, E. (2016) Probabilistic behavioral modeling in building performance simulation – The Brescia eLUX lab. *Energy and Buildings*, **128**, 119–31.

127 Rawlings, J., Coker, P., Doak, J., and Burfoot, B. (2014) Do smart grids offer a new incentive for SME carbon reduction? *Sustainable Cities and Society*, **10**, 245–250.

128 Rawlings, J., Coker, P., Doak, J., and Burfoot, B.(2014) *A Clustering Approach to Support SME Carbon Reduction*, Building Simulation & Optimisation Conference (BSO 14), London, June 2014.

129 Rawlings, J., Coker, P., Doak, A., and Burfoot, B. (2013) *The Need for New Building Energy Models to Support SME Carbon Reduction*, in Proc. 4th TSBE EngD Conference, University of Reading, July 2013.

130 Hsu, D. (2015) Identifying key variables and interactions in statistical models of building energy consumption using regularization. *Energy*, **83**, 144–155.

131 Zhang, Y., O'Neill, Z., Dong, B., and Augenbroe, G. (2015) Comparisons of inverse modeling approaches for predicting building energy performance. *Building and Environment*, **86**, 177–190.

132 Harish, V. and Kumar, A. (2016) Reduced order modeling and parameter identification of a building energy system model through an optimization routine. *Applied Energy*, **162**, 1010–1023.

133 Powell, W.B. and Ryzhov, I.O. (2012) *Optimal Learning*, John Wiley & Sons.

134 Bleil de Souza, C. (2013) Studies into the use of building thermal physics to inform design decision making. *Automation in Construction*, **30**, 81–93.

135 Macdonald, I. and Strachan, P. (2001) Practical application of uncertainty analysis. *Energy and Buildings*, **33**, 219–227.

136 Hopfe, C.J. and Hensen, J.L. (2011) Uncertainty analysis in building performance simulation for design support. *Energy and Buildings*, **43**, 2798–2805.

137 DesignBuilder http://www.designbuilder.co.uk/ (accessed 24 January 2017).

138 JEPlus – An EnergyPlus simulation manager for parametrics http://www.jeplus.org/wiki/doku.php (accessed 24 January 2017).

139 BEOPT – Building Energy OPTimization (https://beopt.nrel.gov/).

140 Apidae labs (https://apidaelabs.com/).

141 ExcaliBEM (https://www.simeb.ca/ExCalibBEM/index_en.php).

142 ANSI/ASHRAE 140:2011 (2011) *Standard Method of Test for the Evaluation of Building Energy Analysis Computer Programs*.

143 ISO 13791:2012 (2012) *Thermal Performance of Buildings – Calculation of Internal Temperatures of a Room in Summer without Mechanical Cooling*, General criteria and validation procedures.

144 Efficiency Valuation Organization (EVO) (2003) *IPMVP New Construction Subcommittee. International Performance Measurement & Verification Protocol: Concepts and Option for Determining Energy Savings in New Construction*, Volume III, Efficiency Valuation Organization (EVO).

145 U.S. Department of Energy Federal Energy Management Program, FEMP. Federal Energy Management Program, M&V Guidelines: Measurement and Verification for Federal Energy Projects Version 3.0. 2008.

146 ASHRAE Guideline 14–2002: Measurement of Energy Demand and Savings.

147 Buildings data hub, Building Performance Institute Europe http://www.buildingsdata.eu/ (accessed 24 January 2017).

148 Building performance database http://energy.gov/eere/buildings/building-performance-database (accessed 24 January 2017).

149 Brown, D., Burns, J., Collis, S. *et al.* (2011) *Applied & Computational Mathematics Challenges for the Design and Control of Dynamic Energy Systems*, Lawrence Livermore National Laboratory (LLNL), Livermore, CA.

150 Samuelson, H.W., Lantz, A., and Reinhart, C.F. (2012) Non-technical barriers to energy model sharing and reuse. *Building and Environment*, **54**, 71–76.

151 Jaffal, I., Inard, C., and Bozonnet, E. (2012) Toward integrated building design: A parametric method for evaluating heating demand. *Applied Thermal Engineering*, **40**, 267–274.

152 Jaffal, I., Inard, C., and Ghiaus, C. (2009) Fast method to predict building heating demand based on the design of experiments. *Energy and Buildings*, **41**, 669–677.

153 Hang, Y., Qu, M., and Ukkusuri, S. (2011) Optimizing the design of a solar cooling system using central composite design techniques. *Energy and Buildings*, **43**, 988–994.

154 Sartori, I., Napolitano, A., and Voss, K. (2012) Net zero energy buildings: A consistent definition framework. *Energy and Buildings*, **48**, 220–232.

155 Marszal, A.J., Heiselberg, P., Bourrelle, J.S. *et al.* (2011) Zero energy building – A review of definitions and calculation methodologies. *Energy and Buildings*, **43**, 971–979.

156 Annunziata, E., Frey, M., and Rizzi, F. (2013) Towards nearly zero-energy buildings: The state-of-art of national regulations in Europe. *Energy*, **57**, 125–133.

157 Building Resistance-Capacitance Modeling (BRCM) http://www.brcm.ethz.ch/doku.php (accessed 24 January 2017).

158 MLE Toolbox – Energy-efficient building automation design, co-simulation and analysis http://mlab.seas.upenn.edu/mlep/ (accessed 24 January 2017).

159 OpenBuild: An Integrated Simulation Environment for Building Control http://la.epfl.ch/openBuild (accessed 24 January 2017).

160 Hamilton, I.G., Summerfield, A.J., Lowe, R. *et al.* (2013) Energy epidemiology: A new approach to end use energy demand research. *Building Research & Information*, **41**, 482–97.

161 Tabula project http://episcope.eu/building-typology/ (accessed 24 January 2017).

162 Commercial Prototype Building Models (https://www.energycodes.gov/development/commercial/prototype_models).

163 Residential Prototype Building Models (https://www.energycodes.gov/development/residential/iecc_models).

164 Pernigotto, G., Prada, A., Gasparella, A., and Hensen, J.L. (2014) *Development of Sets of Simplified Building Models for Building Simulation*.

165 Reinhart, C.F. and Davila, C.C. (2016) Urban building energy modeling – A review of a nascent field. *Building and Environment*, **97**, 196–202.

166 Perez, D. and Robinson, D. (2012) *Urban energy flow modelling: A data-aware approach* Digital Urban Modeling and Simulation, Springer, pp. 200–220.

167 Kämpf, J.H. and Robinson, D. (2007) A simplified thermal model to support analysis of urban resource flows. *Energy and Buildings*, **39**, 445–453.

168 Lee, Y.M., Liu, F., An, L., Jiang, H., Reddy, C., and Horesh, R. *et al.* (2011) *Modeling and Simulation of Building Energy Performance for Portfolios of Public Buildings*. Simulation Conference (WSC), Proceedings of the 2011 Winter: IEEE. p. 915–927.

169 Davila, C.C., Reinhart, C., and Bemis, J. (2016) Modeling Boston: A workflow for the generation of complete urban building energy demand models from existing urban geospatial datasets. *Energy*, **117**, 237–250.

170 Attia, S., Hamdy, M., O'Brien, W., and Carlucci, S. (2013) Assessing gaps and needs for integrating building performance optimization tools in net zero energy buildings design. *Energy and Buildings*, **60**, 110–124.

171 Adhikari, R.S., Aste, N., and Manfren, M. (2012) Optimization concepts in district energy design and management – A case study. *Energy Procedia*, **14**, 1386–1391.

172 Tobias, B., Nicholas, D., Michael, S., and Dirk, N. (2014) *Power Systems 2.0: Designing an Energy Information System for Microgrid Operation*. International Conference on Information Systems (ICIS 2014).

173 Stadler, M., Cardoso, G., Mashayekh, S. *et al.* (2016) Value streams in microgrids: A literature review. *Applied Energy*, **162**, 980–989.

174 Cao, S., Hasan, A., and Sirén, K. (2014) Matching analysis for on-site hybrid renewable energy systems of office buildings with extended indices. *Applied Energy*, **113**, 230–247.

175 Cao, S., Hasan, A., and Sirén, K. (2013) On-site energy matching indices for buildings with energy conversion, storage and hybrid grid connections. *Energy and Buildings*, **64**, 423–438.

176 Lund, H., Marszal, A., and Heiselberg, P. (2011) Zero energy buildings and mismatch compensation factors. *Energy and Buildings*, **43**, 1646–1654.

177 Frontini, F., Manfren, M., and Tagliabue, L.C. (2012) A case study of solar technologies adoption: Criteria for BIPV integration in sensitive built environment. *Energy Procedia*, **30**, 1006–1015.

178 Baetens, R., De Coninck, R., Van Roy, J. *et al.* (2012) Assessing electrical bottlenecks at feeder level for residential net zero-energy buildings by integrated system simulation. *Applied Energy*, **96**, 74–83.

179 Steen, D., Stadler, M., Cardoso, G. *et al.* (2015) Modeling of thermal storage systems in MILP distributed energy resource models. *Applied Energy*, **137**, 782–792.

180 Oldewurtel, F., Parisio, A., Jones, C.N. *et al.* (2012) Use of model predictive control and weather forecasts for energy efficient building climate control. *Energy and Buildings*, **45**, 15–27.

181 Oldewurtel, F., Sturzenegger, D., and Morari, M. (2013) Importance of occupancy information for building climate control. *Applied Energy*, **101**, 521–532.

182 Roels, S. (2014) *The IEA EBC Annex 58-Project on Reliable Building Energy Performance Characterisation Based on Full Scale Dynamic Measurements.* IEA Annex 58 Seminar-Real building energy performance assessment 2014.

183 Strachan, P., Monari, F., Kersken, M., and Heusler, I. (2015) IEA Annex 58: Full-scale empirical validation of detailed thermal simulation programs. *Energy Procedia*, **78**, 3288–3293.

184 Collazos, A., Maréchal, F., and Gähler, C. (2009) Predictive optimal management method for the control of polygeneration systems. *Computers & Chemical Engineering.*, **33**, 1584–1592.

185 Menon, R.P., Paolone, M., and Maréchal, F. (2013) Study of optimal design of polygeneration systems in optimal control strategies. *Energy*, **55**, 134–141.

186 Fazlollahi, S., Mandel, P., Becker, G., and Maréchal, F. (2012) Methods for multi-objective investment and operating optimization of complex energy systems. *Energy*, **45**, 12–22.

187 Nastasi, B. and Lo Basso, G. (2016) Hydrogen to link heat and electricity in the transition towards future Smart Energy Systems. *Energy*, **110**, 5–22.

188 Lythcke-Jørgensen, C., Ensinas, A.V., Münster, M., and Haglind, F. (2016) A methodology for designing flexible multi-generation systems. *Energy*, **110**, 34–54.

189 EN 15232:2012 (2012) *Energy Performance of Buildings – Impact of Building Automation, Controls and Building Management.*

190 Kramer, R., van Schijndel, J., and Schellen, H. (2012) Simplified thermal and hygric building models: A literature review. *Frontiers of Architectural Research*, **1**, 318–325.

191 Kramer, R., van Schijndel, J., and Schellen, H. (2013) Inverse modeling of simplified hygrothermal building models to predict and characterize indoor climates. *Building and Environment*, **68**, 87–99.

192 ISO/DIS 52016-1:2015 (2015) *Energy Performance of Buildings – Calculation of the Energy Needs for Heating and Cooling, Internal Temperatures and Heating and Cooling Load in a Building or Building Zone – Part 1: Calculation Procedures.*

193 ISO/DIS 52017-1:2015 (2015) *Energy Performance of Buildings – Calculation of the Dynamic Thermal Balance in a Building or Building Zone – Part 1: Generic Calculation Procedure.*

194 Masuda, H. and Claridge, D. (2012) *Estimation of Building Parameters Using Simplified Energy Balance Model and Metered Whole Building Energy Use.*

195 Ghiaus, C. (2006) Experimental estimation of building energy performance by robust regression. *Energy and Buildings*, **38**, 582–587.

196 Ghiaus, C. (2006) Equivalence between the load curve and the free-running temperature in energy estimating methods. *Energy and Buildings*, **38**, 429–435.

197 Inard, C., Pfafferott, J., and Ghiaus, C. (2011) Free-running temperature and potential for free cooling by ventilation: A case study. *Energy and Buildings*, **43**, 2705–2711.

198 Liu, G. and Liu, M. (2011) A rapid calibration procedure and case study for simplified simulation models of commonly used HVAC systems. *Building and Environment*, **46**, 409–420.

199 Heo, Y., Choudhary, R., and Augenbroe, G. (2012) Calibration of building energy models for retrofit analysis under uncertainty. *Energy and Buildings*, **47**, 550–560.

200 Haberl, J. and Bou-Saada, T. (1998) Procedures for calibrating hourly simulation models to measured building energy and environmental data. *Journal of solar energy engineering*, **120**, 193–204.

201 Raftery, P., Keane, M., and Costa, A. (2011) Calibrating whole building energy models: Detailed case study using hourly measured data. *Energy and Buildings*, **43**, 3666–3679.

202 Tronchin, L., Manfren, M., and Tagliabue, L.C. (2016) Multi-scale analysis and optimization of building energy performance – Lessons learned from case studies. *Sustainable Cities and Society*, **27**, 296–306.

203 van Schijndel, A.J. and Kramer, R.R. (2014) *Combining Three Main Modeling Methodologies for Building Physics*. Proceedings of the 10th Nordic Symposium on Building Physics (NSB 2014) Lund, Sweden.

204 Adhikari, R.S., Aste, N., and Manfren, M. (2012) Multi-commodity network flow models for dynamic energy management – Smart Grid applications. *Energy Procedia*, **14**, 1374–1379.

205 Manfren, M. (2012) Multi-commodity network flow models for dynamic energy management – Mathematical formulation. *Energy Procedia*, **14**, 1380–1385.

206 Kraning, M., Chu, E., Lavaei, J., and Boyd, S. (2014) Dynamic network energy management via proximal message passing. *Found Trends Optim*, **1**, 73–126.

207 Fattahi, S. and Lavaei, J. *Convex Analysis of Generalized Flow Networks*.

208 Oberkampf, W.L. and Roy, C.J. (2010) *Verification and validation in scientific computing*, Cambridge University Press.

209 Dodson, B., Hammett, P., and Klerx, R. (2014) *Probabilistic Design for Optimization and Robustness for Engineers*, John Wiley & Sons.

210 Boyd, S.P. and Vandenberghe, L. (2004) *Convex Optimization*, Cambridge University Press.

211 Vanderbei, R.J. (2013) *Linear Programming: Foundations and Extensions*, Springer.

212 Powell, W.B. (2007) *Approximate Dynamic Programming: Solving the curses of dimensionality*, John Wiley & Sons.

4

Autonomous Radios and Open Spectrum in Smart Cities

Corey D. Cooke and Adam L. Anderson

Department of Electrical and Computer Engineering, Tennessee Technological University, Cookeville, USA

CHAPTER MENU

Introduction, 99
Candidate Wireless Technologies, 101
PHY and MAC Layer Issues in Cognitive Radio Networks, 105
Frequency Envelope Modulation (FEM), 110
Conclusion, 116

Objectives

- To understand the sensing, communication, and control needs of a smart city.
- To understand the demand posed on radio spectrum by smart devices in a smart city and what technologies are available to meet these needs.
- To become familiar with the basic concepts of cognitive radio (CR) and how it allows for efficient use of radio spectrum in a smart city.
- To consider one solution for CR network organization–frequency envelope modulation (FEM) as a possible means for efficiently communicating spectrum state information for wireless networks with potentially hundreds of nodes.

4.1 Introduction

Smart cities of the future, "tomorrowland" if you will, are not static entities, but organic beings with interconnected parts that exchange goods, services,

---

Smart Cities: Foundations, Principles and Applications, First Edition.
Edited by Houbing Song, Ravi Srinivasan, Tamim Sookoor, and Sabina Jeschke.
© 2017 John Wiley & Sons, Inc. Published 2017 by John Wiley & Sons, Inc.

and information. The world's population has become increasingly more urbanized and is expected to continue to do so in the future [1]; therefore finding ways to optimize the efficiency of interactions within a city is of paramount importance. In the past, cities accomplished this through the use of market economies, manually guided and/or powered transportation, and slow communication networks (ranging from couriers to horseback riders to wireline communications as time progressed). As (i) communication and transportation technologies developed and (ii) the communication and transportation infrastructures of modern cities became increasingly strained, it became clear that cities would have to move beyond the static-decentralized paradigm to that of a dynamic-adaptive paradigm, wherein routing of critical resources across infrastructure networks happens adaptively in real time as needs change [2]. Because the communication and computing resources needed to accomplish this feat need to operate unattended, it is as if these systems have intelligence of their own, hence the name "smart cities" [3].

Smart cities essentially require augmenting a city's existing infrastructure with an advanced command and control network to control the flow of resources. Because this data spans many domains – including but not limited to electric power flow, water/sewage flow, traffic flow, pollution monitoring, and public safety – and because this data has to traverse a network that essentially covers every "nook and cranny" of the city, a ubiquitous network is required with universal connectivity citywide. This network will be composed of an extremely large number of sensors (in the millions or billions per city) [4], which may be wired or wireless, and the supporting data fusion centers that process the data and distill useful information from it using "big data" analytics [2]. Thus it can be seen that a smart city utilizes many different cutting-edge research paradigms of the present era.

This volume of data, however, creates significant technical challenges for the aforementioned command and control network. Some of the requirements this imposes on our network are as follows:

- Spectrum is limited, but the number of wireless sensors is essentially unlimited.
- The network must be flexible and reconfigurable to allow for devices to be added and removed continuously.
- The network must support many different types of devices with varying capabilities.
- The network must be able to reach essentially every corner of the city (ubiquitous connectivity).

The first point is especially prescient; over the last several decades, wireless data rates per device have increased at an exponential rate and are projected to do so *ad infinitum*, as can be seen in Figure 4.1 [5].

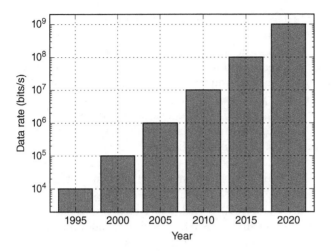

Figure 4.1 Past and projected future wireless data rates in the United States.

In this work we consider several proposed wireless technologies that have the potential to meet these needs and propose a joint physical (PHY)/medium access control (MAC) layer communication scheme that can be applied in conjunction with the aforementioned technologies to fulfill all the requirements set forth in this section. We develop a new modulation called "frequency envelope modulation" (FEM), apply this modulation to the development of a MAC layer network organization algorithm that operates in an unplanned, free spectrum regime, develop theoretical bounds on performance, and demonstrate a practical implementation of said modulation and network organization scheme on a software-defined radio testbed located at Rutgers University.

4.2 Candidate Wireless Technologies

4.2.1 Open Spectrum

Our vision for the future, which we believe will lead to the use of spectrum that will be the most beneficial for the needs of smart cities, is that of "open spectrum" [6]. Currently spectral allocations are rigidly allocated in most countries; in the United States the Federal Communications Commission (FCC) performs this function. In a "fixed" spectrum regime, each spectrum band is allocated for a specific purpose with little room for adaptation, and reassignment is a very difficult and political process. The roots of this policy lie in historical hardware limitations – after the development of the original spark-gap transmitters, it became clear that a way was needed to multiplex many different radio signals at the same time – hence the development of frequency division multiple access, wherein each radio is "hardwired" to a specific frequency or range of carrier frequencies. Due to the analog nature of these radios, these radios could only be designed for a specific purpose, and the regulations that resulted reflected this – hence the adoption of the fixed frequency allocation scheme by the FCC [7]. *Thus the regulatory policies pursued by these bodies are based on technological limitations that are almost a century old.* **As our needs (and technology) change, so should our regulatory regime.** This paradigm is no longer tenable in the twenty-first century. What is needed is a flexible or "open" spectrum policy wherein radios are given more freedom to choose the spectral bands that are available at a given time so that the spectrum can be more fully utilized [8].

With an open spectrum policy in place, the natural technology to exploit this paradigm is that of cognitive radio (CR) [9–11]. Experimental data has shown that at any given time, most spectrum (> 85%) is unoccupied and could be used if the legal limitations were removed [12, 13] . *Therefore an order-of-magnitude improvement in capacity is possible simply by using the spectrum nature has given us.* Due to the proliferation of wireless devices in a smart city, which could number in the billions, **this additional capacity is not only desirable – it becomes a requirement.**

The authors recognize that there will still be the need for some fixed allocations, especially those relating to national security and public safety. This can be seen in Figures 4.2a,b, where exceptions have been made for safety-critical applications such as aircraft landing systems, public safety radio, and GPS. Many other exceptions would doubtlessly need to be made; however, we still want to emphasize that the more open spectrum you have, the more efficiently it can be utilized.

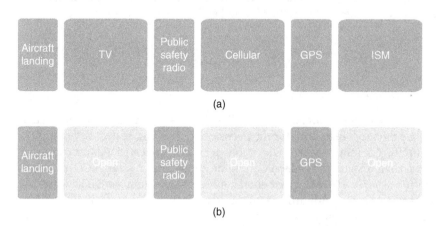

Figure 4.2 Comparison of regulatory schemes for allocating spectrum. (a) Fixed allocation regime. (b) Open spectrum regime. Note that while some safety-critical bands might require fixed allocation, most of the spectrum will be open.

4.2.2 5G Wireless Technologies

One technology that will be a critical part of a smart city's communication infrastructure is mobile cellular networks (MCN). Currently we sit at or are close to so-called 4G (fourth-generation wireless technology) – there is yet no single 5G standard that has been agreed upon [14, 15]. The proposed features of 5G can be summarized with the following slogan: **More, Better, Faster, Cheaper.**

- **More**: More bandwidth, more antennas
- **Better**: Better interference management, better scheduling
- **Faster**: Higher data rates, lower latency
- **Cheaper**: Increased support for low-capacity devices/sensors.

Because bandwidth is very limited at the frequencies for which Long Term Evolution (LTE) was designed, many researchers propose using the 3–300 GHz bands (millimeter wave) as candidate spectrum for 5G systems [16–19]. Millimeter-wave (mmWave) bands contain more than sufficient bandwidth, and the reduced wavelength allows for denser antenna arrays, and it also has unique propagation characteristics that necessitate a new cell deployment strategy.

An additional feature of 5G systems will be the utilization and fusion of data streams coming from heterogeneous networks (HetNets), such as cellular, Wi-Fi, and Bluetooth, in an optimal way that is transparent to the end user [20–22]. This is shown in Figure 4.3, wherein not only the base stations (BSs) but also the devices connecting to the network are heterogeneous in their capabilities.

One new aspect of 5G that has been overlooked by previous wireless standards is that of support for low-capability wireless nodes, such as sensors. Because sensors usually produce data at a low rate and their utility is a function

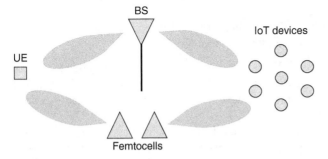

Figure 4.3 Elements of a 5G HetNet. Note that there will exist multiple types of not only base stations (macrocell BSs and femtocells) but also mobile nodes (high-capability UEs, such as a smartphone or tablet, and low-capacity devices, such as sensors) vying for scarce spectrum.

of the number deployed, it is usually desirable to produce large quantities of inexpensive devices rather than smaller quantities of expensive devices. As the world becomes more automated, these devices will be ubiquitous.

4.2.3 Internet of Things (IoT)

The Internet of Things (IoT) is a relatively recent paradigm in the history of networked computing that reflects changes in device usage. Early computer networks were composed of a relatively small number of fixed devices that were usually connected with wired links. For instance, the early iterations of ARPANET, the predecessor to the modern Internet, used large mainframe computers located at major universities as communication nodes and long-distance phone lines served as communication links. Data that flowed over the communication link was the result of a request by a human user at one end of the link to receive data produced by a human at the other end. The personal computing revolution of the 1980s and the broadband revolution of the late 1990s and early 2000s increased the number of nodes and the bandwidth, reliability, and speed of the links, but the basic use case was the same – humans requesting human-generated content. This communication paradigm is commonly referred to as "human-to-human" (H2H) communication in the literature [23].

But what if humans are not at both ends of the communications link? What if:

- A motion sensor is talking to a security system?
- A traffic sensor is talking to a stop light?
- A smart car is talking with another smart car (to prevent a wreck)?

As we think about the future needs of a smart city, it becomes increasingly clear that we no longer have an "Internet of Humans," but rather an "Internet of Things." Here, most communication will be "machine to machine" (M2M) rather than H2H.

The IoT paradigm has many enabling technologies, most of which overlap with each other in function and in terminology [24]:

- **IPV6**: Internet Protocol Version 6. The explosive proliferation of wireless devices will necessitate the adoption of IPV6 [25], which contains expanded address space so that individual devices may be addressed without network address translation (NAT) [26].
- **M2M**: Machine-to-machine communication. Called machine-type communication (MTC) in the MCN literature [27], expanded support of the type of low data rate but consistent communication that machines use will be needed, rather than high data rate but bursty communication characteristic of H2H communication [28].

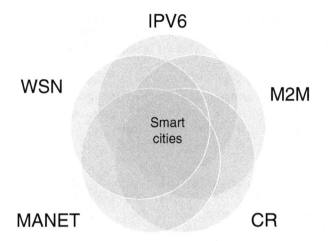

IPV6

WSN

M2M

Smart
cities

MANET

CR

Figure 4.4 Venn diagram showing the overlap of IoT technologies in their applicability to smart cities.

- **WSN**: Wireless sensor networks. Smart cities will be filled to the brim with sensing devices, many of which will be wireless [29, 30].
- **MANET**: Mobile *ad hoc* networks. It will be necessary for wireless sensors to form their own network topologies "on the fly" as sensors are added and removed [31].
- **CR**: Cognitive radio. Wireless sensors will not only need to adapt to needs at the network layer but at the PHY layer as well to find open spectrum and utilize it efficiently. One difference that breaks with the CR tradition is that we propose that the primary user (PU)/secondary user (SU) paradigm are disposed of and instead all nodes are treated as equal "peer nodes" [32].

It should be clear from this description that all of these technologies have a large degree of overlap with one another, as is shown in Figure 4.4.

4.3 PHY and MAC Layer Issues in Cognitive Radio Networks

One proposed solution for the myriad of open spectrum problems over the last decade has been CR [33]. The universally accepted paradigm for cognitive radio networks (CRNs) is that of a PU/SU dichotomy – PUs are incumbent users who have been allocated a spectral band by a regulatory agency and have primary claim on said band in the event of a conflict. However, because frequency bands (as perceived by a mobile receiver) are frequently unoccupied, due to shadowing, transient channel use by the PU, or overly conservative band allotment by the regulatory agency, it is desirable to "recycle" these bands in cases where they are unused so that the full capacity of the spectrum can be realized. In this paradigm, SUs must opportunistically look for and exploit unused bands [34, 35].

The issues relating to resource sharing have been studied extensively in the literature. At the PHY and MAC layers, spectrum sensing [36], spectrum handoff [37], and spectrum management [38] are fundamental issues.

4.3.1 Spectrum Sensing

The first issue that must be addressed for a CRN to function is that of spectrum sensing. Because the SUs do not have priority access to transmit in the spectrum, sensing must be performed to ensure quality of service for PUs. There are two main schemes for how spectrum must be used – spectral "underlay" and "overlay" methods, wherein an SU shares a single channel with a PU and the system is designed to keep interference at a tolerable level, and spectral "interweave," where SUs must look across many channels to find an open one to transmit in [39]. These paradigms are illustrated in Figure 4.5. Like much of the previous work, we also focus on the "interweave" scheme [36].

4.3.1.1 Detection Methods

Many different methods for detecting the presence of a PU are discussed in the literature [36]. There are several considerations in choosing an appropriate detection method – the main trade-off being between specificity and energy efficiency, due to the computational burden of algorithms with higher specificity. At the low end of the specificity scale is energy detection (ED) – which

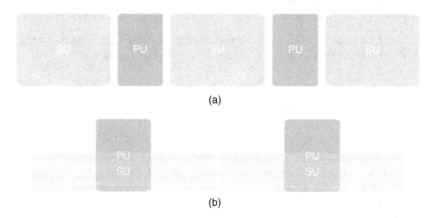

(a)

(b)

Figure 4.5 Comparison of cognitive radio spectrum use paradigms. (a) Interweave paradigm, wherein SUs use unoccupied spectrum. (b) Overlay/underlay paradigms, wherein SUs share the same regions of spectrum with the PU.

is incredibly simple to implement and is the least computationally intensive. In fact, due to its simplicity and reliability, it is widely used in practice [40, 41]. However, ED has no way of determining what type of signal is present and has no way of exploiting signal characteristics to improve detection probabilities for a given signal-to-noise ratio (SNR).

Sensing algorithms with higher specificity take advantage of signal features such as symbol or chip rates, cyclic prefixes, and spatial correlations (particularly in the case of a multiple-input, multiple-output (MIMO) system). Second-order statistics, for example, information gathered from autocorrelation functions or sample covariance matrices, are commonly used to perform feature detection [36]. Cyclostationary-based feature detectors take this one step further to exploit the periodicity of the autocorrelation function present in most types of communication systems. Compressed sensing [42] is another technique that exploits spectral sparsity to achieve sub-Nyquist sample rates, allowing the SU to scan a wide region of spectrum simultaneously.

Other schemes, such as filter banks, multiband sensing, pilot-based detection, matched filter-based detection, and blind detection, are also studied in the literature [36, 43].

4.3.1.2 Cooperative Spectrum Sensing

All of the detection methods discussed in the previous section are implicitly assumed to be operating on multiple independent noncooperative SUs. However, due to the fact that computing and energy resources are scarce on mobile nodes [40, 44–46], cooperative spectrum sensing (CSS) is commonly proposed to alleviate said problems [41, 47, 48].

Spatial diversity is one of the primary motivating factors for CSS. Due to the fact that shadowing and fading can obscure the PU from a given SU, having multiple SUs spread out over a larger geographical region provides diversity to prevent an SU from unintentionally transmitting in a PU band [47]. An example of this is shown in Figure 4.6. Various combining methods, such as soft combining and hard combining, are used to make a decision based on distributed spectrum measurements [36].

Energy efficiency is achieved by minimizing the amount of time an individual SU has to scan – and instead obtaining this information from other nodes. This can be achieved through various means, such as sequential sensing, sleeping, and censoring [36]. Sensing can be further distributed through means of clustering, wherein the head of each cluster of nodes reports back to a central data fusion center.

4.3.1.3 Other Sensing Issues

In addition to the aforementioned issues, recent work has explored issues such as interfering PUs from other CRNs (i.e., a PU who is "supposed" to be far enough away to be considered outside the region of interest but who still

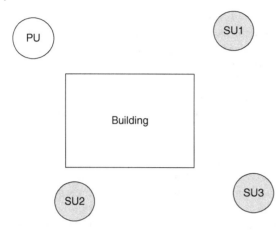

Figure 4.6 Illustration of the need for and advantage of the spatial diversity provided by cooperative spectrum sensing. Note that SU1 has an unobstructed view of the PU, SU2 has a partially obstructed view, and SU3 has a fully obstructed view.

interferes) [49], CRNs with imperfect information [45, 50], CRNs for vehicular *ad hoc* networks (VANETs) [51], CR for smart grid automated metering infrastructure (AMI) [52], and CRN issues in sensor networks [46].

4.3.2 Spectrum Management and Handoff

In addition to the PHY layer issues discussed previously, CRs must effectively organize themselves into a network as appropriate regions of spectrum are found – the spectrum sensing problem of the previous section is purely a PHY layer problem. Once we reach the MAC layer, now we must consider the fact that many SUs must find each other, share spectrum, and communicate, all the while maintaining coexistence with other CRNs and dealing with changing spectral conditions as PUs enter and leave the spectrum.

The aforementioned discussions have focused mainly on issues of forming a CRN. Once a CRN has been formed, it must be maintained as new PUs enter the network and "kick" SUs out of their channels, at which point the SU must find a new channel to transmit in and still maintain a communication link. This problem is referred to as "spectrum handoff" [37]. One example application of spectrum handoff is in cellular HetNets wherein a femtocell may be utilizing the overlay spectrum sharing paradigm described earlier, that is, sharing the same band with a macrocell or microcell BS [53]. In this case there is a serious need for seamless handoff as spectrum conditions change as well as when the user moves. Other recent works have focused on optimizing energy efficiency, delay, and/or spectrum utilization for general cases [54, 55], as well as for applications such as real-time video [56].

Several MAC protocols have been proposed to solve these problems including centralized protocols such as those included in the 802.22 standard [57]

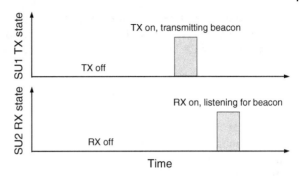

Figure 4.7 Illustration of the rendezvous problem in cognitive radio. Two nodes attempting to discover each other can only rendezvous and exchange control information when the transmitter and receiver communicate in the same time–frequency slot (here they are not).

and decentralized protocols such as hardware-constrained MAC (HC-MAC), decentralized cognitive MAC (DC-MAC), cognitive MAC (C-MAC), and others [57, 58]. In this section we illustrate some of the MAC issues at play.

4.3.3 Rendezvous Problem

One of the most important problems at the MAC layer is the rendezvous problem. When a set of SUs are attempting to form a CRN, they must first become aware of one another and share control information. This process is called neighbor discovery [59] or rendezvous [38, 60].

In most discussions of CRNs, it is assumed that the control information being shared is over a common control channel (CCC) [61]. Nodes engaging in rendezvous will look in a pre-agreed-upon channel or set of channels until they meet at the same time (and frequency, if using multiple channels), at which point they handshake and become aware of one another's presence. This process is shown in Figure 4.7.

However, due to the fact that sensing and transmitting is energy intensive and that all nodes cannot access the CCC simultaneously, the system must be designed such that probabilistically two nodes will almost certainly rendezvous within the desired amount of time. This process is shown in Figure 4.7. This problem is frequently referred to as the "birthday problem" in the literature because it is analogous to finding the probability that two people in a room will share the same birthday [59].

4.3.4 Coexistence

In addition to competing with PUs for spectrum and then having to share said spectrum with other SUs, a CRN must also coexist with other CRNs that are taking advantage of the spectrum that is available. In the literature there are several types of coexistence discussed – vertical coexistence, which refers to the PU/SU coexistence problem that has already been discussed, and horizontal coexistence, which refers to coexistence among CRNs of SUs [61].

Horizontal coexistence can take on two forms – heterogeneous, which is coexistence between CRNs utilizing different protocols, and homogeneous, which is coexistence between CRNs using the same protocol [61].

4.4 Frequency Envelope Modulation (FEM)

We present a novel network organization scheme in this chapter, which is general enough to encompass many different network types, such as the ones shown in Figure 4.8.

We assume that all users are equal – that is, no notion of a "PU" or "SU." A novel modulation scheme, termed FEM [62], is analyzed that allows for the communication of MAC layer control information without the need for a CCC. This scheme breaks the organization problem into two phases: "spectrum sensing," wherein a transceiver node first attempts to find an empty channel to transmit in, and "spectrum recognition," wherein a transceiver node attempts to scan the remaining channels to find the signal intended for it to receive.

Without the presence of a CCC, and in a large open spectrum, the problem of each receiver finding its corresponding transmitter becomes like finding the proverbial "needle in a haystack," especially as the number of nodes and open channels becomes large. We demonstrate that by utilizing collaborative spectrum sensing and knowledge sharing with FEM, we can drastically reduce network configuration time, especially as the number of nodes and open channels increases asymptotically. This is similar to the well-studied "rendezvous problem" in the literature [63–66], but with a few differences:

- The classical treatment of the rendezvous problem usually only studies two nodes at a time [59, 67], whereas we consider an arbitrary number.
- Our scheme does not require precise time-domain synchronization to exchange information.

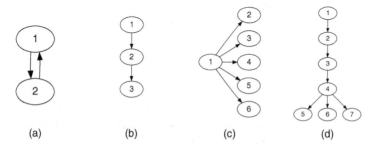

(a)   (b)   (c)   (d)

Figure 4.8 Example network topologies that can be used with FEM. (a) "Normal" full-duplex configuration. (b) Multihop configuration. (c) Multicast configuration. (d) Multihop to multicast configuration.

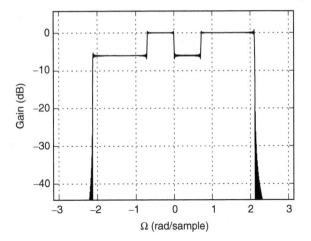

Figure 4.9 Crenel filter with $\gamma = 0.5$. The crenel pattern represented is "001011."

- A "third-party" node can still read information encoded using FEM from other nodes that are communicating with each other.

We first explain the motivation behind FEM and then study its network and PHY layer performance.

The essence of FEM is taking an input time-domain root-raised cosine (RRC) pulse-shaped quadrature amplitude-modulated (QAM) signal and modulating its spectrum in the frequency domain to convey higher-level signaling information. This is accomplished by passing the time-domain signal through a time-varying finite impulse response (FIR) filter whose frequency response at any given time represents a bit pattern in the frequency-domain envelope, resembling the crenels ("valleys") and merlons ("peaks") on the parapets of a castle. This is illustrated in Figure 4.9 and is referred to as crenel filtering.

For simplicity of nomenclature, we refer to a single bit represented in the frequency domain by a crenel or a merlon simply as a crenel. The depth of the crenel affects the error rate of the underlying time-domain signal – the deeper the crenel, the more distortion the time-domain signal suffers. However, the higher the crenels, the more difficult it is to identify the crenel pattern in the frequency domain. Therefore a compromise must be struck to ensure the quality of the signal in both domains.

There are many advantages to transmitting control data in this manner: it obviates the need for a CCC to convey MAC layer information; the data rate is relatively low, so precise timing synchronization is not a burden; it is relatively forgiving of minor frequency offsets, and phase information is irrelevant; and most importantly, the underlying time-domain data stream is still able to be transmitted as normal as long as the crenel filter is designed correctly.

For simulation purposes, nodes attempt to self-configure according to the following rules:

- The spectrum of interest is completely open – no *PU* or *SU*.
- Each node is a data source/sink pair.
- Nodes know the node ID of the node they are supposed to sink from but not source to.
- Each "round" all nodes that have free receivers sense one channel in the spectrum.
- After sensing an open channel, a node occupies that channel.
- A node continues to scan until they find the node they are supposed to sink from, at which point that node is considered *configured*.
- The network is considered *configured* when all nodes are configured.

In the simplest case nodes could use FEM to broadcast their node ID for spectrum recognition [62], but this does not significantly expedite network configuration. As nodes scan random channels, they may not encounter their source node but may still gain information by reading the FEM that other nodes are sending. Nodes can use FEM to continuously broadcast the contents of their spectrum database map so that each node that scanned its spectrum adds this information to its own map. Node organization time can be sped up at an exponential rate as each distributed node contributes knowledge of sensed spectrum to the entire network.

An example spectrogram of a node organization scheme using FEM is shown in Figure 4.10, in which four nodes are attempting to organize in a band of spectrum with four 100 kHz channels. The first two nodes are able to immediately sense and pick an open channel, while the other two nodes are not able to find an open channel until round 3. It can be seen that as new information is gained from scanning, the crenel pattern on each node's spectrum changes. The scanning continues until all nodes have not only found an open channel but also found their "pair node" that they receive data from.

We simulate network configuration time by creating a turn-based game wherein each radio is an agent that acts independently each round using the rules in Section 4.4.1. Based on the results from Section 4.4.2, essentially error-free communication can take place at a crenel rate of roughly 1/1000th the time-domain date rate. Therefore, with a sufficiently low crenel rate, it is reasonable to abstract away the PHY layer and assume that the spectral map is shared in each round where two nodes line up in the same channel.

In our simulations we consider three knowledge-sharing regimes:

- "Uncoordinated" (Unc): No spectrum state information (SSI) is shared. Nodes simply sweep the entire list of channels to find their source node. This can be considered the lower bound on performance.

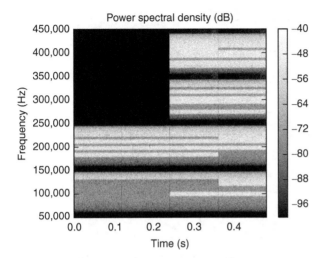

Figure 4.10 Example spectrogram of a FEM organization period with four nodes and four channels where each channel has a bandwidth of 100 kHz. Note that for each round the crenel patterns change as nodes gain new sensing information.

- "Distributed Sharing" (DistSh): Nodes use FEM to share the entire SSI database each time its spectrum is scanned.
- "Centralized Sharing" (CentSh): All nodes share SSI knowledge simultaneously. This can be considered the upper bound on performance.

In the simulation represented in Figure 4.11, the number of nodes and channels is the same, and the number of nodes is varied from 4 to 100 in steps of 4. For each data point in the figure, the network configuration time is averaged over 2000 runs. It can be seen that for "Unc," the configuration time is approximately linear as the number of nodes increases, which follows our intuition. For both FEM-aided "DistSh" and genie-aided "CentSh," it can be seen that the configuration time is roughly logarithmic with the number of nodes – this accords with our intuition that with all the nodes working together, the efficiency should increase exponentially. While "CentSh" obviously outperforms "DistSh," what is encouraging is the fact that both appear to be of a similar asymptotic complexity class.

4.4.2 Physical Layer Performance

The effect of adding FEM on top of the normal time-domain modulation is studied in this section as well as the detection probability of the crenels themselves in the frequency domain.

The bit error rate (BER) of a BPSK signal with FEM added in an AWGN channel as a function of γ and E_b/N_0 is plotted in Figure 4.12a. In Figure 4.12b an inverse crenel filter with the crenel pattern inverted to perform equalization.

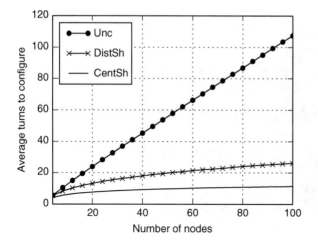

Figure 4.11 Average network configuration time in rounds as a function of the number of nodes, with the number of channels equaling the number of nodes for each data point. The number of nodes was varied from 4 to 100 in steps of 4; 2000 iterations were averaged for each data point in the figure.

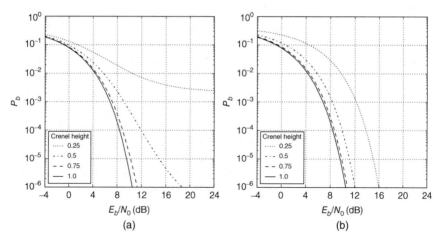

Figure 4.12 Time-domain BER curves of FEM-BPSK as a function of crenel height γ with $L = 12$ (4 of which are training crenels), $K = 2$, and $\beta = 0.25$. (a) Without crenel equalization. (b) With crenel equalization. The $\gamma = 1$ curve corresponds to unimpaired BPSK. Note that in general the equalized version performs better, but for $\gamma \geq 0.75$ the impairment added by FEM is negligible even without equalization.

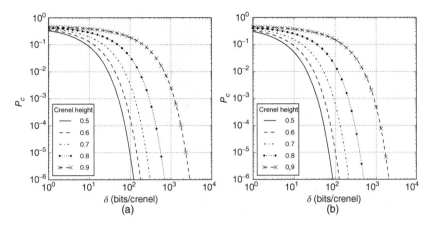

Figure 4.13 FEM crenel detection BER as a function of crenel depth and symbol rate ratio with $\beta = 0$ and $\rho = 1$. (a) $E_s/N_0 = 5$ dB. (b) $E_s/N_0 = \infty$ dB. It can be seen that for acceptable error performance, the crenel rate must be approximately 1/1000th of the time-domain symbol rate.

It can be seen that the equalization does improve performance significantly by removing intersymbol interference (ISI), but for high γ the difference in performance is negligible.

The crenel detection error probability is plotted in Figure 4.13 for (time domain) $E_b/N_0 = 5$ dB and for high SNR. It can be seen that the crenel BER depends very little on the SNR – the most important variables are the symbol rate ratio δ and the crenel depth γ. This follows intuition since noise is flat in the frequency domain, and as long as the signal is above the noise floor, crenel bit "energy" depends simply on the number of crenels per envelope per second and crenel depth.

4.4.3 Experimental Results

To prove the efficacy of FEM and the corresponding rendezvous scheme, we have implemented them on a network of Ettus Research USRP N210 hosted by the ORBIT Lab at Rutgers University [68]. The ORBIT testbed is an open-access testbed available to university researchers anywhere in the world for the development of wireless network protocols. It contains a 20-by-20 grid of nodes, with each node having its own controller PC and a selection of radio interfaces (such as SDRs, Wi-Fi cards, etc). Of this 400-node grid, approximately 20 have USRP N210. We chose to use 15 of these nodes for our FEM experiments, with one additional node serving as a spectrum observer, for a total of 16 nodes.

We chose 15 nodes because it allowed us to display a node's entire SSI database using a reasonable number of crenels without the need for precise timing synchronization. Fifteen nodes mean that with a 4-bit number, you can

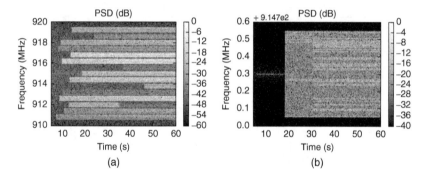

Figure 4.14 Spectrograms for an example FEM organization experiment run on the ORBIT testbed. (a) FEM organization spectrogram. (b) Changing crenel pattern for a single channel (zoomed in).

specify each node's location in the spectrum and reserve one code (all zeros) to represent an empty or unknown channel. This yields 60 crenels in total. With the addition of 4 calibration crenels for empty channel detection and threshold determination on the edges of the spectrum and an 8-crenel CRC, this yields a 72-crenel spectrum, which was realizable to acceptable accuracy using an FIR filter with approximately 500 taps.

The implementation of the organization algorithm was essentially a direct translation of what has been described in previous section with a few modifications to deal with practical issues (power amplifier saturation, frequency-selective fading, collision avoidance, and the hidden-node problem). K-means clustering was used as our crenel detector once the PSD had been obtained and averaged.

The spectrogram of one of these experiments is shown in Figure 4.14a. It can be seen that the spectrum fills up rather quickly as nodes find open spectrum to transmit in. It can be seen in Figure 4.14b that when you zoom in and look at the spectrogram of a single channel, the individual striations on a node's spectrum will change over time as they gain additional information.

This experiment used only 15 nodes, but the spectrum behavior is what would be expected in a larger region of open spectrum within a smart city as innumerable nodes access the spectrum and attempt to rendezvous with their links.

4.5 Conclusion

It is expected that the urbanization of the world's population will continue to grow well into the future. This will lead to an increased density of not only human beings within the cities of the future but also of the necessities required to sustain a modern lifestyle, such as electricity, communications, transportation, water, and sewage disposal. In the past these resources were managed as efficiently as they could be, but as cities become more dense, connected, and full of sensors, it is inevitable that they will also become more "smart." Smart cities are the vision for future cities. In a smart city, sensing, communication, and control networks will be ubiquitous, which has the potential to bring not only great benefits to mankind but also great challenges. Because the sensor and control networks are distributed over large geographic areas, communication networks will bear a considerable burden in relaying all of the required information for the sensing and control loops to operate. Much of this communication infrastructure can be fixed, but much of it will have to be wireless due to mobility considerations as well as the ease of deploying wireless *ad hoc* networks. In addition, smart cities will also be full of smart consumers of information, leading to an increased proliferation of portable media and communication devices such as laptops, smart phones, tablets, etc. All of these devices will be competing for limited spectrum; therefore it is imperative that the spectrum be fully utilized to realize this vision.

Several candidate technologies have been developed in the last few years that have the potential to meet these challenges, such as fifth-generation wireless, the IoT, machine-to-machine communication, wireless sensor networks, mobile *ad hoc* networks, and CR. Many of these technologies have significant overlap with one another.

Due to the flexibility demanded of wireless communication networks in smart cities, CR will be a natural fit to meet these challenges. In fact, as time progresses, it is not unlikely that most if not all radios will become cognitive to varying degrees. CR allows for the flexible and adaptive use of radio spectrum such that empty spectrum can quickly be recycled by the millions of sensors and other wireless devices that will be present in a smart city. There are several new PHY and MAC layer issues that must be addressed in designing a CRN, such as spectrum sensing, spectrum management, spectrum handoff, node discovery, and coexistence, all of which are open areas of research in the present and will be for some time to come.

Open use of spectrum has been discussed in conjunction with the use of CR to further increase the utilization of the spectrum. This goes beyond the common PU/SU paradigms of traditional CR. A novel control information sharing scheme and rendezvous protocol called FEM has been presented, which is particularly suited to the challenges faced in an open spectrum regime.

Final Thoughts

Herein we have discussed the needs and challenges of future smart cities and the critical role wireless devices play in conveying the information needed to operate said cities. We have shown that the open spectrum regime provides a compelling solution to the demands of having a connected city, that CR is the solution to exploit this free spectrum, and that the FEM and network organization scheme provides a way for CRs to effectively form *ad hoc* networks in an efficient manner.

Questions

1 What are three possible systems present in a smart city that require control and communication?

2 Why will spectrum be more crowded in a smart city than normal cities?

3 What are the two distinguishing features that cognitive radio possesses that prior communication networks lacked?

4 How does a secondary user in a cognitive radio network sense and adapt to the presence of primary users of a band?

5 What is frequency envelope modulation, and what aspects of the PHY and MAC layers does it address?

References

1 United Nations (2014) *World's Population Increasingly Urban with More Than Half Living in Urban Areas*, http://www.un.org/en/development/desa/news/population/world-urbanization-prospects-2014.html (accessed 10 December 2016).

2 Zanella, A., Bui, N., Castellani, A., Vangelista, L., and Zorzi, M. (2014) Internet of things for smart cities. *IEEE Internet of Things Journal*, **1** (1), 22–32.

3 Ianuale, N., Schiavon, D., and Capobianco, E. (2016) Smart cities, big data, and communities: reasoning from the viewpoint of attractors. *IEEE Access*, **4**, 41–47.

4 Liu, J., Li, Y., Chen, M., Dong, W., and Jin, D. (2015) Software-defined internet of things for smart urban sensing. *IEEE Communications Magazine*, **53** (9), 55–63.

5 Fettweis, G. (2014) The tactile internet: applications and challenges. *IEEE Vehicular Technology Magazine*, **9** (1), 64–70.

6 Berger, R.J. (2003) Open spectrum: a path to ubiquitous connectivity. *Queue*, **1** (3), 60–68.

7 Benkler, Y. (1997) Overcoming agoraphobia: building the commons of the digitally networked environment. *Harvard Journal of Law and Technology*, **11**, 287–400.

8 Benkler, Y. (2012) Open wireless vs. licensed spectrum: evidence from market adoption. *Harvard Journal of Law and Technology*, **26** (1).

9 Akyildiz, I.F., Lee, W.Y., Vuran, M.C., and Mohanty, S. (2006) NeXt generation/dynamic spectrum access/cognitive radio wireless networks: a survey. *Comput. Networks*, **50** (13), 2127–2159.

10 Popescu, A., Erman, D., Fiedler, M., Popescu, A., and Kouvatsos, D. (2010) *A Middleware Framework for Communication in Cognitive Radio Networks*. Ultra Modern Telecommunications and Control Systems and Workshops (ICUMT), 2010 International Congress on, IEEE, pp. 1162–1171.

11 Wang, C.X., Haider, F., Gao, X., You, X.H., Yang, Y., Yuan, D., Aggoune, H., Haas, H., Fletcher, S., and Hepsaydir, E. (2014) Cellular architecture and key technologies for 5G wireless communication networks. *IEEE Communications Magazine*, **52** (2), 122–130.

12 Microsoft (2016) Microsoft Spectrum Observatory, https://observatory .microsoftspectrum.com (accessed 10 December 2016).

13 Bacchus, R.B., Fertner, A.J., Hood, C.S., Roberson, D. et al. (2008) *Long-Term, Wide-Band Spectral Monitoring in Support of Dynamic Spectrum Access Networks at the IIT Spectrum Observatory*. New Frontiers in Dynamic Spectrum Access Networks, 2008. DySPAN 2008. 3rd IEEE Symposium on, IEEE, pp. 1–10.

14 Osseiran, A., Boccardi, F., Braun, V., Kusume, K., Marsch, P., Maternia, M., Queseth, O., Schellmann, M., Schotten, H., Taoka, H. et al. (2014) Scenarios for 5G mobile and wireless communications: the vision of the METIS project. *IEEE Communications Magazine*, **52** (5), 26–35.

15 Whitacre, J. (2014) Toward 5G: what's changing and how to address design and test challenges? *Microwave Journal*, **57** (5), 94–.

16 Cudak, M., Ghosh, A., Kovarik, T., Ratasuk, R., Thomas, T., Vook, F.W., Moorut, P. et al. (2013) *Moving towards mmWave-based beyond-4G (B-4G) technology*. Vehicular Technology Conference (VTC Spring), 2013 IEEE 77th, IEEE, pp. 1–5.

17 Rappaport, T., Sun, S., Mayzus, R., Zhao, H., Azar, Y., Wang, K., Wong, G., Schulz, J., Samimi, M., and Gutierrez, F. (2013) Millimeter wave mobile communications for 5G cellular: it will work! *IEEE Access*, **1**, 335–349.

18 Roh, W., Seol, J.Y., Park, J., Lee, B., Lee, J., Kim, Y., Cho, J., Cheun, K., and Aryanfar, F. (2014) Millimeter-wave beamforming as an enabling technology for 5G cellular communications: theoretical feasibility and prototype results. *IEEE Communications Magazine*, **52** (2), 106–113.

19 Pi, Z. and Khan, F. (2011) An introduction to millimeter-wave mobile broadband systems. *IEEE Communications Magazine*, **49** (6), 101–107.

20 Ghosh, A., Mangalvedhe, N., Ratasuk, R., Mondal, B., Cudak, M., Visotsky, E., Thomas, T., Andrews, J.G., Xia, P., Jo, H.S. et al. (2012) Heterogeneous cellular networks: from theory to practice. *IEEE Communications Magazine*, **50** (6), 54–64.

21 Li, X., Salleh, R., Aani, A., and Zakaria, O. (2009) *Multi-Network Data Path for 5G Mobile Multimedia.* Communication Software and Networks, 2009. ICCSN'09. International Conference on, IEEE, pp. 583–587.

22 Mehbodniya, A., Kaleem, F., Yen, K.K., and Adachi, F. (2013) *A Novel Wireless Network Access Selection Scheme for Heterogeneous Multimedia Traffic.* Consumer Communications and Networking Conference (CCNC), 2013 IEEE, IEEE, pp. 485–489.

23 Ali, A., Hamouda, W., and Uysal, M. (2015) Next generation M2M cellular networks: challenges and practical considerations. *IEEE Communications Magazine*, **53** (9), 18–24.

24 Gubbi, J., Buyya, R., Marusic, S., and Palaniswami, M. (2013) Internet of things (IoT): a vision, architectural elements, and future directions. *Future Generation Computer Systems*, **29** (7), 1645–1660.

25 Savolainen, T., Soininen, J., and Silverajan, B. (2013) IPv6 addressing strategies for IoT. *IEEE Sensors Journal*, **13** (10), 3511–3519.

26 Forouzan, B.A. (2002) *TCP/IP Protocol Suite*, McGraw-Hill, Inc.

27 Palattella, M., Dohler, M., Grieco, A., Rizzo, G., Torsner, J., Engel, T., and Ladid, L. (2016) Internet of things in the 5G era: enablers, architecture, and business models. *IEEE Journal on Selected Areas in Communications*, **34** (3), 510–527.

28 Zhang, Y., Yu, R., Xie, S., Yao, W., Xiao, Y., and Guizani, M. (2011) Home M2M networks: architectures, standards, and QoS improvement. *IEEE Communications Magazine*, **49** (4), 44–52.

29 Prasath, A., Venuturumilli, A., Ranganathan, A., and Minai, A. (2009) Self-organization of sensor networks with heterogeneous connectivity, in *Sensor Networks*, Signals and Communication Technology (ed. G. Ferrari), Springer, Berlin, Heidelberg, pp. 39–59.

30 Zhang, J., Shan, L., Hu, H., and Yang, Y. (2012) Mobile cellular networks and wireless sensor networks: toward convergence. *IEEE Communications Magazine*, **50** (3), 164–169.

31 Conti, M. and Giordano, S. (2007) Multihop Ad Hoc networking: the theory. *IEEE Communications Magazine*, **45** (4), 78–86.

32 Cooke, C.D. and Anderson, A.L. (2015) *Frequency Envelope Modulation (FEM): A Passive Approach to Universal Spectrum Recognition and Network Self-Configuration.* IEEE GLOBECOM Workshop on 5G and Beyond.

33 Mitola, J. and Maguire, G.Q.J. (1999) Cognitive radio: making software radios more personal. *IEEE Personal Communications*, **6** (4), 13–18.

34 Akyildiz, I., Lee, W.Y., Vuran, M.C., and Mohanty, S. (2008) A survey on spectrum management in cognitive radio networks. *IEEE Communications Magazine*, **46** (4), 40–48.

35 Lu, Y. and Duel-Hallen, A. (2016) Channel-aware spectrum sensing and access for mobile cognitive radio Ad Hoc networks. *IEEE Transactions on Vehicular Technology*, **65** (4), 2471–2480.

36 Axell, E., Leus, G., Larsson, E., and Poor, H. (2012) Spectrum sensing for cognitive radio: state-of-the-art and recent advances. *IEEE Signal Processing Magazine*, **29** (3), 101–116.

37 Zhang, Y. (2009) Spectrum Handoff in Cognitive Radio Networks: Opportunistic and Negotiated Situations. 2009 IEEE International Conference on Communications, pp. 1–6.

38 Lo, B.F. (2011) A survey of common control channel design in cognitive radio networks. *Physical Communication*, **4** (1), 26–39.

39 Mili, M.R., Musavian, L., Hamdi, K.A., and Marvasti, F. (2016) How to increase energy efficiency in cognitive radio networks. *IEEE Transactions on Communications*, **64** (5), 1829–1843.

40 Xiong, G., Kishore, S., and Yener, A. (2013) Spectrum sensing in cognitive radio networks: performance evaluation and optimization. *Physical Communication*, **9**, 171–183.

41 Yang, G., Wang, J., Luo, J., Wen, O.Y., Li, H., Li, Q., and Li, S. (2016) Cooperative spectrum sensing in heterogeneous cognitive radio networks based on normalized energy detection. *IEEE Transactions on Vehicular Technology*, **65** (3), 1452–1463.

42 Jiang, J., Sun, H., Baglee, D., and Poor, H.V. (2016) Achieving autonomous compressive spectrum sensing for cognitive radios. *IEEE Transactions on Vehicular Technology*, **65** (3), 1281–1291.

43 Umar, R. and Sheikh, A.U. (2013) A comparative study of spectrum awareness techniques for cognitive radio oriented wireless networks. *Physical Communication*, **9**, 148–170.

44 Monemian, M., Mahdavi, M., and Omidi, M.J. (2016) Optimum sensor selection based on energy constraints in cooperative spectrum sensing for cognitive radio sensor networks. *IEEE Sensors Journal*, **16** (6), 1829–1841.

45 Zhang, L., Xiao, M., Wu, G., Li, S., and Liang, Y.C. (2016) Energy-efficient cognitive transmission with imperfect spectrum sensing. *IEEE Journal on Selected Areas in Communications*, **34** (5), 1320–1335.

46 Ren, J., Zhang, Y., Zhang, N., Zhang, D., and Shen, X. (2016) Dynamic channel access to improve energy efficiency in cognitive radio sensor networks. *IEEE Transactions on Wireless Communications*, **15** (5), 3143–3156.

47 Akyildiz, I.F., Lo, B.F., and Balakrishnan, R. (2011) Cooperative spectrum sensing in cognitive radio networks: a survey. *Physical Communication*, **4** (1), 40–62.

48 Saber, M.J. and Sadough, S.M.S. (2016) Multiband cooperative spectrum sensing for cognitive radio in the presence of malicious users. *IEEE Communications Letters*, **20** (2), 404–407.

49 Furtado, A., Irio, L., Oliveira, R., Bernardo, L., and Dinis, R. (2016) Spectrum sensing performance in cognitive radio networks with multiple primary users. *IEEE Transactions on Vehicular Technology*, **65** (3), 1564–1574.

50 Liu, B., Li, Z., Si, J., and Zhou, F. (2016) Optimal sensing interval in cognitive radio networks with imperfect spectrum sensing. *IET Communications*, **10** (2), 189–198.

51 Huang, X.L., Wu, J., Li, W., Zhang, Z., Zhu, F., and Wu, M. (2016) Historical spectrum sensing data mining for cognitive radio enabled vehicular Ad-Hoc networks. *IEEE Transactions on Dependable and Secure Computing*, **13** (1), 59–70.

52 Khan, A.A., Rehmani, M.H., and Reisslein, M. (2016) Cognitive radio for smart grids: survey of architectures, spectrum sensing mechanisms, and networking protocols. *IEEE Communications Surveys Tutorials*, **18** (1), 860–898.

53 Al-Dulaimi, A., Anpalagan, A., and Cosmas, J. (2016) Spectrum handoff management in cognitive HetNet systems overlaid with femtocells. *IEEE Systems Journal*, **10** (1), 335–345.

54 Bicen, A.O., Pehlivanoglu, E.B., Galmes, S., and Akan, O.B. (2015) Dedicated radio utilization for spectrum handoff and efficiency in cognitive radio networks. *IEEE Transactions on Wireless Communications*, **14** (9), 5251–5259.

55 Lee, D.J. and Yeo, W.Y. (2015) Channel availability analysis of spectrum handoff in cognitive radio networks. *IEEE Communications Letters*, **19** (3), 435–438.

56 Liu, F., Ma, Y., Zhao, H., and Ding, K. (2015) Evolution handoff strategy for real-time video transmission over practical cognitive radio networks. *China Communications*, **12** (2), 141–154.

57 Hiremath, S., Mishra, A., and Patra, S. (2015) Engineering review of the IEEE 802.22 standard on cognitive radio, in *White Space Communication*, Signals and Communication Technology (eds A.K. Mishra and D.L. Johnson), Springer International Publishing, pp. 1–31.

58 Xiang, J., Zhang, Y., and Skeie, T. (2010) Medium access control protocols in cognitive radio networks. *Wireless Communications and Mobile Computing*, **10** (1), 31–49.

59 McGlynn, M.J. and Borbash, S.A. (2001) *Birthday Protocols for Low Energy Deployment and Flexible Neighbor Discovery in Ad Hoc Wireless Networks*. Proceedings of the 2nd ACM International Symposium on Mobile Ad Hoc Networking and Computing, ACM, New York, NY, USA, MobiHoc '01, pp. 137–145.

60 Bian, K. and Park, J.M. (2011) *Asynchronous Channel Hopping for Establishing Rendezvous in Cognitive Radio Networks*. INFOCOM, 2011 Proceedings IEEE, pp. 236–240.

61 Bian, K., Park, J.M., and Gao, B. (2014) *Cognitive Radio Networks: Medium Access Control for Coexistence of Wireless Systems*, Springer.

62 Anderson, A.L., Witherspoon, C.B., and Papari, B. (2014) *Spectrum Recognition in Large-Scale Cognitive Radio Networks with Spectral Data Mining*. Proceedings of WORLDCOMP.

63 Alonso, G., Kranakis, E., Sawchuk, C., Wattenhofer, R., and Widmayer, P. (2003) Probabilistic protocols for node discovery in Ad Hoc multi-channel broadcast networks, in *Ad-Hoc, Mobile, and Wireless Networks*, Lecture Notes in Computer Science, vol. **2865** (eds S. Pierre, M. Barbeau, and E. Kranakis), Springer, Berlin, Heidelberg, pp. 104–115.

64 Hamida, E.B., Chelius, G., and Fleury, E. (2006) Revisiting Neighbor Discovery with Interferences Consideration. Proceedings of the 3rd ACM International Workshop on Performance Evaluation of Wireless Ad Hoc, Sensor and Ubiquitous Networks, ACM, New York, NY, USA, PE-WASUN '06, pp. 74–81.

65 Borbash, S.A., Ephremides, A., and McGlynn, M.J. (2007) An asynchronous neighbor discovery algorithm for wireless sensor networks. *Ad Hoc Networks*, **5** (7), 998–1016.

66 Silvius, M., MacKenzie, A., and Bostian, C. (2009) Rendezvous MAC Protocols for Use in Cognitive Radio Networks. Military Communications Conference, 2009. MILCOM 2009. IEEE, pp. 1–7.

67 Bian, K. and Park, J.M. (2013) Maximizing rendezvous diversity in rendezvous protocols for decentralized cognitive radio networks. *IEEE Transactions on Mobile Computing*, **12** (7), 1294–1307.

68 Raychaudhuri, D., Seskar, I., Ott, M., Ganu, S., Ramachandran, K., Kremo, H., Siracusa, R., Liu, H., and Singh, M. (2005) *Overview of the ORBIT Radio Grid Testbed for Evaluation of Next-Generation Wireless Network Protocols*. Wireless Communications and Networking Conference, 2005 IEEE, vol. 3, pp. 1664–1669.

5

Mobile Crowd-Sensing for Smart Cities

Chandreyee Chowdhury and Sarbani Roy

Department of Computer Science and Engineering, Jadavpur University, India

CHAPTER MENU

Introduction, 125
Overview of Mobile Crowd-Sensing, 127
Issues and Challenges of Crowd-sensing in Smart Cities, 135
Crowd-sensing Frameworks for Smart City, 144
Conclusion, 149

Objectives

- To become familiar with the concept of mobile crowd-sensing in smart cities.
- To familiarize with the vast set of smart applications on mobile crowd-sensing.
- To be aware of the issues of mobile crowd-sensing.
- To become familiar with the existing frameworks for mobile crowd-sensing.

5.1 Introduction

As more people are moving toward urban areas, cities need to be made smarter to optimally address problems like resource limitation, maintenance of healthy neighborhood, and so on. Many of the problems are dynamic in nature, for example, predicting traffic pattern on a day [1], finding empty space in a parking lot, reporting from disaster situations, or detecting the presence of dangerous pollutants [2]. To measure these in a centralized manner, expensive infrastructure is needed to be maintained. Nowadays, smart mobile phones are

Smart Cities: Foundations, Principles and Applications, First Edition.
Edited by Houbing Song, Ravi Srinivasan, Tamim Sookoor, and Sabina Jeschke.

quite efficient in sensing tasks as well as computing and can also be extended easily to the Internet of Things (IoT), for instance, smartphones, music players, sensor-embedded gaming systems, and in-vehicle sensing devices (GPS, OBD-II) [3]. These devices have many built-in sensors like accelerometer, gyroscope, digital compass, light sensor, Bluetooth as proximity sensor [4], and so on and hence are potentially important sources of sensed data. This trend is even rising as more sensors will be incorporated into the smart devices, for instance, sensors needed for healthcare applications. These devices also come with some computational facilities as well as capability of uploading data to the Internet.

Thus instead of providing complex infrastructure, smart devices from citizens could be utilized for their own benefit. Consequently, mobile crowd-sensing (MCS) is defined in [3] as a category of applications "where individuals with sensing and computing devices collectively share data and extract information to measure and map phenomena of common interest." Here citizens and/or their mobile devices act both as sensors and actuators [5]. This can be viewed as a variant of mobile sensor networks, which rely on people's smartphones utilizing the sensors integrated in these devices. Smart devices are uncontrolled mobile sensors as mobility is determined by the device owners [6].

MCS if utilized effectively can make the cities smarter. The concept of smart city is based on "ideas of supporting infrastructure through the use of data and the importance of deploying processes that respond to that data" [7]. There are many applications of smart city like smart infrastructure, smart home, smart grid, smart transportation, and so on. However, smart citizen closes the loop by participating in sensing and actuation using their smartphones [8]. MCS may link infrastructure to its operations in smart cities [9]. Commercial organizations may be very much interested in collecting mobile sensing data to learn more about customer behavior. Government bodies may also be interested in collecting sensing data to know about road conditions, air pollution, and so on. Citizens may themselves be interested for various reasons like they can take precautions before entering an area. The idea of MCS can be applied in social, environmental, and infrastructure context [3]. However, before deployment of such applications, the desirable properties of such a system in the context of smart city need to be identified along with its associated challenges.

Consequently in this chapter we plan to focus on the scope of crowd-sensing for smart cities. Crowd-sensing literature is reviewed thoroughly in the following section. The challenges of crowd-sensing in the context of smart city are discussed in Section 5.3 followed by a brief overview of existing frameworks in Section 5.4. Finally Section 5.5 concludes with a hint of open issues in this direction of research.

5.2   Overview of Mobile Crowd-Sensing

It is well known that complex tasks can be effectively solved by a group than by an individual (e.g., SETI@Home [10]). In fact, groups perform smarter than the smartest person in the group. Thus involving the intelligence of crowd in solving complex tasks is known as crowdsourcing [6]. Here each member solves a small subtask. With today's smart devices crowdsourcing is even made easier as the users may send information (sensed by the sensors embedded into the device) as well as their opinion about a problem. Crowdsourcing spans from free user-generated content like YouTube to Amazon's mechanical turk where not only creative tasks are outsourced to the crowd, but also small tasks are assigned to the crowd [11]. Crowd-sensing can be viewed as a subset of crowdsourcing where the individuals share sensed information only (not opinion) [12, 13].

This section details the categories of crowd-sensing, architecture, and applications.

5.2.1   Categories of Crowd-sensing

Based on the sensing pattern, crowd-sensing applications can in general be classified into two categories: personal sensing and community sensing. In personal sensing, the aim is to capture information mainly related to the smart device's user, for instance, monitoring user activities or individual's carbon footprint. While in community sensing, data is collected from smartphones of many individuals in order to monitor environmental phenomena around a region, for example, traffic congestion level.

Community sensing can further be subdivided into participatory sensing and opportunistic sensing, depending on the mode of user involvement. In participatory sensing users join the task of sensing, whereas in opportunistic sensing, sensing takes place seamlessly without any user intervention [2]. This is summarized in Figure 5.1. Thus community sensing involving the crowd is known as MCS. Crowd-sensing benefits from the pervasive dominance of smart devices including smartphones, tabs, and so on in order to collect large-scale sensor data [3, 14].

5.2.2   Architecture of Mobile Crowd-sensing

Crowd-sensing mostly follows a pull model where a server (deployed by a government agency or a commercial organization) decides to execute a task and assign it to the smart devices of the crowd. The server usually chooses the devices where the sensing task is to be executed. A typical architecture of crowd-sensing is shown in Figure 5.2. The users may agree to the task

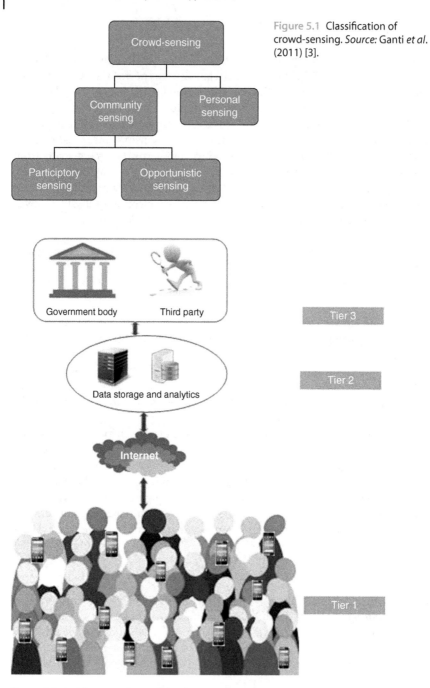

Figure 5.1 Classification of crowd-sensing. *Source:* Ganti *et al.* (2011) [3].

Figure 5.2 Architecture of crowd-sensing.

and start sensing (participatory) or the device may automatically respond (opportunistic) to the request and send the sensed data to the server through the Internet (periodically). Cloud services may be integrated here to store and apply analytics on such huge amount of sensing data from citizens. The information extracted from these data is sent to the task designer, which can be a commercial organization or government agency as shown in Figure 5.2. Thus it can be viewed as a three-tier architecture where in tier 1, the sensing tasks are actually deployed and data is sent to tier 2 that constitutes the server and/or cloud services. This tier is responsible for task assignment, user profiling (deciding about whom to assign the task depending on user credentials), and data analytics to extract necessary information from data. This information is finally sent to the government agency or commercial organization at tier 3. This tier is responsible for designing the task and decision-making. Though for social crowd-sensing applications, tier 1 and 3 devices are the same. Thus tier 2 plays an important role in processing huge data from tier 1 devices and sending output either back to tier 1 or to tier 3, depending on the application. Cloud services could be utilized here for faster task execution.

National Institute of Standards and Technologies (NIST) defines the cloud [15]: *Cloud computing is a model for enabling ubiquitous, convenient, on-demand network access to a shared pool of configurable computing resources (e.g., networks, servers, storage, applications, and services) that can be rapidly provisioned and released with minimal management effort or service provider interaction.* This new computing model is enabled due to the availability of virtually unlimited storage and processing capabilities. In the cloud, these virtualized resources can be leased in an on-demand fashion or as general utilities. Also it can host services to be delivered over the Internet. A cloud service has three distinct characteristics that differentiate it from traditional hosting.

1) On demand - Services on cloud are available on demand.
2) Elastic - The resources or services provided by cloud varies with varying demand.
3) Low (no) headache: Services and resources are mainly managed by the cloud service provider, and the service consumer (at tier 1 and/or tier 3 in Figure 5.2) only needs a computing device with Internet access.

In the crowd-sensing context, not only a vast number of smartphones may provide data to the cloud but also the sensors of a smartphone can generate large data over time, which is called *big data*. The real challenge is to analyze these data to build the knowledge base and ultimately the ability to respond to the world with greater intelligence. Hadoop is an open-source cloud computing environment created and maintained by the Apache project for distributed programming on commodity hardware.

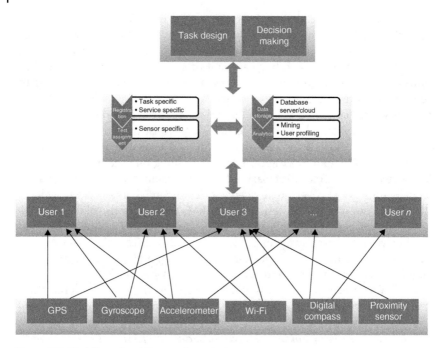

Figure 5.3 Workflow of mobile crowd-sensing.

The tasks performed by the devices lying in the different tiers are summarized in Figure 5.3. As shown in the figure, there may be a feedback from tier 3 to tier 1 through tier 2 about alert. For instance, in [16], crowd-sensing-based real-time public transport information service is implemented, and its front-end Android application, called TrafficInfo, is discussed in detail. The authors also proposed a publish/subscribe (pub/sub) communication model (using Extensible Messaging and Presence Protocol (XMPP)) for crowd-sensing applications for the smart city context. They considered the smartphone users as both publisher and consumer of data (prosumers). A service provider residing between the producer and consumer intercepts crowd-sensed data, processes it, and publishes meaningful information to the interested consumers.

Alternatively, the smart devices of the crowd may also initiate the task by requesting task execution at the server and sending sensed data. For example, while entering a polluted region, a person may proactively send this information to the server. The server may initiate sensing application and assign it to the smart devices around the region, collect data, process it and publish the level of pollution to the government authority, and may warn the citizens moving toward that region if the presence of any dangerous pollutant is detected. Thus tier 2 devices are responsible for task assignment, data storage, and analytics, whereas tier 3 devices are responsible for designing a task and deciding about its outcome. Analysis of users of tier 1 devices is also done by tier 2 devices such as user profiling and task registration.

Crowd-sensing applications for smart cities may be classified into three domains: infrastructure, environment, and social [3]. Applications involving measurement of large-scale phenomena related to public infrastructure fall into the first category. In environmental applications, participatory sensing is used to monitor natural environment like pollution levels in a city. In social applications individuals share sensed information among themselves so tier 1 and tier 3 devices (of Figure 5.2) are the same here. The overall classification of the applications is shown in Figure 5.4. Description of each category is presented below.

5.2.3.1 Applications in Infrastructure

The applications in this category include works related to smart transportation involving route planning and public safety as is summarized in Figure 5.4.

Many works are done on smart transportation as in [16]. In CityPulse [34], 101 smart city application scenarios have been identified including facilitating transportation such as a real-time travel planner or a service predicting public parking space availability. In [26], Singapore's bike sharing system is proposed. Here the idea was to replace short train routes (maximum three stops in the popular train network of the city) by bicycles that may be taken for a rent and parked near the destination [27]. The authors propose that if an individual starts a ride in the system he/she is asked to give his/her destination. Built-in GPS sensors can be utilized for tracking the bicycles and predicting its availability.

Tracking public transport vehicles can also be utilized for predicting arrival times of buses at a bus stop and also availability of seats. This is particularly important for harsh weather conditions. This kind of application mostly relies on accelerometer readings of the smart devices, and it uses a progressive localization technique comparing Wi-Fi SSIDs sensed at different stopping places as in [35]. Applications like Tranquilien [30], Moovit [36], and Tiramisu [37] are also built on similar idea of route planning by predicting the conditions of public vehicles, for instance, crowdedness, arrival times, cleanliness, availability of air conditioning, and so on. In Tranquilien [30], citizens can predict well in advance the comfort of trains in France. It uses optimization algorithms to predict if a person (in a compartment) should be able to find a seat, some chance of obtaining a seat, and standing room only up to 3 days in advance. History data of passengers are fed into the system. There is also option for crowdsourcing where the passengers can share their experience so that correction may be done for any wrong prediction. On the contrary applications like Moovit [36] use data to plan future infrastructure and service provision based on demand. Alternatively, opportunistic sensing may be utilized to identify the crowded routes of

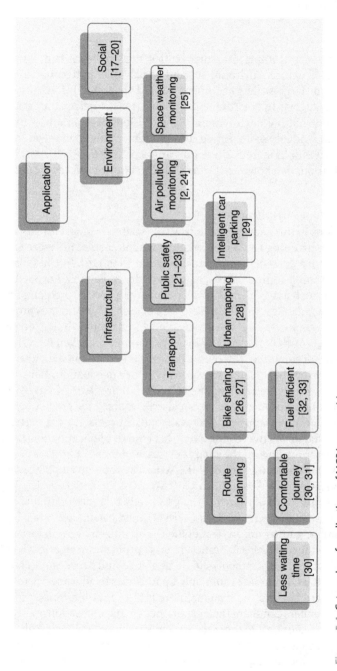

Figure 5.4 Categories of applications of MCS in smart cities.

the city, predict pollution levels, and design better public transport service as in projects like Istanbul in Motion initiated by Vodafone and IBM [31].

Finding fuel-efficient route is another application of crowd-sensing in a smart city that is also environment friendly as it finds the route having the minimum carbon footprint. This is done in [32] where the on-board diagnostic (OBD-II) interface of the cars is utilized to collect sensing data about fuel consumption of a route. Though the members having OBD-II interface may contribute to this participatory sensing event, the members not having the facility may at least get an estimate of fuel-efficient route. In [33] the cameras of windshield-mounted smartphones are used to take images of the road and the environment opportunistically and share it with other cars. Optimal route can be obtained by processing the collection of these images; when combined with fuel consumption data, this application can generate fuel-efficient routes and save fuel consumption. Experiments are conducted in Singapore and Cambridge, USA.

Drivers can benefit from real-time parking data collected from cars equipped with ultrasonic sensors as in [29]. Crowd-sensing can also be used in the mapping of on-street parking spaces to construct legal/illegal parking maps as shown in [28]. A large number of new-generation vehicles possess range finder parking sensors [38]. While in motion, these sensors can be utilized to detect the presence or absence of parked vehicles on the street. The sensor measurements can then be reported to the server along with the vehicle's GPS coordinates to estimate if the reported parking spaces are legal or illegal. Thus parking sensor data can be collected from the crowd to categorize streets into legal and illegal parking spaces with 90% accuracy.

Crowd-sensing can also be used to monitor crowd movement patterns in smart cities especially for detecting dangerous events, such as fights, riots, protests, demonstrations, fires, chemical leaks, stampedes, and high crowd levels, and provide better situational awareness. In [21] Smartphone Augmented Infrastructure Sensing (SAIS) is proposed where information gathered by civilians and officials are collectively utilized to build a dashboard mobile application that provides a sense of situation awareness. Proximity sensors like Bluetooth and GPS can be used for this purpose in addition to *in situ* sensors. All smartphone users should use the SAIS dashboard to produce and consume information about safety events and help each other for better situation awareness. In [22] machine learning techniques are used to process sensor readings to provide meaningful insights to crisis responders. In [23], routing behavior of taxi cabs is analyzed for detecting traffic anomalies, revealing the affected spatial regions and relations between individual road segments and displaying potential alternative routes. The routing trajectory of the taxicabs is stored for offline mining that helps in online anomaly detection. It also investigates the reason for the anomaly from social media.

5.2.3.2 Environmental Applications

This category of MCS applications includes measuring pollution levels in a city, water levels in creeks [24], and monitoring wildlife habitats. Most of these applications combine both sensing and sourcing of the crowd like taking snaps of waterways and trash and sending it to a server for processing. There are few applications that utilize only the sensing capabilities of the crowd as discussed below.

Common Sense [2] provides a prototype deployment for pollution monitoring based on participatory sensing. In this work, specialized handheld air quality sensing devices that communicate with mobile phones (using Bluetooth) are utilized to measure various air pollutants (e.g., CO_2, NO_x). One can utilize microphones on mobile phones to monitor noise levels in an area as well. The devices when used by crowd can help monitor pollution levels across an entire region.

Another type of application includes exploiting mobile phones for space weather monitoring. The Mahali project [25] proposes a revolutionary architecture using smart handheld devices to form a global space weather monitoring network. This involves predicting electron densities in the ionosphere. GPS signals can penetrate ionosphere. Thus data from GPS receivers that have a line of sight to several GPS satellites can be collected by smartphones. This information is sent to a cloud-based environment through the Internet for analysis.

5.2.3.3 Social Applications

In social crowd-sensing, participants share their produced data with each other through a server (tier 2 device in Figure 5.2). Such a database provides better understanding of community-related problems. For example, microblogs [17], which is a universal platform where users can share the information they have sensed about a region (e.g., tourist spots) and also real-time questions about a certain venue, can be entertained. Such a platform can also be utilized for spreading short-term news or advertisements. However this is more related to crowdsourcing. In DietSense [18], individuals take pictures of what they eat and share it within a community to compare their eating habits. This can be utilized for a community of diabetics to watch other diabetics and control their diet or provide suggestions. This again involves both crowd-sensing as well as crowdsourcing.

Crowd-sensing is used in applications like BikeNet [19] where individuals measure location and bike route quality based on parameters like CO_2 content on route, bumpiness of ride, and so on. Data collected is analyzed to obtain the "most" bikeable routes.

In [20], opportunistically captured images and audio clips from smartphones are exploited to link place visits with place categories like store and restaurant. The framework presented in the work combines signals based on location and user trajectories (using Wi-Fi/GPS) and maps it with visual and audio place "hints" mined from opportunistic sensor data. For instance, words spoken by people, text written on signs, or objects recognized in the environment can indicate a particular place. In this way, community of users may get feedback about a place.

Comparison of the crowd-sensing applications is summarized in Figure 5.5. Most of these applications are found to be beneficial to the community. GPS seems to be an indispensable sensor for applications in the infrastructure domain due to its accuracy in specifying location. However GPS is very power-hungry sensor, and it does not work indoor though a person spends most of his/her time indoor. Notably most of the applications designed are for the benefit of a community or society.

5.3 Issues and Challenges of Crowd-sensing in Smart Cities

The abovementioned applications are designed mostly based on the workflow defined in Figure 5.3. Functions like task assignment are done by the servers at tier 2 of the three-tier architecture (Figure 5.2) with the help of task requirements and feedback from tier 3 devices. Task assignment also depends on user profiling that involves pattern and quality of user responsiveness subject to incentives. What incentives to be given to a user for a task depends on the task design done by tier 3 devices. Tier 2 also involves selection of correct data and filtering out private information from it. This filtering along with other processing (localized analytics) can also be done at tier 1 devices in order to minimize data transfer through wireless channels. Analysis of data at tier 1 and/or tier 2 depends on the task design at tier 3 of the system. This is summarized in Figure 5.6.

Though numerous applications of MCS for smart cities have been proposed, the effectiveness of such applications depends on the efficient design of the functionalities mentioned in Figure 5.6. Consequently we briefly discuss about the issues regarding task assignment, user profiling and trustworthiness, design of incentive mechanisms, localized analytics, and security and privacy.

5.3.1 Task Assignment Problem

Few works have been done on task assignment and displaying results [14] in crowd-sensing though there are many challenges to overcome. While crowd-sourcing is aimed to utilize collective intelligence of the crowd to solve complex tasks by breaking them down to smaller tasks, crowd-sensing splits the responsibility of gathering correct information to the crowd. Toward this, a geo-social model of MCS is proposed in [5] that is based on a distributed architecture

Category of applications	Referenced solutions	Type of sensors used	Data access paradigm	Type of sensing	Direct benefit
Infrastructure	BSS Singapore [26]	GPS	Periodic	N/A	Private company
	Tranquilien [30]	N/A	Push	Participatory	Individual
	Moovit [36]	GPS	Push	Participatory	Community
	Tiramisu [37]	GPS	Push	Participatory	Community
	Istanbul in motion [31]	GPS	Periodic	Opportunistic	Society
	GreenGPS [32]	GPS, OBD-II interface	Periodic	Participatory	Individual
	Efficient route finder [33]	OBD-II interface, smartphone cameras	N/A	Opportunistic	Community
	Parking map [28]	GPS, parking sensors	Push	Opportunistic	Society
	SAIS [21]	GPS, Bluetooth	Periodic	Opportunistic	Community, fast responders
Environment	Common sense [2]	Bluetooth, air pollutant sensor	Periodic	Participatory	Community, individual
	Mahali project [25]	GPS, WLAN	Poll	Opportunistic	Society

Figure 5.5 Comparison of mobile crowd-sensing applications.

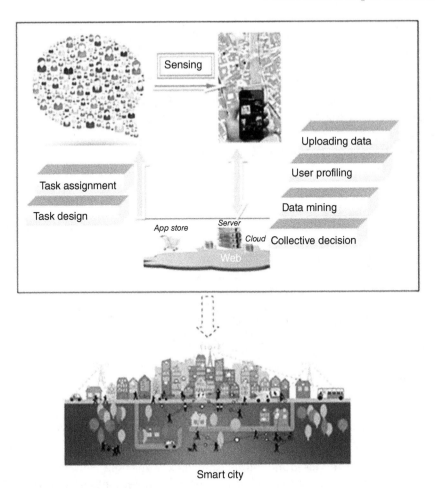

Figure 5.6 Role of mobile crowd-sensing in smart city context.

for task design, assessment, and execution. According to [39], a task can be defined as *a representation of a crowdsensing campaign to gather sensing indicators (considered relevant) at the desired quality level and with the desired coverage.* In this work, the authors associated different parameters with task like acceptance window, that is, how long the task will be available for users to accept; description of the task; duration, that is, the maximum time allowed to the user to finish task execution; and so on. The task life cycle as described in the work is given in Figure 5.7. Any geo-social task remains inactive and hidden from user unless the device reaches the desired area. The task execution is at user's discretion. So this kind of task assignment works only for participatory sensing. A geo-social task can only get completed at the desired region.

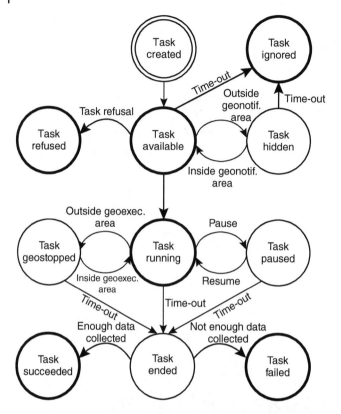

Figure 5.7 Lifecycle of crowd-sensing tasks according to [39]. *Source:* Courtesy of Bellavista, 2015.

A task can also be paused and still get completed in due time. However pause state need not be synchronized with the server; only the relevant states like running, successful, failed, and ignored states are transparently synchronized with the server via the notion of soft state. Tasks are actually assigned to mobile devices through pub/sub communication model in most crowd-sensing applications.

In [5] a framework McSense is proposed for MCS where three task assignment policies are discussed, namely,

- *Random policy* - This policy selects the set of crowd users to employ as a random group of available people in the whole city.
- *Attendance policy* - This policy exploits the knowledge of the citizens who have previously visited the area for significant time. It ranks citizens according to duration of previous visits.
- *Recency Policy* - This policy prefers users who have recently visited the sensing task area.

For each policy, smart devices whose battery level is below a certain threshold, called *battery threshold*, at task starting time are not considered for better task reliability. *Workers ratio*, that is, the percentage of candidate workers that will receive the task assignment, is another parameter considered in the work. For instance, initially for a task, *workers ratio* may be kept higher, but with user profiling, for a particular assignment policy lesser *workers ratio* may suffice later on.

In case of opportunistic sensing, the smartphones register with the server in tier 2. Once registration is complete, data streaming starts following push-or-pull model as discussed in [40]. In most crowd-sensing tasks especially for opportunistic sensing, due to high volume of data, cloud services can be utilized as server in tier 2 to store and effectively control tasks [41].

5.3.2 User Profiling and Trustworthiness

User reliability and categorization of users based on trust are important to ensure correctness of collected data. The phenomenon or thing observed by crowd users may be termed as entities like temperature of a place at a time instant. Users provide observations about entities in crowd-sensing applications with the help of sensors and send it to the server, which is a tier 2 device. The observation of users is noisy and often contradicts with one another. The false observations may outnumber true observations in which case voting or averaging would not extract the truth. However, correlation among entities may be exploited to find out the true observation and hence plot reliable users as is done in [42]. Assessing user reliability degrees is very important as we need observation from a number of reliable users. In [42], the task of truth discovery is formulated as an optimization problem. The correlation between entities is exploited here to find out the truth as well as reliable users. For instance, temperature or pollution level of nearby places should not vary abruptly. Accordingly, the entities are partitioned into disjoint independent sets, and block coordinate descent is applied to solve the problem and find out the user reliability as well as the truth. The sequential design of algorithm is often inefficient to capture temporal correlations. Thus, if data is fed to a cloud environment, parallel implementation of the algorithm on MapReduce [43] could be designed as shown in the work.

User trustworthiness denotes the average reputation of a user over time. An adversary may first build reputation over time, and when it reaches a certain threshold (and hence get recruited for many crowd-sensing tasks), it may start to misbehave, which could be difficult to detect. In [44], user credibility is assessed from sensing activity, reports from review requests, and authorities. Anchor-based voting as in [8] can also be applied to identify trusted users. For voting, history data does not need to be stored but gets reflected in the trust calculation. Anchor nodes are considered to be 100% trusted but their sensing accuracy may vary due to malfunctioning. Anchor nodes are used to vote new users for their trustworthiness.

5.3.3 Incentive Mechanisms

Incentive mechanisms also ensure reliability of users as users tend to provide reliable data for good incentives. MCS system can be crippled if incentive mechanisms do not stimulate human participation. Incentives should be provided to reward cooperation. It can be monetary, gamification, or by other means like providing service, increased social recognition or visibility, and so on [5].

Monetary incentive is the most common form of incentives in crowdsourcing literature as well. It can depend on the number of sensors engaged in a task, power consumed by the sensors [6], and/or data transmission required for the task. Users may be rewarded after each task completion or it may depend on timely completion of a predetermined number of tasks. In [14], an MCS framework is proposed that uses monetary incentives to encourage user participation. Such incentives could be given after completion of each task or only after completing a predefined number of tasks.

In some works, the incentive is not monetary but in terms of service a user would get. For instance, in [17] a credit system is proposed as incentive so that users do not ask for others' participation unless they help others.

Many works are done on gamification of incentives. In [45], two incentive methods are compared experimentally: one is micro-payments where a small amount of money is paid for each task and the other is weighted lottery where the return is two times more than micro-payments if won. It was observed that for the crowdsourcers, weighted lottery was more attractive; however, micro-payments resulted in more productive users because of guaranteed payments. When enough users are not participating in an event, a small amount of monetary incentive may be rewarded to the group in order to attract the idle users of the community for future sensing tasks as in [46]. To stimulate user participation, in [47] a reverse auction-based dynamic pricing incentive mechanism is devised. Here users can participate in sensing tasks and sell their sensing data to a service provider for bid prices claimed by the users themselves. The user with minimum bidding price wins. However if a reliable user loses multiple times, s/he may leave the system. In order to prevent such possibility of starvation, the authors devised a mechanism called virtual participant credit (VPC). The server gives a virtual credit to the participants who lost in the reverse auction as a reward for their participation only. This can only be used for lowering bid price of future rounds thus increasing the winning probability of user for the future auction rounds. Such participation incentive maintains enough active bidders (i.e., desired level of participatory sensing service quality) and stabilizes the incentive cost by keeping the price competitions. The mechanism not only reduces the incentive cost for retaining the same number of participants but also removes the burden of accurate price decision for user data.

In [48] incentive mechanisms are devised based on Stackelberg games and contract theory. However, the mechanisms provided require either the complete information or the prior distributions of users' private types. Thus a priori knowledge is needed. This restriction is removed in [49] by exploiting bidding mechanism. Here (in [49]) the server announces a set of sensing tasks. Crowd users with different available time and sensing costs bid for these tasks. The authors designed mechanisms to optimally schedule the users achieving multiple performance objectives including truthfulness, individual rationality, provable approximation ratios, and computational efficiency simultaneously. The mechanism is shown to work for two underling scenarios:

- when all users arrive at the same time (offline)
- when users appear sequentially (online).

In [50, 51] game theoretic approaches are explored to devise incentive mechanisms though it is assumed that any sensing task can be done instantly. The nature of smartphone users opportunistically occurring in the area of interest is exploited [51]. This online approach is characterized by computational efficiency, individual rationality, and profitability. Here a user, appearing at a certain area of interest, receives available task descriptions from users. Then if it decides (according to the mechanism proposed) to carry out some tasks, it bids for the same. The server may accept the bidding and assign those tasks and payoff.

Summary of incentive mechanisms for MCS is presented in Figure 5.8.

5.3.4 Localized Analytics

An important issue of MCS is to decide about the information processing to be done in tier 1 devices itself. Depending on the type of applications, sensor

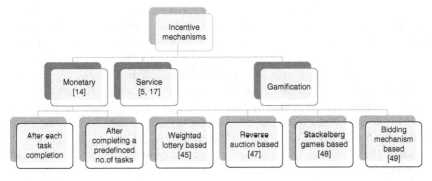

Figure 5.8 Categorization of incentive mechanisms for mobile crowd-sensing applications.

readings need to be preprocessed before sending it to a server. For instance, in pothole detection application as described in [52], the presence of spikes is detected from 3-axis acceleration sensor data to determine potential potholes. Only when potholes are detected, relevant data need to be sent to the server. Local analytics may also help in data aggregation, thus consuming less energy and bandwidth because of the compact representation. Noise and redundant data may also be removed through local analytics. Here the trade-off between energy consumption due to local computation and that due to data transmission is to be taken care of as more local analytics may save size and frequency of data transmission. Two functions of local analytics have been identified in [3], namely:

1) *Data mediation* - It involves filtering of outliers, elimination of noise, or filling in data gaps. For example, due to lack of line of sight, GPS samples acquired may not be accurate. Thus noise elimination or filling in data gaps by extrapolating samples may be required.
2) *Context inference* - Sometimes it is important to detect the context in which the sensor readings are taken. This involves processing of some other sensor readings, for instance, to detect kinetic mode (walking, standing, jogging, running) of humans. Like a person may want to know if many joggers are there at a park so that s/he may go for jogging safely [40]. For this application, only data from joggers who are running is important to the server for analysis.

   The analytics performed for inferring context is mostly application specific. Hence if multiple crowd-sensing applications run on the same device, context inference could be computationally expensive for all of them though many may require readings from the same sensor or similar computation. Moreover, the health of the device in terms of remaining energy and current load should be taken into account before performing mining on the data.

5.3.5 Security and Privacy

MCS calls for many important concerns from citizen's point of view, such as the sharing of personal data (e.g., user location, ambient sound), can raise significant concerns about security and user privacy. As stated in [53], MCS applications potentially collect sensitive sensor data pertaining to individuals that can be used to detect behavioral patterns of individuals. For example, GPS sensor readings can be used to estimate traffic congestion levels and/or anomalies in a given community, but at the same time these can be used to infer private information like movement trajectory of an individual, routes they take during their daily commutes, and home and work locations. In [54], a reputation system based on the Gompertz function is proposed that estimates the trustworthiness of the collected data. Privacy can be preserved if sensing data is processed to make it anonymous and the private part of the data is removed before sending it to the server. However, as predicted in [55], it may cost energy and computation for the crowd-sensing devices. Rather privacy-preserving architecture for crowd-sensing must be designed. AnonySense [56] is such an architecture. It provides a new tasking language for context queries, which will be submitted by the crowd-sensing applications. These tasks are assigned to anonymous nodes, and eventually collected verifiable yet unlinkable reports are fed back to the applications. Consequently, in [57], a privacy-preserving approach for untrusted aggregator is proposed that delinks data from its source in a group of n user's data. Thus it does not hide data and only shows specific aggregation results like conventional privacy-preserving approaches. This protocol shows individual data but achieves "n-source anonymity" in the sense that an aggregator only learns that the source of any particular piece of data is one of n users in a group. Thus any aggregation function can be applied on the data yet maintaining the privacy of individual users in the group. For a large pool of users having varying privacy requirements, authors categorize users in groups having similar privacy requirements.

The issues discussed above are not independent of each other. For instance, task assignment is very important as it decides whether localized analytics is needed by the task. Like for context-sensitive tasks, deciding about the context locally is very important. Moreover, incentives are also task specific and the amount and/or even type of incentive varies with tasks. However, the crowd is attracted by incentive mechanisms. Thus user profiling is driven by incentive mechanisms. Moreover, for tasks requiring localized analytics, it is important to attract trusted users being capable of sensing the context efficiently. Privacy of collected data should be preserved. Thus localized analytics is also a factor in profiling users. This is summarized in Figure 5.9.

These issues maps to the following challenges:

- Strategies should be figured out to attract majority of citizens in terms of incentives (feedback from user profiling to task assignment). It may be money, social recognition, and so on [5].
- Framing the data collection phase in a manner to reduce communication and storage overhead. For instance, text fields could be preferred than sending images and/or videos. Localized analytics may play a key role here.
- Correctness of data should be ensured before assessing them. False positives need to be filtered out. This again depends on proper incentive mechanism.

Figure 5.9 Major issues of crowd-sensing in smart cities and relationship between them.

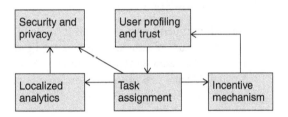

- How to reuse one sensor data for different applications in energy efficient manner [3].
- Ensuring security and privacy of data.
- Efficient data processing and aggregation poses an important challenge [41].

5.4   Crowd-sensing Frameworks for Smart City

Few works are done on designing frameworks for crowd-sensing applications in smart cities. Many of them are based on pub/sub communication paradigm where the tier 3 and tier 1 of Figure 5.2 are the same devices, that is, the smart-phone users are publishing data as well as consuming the services. Few have extended the XMPP protocol [58] in this context as in [16]. Medusa [14] is a programming framework for MCS; it is one of the initial attempts in this domain. It uses a high level XML-based domain-specific programming language called MedScript. By using this language, users can define sensing tasks and workflows for monetary incentives, while the underlying Medusa framework hides the resulting complexities and takes care of task coordination, worker management, incentive assignment, and result collection. Thus task designers like city managers could use this language to take advantage of crowd-sensing. However, most of the frameworks designed later for MCS are specific for application types and take into account one or more challenges mentioned in Section 5.3. Some of the frameworks are detailed in the following subsections.

5.4.1   Here-*n*-Now Framework

Here-*n*-Now is an MCS framework for smart city applications where individuals want to know about a nearby place in terms of noise level, activity level, light intensity, and crowd intensity [40]. For instance, a person wants to select a quiet restaurant for peaceful discussion or a park for jogging where many joggers are there (for safety). Such framework uses opportunistic sensing to collect data from citizens. It correlates mobile sensory information to information from social media in real time. The design of the framework is shown in Figure 5.10. There are two entities in the framework - the crowd users and the cloud data provider. The crowd users are responsible for two functions - data collection and analysis, querying clients. The cloud data provider is responsible for context-aware data processing.

1) Data collection and analysis - This module captures sensory data, performs real-time stream mining on the data, and uploads analyzed information to the cloud. To reduce the communication cost, only analyzed information from each device is sent to the cloud. Resource-aware clustering technique is applied to the collected sensor readings to identify significant changes in

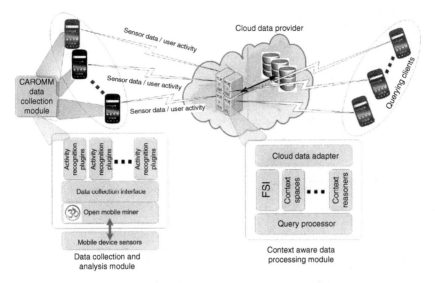

Figure 5.10 Design of Here-*n*-Now framework. *Source:* Reprinted with permission and courtesy of Jayaraman *et al.* (2012) [40].

the situation. This module also includes the ability to plug in mobile activity recognition model. The authors used a neural network-based activity recognition model for recognizing four basic activities - walking, running, sitting, and driving - using the sensed accelerometer data. Thus only if a person is jogging at a park, his/her information would be meaningful for the query mentioned above.

2) Context-aware processing - This module inferences situations using the real-time sensory and activity data collected from mobile users. As shown in Figure 5.10, any context reasoning engine can be integrated into Here-*n*-Now for inference. The authors used fuzzy situation inference (FSI) model, which integrates fuzzy logic into the probabilistic context spaces (CS) model. It combines the benefits of the CS model for supporting pervasive environments along with fuzzy logic to deal with uncertainty associated with human concepts and real-world situations.

Localized analytics for activity recognition is a big advantage in this framework as it only sends meaningful information to the cloud. However, continuous data streaming could be a bottleneck. Moreover, the accuracy of query results depends on availability of active users in desired places.

In [59], a similar framework named silent mobile sensing framework is proposed where the cloud server after statistical analysis publishes data on top of maps for better visualization. In this work, authors also considered data compression before uploading to the cloud server.

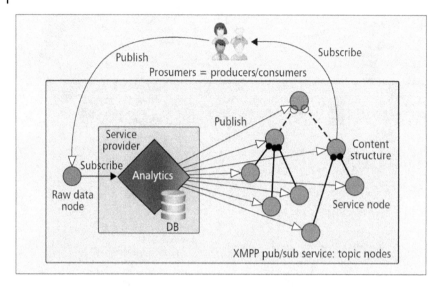

Figure 5.11 Design of XMPP-based mobile crowd-sensing framework. Farkas et al. (2015) [16].

5.4.2 Crowd-sensing Framework based on XMPP

XMPP realizes a pub/sub communication model [16, 60], where publications sent to a node are automatically multicast to the subscribers of that node. In [16], a crowd-sensing framework based on XMPP pub/sub model is proposed. In fact, the pub/sub model fits well with many crowd-sensing applications where the data is collected from crowd and analyzed and service is delivered to the crowd itself, for instance, smart transport service where the user standing at a bus stop may know about availability of seats or current position of the bus. As shown in Figure 5.11, here the crowd is the producer as well as the consumers of data also known as prosumers. However, there is a service provider entity mostly the cloud data provider that can play several roles at the same time like collecting data (consumer role) and storing and analyzing producers' data to offer (service provider role) value added service. Producers as shown in the figure publish raw data to the event nodes. Service providers intercept these data by subscribing to the raw event nodes, analyze it asynchronously, and publish cleaned up information or value-added service to the content nodes. Interested consumers receive the added value/service by subscribing to the appropriate content node(s) in an asynchronous manner.

5.4.3 McSense

In [5], an MCS platform, McSense, is proposed that is based on three-tier architecture as shown in Figure 5.12. Here the three entities are the following:

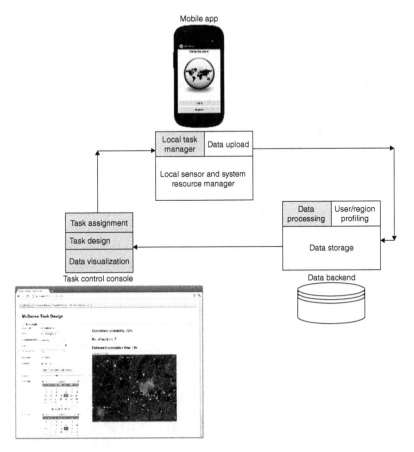

Figure 5.12 Design of McSense - a mobile crowd-sensing platform for smart cities. *Source:* Cardone et al. (2013) [5].

- *McSense mobile app* - This component resides at the smart devices and receives task offers, allows users to accept them, and provides the tools to complete them by seamlessly configuring all needed sensors available on board. The task apps report their collected data to the McSense mobile app. In addition, the McSense mobile app may also collect data to profile users, devices, and regions. Upon completion sensed results and profiling data are uploaded to the McSense data back end, and incentives are received.
- *McSense data back end* - This component resides in some server and receives data from the McSense mobile app. It stores and analyzes sensed data to evaluate task performance and profile technical and social dimensions of users. Task performance includes parameters like task completion time, number of users employed, and so on. However, for many tasks, data from several users

need to be aggregated and collectively analyzed at the back end. Technical dimensions of user devices include number and type of sensors available on the smartphone, available battery, and so on. Social ones include relationship among users, frequency of users visiting the sensing task area, and so on.

- *McSense task control console* - This component may be handled by city managers corresponding to tier 3 in Figure 5.2. It may reside as a web application, and, apart from data visualization, it offers two main functions: task design and task assignment. The task design component takes into account profiles stored in the data back end and task-related data (location, area, and duration) and evaluates the desired task completion time and workers ratio to complete the task with a desired probability. The task assignment component defines the optimal set of workers to carry out the sensing task. For example, assigning a task to a user typically spending much time in an area (as reflected in user profiles) increases its probability of success. The task assignment strategies of McSense are already detailed in Section 5.3.1.

The framework also talks about monetary or service incentives given to users. It is valid for participatory sensing. The authors implemented the framework and were shown to experiment with different types of task assignments. However, the need for localized analytics for better task processing may be considered.

5.4.4 Supporting Framework for Crowd-sensing Apps

In [61], a framework is proposed for supporting crowd-sensing applications in smart cities. The main focus of this work is to help the non-programmers design and deploy crowd-sensing applications. The framework is built on two entities: participant and initiator. Initiators are responsible for developing the application and asking for sensing data from participants. Participants may accept to sense and send data to a server. Both participant and initiator are smartphone users. The framework runs in two parts - configuring and assembling part and mobile crowd-sensing app runtime (CAR) part. In the configuring and assembling part, the app designer specifies the functionalities and constraints through XML-based language, which is used by a predefined campaign model to form the participant module, initiator module, and server side programs for both initiator and participants. The notable functionalities of CAR module include participant recruitment, energy- and cost-efficient sensing data collection, participant profile management, participant experience control, and location reliability verification.

The above mentioned frameworks are compared with respect to the issues mentioned in Section 5.3 and are summarized in Figure 5.13. Frameworks

Referenced framework	Type of incentive	Localized analytics	Type of sensing	Tiers
Here-*n*-Now [40]	Service	Done	Opportunistic	2
Silent mobile sensing framework [59]	Service	Done	Opportunistic	2
Medusa [14]	Monetary/other	NA	Participatory	3
Crowd-sensing framework based on XMPP [16]	Service	NA	Opportunistic	2
Mcsense [5]	Monetary/other	NA	Participatory	3
Supporting framework for crowd-sensing apps [61]	Monetary	NA	Participatory	2

Figure 5.13 Comparison of frameworks proposed in literature.

proposed in [14, 61] are for designers of crowd-sensing applications. These frameworks aim at abstracting the details of the execution and programming burden from users while providing an easy-to-learn user interface for designing applications. However, many works [5, 16, 40] focus on designing a framework for crowd-sensing applications itself. Money seemed to be a common incentive mechanism; however, some works like [40] indicate that the benefit goes to both the contributors and initiators of crowd-sensing applications.

5.5 Conclusion

In this chapter, the literature of MCS for smart cities is reviewed thoroughly including motivation, possible applications, and issues that are key to successful deployment of such services and overview of existing frameworks. Several open issues came out as a result. For instance, the possibility of several crowd-sensing applications running on the same device having overlapping sensing criteria should be taken care of. Trade-off between localized analytics for better privacy versus energy efficiency should be investigated. Most of the existing frameworks discussed address few challenges of MCS for smart cities for some specific type of applications. However, a comprehensive framework taking care of the challenges with the ability to tune in for different application types would be highly beneficial for smart cities.

Final Thoughts

In this chapter the concept of MCS and its applications in smart cities are discussed. The associated issues and their existing solutions are also detailed. The chapter investigates the issues addressed by the existing crowd-sensing frameworks in the smart city context.

Questions

1 Define mobile crowd-sensing and its significance to smart city.

2 Mention the issues associated with mobile crowd-sensing.

3 Describe the importance of incentive mechanism in mobile crowd-sensing.

4 How can privacy be breached in mobile crowd-sensing?

References

1 Hull, B. et al. (2006) *CarTel: A Distributed Mobile Sensor Computing System*. Proceedings of SenSys, pp. 125–138.

2 Dutta, P., Paul, M.A., Kumar, N., Mainwaring, A., Myers, C., Willett, W., and Woodruff, A. (2009) *Demo Abstract: Common Sense: Participatory Urban Sensing Using a Network of Handheld Air Quality Monitors.* Proceedings of ACM SenSys, 2009, pp. 349–350.

3 Ganti, R.K., Fan, Y., and Hui, L. (2011) Mobile crowdsensing: current state and future challenges. *IEEE Communications Magazine*, **49** (11), 32–39.

4 Reality Mining Project, http://realitycommons.media.mit.edu/ (accessed 20 December 2016).

5 Cardone, G., Foschini, L., Bellavista, P., Corradi, A., Borcea, C., Talasila, M., and Curtmola, R. (2013) Fostering participaction in smart cities: a geo-social crowdsensing platform. *IEEE Communications Magazine*, **51** (6), 112–119.

6 Petkovics, A., Simon, V., Gôdor, I., and Børøcz, B. (2015) Crowdsensing solutions in smart cities towards a networked society. *EAI Endorsed Transactions on Internet of Things*, **15** (1), 150600.

7 Gooch, D., Wolff, A., Kortuem, G., and Brown, R. (2015). *Reimagining the Role of Citizens in Smart City Projects*. 1st International Workshop on Smart Cities: People, Technology and Data, September 2015, ACM.

8 Pouryazdan, M. and Kantarci, B. (2016) The Smart Citizen Factor in Trustworthy Smart City Crowdsensing, *IT Professional*, http://nextconlab .academy/PouryazdanKantarci-IT-Professional2016-MinorRevised.pdf (accessed 20 December 2016).

9 Batty, M. et al. (2012) Smart cities of the future. *The European Physical Journal Special Topics*, **214** (1), 481–518.

10 Search for Extraterrestrial Intelligence (SETI) *Crowdsourcing Community*, http://setiathome.ssl.berkeley.edu/ (accessed 15 August 2015).

11 Amazon Mechanical Turk https://www.mturk.com/mturk/welcome (accessed 20 December 2016).

12 Schuurman, D., Baccarne, B., Marez, L.D., and Mechant, P. (2012) Smart ideas for smart cities: investigating crowdsourcing for generating and selecting ideas for ICT innovation in a city context. *Journal of Theoretical and Applied Electronic Commerce Research*, **7** (3), 49–62.

13 Talasila, M., Curtmola, R., and Borcea, C. (2015) Handbook of sensor networking: advanced technologies and applications, in *Mobile Crowd Sensing*, CRC Press.

14 Ra, M.-R. et al. (2012) *Medusa: A Programming Framework for Crowd-Sensing Applications*. Proceedings of the 10th International Conference Mobile Systems, Applications, and Services, 2012, pp. 337–350.

15 Mell, P. and Grance, T. (2011) *The NIST Definition of Cloud Computing*, US National Institute of Science and Technology, http://csrc.nist.gov/publications/nistpubs/800-145/SP800-145.pdf (accessed 20 December 2016).

16 Farkas, K., Fehér, G., Benczúr, A., and Sidló, C. (2015) Crowdsending based public transport information service in smart cities. *IEEE Communications Magazine*, **53** (8), 158–165.

17 Gaonkar, S., Li, J., Choudhury, R.R., Cox, L., and Schmidt, A. (2008) *Micro-Blog: Sharing and Querying Content Through Mobile Phones and Social Participation*. Proceedings of the 6th International Conference on Mobile Systems, Applications, and Services.

18 Reddy, S., Parker, A., Hyman, J., Burke, J., Estrin, D., and Hansen, M. (2007) *Image Browsing, Processing, and Clustering for Participatory Sensing: Lessons from a DietSense Prototype*. Proceedings of the 4th Workshop on Embedded Networked Sensors (EmNets '07), pp. 13–17.

19 Eisenman, S.B., Miluzzo, E., Lane, N.D., Peterson, R.A., Ahn, G.-S., and Campbell, A.T. (2007) *The BikeNet Mobile Sensing System for Cyclist Experience Mapping*. Proceedings of the 5th International Conference on Embedded Networked Sensor Systems (SenSys '07), pp. 87–101.

20 Chon, Y., Lane, N.D., Li, F., Cha, H., and Zhao, F. (2012) *Automatically Characterizing Places with Opportunistic Crowdsensing Using Smartphones*. Proceedings of the 2012 ACM Conference on Ubiquitous Computing (UbiComp '12), pp 481–490.

21 Liao, C.-C., Hou, T.-F., Lin, T.-Y., Cheng, Y.-J., Erbad, A., Hsu, C.-H., and Venkatasubramania, N. (2014) *SAIS: Smartphone Augmented Infrastructure Sensing for Public Safety and Sustainability in Smart Cities*. Proceedings of the 1st International Workshop on Emerging Multimedia Applications and Services for Smart Cities (EMASC '14), pp. 3–8.

22 Radianti, J., Dugdale, J., Gonzalez, J., and Christoffer-Granmo, O. (2014) *Smartphone Sensing Platform for Emergency Management*. Proceedings of 11th International Conference on Information Systems for Crisis Response and Management ISCRAM2014.

23 Pan, B., Zheng, Y., Wilkie, D., and Shahabi, C. (2013) *Crowd Sensing of Traffic Anomalies Based on Human Mobility and Social Media*. Proceedings of the 21st ACM SIGSPATIAL International Conference on Advances in Geographic Information Systems (SIGSPATIAL'13), ACM, pp. 344–353, doi: 10.1145/2525314.2525343.

24 Creek Watch *Explore Your Watershed*, http://creekwatch.researchlabs.ibm .com/ (accessed 20 December 2016).

25 Pankratius, V., Lind, F., Coster, A., Erickson, P., and Semeter, J. (2014) Mobile crowd sensing in space weather monitoring: the mahali project. *Communications Magazine*, **52** (8), 22–28.

26 Shu, J., Chou, M., Liu, Q., Teo, C.P., and Wang, I.L. (2010) Bicycle-Sharing System: Deployment, Utilization and the Value of Re-Distribution. Technical report, National University of Singapore-NUS Business School, Singapore, http://bschool.nus.edu.sg/staff/bizteocp/BS2010.pdf (accessed 27 December 2016).

27 Fricker, C. and Gast, N. (2014) Incentives and redistribution in homogeneous bike-sharing systems with stations of finite capacity. *EURO Journal on Transportation and Logistics*, **5** (3), 261–291.

28 Coric, V. and Gruteser, M. (2013). *Crowdsensing Maps of On-street Parking Spaces*. Proceedings of the 2013 IEEE International Conference on Distributed Computing in Sensor Systems, pp. 115–122.

29 Mathur, S., Jin, T., Kasturirangan, N., Chandrasekaran, J., Xue, W., Gruteser, M., and Trappe, W. (2010) *ParkNet: Drive-By Sensing of Road-Side Parking Statistics*. Proceedings of the 8th International Conference on Mobile Systems, Applications, and Services, pp. 123–136.

30 Tranquilien *Mobile Application*, http://www.tranquilien.com (accessed July 15 2015).

31 Guide to Smart Cities (2013) *The Opportunity for Mobile Operators, GSMA Report*.

32 Ganti, R.K., Pham, N., Ahmadi, H., Nangia, S., and Abdelzaher, T.F. (2010) *GreenGPS: A Participatory Sensing Fuelefficient Maps Application*. Proceedings of the 8th International Conference on Mobile Systems, Applications, and Services, pp. 151–164.

33 Koukoumidis, E., Martonosi, M., and Peh, L.-S. (2012) Leveraging smartphone cameras for collaborative road advisories. *IEEE Transactions on Mobile Computing*, **11** (5), 707–723.

34 Presser, M., Vestergaard, L., and Ganea, S. (2014) *Smart City Use Cases and Requirements*, CityPulse Project Deliverable D2.1, May 2014.

35 Petkovics, Á. and Farkas, K. (2014) *Efficient Event Detection in Public Transport Tracking*. International Conference on Telecommunications and Multimedia (TEMU), pp. 74–79.

36 Moovit *Mobile Application*, http://www.moovitapp.com/ (accessed 18 July 2015).

37 Tiramisu *The Real-Time Bus Tracker*, http://www.tiramisutransit.com/ (accessed 15 August 2015).

38 Brosicke, G., Mayer, O., Eri, R., and Seeger, H. (2001) The automatic parking brake. *ATZ Automobiltechnische Zeitschrift*, **103**, 39–42.

39 Bellavista, P., Corradi, A., Foschini, L., and Ianniello, R. (2015) Scalable and cost-effective assignment of mobile crowdsensing tasks based on profiling trends and prediction: the participact living lab experience. *Sensors Journal*, **15** (8), 18613–18640.

40 Jayaraman, P.P., Sinha, A., Sherchan, W., Krishnaswamy, S., Zaslavsky, A., Haghighi, P.D., Loke, S., and Do, M.T. (2012) *Here-n-Now: A Framework for Context-Aware Mobile Crowdsensing*. Proceedings of the 10th International Conference on Pervasive Computing, pp. 1–4.

41 Antonić, A., Rôzanković, K., Marjanović, M., Pripuzić, K., and Zarko, I.P. (2014) *A Mobile Crowdsensing Ecosystem Enabled by a Cloud-Based Publish/Subscribe Middleware*. Proceedings of FiCloud-2014.

42 Meng, C., Jiang, W., Li, Y., Gao, J., Su, L., Ding, H., and Cheng, Y. (2015) *Truth Discovery on Crowd Sensing of Correlated Entities*. Proceedings of the 13th ACM Conference on Embedded Networked Sensor Systems (SenSys '15), pp. 169–182.

43 Thusoo, A., Sarma, J.S., Jain, N., Shao, Z., Chakka, P., Anthony, S., Liu, H., Wyckoff, P., and Murthy, R. (2009) *Hive: A Warehousing Solution Over a Map-Reduce Framework*. Proceedings of VLDB Endow. 2, August 2, 2009, pp. 1626–1629.

44 Prandi, C., Ferretti, S., Mirri, S., and Salomoni, P. (2015) *Trustworthiness in Crowd-Sensed and Sourced Georeferenced Data*. Proceedings of IEEE International Conference on Pervasive Computing and Communications Workshops (PERCOM), March 2015, pp. 402–407.

45 Rula, J.P., Navda, V., Bustamante, F.E., Bhagwan, R., and Guha, S. (2014) *No One-Size Fits All: Towards a Principled Approach for Incentives in Mobile Crowdsourcing*. Proceedings of the 15th Workshop on Mobile Computing Systems and Applications, p. 3.

46 Sun, J. (2013) *An Incentive Scheme Based on Heterogeneous Belief Values for Crowd Sensing in Mobile Social Networks*. Proceedings of the Global Communications Conference (GLOBECOM), pp. 1717–1722.

47 Lee, J.-S. and Hoh, B. (2010) Dynamic pricing incentive for participatory sensing. *Pervasive and Mobile Computing*, **6** (6), 693–708.

48 Duan, L., Kubo, T., Sugiyama, K., Huang, J., Hasegawa, T., and Walrand, J. (2012) *Incentive Mechanisms for Smartphone Collaboration in Data Acquisition and Distributed Computing*. Proceedings of the 31st IEEE INFOCOM, 2012, pp. 1701–1709.

49 Han, K., Zhang, C., Luo, J., Hu, M., and Veeravalli, B. (2016) Truthful scheduling mechanisms for powering mobile crowdsensing. *IEEE Transactions on Computers*, **65** (1), 294–307.

50 Zhao, D., Li, X.-Y., and Ma, H. (2014) *How to Crowdsource Tasks Truthfully Without Sacrificing Utility: Online Incentive Mechanisms with Budget Constraint*. Proceedings of IEEE INFOCOM, 2014, pp. 1213–1221.

51 Zhang, X., Yang, Z., Zhou, Z., Cai, H., Chen, L., and Li, X. (2014) Free market of crowdsourcing: incentive mechanism design for mobile sensing. *IEEE Transactions on Parallel and Distributed Systems*, **25** (12), 3190–3200.

52 Mohan, P., Padmanabhan, V., and Ramjee, R. (2008) *Nericell: Rich Monitoring of Road and Traffic Conditions Using Mobile Smartphones*. Proceedings of ACM SenSys, 2008, pp. 323–336.

53 Sun, Y., Song, H., Jara, A.J., and Bie, R. (2016) Internet of things and big data analytics for smart and connected communities. *IEEE Access*, **4**, 766–773, doi: 10.1109/ACCESS.2016.2529723.

54 Huang, K.L., Kanhere, S.S., and Hu, W. (2010) *Are You Contributing Trustworthy Data? The Case for a Reputation System in Participatory Sensing*. Proceedings of the 13th ACM International Conference on Modeling, Analysis, and Simulation of Wireless and Mobile Systems (MSWIM'10), pp. 14–22.

55 Guo, B., Wang, Z., Yu, Z., Wang, Y., Yen, N.Y., Huang, R., and Zhou, X. (2015) Mobile crowd sensing and computing: the review of an emerging human-powered sensing paradigm. *ACM Computing Surveys*, **48** (1), 1–31.

56 Cornelius, C., Kapadia, A., Kotz, D., Peebles, D., Shin, M., and Triandopoulos, N. (2008) *AnonySense: Privacy-Aware People-Centric Sensing*. Proceedings of the 6th International Conference on Mobile Systems, Applications, and Services, ACM, pp. 211–224.

57 Zhang, Y., Chen, Q., and Zhong, S. (2016) Privacy-preserving data aggregation in mobile phone sensing. *IEEE Transactions on Information Forensics and Security*, **11** (5), 980–992, doi: 10.1109/TIFS.2016.2515513.

58 Saint-Andre, P. (2011) *Extensible Messaging and Presence Protocol (XMPP): Core*, Internet Engineering Task Force, RFC 6120 (Proposed Standard), http://www.ietf.org/rfc/rfc6120.txt (accessed 20 December 2016).

59 Hariri, F., Daher, G., Sibai, H., Frenn, K., Doniguian, S., and Dawy, Z. (2013). *Towards a Silent Mobile Sensing Framework for Smart Cities*. Wireless World Research Forum (WWRF 30).

60 Eugster, P. et al. (2013) The many faces of publish/subscribe. *ACM Computing Surveys*, **35** (2), 114–131.

61 Wang, J., Wang, Y., and Zhao, J. (2015) Helping Campaign Initiators Create Mobile Crowd Sensing Apps: A Supporting Framework. 2015 IEEE 39th Annual Computer Software and Applications Conference (COMPSAC), vol. 2, pp. 545–552.

6

Wide-Area Monitoring and Control of Smart Energy Cyber-Physical Systems (CPS)

Nilanjan R. Chaudhuri

School of Electrical Engineering and Computer Science, Pennsylvania State University, State College, PA, USA

CHAPTER MENU

Introduction, 155
Challenges and Opportunities, 156
Solutions, 159
Conclusions and Future Direction, 173

Objectives

- To become familiar with wide-area monitoring and control in smart energy CPS.
- To learn about the opportunities and challenges of the stability monitoring using wide-area measurements.
- To appreciate the opportunities and challenges of the oscillation damping application of wide-area control.
- To understand the challenges of wide-area control in the presence of wind farms.
- To know about techniques to solve challenges like latency compensation in wide-area damping control.
- To become familiar with an approach for wide-area damping control using wind farms.

6.1 Introduction

This chapter presents the stabilization aspect of energy cyber-physical systems (CPSs). The focus is on wide-area monitoring and control of smart grids, which

Smart Cities: Foundations, Principles and Applications, First Edition.
Edited by Houbing Song, Ravi Srinivasan, Tamim Sookoor, and Sabina Jeschke.
© 2017 John Wiley & Sons, Inc. Published 2017 by John Wiley & Sons, Inc.

is enabled by remote signals communicated from the phasor measurement units (PMUs) to the control center through a phasor data concentrator (PDC). A global positioning system (GPS) provides timing pulse to correlate the sampled measurements and achieve precise time synchronization. The PDC synchronizes the measurements from all the PMUs with microsecond precision and, under normal condition, sends data once every cycle to the control center. This data is processed to estimate the modal frequency, damping ratio, and mode shapes in near real time, which are the vital indicators of the grid's health and stability. While this information is vital for the operators to assess the health of the system in near real time, the logical next step is to stabilize the grid using remote feedback signals through continuous control action, which is called wide-area control. One of the challenges arise from latency that can adversely impact the performance of closed-loop control system, and what more, it can vary with time. Finally, integration of cyber infrastructure is critical for stabilizing the grid when inverter-interfaced wind farms (WFs) replace conventional generating plants.

The chapter is organized as follows: after this brief introduction, the challenges and opportunities in the areas of wide-area monitoring, wide-area damping control, and the latter in the presence of WFs are discussed. Solutions to these problems based on the author's past research work [1–3] are also presented. Finally, the conclusion and the future direction of research are also summarized.

6.2 Challenges and Opportunities

6.2.1 Wide-Area Monitoring: Damping, Frequency, and Mode-shape Estimation

The 1996 blackout in the Western Electricity Coordinating Council (WECC) system resulted in a lot of research attention in the area of wide-area measurement-based monitoring to avoid or limit the spread of such a catastrophe [4]. Multiple events of line tripping, abrupt load, and generation shedding took place during this blackout that continuously changed the modal behavior of the system [5, 6]. With accurate and near real-time knowledge of system frequency and damping, which are vital indicators of system stress and stability [7, 8], the cascading events could possibly have been averted following appropriate operator intervention. Moreover, relative mode shape estimated in near real time could also provide crucial information about oscillation interaction paths and in turn might become critical for corrective actions like generator and load tripping.

Estimation of damping ratio and frequency from measured signals is widely reported in the literature [6–18]. Although most of the research has been focused on damping and frequency estimation, not many publications exist on mode-shape estimation, especially in near real time [6, 19–27]. One promising approach called the transfer function (TF) approach was proposed in [27] for mode-shape estimation under ring-down/transient response.

In the TF approach the TF between two output signals is estimated, which in effect is the ratio of the numerators of individual output–input TFs. Accurate estimation of numerator is not always straightforward. Moreover, the estimated TF might be noncausal [27] requiring the knowledge of channel delay – which could be an issue for a near real-time realization. The communication approach [25] on the other hand is nonparametric and seems to be more appropriate. However it needs a narrow band filter (NBF) to preprocess the signal as it is suitable for handling only one mode at a time. NBFs need the precise knowledge of center frequency and are typically high-order filters causing delays. For practical applications the modal frequencies can change with operating conditions requiring online adjustment of the center frequency.

Here a phasor approach is presented for estimation of modal frequency, damping, and relative mode shape in near real time during transient conditions. The proposed technique [1] is in line with the communication approach [25] but eliminates its abovementioned drawbacks by using recursive Kalman filtering algorithm along with online correction of modal frequencies, which is described in Section 6.3.1.

6.2.2 Wide-Area Damping Control: Latency Compensation

Advances in phasor measurements and data communication technologies enable utilities to employ remote feedback signals for effective power oscillation damping (POD) controllers [28–31]. Wide-area POD control, which has primarily been a research issue till recent past, is now a practical possibility [32]. Most utilities, although encouraged by this exciting prospect, are still concerned about the consequences of problems in data communication either in the form of unacceptable latency or complete loss of signals in the worst case. Latencies beyond a certain threshold disturb the effective phase compensation provided by the damping controllers leading to potentially unacceptable closed-loop dynamic response.

Remote signals are communicated from the PMUs to the control center through a PDC. A GPS provides precise timing pulse to correlate the sampled measurements and achieve precise time synchronization. The PDC synchronizes the measurements from all the PMUs with microsecond precision and, under normal condition, sends data once every cycle to the control center. In case of congestion in one or more channels, the PDC (e.g., PCU400) waits till it receives data from all the PMUs. Therefore, the total latency is the sum of latency in the most congested channel and the time required for synchronization [33]. Once the PDC receives data from all the channels, it starts sending data to control center at a much faster rate (1 kHz max) until it clears the backlog. During this period the control center could encounter feedback signals with time-varying latency.

State-of-the-art data communication infrastructures ensure that the latencies are generally limited within milliseconds [34]. However, it could be as high as hundreds of milliseconds, or even more under unusual circumstances [34–36]. Latencies less than or equal to tens of milliseconds do not interfere significantly with low-frequency oscillations and can be tolerated without paying attention. However, longer delays affecting the oscillation frequencies, although relatively infrequent, cannot simply be ignored and should be considered during the control design.

Power system researchers, over the years, have come up with different approaches to cater for latency [37, 38]. Use of Padé approximation of nominal delays in the system model is at the heart of many of these techniques. Suitable order of Padé approximation for different delays needs to be selected carefully without introducing appreciable error. Since the early work of Smith [39], several researchers have addressed the issue of fixed [40] and time-varying [41] latencies in the control loops. An infinite dimensional representation of delay was considered in a predictor framework [42, 43]. Such fixed controllers perform "acceptably" over a range of delays but tend to be conservative [37] yielding suboptimal performance under different scenarios.

Conservativeness can be reduced through a gain-scheduling controller if the actual latency is known from accurate time-stamping at both the PMUs and the control center [37]. However, continuous compensation of time-varying latency is hard to achieve as a large number of predesigned controllers might be required with provision for switching from one to other depending on the actual latency.

As a solution to this problem, Section 6.3.4 presents continuous compensation of time-varying latencies using a phasor power POD framework. A phasor POD [44, 45] was proposed and commercialized by ABB [46] and is now used by several utilities. The advantage of this approach is its ability to quickly extract the oscillatory component of the measured feedback signal, whereas a conventional lead–lag-based damping controller might fail due to its washout filtering requirements. The oscillatory component of the measured signal is expressed in phasor form (in a rotating coordinate system) before applying an appropriate phase shift to generate the control signal. Here an adaptive phase-shift algorithm is presented to automatically produce the most appropriate phase shift necessary for different feedback signals and varying operating conditions. To compensate for latency, the rotating coordinates for phasor extraction are adjusted according to the phase change caused by the delay [2]. Phasors are then transformed back to time domain to retrieve the oscillatory component of the original signal out of the delayed signal received.

The North American power system will undergo major changes in the coming decades. The United States currently has 1212 coal-fired power plants with a total capacity of 329.8 GW. The "Clean Power Plan" [47], recently proposed by the Environmental Protection Agency (EPA), necessitates shutting down a lot of these plants to reduce carbon emission. According to the Energy Information Agency (EIA) projection report [48], coal-fired power plants with 90 GW capacity will retire by 2040. Many generators in these coal-fired plants are equipped with power system stabilizers (PSSs), which are indispensable for maintaining stability. This will seriously jeopardize the grid reliability and might lead to widespread blackouts.

Renewable energy resources including doubly fed induction generator (DFIG)-based WFs will replace the retired coal-fired plants. As a result the WFs will also be mandated to contribute toward grid stability. However, it is more likely that the WFs will be interconnected at a location different from the coal-fired plants, which will render this a lot more difficult problem to solve.

In [49, 50], lack of participation of DFIGs in oscillatory modes was highlighted. Moreover, our research shows the lack of adequate observability of the critical inter-area modes in signals locally available at the WFs (Section 6.3.5.6). Thus, the remote feedback signals have significant potential compared with local signals in damping inter-area modes through DFIGs. Past research on WF integration [51] has considered remote feedback signals using dedicated communication channels to stabilize one mode. A realistic power grid with multiple interconnected areas usually has multiple inter-area modes. Control coordination among multiple WFs is necessary to damp these modes, which is demonstrated in Section 6.3.5.6.

6.3 Solutions

In this section some proposed solutions to the abovementioned problems based on the author's past research work [1–3] will be presented.

6.3.1 Phasor Approach

Measured signals during transient condition consist of two components: constant or slowly time-varying average and oscillatory components with one or more modal frequencies. The idea of the phasor approach is to extract the oscillatory component(s) as space phasor(s) in synchronously rotating d–q reference frame(s) [45]. Thus the measured signal $S(t)$ (e.g., bus voltage, line current, real power flow, etc.) can be expressed as

$$S(t) = S_{av}(t) + \text{Re} \sum_{i=1}^{m} \{\vec{S}_{ph(i)} e^{j\omega_i t}\} \tag{6.1}$$

where m is the number of modes in the signal. The space phasor is decomposed into d and q components in the individual rotating reference frame as follows:

$$S(t) = S_{av}(t) + \sum_{i=1}^{m} \{S_{d(i)}(t) \cos \varphi_i(t) - S_{q(i)}(t) \sin \varphi_i(t)\} \tag{6.2}$$

Here, $\varphi_i(t) = \omega_i t$. Note that $\varphi_{0(i)} = \tan^{-1}\left(\frac{S_{q(i)}(t)}{S_{d(i)}(t)}\right)$ is the angle at which each estimated space phasor gets locked with its d–q frame of reference.

Recursive Kalman filter estimation approach is adopted to estimate the parameter vector $\Theta = [S_{av}(t)S_{d(1)}(t) \dots S_{d(m)}(t)S_{q(1)}(t) \dots S_{q(m)}(t)]^{\mathrm{T}}$ following the standard steps [52, 53] shown below:

Step I: Calculate the prediction error:

$$\varepsilon(t) = S(t) - \phi(t)\Theta(t-1) \tag{6.3}$$

where $\phi(t)$ is the regressor expressed as

$$\phi(t) = [1 \cos \varphi_1(t) \dots \cos \varphi_m(t) - \sin \varphi_1(t) \dots - \sin \varphi_m(t)] \tag{6.4}$$

Step II: Compute the Kalman gain vector $K_d(t)$:

$$K_d(t) = \frac{\wp(t-1)\phi^T(t)}{R_2 + \phi(t)\wp(t-1)\phi^T(t)} \tag{6.5}$$

Step III: Update the covariance matrix $\wp(t)$:

$$\wp(t) = [I - K_d(t)\phi(t)]\wp(t-1) + R_1 \tag{6.6}$$

Step IV: Update parameter vector $\Theta(t)$:

$$\Theta(t) = \Theta(t-1) + K_d(t)\varepsilon(t) \tag{6.7}$$

The parameter vector $\Theta(t)$ is initialized with zeros, while the covariance matrix $\wp(t)$ with a high value $(10^4 I)$. R_1 is a diagonal matrix normalized with respect to R_2, which results in R_2 being unity. Choice of R_1 depends on the process noise covariance, which is difficult to know a priori. Hence, R_1 is tuned to attain a proper balance between the filter convergence speed and tranquility in parameter estimates [52, 53]. Here the elements of R_1 were chosen to be 0.5 [52].

The basic concept of phasor extraction is illustrated in Figure 6.1 [1].

A self-initialization method is employed where the estimation algorithms appropriate for ambient conditions are used to obtain the frequency spectrum of the measured data to initialize the frequencies. Changes in frequencies with varying operating condition are tracked online through frequency correction loops for individual modes wherein PI compensators minimize the error between the phase angles in consecutive samples (see the dotted box in Figure 6.1). The frequency corrections are limited to ±0.1 Hz to avoid overlap between adjacent frequencies present in a multimodal signal [1].

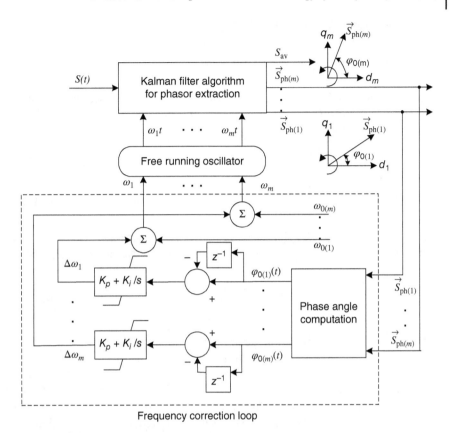

Figure 6.1 Extraction of oscillatory components in phasor approach. *Source:* Chaudhuri and Chaudhuri (2011)[1]. Reproduced with permission of IEEE.

6.3.2 Wide-Area Monitoring: Damping, Frequency, and Mode-Shape Estimation

In phasor approach, the oscillatory component of the measured signal is transformed into a set of phasors containing the amplitudes of individual modes. The ith estimated phasor in time domain can be expressed as

$$|\vec{S}_{\text{ph}(i)}| = \sqrt{S^2_{d(i)}(t) + S^2_{q(i)}(t)} = S_{0(i)}e^{-\frac{\xi_i}{\sqrt{1-\xi_i^2}}\omega_i t} \tag{6.8}$$

where ξ_i is the damping ratio, $S_{0(i)}$ is the amplitude of the phasor at $t = 0$ s, and ω_i is the corrected frequency for the ith mode. Assuming $\chi_i(t) = \frac{\xi_i}{\sqrt{1-\xi_i^2}}t$, natural logarithm on both sides of (6.8) gives

$$\chi_i(t) = -\frac{1}{\omega_i} \ln \left\{ \frac{|\vec{S}_{ph(i)}|}{S_{0(i)}} \right\} = p_1 t + p_0 \tag{6.9}$$

As a result, $\chi(t)$ can be expressed as a linear function of time t. The coefficients p_1 and p_0 can be estimated using a moving window least squares error (LSE) approach as follows:

$$P = (\Gamma^T \Gamma)^{-1} \Gamma^T X(t) \tag{6.10}$$

$$P = \begin{pmatrix} \hat{p}_1 \\ \hat{p}_0 \end{pmatrix}, \Gamma = \begin{bmatrix} 0 & T_s & \cdot \cdot & (M-1)T_s \\ 1 & 1 & \cdot \cdot & 1 \end{bmatrix}^T \tag{6.11}$$

where $X(t) = \begin{bmatrix} \chi(t_1) & \chi(t_2) .. \chi(t_M) \end{bmatrix}^T$ is the measurement vector.

The damping ratio can be calculated in terms of the estimated parameters P as

$$\xi_i = \frac{\hat{p}_1}{\sqrt{1 + \hat{p}_1^2}} \tag{6.12}$$

The mode-shape vector can be normalized with respect to any of the state variables and expressed as relative mode shape as is done throughout the rest of the chapter. For each mode, the relative mode shape is the normalized magnitude and phase shift between the state trajectories in time domain, which can be represented as phasors. During phasor extraction and frequency correction, the estimated phasor $\vec{S}_{ph(i)}$ changes its position dynamically and finally locks itself at an angle $\varphi_{0(i)}$ with the $d-q$ frame. Relative mode shape can thus be estimated using the following steps [1]:

1) Express the speed of the reference generator in phasor form considering the particular mode of interest.
2) Use the output of the frequency correction loop of the reference phasor to estimate the corresponding phasors from other speed signals. This would ensure that the other phasors are locked with respect to the reference.
3) Finally, compute the relative phasor magnitudes and angles with respect to the reference phasor.

6.3.3 Test System: 16-Machine, 5-Area System

A 16-machine, 5-area test system, shown in Figure 6.2, was considered.

All synchronous generators (SGs) were represented by sub-transient models, and eight of them (G1–G8) were equipped with IEEE DC1A excitation systems, while a static excitation system with a PSS (referred to as SG-PSS) was installed at G9. The rest of the SGs were under manual excitation control. A detailed

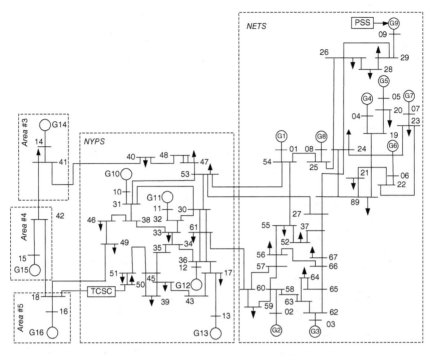

Figure 6.2 Test system: 16-machine, 5-area system with a TCSC.

description of the study system including machine, excitation system, and network parameters can be found in [54]. A Thyristor-Controlled Series Compensation (TCSC) is installed on the tie line connecting buses 18 and 50. There are three critical inter-area modes with nominal frequencies: 0.398 Hz (mode # 1), 0.525 Hz (mode # 2), and 0.623 Hz (mode # 3).

6.3.3.1 Simulation Results

Integrated Gaussian noise is injected at all load buses to simulate random load fluctuations. Monte Carlo simulation based on 100 trials was carried out to investigate the statistical variations. Although both mean and standard deviations of the estimated parameters were monitored for all scenarios, only the results for a particular trial are shown here for clarity. Phase angle difference between voltages at buses 45 and 51 was used for damping estimation, whereas generator speeds were considered for mode-shape estimation.

Out of several scenarios considered, only one representative case following a three-phase fault near bus 54 and line 54–53 outage is presented here. The proposed phasor technique was used for estimating modal frequency, damping ratio, and mode shapes. Results of Monte Carlo simulations with different values of signal-to-noise ratio (SNR) are shown in the form of box plots in

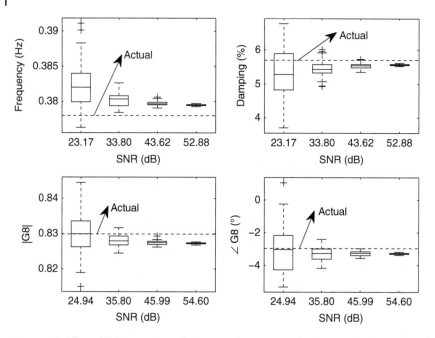

Figure 6.3 Effect of SNR on estimated damping, frequency, and relative Mode shape for mode # 1 after a three-phase fault at $t = 5$ s near bus 54 followed by line 54–53 outage. Box plots show the lower, median, and upper quartiles and extent of the rest of the data and any outliers . *Source:* Chaudhuri and Chaudhuri (2011) [1]. Reproduced with permission of IEEE.

Figure 6.3. The statistical spread in estimated damping, frequency, and relative mode shape is shown in terms of the median, lower, and upper quartiles with their actual values (dotted line) in the backdrop. Estimation is more consistent across different Monte Carlo trials for higher values of SNR. However, in all the cases the estimated quantities are quite close to the respective actual values indicating the effectiveness of the proposed algorithm. Mode-shape magnitude and phase angle of only one generator G8 is shown due to space restrictions. The actual values were obtained from linear analysis of the system.

Further, the dynamic behavior of the test system shown in Figure 6.2 was emulated using an Opal-RT real-time simulator [55]. Analog signals out of the real-time simulator were captured in an oscilloscope and passed on to an oscillation monitor platform for estimation of damping and relative mode shape in near real time. Figure 6.4 shows the damping and relative mode-shape estimation following line 54–53 outage.

The oscilloscope traces of the measured signals are shown here for the phase angle difference between buses 45 and 51 along with generator speed deviations

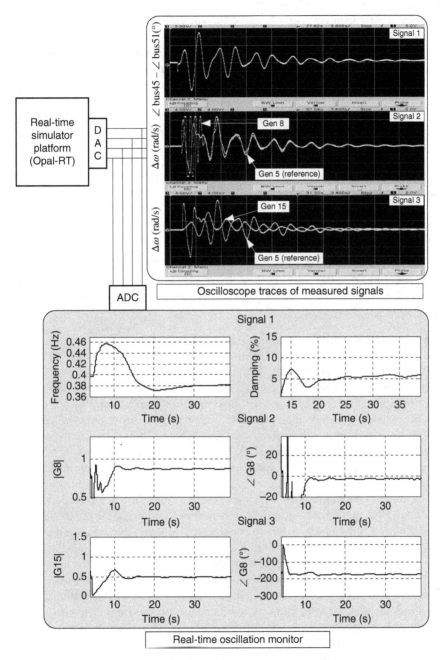

Figure 6.4 Oscilloscope traces of measured signals and corresponding damping and mode-shape estimation with phasor approach for line 54–53 outage . *Source:* Chaudhuri and Chaudhuri (2011)[1]. Reproduced with permission of IEEE.

of generator # 5 (reference), # 8, and # 15. Not a single instance of overrun was detected in the real-time simulator.

Although the presence of multiple modes is visible from the traces of generator speeds, mode #1 is dominant, and oscillation pattern of generator #8 speed deviation shows that it is similar in magnitude and almost in phase with the reference generator. On the other hand, bottom signal shows that oscillation amplitude of the speed of generator #15 is much lesser and almost out of phase with respect to the reference generator. The same information is reflected in the estimated relative mode-shape magnitudes and phase angles (see Figure 6.4). The estimated values were very close to the actual values obtained from linear analysis, as shown in Figure 6.3. Interested readers can refer to [1] for further details.

While this information is vital for the operators to assess the health of the system in near real time, the logical next step is to stabilize the grid using remote feedback signals through continuous control action, which is called wide-area control. In the following section, we will present a case study for latency compensation in wide-area control using the phasor approach.

6.3.4   Wide-Area Damping Control: Latency Compensation

In this section, continuous compensation of time-varying latencies using an adaptive phasor POD (APPOD) framework will be considered. The APPOD is based on the representation of the measured signal $S(t)$ (e.g., bus voltage, line current, real power flow, etc.) as a space phasor, which is estimated using the recursive Kalman filtering technique mentioned in Section 6.3.1. Appropriate phase shift can then be provided by changing the relative position of the $d-q$ reference frame with respect to the space phasor and regenerating a time domain signal in the stationary frame of reference. An approximate idea of the oscillatory frequencies of the system is used as the initial guess ω_0 for phasor extraction.

Changes in frequency with varying operating conditions are tracked online through a frequency correction loop wherein a PI compensator minimizes the error between the phase angles in consecutive samples (see the dotted box in Figure 6.5). The correction is limited to a judiciously chosen range as the POD is not expected to damp modes with higher frequency deviation. The limits can be relaxed to accommodate larger expected deviations in frequency under different operating conditions such as in [56]. However, for multiple modes with frequencies that are very close by, a wider limit on the frequency correction loop might cause an overlap with the neighboring frequencies making the phasor extraction more challenging.

A GPS receiver at the control center time-stamps the signal (with a microsecond precision) received from the PDC. The latency in the communication channel and associated hardware is computed by subtracting the instant of origin at

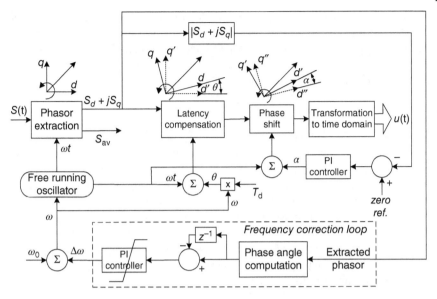

Figure 6.5 Operating principle of an adaptive phasor POD (APPOD). *Source:* Chaudhuri *et al.* (2010) [2]. Reproduced with permission of IEEE.

the PMU from that of arrival at the control center [33]. Thus the state of the art not only provides time-stamped measurements but also the latency associated with each sample [2].

From a phasor point of view, the latency (T_d) introduces a phase lag in the received signal with respect to the original measurement at the PMU location. The phase lag (θ) is compensated by adjusting the rotating d–q frame to d'–q' through an angle θ calculated as the product of phasor angular frequency ω and time delay (T_d). Thus the phasor corresponding to the original PMU signal is retrieved in the new reference frame as follows:

$$\begin{bmatrix} S'_d \\ S'_q \end{bmatrix} = \begin{bmatrix} \cos \omega T_d & -\sin \omega T_d \\ \sin \omega T_d & \cos \omega T_d \end{bmatrix} \begin{bmatrix} S_d \\ S_q \end{bmatrix} \qquad (6.13)$$

To achieve a desired level of damping, an appropriate phase-shift α [45, 46] is introduced by further rotating the reference frame to d''–q'' (see Figure 6.5). The computed phasor magnitude, $|\vec{S}_{ph}(t)| = |S_d(t) + jS_q(t)|$, is driven toward a zero reference (see Figure 6.5) with the error feedback through a PI compensator ($K_p + \frac{K_i}{s}$), which in turn generates the desired phase-shift ($\alpha(t)$) as follows:

$$\alpha(t) = -K_p \times |\vec{S}_{ph}(t)| - K_i \times \int |\vec{S}_{ph}(t)| dt \qquad (6.14)$$

Appropriate phase shifts are provided *adaptively* to individual phasors extracted from a multimodal signal [2]. With appropriate gain this signal constitutes the control input $u(t)$ for the actuators (e.g., TCSC in this case). The gain can be computed from the gain of a lead–lag compensator designed through pole placement technique [57].

6.3.4.1 Simulation Results

Effectiveness of APPOD in continuous latency compensation is investigated in a 16-machine, 5-area test system, shown in Figure 6.2. A TCSC is installed on the tie line connecting the buses 18 and 50 and is used to damp power oscillations with the real power flow in line 16–18 as remote feedback signal. A three-phase fault near bus 54 with subsequent outage of line 53–54 is simulated. The signal latency is assumed to build up at $t = 8$ s from its nominal value ($T_d = 25$ ms) to 800 ms followed by reduction to 500 ms at $t = 12$ s and a further buildup at $t = 16$ s (see Figure 6.6). The continuous adaptation of APPOD in the face of time-varying latency is evident as opposed to the adverse impact when delay compensation is not accounted for.

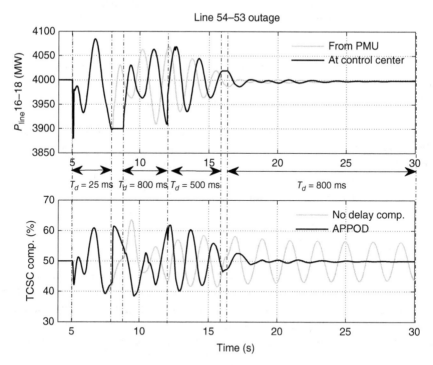

Figure 6.6 System behavior with tie-line 54–53 outage. *Source:* Chaudhuri *et al.* (2010) [2]. Reproduced with permission of IEEE.

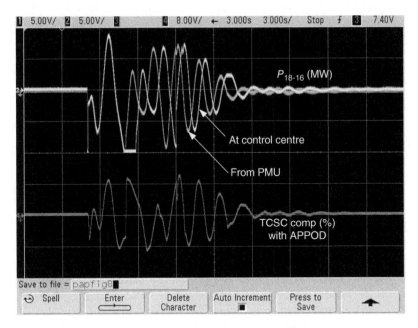

Figure 6.7 Oscilloscope recordings from real-time simulation for the case study shown in Figure 6.6. *Source:* Chaudhuri *et al.* (2010) [2]. Reproduced with permission of IEEE.

To investigate the computation time of APPOD and its possible impact on the overall latency, real-time simulations were carried out on Opal RT simulator [58]. A representative real-time simulation result for the same scenario as in Figure 6.6 is shown in Figure 6.7. For 20-ms sampling time considered for APPOD (based on PMU data rate), there was no instance of overrun for this case study. Identical nature of the closed-loop responses in Figure 6.6 and the oscilloscope traces in Figure 6.7 confirms that the APPOD computations do not introduce any additional latency [2].

6.3.5 Wide-Area Damping Control with Wind Farms

The challenges and opportunities in wide-area damping control in the presence of a large number of inverter-interfaced WFs were described before. Here, the control coordination among different WFs to damp multiple inter-area modes will be demonstrated.

6.3.5.1 DFIG-based Wind Farm Modeling

The overall structure of a DFIG is shown in Figure 6.8a where an aggregated model of the WF is adopted [49, 59].

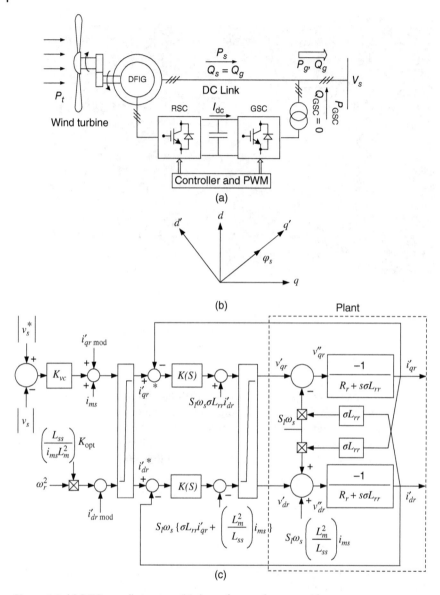

Figure 6.8 (a) DFIG overall structure (b) *d–q*: reference frame used for power system modeling, *d′–q′*: modified reference frame for vector control (c) rotor converter control structure.

Modeling of the DFIG is done in synchronously rotating d–q reference frame [57] with d-axis leading the q-axis as per IEEE convention (see Figure 6.8b). The stator transients of the machine are neglected, the converters are assumed to be ideal, and the DC link dynamics is to be neglected as suggested in [60] – further details can be found in [61]. Besides the standard differential and algebraic equations used to model the generator [61], a two-mass model of the turbine and drive train is considered to take the torsional mode into account. All notations in the following sections are standard – interested readers are referred to [62].

6.3.5.2 Rotor Side Converter (RSC) Control

Standard vector control approach [62] is adopted where the q-axis is aligned with ψ_s (see Figure 6.8b). All notations in the modified reference frame are henceforth denoted with a prime. Therefore,

$$L_{ss}i'_{ds} + L_m i'_{dr} = 0 \Rightarrow i'_{ds} = -\frac{L_m}{L_{ss}}i'_{dr},$$

$$\psi'_{qs} = L_{ss}i'_{qs} + L_m i'_{qr} \tag{6.15}$$

Neglecting R_s and assuming i_{ms} constant, we can write

$$\psi'_{qs} \approx L_m i_{ms} \Rightarrow i'_{qs} = \frac{L_m}{L_{ss}}(i_{ms} - i'_{qr}) \tag{6.16}$$

This results in simplification of v'_{dr} and v'_{qr} as follows:

$$v'_{dr} = -R_r i'_{dr} - \sigma L_{rr}\frac{d(i'_{dr})}{dt} - s_l \omega_s \sigma L_{rr}i'_{qr} - s_l \omega_s \frac{L_m^2}{L_{ss}} i_{ms} \tag{6.17}$$

$$v'_{qr} = -R_r i'_{qr} - \sigma L_{rr}\frac{d(i'_{qr})}{dt} + s_l \omega_s \sigma L_{rr}i'_{dr} \tag{6.18}$$

where $\sigma = (1 - \frac{L_m^2}{L_{ss}L_{rr}})$. As shown in Figure 6.8c "PLANT," the above equations can be rewritten in terms of v''_{dr} and v''_{qr} after isolating the disturbance terms: $s_l \omega_s \sigma L_{rr}i'_{dr}$ and $s_l \omega_s(\sigma L_{rr}i'_{qr} + \frac{L_m}{L_{ss}}i_{ms})$, respectively.

Note that i'_{dr} and i'_{qr} are measurable parameters, whereas i_{ms} can be estimated from (6.16). Therefore the measurable disturbances can be used as feed-forward terms with appropriate signs to achieve decoupling between d and q axes current control loops.

6.3.5.3 Grid Side Converter (GSC) Control

The GSC is assumed to be lossless, that is, the same real power flows through RSC and GSC. On the other hand Q_{GSC} is kept zero to attain minimum converter size.

6.3.5.4 Control Input

The d and q components of the current control loop are modulated for grid stabilization. The control inputs are shown as i'_{drmod} and i'_{qrmod} in Figure 6.8c.

6.3.5.5 Coordinated Control Design

The objective is to ensure the following using a coordinated (simultaneous) approach to calculation of the parameters of multiple controllers:

1) A minimum settling time for all the closed-loop oscillatory modes under possible operating conditions
2) A diagonal controller resulting in decentralized control
3) Fixed structure low-order controller

Analytical solution to the above problem is not straightforward [63]. Hence, heuristic optimization was used here to determine $K(s)$ by solving the above problem. Constraint on closed-loop stability was implicitly imposed through introduction of a high penalty (e.g., 10^6) in the objective function in the case of closed-loop poles on the right half of the s-plane. For further details of the control design principle and methodology, the readers can refer to [64, 65].

6.3.5.6 Simulation Results

The test system shown in Figure 6.2 was considered for the case study along with dedicated, ideal communication channels for feedback control. The effectiveness of the decentralized PSSs at two distant WFs in damping three critical modes using feedback signal communicated from remote locations is demonstrated [3].

Two SGs G9 and G15 were replaced by equivalent DFIG-based WFs. Modal analysis and controller design were performed to achieve a settling time of 15.0 s following coordinated control design [3] described in Section 6.3.5.5.

Linear analysis around a nominal condition shows the presence of three dominant inter-area modes. Modal controllability of DFIG rotor currents for G9 and G15 shows that I'_{drmod} of G9 is the natural choice for modulation as this has the highest magnitudes for mode #1 and #3 (Table 6.1). Additionally, I'_{qrmod} of G15 was selected, in spite of a higher controllability of G9 q-axis rotor current for mode #2 (Table 6.1) in order to distribute the damping duty among the available WFs.

As shown in Table 6.2, magnitudes of modal observability of stator powers of G9 and G15 are significantly reduced when corresponding SGs are replaced by DFIGs. Based on residue angle criterion [66], P_{14-41} and P_{27-37} were chosen as feedback signals for G9 and G15, respectively. Coordinated control design was performed following the approach mentioned in Section 6.3.5.5. For further details, readers are referred to [3].

The dynamic performance of the system following a three-phase self-clearing fault of five-cycle duration near bus 60 is shown in Figure 6.9. The real power

Table 6.1 Modal controllability of DFIG rotor currents [3].

	Modes	1	2	3
G9	I'_{drmod}	1.000	0.277	1.000
	I'_{qrmod}	0.504	1.000	0.014
G15	I'_{drmod}	0.726	0.201	0.727
	I'_{qrmod}	0.303	0.439	0.015

Source: Chaudhuri and Chaudhuri (2013) [3]. Reproduced with permission of IEEE.

Table 6.2 Modal observability of local and remote power flow signals [3].

Mode	#1		#2		#3	
Signals	SG	DFIG	SG	DFIG	SG	DFIG
P_{G9}	0.077	0.003	0.016	0.000	0.031	0.001
P_{G15}	0.284	0.002	0.229	0.001	0.008	0.000
P_{14-41}	0.192	0.197	0.813	1.000	0.015	0.016
P_{27-37}	0.072	0.102	0.028	0.017	0.069	0.075
P_{13-17}	1.000	1.000	0.238	0.453	1.000	1.000

Source: Chaudhuri and Chaudhuri (2013) [3]. Reproduced with permission of IEEE.

flow in the tie-line 54–53 connecting NETS and NYPS shows the presence of three poorly damped modes for DFIGs without PSSs. Modulation of I'_{dr} of G9 and I'_{qr} of G15 resulted in settling of these modes in about 15.0 s. The amplitude of modulation of I'_{dr} is higher as compared with that of I'_{qr}.

6.4 Conclusions and Future Direction

In this chapter a few important aspects of wide-area monitoring and stabilization of the energy CPS are analyzed. The nonidealities introduced by the cyber portion are considered, and an approach for mitigating its impact on wide-area control is presented. Upcoming changes in the physical portion due to the introduction of inverter-interfaced WFs and extracting damping contribution from them in a wide-area framework are also considered.

Although a lot of scholarly research work has already been reported on this topic, the utilities have reacted rather slowly in adopting this technology. In the past decade a lot of utilities have installed PMUs in the whole world,

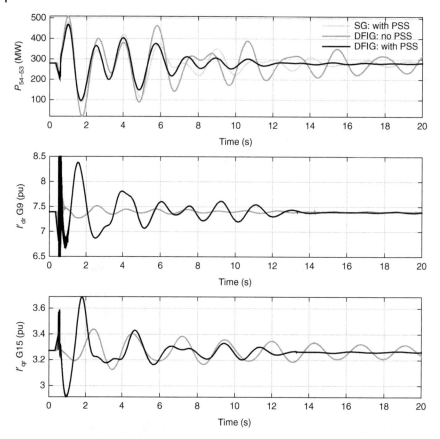

Figure 6.9 Dynamic performance of the system (see Figure 6.2) following a self-clearing three-phase fault near bus 60 for 80 ms. Gray trace: SGs G9 and G15. Dark gray trace: DFIGs G9 and G15 without PSSs. Black trace: DFIGs G9 and G15 with PSSs [3]. *Source:* Chaudhuri and Chaudhuri (2013) [3]. Reproduced with permission of IEEE.

and the energy management systems (EMS) vendors have developed suits of applications for wide-area monitoring, which in turn are slowly making inroads into the control rooms. On the contrary, except a few instances of pilot projects in China, the Nordic Power System, and Hydro-Québec, the wide-area damping control has largely remained unexploited in the utility industry. The future direction of research emphasis should be directed toward realizing the full potential of wide-area signals for grid stabilization through extensive utility adoption.

Final Thoughts

Wide-area monitoring and control has a broad range of applications for smart energy CPS. The goal of this chapter is to familiarize the readers with the opportunities and challenges in the stability-related applications. Solutions to overcome issues like latency in wide-area damping control are described. Finally, wide-area damping control in the presence of WFs is presented.

Questions

1 How wide-area monitoring and control can help stabilize the energy CPS?

2 What are the challenges facing mode-shape estimation using wide-area measurements?

3 What is a phasor-based approach and how can it be used for mode-shape estimation and latency compensation?

4 Why does the local signals of a DFIG-based wind farm have very limited observability of inter-area modes?

References

1 Chaudhuri, N.R. and Chaudhuri, B. (2011) Damping and relative mode-shape estimation in near real-time through phasor approach. *IEEE Transactions on Power Systems*, **26** (1), 364–373.

2 Chaudhuri, N.R., Ray, S., Majumder, R., and Chaudhuri, B. (2010) A new approach to continuous latency compensation with adaptive phasor power oscillation damping controller (POD). *IEEE Transactions on Power Systems*, **25** (2), 939–946.

3 Chaudhuri, N.R. and Chaudhuri, B. (2013) Considerations toward coordinated control of DFIG-based wind farms. *IEEE Transactions on Power Delivery*, **28** (3), 1263–1270.

4 Hauer, J., Trudnowski, D., Rogers, G., Mittelstadt, B., Litzenberger, W., and Johnson, J. (1997) Keeping an eye on power system dynamics. *IEEE Computer Applications in Power*, **10** (4), 50–54.

5 Hauer, A.J.F., Trudnowski, D.J., and DeSteese, J.G. (2007) *A Perspective on WAMS Analysis Tools for Tracking of Oscillatory Dynamics*. Proceedings of IEEE Power Engineering Society General Meeting, 2007, pp. 1–10.

6 Liu, G. and Venkatasubramanian, V. (2008) *Oscillation Monitoring from Ambient PMU Measurements by Frequency Domain Decomposition*. Proceedings of IEEE International Symposium on Circuits and Systems, 2008 (ISCAS 2008), pp. 2821–2824.

7 Trudnowski, D.J., Pierre, J.W., Zhou, N., Hauer, J.F., and Parashar, M. (2008) Performance of three mode-meter block-processing algorithms for automated dynamic stability assessment. *IEEE Transactions on Power Systems*, **23** (2), 680–690.

8 Zhou, N., Trudnowski, D.J., Pierre, J.W., and Mittelstadt, W.A. (2008) Electromechanical mode online estimation using regularized robust RLS methods. *IEEE Transactions on Power Systems*, **23** (4), 1670–1680.

9 Hauer, J.F., Demeure, C.J., and Scharf, L.L. (1990) Initial results in Prony analysis of power system response signals. *IEEE Transactions on Power Systems*, **5** (1), 80–89.

10 Kamwa, I., Grondin, R., Dickinson, E.J., and Fortin, S. (1993) A minimal realization approach to reduced-order modelling and modal analysis for power system response signals. *IEEE Transactions on Power Systems*, **8** (3), 1020–1029.

11 Sanchez-Gasca, J.J. and Chow, J.H. (1999) Performance comparison of three identification methods for the analysis of electromechanical oscillations. *IEEE Transactions on Power Systems*, **14** (3), 995–1002.

12 Liu, G., Quintero, J., and Venkatasubramanian, V. (2007) *Oscillation Monitoring System Based on Wide Area Synchrophasors in Power Systems*. Proceedings of Bulk Power System Dynamics and Control - VII. Revitalizing Operational Reliability, 2007 iREP Symposium, pp. 1–13.

13 Messina, A.R. and Vittal, V. (2006) Nonlinear, non-stationary analysis of interarea oscillations via Hilbert spectral analysis. *IEEE Transactions on Power Systems*, **21** (3), 1234–1241.

14 Hauer, J.F. and Cresap, R.L. (1981) Measurement and modeling of pacific ac intertie response to random load switching. *IEEE Transactions on Power Apparatus and Systems*, **PAS-100** (1), 353–359.

15 Wies, R.W., Pierre, J.W., and Trudnowski, D.J. (2002) Use of ARMA block processing for estimating stationary low-frequency electromechanical modes of power systems. *IEEE Power Engineering Review*, **22** (11), 57.

16 Kamwa, I., Trudel, G., and Gerin-Lajoie, L. (1996) Low-order black-box models for control system design in large power systems. *IEEE Transactions on Power Systems*, **11** (1), 303–311.

17 Wies, R.W., Pierre, J.W., and Trudnowski, D.J. (2004) *Use of Least Mean Squares (LMS) Adaptive Filtering Technique for Estimating Low-Frequency*

Electromechanical Modes in Power Systems. IEEE Power Engineering Society General Meeting, 2004, vol. 2, pp. 1863–1870.

18 Zhou, N., Pierre, J.W., Trudnowski, D.J., and Guttromson, R.T. (2007) Robust RLS methods for online estimation of power system electromechanical modes. *IEEE Transactions on Power Systems*, **22** (3), 1240–1249.

19 Li, W., Gardner, R.M., Dong, J., Lei, W., Tao, X., Yingchen, Z., Liu, Y., Guorui, Z., and Xue, Y. (2009) *Wide Area Synchronized Measurements and Inter-Area Oscillation Study.* Proceedings of IEEE/PES Power Systems Conference and Exposition, 2009. PSCE '09, pp. 1–8.

20 Koessler, R.J., Prabhakara, F.S., and Al-Mubarak, A.H. (2007) *Analysis of Oscillations with Eigenanalysis and Prony Techniques.* IEEE Power Engineering Society General Meeting, 2007, pp. 1–8.

21 Freitas, F.D., Martins, N., and Fernandes, L. (2008) *Reliable Mode-Shapes for Major System Modes Extracted from Concentrated WAMS Measurements Processed by a SIMO Identification Algorithm.* IEEE Power and Energy Society General Meeting - Conversion and Delivery of Electrical Energy in the 21st Century, 2008, pp. 1–8.

22 Anaparthi, K., Chaudhuri, B., Thornhill, N., and Pal, B. (2005) Coherency identification in power systems through principal component analysis. *IEEE Transactions on Power Systems*, **20** (3), 1658–1660.

23 Wilson, D.H., Hay, K., and Rogers, G.J. (2003) *Dynamic Model Verification Using a Continuous Modal Parameter Estimator.* IEEE Power Tech Conference Proceedings, 2003 Bologna, vol. 2, 6 pp.

24 Banejad, M. and Ledwich, G. (2002) *Correlation Based Mode Shape Determination of a Power System.* Proceedings of IEEE International Conference on Acoustics, Speech, and Signal Processing, 2002 (ICASSP '02), vol. 4, pp. IV–3832–IV–3835.

25 Dosiek, L., Trudnowski, D.J., and Pierre, J.W. (2008) *New Algorithms for Mode Shape Estimation Using Measured Data.* IEEE Power and Energy Society General Meeting - Conversion and Delivery of Electrical Energy in the 21st Century, pp. 1–8.

26 Dosiek, L., Pierre, J.W., Trudnowski, D.J., and Zhou, N. (2009) *A Channel Matching Approach for Estimating Electromechanical Mode Shape and Coherence.* IEEE Power and Energy Society General Meeting, pp. 1–8.

27 Zhou, N., Huang, Z., Dosiek, L., Trudnowski, D., and Pierre, J.W. (2009) *Electromechanical Mode Shape Estimation Based on Transfer Function Identification Using PMU Measurements.* IEEE Power and Energy Society General Meeting, pp. 1–7.

28 Kamwa, I., Heniche, A., Trudel, G., Dobrescu, M., Grondin, R., and Lefebvre, D. (2005) *Assessing the Technical Value of FACTS-Based Wide-Area Damping Control Loops.* Proceedings of IEEE Power Engineering Society General Meeting, 2005, vol. 2, pp. 1734–1743.

29 Chow, J., Sanchez-Gasca, J., Ren, H., and Wang, S. (2000) Power system damping controller design-using multiple input signals. *IEEE Control Systems Magazine*, **20** (4), 82–90.

30 Kamwa, I., Grondin, R., and Hebert, Y. (2001) Wide-area measurement based stabilizing control of large power systems-a decentralized/hierarchical approach. *IEEE Transactions on Power Systems*, **16** (1), 136–153.

31 Aboul-Ela, M., Sallam, A., McCalley, J., and Fouad, A. (1996) Damping controller design for power system oscillations using global signals. *IEEE Transactions on Power Systems*, **11** (2), 767–773.

32 Xie, X., Xin, Y., Xiao, J., Wu, J., and Han, Y. (2006) WAMS applications in Chinese power systems. *IEEE Power and Energy Magazine*, **4** (1), 54–63.

33 Korba, P., Segundo, R., Paice, A.D., Berggren, B., and Majumder, R. (2013) *Time delay compensation in power system control*. US Patent 8,497,602, filed May 23, 2008 and filing date Nov 23, 2010, https://www.google.com/patents/US8497602 (accessed 24 January 2017).

34 Stahlhut, J.W., Browne, T.J., Heydt, G.T., and Vittal, V. (2008) Latency viewed as a stochastic process and its impact on wide area power system control signals. *IEEE Transactions on Power Systems*, **23** (1), 84–91.

35 Heydt, G., Liu, C., Phadke, A., and Vittal, V. (2001) Solution for the crisis in electric power supply. *IEEE Computer Applications in Power*, **14** (3), 22–30.

36 Cai, J.Y., Zhenyu, H., Hauer, J., and Martin, K. (2005) *Current Status and Experience of WAMS Implementation in North America*. IEEE/PES Transmission and Distribution Conference and Exhibition: Asia and Pacific, 2005, pp. 1–7.

37 Hongxia, W., Tsakalis, K.S., and Heydt, G.T. (2004) Evaluation of time delay effects to wide-area power system stabilizer design. *IEEE Transactions on Power Systems*, **19** (4), 1935–1941.

38 Dotta, D., e Silva, A.S., and Decker, I.C. (2009) Wide-area measurements-based two-level control design considering signal transmission delay. *IEEE Transactions on Power Systems*, **24** (1), 208–216.

39 Smith, O. (1958) *Feedback Control Systems*, McGraw-Hill, New York.

40 Meinsma, G. and Zwart, H. (2000) On H_∞ control for dead time systems. *IEEE Transactions on Automatic Control*, **45** (2), 272–285.

41 Mirkin, B. and Gutman, P.O. (2008) Robust output-feedback model reference adaptive control of SISO plants with multiple uncertain, time-varying state delays. *IEEE Transactions on Automatic Control*, **53** (10), 2414–2419.

42 Chaudhuri, B., Majumder, R., and Pal, B. (2004) Wide-area measurement-based stabilizing control of power system considering signal transmission delay. *IEEE Transactions on Power Systems*, **19** (4), 1971–1979.

43 Majumder, R., Chaudhuri, B., Pal, B., and Zhong, Q.C. (2005) A unified smith predictor approach for power system damping control design using remote signals. *IEEE Transactions on Control Systems Technology*, **13** (6), 1063–1068.

44 Latorre, H.F. and Angquist, L. (2003) *Analysis of TCSC Providing Damping in the Interconnection Colombia-Ecuador 230 kV*. IEEE Power Engineering Society General Meeting, vol. 4, p. 2366.

45 Angquist, L. and Gama, C. (2001) *Damping Algorithm Based on Phasor Estimation*. IEEE Power Engineering Society Winter Meeting, vol. 3, pp. 1160–1165.

46 Angquist, L. (2003) *Method and a device for damping power oscillations in transmission lines*. US Patent 6 559 561, May 6, 2003.

47 EPA (2015) *Clean Power Plan for Existing Power Plants*, http://www2.epa .gov/cleanpowerplan/clean-power-plan-existing-power-plants (accessed 14 December 2016).

48 eia (2015) *Projection Report*, http://www.eia.gov/analysis/requests/ powerplants/cleanplan/ (accessed 14 December 2016).

49 Sanchez-Gasca, J.J., Miller, N.W., and Price, W.W. (2004) *A Modal Analysis of a Two-Area System with Significant Wind Power Penetration*. IEEE PES Power Systems Conference and Exposition, 2004, vol. 2, pp. 1148–1152.

50 Piwko, R., Miller, N., Girad, R., MacDowell, J., Clark, K., and Murdoch, A. (2010) Generator fault tolerance and grid codes. *IEEE Power and Energy Magazine*, **8** (2), 18–26.

51 Zhixin, M., Lingling, F., Osborn, D., and Yuvarajan, S. (2009) Control of DFIG-based wind generation to improve interarea oscillation damping. *IEEE Transactions on Energy Conversion*, **24** (2), 415–422.

52 Wellstead, P.E. and Zarrop, M.B. (1991) *Self-Tuning Systems: Control and Signal Processing*, John Wiley & Sons.

53 Zima, M., Larsson, M., Korba, P., Rehtanz, C., and Andersson, G. (2005) Design aspects for wide-area monitoring and control systems. *Proceedings of the IEEE*, **93** (5), 980–996.

54 Pal, B. and Chaudhuri, B. (2005) *Robust Control in Power Systems*, Power Electronics and Power Systems, Springer, New York.

55 Majumder, R., Chaudhuri, B., Pal, B.C., and Dufour, C. (2006) *Real Time Dynamic Simulator for Power System Control Applications*. IEEE Power Engineering Society General Meeting, p. 7.

56 Kamwa, I., Grondin, R., and Trudel, G. (2005) IEEE PSS2B versus PSS4B: the limits of performance of modern power system stabilizers. *IEEE Transactions on Power Systems*, **20** (2), 903–915.

57 Kundur, P. (1994) *Power System Stability and Control*, The EPRI Power System Engineering Series, McGraw-Hill, New York, London.

58 Majumder, R., Pal, B.C., Dufour, C., and Korba, P. (2006) Design and real-time implementation of robust FACTS controller for damping inter-area oscillation. *IEEE Transactions on Power Systems*, **21** (2), 809–816.

59 Chaudhuri, N. and Chaudhuri, B. (2011) *Impact of Wind Penetration and HVDC Upgrades on Dynamic Performance of Future Grids*. IEEE Power and Energy Society General Meeting, 2011.

60 Hughes, F.M., Anaya-Lara, O., Jenkins, N., and Strbac, G. (2005) Control of DFIG-based wind generation for power network support. *IEEE Transactions on Power Systems*, **20** (4), 1958–1966.

61 Slootweg, J.G., Polinder, H., and Kling, W.L. (2001) *Dynamic Modelling of a Wind Turbine with Doubly fed Induction Generator*. IEEE Power Engineering Society Summer Meeting, 2001, vol. 1, pp. 644–649.

62 Pena, R., Clare, J.C., and Asher, G.M. (1996) Doubly fed induction generator using back-to-back PWM converters and its application to variable-speed wind-energy generation. *IEE Proceedings on Electric Power Applications*, **143** (3), 231–241.

63 Bhattacharyya, S.P., Chapellat, H., and Keel, L.H. (1995) *Robust Control: The Parametric Approach*, Prentice Hall, Upper Saddle River, NJ, London.

64 Kennedy, J. and Eberhart, R. (1995) Particle Swarm Optimization. IEEE International Conference on Neural Networks, vol. 4, pp. 1942–1948.

65 Chaudhuri, B., Ray, S., and Majumder, R. (2009) Robust low-order controller design for multi-modal power oscillation damping using flexible AC transmission systems devices. *IET Generation, Transmission and Distribution*, **3** (5), 448–459.

66 Ray, S., Chaudhuri, B., and Majumder, R. (2008) *Appropriate Signal Selection for Damping Multi-Modal Oscillations Using Low Order Controllers*. Proceedings of IEEE Power Engineering Society General Meeting, 2008, Pittsburgh, PA.

7

Smart Technologies and Vehicle-to-X (V2X) Infrastructures for Smart Mobility Cities

Bernard Fong[1], Lixin Situ[1], and Alvis C. M. Fong[2]

[1] Faculty of Health and Environmental Sciences, Auckland University of Technology, Auckland
[2] Department of Computer Science, Western Michigan University, Kalamazoo, MI, USA

CHAPTER MENU
Introduction, 181
Data Communications in Smart City Infrastructure, 182
Deployment: An Economic Point of View, 186
Connected Cars, 195
Concluding Remarks, 202

Objectives

- To outline the key elements of intelligent transportation in a smart city.
- To understand how various subsystems are connected within the smart city infrastructure.
- To investigate the economic aspect on smart mobility solutions.
- To elaborate on data processing optimization techniques for enhanced resource utilization.

7.1 Introduction

Smart and assistive technologies interoperate practically in urban infrastructure such that many cities across the world strive to promote technological advances in urbanization and integrate technology to create smart cities through infrastructural enhancement. While this opens up numerous opportunities in smart city infrastructure development, it also entails smooth integration of a wide range of technologies associated with something within the smart city locality as well as connecting with the outside world where

Smart Cities: Foundations, Principles and Applications, First Edition.
Edited by Houbing Song, Ravi Srinivasan, Tamim Sookoor, and Sabina Jeschke.

foundation still heavily depends on legacy systems. An interoperable and sustainable infrastructure is therefore needed for smooth integration.

Advanced infotainment technology together with a fast data connection to the vehicle-to-infrastructure (V2I) network provides elementary features that shape the underlying platform for smart walkability and mobility [1]. To carry a person from one point to another point, namely, an original to an intended destination, a personalized micromobility vehicle can connect individuals to a nearby public transport hub that brings people from a local area together [2]. The broad technological scope that a smart city entails involves almost all aspects of information and communications technology (ICT) in both individual vehicles and the supporting infrastructure. For example, signal processing provides contextual voice controls to help create a safer driving experience and to analyze data traffic flow; wireless communications are necessary for big data delivery in a reliable and secure manner. Social networks and Internet of Things (IoT) support acquisition as well as integration of both structured and unstructured datasets from various sources for city planning and administration. Smart traffic routing and advising services are both essential components of a smart traffic system [3].

To study the technological challenges of integrating these into an efficient smart city infrastructure, this chapter is organized as follows: various aspects of data communications within a smart city are presented followed by the economic aspects of infrastructural deployment. We then take a closer look at the issues as well as the benefits of connecting each and every vehicle in a smart city environment.

7.2 Data Communications in Smart City Infrastructure

The basic building block of a sustainable smart city infrastructure entails utilizing a vast amount of data from a wide variety of sources for different purposes. As such, making good use of the collected data for the enhancement of efficiency and profitability becomes a challenging task for businesses as well as government authorities and agencies. While this opens up numerous opportunities in smart city infrastructure development, it also entails smooth integration of a wide range of technologies associated with something within the smart city locality as well as connecting with the outside world where foundation still heavily depends on legacy systems.

Data from different sources within a smart city proximity can be used for enhancement in many areas. Taking transportation as an example, supports for the day-to-day operation of the city include various pieces of connected infrastructural equipment, fleet management system, and critical automotive components that may have a substantial impact on traffic flow and road utilization efficiency. Further, generalized extensions to public transport systems that

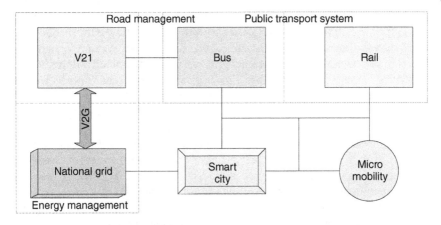

Figure 7.1 Intelligent transport system block diagram.

carry a significant proportion of commuters during peak hours would require data analysis of travel pattern. The analysis of a huge amount of data collected enables better resource planning and utilization that enhances profitability and energy efficiency.

In this simple example of supporting intelligent transportation in a smart city, it can be seen that data processing entails many different types of data collected from many sources of varying nature. Data analytics therefore entails different types, like images that show the condition of real-time traffic (the overall picture entails analysis of video footage from an array of cameras that show vehicle movement in the area), pure numerical data about public transport usage, predefined routing of each and every bus operating in the area, and the most unpredictable movement of pedestrians that exhibits a high degree of randomness. To process such diverse types of data, a communication system must be able to carry data efficiently and securely irrespective of the type and intended destination. A simplified version of the intelligent transportation system is shown in Figure 7.1 that illustrates the basic operation in a smart city context. The overall system within a smart city would involve integration and coordination of various transportation modes that make up the entire city's transport system. The V2X communication network that covers the left half of Figure 7.1 as well as the micromobility subsystem form a vital part of the overall smart city infrastructure in that it provides the vital link to each and every entity that forms the fundamental building blocks in controlling the efficient allocation of various resources within the city.

7.2.1 Data Acquisition

For any communication network, the fundamental part is the data that is transmitted from the source to its intended destination. Data from different sources serve different target audiences and each can take a very different form. Figure 7.1 shows, for example, a case where traffic flow data is collected within the road management subsystem; such data can drive adaptive lighting control to save electricity when street lights can power on and off automatically as vehicles or pedestrians pass by. The same set of data that controls traffic lights can ease congestion at junctions by alerting drivers to traffic jams so that they can either be suggested to take alternative routes or the traffic lights can be made adaptive to enhance traffic flow efficiency.

Traffic flow theory is a traditional methodology widely used by transportation engineers to apprehend the properties of traffic flow at any given time [4]. Early travel demand forecasting processes used in the 1970s [5] relied heavily on manual data analysis due to the limitations of computational power during that time. More recent models such as speed–flow–density relationships, queuing theory, shock waves, and continuum flow models are adopted to enhance traffic flow modeling under nonstationary conditions [6]. Past traffic flow analysis relied on two-dimensional spatial–temporal diagram that shows the trajectory of a specific vehicle through time as it moves from its origin to a destination that may or may not be known [7]. Acquisition of datasets heavily depends on parameters such as mean speeds of individual vehicles, flow, and density.

Moving into the smart city era, intelligent transportation systems entail far more than monitoring traffic flow on certain roads within a close proximity. Automation tasks such as electronic billing of road usage, collection of car parking fees, and electricity consumption of charging an electric vehicle (EV) should all be handled by a single payment platform. This platform then becomes a subsystem. So, intelligent transportation systems must integrate separate subsystems that each perform very important societal functions in such diverse areas as telecommunications, transportation, energy networks, and financial transactions.

One of the key challenges is to assess data flow across the overall system, which has become a key part of smart transportation infrastructure development. The major concerns are the ability to detect and identify events that are early warnings of traffic disruption, which could lead to eventual congestions. The ability to carry out an advance traffic assessment, data collection from multiple streams, integration and prognostication of multiple data streams for knowledge discovery, and decision-making will require a very sophisticated big data analysis platform. All of these outputs will ensure the reliability of the performance of the complex traffic analysis system.

In this simple example of big data analysis in an intelligent transport system, data set that represent traffic flow entails not only the number of vehicles in the region of interest but also parameters such as speed and direction of each vehicle, and any vehicles in nearby regions that may soon enter the region all need to be considered in order to derive a routing map that shows

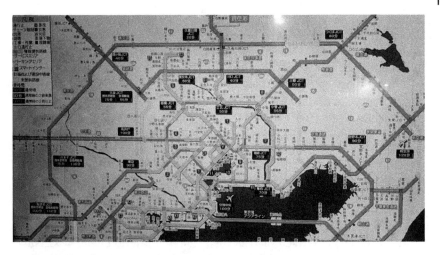

Figure 7.2 Map showing motorists' approximate travel times.

the approximate travel time as shown in Figure 7.2. The accuracy of this system depends on reliable collection of real-time traffic data from many monitoring stations and roadside cameras. The vast amount of data, from different sources of different forms, are transmitted across the city's communication infrastructure for subsequent centralized processing followed by distribution of information throughout the city

The traffic model is even more complicated in the real world since factors such as road works and change in road conditions due to rain or snow can cause significant delays. The dissimilar types of information that are gathered from different sources at the same time pose significant challenges to the communication system for generating such map with real-time instantaneous update.

While the Global Positioning System (GPS) provides important real-time traffic information, it can show the geographical location of every vehicle in the region. The GPS makes traffic flow analysis more predictable when drivers enter their intended destination, giving a certain degree of knowledge on the routing. Additional information such as vehicle type also provides important information on traffic analysis. For example, a motorcycle has a far less contribution to congestion than a 40-ft container truck. More detailed information on how each individual road user, both vehicle drivers and pedestrians, would provide a better overall picture on the actual traffic condition [8].

7.2.2 Traffic Surveillance

Given the diverse types of data that need to be analyzed, a key challenge is to identify and extract essential performance features as described above. Features of data provide crucial means of accurately identifying the onset of any anomalous or changing behavior that might affect traffic flow in parts of the city. This traffic surveillance approach will apply to two-class classification (i.e., smooth vs. congested), extend to multi-class classification where flow reduction occurs while traffic still flows reasonably smoothly at slower speeds, and perform prognostic analysis to quantify the congestion over time.

For dimensional reduction, which is important for discriminating intermittent speed reduction from traffic jams in real time, data analysis would involve exploiting mutual information (MI)-based methods [9]. MI-based feature selection methods can effectively eliminate the difficulties posed by nonlinear high-dimensional datasets by considering higher-order statistics to recognize salient features that are easily masked by noise and data transformations [10]. Traffic analysis taking a prognostic approach [11] can be accomplished by computing the MIIO, that is, MI between the selected input variables and output class labels in computer systems as a method to gradually increase the MI, in order to extract all relevant features for subsequent processing. The salient features necessary for carrying out prognostic analysis can then be computed using the MIIO method with a window estimator, which can determine the most prominent attributes or features that cause congestion.

Intelligent transport systems are special in that they contain unlabeled and partially labeled data. Labeled datasets in transportation systems are complex in nature due to the vast range of data types continuously flowing in from many independent sources. However, using datasets without labels makes diagnostic and prognostic analysis impossible for real-time self-cognizance. An adaptive algorithm for prognostic analysis may therefore become necessary when simultaneously analyzing incoming data from different sources. An adaptive approach can be taken from the hypothesis that if the learning algorithm is allowed to choose the data from which it learns, it will provide more accurate estimation with less training effort.

Since congestion is often not simply binary (stop or go) in traffic flow analysis, it may be necessary to utilize a two-pronged method in order to obtain multi-class probabilities by deconstructing the multi-class problem to yield a series of binary problems [12]. In theory, the data can be collected from both GPS information and roadsides. More than often, traffic control relies on video cameras showing real-time traffic flow at various preselected points. Moving into a smart city environment, route planning with GPS provides prognostic significance of specific traffic patterns, allowing early prediction of where congestions are likely to occur.

7.3 Deployment: An Economic Point of View

The development of an underlying infrastructure that forms a smart city heavily relies on the potential profitability that can be generated directly or indirectly from the vast amount of data that flows across the smart city. This involves making appropriate use of data to maximize business incentives of smart systems and optimization of return on investment (ROI). Keeping up with technological advances while maximizing profitability is particularly important in the fast-changing environment when serving consumers through analyzing market information as an integral part of a smart city infrastructure. There is also a special need to address technological changes as everything associated with public transport becomes smarter. In this section, we will take a brief look at selected case studies in telemedicine, green city, intelligent transportation, and autonomous vehicles. We observe how they are linked to the overall smart city infrastructure through a reliable and efficient communication system and what economic considerations are there when different entities with different economic interests share the bulk of collected data.

There are different implications when dealing with cross-platform data analytics as data takes different forms and types. Car parks can improve utilization efficiency through sensors that track when a vehicle leaves and alert nearby drivers to nearby empty spots. Stores in shopping malls can use BI technology to send shoppers promotional coupons to their smartphones based on certain demographic parameters.

7.3.1 Detecting Abnormal Events

The worst-case scenario from the business point of view is to deal with the unexpected. It is therefore highly desirable to carry out anomaly detection so that the chance of having to deal with awkward situations can be minimized, be it a car park that has highly fluctuating utilization rates or bus operators that have long queues for certain routes while empty buses run on other routes. Using statistical process control (SPC) to monitor multivariate and auto-correlated data can provide a big data analytics solution in such situation [13]. SPC can also be used for business activity monitoring and traffic surveillance. However, most traditional SPC detection methods depend on distributional assumptions on incoming data and therefore impose certain limitations on anomaly detection [14]. In intelligent transportation systems, similar to problems in business activity monitoring and traffic surveillance, data is more likely to be auto-correlated and multivariate with mixed continuous and discrete measurements [15].

To address such issues in data analytics, an integrated approach for modeling and monitoring of complex multivariate data should entail:

- Baseline characterization and correlation analysis of traffic and environmental conditions
- Followed by classification-based flow monitoring methods for SPC monitoring of forecasting residuals.

In many cases, there may be a loss of data in the system, for example, the failure of a set of traffic lights that no longer provides real-time traffic information, yet this may or may not have a substantial impact on traffic around the city. The nature or type of missing data can provide clues to identify the underlying problems that are often discovered when the reasons for the absence of the data are probed.

Data integrity issues may also arise from masking events, measurement errors, and reporting errors resulting from various types of uncertainty, aggregation, structural changes in data generation mechanisms [16], and possibly technical limitations such as insufficient bandwidth or computational power to handle the amount of incoming data. Issues of merging disparate data structures, which are created under varying traffic conditions, are a common problem with many routine operational and administrative datasets in most public transport systems. Such data can be evaluated by adapting data imputation methods and through the formulation of spatial and temporal interpolation techniques related to traffic monitoring systems [17].

7.3.2 Network Failure

The communication network backbone system plays a vital role in smart city infrastructure maintenance. More specifically, many component or system failures can be made predictable through analysis of certain critical components to detect any deviation from normality that eventually leads to an impending failure [18]. Fusion prognostic approach combining physics-of-failure (PoF) and data-driven prognostics facilitates the assessment and prediction of reliability of individual subsystems that form the overall smart city infrastructure under their specific actual operating conditions. This leads to the possibility of assessing the actual degradation of different components that can provide useful information on remaining useful life (RUL) estimation.

The PoF approach uses knowledge of a hardware failure mechanisms and life cycle loadings. Summarized in the flowchart in Figure 7.3 that takes a diagnostics and prognostic approach, the first step for the PoF-based prognostics is to identify the hardware architecture of a subsystem. Physical connections and the functional relationship between system components will be identified. Next, application loads during the life cycle can be evaluated. The system may experience various loading conditions, including thermal, chemical, electrical, and mechanical loads, and even radiation from the environment. The final step is to conduct failure modes, mechanisms, and effect analysis (FMMEA) with the primary objective of identifying and prioritizing potential failure mechanisms. FMMEA is normally performed in conjunction with stress analysis to acquire knowledge of stress levels at a specific location for a given application load. These analyses yield the statistically likely time to failure for a dominant failure mechanism at failure sites based on PoF models.

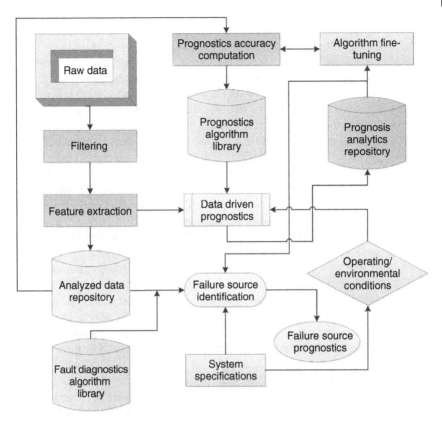

Figure 7.3 Network failure analysis.

To illustrate how data analytics is applied to the derivation of maintenance schedule, we take a look at an example where roadside light-emitting diode (LED) display signboard is subjected to continuous use in an outdoor environment.

The data-driven technology will be implemented to carry out a precursor parameter monitoring for the detection of LED degradation [19]. A precursor parameter is defined as a parameter that indicates incipient faults or damages before a failure occurs. As shown in Figure 7.3, the first step is to identify the correlation between precursor parameters and potential failure modes. This is followed by accelerated testing procedures and loading conditions that reflect the environmental and operating loads in the field. The last step is to develop algorithms that enable the assessment of the reliability of LEDs. All these steps require a precursor parameter monitoring system that enables precursor parameter sensing, data acquisition, and signal processing.

The concept of micromobility provides flexibility that occupies less space on the road. These vehicles have common features such as low power consumption, short-range, and short time battery recharge, making them particularly suited for commuters in a confined area of locality. As illustrated in Figure 7.4, a smart mobility vehicle has a small size that facilitates ease of driving and parking whose maneuverability is almost as malleable as a motorcycle.

The network serves many purposes in micromobility management, not only serving the vehicle user but processing information that is necessary for overall planning of various resources to optimize utilization. In Figure 7.5, micromobility management is accomplished by combining both the PoF (riding on the same PoF approach as described in the previous section on infrastructure reliability analysis) and the data-driven methodology such that data analysis benefits from the merits of each. The advantage of the PoF method is its ability to isolate the root cause of congestion that contributes to slower traffic. The data-driven approach is advantageous as it is capable of detecting intermittent flow behavior that can address the complexity of the analysis by extracting relevant information of environmental and traffic data. The first step in the process is to determine the set of parameters that can be monitored *in situ*. Identification of the parameters for monitoring can be aided by anomaly detection. It is carried out by comparing the monitored data with a baseline of free flow, that is, when every vehicle can travel at the speed limit while binding to the two-second rule. This baseline is effectively the ideal driving condition that consists of a set of parameter data that best represents the possible variations of the normal operation of the entire traffic control system. When an anomaly is detected, the parameters that contribute significantly to the observed anomaly are isolated. The isolated parameters help determine the PoF models most relevant to slower traffic, which in turn provide information such as maximum number of vehicles that can pass through a given road junction. Assessment of the remaining capacity of that road junction can then be computed using data-driven techniques such as neural network, multistep-ahead prediction, and support vector regression.

The data-driven estimate is obtained by trending the critical parameters to calculate the time until the junction reaches saturation so that other vehicles can be advised to take alternative routes, while the PoF estimate is obtained using the relevant PoF models identified in the above paragraph. By the fusion methodology, a number of advantages can be obtained, such as the determination of root cause of speed reduction, detection of intermittent congestion, and reassignment of traffic signal sequence.

Vehicles that are parked are not being utilized. They can therefore be made more efficient through a sharing scheme. An efficient smart city should facilitate a service that allows registered users to share a vehicle and parking space to

Figure 7.4 Micromobility vehicle. (a) Personalized flexibility. (b) Assistive disability access.

(a)

(b)

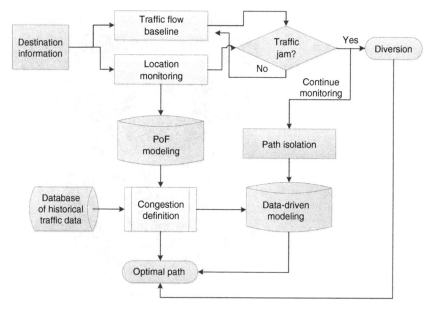

Figure 7.5 Micromobility management network through fusion prognostic operation.

alleviate traffic congestion and improve the utilization efficiency of resources. The system allows users to pick up and return a rental vehicle at different service locations throughout the smart city.

Figure 7.6 shows a trial implementation of an electrically assisted bicycle-sharing service system that allows subscribed members to rent and return the bicycles at different locations. The operation of such system relies on a communications network that supports optimal usage of vehicles and parking space. The most challenging part of making the system effective is managing the automated booking system through big data analytics processing. Since advance booking is not required, the system allows instant booking as a user is ready to pick up a bicycle and go. There is no way of forecasting bicycle demands prior to receiving a booking. Demand estimation modeling is necessary to make prediction based on past history [20]. Data analytics based on computational modeling of known events of occurrence can be applied to the prediction and modeling of demands. Through collection of relevant data, model-driven prognostics can assist with forecasting the possible implications of the rate of demand change and the concentration of available bicycles across different service stations. Prognostic prediction can also reduce the impact of demand change due to different times of the day. To address this challenge, efficient prediction and early warning of an availability shortage in certain service stations of the system through real-time booking data collection, usage monitoring, and management through prognostic analysis during peak hours

(a)

(b)

Figure 7.6 Vehicle sharing for enhanced resource utilization efficiency. (a) Electric bicycle service location. (b) Automated booking system.

are necessary. The most common existing methods for availability shortage detection include spatial and temporal surveillance [21].

In reality, the booking data are likely to be heterogeneous and correlated and often exhibit seasonal patterns over time. Efficient detection and prognostic methods under these situations are necessary for system management. Effective mitigation strategies and bicycle management entail diagnostics, prognostics, and forecasting methods. Efficient spatiotemporal surveillance algorithms will be part of the key infrastructural elements for monitoring multiple streams of data, including booking and actual bicycle return statistics and the planned route of each bicycle in the system. The effectiveness of bicycle monitoring system encounters multiple testing issues at several levels. There are multiple data sources and within each data source there are usually multiple series, and many of the series are further broken down into subseries, for example, by age-group of users. All data series are monitored in a univariate fashion, and finally, multiple detection algorithms are applied to each series. Each of the mainstream methods for handling multiple data streams has its limitations [22]. For instance, the commonly used Bonferroni method is considered overconservative, false discovery rate (FDR) corrections depend on the number of hypotheses and are problematic with too few hypotheses, and Bayesian methods are sensitive to the choice of prior and it is unclear how to choose a prior. As mentioned, the booking data are likely to be heterogeneous and correlated and often exhibit seasonal patterns over time. Surveillance approaches should be developed toward faster detection by monitoring real-time booking such as sudden increase in utilization of the automated booking terminals. Other correlated indicators from special events and internet search of bicycle availability in addition to monitoring confirmed that bookings should also be considered. It is needed to well recognize and handle the multiplicity problems as their effects are very visible in bicycle usage surveillance.

7.3.4 V2X Network Integration and Interoperability

In the context of the smart city, interoperability with legacy networks and infrastructures is perhaps the greatest challenge that network planning has to deal with. This is significant as it entails more than full compliance with various networking standards. The legacy networks are not an issue by themselves, yet linking them with the smart city network backbone may entail validation of technology prior to network integration, and this may pose problems.

When dealing with integration and interoperability, there is even more than integrating harmoniously between a smart city and the outside world. Smart city infrastructure should be designed such that vehicles ranging from the smallest personal mobility vehicle, being a slow single-person transportation mode as shown in Figure 7.7 that enables a user to ride safely alongside the pedestrians, share the same space with all other types of vehicles. These

Figure 7.7 Personal mobility vehicles.

personal mobility vehicles are literally designed and customized for personal usage. They are optimized for efficient usage in a smart city environment. Such transportation system has to allow safe passage for the right vehicle type in the right place.

7.4 Connected Cars

Similar to the case of public transport, road safety and utilization efficiency heavily depend on collecting and analyzing data of traffic flow. Autonomous driving vehicles offer an effective solution for intelligent traffic control as pre-programming routing allows smoother traffic flow control. Ambulance traveling across metropolitan cities has long been held up by traffic congestion [23]. Optimizing the travel route for mission critical emergency support vehicles has to be thoroughly addressed in a smart city.

7.4.1 Multi-Hop Communication in V2X

With the widespread adoption of V2X communication, smart cities are surrounded by various highly localized vehicular networks such as vehicle

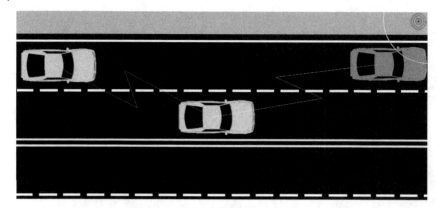

Figure 7.8 Data coverage extension with multi-hop V2X.

to vehicle (V2V), vehicle to home (V2H), and V2I. The proliferation of these communication-capable motor vehicles forms a broader V2X communication system, in that each vehicle as a node with its unique IDs constructs a network across diverse platforms and shares data for each intended purpose [24]. In the V2X context, the most important attribute of managing a network (or subnetwork) is constructing and maintaining network resources. Since momentous computational power is necessary to transmit and receive communication packets simultaneously, each vehicle in the V2X system may have certain limitations to share their resources by relaying packets destined for other nodes. However, if the majority of nodes in a network are unable to relay packets, as in the case of inter-vehicular communication, multi-hop communication may not be feasible and vehicular system be rendered ineffective. This problem needs to be thoroughly addressed in rural areas where cellular coverage may not be available.

In the case of Figure 7.8, only the front red car is under cellular coverage. The other two cars further back are too far away from the cellular base station and therefore unable to send any callout signal unless a signal extension path is made possible by relaying through nearby vehicles traveling in the same direction.

To model such situation, the multi-hop network illustrated in Figure 7.8 is supported by a compensation scheme for each vehicle participating in multi-hop forwarding. Data packets are relayed from V2V through short-range V2V direct communication so that communication is made possible even when a vehicle is out of cellular coverage. Awarding compensation for participating in forwarding packets implies that participants contribute their resource to all other vehicles in the V2V subnetwork. This is equivalent to a structure in which nodes within the network regulate the amount of resources made available for the forwarding service by a given node based on a free and

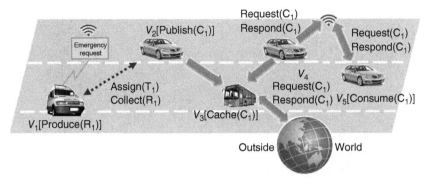

Figure 7.9 Multi-hop network management.

fair policy. To illustrate a case, consider an emergency vehicle initiating an emergency request within a multi-hop V2V network such that each vehicle v that participates in providing resources announces its participation $p(v)$ for the data packet forwarding service. A network management system (NMS) then communicates with vehicles that want to use multi-hop communication and decides whose resources will be used. In the example shown in Figure 7.9, vehicle v_1 decides to use the path $P = \{(v_1, v_2), (v_2, v_3), (v_3, v_4)\}$ to send data to v_4; $p(v_2)$ and $p(v_3)$ are assigned to v_2 and v_3, respectively. The prices of the nodes are renewed during the beginning of each round, and the paths are valid until the end of the round.

The main challenge here is to find the best routing for a data packet $v \in V$, $p(v)$ when v is outside cellular coverage, being assigned with full network resources $f(v)$ to each $v \in V$ when all requests $r \in R$ use the shortest paths to reach the nearest base station. This problem is addressed in two stages: (i) to find the optimal data transmission path and (ii) to compute the required network resources for successful transmission of all data packets to the nearest base station.

Computation: Maximum Available Bandwidth

INSTANCE: Graph $G = (V, E)$, requests R, source $s(r) \in V$, destination $d(r) \in V$, for each $r \in R$.
QUESTION: Find forwarding prices $p(v)$ of all nodes $v \in V$ with maximum network resources $f(v)$ when all requests $r \in R$ uses the shortest paths, where path length = sum of the bandwidth across path P.

To reduce the impact of mobility on topology control, Dijkstra's algorithm [25] and *ad hoc* on-demand distance vector routing [26] are used. The objective is to overflow the subnetwork with packets and trace return packets to find the shortest routing paths [27]. Since each node within the same subnetwork

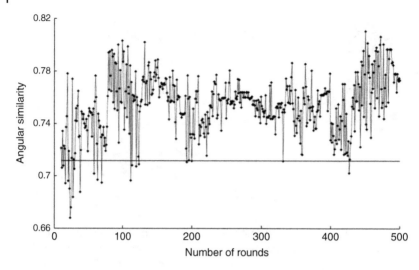

Figure 7.10 Angular similarity performance evaluation of prognostics and network health management.

may have its own objective function, it is quite possible that the result will yield multiple optimal paths. In the simplest case, the path that maximizes the hops of v_1 may minimize the hops of v_2 and vice versa. So, to address multi-objective optimization such as in the case of computing the maximum available bandwidth above, a Pareto-optimal solution, which is a solution in which there is no feasible solution that improves one objective while the other objectives suffer degradation, can be deployed [28].

To validate NMS effectiveness in its most primitive form, a uniform fixed resource allocation policy is set as the baseline. The Pareto-optimal solution of the centralized prognostics and network health management (PHM) scheme [29] uses multiple iterations such that crossover and mutation are evaluated with the mean values of 10 different random connected vehicles. Angular similarity against the number of rounds of packets as a performance evaluation metric is shown in Figure 7.10. It compares the efficiency of an NMS to a pre-determined baseline for network optimization. A baseline is drawn at 0.707 $(1/\sqrt{2})$ [30]. Figure 7.10 shows that the network management scheme becomes significantly more efficient as the number of rounds of packets exceeds 100.

7.4.2 Green V2X Communications in Smart Cities

Smart technologies and V2X together provide a green environment through efficient use of energy. Optimizing energy consumption cannot be made possible without the support of a reliable green communication infrastructure [31]. In a smart city infrastructure, the two most relevant topics of interests are

Table 7.1 Mainstream EV charging specifications (220 V market).

Model	Single phase		Three phase
	IEC model	GB model	IEC model
Input power	220 VAC, 32 A	220 VAC, 32 A	380 VAC, 32 A
Output voltage and currents	Mode 1: 220 V 10 A Mode 2: 220 V 32 A	Mode 1: 220 V 10 A Mode 2: 220 V 32 A	Mode 1: 220 V 10 A Mode 2: 220 V 32 A Mode 3: 380 V 32 A
Output sockets	BS 1363 IEC Type 1	CPCS-CCC GB 20234	BS 1363 IEC Type 1
Charging protocol	PWM ready to support IEC 61851	PWM ready to support GB 18487	PWM ready to support IEC 61851
Safety compliance	IEC 61851-1, IEC 61851-22	GB 18487.1, GB 18487.2	IEC 61851-1, IEC 61851-22
Environment rating	IP54		
Internal control	Modular PCBA system with pilot duty communication and networking capabilities using embedded system CPU		
Cable/socket lock	Interlock of IEC/GB type		

grid-integrated vehicle (GIV) and V2H communications for optimal energy utilization [32].

GIVs integrate into the smart city infrastructure by connecting each vehicle into the city's energy management network so that data can be gathered for efficiency and capacity planning of the grid. The network connects each vehicle to entities such as fleet management/operation companies, power utilities, and manufacturers of both EVs and charging stations. All these are the elementary building blocks of the grid management system. Data related to individual EV's use profile and battery's RUL are of particular interest for evaluating load management.

Load management is a particularly important topic in a green smart city environment since it optimizes electricity optimization in managing a collective type of EV chargers that comply different charging standards. Table 7.1 shows a number of different charging standards for 220 V (mainly Europe and Austrasia). These chargers utilize both single-phase and three-phase power supplies.

Integrated vehicle-to-grid (V2G) systems facilitate the reduction of peak-power demands. One of the key features of V2G is the ability to balance load such that a plug-in EV can charge overnight when electricity demand is at the lowest of the day and releases its remaining charge back to the grid for redistribution during the day [33]. Connected cars can also be initiated to dynamically return electricity to meet sudden demands for power [34]. The vehicle communicates with the load management system for energy optimization through the following daily cycle:

1) Plug-in connected car charges its battery at night.
2) Battery fully charged by the time the driver awakes.
3) Driver receives information about optimal routing as journey commences.
4) Preassigned route brings the vehicle to the nearest motorway with planned charging capacity.
5) Inductive road lane markers and electrified strips allow battery recharges via electrified motorways to compensate for depletion when the vehicle was driven from origin to the motorway.
6) Vehicle leaves motorway with fully charged battery.
7) Vehicle parked during the day can return electricity to the grid on-demand upon receiving request.
8) Driver returns home by repeating (2)–(6).

To facilitate Step (7), it is necessary to determine the amount of remaining charge necessary for completing the journey. A data-driven model can be constructed by analyzing either past routes, such as in the case of a commuter whose route between home and workplace is almost the same throughout all workdays or from user preprogrammed route through in-vehicle GPS. The main objective is to gather information about anticipated power consumption before the vehicle reaches the next charging point within supported range so that a good understanding about the usage profile of a particular vehicle can be computed.

7.4.3 Vehicular Communications Infrastructure Reliability

The challenges associated with reliable data communication with vehicles are mainly due to outdoor environmental conditions [35] and traveling speed [36]. Although the latter is a controllable parameter, it is practically infeasible to adjust the vehicle's traveling speed to trade off for better communications. For this reason, any wireless communication system that supports vehicular communications should allow continuous connection as the vehicle travels at a realistic speed, which may be in excess of 100 mph (160 km/h) in many countries. This is in fact one of the main advantages of advancing from HSDPA (3G) to LTE (4G) networks since LTE utilizes orthogonal frequency division multiplexing (OFDM) transmission scheme with better robustness to frequency-selective fading [37]. Generally, 4G is said to be able to maintain connection at normal vehicle traveling speeds with adequately fast handover from nearby base stations. LTE infrastructure is designed for deployment in a wide variety of frequency bands around 700 MHz to 2.6 GHz. LTE offers scalable bandwidths, from less than 5 MHz up to 20 MHz, together with support for both frequency division duplex (FDD) paired and time division duplex (TDD) unpaired spectrum.

In rural areas where base stations are far apart and users are widely scattered, there is generally no obstruction to wireless propagation so long as the terrain is flat, that is, signal undergoes free-space attenuation. However, issues such as scattering and multipath can be significant degradation factors to consider [38]. Furthermore, in vast metropolitan cities, IP addressing and user mobility across various network segments will become increasingly important. To make best use of the available wireless spectrum, systems can utilize both sector and frequency reuse to decrease adjacent and co-channel interference [39].

In the process of shaping a new smart city infrastructure, a brand new LTE network can be installed under the assumption that no existing legacy 2G/3G network needs to be considered for the purpose of network planning. Nonetheless, it is important to be looking forward to future planning where 5G is on the horizon. Utilizing a far higher carrier frequency than the current 4G LTE system would imply a very different set of signal propagation characteristics to be considered [40]. Most notable is the effect of rain-induced attenuation and depolarization when the vehicle is located several miles away from the base station [41]. The extent of radio link performance degradation is measured by cross polarization diversity (XPD) that is determined by the degree of coupling between signals of orthogonal polarization. XPD typically results in a 10% reduction in coverage due to cell-to-cell interference. The use of carrier frequencies way above 10 GHz has an effective trade-off for higher frequency reuse and capacity being offered due to higher spectral efficiency versus signal degradation due to rain.

To deal with a number of factors that influence various transmission reliability-related parameters, an adequate system operating or often known as the fade margin should be set to maintain communication link availability [42]. The fade margin, which measures the difference between the received signal strength and the signal minimum level, is maximized to provide maximum link reliability without increasing the transmission power.

7.4.4 Business Intelligence in Connected Cars

Navigation, telematics, and infotainment are the major applications of the connected car market [43]. The growth trend seen in the navigation segment is majorly due to the vast potential of realizing profit generation through integrating business intelligence (BI) technology into connected cars. As mentioned in the beginning of this chapter, BI opens up vast opportunities for businesses. We conclude this chapter by taking a brief look at what roles connected cars play in the context of BI, which relies on a secure communication network that strikes an acceptable balance between privacy and the type/amount of information that can be used [44]. The system relies heavily on both secured information communication and subsequent data analytics to extract relevant information. Associative classification technique applied to BI provides a new horizon for better matching between service providers and consumers in the fast-changing technology-driven market place [45].

Telematics for connected cars took off two decades ago when GM piloted the OnStar program in 1995 [46]. This market segment will certainly undergo a significant boost when the EU adopts compulsory installation of Automatic Emergency Call System (eCalls) (PE-COS 77/14) in 2017, while the US NHTSA also announced its intention to mandate V2V technology in new cars to induce collective action. Such regulations across the EU and USA drive a rapid expansion in both business opportunities and feature enhancements in the telematics and connected cars market. To enhance profitability, there are many complex steps requiring various inputs, algorithms, analysis strategies, and implementation decisions [47]. Modeling and drawing inference from a continuous flow of incoming data stream requires extraction of meaningful information from a set of observations. As there are practical problems with handling a large amount of data points collected from each vehicle, using data reduction methods as preprocessing is essential. Currently, sophisticated data reduction techniques are usually classification-based approaches [48]. These approaches work with a specified classification model and the reduced dataset is obtained according to classification accuracy. However, these data reduction methods often distort the original data distribution, making them inept for other classification models and other pattern recognition tasks. These methods are both computationally and memory intensive for feature extraction. The purpose of feature extraction is to yield representation that makes the classifier and data analysis effective and accurate.

Data received from each vehicle will be free to collect as many features as possible. The collected data will generally be in the form of ordinal data, also mixed with some nominal data, that is, the connector surface is "smooth" or "rough" and the residual is "clean," "dark," or "corroded." One way to handle nominal data is based on mapping the nominal data into a continuous domain [49].

7.5 Concluding Remarks

This chapter describes various types of wireless communication systems interconnected to form an overall intelligent transportation infrastructure for smart city development. The vital role of the intelligent transportation system in the V2X communication network is to support smoother traffic flow.

Data acquisition is another key part of the smart city framework as a series of information is required for efficient planning and resource allocation. We also studied the economic aspect on mainstream IoT solutions by focusing on network failure management and to analyze data coverage extension with multi-hop V2X. Finally, the chapter ends with a discussion on vehicular communications infrastructure interoperability and reliability such that the network can be made scalable to cater for future expansion in terms of both increased traffic volume and area of coverage.

Final Thoughts

In this chapter we discussed various aspects of smart mobility technology that form the basic building blocks of a smart city. The V2X infrastructure plays a vital role in connecting individual vehicles that travel across the smart city. Each vehicle may have different information set to transmit and take different routes from its origin to intended destination. All they exhibit in common is that they share essentially the same resources with the aim of reaching the destination safely in the shortest time with the least amount of energy consumption. All these can only be optimized by maintaining a reliable wireless communication link across the V2X network. At the same time, the diverse range of data collected from these vehicles contributes to the more efficient utilization of resources like car parking and traffic signal timing. These also lead to vast business opportunities through efficient deployment of BI technology in related business sectors, making the initial investment of building a comprehensive smart city network infrastructure economically worthwhile.

Questions

Although V2X technology provides an effective means for connecting individual vehicles to the broader smart city infrastructure, a number of major challenges still need to be addressed:

1 How to adopt surveillance methods to monitor real-time traffic condition in a dynamic smart city environment while striking a balance between privacy and data availability?

2 How to utilize data acquired from various sources such as in-vehicle cameras, road sensors, and other traffic-related indicators to improve efficiency of traffic flow?

3 How to grasp a better understanding about the behavioral properties of drivers in a dynamic environment?

4   How to structure a more realistic and accurate stochastic simulation model to facilitate connection between different modes of transport?

5   How to ensure infrastructure scalability to cope with future expansion and feature enhancement when keeping up with technological advances?

6   How to synergize the isolated research areas in road surveillance anonymously, simulation modeling, seasonal traffic pattern behavior, and traffic modeling?

7   How to enhance road safety within a smart city through critical risk analysis that effectively supports driving suitability evaluation of medical and nonmedical intervention strategies?

8   How to evaluate the sensitivity and robustness of different surveillance methods from the data generated from different parts of the smart city?

9   How to implement and validate the developed traffic analysis algorithms, models, and approaches through big data analytics knowledge, field experiments, and test-bed simulator systems?

10   How to optimize reliable continuous data communication for every vehicle throughout the smart city with the minimum number of base stations and access points?

References

1 Baccarelli, E., Cordeschi, N., Mei, A. *et al.* (2016) Energy-efficient dynamic traffic offloading and reconfiguration of networked data centers for big data stream mobile computing: review, challenges, and a case study. *IEEE Network*, **30** (2), 54–61.

2 Tracy, A.J., Su, P., Sadek, A.W. *et al.* (2011) Assessing the impact of the built environment on travel behavior: A case study of Buffalo, New York. *Transportation*, **38** (4), 663–678.

3 Djahel, S., Doolan, R., Muntean, G.M. *et al.* (2015) A communications-oriented perspective on traffic management systems for smart cities: Challenges and innovative approaches. *IEEE Communications Surveys and Tutorials*, **17** (1), 125–151.

4 Hall, F.L. and Motgomery, F.O. (1993) Investigation of an alternative interpretation of the speed-flow relationship for UK motorways. *Traffic Engineering and Control*, **34** (9), 420–425.

5 McFadden, D. (1974) The measurement of urban travel demand. *Journal of Public Economics*, **3** (4), 303–328.

6 Van Woensel, T. and Vandaele, N. (2007) Modeling traffic flows with queueing models: A review. *Asia Pacific Journal of Operational Research*, **24** (04), 435–461.

7 Levinson, D., Liu, H., Garrison, W. *et al.* (2011) *Fundamentals of Transportation*, CreateSpace Independent Publishing Platform.

8 Schäfer, R.P., Thiessenhusen, K.U., Brockfeld, E., *et al.* (2002) A traffic information system by means of real-time floating-car data. ITS World Congress Jan 2002.

9 Park, B., Won, Y.J., Choi, M.J. *et al.* (2008) Empirical analysis of application-level traffic classification using supervised machine learning, in *Challenges for Next Generation Network Operations and Service Management* (eds Y. Ma, D. Choi, and S. Ata), Springer, Berlin Heidelberg.

10 Fayyad, U., Piatetsky-Shapiro, G., and Smyth, P. (1996) From data mining to knowledge discovery in databases. *AI Magazine*, **17** (3), 37.

11 Lau, D.K. and Fong, B. (2011) Prognostics and health management. *Microelectronics Reliability*, **51** (2), 253–254.

12 Klijnsma, T. (2013) Preventing line congestion at Aebi Schmidt Nederland.

13 Montgomery, D.C. and Woodall, W.H. (1999) Research issues and ideas in statistical process control. *Journal of Quality Technology*, **31** (4), 376–387.

14 Shyu, M.L., Chen, S.C., Sarinnapakorn, K. *et al.* (2003) *A novel anomaly detection scheme based on principal component classifier*, University of Miami Coral Gables.

15 Steele, F. (2008) Multilevel models for longitudinal data. *Journal of the Royal Statistical Society, Series A*, **171** (1), 5–19.

16 Carroll, R.J., Ruppert, D., Stefanski, L.A. *et al.* (2006) *Measurement error in nonlinear models: a modern perspective*, CRC Press.

17 Fong, B., Rapajic, P.B., Hong, G.Y. *et al.* (2003a) On performance of an equalization algorithm based on space and time diversity for wireless multimedia services to home users. *IEEE Transactions on Consumer Electronics*, **49** (3), 597–601.

18 Fong, B. and Li, C.K. (2012) Methods for assessing product reliability: Looking for enhancements by adopting condition-based monitoring. *IEEE Consumer Electronics Magazine*, **1** (1), 43–48.

19 Fong, B., Ansari, N., and Fong, A.C.M. (2012a) Prognostics and health management for wireless telemedicine networks. *IEEE Wireless Communications*, **19** (5), 83–89.

20 Zhou, X. and Mahmassani, H.S. (2007) A structural state space model for real-time traffic origin–destination demand estimation and prediction in a day-to-day learning framework. *Transportation Research Part B: Methodological*, **41** (8), 823–840.

21 Dollár, P., Rabaud, V., Cottrell, G., *et al.* (2005) Behavior Recognition via Sparse Spatio-Temporal Features. Conf. Proc. 2nd Joint IEEE International Workshop on Visual Surveillance and Performance Evaluation of Tracking and Surveillance, October 2005, 65–72.

22 Huang, L., Guo, T., Zalkikar, J.N. *et al.* (2014) A review of statistical methods for safety surveillance. *Therapeutic Innovation and Regulatory Science*, **48** (1), 98–108.

23 Budge, S., Ingolfsson, A., and Zerom, D. (2010) Empirical analysis of ambulance travel times: The case of Calgary emergency medical services. *Management Science*, **56** (4), 716–723.

24 Han, T. and Ansari, N. (2014) Offloading mobile traffic via green content broker. *IEEE Internet of Things Journal*, **1** (2), 161–170.

25 Nishiyama, H., Ngo, T., Ansari, N. *et al.* (2012) On minimizing the impact of mobility on topology control in mobile *ad hoc* networks. *IEEE Transactions on Wireless Communications*, **11** (3), 1158–1166.

26 Liu, W., Nishiyama, H., Ansari, N. *et al.* (2013) Cluster-based certificate revocation with vindication capability for mobile *ad hoc* networks. *IEEE Transactions on Parallel and Distributed Systems*, **24** (2), 239–249.

27 Fong, B., Fong, A.C.M., and Li, C.K. (2011) *Telemedicine Technologies: Information Technologies in Medicine and Telehealth*, Wiley.

28 Zitzler, E. and Thiele, L. (1999) Multi-objective evolutionary algorithms: a comparative case study and the strength Prato approach. *IEEE Transactions on Evolutionary Computation*, **3** (3), 257–271.

29 Fong, B., Ansari, N., and Fong, A.C.M. (2012b) Prognostics and health management for wireless telemedicine networks. *IEEE Wireless Communications*, **19** (5), 83–89.

30 Apaydin, T. and Ferhatosmanoglu, H. (2006) Access structures for angular similarity queries. *IEEE Transactions on Knowledge and Data Engineering*, **18** (11), 1512–1525.

31 Khaleel, H.R. (2015) *Telemedicine: Emerging Technologies, Applications and Impact on Health Care Outcomes*, Nova Science Publishers, NY.

32 Higuchi, S. (2014) World movement on development and diffusion for next-generation vehicles. *Journal of Automotive Safety and Energy*, **5** (2), 107–120.

33 Sousa, T., Morais, H., Vale, Z. *et al.* (2012) Intelligent energy resource management considering vehicle-to-grid: A simulated annealing approach. *IEEE Transactions on Smart Grid*, **3** (1), 535–542.

34 Lund, H. and Kempton, W. (2008) Integration of renewable energy into the transport and electricity sectors through V2G. *Energy Policy*, **36** (9), 3578–3587.

35 Fong, B., Rapajic, P.B., Hong, G.Y. *et al.* (2003b) Factors causing uncertainties in outdoor wireless wearable communications. *IEEE Pervasive Computing*, **2** (2), 16–19.

36 Wang, Z., Liu, L., Zhou, M. *et al.* (2008) A position-based clustering technique for *ad hoc* intervehicle communication. *IEEE Transactions on Systems, Man, and Cybernetics Part C: Applications and Reviews*, **38** (2), 201–208.

37 Han, T. and Ansari, N. (2015) RADIATE: Radio over fiber as an antenna extender for high-speed train communications. *IEEE Wireless Communications*, **22** (1), 130–137.

38 Fong, B., Fong, A.C.M., and Hong, G.Y. (2005) On the performance of telemedicine system using 17-GHz orthogonally polarized microwave links under the influence of heavy rainfall. *IEEE Transactions on Information Technology in Biomedicine*, **9** (3), 424–429.

39 Fong, B., Ansari, N., Fong, A.C.M. *et al.* (2004) On the scalability of fixed broadband wireless access network deployment. *IEEE Communications Magazine*, **42** (9), 12–18.

40 Hong, W., Baek, K.H., Lee, Y. *et al.* (2014) Study and prototyping of practically large-scale mmWave antenna systems for 5G cellular devices. *IEEE Communications Magazine*, **52** (9), 63–69.

41 Fong, B., Rapajic, P.B., Hong, G.Y. *et al.* (2003c) The effect of rain attenuation on orthogonally polarized LMDS systems in tropical rain regions. *IEEE Antennas and Wireless Propagation Letters*, **2** (1), 66–67.

42 Fong, B., Rapajic, P.B., Fong, A.C.M. *et al.* (2003d) Polarization of received signals for wideband wireless communications in a heavy rainfall region. *IEEE Communications Letters*, **7** (1), 13–14.

43 Fong B, Wong ECC, Situ L, *et al.* (2016). Interoperability Optimization and Service Enhancement in Vehicle Onboard Infotainment Systems. Conf. Proc. 34th IEEE International Conference on Consumer Electronics, January 2016.

44 Hong, J., Shin, J., and Lee, D. (2016) Strategic management of next-generation connected life: Focusing on smart key and car–home connectivity. *Technological Forecasting and Social Change*, **103**, 11–20.

45 Do, T.D., Hui, S.C., and Fong, A.C. (2005) *Artificial immune system for associative classification* Advances in Natural Computation, Springer, Berlin, Heidelberg, pp. 849–858.

46 Barabba, V., Huber, C., Cooke, F. *et al.* (2002) A multimethod approach for creating new business models: The General Motors OnStar project. *Interfaces*, **32** (1), 20–34.

47 Kamburugamuve, S., Fox, G., Leake, D., *et al.* (2013) Survey of Distributed Stream Processing for Large Stream Sources. Technical report.

48 Tsai, C.F., Eberle, W., and Chu, C.Y. (2013) Genetic algorithms in feature and instance selection. *Knowledge-Based Systems*, **39**, 240–247.

49 Tominski, C., Abello, J., and Schumann, H. (2004) Axes-Based Visualizations with Radial Layouts. Conf. Proc. ACM Symposium on Applied Computing, pp. 1242–1247.

8

Smart Ecology of Cities: Integrating Development Impacts on Ecosystem Services for Land Parcels

Marc Morrison[1], Ravi S. Srinivasan[1], and Cynnamon Dobbs[2]

[1] *M.E. Rinker, Sr. School of Building Construction Management, University of Florida, Gainesville, FL, USA*
[2] *Departamento de Ecosistemas y Medio Ambiente, Pontifical Catholic University of Chile, Santiago, Chile*

CHAPTER MENU

Introduction, 209
Need for Smart Ecology of Cities, 212
Ecosystem Service Modeling (CO_2 Sequestration, PM10 Filtration, Drainage), 214
Methodology, 219
Implementation of Development Impacts in Dynamic-SIM Platform, 231
Discussion (Assumptions, Limitations, and Future Work), 234
Conclusion, 235

Objectives

- To understand ecology in the context of the smart city.
- To assess the state of accounting and model building in the realm of ecosystem service provision and target where more research needs to be done.
- To develop specific methodologies for accounting for ecosystem service provision using GIS technology at a large scale.
- To use a case study city/county in order to demonstrate the ability of current knowledge and tools to predict ecosystem service provision in an urban context.

8.1 Introduction

As the realities of climate change become more pressing, the availability of resources from far afield of city centers is destined to become more constrained.

Smart Cities: Foundations, Principles and Applications, First Edition.
Edited by Houbing Song, Ravi Srinivasan, Tamim Sookoor, and Sabina Jeschke.
© 2017 John Wiley & Sons, Inc. Published 2017 by John Wiley & Sons, Inc.

Increases in the frequency of severe weather events will mean that cities need to redact their focus toward accounting, preserving, and fully utilizing the resources within their immediate boundaries and hinterlands if they wish to thrive under these new pressures [1]. While this fact has implications for virtually every sector of city operation, whether power generation, waste management, or agriculture/food provision, no sector is better poised to provide effective tools for confronting these challenges than the building sector.

Designs that are truly sustainable have been identified in academia, and the popular press [2], as a potential solution to many of the social, economic, environmental, and institutional ills that plague society. In light of this, practitioners of the built arts and sciences have pushed in recent decades to try and implement sustainable initiatives and assessment methods in cities at a variety of scales (building, site, neighborhood, and region). Building construction professionals have developed sustainability progress indicators that attempt to identify, measure, and evaluate different choices in regard to sustainability [3]. Tangible results of this trend can be seen in the development of a variety of rating systems, certification schemes, life cycle assessment (LCA)-based tools, technical guidelines, assessment frameworks, checklists, and suites of modeling software increasingly being adopted in recent years. However, these are currently lacking in that they focus too heavily on the building scale, too narrowly define impact as chiefly current energy and water consumption, and neglect important implications of deep ecological concepts such as embodied energy and ecosystem service provision.

In order to contextualize the work that will follow in this chapter, an understanding of the state of building rating systems and sustainability impact assessment techniques in use by industry is essential. From the ecological side, growth in environmental building assessment is ironically much like a tree in structure. As building professionals demand tools of assessment, a dependable trunk of products has been developed, which branches off to reflect differing underlying motives and specializations of tools and methods. Some focus more on comparative financial gains, others on aspects of human health, while still more narrowly in the advantages of environmental stewardship. Demand for these disparate functions has been satisfied by different tools that lack the comprehensive, all-encompassing functionality needed to holistically capture the impacts of development. Examples of this can be seen in tools that satisfy environmental aspects but inadequately perform the equally important task of characterizing economic or social considerations. Examples of this tendency can be seen in the BREEAM, BEPAC, LEED, and HK-BEAM systems [4]. Research has shown that environmental issues are increasingly interconnected with social and economic aspects of society; therefore, to urgently address these challenges, more nuances must be added to metrics that inform design choices [4, 5].

Currently, there are over 20 major building impact assessment systems, and these come in the form of assessment and rating tools. Of these, some extend analysis exclusively to water and energy consumption or aspects of transportation and site management, while a scarce few address embodied energy or ecosystem services. The very existence of these tools demonstrates that building professionals are paying more attention to sustainability assessments; however, the aforementioned trend is an obstacle to reaching the ultimate goal of developing a sustainable building sector. While these rating systems have significantly promoted a sustainable development market, their process has neglected their potential as tools to facilitate sustainable design. They do not provide a balance approach leading to sustainable development [6]. Furthermore, these rating systems are disintegrated, meaning that they do not include all data sources for diligent and comprehensive assessment in one integrated, easy-to-use platform. Ratings systems such as these have, unfortunately, no ecologically derived benchmark for the determination of built environment impacts on the provision of ecosystem services. Aspects regarding the location of impacted areas, that is, whether impacts are affecting an already developed cityscape or an ecologically sensitive wetland that services the drinking resources for a region are entirely ignored. These rating systems only focus on a limited number of building components, often neglecting components that can present a significant amount of impact. Another aspect of these rating systems is their inherently static nature in regard to aspects of building use. Innovations in technology have had a sharp impact on operational energy use, which is slated to decrease significantly due to success in implementation of net zero energy benchmarks. Therefore, these tools need to move toward concentrating on aspects that actually have a proportionally high level of impact such as how buildings connect to the larger networks that make up cities, especially as smart cities are being built from the ground up, that is, cities designed to accommodate some of the most advanced buildings with transportation infrastructure technologies and more.

Of the several rating systems identified to address building-level sustainability assessments, a few major entities offer methods that can be applied on a significantly broader urban scale. Neighborhood sustainability assessment systems (NSAs) are primarily carried out using five of the following frameworks with regionally based popularity [6]: LEED-ND: North America [7]; BREEAM Communities: Europe [8]; DGNB-NSQ: Europe [9]; CASBEE-UD: East Asia [10]; and Pearl Community for Estidama: West Asia, Middle East [11].

Much like assessments on the building scale, these possess a variety of issues. Among them is that these systems often fail to consider the systematically complex and dynamic relationships that inform unique territorial contexts. Attempts are made to be applicable everywhere, not diligently through the inclusion of contextual assessment information but through an attempt to be as generalized as possible [12] by relying on high-level rule-based measurement systems that provide overwhelmingly similar guidelines regardless of contextual characteristics [13] Ecosystems services are also as neglected in this scale of analysis as in the building scale. A common weakness of the assessment methods for both scales is the fact that progress in characterizing building impacts is made in a silo, where advances and achievements made cannot be tapped to create a better overall picture of a region's sustainability. Progress on this method should include a multidisciplinary approach through the connection of academics and professionals to cooperatively share useful information.

Solutions to these problems can only come from the development of a new model that is explicitly developed to include these aspects particularly focusing on the impact on the ecology of the neighborhood. Ideally this model is integrated to virtually simulate, in real time, overall sustainability across multiple scales. The Dynamic-Sustainability Information Modeling (Dynamic-SIM) platform offers the necessary structure to map building and land parcels in a seamless fashion [14]. Inclusion of ecosystem service provision considerations would be added in a way that allows for the comparison of design alternatives. Building components would not just be considered by weight in the demolition processes but also represented in the form of embodied energy on a life cycle basis. This embodied energy is a measure of the amount of energy needed to form the materials, build the structures, and maintain them over the course of their life cycles. This comes from the materials directly used in construction, the energy for transporting these materials, the significant distances to construction sites from where the raw materials are refined and produced, the energy used in the assembly process by both humans and machines, recurring energy amounts for the replacement of certain building components, and finally the energy needed to demolish/deconstruct or recycle buildings at the end of their lives [15]. The metrics of assessment available through this framework capture these costs in a much more extensive manner.

8.2 Need for Smart Ecology of Cities

There has already been work in the academic field to place the benefits of ecosystem services in the context of the built environment, but these have seen limited application in the building, planning, or real estate sector. Only 35 companies explicitly mention ecosystem service accounting as a practice, and the majority of these companies are, ironically, related to natural resource extraction [16]. Through this research, as discussed in this chapter, evaluations regarding the research direction in the field to solve the conundrum of ecosystem service management in an urban context are clarified. Questions such as the following have noted importance to the development of a means to capture the value of green space and ecosystem services in the decision-making process [17]:

> What are appropriate indicators and typologies for the comparative assessment, monitoring, and prediction of the state and trends of urban green spaces and their ecosystem services?
>
> What are the ecosystem services provided by urban green spaces, and how can these services be quantified?

The reason for this gulf in the application of this kind of accounting in decision-making is because the concept of ecosystem services in the urban context has been largely unrecognized until recently with the push for interdisciplinary progress in the field of urban planning. The publication of "Millennium Ecosystem Assessment" by the United Nations (UN) was crucial in this transition and serves as a foundation for this branch of study [18].

The research study discussed in this chapter is vital for creating a balance between city impacts and the surrounding environment because of the complexity of social and economic drivers involved in shaping urban form and smart cities, in particular. Services offered by urban ecosystems can alleviate the negative effects of city development. State-of-the-art assessments suggest a cascade of benefits to be leveraged when ecosystem services are put in tangible, comparable monetary and social terms [19]. This study touches on the importance of framework development in regard to ecosystem services, because indicator development is necessary to quantify associated benefits. Indicators in this context describe an ecosystem processes or components that provide a given service. Once identified, performance is benchmarked to determine the benefits to be gained and the best way to leverage these benefits. Valuation of these services falls in three domains: ecological, sociocultural, and economic. Additionally, they are understood as having use value or "nonuse" value [20]. Conversion of certain land uses (e.g., tropical forests, wetlands, mangroves, etc.) to economically intensive uses, whether in the urban or peri-urban contexts, is shown to actually offer less net worth compared with the areas closer to an intact and undisturbed state [21]. Research in this field has been aimed at determining ecosystem service values generally, but it is worth noting that cities will have to take it upon themselves to micromanage ecosystem services on a local scale in the coming future; therefore, the planning implications of this realization are pertinent.

There is significant evidence that the availability and type of ecosystem services provided in an urban environment is influenced heavily by the density and level of urbanization existing in an area [22]. Tratalos *et al.* uses a framework to map ecosystem services, on a plot-by-plot basis, to quantify storm water runoff, maximum temperature, and carbon sequestration services provided by urban green spaces [23]. Within the local context, the framework developed by Dobbs *et al.* quantifies indicators from 12 different ecosystem services that occur in urban forests and offers benefits to human well-being in the Gainesville area [24]. These methods are not only seen in the regional context, but it is also seen in targeted consideration at the building scale. Olgyay and Herdt used a unique method combining embodied energy, carrying capacity, and building footprint/construction information to develop two indexes of sustainability used to quantify ecosystem services at the building scale [25]. Their work sets the ideal of "restorative architecture" at the heart of what it means to build green. They argue that traditional buildings exceed the carrying capacity of project sites as a matter of course and green buildings often get close to breaking even, while truly restorative architecture actually provide and enrich project sites. Soil chemistry remediation, increases in moisture availability, and changes in biotic materials are just a few ways they envision architecture and thoughtful design producing increases in the vitality of building sites. Through consideration of the nuanced nature of ecosystem services, ecosystem disservices have also seen increased attention recently. As a burgeoning concept, ecosystem disservices saw little concrete consensus in definition until recently. Some working definitions include "disturbed or missing services as consequences of loss of biodiversity" [26] or "negative effects of climate change" [27], but the most appropriate explanation, forming a dichotomy with the anthropocentric definition of ecosystem services, is "functions of ecosystems that are perceived as negative for human well-being" provided by Lyytimäki and Sipilä [28].

On modeling and mapping, current maps and models often include either land cover or land use while failing to relate this to any other spatially integrated measures of land use function. Challenges to this stem from differing scale factors when quantifying ecosystem goods and services (EGS), particularly when focusing on a city or regional scale. This modifies the kind of approaches that can be used where some studies are able to have extremely specific, firsthand collected datasets for integration, while others use more broad datasets with a heavier reliance on literature and underlying theories [29].

8.3 Ecosystem Service Modeling (CO_2 Sequestration, PM10 Filtration, Drainage)

8.3.1 Overview of Ecosystem Services in Urban Contexts

Urban ecosystem services are often broken down into four distinct categories, based on their functions in regard to humans: provisioning, regulating, supporting, and cultural and amenity services. Provisioning services regard those that end in a material product that humans can take directly from ecosystems. Regulating services are those that are gained from ecosystem processes, for example, water purification, and climate regulation or erosion control. Supporting services are linked most closely with habitat and genetic diversity in a given ecosystem and are the ecological functions that underlie the production of other ecosystem services. Cultural services are often the hardest to ascribe value to, due to their intangible nature, but these include the emotional dimensions of spirituality and appreciation that comes from being immersed in different ecosystems. As a new field of study, the implementation of ecosystem service accounting in the urban planning decision-making process has its fair share of challenges, as evident by numerous sources [30–32]. However, many researchers are making strides in developing more concrete and reproducible frameworks for the assessment of these services in a useful manner for the decision-making process [33]. Cities rely on ecosystems at numerous temporal and spatial scales, for example, the food systems that provide for modern city inhabitants even though they can often be found hundreds or thousands of miles away from city centers. Meanwhile, city residents directly get the benefits of services like air purification, noise reduction, or urban cooling, which are derived from the green infrastructure and cityscape immediately around them.

Of the 20 or more potential ecosystem services available for study, the ones focused on in our study were the services of carbon sequestration/storage, PM10 filtration, and drainage potential. These services were chosen based on the availability of relevant datasets and backing methodologies for accounting. The carbon sequestration and storage service is performed by the trees in an urban context through converting excess carbon, absorbed in the process of photosynthesis, into biomass for the trees' physical growth. This acts as a small sink for the greenhouse gases produced through urban living. PM10 filtration is also a function of the plants in an urban ecosystem that absorbs pollutants in the air by acting as sinks; in this way, air quality is improved for city inhabitants. Drainage potential refers to the fact that as cities develop the amount of impervious surfaces that cover their extant increases. When this happens water from storm events does not have time to be absorbed by the ground; it is instead rapidly shuttled across these surfaces turning what should be mild weather occurrences into notable flooding events taking along excess nutrients with it that contribute to issues of water quality. More porous environments in cities can help mitigate these impacts.

8.3.2 CO_2 Sequestration

In order to contextualize the work that will follow, there must be an understanding of the application of this kind of methodology in previous studies, so that lessons can be learned and applied to this work. According to Martinez-Harms and Balvanera, secondary data such as land cover, remotely sensed, and topological data are used by the majority of ecosystem service modeling studies, with most of these studies focusing on a regional scale [34]. However, there has been a marked rise in the aspirations for this community of scientists to increase the applicability of these methods through the standardization of modeling efforts [35]. Computer modeling offers a viable way to achieve this goal because of the ability to consolidate all relevant data in one location, where it can be accessed globally. Application development has come in a variety of forms with different focuses. Douglas-fir simulator (DFSIM) is a modeling tool developed with the explicit purpose of demonstrating different outcomes for development/management options of coastal Douglas-fir populations [36]. It has been extended to include economic considerations, but, due to the output form and the homogenous input tree type, it could easily be extended to consider the carbon sequestration of coastal Douglas-fir trees. Modeling has also been used to account for multiple ecosystem services in the same model. MIMES was formed to account for stakeholder inputs and biophysical datasets to produce an economic valuation of ecosystem services. Ecosystem services are interconnected in their provision and valuation due to the complexities of ecosystems and economies; therefore the model has a focus on both spatial and temporal scales. Another aspect that this model specifically looks at is the demographic implications of the provision of these ecosystem services [37].

For this study, to determine the carbon sequestration potential of areas that are being lost to development, it is important to note the current carbon storage that is in place within the given area of Alachua County in order to inform management decisions. The US Department of Agriculture developed a technical document with factors for converting tree volumes in cubic feet to carbon storage in pounds. This is broken down by region (Southeast, South Central, Northeast, Mid-Atlantic, North Central, Central, Rocky mountain, and Pacific Coast) as well as hard- and soft-wooded forest type (pines, oak–hickory, oak–pine, bottomland, and hardwoods). With this and data regarding tree cover, land cover type, and the associated parcel size, the total carbon storage for a parcel can be determined. Additionally, the average size of the trees must also be determined for all the parcels, which can be done by including LiDAR datasets in the analysis.

While not explicitly a model, the work of Raciti *et al.* is notable due to the use of remote sensing methods to determine carbon sequestration from forests over a large geographic extent, namely, the Boston Metropolitan Area [38]. The combination of both LiDAR datasets and 4-band QuickBird imagery allows for the creation of multiple maps displaying ecosystem service provision, specifically carbon sequestration, and then compared the datasets created by their methods to those using the National Land Cover Database (NLCD). Tree canopy height was determined through a combination of LiDAR multipoint data combined with normalized difference vegetation index (NDVI), a measure based on the amount of infrared light reflected by plants, which can show the health and density of plants in a given area. NDVI is the main measure of tree canopy using multispectral remote sensing methods. As NDVI is based on the different recorded wavelengths, a simple equation is needed to calculate the difference in the near-infrared and red color bands for each pixel. The result is a value between 0 and 1 that corresponds to different densities and health of vegetation.

This can be expressed as $NDVI = \frac{NIR-RED}{NIR+RED}$, where NIR = near infrared (Band 4), RED = Red band (Band 3).

Nelson *et al.* provide a state-of-the-art assessment of multiecosystem service accounting noting that there is a marked increase in the number of models in development that perform this task [39]. Noteworthy are EcoMetrix, Integrated Valuation of Ecosystem Services and Tradeoffs (InVEST), and Artificial Intelligence for Ecosystem Services (ARIES), which link terrestrial ecosystem service provision, economic valuation, and comparative trade-offs all within the same framework. Included in all of these approaches is the ecosystem service of carbon sequestration, which highlights its state of relative interest and ease in quantifying compared with other services such as those aesthetic or recreational in nature. This difficulty in quantification is due to the fact that these intangible qualities depend on social factors, and their value varies based on the ideals of the prevailing culture.

8.3.3 PM10 Filtration

Filtration of PM10 particles has obtained little focus in the predictive ecosystem modeling community; however, its economic and health effects on individuals have been extensively recognized in literature [40–43]. Most work has been focus on determining the ambient levels of PM10 particles rather than on the supply of the filtration ecosystem service, with the exception of the I-TREE tool. I-TREE is self-described as "tools for assessing and managing community forests" and is a suite that provides urban forestry analysis and benefit assessments through the application of six specific tools. Ecosystem services are included in the form of carbon sequestration, PM10 removal, and other greenhouse gas removal that can be quantified while also determining the health and density of tree cover. Others have used this tool as a foundation for quantifying several services [44]. Their study focused on developing a distributed air pollutant modeling framework for dry deposition. The explicit goal of the study was to produce a model that could be used by city managers to determine the best locations for planting future urban forest amenities and facilitate the estimation of temperature, leaf area index (LAI), and air pollutant concentrations in a spatially explicit way within the context of the City of Baltimore.

Drainage, as an ecosystem service, is often defined as a combination of many different benefits. When the conditions of proper drainage are present, runoff can be decreased, which has a direct impact on reducing the erosion effects on landscapes. Vegetated land reduces the frequency of flooding events caused by erosion, and land can be cleansed of fertilizers and dust particles, preventing the buildup of these drivers of eutrophication in surrounding bodies of water [45]. These are just some of the benefits of this service, which can additionally be extended to direct effects on the human health of surrounding individuals [46]. Traditionally, curve number (CN) analysis has a history of application in regard to the proper management of agricultural land or water quality, but as the impacts of drainage in the urban context have been known, their application has been explored in the context of the built environment and cityscape. Urban application can be seen for open space, impervious spaces, streets, roads, and urban districts.

The complicated nature of this interaction can be seen in the work of Booth *et al.* who determined that the conversion from forest to pasture or grassland can have pronounced effects on water quality, even when the amount of new impervious surfaces is kept relatively low [47]. This highlights the importance of accounting for what is lost when sprawled urbanization patterns lead to the loss of traditionally forested areas. While not explicitly applied in the context of urban-scale analysis, the ArcCN tool is a very sound starting example for modeling CNs using GIS technology [48]. The model uses 2013 land cover and soil hydrologic group information to simplify the traditional process developed by the National Resources Conservation Service (NRCS) [49]. However, the spatial nuances across large areas are often lost through this application, a problem that ArcCN hopes to solve in the future. One of the main issues with the application, in regard to our study, is the fact that the method is applied at a very coarse scale.

Our study develops a methodology to integrate development impacts with land parcels. Within the smart cities concept, this method is the result of technological innovations and, as such, becomes more accurate and useful when more data are available. While smart city progress has focused on the growth of integrated networks of buildings, devices, and infrastructure, it is important that we leverage this same kind of thinking in developing means to quantify and integrate ecosystem services considerations for smart city growth as well. This methodology was built with the goal of demonstrating an applied example of the potential already offered by existing techniques.

Section 8.4 provides details on the methodology used to calculate carbon sequestration, drainage, and PM10 filtration for Alachua County land parcels. Section 8.5 discusses the implementation of development impacts in the Dynamic-SIM platform. Section 8.6 details the assumptions and limitations that are inherent to the method that was developed, as well as speculating on potential extensions to the model that can be pursued in the future. The work is concluded with an evaluation of the potential implications that the methodology can have on policy planning and public/private city development processes.

8.4 Methodology

The study area is Alachua County for coarser analysis, while the finer-scale analysis was done in the largest city in the county, the City of Gainesville. The county spans 969 mi^2 with a total population of 247,336 people as of the 2010 census [50]. The City of Gainesville covers a geographic area of 61 mi^2 and has a population of 128,460. Parcel designations are provided by the Alachua County Property Appraisers office, which maintains GIS datasets on the parcels that make up Alachua County and the Gainesville area.

8.4.1 Carbon Sequestration

The first step in the integration of ecosystems services in the context of this specific model was to combine information of land cover from the state of Florida with the parcel data that the model are based on, which includes economic information from the Alachua County property appraiser's office. A variety of organizations offer information regarding land cover, and these differed greatly in their specificity as well as the supporting information these were provided with. This data can include detailed information regarding plant species for a given area, an important factor in the ecosystem services that an area renders. As part of this exercise, we look into several land cover datasets that might inform our methodology. As a first step the Florida Geographic Data Library (FGDL) was selected, created by the University of Florida GEOPlan center based on the 2014 Parcel dataset produced earlier in the year. The institution consolidated 99 detailed land cover categories into 15 generalized land cover categories. The categories that they chose were Residential, Vacant residential, Vacant nonresidential, Retail/Office, Industrial, Agricultural, Recreation, Other, Public/semi-public, Mining, Row, Water, Centrally Assessed, and Acreage not zoned for agriculture.

Secondly, Landsat TM satellite imagery was selected to establish a land cover classification by the Florida Fish and Wildlife Conservation (FWC) Commission. As useful as this dataset is in general, the nature of the work in developing the model calls for a more detailed and fine-grain assessment of land cover, specifically such that it informs on the type of habitat that exists on each given parcel of land.

In order to address the need of finer-scale analysis, we used data provided by the Cooperative Land Cover (CLC) map created by the partnership of the Florida Fish and Wildlife Conservation Commission and the Florida Natural

Figure 8.1 FNAI CO-OP land cover.

Area Inventory (FNAI). This level of detail informs on tree cover as well as the species of plants, which has implications for the ecosystem services of drainage and PM10 filtration [51].

Formed from the combination of 37 existing mapping sources as well as 8 maps with high-resolution aerial photography available, these sources were vetted and corrected for consistency and formatted before being run through the Florida Land Cover Classification System. After consolidation, over 7.4 million acres of land was able to be classified with the remainder being added through inclusion of datasets already held and managed by Florida water districts [52]. After the consolidation process was applied, 190 hierarchical classes spanning and 39.5 million acres were condensed into a color-coded legend of 44 classes (Figure 8.1).

After consulting GIS experts, it was determined that the first step for combining the land cover dataset with the parcel data is to use the Raster Calculator tool. This will deliver an output raster that combines the parcel number with the land use designation (Figure 8.2).

Due to the nature of the datasets and the sheer volume of the data being handled, ArcPy methods were also employed. ArcPy is the Python programming language extension of ArcGIS 10 allowing geoprocessing to be automated and even looped, increasing the speed of analysis [53]. ArcPy is the specific package with the functions, classes, and modules that are related to the process of scripting in ArcGIS. The ArcPy module is responsible for importing the geographic

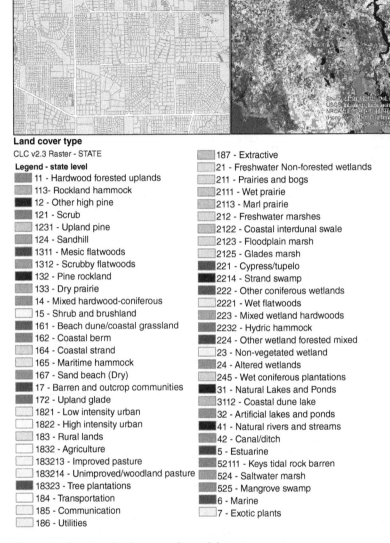

Land cover type

CLC v2.3 Raster - STATE

Legend - state level

- 11 - Hardwood forested uplands
- 113- Rockland hammock
- 12 - Other high pine
- 121 - Scrub
- 1231 - Upland pine
- 124 - Sandhill
- 1311 - Mesic flatwoods
- 1312 - Scrubby flatwoods
- 132 - Pine rockland
- 133 - Dry prairie
- 14 - Mixed hardwood-coniferous
- 15 - Shrub and brushland
- 161 - Beach dune/coastal grassland
- 162 - Coastal berm
- 164 - Coastal strand
- 165 - Maritime hammock
- 167 - Sand beach (Dry)
- 17 - Barren and outcrop communities
- 172 - Upland glade
- 1821 - Low intensity urban
- 1822 - High intensity urban
- 183 - Rural lands
- 1832 - Agriculture
- 183213 - Improved pasture
- 183214 - Unimproved/woodland pasture
- 18323 - Tree plantations
- 184 - Transportation
- 185 - Communication
- 186 - Utilities
- 187 - Extractive
- 21 - Freshwater Non-forested wetlands
- 211 - Prairies and bogs
- 2111 - Wet prairie
- 2113 - Marl prairie
- 212 - Freshwater marshes
- 2122 - Coastal interdunal swale
- 2123 - Floodplain marsh
- 2125 - Glades marsh
- 221 - Cypress/tupelo
- 2214 - Strand swamp
- 222 - Other coniferous wetlands
- 2221 - Wet flatwoods
- 223 - Mixed wetland hardwoods
- 2232 - Hydric hammock
- 224 - Other wetland forested mixed
- 23 - Non-vegetated wetland
- 24 - Altered wetlands
- 245 - Wet coniferous plantations
- 31 - Natural Lakes and Ponds
- 3112 - Coastal dune lake
- 32 - Artificial lakes and ponds
- 41 - Natural rivers and streams
- 42 - Canal/ditch
- 5 - Estuarine
- 52111 - Keys tidal rock barren
- 524 - Saltwater marsh
- 525 - Mangrove swamp
- 6 - Marine
- 7 - Exotic plants

Figure 8.2 Capstone land cover and parcel dataset.

data in a way that can be understood in the coding language. Additionally, NumPy, a module created for scientific analysis in the context of Python, was also used. Creation of *n*-dimensional arrays, broadcasting functions, interoperability with other coding languages, and aspects of linear algebra are just a few of the features that this module possesses [54]. Specifically, NumPy was used to create a container, known as a raster array, which we are then able to

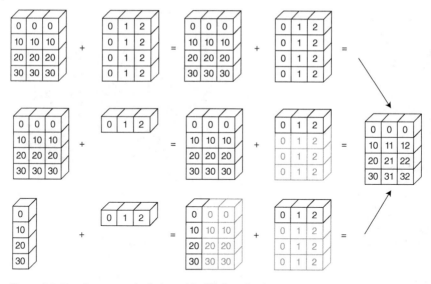

Figure 8.3 NumPy raster calculation with different sized arrays.

populate with the raster that we made in the first step. The "physical" structure of the array by default is a kind of data cluster, represented very much like a cube (Figure 8.3).

In the code used in this study the term "land_use" refers to the ecological land cover for each parcel. These classifications vary between upland, wetland, and exotic designations with specificity as to the type of vegetation that can be found there, for example, Hardwood Forested Uplands. Using the code string "valueList = rasArray.flatten().tolist()," the information found in this raster array can be flattened and structured as a linear list rather than a cube. This increases the speed of the looping process.

Following the importation of the modules, two dictionaries, D and R, are created. A "for" function is used to loop the process of designation so that each entity in the valuelist is looked at one by one and placed in the appropriate dictionary. Dictionary "D" is created to be populated by the parcel's ID and dictionary "R" houses the landuse_id. A separate variable, known as "A," is operated on for the looping procedure as defined by the following code "A = str(parcel_value).rjust(8,'0')," which instructs the script to read each string value starting from the eighth digit from the right until the end of the string as the value for each parcel. We then instruct the program to read from the first to the third digits as the landuse_id and from the third digit to the end as the parcel_id. Following this we specify that if the code finds the parcel_id in D, then it will augment the values that are found in order to assign and keep the designations based on the landuse_id that can be found in the original raster, effectively assigning it to the parcel data. Following this, a loop was created that would iterate the items in the "D" dictionary based on the number of pixels belonging to each parcel and what the associated land use was. In this way, the "R" dictionary is populated with information regarding the predominant land use for each parcel as well as what percentage of the parcel area this land use covers. This is presented as an expression (Parcel_id, Landuse_id, Pixel_numbers, Percentage). This is then printed out to a CSV excel spreadsheet where it can be joined to the original Alachua parcel file based on the parcel_id field.

Above is the first methodology used to incorporate the land cover dataset information into the Alachua County parcel data. Unfortunately, applying this method requires expert knowledge in GIS and ArcPy operation. The second method, more simple, was based on the Florida land cover map created by the Florida Cooperative Fish and Wildlife Research Unit, known as the Florida Gap Project. This classification system is based on the National Vegetation Classification System (NVCS) and was determined using Landsat TM satellite imagery. Instead of processing the data through another ArcPy script, geoprocessing techniques in Arcmap were used. The following steps were applied, and this methodology became the basis for the determination of the other ecosystem services in the area of study:

1) Raster data loaded into Arcmap
2) Zonal statistics tool (MAJORITY) applied, Alachua parcels as zone layer and land cover as raster value layer
3) Conversion of parcels to points
4) Extract features by point tool applied using parcels as points and zonal statistic results as value raster
5) Join applied based on parcel ID in attribute table with original Alachua parcel layer
6) Data imported to original Alachua parcel layer through creation of new field and population with results of spatial join using matching parcel IDs
7) Adding descriptor to zonal statistic result records of land cover type.

Originally, the intent was to have a level of detail in parcel data specific enough to calculate the carbon sequestration capabilities for each ecosystem type; however due to time constraints, averages of tree carbon storage in urban areas and communities were used to determine the total amount of carbon stored on parcels as well as their potential sequestration. Averages for whole urban tree storage were about 7.69 kg C/m^2 and about 0.28 kg C/m^2 of annual carbon sequestration [55]. Using the ArcGIS field calculator tool, this can be determined for parcels using the expression [Shape_Area] * [NDVI_1] * 7.69 = kg C Stored by parcel vegetation and [Shape_Area] * [NDVI_1] * 0.28 = kg C sequestered annually by vegetated parcel (Figure 8.4).

Alachua county carbon sequestration

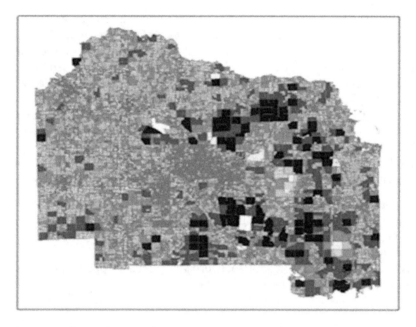

Carbon sequestration (Kg C/year)

Alachua_parcels

Carbon_sequestration kg C/year

- -4681.141602 - 6755.987793
- 6755.987794 - 22382.701172
- 22382.701173 - 52042.230469
- 52042.230470 - 108567.117188
- 108567.117188 - 244246.625000
- 244246.625001 - 484465.343750
- 484465.343751 - 1026675.937500

Figure 8.4 Alachua County carbon sequestration.

8.4.2 Drainage

Another ecosystem service lost through conversion of forests to impervious surfaces is the service of effective drainage. Drainage potential is a function of the type of soil in an area, how much tree cover there is, the quality of the land, and how impervious the surfaces are. In keeping with the framework of Dobbs *et al.* [24], determination of CN for parcels in Alachua County is essential.

Adding the hydrologic soil groups to the appropriate parcels is a necessary process to calculate CNs using a eight-step GIS process:

1) Obtain and load soil survey data for Alachua County into ArcGIS.
2) Convert soil survey polygon data into raster data.
3) Perform zonal statistics.

4) Convert parcels to points to extract results of zonal statistic operation.
5) Spatially join the resulting dataset from step 4 to the original Alachua parcel layer.
6) Create new field for "Hydrologic Code" in Alachua County Parcel attribute table.
7) Populate this new field with zonal statistic results using field calculator.
8) Use field calculator to change each zonal statistic number to its corresponding soil type.

The US Department of Agriculture Natural Resources Conservation Service is responsible for maintaining and organizing the information obtained through the cooperative soil survey. This database has soil maps and data available for over 95% of US counties and offers these datasets online through their web soil survey service.

Soil type and characteristics are organized through a classification system known as soil hydrologic groups or HSG. The groups are based on the first four letters of the alphabet with the possibility of a mix of wet soil types, for example, A, D, or A/D soils. Designation of soil type is a matter of the measured rainfall, runoff, and infiltrometer datasets created for a region. Assumptions are made that soils found in the same climatic regions with similar depths will share many of the same behavior regarding water runoff [56]. Soils "A" are characterized by low runoff potential where the porous nature of the soil allows water to pour through very easily due to the sand and gravel content. Soil "B" has only moderately low runoff potential given they are more likely to have between 10 and 20% clay and mixtures of loam, silt loam, silt, or sandy clay. Soil "C" has moderately high runoff potential because water is typically restricted in its impedance. There is typically less than 50% sand in this soil type, and the amount of clay is also higher than in the two previously mentioned soil classes. Soils of group "D" have the highest runoff potential when inundated with water due to the high concentration of clay (over 40%) and dearth of sand (less than 50%). Due to the complex and varied nature of soil types across a geographic extant, the best method to determine the designation for ecosystem accounting is to give a hydrologic code designation to each parcel based on what type of soil takes up the majority of the area. This can be done once the polygon soil survey dataset has been converted to raster for use as an input in the zonal statistics tool. The zonal statistics tool produces an output using a given statistic, such as majority, maximum, mean, range, and so on, with one layer to define the zone of operation and another providing the values for calculation. In this case the layer that acts as the zone is the Alachua County Parcel layer file and the value raster is the hydrologic group raster (Figure 8.5).

In the zonal statistic results, designations from 1 to 8 represent the original soil codes with the inclusion of one that represents "Null" values, where there were no data available originally (Figure 8.6). These data have to be

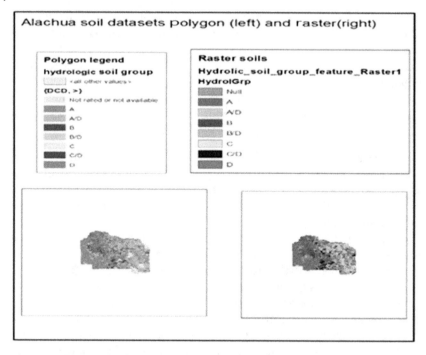

Figure 8.5 Raster and polygon soil data.

meaningfully connected to the discrete parcel data using the feature to point tool. This tool creates a feature class containing points generated from the representative locations of input features. In this case the representative point is the middle of the polygons for the parcel data. Once this feature class has been created, the data can be taken from the zonal statistic using the spatial join tool. Spatial join is an operation that lets you join attributes from one feature to another based on their spatial relationship.

Designations are made for the target feature corresponding to the Alachua parcels joined to the zonal statistic resulting layer. Once this is done, the attribute table for each parcel as a point has a zonal statistic result. This is important to note because in order to perform a joining operation based on attribute tables, there must be at least one unique field in common between the two tables that are to be joined. In this instance we used the unique field for the parcel ID numbers. The join in its current state relies on the presence of the Alachua point feature attribute table to display the zonal statistic result; the data have not actually been added to the Alachua parcels yet. In order to do this, we must create a new field, named "Hydrologic Code," in the Alachua County parcel attribute table and then use the field calculator to populate it with the designation from the zonal statistic column.

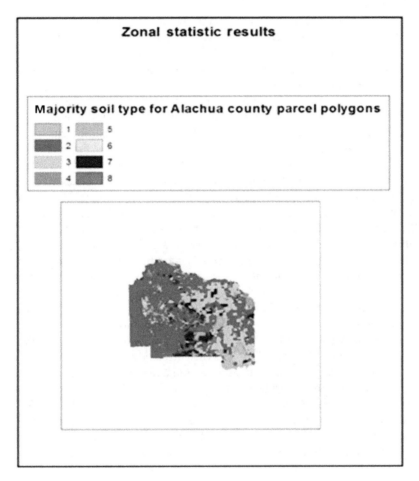

Figure 8.6 Zonal statistics result.

The field calculator is a tool that allows you to input values into an attribute table through the use of mathematical syntax. Calculations can be based on other fields in the table, such as the operation that we used, which specified "Hydrologic Code = Zonal statistic." After this operation the corresponding zonal code is duplicated in the hydrologic column, and the join to the zonal statistic layer file can be removed. A selection by attribute is performed to select each zonal statistic result in turn, that is, 1–8. From the soil data we know that each number represents a given soil type, so once every instance of a number is selected, we can then use the field calculator to only operate on the selected records to change their designation from the largely meaningless numbers 1–8 to the letter designations for HSG. Once this process is finished, the incorporation of the soil hydrologic group to the Alachua County parcel data is done.

However, as previously mentioned, CNs are not a consequence singularly of the HSG of a given piece of land; it is also dictated by the land use of the area as well as the amount of cover and condition seen in the landscape. Included in the work of Schiariti is a section on the proper selection of CNs in light of different ground cover conditions, specifying that "It is crucial to use the CN value that best mimics the Ground Cover Type and Hydrologic Condition." While we have primarily been operating in and using the Alachua County parcel files, the nature of the ground cover classification for CN calls for the use of a coarser land cover designation than the ground cover dataset we have been trying to incorporate into the model so far. Additionally, the ground condition will be determined based on the amount of vegetation present in the area, so we will be using the NDVI method to determine the LAI of each parcel. This is an important step because it is also an underlying dataset for the determination of PM10 filtration. Each parcel will have to be given a TR-55 runoff CN. The literature provides a description of the cover type and hydrologic condition, and then a corresponding CN is provided based on the soil hydrologic group present. For cultivated agricultural land, the specification of conditions is more detailed than is available through the generalized land use shapefile; it calls for treatment information as to whether fields are left fallow, legumes are grown on them, there is row agriculture present, and so on. So for agricultural sites, a generalization will have to be made about their ground condition and treatment. A designation of "poor" or "good" is then given based on conditions such as the amount of vegetation covering the ground and how dense this is. After which, HSG is taken into consideration to provide a specific CN (Figure 8.7).

Designations for urban land are less nuanced than for agricultural land. The method does away with the categorization of land as "poor" or "good" and gives generalized CNs based on the type of impervious surface it is and how much grass cover open spaces have. For residential areas, this generalization is extended based on the average lot size for a given residential district. Once this CN is determined, in order to determine the specific amount of runoff that will occur given a certain rain event, one must apply the SCS runoff equation found in Schiariti's work.

Once the NDVI-derived LAI data have been applied to each parcel, there can be a designation of parcel quality based on land use type. Following this, an SQL syntax query can be used to select parcels of a certain type, that is, ones that are open space and HSG group B and have 50% LAI value to assign the unique CNs to each specific context.

8.4.3 PM10 Filtration

Conceptually, the service of PM10 filtration is simply a factor of the amount of filtration for each land cover type and the amount of land cover by this type.

Chapter 2	Estimating Runoff	Technical Release 55
		Urban Hydrology for Small Watersheds

Table 2-2b   Runoff curve numbers for cultivated agricultural lands [1]

	Cover description		Curve numbers for hydrologic soil group			
Cover type	Treatment [2]	Hydrologic condition [3]	A	B	C	D
Fallow	Bare soil	—	77	86	91	94
	Crop residue cover (CR)	Poor	76	85	90	93
		Good	74	83	88	90
Row crops	Straight row (SR)	Poor	72	81	88	91
		Good	67	78	85	89
	SR + CR	Poor	71	80	87	90
		Good	64	75	82	85
	Contoured (C)	Poor	70	79	84	88
		Good	65	75	82	86
	C + CR	Poor	69	78	83	87
		Good	64	74	81	85
	Contoured & terraced (C&T)	Poor	66	74	80	82
		Good	62	71	78	81
	C&T+ CR	Poor	65	73	79	81
		Good	61	70	77	80
Small grain	SR	Poor	65	76	84	88
		Good	63	75	83	87
	SR + CR	Poor	64	75	83	86
		Good	60	72	80	84
	C	Poor	63	74	82	85
		Good	61	73	81	84
	C + CR	Poor	62	73	81	84
		Good	60	72	80	83
	C&T	Poor	61	72	79	82
		Good	59	70	78	81
	C&T+ CR	Poor	60	71	78	81
		Good	58	69	77	80
Close-seeded or broadcast legumes or rotation meadow	SR	Poor	66	77	85	89
		Good	58	72	81	85
	C	Poor	64	75	83	85
		Good	55	69	78	83
	C&T	Poor	63	73	80	83
		Good	51	67	76	80

[1] Average runoff condition, and $I_a = 0.2S$.

[2] Crop residue cover applies only if residue is on at least 5% of the surface throughout the year.

[3] Hydraulic condition is based on combination factors that affect infiltration and runoff, including (a) density and canopy of vegetative areas, (b) amount of year-round cover, (c) amount of grass or close-seeded legumes, (d) percent of residue cover on the land surface (good ≥ 20%), and (e) degree of surface roughness.

Poor: Factors impair infiltration and tend to increase runoff.

Good: Factors encourage average and better than average infiltration and tend to decrease runoff.

Figure 8.7 Chart detailing CN designation for agricultural land.

However, this is complicated by the fact that different tree species have different rates of PM10 filtration; therefore, in order to obtain a more accurate model, the differing rates would ideally be considered on a parcel by parcel basis [57]. The basic equation that governs the filtration amount is expressed as follows:

$$\text{Deposited amount } (g/m^2) = \text{LAI} \times V_d \times C \times T$$

where

LAI = Leaf area index

V_d = Deposition velocity

C = Air concentration of pollutant

T = Time

This expresses the general rules that dictate the amount of deposition of PM10 onto the surfaces of vegetation, but the difficulties of modeling this relationship at a regional scale has been noted in literature. Specifically, "Regional removal of pollutants by deposition on vegetation in urban areas has been calculated from reported deposition velocities and averaged concentrations together with measured or estimated vegetation surface. Due to large variability of both vegetation surface and air pollutant concentrations [over regional distances], averaging problems are common" [57]. A study has been done to determine the amount of average PM10 filtration for the city of Gainesville, and we extend the results of this study to account for the county. The reason this is acceptable is due to the nature of the city of Gainesville itself. Although it does have its representative share of urban areas, it is also home to agricultural sites and numerous conservation areas and possesses a distribution of tree species we feel that effectively captures the biodiversity of the surrounding study area that it is embedded in [58]. It was determined that the average annual pollution removal per square meter of tree cover is 1.52446 g/m^2 in the area of study [59, 60].

A LAI for Alachua County can be produced by using the MODIS13Q1 dataset. The potential use of EROS Moderate Resolution Imaging Spectrometer (eMODIS) and MODIS15A1 LAI datasets is also explored; however, the use of these was disqualified for a variety of reasons. These reasons included discordance between the spatial resolution that would be best for the study and the resolution that we could obtain from either of these as well as issues converting the datasets into a usable form through the application of the image window tool in ArcGIS.

MODIS13Q1 is a Global MODIS index with a temporal resolution of 16 days and a spatial resolution of 250 m. The dataset in its original form carries an unknown datum on a custom sinusoidal grid and, therefore, must be projected using the Project Raster Tool before it can be analyzed because there is inherent warp in its representation in its original form. Once this conversion has been made, a scaling factor has to be applied to the provided map in order to convert it into a form that has the information that we need. This can be done through the application of map algebra tools, specifically the raster calculator.

By multiplying each cell by 0.0001, the valid ranges of $-2000-9985$ are converted to a ranges of $-0.2-0.9985$, representing the proportion of each cell that is covered in vegetation. There are admittedly errors present in this method, as demonstrated by the presence of a negative 20% value, but the vast majority of the values are valid. After this conversion has been made, the process followed is much the same as that for the HSG determination, specifically steps 3–8. The difference here is that this only gives us the maximum area of leaf cover for each parcel; in order to calculate the LAI, we then have to divide this resulting proportion for each parcel by the total parcel area. This uses the field calculator to populate a field with the result of the required division for every record in the attribute table. Once this is complete, this value is then multiplied by the total parcel size as well as $1.52446 \, g/m^2$ to determine how many grams of PM10 each parcel filters on an annual basis.

8.5 Implementation of Development Impacts in Dynamic-SIM Platform

The Dynamic-SIM platform has the explicit goal of benchmarking buildings on an ecological basis, allowing for the analysis of different design options. Geared for use by building engineers, architects, and planners, the popular Building Information Modeling (BIM) platform is the foundation for the model, with the addition of imported ArcGIS formats to determine land development impacts. BIM uses an Industry Foundation Class (IFC) for data management, which promotes interoperability. The cloud accessible nature of this framework will hopefully lead to a revolution on how fast, inexpensive, and effective we can collect, share, and collectively analyze the data produced by this framework. This cloud-based focus will streamline the development of new and effective tools by crowdsourcing the process to researchers at a variety of institutions for use in their independent studies, with successful developments being speedily available to everyday practitioners for immediate use. This would increase the amount of building data we collectively have access to and can help us begin to aggregate impacts in a meaningful way. Already, preliminary 3D heat transfer can be projected and modeled so that different configurations of building orientation and fenestration can be tested [61]. Recently, Agdas and Srinivasan tested parallel computing methods to improve time in developing results from EnergyPlus energy simulation software while also offering the opportunity to develop scenarios for different design considerations, in platform [62]. Real-time accounting for building energy use can be performed in platform as well [63]. As the use of the building changes over time, the model will be able to track in detail what kind of effects this has on the building's overall energy use in real time while keeping a historical record. More importantly, this platform allows for seamless transfer from city to neighborhood to building levels of analysis (Figure 8.8).

Where BIM focuses on the direct effects of building development, Dynamic-SIM extends to account for the impact of building materials taken from far afield of the area being modeled and neighborhood-scale impacts

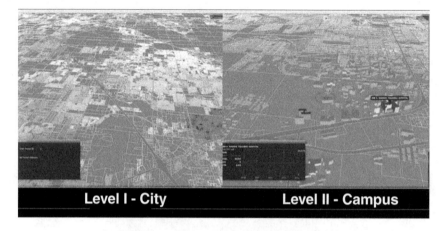

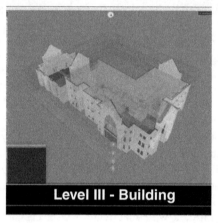

Figure 8.8 Dynamic-SIM platform.

to the accounting process. To facilitate widespread use and development, a focus will be placed on developing along an open source line, where the tools being integrated are free and open for use by the global community. Complete and meaningful environmental analysis of building impacts is one of the main focuses of the Dynamic-SIM model, and that is where the work of this chapter comes in.

Dynamic-SIM is not a fully realized tool; it is largely still in development. Currently, the platform serves as a benchmarking tool, where the performance of buildings can be determined using meter information as well as real-time sensors and then compared with one another over time. Land development impacts are the newest and least developed aspect of the model. At the moment, ecosystem services can be determined and represented, but there is not a solid means of comparing the results with each other in a normalized fashion. At the building level in the future, the model will be able to simulate different design choices for a given building, such as fenestration patterns, orientation, or material choices, and project how these choices will affect the use of energy in the building. There are even efforts being made to quantify the effects of acoustical comfort on building inhabitants, and the final model would be able to simulate how different alternative choices would affect these considerations before building's designs have been finalized and locked in. Moreover, building's site location has been integrated into the model by considering (i) connectivity to existing urban infrastructure, (ii) integration within current neighborhood, (iii) connectivity to transportation network, and (iv) environmental and agricultural land status as developed by Komeily and Srinivasan [64].

At the neighborhood scale, the final model would be able to account for the ecosystem services provision of different development patterns based on projected changes in the landscape in order to provide information relevant to choices made in future development patterns. Additionally, this benchmarking would provide targets in offsetting certain development choices through the implementation of green infrastructure and could even serve as a demonstration tool for the benefits of this kind of development to nonprofessionals. The model could be slowly built over time for a particular city, and neighborhood performance could be compared by policymakers in order to work at the weakest points of a city rather than funneling resources into neighborhoods that are already performing to high standards. Changes in individual buildings or a given neighborhood can be compared over large time frames to determine what works most effectively and what systems are not performing as they should.

While a set ranking system does not exist, inputs at a variety of scales can be relatively compared. At the building level, comparisons can be made based on how much CO_2 is produced to provide the energy it consumes or the amount of water it consumes. At the neighborhood scale, lynchpin neighborhoods can be identified through the amount of services they offer in general and to the city as a whole. This can come in the form of the amount of PM10 particles trees in a given area siphon out of the air, how much carbon that area sinks, or how much it contributes to groundwater recharge through the permeability of its soils. Gamification has recently been explored as potentially being a tool to support sustainability. Patterns of excess energy use have been shown to drop when neighborhood residents are able to see how they stack up in regard to their neighbor's energy use, and, in the hands of the proper planning agency, this tool can be used to develop programs such as those aimed to gamify positive environmental behaviors.

As of now, the system does not have a clear-cut rating system to date. In the future, however, one could be developed much like the LEED system where ratings are given to various buildings based on their projected impacts and how they perform in relation to land parcels of similar form that are untampered by development. A variety of schema can be applied in judging the neighborhood but what is most important is for a standard to be instituted that accurately identifies the most egregious offenders and provides a means for keeping track of in what way different interventions effect the site.

8.6 Discussion (Assumptions, Limitations, and Future Work)

George Box is often quoted as saying "Essentially, all models are wrong, but some are useful" [65]. Wisdom such as this cannot be overstated, and it helps to have this in mind when considering the assumptions and limitations of ecosystem service modeling in its current state. For carbon sequestration and storage, an assumption is made that the processes of urban trees can be applied to trees in a more rural context. Additionally, the application of national averages may smooth over the nuances that would be present if looking at the provision of these ecosystem services from forest types that were only present in the area of study. However, the averages were made with direct consideration of the foliage of the Gainesville area, so the values still hold validity. Assumptions were made that the leaf coverage for each parcel comes only from trees rather than other vegetation types such as shrubs or grasses. For agricultural drainage areas, vegetation type and treatment was universally assumed to be row crops planted in straight rows as there is little reason to plant in a terraced or contoured fashion based on the topographical layout of Alachua County.

In future, and final, iterations of the model designations of "good" and "poor" should be applied to the model results to generate the specific CN value for each parcel. For PM10 filtration, assumptions can be seen in the form of the application of average flux values for the City of Gainesville, applied to the entire county, where ideally there would be more nuance added to the analysis based on the different species of trees found in each parcel. In the future, the class name for each dominant land cover can be used to more thoroughly research what the predominant species are as well as what their specific rates of dry deposition, that is, filtration, would be. There are already studies in the literature that have analyzed the pollution filtration rates of many of the common species in Alachua County, which would be useful for inclusion in future iterations of the model [53]. Universally, across all ecosystem service methods, assumptions were made in reference to the use of the zonal statistic tool. This is due to the fact that the use of the tool introduces a certain aspect of subjectivity to a study as the GIS operator must decide which operation will yield the most accurate and appropriate information, such as the choice to use either the MAJORITY function or the MEAN function in an operation.

With the development of the Dynamic-SIM modeling platform, we hope to nudge the field away from this overall trend by offering decision-makers a viable tool for EGS accounting that can be used in concert with rating systems and other means of sustainability assessment [60–62].

8.7 Conclusion

The looming challenges that cities face in the wake of the reality of climate change calls for a fundamental rethinking of what tools will best equip us to implement the solutions needed to ensure a stable homeostatic relationship with nature in the future while still carrying out the activities that make us indelibly human. Currently we are under prepared, but there have recently been great strides in technology, as well as mindset that demonstrate the ability of the building sector to adapt. The continued development of Dynamic-SIM is one more step in that progression. Increasing the number of tools that account for deep ecological concepts, such as EGS, is essential to the inclusion of these considerations in effective city management. Carbon sequestration, PM10 filtration, and drainage are important ecosystem services; future work should build on this to include some of the more intangible but equally important ecosystem services such as aesthetic or recreational value especially with how important natural areas, conservation areas, and bike trails are to the Alachua County economy.

Another consideration is the potential use of this application to inform regulatory standards and codify the system of ecosystem accounting and provision. Models such as these can be used to replace or improve the traditional tools of cost–benefit analysis process, such as the *pro forma* analysis. When thinking of the cost or benefit of development, currently the loss or provision of ecosystem services is plainly not on the table for discussion largely because developers and interested parties do not have broad access to the tools that would allow for this kind of analysis. In keeping with the goal of fully accounting for the costs and benefits of a given decision, these externalities must be accounted for in order to reflect the true market value of land development in a way that guides the efficient operation of the prevailing free market system.

With the rise of the field of environmental law, fees can now be leveraged by city municipalities for a variety of development impacts, ranging from transportation impact fees, water and sewage impact fees, and so on. These are used to offset some of the increases in costs that can be incurred by cities when sprawled development increases the distance that utility services have to travel to customers or the increased costs that come from operating progressively longer and more complicated public transportation routes and facilities. The results of this study demonstrate how the "tragedy of the commons" concept can be directly applied to the land development and city resource management as well. While a private owner of land does receive a certain amount of monetary benefit from developing land, the greater community losses the PM10 filtration benefit that this land was providing, and the world at large loses the carbon sink that was provided by that space as well. Through the quantification of exactly how much service capacity we are losing through this trade-off, we may be able to demonstrate this to the land owner in a way that makes them reconsider how that land is developed or whether it is actually worth developing at all. If that fails, then the results from this model can be used to develop appropriate impact fees, and if individuals do choose to go ahead with development, impact fees can be leveraged that offset the increases to asthma-related health effects for city occupants and potentially slow down or discourage certain forms of development or development in general.

Final Thoughts

In this chapter we discussed the state of building- and urban-scale sustainability assessment systems; we then moved to the discussion of what questions the concept of smart ecology for cities addresses, and we developed and described a novel means of accounting for urban ecosystem services using a case study site. Finally, we discussed how these functions can be integrated into a larger platform, such as a system known as Dynamic-SIM, and used to make policy decisions by public and private planning entities as well as land developers of all kinds.

Questions

1 Discuss the importance of ecology and smart city development.

2 What are EGS? Discuss five EGS using an example location (city or neighborhood).

3 Discuss the various accounting models used to measure ecosystem services.

References

1 Evans, P.J. (2011) Resilience, ecology, and adaptation in the experimental city. *Transactions of the Institute or British Geographers*, **36** (2), 223–237.
2 Revkin, C. A. (2013). Scientists Propose a New Architecture for Sustainable Development. *New York Times: The Opinion Pages*

3 Pope, J., Annandale, D., and Morrison-Saunders, A. (2004) Conceptualizing sustainability assessment. *Environmental Impact Assessment Review*, **24**, 595–616.

4 Ding, G.K.C. (2007) Sustainable construction – the role of environmental assessment tools. *Journal of Environmental Management*, **86** (2008), 451–464.

5 Holling, S.C. (2001) Understanding the complexity of economic, ecological, and social systems. *Ecosystems*, **4** (5), 390–405.

6 Komeily, A. and Srinivasan, R.S. (2015) A need for balanced approach to neighborhood sustainability assessments: A critical review and analysis. *Sustainable Cities and Society*, **18**, 32–43.

7 USGBC. *LEED-ND: North America*. Retrieved from: http://www.usgbc.org/leed.

8 BRE GROUP. *BREEAM Communities: Europe*. Retrieved from: http://www.breeam.org/.

9 Sustainable Building Alliance. *DGNB-NSQ: Europe*. Retrieved from: http://www.sballiance.org/our-work/libraries/deutsche-gesellschaft-fur-nachhaltiges-bauen-dgnb-or-german-sustainable-building-council/

10 JaGBC. *CASBEE-UD: East Asia*. Retrieved from: http://www.ibec.or.jp/CASBEE/english/overviewE.htm

11 ADUPC. *Pearl Community for Estidama: West Asia, Middle East*. Retrieved from: http://estidama.upc.gov.ae/pearl-rating-system-v10/pearl-community-rating-system.aspx

12 Xing, Y., Horner, R., El-Haram, M., and Bebbington, J. (2009) A framework model for assessing sustainability impacts of urban development. *Accounting Forum*, **33** (3), 209–224.

13 Komeily, A. and Srinivasan, R.S. (2016) What is neighborhood context and why does it matter in sustainability assessment? *Procedia Engineering*, **145**, 876–883.

14 Srinivasan, R.S., Kibert, C.J., Fishwick, P. *et al.* (2015) Dynamic–Building Information Modeling (Dynamic-BIM): An Interactive Platform for Building Energy Engineering Education, in *Transforming Engineering Education through Innovative Computer Mediated Learning Technologies* (eds H. Dib, R. Fruchter, K. Lin *et al.*), ASCE.

15 Srinivasan, R.S., Ingwerson, W., Trucco, C. *et al.* (2014) Comparing energy-based indicators used in life cycle assessment tools for buildings. *Building and Environment*, **79**, 138–151.

16 BSR (2013) Private Sector Uptake of Ecosystem Services Concepts and Frameworks: The Current State of Play.

17 James, P., Tzoulas, K., Adams, M.D. *et al.* (2009) Towards an integrated understanding of Green space in the European built environment. *Urban Forestry & Urban Greening*, **8** (2), 65–75.

18 United Nations Millennium Ecosystem Assessment (2005) *Ecosystems and Human Well-being: Biodiversity Synthesis*, World Resources Institute, Washington, DC.

19 Groot, D.R.S., Alkemade, R., Braat, L. *et al.* (2009) Challenges in integrating the concept of ecosystem services and values in landscape planning, management, and decision making. *Ecological complexity*, **7**, 260–272.

20 MA (2003) *Ecosystems and Human Well-being: A Framework for Assessment*, Island Press, Washington.

21 Balmford, A., Bruner, A., Cooper, P. *et al.* (2002) Economic reasons for conserving wild nature. *Science*, **297**, 950–953.

22 Whitford, V., Ennos, A.R., and Handley, J.F. (2001) City form and natural process—indicators for the ecological performance on urban areas and their application to Merseyside, UK. *Landscape Urban Planing*, **57**, 91–103.

23 Tratalos, J., Fuller, A.R., Warren, H.P. *et al.* (2007) Urban form, biodiversity potential and ecosystem services. *Landscape and Urban Planning*, **83** (4), 308–317.

24 Dobbs, C., Escobedo, F.J., and Zipperer, W.C. (2010) A framework for developing urban forest ecosystem services and goods indicators. *Landscape and Urban Planning*, **99**, 196–206.

25 Olgyay, V. and Herdt, J. (2004) The application of ecosystems services criteria for green building assessment. *Solar Energy*, **77**, 389–398.

26 Chapin, S.F., Zavaleta, S.E., Eviner, T.V. *et al.* (2000) Consequences of changing biodiversity. *Nature*, **405**, 234–242.

27 Balford, A. and Bond, W. (2005) Trends in the state of nature and their implications for human well-being. *Ecology Letters*, **8**, 1218–1234.

28 Lyytimäki, J. and Sipilä, M. (2009) Hopping on one leg – the challenge of ecosystem disservices for urban green management. *Urban Forestry & Urban Greening*, **8**, 209–315.

29 Groot, S.R., Alkemade, R., Braat, L. *et al.* (2010) Challenges in integrating the concept of ecosystem services and values in landscape planning, management, and decision making. *Ecological Complexity*, **7** (3), 260–272.

30 Baggethun, G.E. and Barton, N.D. (2013) Classifying and valuing ecosystem services for urban planning. *Ecological Economics*, **86**, 235–245.

31 Troy, A. and Wilson, M.A. (2006) Mapping ecosystem services: Practical challenges and opportunities in linking GIS and value transfer. *Ecological Economics*, **60**, 435–449.

32 Elmqvist, T., Fragkias, M., Goodness, J. *et al.* (2013) *Urbanization, Biodiversity and Ecosystem Services: Challenges and Opportunities*, SpringerOpen.

33 Ahern, F.J., Cilliers, S., and Niemela, J. (2014) The concept of ecosystem services in adaptive urban planning and design: A framework for supporting innovation. *Landscape and Urban Planning*, **125**, 254–259 Special Issue: Actionable Urban Ecology in China and the World..

34 Martinez-Harms, J.M. and Balvanera, P. (2012) Methods for mapping ecosystem services supply: A review. *International Journal of Biodiversity Science, Ecosystem Services & Management*, **8**, 17–25.

35 Crossman, D.N., Burkhard, B., Nedkov, S. *et al.* (2013) A blueprint for mapping and modelling ecosystem services. *Ecosystem Services*, **4**, 4–14.

36 Curtis, R.O., Clendenen, G.W., and DeMars, D.J. (1981) *A New Stand Simulator for Coastal Douglas-fir: DFSIM User's Guide. USDA Forest Service, Pacific Northwest Forest and Range Experiment Station, Portland, OR.* General Technical Report PNW-GTR-128. 79 p.

37 Boumans, R. (2011) *Overview and Background of the Multi-Scale Integrated Model of Ecosystem Services (MIMES)*, Affordable Futures, ESP, Wageningen.

38 Raciti, M.S., Hutyra, R.L., and Newell, J.D. (2014) Mapping carbon storage in urban trees with multi-source remote sensing data: Relationships between biomass, land use, and demographics in Boston neighborhoods. *Science of The Total Environment*, **500–501**, 72–83.

39 Nelson, J.E. and Daily, C.G. (2010) Modelling ecosystem services in terrestrial systems. *F1000 Biology Reports*, **2**, 53.

40 Sanesi, G., Gallis, C., and Kasperidus, D.H. (2010) Urban forests and their ecosystem services in relation to human health, in *Forests, Trees and Human Health* (eds K. Nilsson, M. Sangster, C. Gallis *et al.*), Springer, Netherlands, pp. 23–40.

41 Pope, A.C., Dockery, W.D., Spengler, D.J., and Raizenne, M.E. (1991) Respiratory health and PM10 pollution: A daily time series analysis. *American Review of Respiratory Disease*, **144** (3_pt_1), 668–674.

42 Pope, C.A. (2000) Epidemiology of fine particulate air pollution and human health: Biogenic mechanisms and who's at risk? *Environmental Health perspectives*, **108** (Suppl 4), 713–723.

43 Oberdorster, G., Maynard, A., Donaldson, K. *et al.* (2005) Principles for characterizing the potential human health effects from exposure to nanomaterials: Elements of a screening strategy. *Particle and Fibre Toxicology*, **2**, 8.

44 Hirabayashi, S., Kroll, N.C., and Nowak, J.D. (2002) Development of a distributed air pollutant dry deposition modeling framework. *Environmental Pollution*, **171**, 9–17.

45 Brezonik, P.L. and Stadelmann, T.H. (2002) Analysis and predictive models of storm water runoff volumes, load, and pollutant concentrations from watersheds in the twin cities metropolitan area, Minnesota, USA. *Water Res*, **36** (7), 1743–1757.

46 Abrahams, P.W. (2002) Review: Soils: Their implications to human health. *The Science of the Total Environment*, **291**, 1–32.

47 Booth, D.B., Hartley, D., and Jackson, R. (2002) Forest cover, impervious surface area, and the mitigation of stormwater impacts. *JAWRA*

Journal of the American Water Resources Association, **38**, 835–845. doi: 10.1111/j.1752-1688.2002.tb01000.x

48 Zhan, X. and Huang, L.M. (2004) ArcCN-Runoff: An Arcgis tool for generating curve number and run off maps. *Environmental Modelling & Software*, **19**, 875–879.

49 USDA NRCS (2013). Web Soil Survey. December 2013.

50 United States Census Bureau Population Division (2015) "Alachua County, FL PeopleQuickFacts" *United States Census 2010*. Washington: US Census Bureau, August 5, 2015. Web. August 25. Retrieved from: http://quickfacts .census.gov/qfd/states/12/12001.html

51 FNAI. "Florida Natural Areas Inventory". Retrieved from: http://www.fnai .org/landcover.cfm.

52 Florida's Wildlife Legacy Initiative Project 08009 (2010) "Development of a Cooperative Land Cover Map: Final Report". July 2010.

53 Flater D, Prince G. 2011. *Python-Getting Started*. Esri International User Conference: Technical Workshops, San Diego, CA.

54 SciPy.org. 2013. Numpy. Retrieved from: www.numpy.org.

55 Nowak, J.D., Greenfield, J.E., Hoehn, E.R., and Lapoint, E. (2013) Carbon storage and sequestration by trees in urban and community areas of the united states. *Environmental Pollution*, **178**, 229–236.

56 USDA NRCS (2007) Hydrologic Soil GroupsChapter 7, in *Part 630 Hydrology National Engineering Handbook*, USDA NRCS.

57 Freer-Smith, H.P., Khatib-El, A.A., and Taylor, G. (2004) Capture of particulate pollution by trees: Comparison of species typical of semi-arid areas (*Ficus nitida* and *Eucalyptus globulus*) with European and North American species. *Water, Air, and Soil Pollution*, **155**, 173–187.

58 Bhardwaj, B., Walton, S., Escobedo, F., and 2012 FOR4090 Students (2013) *Assessment of University of Florida-Main Campus' Urban Forest Structure and Ecosystem Services (January 2013)*, University of Florida, IFAS School of Forest Resources and Conservation.

59 Escobedo F. *Personal Correspondence through Email*, 2015.

60 Hagan, D., Dobbs, C., Escobedo, F. *et al.* (2011) *Urban Soils in Gainesville, Florida, and Their Implications for Environmental Quality and Management*, EDIS: University of Florida IFAS Extension.

61 Srinivasan, R.S., Thakur, S., Parmar, M., and Akhmed, I. (2014) *Towards the Implementation of a 3D Heat Transfer Analysis in Dynamic-BIM (Dynamic Building Information Modeling) Workbench*. Proceedings of the 2014 Winter Simulation Conference: 3324–3235

62 Agdas, D. and Srinivasan, R.S. (2014). *Building Energy Simulation and Parallel Computing: Opportunities and Challenges*. Proceedings of the Winter Simulation Conference 2014.

63 Srinivasan, R., Kibert, C., Thakur, S., Ahmed, I., and Fishwick, P. (2012) *Preliminary Research in Dynamic-BIM (D-BIM) Workbench Development.* Proceedings of the Winter Simulation Conference, 2012.

64 Komeily, A. and Srinivasan, R. (2015) *GIS-Based Decision Support System for Smart Project Location.* In 2015 International Workshop on Computing in Civil Engineering.

65 Box, G.E.P. and Draper, N.R. (1987) *Empirical Model Building and Response Surfaces*, John Wiley & Sons, New York, NY.

9

Data-Driven Modeling, Control, and Tools for Smart Cities

Madhur Behl[1] and Rahul Mangharam[2]

[1] Department of Computer Science, University of Virginia, Charlottesville, VA, USA
[2] Electrical and Systems Engineering, University of Pennsylvania, Philadelphia, PA, USA

CHAPTER MENU

Introduction, 245
Related Work, 248
Problem Definition, 250
Data-Driven Demand Response, 252
DR Synthesis with Regression Trees, 254
The Case for Using Regression Trees for Demand Response, 259
DR-Advisor: Toolbox Design, 261
Case Study, 263
Final Thoughts, 271

Objectives

On the surface, DR may seem simple. Reduce your power when asked to and get paid. However, in practice, one of the biggest challenges with end user DR for large-scale consumers of electricity is the following: *Upon receiving the notification for a DR event, what actions must the end user take in order to achieve an adequate and a sustained DR curtailment?*

This is a hard question to answer because of the following reasons:

1) *Modeling Complexity and Heterogeneity*: Unlike the automobile or the aircraft industry, each building is designed and used in a different way, and therefore, it must be uniquely modeled. Learning predictive models of building's dynamics using first principles based approaches (e.g., with EnergyPlus [9]) is very cost and time prohibitive and requires retrofitting the building with several sensors [10]; the user expertise, time, and associated sensor costs required to develop a model of a single building are very high.

Smart Cities: Foundations, Principles and Applications, First Edition.
Edited by Houbing Song, Ravi Srinivasan, Tamim Sookoor, and Sabina Jeschke.
© 2017 John Wiley & Sons, Inc. Published 2017 by John Wiley & Sons, Inc.

This is because usually a building modeling domain expert typically uses a software tool to create the geometry of a building from the building design and equipment layout plans and add detailed information about material properties, equipment, and operational schedules. There is always a gap between the modeled and the real building, and the domain expert must then manually tune the model to match the measured data from the building [11].

2) *Limitations of Rule-Based DR*: The building's operating conditions, internal thermal disturbances and environmental conditions must all be taken into account to make appropriate DR control decisions, which is not possible with using rule-based and predetermined DR strategies since they do not account for the state of the building but are instead based on best practices and rules of thumb. As shown in Figure 9.1a, the performance of a rule-based DR strategy is inconsistent and can lead to reduced amount of curtailment, which could result in penalties to the end user. In our work, we show how a data-driven DR algorithm outperforms a rule-based strategy by 17% while accounting for thermal comfort. Rule-based DR strategies have the advantage of being simple, but they do not account for the state of the building and weather conditions during a DR event. Despite this lack of predictability, rule-based DR strategies account for the majority of DR approaches.

3) *Control Complexity and Scalability*: Upon receiving a notification for a DR event, the building's facilities manager must determine an appropriate DR strategy to achieve the required load curtailment. These control strategies can include adjusting zone temperature set points, supply air temperature, and chilled water temperature set point, dimming or turning off lights, decreasing duct static pressure set points and restricting the supply fan operation. In a large building, it is difficult to assess the effect of one control action on other sub systems and on the building's overall power consumption because the building sub systems are tightly coupled. Consider the case of the University of Pennsylvania's campus, which has over a hundred different buildings and centralized chiller plants. In order to perform campus-wide DR, the facilities manager must account for several hundred thousand set points and their impact on the different buildings. Therefore, it is extremely difficult for a human operator to accurately gauge the building's or a campus's response.

4) *Interpretability of Modeling and Control*: Predictive models for buildings, regardless how sophisticated, lose their effectiveness unless they can be interpreted by human experts and facilities managers in the field. For example artificial neural networks (ANN) obscure physical control knobs and interactions and hence are difficult to interpret by building facilities managers. Therefore, the required solution must be transparent, human centric, and highly interpretable.

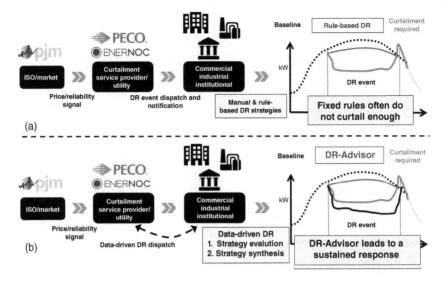

Figure 9.1 Majority of DR today is manual and rule based. (a) The fixed rule based DR is inconsistent and could under perform compared with the required curtailment, resulting in DR penalties. (b) Using data-driven models, DR-Advisor uses DR strategy evaluation and DR strategy synthesis for a sustained and sufficient curtailment.

The goal with data-driven methods for energy systems is to make the best of both worlds, that is simplicity of rule-based approaches and the predictive capability of model-based strategies, but without the expense of first principles or gray-box model development.

In this chapter, we present a method called DR-Advisor (Demand Response-Advisor), which acts as a recommender system for the building's facilities manager and provides the power consumption prediction and control actions for meeting the required load curtailment and maximizing the economic reward. Using historical meter and weather data along with set point and schedule information, DR-Advisor builds a family of interpretable regression trees to learn non parametric data-driven models for predicting the power consumption of the building (Figure 9.2). DR-Advisor can be used for real-time DR baseline prediction, strategy evaluation, and control synthesis, without having to learn first principles based models of the building.

9.1 Introduction

In 2013, a report by the US National Climate Assessment provided evidence that the most recent decade was the nation's warmest on record [1], and experts predict that temperatures are only going to rise. In fact, for the past three years (2014-2016), every year has become the hottest year on record

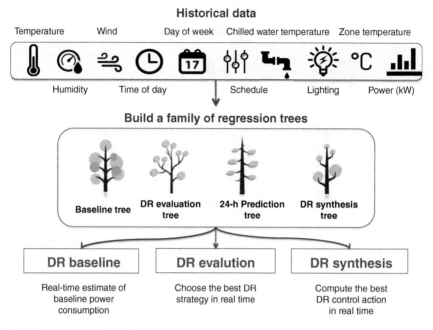

Figure 9.2 DR-Advisor architecture.

since the beginning of weather recording in 1880 [2]. Heat waves in summer and polar vortexes in winter are growing longer and pose increasing challenges to an already over-stressed electric grid.

Furthermore, with the increasing penetration of renewable generation, the electricity grid is also experiencing a shift from predictable and dispatchable electricity generation to variable and non-dispatchable generation. This adds another level of uncertainty and volatility to the electricity grid as the relative proportion of variable generation versus traditional dispatchable generation increases. The organized electricity markets across the world all use some variant of real-time price for wholesale electricity. The real-time electricity market at PJM, one of the world's largest independent system operators (ISO), is a spot market where electricity prices are calculated at 5-min intervals based on the grid operating conditions. The volatility due to the mismatch between electricity generation and supply further leads to volatility in the wholesale price of electricity. For example, the polar vortex triggered extreme weather events in the United States in January 2014, which caused many electricity customers to experience increased costs. Parts of the PJM electricity grid experienced a 86 fold increase in the price of electricity from $31/MW h to $2,680/MW h in a matter of a few minutes [3]. Similarly, the summer price spiked 32 fold from an average of $25/MW h to $800/MW h in July of 2015. Such events show

how unforeseen and uncontrollable circumstances can greatly affect electricity prices that impact ISOs, suppliers, and customers. Energy industry experts are now considering the concept that extreme weather, more renewables, and resultant electricity price volatility could become the new norm.

Across the United States, electric utilities and ISOs are devoting increasing attention and resources to demand response (DR) [4]. DR is considered as a reliable means of mitigating the uncertainty and volatility of renewable generation and extreme weather conditions and improving the grid's efficiency and reliability. The potential DR resource contribution from all U.S. DR programs is estimated to be nearly 72,000 megawatts (MW), or about 9.2% of U.S. peak demand [5], making DR the largest virtual generator in the US national grid. The annual revenue to end users from DR markets with PJM ISO alone is more than $700 million [6]. Global DR revenue is expected to reach nearly $40 billion from 2014 to 2020 [7].

The volatility in real-time electricity prices poses the biggest operational and financial risk for large-scale end users of electricity such as large commercial buildings, industries, and institutions [8], often referred to as *C/I/I* consumers. In order to shield themselves from the volatility and risk of high prices, such consumers must be more flexible in their electricity demand. Consequently, large *C/I/I* customers are increasingly looking to DR programs to help manage their electricity costs.

DR programs involve a voluntary response of a building to a price signal or a load curtailment request from the utility or the curtailment service provider (CSP). Upon successfully meeting the required curtailment level, the end users are financially rewarded but may also incur penalties for under performing and not meeting a required level of load curtailment.

9.1.1 Contributions

This work has the following data-driven contributions:

1) *DR Baseline Prediction:* We demonstrate the benefit of using regression trees based approaches for estimating the DR baseline power consumption. Using regression tree-based algorithms eliminates the cost of time and effort required to build and tune first principles based models of buildings for DR. DR-Advisor achieves a prediction accuracy of 92.8%—98.9% for baseline estimates of eight buildings on the Penn campus.

2) *DR Strategy Evaluation:* We present an approach for building auto-regressive trees and apply it for DR strategy evaluation. Our models takes into account the state of the building and weather forecasts to help choose the best DR strategy among several predetermined strategies.

3) *DR Control Synthesis:* We introduce a novel model-based control with regression trees (mbCRT) algorithm to enable control with regression trees and use it for real-time DR synthesis. Using the mbCRT algorithm,

we can optimally trade off thermal comfort inside the building against the amount of load curtailment. While regression trees are a popular choice for predictive models, this is the first time regression tree based algorithms have been used for controller synthesis with applications in DR. Our synthesis algorithm outperforms rule-based DR strategy by 17% while maintaining bounds on thermal comfort inside the building.

9.1.2 Experimental Validation and Evaluation

We evaluate the performance of DR-Advisor using a mix of real data from eight buildings on the campus of the University of Pennsylvania, Philadelphia, United States, and data sets from a virtual building test-bed for the Department of Energy's (DoE) large commercial reference building. We also compare the performance of DR-Advisor against other data-driven methods using a bench marking data set from the American Society of Heating, Refrigerating and Air-Conditioning Engineers (ASHRAE) Great Energy Predictor Shootout Challenge.

This chapter is organized as follows: In Section 9.2, a detailed survey of related work is presented. Section 9.3 describes the challenges with DR. In Section 9.4, we present how data-driven algorithms can be used for the problems associated with DR. Section 9.5 presents a new algorithm to perform control with regression trees for synthesizing DR strategies. Section 9.7 describes the MATLAB-based DR-Advisor toolbox. Section 9.8 presents a comprehensive case study with DR-Advisor using data from several real buildings. We conclude this chapter in Section 9.9 with a summary of our results and a discussion about future directions.

9.2 Related Work

There is a vast amount of literature [12–15] that addresses the problem DR under different pricing schemes. However, the majority of approaches so far have focused either on rule-based approaches for curtailment or on model-based approaches, such as the one described in [13], in which model predictive control is used for DR based on a gray-box model of a building. Auslander *et al.* [12] use a high-fidelity physics-based model of the building to solve a problem similar to the DR evaluation problem. Van Staden *et al.* [15] use model predictive control for closed-loop optimal control strategy for load shifting in a plant that is charged for electricity on both time-of-use and peak demand pricing. One of the seminal studies of application of model predictive control on real buildings for DR and energy-efficiency operation came from the OptiControl project [10]. After several years of work on using gray-box and white-box models for DR control design, the authors state that the usefulness of any model-based controller must be measured by not only its benefits and savings but also its incurred costs, such as the necessary hardware and software and the system's design, implementation, and maintenance effort. They further conclude that the biggest hurdle to mass adoption of intelligent building control is the cost and effort required to capture accurate dynamical models of the buildings. Since DR-Advisor only learns an aggregate building-level model and combined with the fact that weather forecasts from third-party vendors are expected to become cheaper, there is little to no additional sensor cost of implementing the DR-Advisor recommendation system in large buildings. The difficulties in identifying models for buildings are also highlighted in [16]. The authors observe that while model creation is mentioned only marginally in majority of the academic works dealing with model predictive control, these usually assume that the model of the system is either perfectly known or found in the literature, the task is much more complicated and time consuming in the case of a real application, and sometimes, it can be even more complex and involved than the controller design itself. There are ongoing efforts to make tuning and identifying white-box models of buildings more autonomous [11].

There is recent work that has explored the aspects of modeling, implementation, and implications of DR buildings [17–20]; however, their focus has mainly been on the residential sector. Dupont *et al.* [19] show that in general DR contributes to a lower cost, higher reliability, and lower emission level of power system operation and highlight the societal value of DR. In [20] the authors studied the short- and long-term effects of DR on residential electricity consumers through an elaborate empirical study. A reduced-order physics-based, gray-box modeling technique for simulating residential electric demand is presented in [18]. The ability to determine the correct response for large commercial buildings (from DR evaluation or DR synthesis) on a fast time scale (1–5 min) using purely data-driven methods makes both our approach and tool novel.

Several machine learning and data-driven approaches have also been utilized before for forecasting electricity load. We already compared the performance of DR-Advisor against several data-driven methods in Section 9.8.3. In [21], seven different machine learning algorithms were applied to a residential dataset with the objective of determining which techniques are most successful for predicting next hour residential building consumption. Kialashaki and Reisel [22] use ANN and regression models for modeling the energy demand of the residential sector in the United States. A forecasting method for cooling and electricity load demand is presented in [23], while a statistical analysis of the impact of weather on peak electricity demand using actual meteorological data is presented in [24]. In [25] a software architecture using parallel computing is presented to support data-driven DR optimization. The shortcoming of work in this area is twofold: First, the time scales at which the forecasts are generated range from 15–20 min to hourly forecast, which is too coarse grained for

DR events and for real-time price changes. Second, the focus of these methods is only on load forecasting but not on control synthesis, whereas the mbCRT algorithm presented in this chapter enables the use of regression trees for control synthesis for the very first time.

9.3 Problem Definition

The timeline of a DR event is shown in Figure 9.3. An *event notification* is issued by the utility/CSP at the notification time (~30 min). The time by which the reduction must be achieved is the *reduction deadline*. The main period during which the demand needs to be curtailed is the *sustained response period* (1–6 h). The end of the response period is when the main curtailment is released. The normal operation is gradually resumed during the *recovery period*. The DR event ends at the end of the recovery period.

The key to answering the question of what actions to take to achieve a significant DR curtailment upon receiving a notification lies in making accurate predictions about the power consumption response of the building. Specifically, it involves solving the three challenging problems of end-user DR, which are described next.

9.3.1 DR Baseline Prediction

The DR baseline is an estimate of the electricity that would have been consumed by a customer in the absence of a DR event (as shown in Figure 9.3) The measurement and verification of the DR baseline is the most critical component of any DR program since the amount of DR curtailment and any associated financial reward can only be determined with respect to the

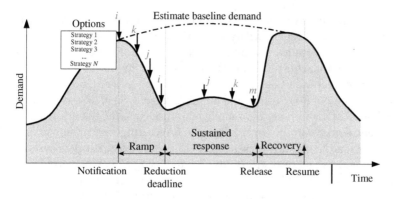

Figure 9.3 Example of a demand response timeline.

baseline estimate. The goal is to learn a predictive model $g()$ that relates the baseline power consumption estimate $\hat{Y}_{base}$ to the forecast of the weather conditions and building schedule for the duration of the DR event, that is, $\hat{Y}_{base} = g(\text{weather, schedule})$.

9.3.2 DR Strategy Evaluation

Most DR today is manual and conducted using fixed rules and predetermined curtailment strategies based on recommended guidelines, experience, and best practices. During a DR event, the building's facilities manager must choose a single strategy among several predetermined strategies to achieve the required power curtailment. Each strategy includes adjusting several control knobs such as temperature set points and lighting levels and temporarily switching off equipment and plug loads to different levels across different time intervals.

As only one strategy can be used at a time, the question then is, *how to choose the DR strategy from a predetermined set of strategies that leads to the largest load curtailment?*

Instead of predicting the baseline power consumption $\hat{Y}_{base}$, in this case we want the ability to predict the actual response of the building $\hat{Y}_{kW}$ due to any given strategy. For example, in Figure 9.3, there are N different strategies available to choose from. DR-Advisor predicts the power consumption of the building due to each strategy and chooses the DR strategy ($\in \{i, j, \ldots, k, \ldots N\}$), which leads to the largest load curtailment subject to the constraints on the thermal comfort and set points. The resulting strategy could be a combination of switching between the available set of strategies.

9.3.3 DR Strategy Synthesis

Instead of choosing a DR strategy from a predetermined set of strategies, a harder challenge is to synthesize new DR strategies and obtain optimal operating points for the different control variables. We can cast this problem as an optimization over the set of control variables, $\mathbb{X}_c$, such that

$$
\begin{aligned}
&\underset{\mathbb{X}_c}{\text{minimize}} \quad f(\hat{Y}_{kW}) \\
&\text{subject to} \quad \hat{Y}_{kW} = h(\mathbb{X}_c) \\
&\qquad\qquad\quad \mathbb{X}_c \in \mathbb{X}_{safe}
\end{aligned}
\tag{9.1}
$$

We want to minimize the predicted power response of the building $\hat{Y}_{kW}$, subject to a predictive model that relates the response to the control variables and subject to the constraints on the control variables.

Unlike rule-based DR, which does not account for building state and external factors, in DR synthesis the optimal control actions are derived based on the current state of the building, forecast of outside weather, and electricity prices.

9.4 Data-Driven Demand Response

Our goal is to find data-driven functional models that relate the value of the response variable, say, power consumption, $\hat{Y}_{kW}$, to the values of the predictor variables or features $[X_1, X_2, \ldots, X_m]$, which can include weather data, set-point information, and building schedules. When the data has lots of features, as is the case in large buildings, which interact in complicated, nonlinear ways, assembling a single global model, such as linear or polynomial regression, can be difficult and lead to poor response predictions. An approach to nonlinear regression is to partition the data space into smaller regions, where the inter- actions are more manageable. We then partition the partitions again – this is called recursive partitioning – until finally we get to chunks of the data space, which are so tame that we can fit simple models to them. Regression tree is an example of an algorithm that belongs to the class of recursive partition- ing algorithms. The seminal algorithm for learning regression trees is CART as described in [26].

Regression tree-based approaches are our choice of data-driven models for DR-Advisor. The primary reason for this modeling choice is that regression trees are highly interpretable by design. Interpretability is a fundamental desir- able quality in any predictive model. Complex predictive models like neural networks and support vector regression go through a long calculation routine and involve too many factors. It is not easy for a human engineer to judge if the operation/decision is correct or not or how it was generated in the first place. Building operators are used to operating a system with fixed logic and rules. They tend to prefer models that are more transparent, where it is clear exactly which factors were used to make a particular prediction. At each node in a regression tree a simple, if this – then that, human-readable plain text rule is applied to generate a prediction at the leaves, which anyone can easily understand and interpret. Making machine learning algorithms more inter- pretable is an active area of research [27], one that is essential for incorporating human-centric models in DR for energy systems.

9.4.1 Data Description

In order to build regression trees that can predict the power consumption of the building, we need to train on time-stamped historical data. As shown in Figure 9.2, the data that we use can be divided into three different categories:

1) *Weather Data*: It includes measurements of the outside dry-bulb and wet-bulb air temperature, relative humidity, wind characteristics, and solar irradiation at the building site.
2) *Schedule Data*: We create *proxy* variables that correlate with repeated pat- terns of electricity consumption, for example, due to occupancy or equip- ment schedules. *Day of Week* is a categorical predictor that takes values from

1 to 7 depending on the day of the week. This variable can capture any power consumption patterns that occur on specific days of the week. For instance, there could a big auditorium in an office building that is only used on certain days. Likewise, *Time of Day* is quite an important predictor of power consumption as it can adequately capture daily patterns of occupancy, lighting, and appliance use without directly measuring any one of them. Besides using proxy schedule predictors, actual building equipment schedules can also be used as training data for building the trees.

3) *Building Data*: The state of the building is required for DR strategy evaluation and synthesis. This includes (i) chilled water supply temperature, (ii) hot water supply temperature, (iii) zone air temperature, (iv) supply air temperature, and (v) lighting levels.

9.4.2 Data-Driven DR Baseline

DR-Advisor uses a mix of several algorithms to learn a reliable baseline prediction model. For each algorithm, we train the model on historical power consumption data and then validate the predictive capability of the model against a test dataset which the model has never seen before. In addition to building a single regression tree (SRT), we also learn cross-validated regression trees (CV-RTs), boosted regression trees (BRTs), and random forests (RFs). The ensemble methods like BRT and RF help in reducing any over-fitting over the training data. They achieve this by combining the predictions of several base estimators built with a given learning algorithm in order to improve generalizability and robustness over a single estimator. For a more comprehensive review of RFs, we refer the reader to [28]. A BRT model is an additive regression model in which individual terms are simple trees, fitted in a forward stage-wise manner [29].

9.4.3 Data-Driven DR Evaluation

The regression tree models for DR evaluation are similar to the models used for DR baseline estimation except for two key differences: First, instead of only using weather and proxy variables as the training features, in DR evaluation, we also train on set-point schedules and data from the building itself to capture the influence of the state of the building on its power consumption. Second, in order to predict the power consumption of the building for the entire length of the DR event, we use the notion of auto-regressive trees. An auto-regressive tree model is a regular regression tree except that the lagged values of the response variable are also predictor variables for the regression tree, that is, the tree structure is learned to approximate the following function:

$$\hat{Y}_{\text{kW}}(t) = f([X_1, X_2, \dots, X_m, Y_{\text{kW}}(t-1), \dots, Y_{\text{kW}}(t-\delta)]) \qquad (9.2)$$

where the predicted power consumption response $\hat{Y}_{\text{kW}}$ at time t depends on previous values of the response itself $[Y_{\text{kW}}(t-1), \dots, Y_{\text{kW}}(t-\delta)]$ and δ is the order of the auto-regression. This allows us to make finite horizon predictions of power consumption for the building. At the beginning of the DR event, we use the auto-regressive tree for predicting the response of the building due to each rule-based strategy and choose the one that performs the best over the predicted horizon. The prediction and strategy evaluation is recomputed periodically throughout the event.

9.5 DR Synthesis with Regression Trees

The data-driven methods described so far use the forecast of features to obtain building power consumption predictions for DR baseline and DR strategy evaluation. In this section, we extend the theory of regression trees to solve the DR synthesis problem described earlier in Section 9.3.3. This is our primary contribution.

Recall that the objective of learning a regression tree is to learn a model f for predicting the response Y with the values of the predictor variables or features $X_1, X_2, \ldots, X_m$; that is, $Y = f([X_1, X_2, \ldots, X_m])$. Given a forecast of the features $\hat{X}_1, \hat{X}_2, \ldots, \hat{X}_m$, we can predict the response $\hat{Y}$. Now consider the case where a subset, $\mathbb{X}_c \subset \mathbb{X}$, of the set of features/variables $\mathbb{X}$'s are manipulated variables, that is, we can change their values in order to drive the response ($\hat{Y}$) toward a certain value. In the case of buildings, the set of variables can be separated into disturbance (or non-manipulated) variables like outside air temperature, humidity, and wind, while the controllable (or manipulated) variables would be the temperature and lighting set points within the building. Our goal is to modify the regression trees and make them suitable for synthesizing the optimal values of the control variables in real time.

9.5.1 Model-Based Control with Regression Trees

The key idea in enabling control synthesis for regression trees is in the separation of features/variables into manipulated and non-manipulated features. Let $\mathbb{X}_c \subset \mathbb{X}$ denote the set of manipulated variables and $\mathbb{X}_d \subset \mathbb{X}$ denote the set of disturbance/non-manipulated variables such that $\mathbb{X}_c \cup \mathbb{X}_d \equiv \mathbb{X}$. Using this separation of variables, we build upon the idea of simple model-based regression trees [30, 31] to *mbCRT*.

Figure 9.4 shows an example of how manipulated and non-manipulated features can get distributed at different depths of model-based regression tree, which uses a linear regression function in the leaves of the tree:

$$\hat{Y}_{\text{Ri}} = \beta_{0,i} + \beta_i^T \mathbb{X} \tag{9.3}$$

where $\hat{Y}_{\text{Ri}}$ is the predicted response in region R_i of the tree using all the features $\mathbb{X}$. In such a tree the prediction can only be obtained if the values of all

Figure 9.4 Example of a regression tree with linear regression model in leaves. Not suitable for control due to the mixed order of the controllable X_c (solid blue) and uncontrollable X_d features.

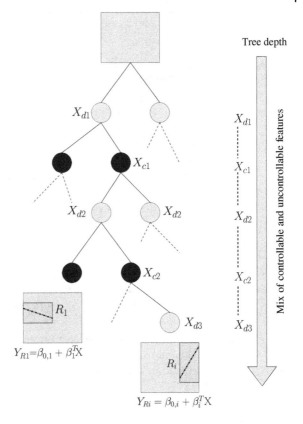

the features X's are known, including the values of the control variables X_{ci}'s. Since the manipulated and non-manipulated variables appear in a mixed order in the tree depth, we cannot use this tree for control synthesis. This is because the value of the control variables X_{ci}'s is unknown, one cannot navigate to any single region using the forecasts of disturbances alone.

The mbCRT algorithm avoids this problem using a simple but clever idea. We still partition the entire data space into regions using CART algorithm, but the top part of the regression tree is learned only on the non-manipulated features $\mathbb{X}_d$ or disturbances as opposed to all the features $\mathbb{X}$ (Figure 9.5) In every region at the leaves of the "disturbance" tree, a linear model is fit but only on the control variables $\mathbb{X}_c$:

$$Y_{Ri} = \beta_{0,i} + \beta_i^T \mathbb{X}_c \tag{9.4}$$

Separation of variables allows us to use the forecast of the disturbances $\hat{\mathbb{X}}_d$ to navigate to the appropriate region R_i and use the linear regression model ($Y_{Ri} = \beta_{0,i} + \beta_i^T \mathbb{X}_c$) with only the control/manipulated features in it as the valid prediction model for that time step.

Algorithm 1 mbCRT: model based control with regression trees

1: DESIGN TIME
2: **procedure** MODEL TRAINING
3:     *Separation of Variables*
4:     *Set* $\mathbb{X}_c \leftarrow$ non-manipulated features
5:     *Set* $\mathbb{X}_d \leftarrow$ manipulated features
6:     Build the power prediction tree T_{kW} with $\mathbb{X}_d$
7:     **for all** Regions R_i at the leaves of T_{kW} **do**
8:         Fit linear model $k\hat{W}_{Ri} = \beta_{0,i} + \beta_i^T \mathbb{X}_c$
9:         Build q temperature trees $T1, T2 \cdots Tq$ with $\mathbb{X}_d$
10:     **end for**
11:     **for all** Regions R_i at the leaves of Ti **do**
12:         Fit linear model $\hat{Ti} = \beta_{0,i} + \beta_i^T \mathbb{X}_c$
13:     **end for**
14: **end procedure**
15: RUN TIME
16: **procedure** CONTROL SYNTHESIS
17:     At time t obtain forecast $\hat{\mathbb{X}}_d(t+1)$ of disturbances $\hat{X}_{d1}(t+1)$, $\hat{X}_{d2}(t+1), \cdots$
18:     Using $\hat{\mathbb{X}}_d(t+1)$ determine the leaf and region R_{rt} for each tree.
19:     Obtain the linear model at the leaf of each tree.
20:     Solve optimization in Eq 9.5 for optimal control action $\mathbb{X}_c^*(t)$
21: **end procedure**

9.5.2 DR Synthesis Optimization

In the case of DR synthesis for buildings, the response variable is power consumption; the objective function can denote the financial reward of minimizing the power consumption during the DR event. However, the curtailment must not result in high levels of discomfort for the building occupants. In order to account for thermal comfort, in addition to learning the tree for power consumption forecast, we can also learn different trees to predict the temperature of different zones in the building. As shown in Figure 9.6 and Algorithm 1, at each time step during the DR event, a forecast of the non-manipulated variables is used by each tree to navigate to the appropriate leaf node. For the power forecast tree, the linear model at the leaf node relates the predicted power consumption of the building to the manipulated/control variables, that is, $k\hat{W} = \beta_{0,i} + \beta_i^T \mathbb{X}_c$.

$$Y = f(X_1, X_2, ..., X_m)$$

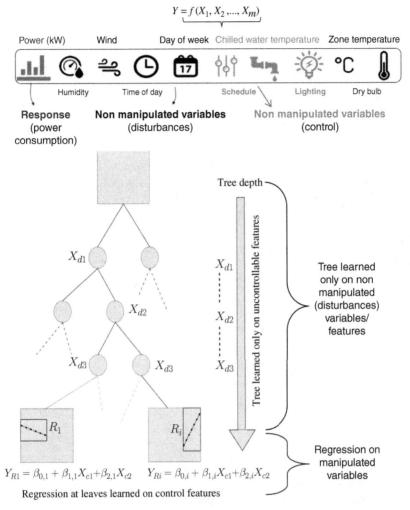

Figure 9.5 Example of a tree structure obtained using the mbCRT algorithm. The separation of variables allows using the linear model in the leaf to use only control variables.

Similarly, for each zone $1, 2, \ldots q$, a tree is built whose response variable is the zone temperature Ti. The linear model at the leaf node of each of the zone temperature tree relates the predicted zone temperature to the manipulated variables $\hat{T}i = \alpha_{0,i} + \beta_j^T \mathbb{X}_c$. Therefore, at every time step, based on the forecast of the non-manipulated variables, we obtain $q + 1$ linear models between the power consumption and q zone temperatures and the manipulated variables.

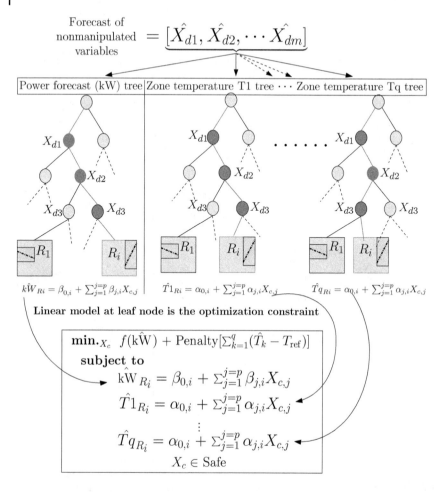

Figure 9.6 DR synthesis with thermal comfort constraints. Each tree is responsible for contributing one constraint to the demand response optimization.

We can then solve the following DR synthesis optimization problem to obtain the values of the manipulated variables $\mathbb{X}_c$:

$$\underset{\mathbb{X}_c}{\text{minimize}}\, f(k\hat{W}) + \text{Penalty}\left[\sum_{k=1}^{q}(\hat{T}_k - T_{\text{ref}})\right]$$

subject to

$$k\hat{W} = \beta_{0,i} + \beta_i^T \mathbb{X}_c$$

$$\hat{T}1 = \alpha_{0,1} + \beta_1^T \mathbb{X}_c \tag{9.5}$$

$$\cdots$$

$$\hat{T}d = \alpha_{0,q} + \beta_q^T \mathbb{X}_c$$

$$\mathbb{X}_c \in \mathbb{X}_{\text{safe}}$$

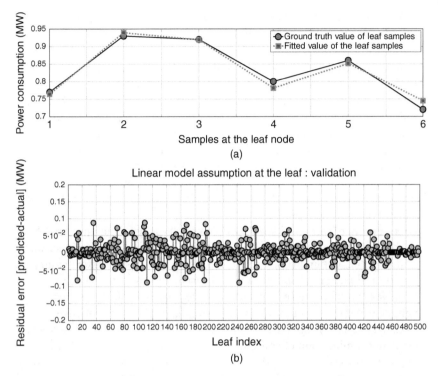

Figure 9.7 Linear model assumption at the leaves. (a) The comparison between fitted values and ground truth values of power consumption for one of the leaves in the power consumption prediction tree. (b) The residual error between fitted and actual power consumption values for all the leaf nodes of the tree.

The linear model between the response variable Y_{Ri} and the control features X_c is assumed for computational simplicity. Other models could also be used at the leaves as long as they adhere to the separation of variables principle. Figure 9.7 shows that the linear model assumption in the leaves of the tree is a valid assumption.

The intuition behind the mbCRT algorithm 1 is that at run time t, we use the forecast $\hat{X}_d(t+1)$ of the disturbance features to determine the region of the *uncontrollable* tree and hence the linear model to be used for the control. We then solve the simple linear program corresponding to that region to obtain the optimal values of the control variables.

The mbCRT algorithm is the first ever algorithm that allows the use of regression trees for control synthesis.

9.6 The Case for Using Regression Trees for Demand Response

Trees share the advantage of being a simple approach, much like other data-driven approaches. However, they offer several other advantages in addition to being interpretable, which make them suitable for solving the challenges of DR discussed in Section 9.4. We list some of these advantages here:

1) *Fast Computation Times*: Trees require very low computation power, both running time and storage requirements. With N observations and p predictors, trees require $pN \log N$ operations for an initial sort for each predictor and typically another $pN \log N$ operations for the split computations. If the splits occurred near the edges of the predictor ranges, this number can increase to $N^2 p$. Once the tree is built, the time to make predictions is extremely fast since obtaining a response prediction is simply a matter of traversing the tree with fixed rules at every node. For fast DR, where the price of electricity could change several times within a few minutes, trees can provide very fast predictions.

2) *Handle a Lot of Data and Variables*: Trees can easily handle the case where the data has lots of features that interact in complicated and nonlinear ways. In the context of buildings, a mix of weather data, schedule information, set points, and power consumption data is used, and the number of predictor variables can increase very quickly. A large number of features and a large volume of data can become too overwhelming for global models, like regression, to adequately explain. For trees, the predictor variables themselves can be of any combination of continuous, discrete, and categorical variables.

3) *Handle Missing Data*: Sometimes, data has missing predictor values in some or all of the predictor variables. This is especially true for buildings, where sensor data streams fail frequently due to faulty sensors or faulty communication links. One approach is to discard any observation with some missing values, but this could lead to serious depletion of the training set. Alternatively, the missing values could be imputed (filled in) with, say, the mean of that predictor over the non-missing observations. For tree-based models, there are two better approaches. The first is applicable to categorical predictors: we simply make a new category for "missing." From this we might discover that observations with missing values for some measurement behave differently than those with non-missing values. The second more general approach is the construction of surrogate variables. When considering a predictor for a split, we use only the observations for which that predictor is not missing. Having chosen the best (primary) predictor and split point, we build a list of surrogate predictors and split points. The first surrogate is the predictor and corresponding split point that best mimics the split of the training data achieved by the primary split. The second surrogate is the predictor and corresponding split point that does second best, and so on. When sending observations down the tree either in the training phase or during prediction, we use the surrogate splits in order, if the primary splitting predictor is missing.

4) *Robust to Outliers*: Tree-based models are generally not affected by outliers but regression-based models. The intuitive reasoning behind this is that during the construction of the tree, the region of the data with outliers is likely to be partitioned in a separate region.

9.7 DR-Advisor: Toolbox Design

The algorithms described thus far have been implemented into a MATLAB-based tool called DR-Advisor. We have also developed a graphical user interface (GUI) for the tool (Figure 9.8) to make it user-friendly.

Starting from just building power consumption and temperature data, the user can leverage all the features of DR-Advisor and use it to solve the different DR challenges. The toolbox design follows a simple and efficient workflow as shown in Figure 9.9. Each step in the workflow is associated with a specific tab in the GUI. The workflow is divided into the following steps:

1) *Upload Data*: When the toolbox loads, the Input tab of the GUI (Figure 9.8) is displayed. Here the user can upload and specify any sensor data from the building, which could be correlated with the power consumption. This includes historical power consumption data, any known building operation schedules, and zone temperature data. The tool is also equipped with the capability to pull historical weather data for a building location from the web. The user can also specify or upload electricity pricing or utility tariff data. Once the upload process is complete, the data structure for learning

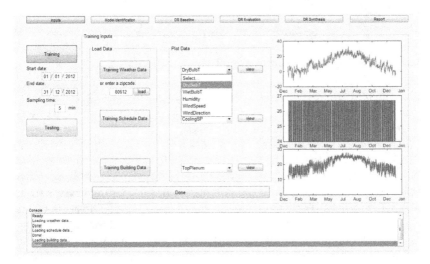

Figure 9.8 Screenshot of the DR-Advisor MATLAB-Based GUI.

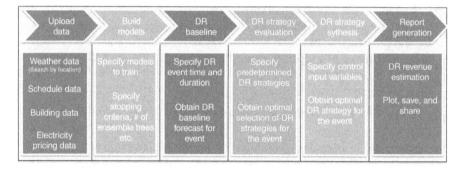

Figure 9.9 DR-Advisor workflow.

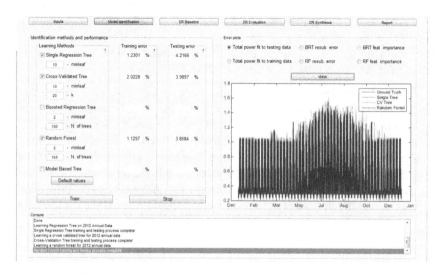

Figure 9.10 DR-Advisor model identification tab.

the different tree-based models is created internally. The GUI also has a small console that is used to display progress, completion, and alert messages for each action in the upload process.

2) *Build Models*: In the next step of the workflow, the user can specify which tree-based models should be learned as shown in Figure 9.10. These include SRT, CV-RT, RF, BRT, and M5 model-based regression tree (M5). For each method the user may change the parameters of the training process from the default values. These parameters include the stopping criteria in terms of MinLeaf or the number of trees in the ensemble and the value for the number of folds in cross-validation. After the models have been trained, the normalized root mean square value for each method on the test data is displayed. The user can also visualize and compare the predicted output versus

the ground truth data for the different methods. For the ensemble methods, the convergence of the resubstitution error and the feature importance plots can also be viewed.

3) *DR Baseline*: In the DR baseline tab, the user can specify the start and end times for a DR event, and DR-Advisor generates the baseline prediction for that duration using the methods selected during the model identification. The user can also specify if the baseline uses only weather data or it uses weather plus building schedule data.

4) *DR Strategy Evaluation*: In this step, the user first has to specify the pre-determined DR strategies, which need to be evaluated during the DR event. The user can choose different control variables and specify their value for the duration of the DR event. A group of such control variables constitute the DR strategy. The user may specify several DR strategies, in which different combinations of the control variables take different values. Upon executing the DR evaluation process, DR-Advisor is capable of selecting the best set of strategies for the DR event based on load curtailment.

5) *DR Strategy Synthesis*: For DR synthesis, two inputs are required: the user needs to provide an electricity/DR rate structure, and the user needs to specify which of the variables are the control variables. DR-Advisor then uses the mbCRT (Section 9.5.1) algorithm to synthesize and recommend a DR strategy for the DR event by assigning suitable values to the control inputs.

6) *Report Generation*: Facilities managers need to log reports of the building's operation during the DR event. DR-Advisor can generate summarized reports of how much load for curtailed and the estimated revenue earned from the DR event. The report also includes plots of what control actions were recommended by DR-Advisor and the comparison between the estimated baseline power consumption and the actual load during the event.

9.8 Case Study

DR-Advisor has been developed into a MATLAB toolbox available at http://mlab.seas.upenn.edu/dr-advisor/. In this section, we present a comprehensive case study to show how DR-Advisor can be used to address all the aforementioned DR challenges (Section 9.4), and we compare the performance of our tool with other data-driven methods.

9.8.1 Building Description

We use historical weather and power consumption data from eight buildings on the Penn campus (Figure 9.11). These buildings are a mix of scientific research labs, administrative buildings, office buildings with lecture halls, and

Figure 9.11 Eight different buildings on the Penn campus were modeled with DR-Advisor.

Table 9.1 Model validation with Penn data.

Building name	Total area (sq-ft)	Floors	Accuracy (%)
LRSM	92,507	6	94.52
College Hall	110,266	6	96.40
Annenberg Center	107,200	5	93.75
Clinical Research Building	204,211	8	98.91
David Rittenhouse Labs	243,484	6	97.91
Huntsman Hall	320,000	9	95.03
Vance Hall	106,506	7	92.83
Goddard Labs	44,127	10	95.07

biomedical research facilities. The total floor area of the eight buildings is over 1.2 million square feet spanned across. The size of each building is shown in Table 9.1.

We also use the DoE Commercial Reference Building (DoE CRB) simulated in EnergyPlus [32] as the virtual test-bed building. This is a large 12-story office building consisting of 73 zones with a total area of 500,000 sq ft. There are 2397 people in the building during peak occupancy. During peak load conditions, the building can consume up to 1.6 MW of power. For the simulation of the DoE CRB building, we use actual meteorological year data from Chicago for the years 2012 and 2013. On July 17, 2013, there was a DR event on the PJM ISO grid from 15:00 to 16:00 h. We simulated the DR event for the same interval for the virtual test-bed building.

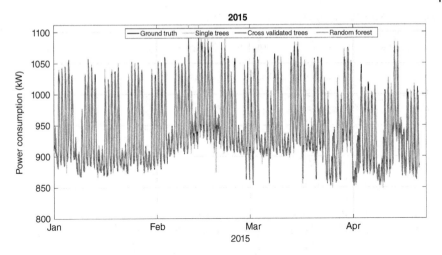

Figure 9.12 Model validation for the clinical research building at Penn.

9.8.2 Model Validation

For each of the Penn buildings, multiple regression trees were trained on weather and power consumption data from August 2013 to December 2014. Only the weather forecasts and proxy variables were used to train the models. We then use the DR-Advisor to predict the power consumption in the test period, that is, for several months in 2015. The predictions are obtained for each hour, making it equivalent to baseline power consumption estimate. The predictions on the test set are compared with the actual power consumption of the building during the test-set period. One such comparison for the clinical reference building is shown in Figure 9.12. The following algorithms were evaluated: SRT, k-fold cross-validated (CV) trees, BRTs, and RFs. Our chosen metric of prediction accuracy is the one minus the normalized root mean square error (NRMSE). NRMSE is the RMSE divided by the mean of the data. The accuracy of the model of all the eight buildings is summarized in Table 9.1. We notice that DR-Advisor performs quite well and the accuracy of the baseline model is between 92.8% and 98.9% for all the buildings.

9.8.3 Energy Prediction Benchmarking

We compare the performance of DR-Advisor with other data-driven method using a benchmarking dataset from the ASHRAE Great Energy Predictor Shootout Challenge [33]. The goal of the ASHRAE challenge was to explore and evaluate data-driven models that may not have such a strong physical basis yet that perform well at prediction. The competition attracted ~ 150 entrants, who attempted to predict the unseen power loads from weather and solar radiation data using a variety of approaches. In addition to predicting the

Table 9.2 ASHRAE energy prediction competition results

ASHRAE Team ID	WBE CV	CHW CV	HW CV	Average CV
9	10.36	13.02	15.24	12.87
DR-Advisor	**11.72**	**14.88**	**28.13**	**18.24**
6	11.78	12.97	30.63	18.46
3	12.79	12.78	30.98	18.85
2	11.89	13.69	31.65	19.08
7	13.81	13.63	30.57	19.34

hourly whole building electricity consumption (WBE, kW), both the hourly chilled water (CHW, millions of Btu/hr) and hot water consumption (HW, millions of Btu/hr) of the building were also required to be prediction outputs. Four months of training data with the following features was provided: 1) Outside temperature (°F) 2) Wind speed (mph) 3) Humidity ratio (water/dry air) 4) Solar flux (W/m^2) In addition to these training features, we added three proxy variables of our own: hour of day, IsWeekend, and IsHoliday to account for correlation of the building outputs with schedule.

Finally, we use different ensemble methods within DR-Advisor to learn models for predicting the three different building attributes. In the actual competition, the winners were selected based on the accuracy of all predictions as measured by the NRMSE, also referred to as the coefficient of variation statistic (CV). The smaller the value of CV, the better the prediction accuracy. ASHRAE released the results of the competition for the top 19 entries that they received. In Table 9.2, we list the performance of the top five winners of the competition and compare our results with them. It can be seen from Table 9.2 that the RF implementation in the DR-Advisor tool ranks second in terms of WBE CV and the overall average CV. The winner of the competition was an entry from Mackay [34], which used a particular form of Bayesian modeling using neural networks.

The result we obtain clearly demonstrates that the regression tree-based approach within DR-Advisor can generate predictive performance that is comparable with the ASHRAE competition winners. Furthermore, since regression trees are much more interpretable than neural networks, their use for building electricity prediction is, indeed, very promising.

9.8.4 DR Evaluation

We test the performance of three different rule-based strategies shown in Figure 9.13. Each strategy determines the set-point schedules for chiller water, zone temperature, and lighting during the DR event. These strategies were

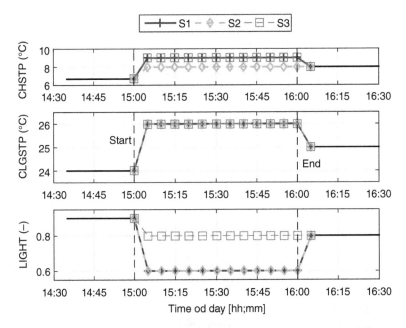

Figure 9.13 Rule-based strategies used in DR evaluation. CHSTP denotes chiller set point and CLGSTP denotes zone cooling temperature set point.

derived on the basis of automated DR guidelines provided by Siemens [35]. Chiller water set point is same in Strategy 1 (S1) and Strategy 3 (S3), higher than that in Strategy 2 (S2). Lighting level in S3 is higher than in S1 and S2.

We use auto-regressive trees (Section 9.4.3) with order $\delta = 6$ to predict the power consumption for the entire duration (1 h) at the start of DR event. In addition to learning the tree for power consumption, additional auto-regressive trees are also built for predicting the zone temperatures of the building. At every time step, first the zone temperatures are predicted using the trees for temperature prediction. Then the power tree uses this temperature forecast along with lagged power consumption values to predict the power consumption recursively until the end of the prediction horizon.

Figure 9.14 shows the power consumption prediction using the auto-regressive trees and the ground truth obtained by simulation of the DoE CRB virtual test-bed for each rule-based strategy. Based on the predicted response, in this case DR-Advisor chooses to deploy S1, since it leads to the least amount of electricity consumption. The predicted response due to the chosen strategy aligns well with the ground truth power consumption of the building due to the same strategy, showing that DR strategy evaluation prediction of DR-Advisor is reliable and can be used to choose the best rule-based strategy from a set of predetermined rule-based DR strategies.

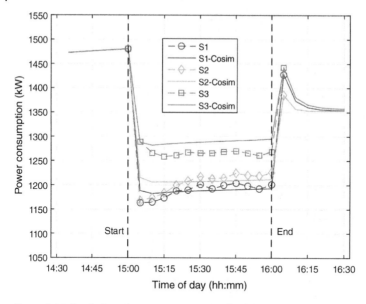

Figure 9.14 Prediction of power consumption for three strategies. DR evaluation shows that Strategy 1 (S1) leads to maximum power curtailment.

9.8.5 DR Synthesis

We now evaluate the performance of the mbCRT (Section 9.5.1) algorithm for real-time DR synthesis. Similar to DR evaluation, the regression tree is trained on weather, proxy features, set-point schedules, and data from the building. We first partition the set of features into manipulated features (or control inputs) and non-manipulated features (or disturbances). There are three control inputs to the system: the chilled water set point, zone air temperature set point, and lighting levels. At design time, the model-based tree built (Algorithm 1) has 369 leaves, and each of them has a linear regression model fitted over the control inputs with the response variable being the power consumption of the building.

In addition to learning the power consumption prediction tree, 19 additional model-based trees were also built for predicting the different zone temperatures inside the building. When the DR event commences, at every time step (every 5 min), DR-Advisor uses the mbCRT algorithm to determine which leaf and therefore which linear regression model will be used for that time step to solve the linear program (Eq. 9.5) and determine the optimal values of the control inputs to meet a sustained response while maintaining thermal comfort.

Figure 9.15 shows the power consumption profile of the building using DR-Advisor for the DR event. We can see that using the mbCRT algorithm, we are able to achieve a sustained curtailed response of 380 kW over a period of 1 h as compared with the baseline power consumption estimate. Also shown

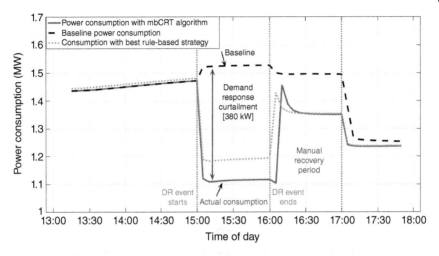

Figure 9.15 DR synthesis using the mbCRT algorithm for July 17, 2013. A curtailment of 380 kW is sustained during the DR event period.

in the figure is the comparison between the best rule-based fixed strategy, which leads to the most curtailment in Section 9.8.4. In this case the DR strategy synthesis outperforms the best rule-based strategy (from Section 9.8.4, Figure 9.14) by achieving a 17% higher curtailment while maintaining thermal comfort. The rule-based strategy does not directly account for any effect on thermal comfort. The DR strategy synthesized by DR-Advisor is shown in Figure 9.16. We can see in Figure 9.17 how the mbCRT algorithm is able to maintain the zone temperatures inside the building within the specified comfort bounds. These results demonstrate the benefit of synthesizing optimal

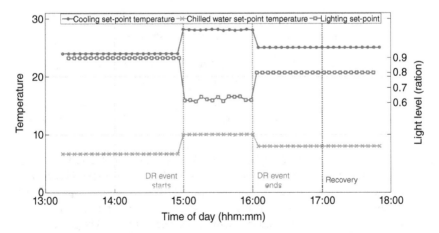

Figure 9.16 Optimal DR strategy as determined by the mbCRT algorithm.

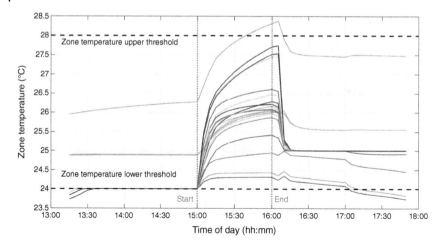

Figure 9.17 The mbCRT algorithm maintains the zone temperatures within the specified comfort bounds during the DR event.

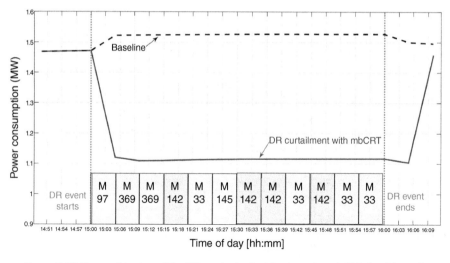

Figure 9.18 Zoomed-in view of the DR synthesis showing how the mbCRT algorithm selects the appropriate linear model for each time step based on the forecast of the disturbances.

DR strategies as opposed to relying on fixed rules and predetermined strategies, which do not account for any guarantees on thermal comfort. Figure 9.18 shows a close of view of the curtailed response. The leaf node that is being used for the power consumption constraint at every time step is also shown in the plot. We can see that the model switches several times during the event, based on the forecast of disturbances.

These results show the effectiveness of the mbCRT algorithm to synthesize DR actions in real time while utilizing a simple data-driven tree-based model.

9.8.5.1 Revenue from Demand Response

We use Con Edison utility company's commercial DR tariff structure [36] to estimate the financial reward obtained due to the curtailment achieved by the DR-Advisor for our Chicago-based DoE CRB. The utility provides a $25/kW per month as a reservation incentive to participate in the real-time DR program for summer. In addition to that, a payment of $1 per kW h of energy curtailed is also paid. For our test-bed, the peak load curtailed is 380 kW. If we consider ~ 5 such events per month for 4 months, this amounts to a revenue of $\sim$ $45,600 for participating in DR, which is 37.9% of the energy bill of the building for the same duration ($120,317). This is a significant amount, especially since using DR-Advisor does not require an investment in building complex modeling or installing sensor retrofits to a building.

9.9 Final Thoughts

We present a data-driven approach for modeling and control for DR of large-scale energy systems, which are inherently messy to model using first principles-based methods. We show how regression tree-based methods are well suited to address the challenges associated with DR for large *C/I/I* consumers while being simple and interpretable. We have incorporated all our methods into the DR-Advisor tool – http://mlab.seas.upenn.edu/dr-advisor/.

DR-Advisor achieves a prediction accuracy of **92.8 98.9%** for eight buildings on the University of Pennsylvania campus. We compare the performance of DR-Advisor on a benchmarking dataset from ASHRAE's energy predictor challenge and rank second among the winners of that competition. We show how DR-Advisor can select the best rule-based DR strategy, which leads to the most amount of curtailment, from a set of several rule-based strategies. We presented an mbCRT algorithm that enables control synthesis using regression tree-based structures for the first time. Using the mbCRT algorithm, DR-Advisor can achieve a sustained curtailment of **380 kW** during a DR event. Using a real tariff structure, we estimate a revenue of $\sim$ **$45,600** for the DoE reference building over one summer, which is **37.9%** of the summer energy bill for the building. The mbCRT algorithm outperforms even the best rule-based strategy by **17%**. DR-Advisor bypasses cost- and time-prohibitive process of building high-fidelity models of buildings that use gray-box and white-box modeling approaches while still being suitable for control design. These advantages combined with the fact that the tree-based methods achieve high prediction accuracy make DR-Advisor an alluring tool for evaluating and planning DR curtailment responses for large-scale energy systems.

Questions

1 Are data-driven models suitable for predictive modeling for large commercial buildings?

2 Is there an alternative to using traditional MPC based algorithms for building operation and demand response?

3 Upon receiving a demand response request to curtail, how does the facilities manager decide what actions to take to temporarily shed aggregate electricity demand?

4 What is separation of variables and the model-based control with regression trees (mbCRT) algorithm?

5 Are their any 'human/operator-in-the-loop' building automation systems?

References

1 Melillo, J.M., Richmond, T., and Yohe, G.W. (2014) *Climate Change Impacts in the United States: The Third National Climate Assessment*, U.S. Global Change Research Program, 841 pp.

2 NOAA National Centers for Environmental Information (2015) State of the Climate: Global Analysis for August 2015, http://www.ncdc.noaa.gov/sotc/global/201508 (retrieved 15 October 2015).

3 Michael, J. and Kormos, P.I. (2014) *PJM Response to Consumer Reports on 2014 Winter Pricing*.

4 Goldman, C. (2010) *Coordination of Energy Efficiency and Demand Response*, Lawrence Berkeley National Laboratory.

5 Federal Energy Regulatory Commission et al. (2008) *Assessment of Demand Response and Advanced Metering*.

6 PJM Interconnection (2014) *Demand Response Operations Markets Activity Report*.

7 Navigant Research (2015) *Demand Response for Commercial & Industrial Markets. Market Players and Dynamics, Key Technologies, Competitive Overview, and Global Market Forecasts*.

8 Mulhall, R.A. and Bryson, J.R. (2014) Energy price risk and the sustainability of demand side supply chains. *Applied Energy*, **123**, 327–334.

9 Crawley, D.B., Lawrie, L.K. et al. (2001) EnergyPlus: creating a new-generation building energy simulation program. *Energy and Buildings*, **33** (4), 319–331.

10 Sturzenegger, D., Gyalistras, D., Morari, M., and Smith, R.S. (2016) Model predictive climate control of a Swiss office building: implementation, results, and cost-benefit analysis. *IEEE Transactions on Control Systems Technology*, **24** (1), 1–12.

11 New, J.R., Sanyal, J., Bhandari, M., and Shrestha, S. (2012) *Autotune e+ Building Energy Models*. Proceedings of the 5th National SimBuild of IBPSA-USA.

12 Auslander, D., Caramagno, D., Culler, D., Jones, T., Krioukov, A., Sankur, M., Taneja, J., Trager, J., Kiliccote, S., Yin, R. et al. *Deep Demand Response: The Case Study of the CITRIS Building at the University of California-Berkeley*.

13 Oldewurtel, F., Sturzenegger, D., Andersson, G., Morari, M., and Smith, R.S. (2013) *Towards a Standardized Building Assessment for Demand Response*. Decision and Control (CDC), 2013 IEEE 52nd Annual Conference on, IEEE, pp. 7083–7088.

14 Xu, P., Haves, P., Piette, M.A., and Braun, J. (2004) *Peak Demand Reduction from Pre-Cooling with Zone Temperature Reset in An Office Building*, Lawrence Berkeley National Laboratory.

15 Van Staden, A.J., Zhang, J., and Xia, X. (2011) A model predictive control strategy for load shifting in a water pumping scheme with maximum demand charges. *Applied Energy*, **88** (12), 4785–4794.

16 Žáčeková, E., Váňa, Z., and Cigler, J. (2014) Towards the real-life implementation of MPC for an office building: identification issues. *Applied Energy*, **135**, 53–62.

17 Kialashaki, A. and Reisel, J.R. (2013) Modeling of the energy demand of the residential sector in the united states using regression models and artificial neural networks. *Applied Energy*, **108**, 271–280, doi: 10.1016/j.apenergy.2013.03.034.

18 Muratori, M., Roberts, M.C., Sioshansi, R., Marano, V., and Rizzoni, G. (2013) A highly resolved modeling technique to simulate residential power demand. *Applied Energy*, **107**, 465–473, doi: 10.1016/j.apenergy.2013.02.057.

19 Dupont, B., Dietrich, K., Jonghe, C.D., Ramos, A., and Belmans, R. (2014) Impact of residential demand response on power system operation: a Belgian case study. *Applied Energy*, **122**, 1–10, doi: 10.1016/j.apenergy.2014.02.022.

20 Bartusch, C. and Alvehag, K. (2014) Further exploring the potential of residential demand response programs in electricity distribution. *Applied Energy*, **125**, 39–59, doi: 10.1016/j.apenergy.2014.03.054.

21 Edwards, R.E., New, J., and Parker, L.E. (2012) Predicting future hourly residential electrical consumption: a machine learning case study. *Energy and Buildings*, **49**, 591–603.

22 Kialashaki, A. and Reisel, J.R. (2013) Modeling of the energy demand of the residential sector in the united states using regression models and artificial neural networks. *Applied Energy*, **108**, 271–280.

23 Vaghefi, A., Jafari, M., Bisse, E., Lu, Y., and Brouwer, J. (2014) Modeling and forecasting of cooling and electricity load demand. *Applied Energy*, **136**, 186–196.

24 Hong, T., Chang, W.K., and Lin, H.W. (2013) A fresh look at weather impact on peak electricity demand and energy use of buildings using 30-year actual weather data. *Applied Energy*, **111**, 333–350.

25 Yin, W., Simmhan, Y., and Prasanna, V.K. (2012) *Scalable Regression Tree Learning on Hadoop Using Openplanet*. Proceedings of 3rd International Workshop on MapReduce and its Applications Date, ACM, pp. 57–64.

26 Breiman, L., Friedman, J., Stone, C.J., and Olshen, R.A. (1984) *Classification and Regression Trees*, CRC Press.

27 Giraud-Carrier, C. (1998) *Beyond Predictive Accuracy: What?* Proceedings of the ECML-98 Workshop on Upgrading Learning to Meta-Level: Model Selection and Data Transformation, pp. 78–85.

28 Breiman, L. (2001) Random forests. *Machine Learning*, **45** (1), 5–32.

29 Elith, J., Leathwick, J.R., and Hastie, T. (2008) A working guide to boosted regression trees. *Journal of Animal Ecology*, **77** (4), 802–813.

30 Quinlan, J.R. et al. (1992) *Learning with Continuous Classes*. 5th Australian Joint Conference on Artificial Intelligence, vol. 92, Singapore, pp. 343–348.

31 Friedman, J.H. (1991) Multivariate adaptive regression splines. *The Annals of Statistics*, **19** (1), 1–67.

32 Deru, M., Field, K., Studer, D. et al. (2010) *U.S. Department of Energy Commercial Reference Building Models of the National Building Stock*.

33 Kreider, J.F. and Haberl, J.S. (1994) Predicting Hourly Building Energy Use: The Great Energy Predictor Shootout–Overview and Discussion of Results. Tech. Rep. American Society of Heating, Refrigerating and Air-Conditioning Engineers, Inc., Atlanta, GA.

34 MacKay, D.J. et al. (1994) Bayesian nonlinear modeling for the prediction competition. *ASHRAE Transactions*, **100** (2), 1053–1062.

35 Siemens (2011) *Automated Demand Response Using OpenADR: Application Guide*.

36 Con Edison *Demand Response Programs Details*, https://www.coned.com/en/save-money/rebates-incentives-tax-credits/demand-management-incentives.

10

Bringing Named Data Networks into Smart Cities

Syed Hassan Ahmed[1], Safdar Hussain Bouk[1], Dongkyun Kim[1], and Mahasweta Sarkar[2]

[1] School of Computer Science and Engineering, Kyungpook National University, Daegu, Korea
[2] Electrical & Computer Engineering Department, San Diego State University, San Diego, CA, USA

CHAPTER MENU
Introduction, 275
Future Internet Architectures, 278
Named Data Networking (NDN), 282
NDN-based Application Scenarios for Smart Cities, 285
Future Aspects of NDN in Smart Cities, 297
Conclusion, 303

Objectives

- To provide an overview of the Internet legacy and its role in envisioning a smart city.
- To become aware of technology trends from traditional Internet to Future Internet.
- To become familiar with Named Data Networking.
- To discuss a set of Future Internet applications that can be considered as an integral part of any city to be referred as smart city.

10.1 Introduction

According to recent studies, it has been stated that by the year 2050, we will be having around 9 billion humans on the Earth, and almost 70% of them would be living in urban areas. This steady and consistent growth in the world

Smart Cities: Foundations, Principles and Applications, First Edition.
Edited by Houbing Song, Ravi Srinivasan, Tamim Sookoor, and Sabina Jeschke.
© 2017 John Wiley & Sons, Inc. Published 2017 by John Wiley & Sons, Inc.

population is alarming and indicates that we need to redesign our cities in terms of constructions, roads, medical assistance, and, on top of everything, our information and communication technology (ICT). For the past few decades, we have witnessed that the ICT services have been improving the lifestyle of laymen, for example, various medical advancements, communication advancements, and traveling facilities. Later on, we have seen rapid agricultural advancements. These are the basic needs of any human on Earth. Here it is worth mentioning that the communication technologies have been actively and passively serving the humanity from the day they came into being. For instance, we take examples of patient monitoring systems, surgery equipment, in-body sensors, and body area networks (BANETs). Also, we have seen noticeable work done by wireless sensor networks (WSNs), and the applications of the WSNs are unlimited. Similarly, the recent research in vehicular *ad hoc* networks (VANETs) has enabled us to have at least three types of services including safety, traffic control, and user applications. All these services are providing secure driving experience, less traffic congestion, and various entertainment applications, respectively, on roads. In short, the wired and wireless communication technologies have been improving our lifestyles, and today we are connected to the world, regardless of location, and altitude and depth on/off the Earth.

However, in the near future, we can assume that the current technologies may be insufficient to entertain the massive increasing demands of the users and consumers seeking the ICT services mentioned above or to be expected in the near future. Recently, the researchers and industry personnel have identified that "smart cities" is the potential solution. The smart cities concept is basically the emergence of the advancements that have been made in various fields. From the recent literature, it is hard to find the exact definition of this term. Furthermore, we have seen that various technologies are making our lives smarter than we have ever imagined before. Therefore, we believe that if we merge those technologies and enable them to communicate with each other for the benefits of citizens and government-oriented departments, we expect that future cities should be able to provide smart services to the humans living in that particular city. For instance, we want purified drinking water, and the level of pollution can be detected by deploying underwater sensor networks for pollution monitoring, and actuators can be designed similar to the filters in our houses. So whenever pollution is detected, the filtering process may start, and, as a result, we will not require individual filters to be installed in our house. Similarly, autonomous driving will be the next revolution in field automation. For example, autonomous ambulances and live monitoring of the patient in the absence of the doctor could be beneficial to the laymen as well.

Smart cities are expected to provide high quality living standards to all their citizens with main focus on a few key aspects such as smarter environment, smarter mobility, smarter connectivity, and smarter governance. When we say "smart," we mean that security, privacy, robustness of the system, and availability of the services should be persistent. The main goal is to build a business-competitive and attractive environment by leveraging on the human capital of the city. For that reason, we state that the ICT builds a foundation of any city to be smart. Since, ICT can provide intelligent transportation systems (ITS), environmental monitoring, efficient utilization of energy sources, healthcare, public security, and e-commerce, today we are able to connect various devices and services using cloud computing (CC), Internet of Things (IoT), the Worldwide Interoperability for Microwave Access (WiMAX), and the Wi-Fi. Moreover, if we look into the revolutions in cellular networks, we have witnessed obvious advancements in the form of third generation (3G), Long-Term Evolution (LTE), and LTE Advanced (LTE-A). Beyond a shadow of a doubt, all these services have been enabling various service providers to make commercial products and providing consumers with remarkable on-move connectivity.

Nevertheless, from a technical point of view, a radical change has also been observed in the use of digital resources. For example, nowadays, users are interested in sharing the contents between connected devices rather than just being interconnected w.r.t remote devices. Also, the main purpose of today's connected devices is to share the Data. However, the current IP-based communications have increased latency in the content retrieval process due to its host-centric nature of communication. Although we have a variety of security protocols in almost every networking paradigm, we still receive spam. One reason is that while keeping our medium and host secure, we neglect the content integrity, and as a result, despite the fact that we have good speeds of downloading, sometimes we get stuck in the retrieval process. These features will affect connection-oriented services that are expected to be a part of future smart cities. At the time of writing of this chapter, we argue that a new and changed perspective of ICT is required to make our future cities more robust, reliably connected, and support mobile applications. For instance, we have few recent works that focus on putting into practice content-centric approaches (i.e., Information-Centric Networks (ICNs)). To date, Palo Alto Research Center (PARC) proposed a promising Future Internet architecture named as CCN. The communication in CCN has been shifted from host centric to Data centric. In CCN, we have names for the contents instead of end-to-end devices. In later stages, the researchers from the University of California, Los Angeles (UCLA) in collaboration with Van Jacobson (the founder of CCN) proposed NDN. NDN further solves different issues faced by the previous ICN architectures and is considered the latest and reliable architecture with active project crew that provides up-to-date debugs and documentations regarding the NDN implementation.

In this chapter, we briefly describe the recently proposed Future Internet architectures followed by insight and discussion on NDN. In addition, we also describe the possible applicability of NDN in smart cities and its potentials. Before the conclusions, we also provide variant application scenarios for NDN-enabled smart cities and future research road map for researchers.

10.2 Future Internet Architectures

Recently, the researchers have put some efforts and have proposed preliminary architectures for Future Internet. In this subsection, we will put some light on some known ones as follows:

10.2.1 Data-Oriented Network Architecture (DONA)

The Data-Oriented Network Architecture (DONA), devised by UC Berkeley, is claimed to be one of the architectures that provide complete ICN solution [1]. In DONA, the type of naming is replaced from URL-based hierarchical naming to flat naming or hash-based naming. The hash-based naming in DONA helps consumer to verify the origin of the content as well as its integrity. The naming in DONA is based on the mapping between the content and principal's or publisher's name. The flat namespace of NCOs is in the form P:L, where P uniquely specifies the principal field globally and is the publisher's public key's cryptographic hash and L specifies the unique label of NCO. The naming granularity is the function of the principal, for instance, the principal can name the entire stream of a video or an individual chunk within that video. The names are globally unique, flat, self-sufficient, and location and application independent. As the names are hash based, so the consumer requires external mechanisms (e.g., search engines) to provide name against the request for that content or in simple words the mapping between the human-readable names and hash-based names, because it is difficult for a human to remember hash-based names.

In DONA, there are specialized servers called resolution handlers (RHs) that provide the name resolution. At least one logical RH exists at each autonomous system (AS) [2]. These RHs are divided into hierarchies for resolving names from top to bottom such as current inter-domain routing, as it can be seen in Figure 10.1. To advertise an NCO to the network, a principal (publisher) first contacts its local RH and forwards a REGISTER message along with the name of content. The local RH creates mapping to the principal. The local RH then broadcasts this registration information to all the neighboring and parent RHs, asking them to store a mapping between the address of RH that forwarded that information and NCO's name. So these registrations are followed up to the top hierarchy, that is, tier-1 and whole network become aware of that mapping. To find an item in the network, a subscriber forwards a FIND message to its local RH, which passes that message to upper tier until a mapping for the request is found. Pointers are followed in request to reach the publisher. When a publisher

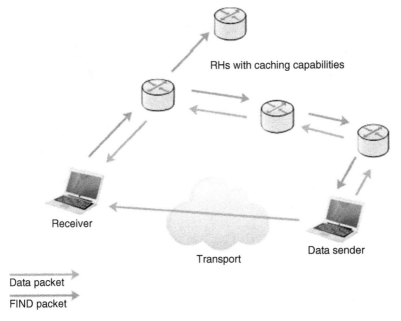

RHs with caching capabilities

Receiver

Transport

Data sender

Data packet

FIND packet

Figure 10.1 Data-oriented network architecture (DONA): overview.

is reached, the reverse path order is used by the publisher toward the subscriber to deliver the content.

The subscriber mobility in DONA is handled simply by just resending the requests for content from a new location to new RH [3]. DONA follows out-of-the-band delivery of content that requires to reestablish a session with either the same publisher or a new publisher once the consumer moves from one RH to another RH. So the process is complicated. To handle publisher's mobility, DONA supports early binding for subscriber mobility in which the binding between the locator and publisher's identifier is created when publisher registers itself. So when a publisher moves, it simply registers itself with a new RH. It is only simple to serve requests under new RH, but requests under previous RH have to be resent or require a mechanism such as Mobile IP. This is one of the limitations of early binding in DONA [4].

10.2.2 Network of Information (NetInf)

Network of Information (NetInf) is a continuation of the EU Framework 7 Program-funded projects, Architecture and Design for the Future Internet (4WARD), and Scalable and Adaptive Internet Solutions (SAIL) [5]. The area of focus of SAIL project contains issues related to network transport, while 4WARD is related to naming and searching content. The NetInf names are flattish; they can reserve a hierarchical naming scheme and can also obtain the

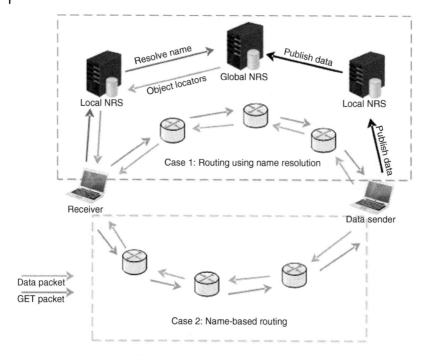

Figure 10.2 Network of Information (NetInf) overview.

hash-based form [6]. The names are in the form ns://A/L hierarchical fashion, where A helps in global name resolution and L helps in local name resolution. Each of these parts can be normal string in the form of URI or it can be a hash.

Name resolution and Data routing in NetInf can be performed in either coupled mode or decoupled mode, as illustrated in Figure 10.2. In decoupled mode, an entity called Name Resolution Systems (NRS) holds the mappings between the NCO names and locators to reach the NCOs. The NRS works in DHT fashion, and each of these NRS handles L part resolution if local and A part resolution if global. To advertise the content, a publisher sends the Publish message to its local NRS along with the locator. The local NRS makes a Bloom filter of the L part that belongs to same authority A and propagates the Publish message to upper-tier global NRS. Global NRS saves the mappings corresponded by local NRS. The subscriber can get the NCO by sending a Get message to its local NRS, which in return negotiates with global NRS for the locator of the NCO. The global NRS returns the locator information to local NRS, which delivers it to the subscriber. The subscriber queries the publisher for the NCO, and the publisher acknowledges the subscriber with the requested NCO [7].

In coupled mode, a routing protocol advertises the name of NCO to CRs. To get an NCO, a subscriber sends request message to its local CR, which propagates it to other CRs hop by hop. When there exists a match for the requested NCO, reverse path is utilized to deliver that NCO to the subscriber [8]. The request messages are aggregated at CRs to track back the subscriber for delivery of the content.

The subscriber mobility in NetInf is handled easily by resending the requests for an NCO under new local NRS. Publisher mobility is a bit difficult because when the publisher moves from one local NRS to a new one, the NR service needs to be updated right from the global NR to local NR [9].

10.2.3 Publish Subscribe Internet Technology (PURSUIT)

The Publish Subscribe Internet Technology (PURSUIT) is an extension of Publish Subscribe Internet Routing Paradigm (PSIRP) [10]. Both of these architectures provide a clean-slate approach to replace the current IP architecture. Both PURSUIT and PSIRP are funded by EU Framework 7 Program. In PURSUIT, the NCO is named by a flat naming scheme through concatenating two unique identifiers called scopeID and rendezvousID. The rendezvousID specifies the actual identity of the NCO, while the scopeID specifies the group to which an NCO belongs. Each NCO must belong to at least one scope, while an NCO can belong to multiple scopes with different rendezvous IDs [11]. The scopeID helps to define access boundaries of an NCO, for instance, a publisher can publish a photograph under "family" scope and "friends" scope, having distinct rights for each scope.

In PURSUIT, different rendezvous nodes (RNs) are implemented that combinely form Rendezvous Network (RENE) for name resolution in hierarchical distributed hash table (DHT) manner. As depicted in Figure 10.3, to advertise content to network, a publisher sends Publish message to its local RN, which disseminates it to upper-tier RENE. When a subscriber puts a request against an NCO, it sends a Subscribe message to its local RN that forwards it to upper-tier RENE, and it forwards it to RENE, which holds the binding. A route is constructed from publisher to subscriber when RN asks Topology Manager (TM) node to do it. TM node sends a message to publisher called Start Publish, which contains route. Publisher establishes route by exploiting this information and uses Forwarding Nodes (FNs) to forward NCO to the subscriber.

The subscriber mobility in PURSUIT is easy to handle [12]. Subscriber just moves to another network and resends the requests for content under new RN. Handling publisher's mobility is quite difficult [13]. Because when the publisher moves, then the topology information has to be resubmitted and changed from lower to upper-tier RENE.

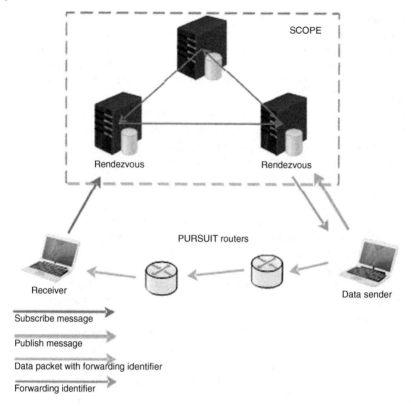

Figure 10.3 PURSUIT overview.

10.3 Named Data Networking (NDN)

The NDN is providing its part to further enhance CCN project and is funded by the US Future Internet Architecture program. The notion of NDN is to transform the existing shape of Internet protocol stack by replacing the narrow thin waist with named Data, and under this waist different technologies for connectivity can be used, such as IP. A layer called strategy layer provides the mediation between the underlying technologies and the named Data layer.

The naming scheme used in NDN is human friendly, hierarchical, and resembles URLs; an example of such name can be /ait.asia/home/index.html. It is not obvious that NDN names must be human-readable or there must exist DNS or IP address in name; rather it can be a hash of a string. In NDN, the names of the NCOs are matched on the basis of longest prefix match, for example, /ait.asia/home/index.html. In such case, the name can be matched with the start and then further can be explored for any piece of information, for example, /ait.asia/home/index.html/v1/s1, which means segment-1 from version-1 of that file. After that the subscriber can apply a direct function by asking the next segment, that is, /ais.asia/home/index.html/v1/s2 or can go with the next sibling under that hierarchy. So it depends upon the exploration of the prefix matching with longest prefix match mechanism.

Another good thing about NDN naming is that the subscriber can ask for a content that has not yet generated. For this, a publisher advertises to network the prefix that it can provide a content with such prefix so anyone interested can ask for it. This helps applications where the content is generated dynamically, and its complete name prefix is not known in advance, such as dynamic or live video generation.

In NDN, there are two types of messages used for requesting and routing information, one is Interest message that a subscriber issues for a certain content and in return publisher provides Data to that subscriber. There are two Data structures (routing tables): FIB, PIT and a CS maintained at each content router (CR) in a hop-by-hop fashion for sending interests and replying with content. The FIB maps the interests received to forward them to interface(s) for publishers or other CRs having content in their caches. The PIT keeps track of the interests for which content is expected to be arrived. Lastly, the CS acts as a cache for the CR to keep the replica of the content that went through that CR.

The order of importance among FIB, PIT, and CS is that CS reserves the highest priority and then PIT and lastly comes FIB in the priority list. When an interest is received for a content, initially CR checks its PIT whether the interest message for the same content has been received earlier; if so the interface over which interest was received is kept as a record in an entry in PIT keeping back track of the interface so that the multicast delivery can be performed for multiple subscribers asking for the same content. If not found in PIT, the second step is to check it into the CS; if found there then it is returned on the interface at which the interest was received, and interest is deleted. If content is not in CS and there is no entry in PIT, then CR makes an entry of it in PIT and forwards it to other CRs by creating an entry in FIB for this interest as depicted in Figure 10.4b. However, CCN has a different philosophy of treating incoming Interest packet, and it can be seen in Figure 10.4a.

Handling subscriber mobility in NDN is quite similar to that of other ICN approaches in which the subscriber reissues interest messages from new location, but the content against old interest messages are delivered to old CR. The publisher mobility is difficult to handle because the FIB entries have to be restated when a publisher moves from one CR to a new one. It becomes harder to handle publisher's mobility in very dynamic networks such as mobile ad hoc network (MANET). For this purpose, NDN employs Listen First, Broadcast Later (LFBL) protocol. In LFBL, a subscriber floods the interest message to all publishers. Any publisher having content against the interest checks the medium whether any other publisher has already replied with the content or not; if not then it sends the content to the subscriber.

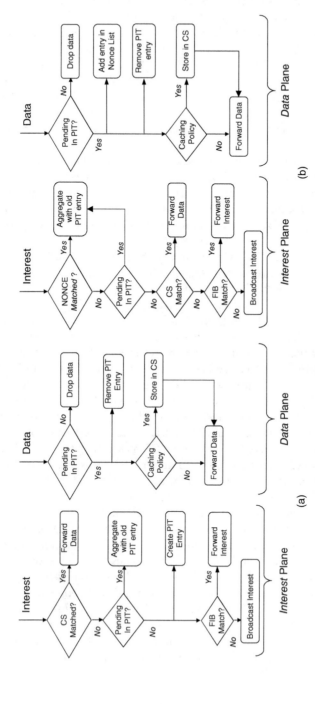

Figure 10.4 Basic operational perspectives of CCN versus NDN. (a) Content-Centric Networking (CCN). (b) Named Data Networking (NDN).

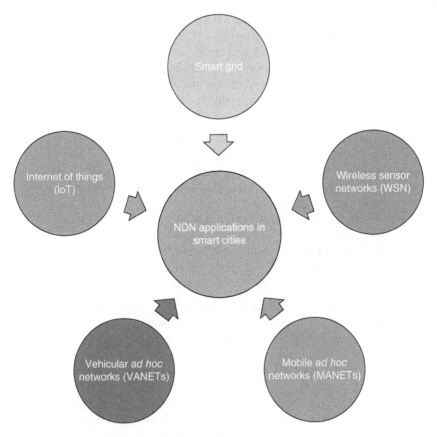

Figure 10.5 **NDN potentials and applications.**

10.4 NDN-based Application Scenarios for Smart Cities

As aforementioned, the NDN has been proved to be one of the most widely explored architectures of the Future Internet. The reason behind this research plethora is the support of NDN for various applications. In this section, we will focus on varying application perspectives that contribute to build a smart city (Figure 10.5).

10.4.1 NDN in IoT for Smart Cities

It took us a while to figure out the exact definition of an IoT system; currently we define IoT as the collection of small battery-operated devices having sensing, computation, and communication capabilities and are attached with the objects (i.e., anything). IoT enables those devices to connect with each other via

Internet. These connections between devices make them smarter to exchange and process the information to/from other devices autonomously or with partial intervention of the humans. Rather than a point-to-point communication scenario, the massive size of network in terms of numerous devices does have a focal point on Data. Thus, the smart applications for IoT argue for the contextual type of information generated either reactively or proactively by those devices [14]. Here we state the following key issues with such type of networks:

- Naming and addressing each device in dynamic fashion
- Protocols and algorithms to ensure energy efficiency
- Self-organization and network management
- Interoperability standards and network scalability
- Network and service discovery
- Cloud connections and their computation
- Minimum latency in real-time communications
- Dynamic partitioning and merging of the network
- Scalable security solutions
- Mechanisms for a push-based communication.

As aforementioned, IoT has been recently researched due to its support to a huge collection of applications. Hence, the Future Internet architecture, that is, NDN, has also shown some potential and is worth to be investigated in the IoT. For example, in [15], the authors proposed ICN architecture for IoT. In the context of this work, the IoT contents are addressed using the names, and the concept as a home automation system was implemented. Also, the implementation in this work used the push-based communication mechanism.

Similarly, in [16], the authors used push-based communication methodology for an IoT traffic using Future Internet architecture. Initially, the authors described the subscription scenario where subscription is done by sending an Interest message that uses the hierarchical name for the desired content without creating the PIT entry for each interest forwarded. The reason for not creating the PIT entry is that no Data is immediately expected for this interest. However, the authors mention that PIT entry creation will avoid the Interest looping.

Another proposal that analyzes the application of NDN in IoT is presented in [17]. It suggests that complete ICN mechanism cannot be implemented on IoT nodes because of their power, sensing, processing, and memory constraints. Therefore, some functionalities, that is, security and caching options, are delegated to the third-party trusted nodes. Furthermore, the optimal caching can be achieved by only storing the latest sensed information that is achieved by using the counter or sequence numbers. In [18], experiments were performed with NDN implementation on IoT deployment in multiple office buildings.

Precisely, most of the IoT proposals just focused on the simple scenarios where the Interest is used to subscribe for Data and the Data is sent for that subscription period. Generally, there is periodic sensing in the IoT application, but the Data may also be generated in response to the event detected by the sensors; this is called unsolicited emergency notification [19]. The application requires this notification to be pushed in the network, either in unicast or multicast manner. However, there should be CCN proposals that implement the unsolicited emergency message communication support along with the solicited Data communication in IoT. There are several other issues that must be addressed when adopting CCN in IoT, and the authors are suggested to explicitly refer the related work.

10.4.2 NDN in Smart Grid for Smart Cities

Still the research community has diverse definitions of a smart grid; however, in the context of this chapter, we define it as follows: smart grid integrates advance communication, control and automation, computer-based technology, and systems that manages, regulates, and brings the responsive and resilient utility electricity network. The grid connects the power generation sources and manages the electricity demand in a reliable, sustainable, and economic manner. Balanced demand-based supply of electricity is one of the main objectives of smart grid because most of the electricity generation relies on fossil fuels that increase the amount of harmful gases in the environment. The smart grid is a system of systems consisting of many components (refer Figure 10.6), including, but not limited to, the following:

- Smart meters: The utility meters enabled with communication technology to connect energy consumers and the providers to automate billing, regulate demand, and detect faults to speedy recovery.
- Smart electricity generation: There are several power generation sources ranging from renewable to the one that consume fossil fuels. Smart electricity generation system optimally generates electricity to meet the demand with minimum cost and carbon emission.
- Smart power distribution: The power distribution system connects the power generation sources with the consumers through distribution lines

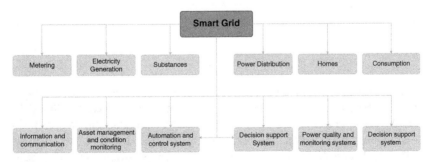

Figure 10.6 Components of the smart grid.

and smart substations. It has self-optimizing, self-healing, and self-balancing capabilities to automatically predict and detect power failures in real time.

- Smart substations: It controls and monitors the critical and noncritical operational Data, that is, battery status, transformer status, breaker information, power factor performance, security, and so on.
- Information and communication technologies: Provide means for all the components to interact with each other to conserve energy by efficiently utilizing, distributing, and generating the electricity.

Recently, ICN architectures and their effectiveness have been investigated in the smart grid systems to investigate their feasibility and effectiveness. For instance, Katsaros *et al.* [20] promote the use of ICN in smart grid applications and suggest the use of publish-/subscribe-like communication to ease the smart grid control with simple and secure Data sharing. The smart grid also uses many-to-many Data communication approach between devices and applications; therefore, ICN is envisioned to be a proper communication architecture for smart grids in future. Currently, the ICN framework is used as an overlay on smart grid communication to enable seamless and robust communication.

Similarly, in [21], the authors implemented the ICN-based communication infrastructure, called C-DAX, to support Data communication in the smart grid. They proposed the C-DAX architecture, and its components and plan have fully functional lab demonstration as well as porting the implementation to the actual smart grid system in the Netherlands.

10.4.3 NDN in WSN for Smart Cities

Wireless Sensor Networks (WSN) is an integral part of IoT as the collection of large number of small battery-operated devices capable of sensing and communicating (see Figure 10.7). The WSN consists of inexpensive and large number of devices spread or installed in the sensing area that monitor the environmental or physical parameters, that is, humidity, temperature, pressure, vital signs, water salinity, soil moisture, and so on. A typical tiny sensor node consists of the following components connected as a single component:

- Communication
- Computing
- Sensing
- Power source
- Actuation.

The operations that consume the battery power are wireless communication, sensing, processing, and listening. Therefore, a node must have to efficiently schedule its operations. Along with the longer lifetime, the WSN must self-configure and self-organize itself due to dynamic network architecture resulted by the node failures [22]. Several routing solutions and proposals

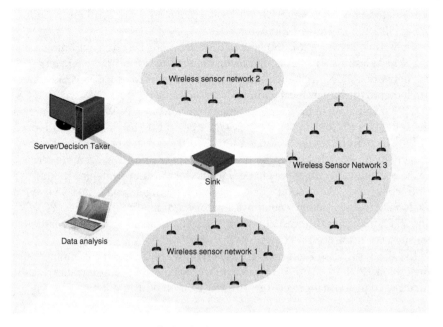

Figure 10.7 Generic overview of WSN deployment.

have been presented in past to achieve energy efficiency, self-configuration, and self-organization. Researchers in the area of WSN may further review [23, 24].

Recently, the Future Internet architectures have been investigated in WSN. In [25], Rawat *et al.* implemented the CCN-named Data communication stack in Contiki [contiki] operating system. Contiki is one of the operating systems for WSNs and embedded systems that contain resource-constrained devices. The implementation considers the hierarchical naming (similar to CCN) consisting of the name prefix followed by the content attributes as follows:

> Prefix:
> */Temperature/halinria*
> Content Attributes:
> */Temperature/halinria/Bld1/Floor2/Office1*
> */Temperature/halinria/Bld1/Floor1/Office12*

The implementation uses Interest and Data messages of 102 bytes to match the IEEE 802.15.4 frame that is 127 bytes long (127−25 (bytes MAC header) =102 bytes). Processing steps of these messages are modified to suit the processing capabilities of the WNS nodes. CCN PIT, FIB, and CS are also implemented accordingly. Moreover, the Future Internet implementation in Contiki is evaluated through simulations and real deployment using synthetic monitoring applications for varying network sizes.

Singh and Sharma [26] implemented the content-centric environment for WSN in Wieslib [27]; that is, library of algorithms form heterogeneous sensor networks. CCN implementation for WSN is named as CCN-WSN, which implements the hierarchical naming, Interest message, Data message, PIT, CS, and FIB. Evaluation of this flexible implementation of CCN-WSN demonstrates the suitability of CCN in WSN. The CCN architecture for WSN was proposed in [28]. The architecture is divided into two tiers; first tier manages the heterogeneity devices that comprise the WSN (sensor node, sink, remote server). CCN is enhanced with some changes to the forwarding strategies to improve Data collection. The second tier is the modified, lightweight, and shortened CCN forwarding strategy encompassing forwarding Data structures (FIB, CS, and PIT), messages (Interest and Data message), transmission, and techniques for message retransmission.

The above discussed solutions are the preliminary CCN implementation and their validations in WSN. However, a more serious attention of researchers is required to propose more energy-efficient CCN-based solutions for WSNs.

10.4.4 NDN in MANETs for Smart Cities

MANET is an autonomous, infrastructure less, self-healing, self-organizing, dynamic topology, and multi-hop nature network of mostly battery-operated nodes. Varying applications are precisely presented in Figure 10.8. These characteristics of the network make Data delivery between source and destination node more challenging. Due to its multi-hop nature and dynamic network topology, Data dissemination is with less control overhead to conserve energy and more reliability. There have been a lot of research to resolve these issues; however, the issues are still pending and the research is still ongoing. There is a lot of literature and books available on the topic, and the readers can go through the following very recent articles to get the quick information about the topic [29, 30].

After the invention of a new future emerging network paradigm, which substitutes the traditional IP-based, location-dependent, and host-centric networks with the name-based, location-independent, and Data-centric network, researchers investigated those in MANETs. For instance, in [31], the authors proposed the NDN-based new Data forwarding scheme for MANETs named LFBL. LFBL uses three-way message exchange (announce the name prefixes, forward Interest, and return Data) that is supported by NDN. Initially, a node disseminates the network-wise request carrying the name of the Data requested by the application. Any node with the requested named Data sends a response packet, backward to the requesting node. Next, the Data receiving node sends the acknowledgement (positive or negative)

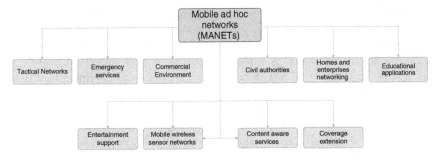

Figure 10.8 Different applications and services supported by MANETs.

packet. Similarly, Mahmood and Manivannan [32] implemented the CCN on laptops running Linux OS creating on-demand MANET. The CCN Data structures that are implemented include CS, MetaData registry, and Interest table. The large MANET for emergency and tactical scenario with hierarchical architecture, group mobility, and operation is considered in this work. Initially, the publisher disseminates the meta-content information that is disseminated through gateway nodes at the upper layer of the hierarchy in each group and recorded at each node's registry in each group. The node sends Interest to gateway node and the gateway replies with the matching content if it has the copy. Otherwise, the Interest is sent to other gateways to find the content, which makes these gateways responsible to publish and deliver Data. Furthermore, in [33], the authors pointed out the fundamental approaches to identify the fundamental points in the design of content-centric MANET (CCM). The performance and efficiency of the identified design is evaluated through modeling. Announcement of content availability, sending query to find the content, and fetching the content are three major operations that are considered in the modeling space. There are three schemes that are identified in the paper to retrieve content in MANET. The first is reactive flooding, where the requesting node floods the requested message to find the content, service, or location of the content or service provider node. In this case there is no announcement; just a node floods the query message and the content provider replies with Data or service in a unicast manner to the requesting node. The second proposed design is the proactive flooding, where each node periodically floods the resource or content availability in MANET. The requesting node just listens to the periodic announcements, and then the request and content delivery are performed in a unicast manner. The third and the last CCN design for MANET uses Geographic Hash Tables (GHT). This design assigns a key to each resource in the network, and the hash operation through this key may provide a pair of two-dimensional (x, y) coordinates. The node first announces the pairs (resource, host) to the nodes closest to the resource key. The requester first computes the hash of the resource that it is interested in, which provides the location of node(s) holding the pair of (resource, host). The query is forwarded in a unicast manner to these nodes through GPSR protocol. The content fetch operation is performed once the resource host information is retrieved. The fetch and content forwarding are unicast operations. Content availability, latency, and overhead cost are modeled by the authors.

In [34], Content-centric fasHion mANET (CHANET) has been proposed. The CHANET architecture provides detailed processing of the broadcast nature Interest and Data messages and many features for 802.11 based MANET including:

- Hierarchical naming
- Content segmentation and reassembly (content divided into chunks)
- Content advertisement (periodically by fixed providers, e.g., access point)
- Content discovery (interest message) and delivery (Data message)
- Retransmission request (Int-Ack message).

CHANET brings simplicity and robustness by using the broadcast nature of CCN messages. Nodes overhear the broadcasted messages and defer time to reduce the number of collisions. The Interest and retransmission request message forwarding decision is made by the node itself that receives the messages. It indirectly provides the retransmission request (embedding acknowledgments in the Interest packets) and sequence control mechanism. The consumer-provider mobility is inherently implemented and supported by the scheme.

Meisel *et al.* [35] propose the CCN scheme for MANETs that avoids the message loss in MANETs. The scheme uses the neighborhood information and defers time to achieve this goal. Detection of neighbors is done by periodic overhearing of the communication activities of the neighbors. If a node does not hear any periodic advertisement or any communication activity, the link to that neighbor is considered broken. The Interest and Data messages may be broadcasted or unicasted. In the case of broadcasted messages, time-to-live (TTL) value is used; however, TTL is not considered for the unicast messages. If a node has any content, it is advertised by that node. Intermediate nodes that receive these content advertisements may record all the paths toward the content source in FIB. Another Interest flooding control scheme, called neighborhood-aware interest forwarding (NAIF), for NDN-enabled MANETs has been proposed in [36]. NAIF initially prunes the ineligible forwarders because it selects the potential forwarders from the forwarder set based on Data retrieval rate and its distance to the content provider.

Cianci *et al.* [37] and [38] implemented the content-centric design for MANET's one handheld device called SCALE. The authors use nodeID to reflect its geographic coordinates, content name, and several other parameters to achieve content communication. The demo supports three functions:

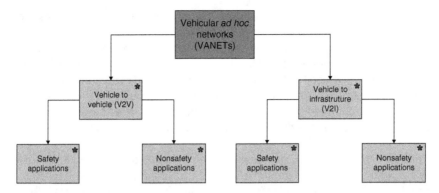

Figure 10.9 Application overview of VANETs.

publish, subscribe, and query the content. The Data in cache is indexed based on the content instead of content name.

As most of the MANET nodes are battery operated, mobile, and communicate without infrastructure support, the CCN solution must be efficient in content discovery, reliably forwarding messages, and support security. The proposed solutions just focus on the application and testing of CCN architecture, and more solutions are required to support reliability and efficiency of CCN messages and alleviate bandwidth congestion resulting from the broadcast nature of network.

10.4.5 NDN in VANETs for Smart Cities

VANETs are reality of the near future and play a vital role in everyday life by providing services, for example, safety and comfort of passengers on the drive and infotainment. VANETs support various sets of applications as shown in Figure 10.9. The vehicular network comprises vehicles with communication capabilities. One of the most prominent characteristics of VANET is its highly dynamic topology that makes it challenging to propose any efficient and promising communication solution. Along with that it is also proved by many researchers that TCP/IP communication protocol stack is inefficient for mobile networks. This is the reason that a separate protocol stack for VANETs called the "wireless access in vehicular environments" (WAVE) [39] has been proposed. WAVE supports Data exchange without the TCP/IP overhead, through WAVE short message protocol (WSMP) that was designed for safety critical and control messages.

Recently, there has been a lot of research with the intention to reliably communicate emergency information, traffic status, vehicle sensory Data, and infotainment information in a nonhost-centric manner by using information-centric communication mechanism. Content-centric approaches are briefly discussed and summarized in [38]. Here, we discuss and summarize the very recent NDN-based schemes for VANETs.

The very first proposal in [40] employs named Data communication mechanism to collect vehicles' sensory Data from the mobile vehicular network. This information may be used by the manufacturing or related companies to provide vehicle management, safety, and alert information to the drivers or owners of the vehicles or companies. A hierarchical naming scheme with company, type of information, country, state, and so on is used to request this sensory information.

Since, NDN uses hierarchical naming with general components to identify the contents. The formal naming scheme for vehicular networks to represent spatial and temporal vehicular information has been proposed in [41]. This naming scheme is used by vehicular applications to communicate contents between vehicles and the infrastructure. The vehicular network information is identified by the following naming scheme:

$$/traffic/geographic\_scope/temporal\_scope/data\_type/NONCE$$

TalebiFard *et al.* [42] proposed a content-centric communication scheme for autonomous vehicles, named "CarSpeak." CarSpeak-enabled autonomous vehicles communicate sensory information from neighboring cars as well as the sensor installed on the static infrastructures, for example, RSU, road side buildings, and so on using the content-centric Interest–Data mechanism.

Due to the broadcast nature of the wireless faces, collision of Interest and Data messages is inevitable. Amadeo *et al.* [43] introduced different timers to avoid the NDN message collisions in an NDN-enabled vehicle-to-vehicle (V2V) multi-hop highway communication network. The timers include collision-avoidance timer, pushing timer, NDN-layer retransmission, and application retransmission timer. In collision-avoidance timer, a node randomly delays Interest of Data transmission between 0 and 2 ms. Push type messages are scheduled based on the pushing timer that is computed based on the transmitter-receiver distance, maximum transmission range, and minimum next-hop delay. NDN-layer retransmission timer (50 ms) and application retransmission timer are used to schedule the retransmission of NDN messages over the lossy wireless network.

A hierarchical Bloom-filter routing (HBFR) has been proposed for CCN-enabled VANET in [44]. HBFR identifies each chunk of the content object with hierarchal name, for example, /Category/Service-Name/Additional-Info/. The category shows the type of content based on its size, type, popularity, and so on. The service name identifies the Data services provided by the node(s) based on time sensitivity of the content. Additional-Info consists of any additional information about the content. The proposed HBFR framework adaptively performs reactive and proactive content discovery based on content characteristics. It uses Bloom filters to announce the popular prefixes, and the announcement is restricted within a geographical region. The network is divided into hierarchical geographic regions to restrict the distribution of messages to reduce overhead. Vehicles in a single region form a cluster, and every vehicle in that region is a member of that cluster. The contents' advertisement by a vehicle includes its region information, and it adds the content prefixes using Bloom filters in the content advertisements. These Bloom-filter-based advertisements are shared between regions to get full view of the contents and their respective regions to easily discover and communicate the desired content.

Amadeo *et al.* [45] proposed the content segmentation–reassembly and reliable content delivery scheme by scheduling the retransmission of Interests for the lost Data messages. For faster recovery of the lost Data messages, the Interest is retransmitted depending on the dynamic round-trip time (RTT) of Interest–Data exchange. Average RTT over multi-hop path is estimated as weighted moving average of RTT samples, and based on this information, retransmission time out is scheduled.

A last encounter content routing (LER) in [46] is an opportunistic geo-inspired routing scheme for CCN-enabled VANET. Each vehicle that runs LER maintains two tables named content list and last encounter list (LEL). The content list contains the list of all contents that the vehicle holds and is shared within 1-hop neighbors. LEL enlists the content information received from the neighboring vehicles and enlists the provider ID, encounter position, and time. A content provided by multiple vehicles have multiple entries in the LEL. The Interests are forwarded toward the location provided in the message.

The results of the vehicular NDN implementation over vehicles have been presented in [47]. The implementation uses various wireless faces, that is, Wi-Fi, WiMAX, IEEE802.11p, and so on, and the messages are transmitted over all the available faces to avoid the disruptions caused by the intermittent connectivity. The experiments have been performed when the vehicular network had no mobility, vehicles moved in platoon, and vehicles moved around the UCLA campus.

In [48], authors proposed the forwarding scheme that fetches Data from a plethora of providers using the digital map information. Navigo binds the NDN Data names to the producers' geographic area(s). It uses the shortest path algorithm to forward Interests to the geographic area of the potential provider. The authors also claim the application of adaptive best Data provider discovery and selection scheme from multiple geographic areas.

A RobUst Interest Forwarder Selection (RUFS) scheme for VANET has been proposed in [49]. RUFS mitigates the interest broadcast storm by selecting the suitable next-hop forwarder vehicle. In RUFS, each vehicle shares its satisfied interests' statistics with the neighboring nodes. This information is managed in the neighbors satisfied list (NSL).

The traffic violation ticketing (TVT) application for the CCN-enabled vehicular networks has been proposed in [50]. It discusses the utilization of CCN Interest and Data messages used by the cop vehicles to issue tickets to the vehicles that commit violations. Additional Data structures are maintained to achieve this application perspective. The same authors extended their work one step ahead and evaluated their smart traffic ticketing architecture with the support of NDN and named it as *"SmartCop"* [51].

Since naming any content also plays a vital role in searching latency and does affect the overall performance, we do need research efforts in this context. For example, the hierarchical and hash-based content naming scheme for vehicular networks is proposed and evaluated in [52]. This new naming scheme encompasses provider's identity, different components that represent the content attributes, and the spatiotemporal resolution of contents. A small hash component is also part of the name that helps to precisely identify the content. Along with that a compact trie-based name management scheme has been adapted to manage the content name to speedy search, delete, and add the name in the name prefix tables. Analysis shows that the proposed name management scheme is more suitable for the variable length name prefix management in CNN.

10.4.6 NDN in Climate Data Communications

In addition to the aforementioned applications, the researchers also explored the use of NDN in Climatic Data Communications. Similar to others, the NDN has also been foreseen as a potential solution in Data-intensive field such as climate science. Shannigrahi *et al.* [53] discussed the similarities and differences between the climate and high energy physics (HEP) use cases, along with specific issues that HEP faces and will face during LHC Run2 and beyond, which can be addressed by NDN.

The researchers have successfully tested NDN in the climate application domain [54]. To handle the various naming schemes used in climate applications, they designed and implemented translators that take existing names with arbitrary structure (produced by climate models or home grown) and translated them into NDN-compliant names. Depending on the original name structure, the translation can be fairly direct (e.g., Data that comply with the "Data Reference Syntax" from the Coupled Model Intercomparison Project) or complex (from home-grown naming schemes that require the analysis of metadata embedded in the dataset or even user feedback in order to construct proper NDN names).

Likewise, there are several features of NDN that can be beneficial to the HEP computing use case. For example, the Data sources publish new content to the network following an agreed-upon naming scheme. Data delivery is always performed in a pull mode, driven by the consumer issuing interest packets.

Intermediate nodes in the network dynamically cache Data based on content popularity, ready to satisfy subsequent interests directly from the cache, thus lowering the load on servers with popular content. Combining this with the pull mode results in a multi-cast like Data delivery, possibly optimizing both the network utilization and server load. Similarly, the use of multiple Data sources simultaneously, as well as the native use of multiple paths between client and Data source, provide for robust fail-over in the case of network segment, node, or end-site failure.

HEP experiments using the Worldwide LHC Computing Grid (WLCG) have well-developed, hierarchical naming schemes in use, which already fit the NDN approach well. We can take this logical file name structure as a starting point for investigating the benefits of using NDN as the Data distribution and access network for HEP Data processing. In short, all these are active research areas today, where caching as well as forwarding strategies, naming schemes, multi-sourcing, and multipath forwarding need to be investigated from not only the network but also the application perspective.

10.5   Future Aspects of NDN in Smart Cities

There are different categories of future aspects in NDN that must be addressed; however, we divide them as per their related NDN components, including content (chunking, discovery, manifest, multihoming, etc.), naming (hierarchical, hash based, attribute based, name management scheme, etc.), Data Structures (PIT management, FIB management, etc.), NDN message forwarding, content caching, dynamic network topology (provider mobility, consumer mobility, effects of dynamic topology on content forwarding, etc.), and security and privacy. For instance, NDN research directions are briefly discussed in this section.

10.5.1   NDN Content/Data

Content can be any binary object or binary stream, such as a file of any type, audio stream, and video stream. In general, most of the research is focused on how to name a content and efficiently forward the content in NDN. There are some basics about the content that should be addressed, for example, chunking, manifest, and multihoming. Content chunking divides a large content into multiple small identifiable pieces to efficiently forward the content from the consumer to the provider. It also provides a means to evenly disperse the content in the distributed network caches. There is no NDN standard to divide a content into chunks of universally equal size. Some of the research works assume that a content chunk is equal to the size of an MTU. The fact is that the MTU size may vary based on the face type. Therefore, if a node receives multiple chunks of the content that do not fit the outgoing face, then the chunks require rearrangement. It also poses additional issues that need to be resolved for the NDN transport layer. These issues are enlisted, but not limited to, as follows:

- Synchronization between different chunk providers or chunk sources
- Receiver-driven synchronization
- Cache management policies for chunks
- Chunk-level security checks
- Chunk-based or content-based identification
- Content or chunk-based authentication
- Content manifest.

The above said issues have influence and pose complexity to many other aspects that affect efficient content communication in NDN.

10.5.2 Naming Content/Data in NDN

Content object retrieval in NDN usually involves two stages: discovery and delivery of the content using Interest and Data messages. The content is discovered through its name that uniquely identifies it in the network. This is the reason that a content name is a mandatory TLV in the Interest and Data message. Interest message with a content name is used to discover content. The content delivery involves the rules to route contents on the network, and it also uses the content name to take routing decisions.

There are different ways to represent the name, including hierarchical, flat, and attribute-based "names." The pros and cons, management, and respective issues related to each of these schemes are yet to be explored. However, here we will focus on the future directions concerning the hierarchical naming scheme used by the NDN architecture.

Due to the hierarchical nature of the naming scheme used in NDN, names can easily be aggregated and naturally have the longest prefix matching feature. Along with these inherent features, there are still some issues that need to be resolved. Following are the future research directions, but not limited to, that must be addressed:

- The evaluation to measure the effectiveness and scalability to Internet level is yet required for hierarchical names.
- Security information is not part of the NDN naming because it assumes explicit security provided by the content itself. Therefore, the hybrid schemes are required for NDN naming to provide security information within the name.
- Longest prefix matching on name string is time consuming and requires feasible solutions to minimize the search time.
- Efficient Data structures are required to optimize the memory usage for hierarchical names with varying prefix sizes.

- Related to the previous point, the Data structure should also be effective enough to perform speedy management of hierarchical names such as adding and deleting the name prefixes because the arrival rate of the names can be much higher in the global scale Internet.
- Number, position, and size of the hierarchical name components are not fixed, and inferring the name components is still application dependent. Hence, there is a need to standardize naming structure and inference of the rules.
- There are some applications, for example, vehicular networks, WSNs, and so on where spatial and temporal scope contents are required. There are few schemes that discuss these issues; however, more research in this aspect is required.
- Hardware-level implementation and management of hierarchical names is still a pending issue.

10.5.3 NDN Data Structures

The name prefixes are managed in PIT and FIB to forward Interest and Data messages in the network. NDN keeps track of many parameters, that is, NONCE list, dead NONCE list, timers, and so on to avoid routing loops. The incoming name plus face information and outgoing face plus prefix information are stored in PIT and FIB tables of NDN. These tables are refreshed and stale information is removed. The size of these tables and Data delivery rate are directly related to the refreshing duration. Therefore, it is required to test this effect in varying network scenarios, for example, wired, wireless, and dynamic topology networks. The size and look-up efficiency of these Data structures is still an open issue in NDN.

10.5.4 NDN Message Forwarding

NDN uses FIB and PIT to forward Interest and Data messages in the network. Basically, most of the Interest messages are forwarded through longest prefix matching within FIB. The Interest message is forwarded to face that is associated with the matching FIB entry. In the case of multiple outgoing faces, the forwarding strategy selects the most stable (Note: The description of most suitable depends on which parameter is being used to rank the outgoing face as most suitable, for example, Interest satisfaction rate, most recent activity over face.). However, there may be the possibility that a subset of the faces may be selected to forward interest message. Therefore, an efficient forwarding mechanism is required that not only selects outgoing face on FIB match but also considers network and neighborhood parameters.

Moreover, the Interest is forwarded between the consumer and the provider node through multiple paths, and the consumer may receive multiple copies of the Interest message at varying time instances. Normally, a provider or a caching node in NDN replies with Data to the first Interest message that it receives, and the following received copies of the Interest message are dropped by the provider node. In this case, there may be the possibility that the down-stream direction in that path may not be suitable to forward Data message. Therefore, a provider node must hold the Data message and select that best path based on the path information received through multiple copies of the Interest message from different paths. The duration to hold the Data message and selection of the suitable path(s) is still an open issue and a possible future research direction.

The Interest overhead is one of the issues that should be mitigated in NDN [55]. An Interest message is forwarded by each node that receives the Interest message and has no PIT entry. This may lead to a very high Interest overhead, and that can be minimized by limiting the number of Interest forwarders to be investigated [56]. Additionally, the new transport modes should be examined that may include anycast (any to any), multicast (i.e., one to many), and concast (i.e., many to one).

Likewise, distributed caching in NDN may also pose a situation where multi-provider for a single Interest is possible. In this case, the consumer must select the most suitable provider among the subset [57]. When an interest is issued for a content object that can be satisfied by the subset of providers, the synchronization between that subset of providers is necessary to increase the content transfer efficiency.

10.5.5 Content Discovery in NDN

NDN uses content name in the Interest message to discover the content in the network. As previously discussed, the interest messages are routed based on the longest prefix matching in content name tables, that is, PIT and FIB, which are maintained at each NDN-enabled node in the network. The provider node or any intermediate node that has the matching copy of the content in its cache sends the Data message to the consumer node following the reverse path. There is no location information available for the desired content provider in the network. One of the solutions to cope this problem is to maintain additional information regarding the content providers in the Data structure(s) based on the previously satisfied Interests. The next Interests demanding the same contents are forwarded toward those providers. However, if a node has no provider information available, then it follows the conventional CCN forwarding mechanism.

Our discussions so far have provided the base to explore the topic to answer the following questions:

- How contents can be discovered with minimum overhead?
- How to reduce content discovery delay?

- What are the efficient methods to manage neighborhood or network-wide content information on a node?
- What is the effect of content discovery on content communication efficiency in NDN?
- Is content discovery feasible in highly mobile or highly dynamic topology networks?

10.5.6 NDN in Dynamic Network Topology

Dynamic network topology is defined as a network topology that varies over time due to node mobility, node failures, link failures, and so on. There has been a lot of research to handle network dynamicity in the past, and many solutions have been proposed for wired and wireless networks. It is argued that consumer mobility is inherently handled by most of the ICN architectures (especially CCN/NDN) because they use pull-based or consumer-driven scheme. Previously cited work mostly focuses on the consumer and/or provider mobility, where they try to achieve higher Interest satisfaction and delivery rate and minimize the latency. It is claimed in architecture documentations that NDN inherently handles the mobility by detaching the location information binding with the content. However, the receiver and provider's mobility may have a high impact on the Interest and Data message transmission. Another side effect of dynamic topology other than the Data-Interest message loss is that it leaves routing information traces in the tables at intermediate nodes, which may increase the communication delay due to high lookup cost. The issues related to provider mobility that are identified and need to be addressed are locating provider node all the time and maintaining the connectivity until complete content is received. In the absence of position information and route updates in NDN, it requires new mechanism(s) to cope with these issues. Another question that is waiting for an answer is the examination and the benefits of caching in dynamic topology networks. Additionally, the new schemes are required to ensure intactness of content in the presence of mobility.

10.5.7 Content Caching in NDN

Content caching is not a new topic in communications networks and has been intensively investigated in the past. Most of the studies are focused on caching in web applications and peer-to-peer networks. There are different caching schemes, that is, least frequently/recently used (LFU/LRU), most frequently/recently used (MFU/MRU), leave copy down (LCD), popularity-based caching, and so on, that are investigated in CCN. It is argued and serious concerns are discussed that extensive use of caching in CCN will not have considerable benefits. Therefore, more investigations are required to explore caching based on various communication patterns in CCN. Caching may pose more challenges on efficient cache size utilization when the network topology is unpredictable, and there is need to focus some research on investigating issues. The information popularity is another issue that must be investigated that how dynamic content popularity is decided.

To summarize the discussion, solicitation of in-network caching and replication schemes for CCN requires new paradigms that jointly investigate routing, forwarding, and cache management optimization, that is, effect of cache locations on routing decisions, cache contention for varying information nature, and so on.

10.5.8 Security and Privacy

CCN claims to secure the content rather than the connection. In other words, instead of securing the connection between the sender and receiver, CCN inherently secures the content by sending security-related information along with the content itself. Now the question is, what is content security, or what are the security aspects related to the content in CCN?

CCN architecture requires public key cryptography (PKC) to bind a public key with content name. The PKC is also used by most of the ICN architectures to provide security and privacy features. Main focus of CCN is to secure the content object (either as whole content object or each individual chunk of the content object). However, it poses the following questions: Is it necessary to digitally sign and encrypt every content and how to decide which content should be encrypted or signed? To secure the content, a producer or generator of the content digitally signs the content, and the publisher's information is also provided within. This avoids third-party dependency. The consumer uses this supplementary information with the content object to detect the content integrity that the content object is tempered by any node during communication between producer to the consumer, and it also assures that the content is produced by the trusted publisher. The hierarchical names are human readable and can easily identify the publisher of the content. This can easily bind the name with the publisher. However, there must be some mechanism to verify whether the key associated with the name really belongs to the publisher or not.

As most of the ICN architectures, including CCN, rely on the PKC to provide security, then the paramount question is that who will be responsible for creating, distributing, and revoking these keys. In addition to that, if CCN relies on the trusted entities for name verification, then the key management can become a prime issue in CCN.

In mobile networks, especially *ad hoc* nature networks, the security and resolution of several security attacks is still a challenging issue. CCN claims and focuses on securing the content using public key encryption rather than the connection, which means that it promises that the content is same as it asserts.

This may raise more privacy issues because the content name is advertised in the Interest messages. In infrastructure-supported wireless networks, the privacy and trust are alleviated due to various points of management in the network, for example, access routers. However, pure *ad hoc* and mobile networks without any infrastructure support require robust solutions to handle the privacy and security threats. The solutions can be more challenging in constrained device scenarios, for example, limited processing power, small memory, and limited bandwidth.

There are several security attacks that have been identified for CCN/NDN, and they require robust solutions to prevent those attacks, for example, denial of service (DoS), distributed DoS (DDoS), sniffing and watchlist, blackhole, flooding, and content pollution.

10.5.9 Evaluation Methods

Currently, many researchers are pursuing and publishing their solutions for CCN, and they evaluate their schemes through simulations and theoretical and empirical evaluations. There are only few schemes that have been empirically evaluated due to time, budget, access, and other limitations. However, most of the proposals have been proposed for a varying range of network scenarios (IoT, VANET, WSN, MANET, etc.) and simulated in the freely available or open-source simulators.

Most of the solutions are evaluated through simulations, and, along with that to keep fairness in evaluation, there should be standard or baseline network scenarios to be considered in the simulations. In addition to that, traffic load, content popularity, and different other metrics are also explored by the authors. To keep the fairness in simulation evaluation for the proposed schemes, it is necessary to have baseline scenarios and standard parameters. There may be additional challenges in CNN that should be explored because an active, ongoing research on the topic is being pursued around the globe.

10.6 Conclusion

Recently, it has been noticed that people are more interested to live in urban areas due to the more healthy and secure environmental technologies. Moreover, the connectivity among devices and oversees plays a vital role in transforming a smarter life in cities. On the other hand, the current IP-based Internet is facing a lot of delays and security issues due to the massive traffic and demand from the end-user. Therefore, we expect that in future smart cities, the current Internet will not be enough to handle the demands. Hence, we need to integrate the concepts of Future Internet into smart cities and check the feasibility of these emerging technologies.

In this chapter, therefore, we have enlightened the promising Future Internet architectures and their applications in smart cities. We also summarize the current advancements in the field of CNN and named Data networks that are relevant to the smart cities. We expect that our chapter will contribute and encourage our readers pursue the challenges and research road map provided in the end of the chapter.

Final Thoughts

In this chapter, we first discussed the legacy of the Internet architectures and the relevant recent developments. Furthermore, we introduced the Future Internet architectures and provided a brief introduction to the groundbreaking technologies. In addition, we also focused on the main objective of this chapter, that is, to provide an insight on NDN and its applicability in various applications that can be useful in building a smart city. In the end, we also enlist few of the existing challenges and research areas for NDN research community. We expect that our chapter will make our readers familiar with the basics of NDN and its applications.

Questions

1   What is the difference between IP-based and Named Data Networking?

2   What does DONA stand for and what are the naming characteristics of DONA in general?

3   What are the basic operational differences between content-centric and Named Data Networks? You may draw a flowchart diagram that shows the differences?

4   Describe any application of the Vehicular Named Data Networking for smart cities?

5   What are the existing challenges or future aspects of message packet forwarding in NDN?

References

1 Mazières, D., Kaminsky, M., Kaashoek, M.F., and Witchel, E. (1999) *Separating Key Management from File System Security*. Proceedings of SOSP '99, December 1999, Charleston, SC, USA, pp. 124–139.

2 Moskowitz, R. and Nikander, P. (2006) *Host Identity Protocol Architecture RFC 4423*, IETF.

3 Xylomenos, G., Ververidis, C.N., Siris, V.A., Fotiou, N., Tsilopoulos, C., Vasilakos, X., Katsaros, K.V., and Polyzos, G.C. (2014) A survey of information-centric networking research. *IEEE Communications Surveys & Tutorials*, **16** (2), 1024–1049.

4 Ghodsi, A. *et al.* (2011) *Naming in Content-Oriented Architectures*. Proceedings of ACM SIGCOMM Workshop Information-Centric Networking, August 2011, Toronto, Canada.

5 Dannewitz, C. *et al.* (2010) *Secure Naming for A Network of Information*. Proceedings of the 13th IEEE Global Internet Symposium '10, March 2010, San Diego, CA, USA.

6 D'Ambrosio, M., Dannewitz, C., Karl, H., and Vercellone, V. (2011) *MDHT: A Hierarchical Name Resolution Service for Information-Centric Networks*. Proceedings ACM SIGCOMM Workshop on Information-Centric Networking, ACM, New York, NY, USA, pp. 7–12.

7 Dannewitz, C., D'Ambrosio, M., Karl, H., and Vercellone, V. (2013) Hierarchical DHT-based name resolution for information-centric networks. *Computer Communications*, **36** (7), 736–749.

8 Eriksson, A. and Ohlman, B. (2007) *Dynamic Internetworking Based on Late Locator Construction*. 10th IEEE Global Internet Symposium.

9 Kutscher, D. *et al.* (2012) *Content Delivery and Operations, Deliverable*. SAIL 7th FP EU-Funded Project, May 2012.

10 Ain, M. *et al.* (2009) *D2.3 - Architecture Definition, Component Descriptions, and Requirements Deliverable*. PSIRP 7th FP EU-Funded Project, February 2009.

11 Bloom, B.H. (1970) Space/time trade-offs in hash coding with allowable errors. *ACM Communications*, **13** (7), 422–426.

12 Miller, V.S. (1985) *Use of Elliptic Curves in Cryptography*. Proceedings of CRYPTO '85: The Advances in Cryptology, August 1985.

13 Lagutin, D. (2008) *Redesigning internet - the packet level authentication architecture. Licentiate's thesis*. Helsinki University of Technology, Finland.

14 Miorandi, D., Sicari, S., De Pellegrini, F., and Chlamtac, I. (2012) Internet of things: vision, applications and research challenges. *Ad Hoc Networks*, **10** (7), 1497–1516, doi: 10.1016/j.adhoc.2012.02.016. ISSN: 1570-8705.

15 Waltari, O.K. (2013) *Content-centric networking in the internet of things. MSc thesis*. Department of Computer Science, University of Helsinki, http://hdl.handle.net/10138/42303 (accessed 16 December 2016).

16 Francois, J., Cholez, T., and Engel, T. (2013) *CCN Traffic Optimization for IoT*. 2013 4th International Conference on the Network of the Future (NOF), October 23–25, 2013, pp. 1–5.

17 Quevedo, J., Corujo, D., and Aguiar, R. (2014) *A Case for ICN Usage in IoT Environments*. 2014 IEEE Global Communications Conference (GLOBE-COM), pp. 2770–2775.

18 Amadeo, M., Campolo, C., and Molinaro, A. (2014) *Multi-Source Data Retrieval in IoT via Named Data Networking*. Proceedings of the 1st International Conference on Information-Centric Networking (ICN '14). ACM, New York, NY, USA, pp. 67–76.

19 Shang, W., Bannis, A., Liang, T., Wang, Z., Yu, Y., Afanasyev, A., Thompson, J., Burke, J., Zhang, B., and Zhang, L. (2016) *Named Data Networking of Things*. Proceedings of the 1st IEEE International Conference on Internet-of-Things Design and Implementation, April 4–8, 2016, Berlin, Germany.

20 Katsaros, K., Chai, W., Wang, N., Pavlou, G., Bontius, H., and Paolone, M. (2014) Information-centric networking for machine-to-machine data delivery: a case study in smart grid applications. *IEEE Network*, **28** (3), 58–64.

21 Yu, K., Zhu, L., Wen, Z., Mohammad, A., Zhou, Z., and Sato, T. (2014) *CCN-AMI: Performance Evaluation of Content-Centric Networking Approach for Advanced Metering Infrastructure in Smart Grid*. Applied Measurements for Power Systems Proceedings (AMPS), 2014 IEEE International Workshop on, September 24-26, 2014, pp. 1–6.

22 Chai, W.K., Katsaros, K.V., Strobbe, M., Romano, P., Ge, C., Develder, C., Pavlou, G., and Wang, N. (2015) *Enabling Smart Grid Applications with ICN*. 2nd ACM Conference on Information-Centric Networking (ICN 2015), September 30–October 2, 2015, pp. 207–208.

23 Rault, T., Bouabdallah, A., and Challal, Y. (2014) Energy efficiency in wireless sensor networks: a top-down survey. *Computer Networks*, **67**, 104–122.

24 Kafi, M.A., Djenouri, D., Ben-Othman, J., and Badache, N. (2014) Congestion control protocols in wireless sensor networks: a survey. *IEEE Communications Surveys & Tutorials*, **16** (3), 1369–1390.

25 Rawat, P., Singh, K.D., Chaouchi, H., and Bonnin, J.M. (2013, 2014) Wireless sensor networks: a survey on recent developments and potential synergies. *The Journal of Supercomputing*, **68** (1), 1–48.

26 Singh, S.P. and Sharma, S.C. (2015) A survey on cluster based routing protocols in wireless sensor networks. *Procedia Computer Science*, **45**, 687–695.

27 Butun, I., Morgera, S.D., and Sankar, R. (2014) A survey of intrusion detection systems in wireless sensor networks. *IEEE Communications Surveys & Tutorials*, **16** (1), 266–282.

28 Saadallah, B., Lahmadi, A., and Festor, O. (2012) *CCNx for Contiki: Implementation Details*. Technical Report RT-0432, INRIA, p. 52.

29 Burke, J., Gasti, P., Nathan, N., and Tsudik, G. (2014) *Secure Sensing Over Named Data Networking*. Proceedings of the 13th IEEE International Symposium on Network Computing and Applications (NCA).

30 Baumgartner, T., Chatzigiannakis, I., Fekete, S.P., Koninis, C., Kröller, A., and Pyrgelis, A. (2010) Wiselib: a generic algorithm library for heterogeneous sensor networks, in *Proceedings of the 7th European Conference on Wireless Sensor Networks (EWSN'10)* (eds J. Sá Silva, B. Krishnamachari, and F. Boavida), Springer-Verlag, Berlin, Heidelberg, pp. 162–177.

31 Dorronsoro, B., Ruiz, P., Danoy, G., Pigne, Y., and Bouvry, P. (2014) *Evolutionary Algorithms for Mobile Ad Hoc Networks*, John Wiley & Sons, Inc.

32 Mahmood, B.A. and Manivannan, D. (2015) Position based and hybrid routing protocols for mobile Ad Hoc networks: a survey. *Wireless Personal Communications*, **83** (2), 1009–1033.

33 Reina, D.G., Askalani, M., Toral, S.L., Barrero, F., Asimakopoulou, E., and Bessis, N. (2015) A survey on multihop Ad Hoc networks for disaster response scenarios. *International Journal of Distributed Sensor Networks*, **2015**, 1–16.

34 Attia, R., Rizk, R., and Ali, H.A. (2015) Internet connectivity for mobile ad hoc network: a survey based study. *Wireless Networks*, **21** (7), 2369–2394.

35 Meisel, M., Pappas, V., and Zhang, L. (2010) *Ad Hoc Networking via Named Data*. Proceedings of the 5th ACM International Workshop on Mobility in the Evolving Internet Architecture (MobiArch '10), ACM, New York, NY, USA, pp. 3–8.

36 Oh, S.Y., Lau, D., and Gerla, M. (2010) *Content Centric Networking in Tactical and Emergency MANETs*. IFIP Wireless Days (WD), 2010, October 20–22, 2010, pp. 1–5.

37 Cianci, I., Grieco, L.A., and Boggia, G. (2012) *CCN - Java Opensource Kit EmulatoR for Wireless Ad Hoc Networks*. Proceedings of the 7th International Conference on Future Internet Technologies (CFI '12), ACM, New York, NY, USA, pp. 7–12.

38 Bouk, S.H., Ahmed, S.H., Kim, D., and Song, H. (2017) *Named-Data-Networking-Based ITS for Smart Cities*. IEEE Communications Magazine, 2017, January 105–111, **55**, 1.

39 Yu, Y.-T., Dilmaghani, R.B., Calo, S., Sanadidi, M.Y., and Gerla, M. (2013) *Interest Propagation in Named Data Manets*. International Conference on Computing, Networking and Communications (ICNC), 2013, January 28–31, 2013, pp. 1118–1122.

40 Varvello, M., Schurgot, M., Esteban, J., Greenwald, L., Guo, Y., Smith, M., Stott, D., and Wang, L. (2013) *SCALE: A Content-Centric MANET*. Computer Communications Workshops (INFOCOM WKSHPS), 2013 IEEE Conference on, April 14–19, 2013, pp. 29–30.

41 Detti, A., Tassetto, D., Melazzi, N.B., and Fedi, F. (2015) Exploiting content centric networking to develop topic-based, publish–subscribe MANET systems. *Ad Hoc Networks*, **24**, Part B, 115–133.

42 TalebiFard, P., Leung, V.C.M., Amadeo, M., Campolo, C., and Molinaro, A. (2015) Information-centric networking for VANETs, in *Vehicular Ad Hoc Networks*, Chapter 17 (eds C. Campolo, A. Molinaro, and R. Scopigno), Springer-Verlag, pp. 503–524.

43 Amadeo, M., Campolo, C., and Molinaro, A. (2012) *Content-Centric Vehicular Networking: An Evaluation Study*. 3rd International Conference on the Network of the Future (NOF), 2012, November 21–23, 2012, pp. 1–5.

44 Yu, Y.-T., Li, X., Gerla, M., and Sanadidi, M.Y. (2013) *Scalable VANET Content Routing Using Hierarchical Bloom Filters*. 9th International Wireless Communications and Mobile Computing Conference (IWCMC), 2013, July 1–5, 2013, pp. 1629–1634.

45 Amadeo, M., Campolo, C., and Molinaro, A. (2013) *Design and Analysis of a Transport-Level Solution for Content-Centric VANETs*. Proceedings of the IEEE International Conference on Communications Workshops (ICC), 2013, June 9–13, 2013, pp. 532–537.

46 Yu, Y.-T., Li, Y., Ma, X., Shang, W., Sanadidi, M.Y., and Gerla, M. (2013) *Scalable Opportunistic VANET Content Routing with Encounter Information*. Network Protocols (ICNP), 2013 21st IEEE International Conference on, October 7–10, 2013, pp. 1–6.

47 Grassi, G., Pesavento, D., Pau, G., Vuyyuru, R., Wakikawa, R., and Zhang, L. (2014) *VANET via Named Data Networking*. IEEE Conference on Computer Communications Workshops (INFOCOM WKSHPS), 2014, 27 April- 2 May 2014, pp. 410–415.

48 Grassi, G., Pesavento, D., Pau, G., Zhang, L., and Fdida, S. (2015) *Navigo: Interest Forwarding by Geolocations in Vehicular Named Data Networking*. IEEE 16th International Symposium on "A World of Wireless, Mobile and Multimedia Networks" (WoWMoM), June 2015, pp. 1–10.

49 Ahmed, S.H., Bouk, S.H., and Kim, D. (2015) RUFS: RobUst forwarder selection in vehicular content-centric networks. *IEEE Communications Letters*, **19** (9), 1616–1619.

50 Ahmed, S.H., Yaqub, M.A., Bouk, S.H., and Kim, D. (2015) *Towards Content-Centric Traffic Ticketing in VANETs: An Application Perspective*. Ubiquitous and Future Networks (ICUFN), 2015 7th International Conference on, July 7–10, 2015, pp. 237–239.

51 Ahmed, S.H., Yaqub, M.A., Bouk, S.H., and Kim, D. (2016) SmartCop: enabling smart traffic violations ticketing in vehicular named data networks. *Mobile Information Systems*, **2016**, 1–12, Article ID 1353290, doi: 10.1155/2016/1353290.

52 Bouk, S.H., Ahmed, S.H., and Kim, D. (2015) Hierarchical and hash based naming with compact trie name management scheme for vehicular content centric networks. *Computer Communications*, **71**, 73–83.

53 Shannigrahi, S., Barczuk, A., Papadopoulos, C., Sim, A., Monga, I., Newman, H., Wu, J., and Yeh, E. (2015) *Named Data Networking in Climate Research and HEP Applications.* Proceedings of the 21st International Conference on Computing in High Energy and Nuclear Physics (CHEP2015), April 2015, Okinawa, Japan.

54 Olschanowsky, C. *et al.* (2014) *Supporting Climate Research Using Named Data Networking.* LANMAN.

55 Ahmed, S.H., Bouk, S.H., Yaqub, M.A., Kim, D., and Gerla, M. (2016) *CONET: Controlled Data Packets Propagation in Vehicular Named Data Networks.* Proceedings of the 13th IEEE Annual Consumer Communications & Networking Conference (CCNC), Las Vegas, NV, pp. 620–625.

56 Bouk, S.H., Ahmed, S.H., Yaqub, M.A., Kim, D., and Gerla, M. (2016) DPEL: dynamic PIT entry lifetime in vehicular named data networks. *IEEE Communications Letters*, **20** (2), 336–339.

57 Ahmed, S.H., Bouk, S.H., Yaqub, M.A., Kim, D., Song, H., and Lloret, J. (2016) CODIE: controlled data and interest evaluation in vehicular named data networks. *IEEE Transactions on Vehicular Technology*, **65** (6), 3954–3963.

11

Human Context Sensing in Smart Cities
Juhi Ranjan, Erin Griffiths, and Kamin Whitehouse

University of Virginia, Charlottesville, USA

CHAPTER MENU
Introduction, 311 Human Context Types, 312 Sensing Technologies, 317 Conclusion, 331

Objectives

- To become familiar with the four main types of human context sensing
- To become familiar with the applications and measurable parameters of each type of context
- To become familiar with the technologies that support the measurement and sensing of different human contexts
- To become familiar with the strengths and weaknesses of the technologies currently used in human context sensing

11.1 Introduction

The complex systems that make up any city all revolve around one common aspect: the people. Today, 54% of the human population live in urban areas, and by 2050 that number is expected to increase to 66% [1]. In the future, a smart city will have applications, both individual and societal, that are tailored to the people living within them and help to improve healthcare, transportation, energy use, and the lives of people within the city. However, in order to include the city's people in these applications, such applications must be able to sense and monitor the people themselves through *human context sensing*. Human

Smart Cities: Foundations, Principles and Applications, First Edition.
Edited by Houbing Song, Ravi Srinivasan, Tamim Sookoor, and Sabina Jeschke.

context sensing uses different sensing technologies to monitor an individual's human context: their state, physiologically and emotionally, and information about their interactions with the world around them. Only by understanding this context on an individual level can applications be tailored to the people within a city to provide the most personalized care, feedback, and control for the individuals and the city around them.

Human context sensing is structured into four categories that holistically capture the user's condition and their interaction with the world around them. These categories are the user's emotional state (emotive sensing), the user's physiological state (physiological sensing), the user's motor functions and skills (functional sensing), and the user's physical presence and location in their environment (location sensing). The four categories of human context sensing sense different, but interrelated, aspects of a person's context. Emotive sensing detects the emotional state of the user, including things such as stress, happiness, engagement, and general emotional intelligence. Physiological sensing is similar to emotive sensing in the changes it might detect, such as increased heart rate, except it looks for anomalies that indicate health problems instead of emotional state. Functional sensing is the detection of the state of a user's activity and ability. The huge variety of motor skills performed by humans makes this sensing highly variable, with solutions often presented for specific applications such as detecting eating, walking, and daily activity routines. Presence and location sensing detects if a person is actually present within a given space or identifies where they are located in reference to other objects. This sensing gives an application a wider perspective of the user in the context of the environment, including how they interact with objects and other people.

Like the sensing categories themselves, the technologies that perform the sensing often overlap, and many of the same technologies have been used to sense each piece of human context. One common technology example would be video and audio – a technology that has both been used to accurately sense a person's location as well as read their facial features and sense their emotional state. The four most common types of technology used to perform context sensing are video and audio, wearables, smartphones, and sensing from the environment. Each performs differently depending on the type of context being sensed. Additionally, each has practical uses, costs, and privacy implications for use in a smart city. In this chapter we will discuss the different types of human context sensing and the technologies being researched to perform this sensing.

11.2 Human Context Types

The pieces of human context that a smart city and its applications must leverage fall roughly into four categories: physiological, emotive, functional, and location. Each context provides a different benefit for smart applications, affecting a variety of different industries. Below we describe the four types of human context, their potential impact, and what must be monitored or sensed to infer each context.

11.2.1 Physiological Context

Physiology is the science of life, and sensing physiological context involves measuring biological signs that are vital for sustaining a healthy life [2]. Tracking vital signs is critical to the detection, prevention, and treatment of diseases. Timely detection of health issues has a huge potential to save healthcare expenses; in the United States alone, it is estimated to be able to save $16 billion in healthcare costs annually [3]. The state of human health can be expressed by a variety of interdependent physiological signals. Some physiological signals require precise settings and sensitive equipments, while others are easier to measure. Developing physiological context-sensing devices includes determining what physiological signals need to be measured based on the ease of measurement and importance of the signal in diagnosing different health issues.

Smart health applications monitor specific body functions to diagnose if a person is either likely to develop or currently ailing from a certain disorder. These applications differ in their mode of operation. Some applications monitor a person passively, where they opportunistically sense physical parameters when they can [4–8]. Other applications require active participation from its users to diagnose medical issues [9, 10]. Another way to classify applications that use physiological context is based on the different aspects of an individual's quality of healthcare that they cater to, such as (i) systems to improve accessibility to advanced healthcare from any location, (ii) systems to integrate health monitoring with people's lifestyles in order to detect medical problems early, and (iii) systems that use social and behavioral models in developing and enforcing healthy life practices. The use of alternative healthcare delivery methods can augment the existing facilities and reduce demand on healthcare infrastructure [7, 11–13]. The ability to detect deterioration in physiological signs conveniently in people's homes can lead to early detection of diseases, which ultimately reduces healthcare cost [14, 15]. Finally, advanced healthcare systems can use social media and other socially influencing factors to enable good health values by establishing social norms [16].

Most applications that use physiological contexts quantify physical health of a person with two types of parameters: vital signs, such as respiratory rate, and life-sustaining activities such as sleep. Medical problems manifest into unique symptoms, which can be changes in vital signs, such as increased heart rate and blood pressure, or changes in life-sustaining activities, such as change in respiration rate, or changes in both types of parameters. Body temperature and blood pressure are two of the most commonly measured physiological parameters. Medical investigations have established that the most important physiological parameters are those that measure the state of the heart and the respiratory system of the body. Some of the measured parameters are ECG (variation of electrical heart vector, heart work rate), pulse (heart work rate), respiratory rate, respiratory volume (minute respiratory volume gauge), body temperature, blood pressure (systolic and diastolic blood pressure), and heart rate.

11.2.2 Emotive Context

Sensing the emotive context of an individual has far-reaching consequences beyond just determining if someone is happy or sad. At its most basic, emotive context can be used to help understand and diagnose physiological context – an application should respond very differently to fast breathing when someone is afraid as opposed to when they're excited. However, future applications that understand emotive context can affect larger industries. Online advertising, a $58.61 billion industry in the United States, which has continued to look for better ways to personalize their ads through search histories and personal information, may be able to leverage our emotions in ad selection – for better or for worse [17]. In the $7 billion tutoring industry in the United States, a tutor's responsiveness to a student's frustration or confusion helps promote learning, but high prices often prevent those of limited means from engaging these personalized services [18]. Tutoring applications that provide emotional responsive services through emotive context sensing could provide these services more cheaply in the future [19].

Research on applications that can leverage emotive context sensing is being performed today. For a city, applications are being created that identify angry or distracted drivers and reroute them to less-trafficked areas, improving driving safety in the entire city [20–22]. At an individual level, knowledge of a person's emotional state can provide more beneficial and productive responses from the smart applications around them. For example, there exists a movie recommender that measures an emotional response to watching videos and provides more meaningful recommendations for future films [23, 24]. DRESS responds to an elderly person's frustration by leading them through a soothing activity, decreasing frustration, and allowing them to complete the task of dressing with only the smart application for assistance [25]. By sensing when players are bored and increasing difficulty, or decreasing difficulty and offering tutorials when frustration is sensed, new video games research can better customize the experience to individuals [26]. From a health perspective, there exists an application that monitors a person's emotional state for signs of depression and can be employed when they begin a new medication where depression is a potential side effect [27]. For smart applications like these to incorporate this emotive information into their operation, sensing systems must be created to sense and identify emotion.

We exhibit emotions in roughly two ways: externally, where other humans can read and interpret our emotions, and internally, where the physiological state of our bodies can be a response to an emotion rather than physical health. Smart applications can monitor both these aspects to assess our emotional states. We communicate emotions to others through facial actions, such as frowning, smiling, or raising an eyebrow; through body language, such as leaning away or toward a speaker; through gestures, such as touching another's arm or shoulder and offering a hug; and through our voice, as pitch, volume, and word choice changes to reflect emotions. This communication is external, a way of communicating our emotions to those around us. We also exhibit the effect of emotion internally with increased heart or respiratory rate, changes in blood pressure, and changes in skin conductance, called the galvanic skin response, that reflects physiologic arousal. Smart applications sense these emotional states by monitoring one or more of these exhibited signs and inferring the emotion that underlies them. Usually, this becomes more difficult with internal signs, since these can be attributed to physical as well as emotional. For example, a raised heart rate may be due to either exercise, emotional arousal, or both.

11.2.3 Functional Context

Functional human context refers to information that can help assess activities of daily living (ADL) that individuals perform, such as how many hours a person watches TV or how often a mother and daughter cook together [28]. As people grow older, they may need assistance in performing some or all ADLs. Living in a single bedroom apartment at an assisted living facility can be as expensive as $45K per year [29]. Therefore, there is a growing emphasis on supporting seniors living at home with appropriate in-home care. Knowing the functional context of individuals can help a city plan infrastructure to support the lifestyle of its citizens. For example, if a city knows neighborhoods with higher number of mobility-impaired persons, it can plan to make sure these locations have sufficient accessibility options. Typically, sensing a person's functional abilities involves a trained personnel evaluating the person based on their performance in clinical tests, which is expensive. Technology can help reduce the cost and increase the ubiquity of functional skill assessment [30]; once the technological infrastructure for performing this assessment is established, the cost of making additional measurements over time becomes negligible.

There are many applications where knowing the functional ability of an individual can assist the individual in leading a more fulfilling life. Some of the compelling applications that depend on sensing functional ability are assisted living and care for the elderly, sensing context for diagnosing medical disorders, personal energy accountability, and improvement in personal skills, such as games. Each of these applications focus on sensing a different aspect of functional ability, such as what activity a person did (Did a person cook today?), how well did the person do the activity (Can a person dress without help of another person?), what instruments were used to perform the activity (What appliances did a person use to cook?), and how well an activity was performed (Can the person eat food without spilling?). Knowing the functional abilities and disabilities of a person can also help predict other potential health challenges, such as cognitive decline [31, 32]. It is important to diagnose the exact extent of functional disability, because the cost of care typically depends on the type of care being provided. For example, the cost of helping a person with eating food is more expensive than the cost of assisting them bathe. Since comprehensive care is extremely expensive and therefore impractical for a large population, functional assessment can help city manage its resources by providing directed care.

Assessing the functional behavior of individual involves measuring their ability to perform two types of self-care activities: basic ADL and instrumental ADL [33]. Basic ADL refers to the fundamental activities that a person needs to lead a healthy and hygienic lifestyle. It consists of activities such as functional mobility, bathing and showering, self-feeding, personal hygiene, and toilet hygiene. Instrumental ADL refers to activities that allow an individual to live independently in a community. It consists of activities such as housework, cooking, financial management, health management, and shopping. Doctors typically assess persons with functional disabilities on their capability to perform both basic and instrumental ADLs in order to prescribe the appropriate level of care for them.

11.2.4 Location Context

By knowing where a person is and what devices they could interact with at that location, a smart building, city, or application can understand how that individual fits into the community. This information could be used to better route individuals in cars through a city safely – potentially decreasing the $300 billion spent on traffic crashes [34]. Outside of a car, smart applications and cities could also track individuals in order to support their needs in a similar manner. Given the information of location and an understanding of how individuals interact with the environment through it, applications can use that information to prompt these interactions to change. A dramatic effect of this could be through a better understanding of how we use and waste energy. Motion sensors that control lights in commercial buildings are already being used today, responding to the presence of humans to save energy. Heating and cooling can be controlled by human location in a similar manner, and research is being performed to make this more accurate without necessary waving of hands at the motion sensor [35, 36]. Ranjan *et al.* assign energy use to individuals based solely on their location in a home, knowing that the person closest to the appliance when it is used is likely the one to have used it [37]. With this, utility bills could be apportioned to those who actually used the electricity, and wasteful users can be identified and encouraged to change their behavior.

While locating individuals outdoors has been largely solved by GPS in recent years, locating individuals indoors is still an area of research. It can roughly be divided into two thrusts: exact location and presence. When finding the exact location of an individual, the layout or floor plan around them is implicit in the design of the sensing system. For instance, smart applications that assist hospital workers in locating patients must know the layout of the hospital to provide meaningful location information. Knowing that a patient has left their room to visit the cafeteria may prompt a different response than knowing they are located in the hallway or an incorrect patient room. In some cases the exact location of the individual as it relates to the floor plan doesn't matter, only the objects in the environment they are interacting with. This type of research focuses more on whether a person is in the presence of a particular object and their proximity to it. In a mall, knowing exactly where a person is doesn't matter to an advertising company, but how long they spent in proximity to their advertisement does. Applications may also care about the presence of another human, rather than an object. Knowing how often people are in the same room and how often they are in proximity to each other provides information on their health and interaction with others.

11.3 Sensing Technologies

The technologies used to perform human context sensing are being researched extensively today. The most common technologies being used and researched are video and audio, wearables, smartphones, and sensing from the environment. Each technology has advantages and disadvantages in both sensing the four types of human context, as well as the technology's use, practicality, cost, and privacy invasiveness.

11.3.1 Video and Audio

Video and audio are often the most information-rich way to sense human context information. Cameras are quickly becoming the most ubiquitous sensor in the world today, with decrease in size and cost and an increase in the recording capabilities and ease of use. Millions of people have access to a camera through a mobile phone or computer, and security cameras litter the environment around us. Due to these factors, cameras are an ideal sensor to leverage for human context sensing (Figure 11.1). Any human context that can be seen or heard can be recorded, processed, and identified using video and audio. Processing the video in all these cases is often the largest portion of research work. Applications that use video for sensing must deal with automatic detection of the object

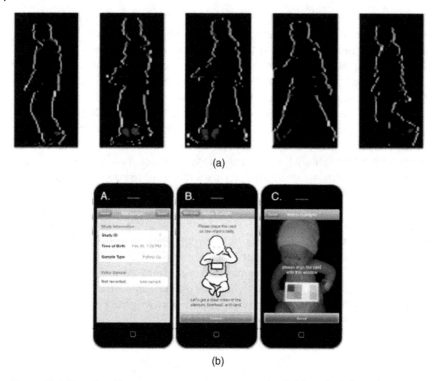

Figure 11.1 Functional human context such as a person's gait (a) or physiological context such as jaundice indications from skin color in babies (b) is sensed through video-based technologies [41, 42]. *Source:* Greef *et al.* (2014) [41]. Reproduced with permission of Proceedings of the 2014 ACM International Joint Conference on Pervasive and Ubiquitous Computing & Kale *et al.* (2003) [42]. Reproduced with permission of Springer.

of interest and filtering out any noise such as changing light sources or movement in the video. The placement and processing of the video changes greatly depending on the human context information to be sensed, and some areas of human context are more difficult to sense through video than others.

Physiological sensing is often most difficult using video and audio. Many indicators of our health are internal, and we as humans cannot often identify health signs in others by eye. Some signs, however, are obvious in video such as the weight gain or loss of an individual and the outward signs of unhealthy sleep, such as tossing and turning monitored by an infrared-capable camera [38, 39]. Medical nurses often detect respiration rates by watching a person's chest rise and fall – a task easily accomplished by a camera. Some physiological signs that cannot be detected by the human eye can be sensed in video. One piece of work in particular monitors heart rate through a computer camera by detecting the minute changes in skin color as blood is pumped through the face [40]. Slight skin color changes that indicate health, such as jaundice in babies, can also be detected through video monitoring before it is visible to the unpracticed human eye for at-home testing [41]. However, many of these signs require absolute stillness to detect. For example, heart rate detection in the face can be disrupted by head movement to music.

Emotive human context is often most easily sensed through video or audio because humans communicate emotions through the eyes and ears of those around them. Facial expressions, one of the first and most obvious ways we communicate emotions, have been monitored by cameras since the 1990s. Applications that do this must segment facial expression (i.e., determine when an emotional facial expression is occurring) and classify that expression as an emotion based on what facial features indicate certain emotions [43]. Emotions communicated through body language are sensed in a similar manner, though current work augments pure video with depth sensors (such as those used by Microsoft's Kinect gaming device) to detect movements and gestures indicative of emotion [44]. Emotion is also commonly communicated through vocal patterns, such as a raised voice indicating excitement or anger, and audio-based applications detect emotion from speech features such as pitch, speech rate, and pauses [45]. In most video- and audio-based emotive detection, machine learning is used to train a classifier to classify emotional states – requiring a large number of training examples of each emotion before performing detection.

Functional human context is often easily recorded in video, but often difficult to detect or classify automatically. Current research focuses on simple, easily identified activities such as walking, running, and eating. Many such activities are monitored for health reasons, such as monitoring how ambulatory a person is. Gait analysis is a common functional assessment through video, where the identity of a person can be detected through the motion of their legs [42]. Additionally, gait analysis is used to predict overall health and identify warning signs of potential future falls in the elderly [46]. Each functional activity may have its own visual markers that need to be identified to accurately detect the activity, such as being able to identify a spoon to recognize an eating activity. Because of this, current research creates a new vision-based algorithm for each new functional human context-based application, and few techniques generalize well to any large subset of possible applications.

Location information on humans is easily obtained from video data. Often motion detection algorithms, such as detecting the change in pixels of a video over time or performing background subtraction, are used to identify people in video data. By understanding where the camera is in the environment and what its viewing range covers, the exact location of a person can be inferred [47]. One of the main problems with this approach is occlusion, when one person blocks the view of another and applications mistakenly detect only one occupant. Approaches to counteract this often include using multiple cameras monitoring the same space from different angles [48]. Video-based localization can also detect when a person approaches an object in the environment by identifying and tracking that object. By using targeted microphones, audio information can also be used to localize people in an environment as they speak or otherwise make noise. Additionally, audio and video sensing systems can be combined to increase the accuracy of the detected location [49].

While video and audio can be used to sense many aspects of each of the four human contexts, it has two major drawbacks: processing and privacy. Because video records a huge amount of information unrelated to the task of human context sensing (background objects and colors, animals, lighting changes, etc.), all algorithms that process that data must be able to accurately filter out this unrelated data. In physiological sensing this may include the motion of the person themselves in order to sense heart or respiratory rate. In location sensing this may include filtering out other moving objects or people who are not the specific human who needs to be located. As future work in computer vision continues, faster and more accurate algorithms may make video-based human context sensing more accurate and able to leverage the information-rich sensor for a larger variety of human context sensing. However, the largest drawback in human context sensing from video is the notion of privacy. Video and audio recording is considered one of the most privacy-invasive sensing technologies today [50]. This often prevents video from being used for human context sensing in homes, where 68% of our lives are spent [51], unless concerns of health or safety rise above concerns for privacy. Because of this, current research often looks to other less privacy-invasive sensing system to perform human context sensing.

11.3.2 Wearables

Wearable sensing technology and devices are clothing, accessories, and shoes that have processing and sensing components embedded in them. There has been a surge in the number of wearable devices available in today's market. The biggest advantage of using wearables to sense a person's context is its ability to permeate in a person's daily activities. The availability of low power sensors, such as temperature, accelerometer, gyroscope, magnetometer, microphone, and so on, makes it possible to obtain rich data about a person's activities. The biggest challenge in using wearable sensing for continuous monitoring is its limited battery life. The size and weight of the battery is limited in order for wearable devices to be small and light enough to be inconspicuous during daily wear. Despite this shortcoming, wearables are a very exciting concept, as it allows regular accessories and objects to be transformed into smarter things that can potentially add value to a person's life.

Wearable devices typically use contact-based sensing methods to detect physiological signs, such as heart rate, blood pressure, stress, and so on. Body functions are measured using specialized sensors onboard, capable of sensing different vital signs. For example, many smart watches use pulse oximetry to measure a person's heart rate. Pulse oximetry is a noninvasive technique that measures the oxygen saturation of blood, which pulsates at the rate at which the heart pumps the blood. The wrist-worn devices shine two different wavelengths of light through the tissue and measure how it's attenuated [52]. Measuring blood glucose is essential for an effective diabetic management plan. Wearable blood glucose sensors make it easy to obtain this information at regular intervals. While basic vital signs are monitored with sensors that measure factors that are directly correlated to the event being measured, such as heart rate monitoring to detect heart problems, other systems use indirect sensing techniques to infer certain physical conditions, such as epilepsy. A combination of data from electrodermal activity (EDA) sensor and accelerometer sensor embedded in wearable wrist is used to detect episodes of epilepsy [53]. Many wearable devices are also used for discovering more information about one's lifestyles. Tracking sleep habits is one such application, and many wearable devices also act as a personal sleep tracker and can monitor the sleep quality of its wearer. New wearable products are being investigated, which are tightly coupled with objects commonly used in day-to-day lives. For example, a new product called the smart vest has been created that can track its wearer's biometric data such as heart rate, breathing, steps, pace, and calories [54]. Another smart shirt can help alleviate back pain by monitoring posture and helping its wearer improve the way they sit [55].

The emotional state of a person can often be detected by monitoring changes in certain physiological signs. The main difference with physiological context sensing is that these devices have algorithms that fuse data from multiple physiological sensing to determine a high level emotional state. Based on the detected emotion, a system can decide if it wants to initiate certain remediations to ensure that the user is comfortable. For example, wearables can attach to undergarments and detect the user's physical activity and state of mind by tracking their breathing pattern [56]. This can be used to determine emotional states during driving, such as if a person is relaxed during peaceful stretches on the road, or dull and bored, or tensed in traffic. Based on the detected state of mind, the app can suggest the user to take a break or perform deep breathing to get calm. Another wrist-wearable device tracks a person's mood using blood pressure sensing and heart rate variability and can differentiate between good stress and bad stress [57]. Another wearable device is in the form of glasses, where the glasses are augmented with microphone, camera, and other sensors, which can detect the emotion of a person in the field of view [58]. Some wearable devices not only detect emotions but also share the inference with others. For example, a butterfly pin called MoodWings flutters when your stress level rises and a scarf that heats up when you're sad and plays cheerful music when you're happy [59]. Some wearables are used to report emotional information periodically to caregivers of people who need attention. For example, autistic children often don't realize when they are getting stressed, and their tension may build to a point that it leads to a meltdown. By monitoring the wearer's stress level, measured by a galvanic skin conductance sensor, a potential outburst can be anticipated and prevented by an experienced caregiver [60]. A wearable sleeve can detect the emotion of affection from others through pressure sensors that detect when the sleeve is touched or squeezed [61].

Wearables are also highly informative about their wearer's functional activities, such as eating, cooking, bathing, and so on. The use of different sensors, such as microphone, camera, accelerometer, and other inertial sensors, can be used with complex inference algorithms to determine the activity state of the user (Figure 11.2). Wearables are especially becoming very popular in detecting and monitoring the ambulatory aspects of functional health such as walking, running, climbing stairs, lying down, sitting, and so on. Accelerometers are one

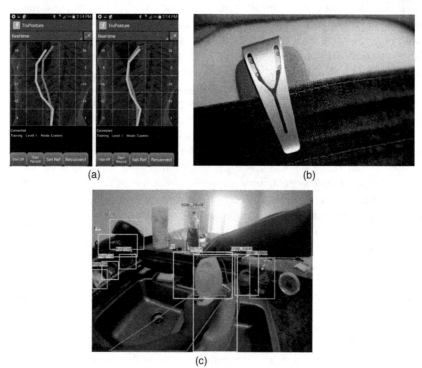

(a) (b)

(c)

Figure 11.2 Wearable sensing technologies: (a) Detecting back posture using sensors embedded in a smart shirt. (b) Determining physical activities and respiratory pattern using device clipped to clothes. (c) First person views using wearable camera. *Source:* Adapted from TruPosture (2015), Spire (2015), and Pirsiavash (2012) [55, 56, 62]. Reproduced with permission of IEEE.

of the most popularly used sensors for ambulation detection. Various wearable devices that are commercially available use at least one triaxial accelerometer, which is placed either at the wrist or attached at the hips [63]. Some activities are relatively easy to detect with a single sensor only, such as sitting, standing, and walking, and the sensor placement is robust to different locations on the body. However, for complex activities, such as climbing up and down stairs, the sensor placement becomes crucial for high accuracy. While leg activities can be recognized by sensors placed on legs as well as upper body, hand activities such as typing on keyboard can be detected accurately only with sensors worn on the wrist. There again, since the two hands can move freely and independently of each other, wearing the sensor on one hand is not sufficient to determine activities performed by the other hand [64]. Unintentional falls while walking is a common cause of injury among the elderly. Wearable devices worn by the elderly can detect when they fall using algorithms that monitor for its characteristic signal in the devices. Data from these sensors can indicate when a significant impact co-occurs with a change in orientation to determine if a person has fallen [65]. Some of the non-ambulation activities, such as cooking, eating, and so on, are also important from the perspective of remote monitoring if the elderly are able to live comfortably in their own homes. However, these functional activities are harder to detect. One way to detect eating and drinking is to recognize the associated arm gestures with these activities using inertial sensing devices worn on the wrist [66]. Another device fitted with a microphone, worn in ear like a Bluetooth headset, can detect additional information about the eating habits of a person, such as when the person is eating chips [67]. Detecting more complex activities such as making coffee, washing dishes, and so on, requires more intrusive technology. Wearable camera systems have been proposed, which can analyze the scene to determine what objects are being used and what activities are being performed [62].

Wearables are very useful in locating the position of people indoors and outdoors. This is because wearables have systems that can transmit and receive information wirelessly, which can be used to locate the device and therefore its wearer. Locating a person outdoors has now become very common because of the GPS sensor in fitness trackers and other wearables. In fact by clustering and modeling the GPS locations of a person, it may even be possible to predict their future locations [68]. Some applications require more than just coordinates of a person's location. To create more contextual utilities, applications often require what is referred to as the "logical location" – the store, business, or point of interest that a person is at. The use of wearable cameras with GPS sensor can determine what objects a person is paying attention to in a physical location and therefore identify the logical location [69]. While GPS remains the primary sensor for locating a person outdoor, locating a person indoors remains a challenge. A GPS cannot be used indoors since they require a clear line of sight to the sky. A wide range of technologies are currently in the works for indoor localization. While wireless-based technologies, such as Wi-Fi, Bluetooth, infrared, ultrasound, and so on are more commonly explored, other methods such as pedestrian dead reckoning use inertial sensors in wearable devices. Another way to locate indoors is to determine the proximity of people to objects that have a fixed location. This is done typically using RFID and magnetic beaconing methods.

The biggest advantage of using wearables for human context sensing is that since they are present on the wearer's person, they are able to capture the sounds, pictures, physiology, and gestures of the person as they perform different activities. While most commercially available applications of wearable devices are used for physical health quantification, they are used less for the other three contexts. Part of the reason is that some of these contexts require significant amount of training data due to the complex and subjective inference required. While research is ongoing to solve these challenges, one of the biggest bottlenecks in using wearable technology remains the battery life of the device. A limited battery life restricts the use of information-rich sensors such as microphone and camera.

11.3.3 Smartphones

Smartphones are the norm in cell phones these days, with a projected 2459 million users worldwide by 2019 [70]. There are many reasons why the smartphone technology has been largely successful. In addition to making phone calls, a smartphone can act as a navigator to give directions to a location and a pedometer to count a person's steps and detect a person's social interactions. The applications for a smartphone are endless. The flexibility of a smartphone can be attributed to the various sensors on board and powerful processors. The fact that humans tend to keep their phones in their proximity or on their person allows application developers to develop contextual systems that sense what its user is doing and respond to it. Some of the popular sensors on a smartphone are light, proximity, and inertial sensors such as accelerometer, gyroscope, magnetometer, barometer, microphone, Bluetooth, and Wi-Fi radio. Fusion of these sensors can help detect complex activities and cancel out potential errors caused by a single sensor. With new advances, smartphones are expected to be loaded with more features such as gaze tracking, where the phone can detect when the person is looking at the screen. Other possible advancements include the ability to transform a smartphone into a medical device, where attached peripherals can perform more specialized medical monitoring functions. Another future direction in this research is the synergistic fusion of smartphones with smart environmental sensors to provide more advanced applications. For example, the smartphone could communicate with the microwave and fridge and deduce the number of calories a person is consuming and combine this information with the amount of calories the person is burning daily to give a net calorie balance.

Most of the physiological signs of a person are challenging to monitor implicitly using a smartphone alone, since these devices are not typically in contact with the body of its user. Smartphones are typically used in mainly three ways as medical devices: (i) implicitly, where the smartphone automatically detects and monitors certain conditions with its user; (ii) explicitly, where a person places the phone in an appropriate position for measurement; and (iii) in conjunction with other peripheral devices. Some applications of explicit smartphone sensing are heart rate monitoring, lung function monitoring, and so on. To detect heart rate, researchers propose the use of the camera in the smartphone. In this technique, a person places their finger over a camera and waits for a few seconds. As the blood pumps in, it obstructs the light coming through the finger momentarily. The person's heart rate can be estimated by detecting the rate at which the light gets obstructed. Another application of explicit sensing is that of measuring lung function or spirometry. Spirometry is a method to assess the condition of lungs in which a person exhales into a tube-like device, which measures the instantaneous flow and cumulative exhaled volume of air. Researchers propose a method in which a person holds a smartphone at one arm's distance from the mouth and blows full lung capacity toward the phone [10]. The in-built microphone of the smartphone detects the flow rate using sound frequency features and determines the flow and volume of the air exhaled. One of the few implicit monitoring applications of smartphones include footstep count, in which an accelerometer is continuously sampled at a low frequency [71]. Another exciting application of implicit sensing is in detecting sleep apnea. The proposed system senses for modulated ultrasonic waves transmitted by a phone to detect the movement of the chest of a person. Once the system establishes a breathing pattern, it can then register every time a person changes breathing pattern or stops breathing [72]. In sensing using peripheral devices, one proposed system looks at determining the refractive error in human eye. The system consists of a near-eye optical probe connected to a smartphone. The system substitutes mechanically, moving parts with moving patterns on a digital screen, to determine problems with the human eye [73].

Smartphones can also be used for emotive tracking in implicit and explicit ways. In implicit method, a smartphone samples its sensors at regular intervals to determine if it can detect the emotional state of its user. For example, researchers have proposed a mood-detecting techniques that mines a person's usage of their smartphone to determine their mood [74]. This is based on the intuition that people use their smartphone differently when they are in different mood states. The proposed system leverages these patterns by learning about its user and associating smartphone usage patterns with certain moods. Another implicit method continuously detects a person's stress levels in their voice using the microphone in the smartphone [75]. One of the main challenges in this type of system is the presence of diverse ambient noise that the microphone picks up as well. In explicit method, the smartphone knows when it's being used and therefore has higher chances of capturing a person's facial expression or voice. For example, researchers have explored the use of gaze tracking in smartphone to determine user engagement with the contents of the phone [76]. The small size of the phone screen and the quality of the front facing camera make traditional gaze tracking algorithms ineffective for use with smartphones as is. Researchers have tested various features of dwell time selections and gaze gestures to determine what is effective for a small screen. Facial recognition is another way of detecting the emotional state of a person such as surprise, fear, sadness, anger, disgust, and happiness. Traditionally, this has been done using computer vision and requires high computing resources. With the new high computing resources packed in smartphones, researchers have demonstrated how smartphones can be used for face detection, facial feature tracking, and classification of facial expressions [77, 78]. Smartphones are also useful for detecting social group emotions. By using Bluetooth to discover the proximity of other people, microphones, and accelerometer, systems are able to determine the patterns of interaction and the correlation of emotions with places, groups, and activity [79].

Smartphones have also been widely studied as a means for detecting the functional activities performed by their users. The onboard accelerometer is one of the most commonly used sensors for activity recognition. For example, in a system proposed by researchers [80], the accelerometer is used with kernel discriminant analysis to determine if the person is walking, running, climbing up stairs, or climbing down stairs. The inaccuracies of smartphones in detecting complex activities stem from the fact that users often carry their phones in different pockets and often remove it from their person. A proposed system claims that if smartphones are worn by their users in an arm holster, it can be used to detect 10;12 different types of upper body exercises performed by the user [81]. In addition to ambulatory activities, such as cycling, lying down, walking, and so on, it is also important to detect other complex activities, such as eating, reading, cooking, and so on. However, to detect these activities, the use of smartphone alone is not sufficient. Smartphones can be augmented with more on-body sensors to be able to detect these activities. For example, a proposed system to detect activities [82] uses five on-body accelerometer and microphone sensors to detect some functional activities such as watching TV, cleaning house, and so on. Another approach is to augment the objects in the environment, which are used for different activities, with sensors. For example, a proposed system [83] uses smartphones carried by people that are augmented with a magnetic beacon transmitter, which is unique for every person, and receivers installed on objects, such as refrigerator, microwave, stove, and so on. When the object detects that it is being used, it detects who is using it by detecting the strongest magnetic beacon it receives. It is possible to detect what activity a person is doing by detecting what object the person is using.

Smartphones can locate people indoors as well as outdoors. Most of the localization techniques are standard with smartphones as they are with wearable devices. To locate people outdoors there are a couple of different methods. The use of GPS sensor in the phone is one of the most common ways. However, GPS sensor only provides coordinates of a person's location. GPS coordinates need to be translated into more meaningful data for other applications to make use of it. For example, researchers have developed a system that takes as input GPS data streams generated by users' phones and creates a searchable database of locations and activities. To do this they use an external database of businesses and their locations, along with possible activities, and match this data with the location of the person [84]. The biggest drawback of using GPS for tracking is that it drains the battery of the smartphone very fast. Another way to locate a person outdoors is using Wi-Fi radio. In this technique, a person's proximity to a tagged and known Wi-Fi AP in the city is used to locate where the person is [85]. Of course, this technique is prone to failure in regions where there aren't many Wi-Fi APs, such as in rural places. Another relative energy-efficient method of tracking outdoors is using cell tower triangulation. However, the biggest drawback of this method is its large granularity of tracking that may be imprecise for certain applications. Tracking a person indoors is perhaps more challenging than tracking them outdoors. While there are methods to track a person indoors using Wi-Fi fingerprinting, and other radio-based methods, there are new techniques that automatically learn the characteristics of an indoor floor plan and can determine where the person is located [86]. The key intuition behind the fusion approach is that certain locations of indoor environment pose a distinct signature for the various sensors on a smartphone. For example, an elevator has a unique accelerometer pattern, corridors may have a distinct set of visible Wi-Fi AP and signal strengths, and so on. Each of these observations can help pinpoint the current location of a person without the need to set up infrastructural support or additional sensors.

Smartphones are versatile devices. The processing power and unique array of onboard sensors make it useful in all aspects of human context sensing. However, the key bottleneck of smartphone technology remains its limited battery life. While battery technology has improved in the latest smartphone models, so has the number of high-powered components. Bigger screens and high power sensors sampled at a fast frequency in the background are some of the reasons that attribute to limited battery life. Other than technology, another factor that limits the utility of smartphones in human context sensing is the fact that phones do not remain on their users' person all the time, especially when people are at home or are charging the phone. In addition to these issues, there remains an unsolved problem of privacy and security due to a smartphone's constant or near-constant connection to the Internet. Just as smartphones have access to their user's lifestyle by means of various sensors, so do hackers, who can use this information for malicious purposes. As such application developers need to ensure that privacy and security are features that are built in to the applications, while smartphone developers need to build operating systems that are resilient to hacking attempts.

11.3.4 Environment

The last major technology type to be used for human context sensing isn't so much a technology as a sensing location. The above sensing technologies, such as video and audio, wearables, and smartphones, have two major disadvantages to widespread, practical use: privacy concerns and the requirement of a carried device. Many people are concerned about the invasion of privacy implicit in video and audio recording, as information beyond that required for the application, sometimes sensitive or personal information, can be recorded and stored in a way that could be hackable or seen by others. Wearables and smartphones have issues with practicality since they must be carried by the user being sensed for the device to work. This means that whenever someone forgets to grab their cell phone or put on their wearable device, they collect no data for the duration. If the user gets bored with the device or finds that it no longer suits their fashion or lifestyle, that data can be lost indefinitely. To deal with these two issues, environment-based sensing has been introduced for many human-based context-sensing applications. This type of sensing places sensors in the environment that a user will interact with – chairs, rooms, walls, appliances, and so on – and senses a person's context from their interactions with those objects (Figure 11.3).

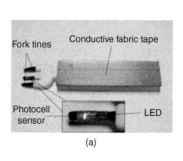
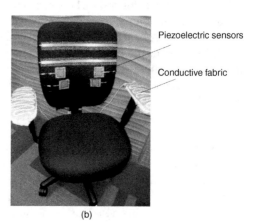

Figure 11.3 A variety of sensors embedded in the environment can sense the functional context of a child's eating habits (a) or physiological context such as heart rate and respiratory rate when a person is sitting (b). Griffiths (2014) [8]. Reproduced with permission of Proceedings of the 2014 ACM International Joint Conference on Pervasive and Ubiquitous Computing, ACM and Kadomura (2013) [87]. Reproduced with permission of Human Factors in Computing Systems, ACM.

Physiological sensing is the hardest to perform from the environment, since so many of our physiological signs present below the surface of our skin. Sensing these signs often requires a firm, steady contact with the environmental object in order to obtain a good measurement – think of the heart rate monitors seen in exercise equipments. In order to obtain a heart rate, a person must have a firm, full contact and relatively unmoving interaction with the sensor. In the environment around us, the objects that we interact with most commonly in this manner are objects we either sit or lay on, including beds, chairs, car seats, and toilets. Because of this, these are the commonly used objects for physiological sensing from the environment. The solid, consistent contact of a toilet seat provides a way to detect blood pressure using ECG and PPG to detect the pulse arrival time and estimate the blood pressure of an individual [88]. The HealthChair senses heart rate and respiratory rate through ECG and pressure sensors attached to a standard office chair. It detects heart rate whenever an occupant uses both arms of the chair where the ECG is attached and respiratory rate whenever they press against the pressure sensors on the backrest [8]. Pressure sensors in beds are used to detect sleep time, quality, snoring, apnea, heart rate, respiratory rate, and body position [89].

Like physiological sensing, emotive sensing is difficult to perform from the environment. Without the benefit of sight previously provided by the video sensor, emotion must be sensed through either physiological signs such as heart or respiratory rate or by attributing emotion to a person's interaction with an object. The sensing technology used to collect emotive data is similar to physiological sensors and those found in wearables but are embedded into the environment. To be able to attribute emotion to this data, the sensors are often embedded in specific use objects, such as a computer mouse and keyboard, to sense a person's emotions as they perform an activity with that object. Work on the emotion mouse prototypes the use of temperature sensors, a galvanic skin response sensor, and a heart rate sensor to determine a user's emotions as they interact with a computer program [90]. Pressure-sensing keyboards have also been explored to sense user frustration with a computer program – the more frustrated you are, the harder you press each key [91]. Emotive sensing from the environment that does not include physiological information tends to fall into the category of posture recognition. Our posture, especially in chairs, tends to indicate emotions such as engagement or how interested we are in an activity. By using a pressure sensor array attached to the seat and back of a chair, Mota and Picard detect the interest level of a learner with the computer program they are using [92].

Functional sensing from the environment falls into two main categories: a person's skill in interacting with a specific object and the perceived activity from interactions with many instrumented objects. The first, interacting with a specific object, is used when that object has only one specific use. This includes things like spoons and forks, coffee makers, showers, and so on. that are unlikely to be used for a task other than their common one (a shower is used very infrequently to wash clothes rather than people). Work in this area embeds sensors into these objects to detect when and how a person interacts with them. One example includes the Sensing Fork, an instrumented fork that, combined with a tablet or phone for feedback, senses a child's interaction with the fork and their food and encourages them to eat [87]. Food color is determined through a single pixel RGB sensor, and contact with food or contact with a child's mouth is sensed through conductance – a circuit is made between the tines of a fork through the food when food is touched or between the tines and the handle through the grip and mouth of the child when it's being eaten. Looking at more complex activities, such as making dinner, requires contact with many objects in the environment. When we cook dinner, for example, we might interact with cupboards, the refrigerator, the stove, the sink, the trashcan, the dishwasher, the oven, and the microwave. However, we might contact many of these for just washing dishes or eating snacks. The environment-based solution to detecting and differentiating these activities involves instrumenting everything in the environment and then learning to recognize what interactions are a specific activity through rule-based or machine learning classification algorithms [93].

Location sensing is the best fit for environment sensing as this human context sensing is about localizing the person within the environment. A sensor for location must know of the existence of the environment in order to find a person's location, and sensors placed at specific points in the environment know their own location and can map a person within their knowledge of the environment. Technologies, algorithms, and sensors themselves vary widely in research efforts to perform this localization. Some examples of current technologies can be found in [36], which includes systems that sense from the environment, both with and without carried devices. To highlight the systems that require no user participation, separating them from the wearable and smartphone solutions above, we discuss only systems that do not require carried devices. The intuitive example of this is the SmartFloor, where the entire floor of a space is covered with pressure sensors to perform localization anywhere in that space [94]. However, the high installation cost makes this and similar solutions too expensive for practical use. Other systems look at a more limited understanding of location, such as the Doorjamb system, which localizes a person to a room by monitoring only the doorways in a home with ultrasound sensors to detect when a person moves from room to room [35]. Lastly, some systems look at using the existing infrastructure in a building to perform localization and decrease cost. The WiTrack system uses the wireless signals sent out by Wi-Fi, signals that change significantly when they hit a human body, to localize people and even track some of their leg and arm movements [95].

While sensing from the environment has the benefits of not requiring users to carry devices and being minimally privacy invasive compared to cameras, it has a number of detractors. The first and most cost impacting is the huge number of sensors required. The entire environment that a person goes through, including home, work, car, and so on, must be instrumented to perform anything close to the continuous sensing possible with wearable objects. This can be very costly, especially if the instrumentation is meant to monitor only one person. Amortizing the costs by monitoring multiple people from the environmental infrastructure causes its own problem, mainly determining the identity of the specific person being sensed. With video monitoring, this can be done using facial recognition or similar techniques. With environmental sensing, weaker identifying characteristics must be used, such as a person's weight in the SmartFloor or height in Doorjamb. Often, these systems are unable to determine the identity of the person they sense. Because of this, environmental sensing is commonly limited to applications where the identity is known, such as a personal office chair for the HealthChair, or where the identity is unimportant. Additionally, applications that support opportunistic sensing, where an occasional measurement is all that is needed, often use environmental technologies. The HealthChair operates on this principle, where occasional health readings when a person interacts with the chair still provides a historic day-by-day or week-by-week set of health measurements.

11.4 Conclusion

Around the world, cities are adopting smart sensor technologies to improve the lives of their citizens. While some goals for smart cities are generic, such as conserving water, reducing electricity consumption, and eliminating traffic jams, other goals cater very specifically to the requirements of people, such as planning assistive facilities for senior citizens, deciding new bus stops, and so on. Knowing the abilities, lifestyles, and impending problems faced by its citizens can help a smart city adapt its systems appropriately. Therefore, for a city to function efficiently for its citizens, sensing the context of its citizens' lives remains one of its most crucial endeavors. In this chapter, we have discussed human context sensing as four different categories that aim to holistically capture a person's state and their interactions with the world: physiological sensing, emotive sensing, functional sensing, and location sensing. Together, these facets capture the mental states, physical body conditions, lifestyles, and location of individuals. The goals and applications for each category unify in improving the quality of life of an individual by monitoring different aspects of life that can help the city provide an individual with the right level of assistance and facilities.

We have also discussed the impact of the four main technological thrusts in each category of human context sensing: video and audio, wearables, smartphones, and environmental sensing. Each type of technology has a unique set of sensing abilities as well as constraints. For example, while audio and video sensing technologies have rich information content, they are hard to process automatically and also pose privacy concerns. The smartphone technology is pervasive and has a full suite of sensors that can sense many aspects of a person's life. But a limited battery life and the fact the people don't always carry their phones on their person make it unreliable for many applications. This is where wearable devices have an advantage, since these devices are present on a person's body most of the time. However, wearable devices suffer mainly from short battery life and limited processing capabilities. Environmental sensors have the advantage that they don't require users to carry devices with them in order to guess their context. However, this works well only when there is a single person in a given space. With multiple people sharing the same environment, such as roommates, it may become necessary to use personal devices along with environmental sensors to disambiguate each person's activities in the space.

One of the key takeaways from our discussion is that all categories of human context sensing are intertwined, as are the technologies that support them. This is because the state of the mind and the body affect the lifestyles that people live and the places they choose to spend time at. Conversely, people's lifestyles affect the state of their mind and body. The various human context technologies are not isolated from each other either. Developments in each of these sectors advance the capabilities of others. For example, smartphone technology is highly dependent on advances in audio and video processing, as well as environmental sensing.

Final Thoughts

In this chapter we discussed the concept of human context sensing, the definitions of the four main types of human contexts, and the current technological sensing mechanisms. We discuss in detail the definition, applications, and measurable parameters of each of the four types of human context. And finally, we describe and critique technological methods that can be used to sense the different contexts. We believe that in the future smart cities will have several applications, which will be tailored to the people living in them, and in creating such applications, human context sensing will be the key to making critical design decisions.

Questions

1   What are the main types of human context sensing?

2   What are some of the contexts that are sensed for: (a) an individual and (b) groups of individuals?

3 What are the main technologies used in sensing human context, and how are they interdependent?

4 Describe strategies that can mitigate the main limitations of using video-based technology in sensing human context.

5 Describe two smart city applications that are dependent on multiple human contexts.

References

1 United Nations Department of Economic and Social Affairs (2014) *World's Population Increasing Urban*, https://www.un.org/development/desa/en/news/population/world-urbanization-prospects.html (accessed 16 December 2016).

2 Lukowicz, P., Anliker, U., Ward, J., Tröster, G., Hirt, E., and Neufelt, C. (2002) *AMON: A Wearable Medical Computer for High Risk Patients*. Proceedings of the 6th International Symposium on Wearable Computers, IEEE, p. 0133.

3 National Center for Chronic Disease Prevention and Health Promotion (2009) *The Power of Prevention: Chronic disease...the Public Health Challenge of the 21st Century*, http://www.cdc.gov/chronicdisease/pdf/2009-Power-of-Prevention.pdf (accessed 16 December 2016).

4 Baek, H.J., Chung, G.S., Kim, K.K., and Park, K.S. (2012) A smart health monitoring chair for nonintrusive measurement of biological signals. *IEEE Transactions on Information Technology in Biomedicine*, **16** (1), 150;158.

5 Grajales, L. and Nicolaescu, I.V. (2006) *Wearable Multisensor Heart Rate Monitor*. Wearable and Implantable Body Sensor Networks, 2006. BSN 2006. International Workshop on, IEEE, pp. 154;157.

6 Jovanov, E., Donnel, A.O., Morgan, A., Priddy, B., and Hormigo, R. (2002) *Prolonged Telemetric Monitoring of Heart Rate Variability Using Wireless Intelligent Sensors and a Mobile Gateway*. Engineering in Medicine and Biology, 2002. 24th Annual Conference and the Annual Fall Meeting of the Biomedical Engineering Society EMBS/BMES Conference, 2002. Proceedings of the Second Joint, vol. 3, IEEE, pp. 1875;1876.

7 Jovanov, E., Raskovic, D., Price, J., Krishnamurthy, A., Chapman, J., and Moore, A. (2001) Patient monitoring using personal area networks of wireless intelligent sensors. *Biomedical Sciences Instrumentation*, **37**, 373;378.

8 Griffiths, E., Saponas, T.S., and Brush, A. (2014) *Health Chair: Implicitly Sensing Heart and Respiratory Rate*. Proceedings of the 2014 ACM International Joint Conference on Pervasive and Ubiquitous Computing, ACM, pp. 661–671.

9 Larson, E.C., Goel, M., Boriello, G., Heltshe, S., Rosenfeld, M., and Patel, S.N. (2012) *SpiroSmart: Using a Microphone to Measure Lung Function on a Mobile Phone.* Proceedings of the 2012 ACM Conference on Ubiquitous Computing, ACM, pp. 280–289.

10 Gupta, S., Chang, P., Anyigbo, N., and Sabharwal, A. (2011) *mobileSpiro: Accurate Mobile Spirometry for Self-Management of Asthma.* Proceedings of the 1st ACM Workshop on Mobile Systems, Applications, and Services for Healthcare, ACM, p. 1.

11 Dell, N., Francis, I., Sheppard, H., Simbi, R., and Borriello, G. (2014) *Field Evaluation of A Camera-Based Mobile Health System in Low-Resource Settings.* Proceedings of the 16th International Conference on Human-Computer Interaction with Mobile Devices & Services, ACM, pp. 33–42.

12 Otto, C., Milenkovic, A., Sanders, C., and Jovanov, E. (2006) System architecture of a wireless body area sensor network for ubiquitous health monitoring. *Journal of Mobile Multimedia*, **1** (4), 307–326.

13 Khusainov, R., Azzi, D., Achumba, I.E., and Bersch, S.D. (2013) Real-time human ambulation, activity, and physiological monitoring: taxonomy of issues, techniques, applications, challenges and limitations. *Sensors*, **13** (10), 12 852–12 902.

14 Agu, E., Pedersen, P., Strong, D., Tulu, B., He, Q., Wang, L., and Li, Y. (2013) *The Smartphone as a Medical Device: Assessing Enablers, Benefits and Challenges.* Internet-of-Things Networking and Control (IoT-NC), 2013 IEEE International Workshop of, IEEE, pp. 48–52.

15 Yilmaz, T., Foster, R., and Hao, Y. (2010) Detecting vital signs with wearable wireless sensors. *Sensors*, **10** (12), 10 837–10 862.

16 Munson, S.A., Cavusoglu, H., Frisch, L., and Fels, S. (2013) Sociotechnical challenges and progress in using social media for health. *Journal of Medical Internet Research*, **15** (10), e226.

17 eMarketer (2015) *US Digital Ad Spending Will Approach $60 Billion This Year, with Retailers Leading the Way*, http://www.emarketer.com/Article/US-Digital-Ad-Spending-Will-Approach-60-Billion-This-Year-with-Retailers-Leading-Way/1012497 (accessed 16 December 2016).

18 Alina Dizik, B. (2014) *Does Your Child Really Need a Private Tutor?* http://www.bbc.com/capital/story/20131016-the-global-tutoring-economy (accessed 16 December 2016).

19 Spaulding, S. and Breazeal, C. (2015) *Affect and Inference in Bayesian Knowledge Tracing with a Robot Tutor.* Proceedings of the 10th Annual ACM/IEEE International Conference on Human-Robot Interaction Extended Abstracts, ACM, pp. 219–220.

20 You, C.W., Lane, N.D., Chen, F., Wang, R., Chen, Z., Bao, T.J., Montes-de Oca, M., Cheng, Y., Lin, M., Torresani, L. et al. (2013) *CarSafe App: Alerting Drowsy and Distracted Drivers Using Dual Cameras on*

Smartphones. Proceeding of the 11th Annual International Conference on Mobile Systems, Applications, and Services, ACM, pp. 13–26.

21 Healey, J. and Picard, R.W. (2005) Detecting stress during real-world driving tasks using physiological sensors. *IEEE Transactions on Intelligent Transportation Systems*, **6** (2), 156–166.

22 Sathyanarayana, A., Nageswaren, S., Ghasemzadeh, H., Jafari, R., and Hansen, J.H. (2008) *Body Sensor Networks for Driver Distraction Identification*. Vehicular Electronics and Safety, 2008. ICVES 2008. IEEE International Conference on, IEEE, pp. 120–125.

23 Bao, X., Fan, S., Varshavsky, A., Li, K., and Choudhury, R.R. (2013) *Your Reactions Suggest You Liked the Movie: Automatic Content Rating via Reaction Sensing*. Proceedings of the 2013 ACM International Joint Conference on Pervasive and Ubiquitous Computing, ACM, pp. 197–206.

24 Obrador, P. (2003) *Video indexing based on viewers' behavior and emotion feedback*. US Patent 6,585,521.

25 Mahoney, D.F., Burleson, W., Lozano, C., Ravishankar, V., and Mahoney, E.L. (2015) Prototype development of a responsive emotive sensing system (dress) to aid older persons with dementia to dress independently. *Gerontechnology*, **13** (3), 345–358.

26 Wu, D., Courtney, C.G., Lance, B.J., Narayanan, S.S., Dawson, M.E., Oie, K.S., and Parsons, T.D. (2010) Optimal arousal identification and classification for affective computing using physiological signals: virtual reality stroop task. *IEEE Transactions on Affective Computing*, **1** (2), 109–118.

27 Alghowinem, S., Goecke, R., Wagner, M., Epps, J., Breakspear, M., and Parker, G. (2013) *Detecting Depression: A Comparison Between Spontaneous and Read Speech*. Acoustics, Speech and Signal Processing (ICASSP), 2013 IEEE International Conference on, IEEE, pp. 7547–7551.

28 Foti, D. and Kanazawa, L. (2008) Activities of daily living, in *Pedretti's Occupational Therapy: Practice Skills for Physical Dysfunction*, vol. **6** (eds H.M. Pendleton and W. Schultz-Krohn), Mosby Elsevier, St. Louis, MS, pp. 146–194.

29 BankRate (2013) *5 Ways to Cover Assisted Living Expenses*, http://www .bankrate.com/finance/insurance/paying-for-assisted-living-1.aspx (accessed 16 December 2016).

30 Ranjan, J. and Whitehouse, K. (2015) Rethinking the fusion of technology and clinical practices in functional behavior analysis for the elderly, in *Human Behavior Understanding*, Springer, pp. 52–65.

31 Rajan, K.B., Hebert, L.E., Scherr, P.A., de Leon, C.F.M., and Evans, D.A. (2013) Disability in basic and instrumental activities of daily living is associated with faster rate of decline in cognitive function of older adults. *The Journals of Gerontology Series A: Biological Sciences and Medical Sciences*, **68** (5), 624–630.

32 Covinsky, K.E., Palmer, R.M., Fortinsky, R.H., Counsell, S.R., Stewart, A.L., Kresevic, D., Burant, C.J., and Landefeld, C.S. (2003) Loss of independence in activities of daily living in older adults hospitalized with medical illnesses: increased vulnerability with age. *Journal of the American Geriatrics Society*, **51** (4), 451–458.

33 Thomas, V.S., Rockwood, K., and McDowell, I. (1998) Multidimensionality in instrumental and basic activities of daily living. *Journal of Clinical Epidemiology*, **51** (4), 315–321.

34 AAA (2011) *AAA Study Finds Costs Associated with Traffic Crashes are More than Three Times Greater than Congestion Costs*, http://newsroom .aaa.com/2011/11/aaa-study-finds-costs-associated-with-traffic-crashes-are-more-than-three-times-greater-than-congestion-costs/ (accessed 16 December 2016).

35 Hnat, T.W., Griffiths, E., Dawson, R., and Whitehouse, K. (2012) *Doorjamb: Unobtrusive Room-Level Tracking of People in Homes Using Doorway Sensors*. Proceedings of the 10th ACM Conference on Embedded Network Sensor Systems, ACM, pp. 309–322.

36 Lymberopoulos, D., Liu, J., Yang, X., Choudhury, R.R., Handziski, V., and Sen, S. (2015) *A Realistic Evaluation and Comparison of Indoor Location Technologies: Experiences and Lessons Learned*. Proceedings of the 14th International Conference on Information Processing in Sensor Networks, ACM, pp. 178–189.

37 Ranjan, J., Griffiths, E., and Whitehouse, K. (2014) *Discerning Electrical and Water Usage by Individuals in Homes*. Proceedings of the 1st ACM Conference on Embedded Systems for Energy-Efficient Buildings, ACM, pp. 20–29.

38 Konofal, E., Lecendreux, M., Bouvard, M.P., and Mouren-Simeoni, M.C. (2001) High levels of nocturnal activity in children with attention-deficit hyperactivity disorder: a video analysis. *Psychiatry and Clinical Neurosciences*, **55** (2), 97–103.

39 Heinrich, A., Aubert, X., and de Haan, G. (2013) *Body Movement Analysis During Sleep Based on Video Motion Estimation*. e-Health Networking, Applications & Services (Healthcom), 2013 IEEE 15th International Conference on, IEEE, pp. 539–543.

40 Poh, M.Z., McDuff, D.J., and Picard, R.W. (2011) Advancements in non-contact, multiparameter physiological measurements using a webcam. *IEEE Transactions on Biomedical Engineering*, **58** (1), 7–11.

41 de Greef, L., Goel, M., Seo, M.J., Larson, E.C., Stout, J.W., Taylor, J.A., and Patel, S.N. (2014) *BiliCam: Using Mobile Phones to Monitor Newborn Jaundice*. Proceedings of the 2014 ACM International Joint Conference on Pervasive and Ubiquitous Computing, ACM, pp. 331–342.

42 Kale, A., Cuntoor, N., Yegnanarayana, B., Rajagopalan, A., and Chellappa, R. (2003) Gait analysis for human identification, in *Audio-and Video-Based Biometric Person Authentication*, Springer, pp 1058–1058.

43 Cohen, I., Sebe, N., Garg, A., Chen, L.S., and Huang, T.S. (2003) Facial expression recognition from video sequences: temporal and static modeling. *Computer Vision and Image Understanding*, **91** (1), 160–187.

44 Kapur, A., Kapur, A., Virji-Babul, N., Tzanetakis, G., and Driessen, P.F. (2005) Gesture-based affective computing on motion capture data, in *Affective Computing and Intelligent Interaction*, Springer, pp. 1–7.

45 Petrushin, V. (1999) *Emotion in Speech: Recognition and Application to Call Centers*. Proceedings of Artificial Neural Networks in Engineering, vol. 710.

46 Davis, R.B., Ounpuu, S., Tyburski, D., and Gage, J.R. (1991) A gait analysis data collection and reduction technique. *Human Movement Science*, **10** (5), 575–587.

47 Ghidary, S.S., Nakata, Y., Takamori, T., and Hattori, M. (2000) *Human Detection and Localization at Indoor Environment by Home Robot*. Systems, Man, and Cybernetics, 2000 IEEE International Conference on, vol. 2, IEEE, pp. 1360–1365.

48 Yu, C.R., Wu, C.L., Lu, C.H., and Fu, L.C. (2006) *Human Localization via Multi-Cameras and Floor Sensors in Smart Home*. Systems, Man and Cybernetics, 2006. SMC'06. IEEE International Conference on, vol. 5, IEEE, pp. 3822–3827.

49 Lo, D. (2007) *Multimodal Human Localization Using Bayesian Network Sensor Fusion*.

50 Oulasvirta, A., Pihlajamaa, A., Perkiö, J., Ray, D., Vähäkangas, T., Hasu, T., Vainio, N., and Myllymäki, P. (2012) *Long-Term Effects of Ubiquitous Surveillance in the Home*. Proceedings of the 2012 ACM Conference on Ubiquitous Computing, ACM, pp. 41–50.

51 Nelson, W.C., Ott, W.R., Robinson, J.P., Tsang, A.M., Switzer, P., Behar, J.V., Hern, S.C., and Engelmann, W.H. (2001) The National Human Activity Pattern Survey (NHAPS): a resource for assessing exposure to environmental pollutants. *Journal of Exposure Analysis and Environmental Epidemiology*, **11** (3), 231–252.

52 Lebak, J., Yao, J., and Warren, S. (2003) *Implementation of a Standards-Based Pulse Oximeter on a Wearable, Embedded Platform*. Engineering in Medicine and Biology Society, 2003. Proceedings of the 25th Annual International Conference of the IEEE, vol. 4, pp. 3196–3198.

53 Poh, M.Z., Loddenkemper, T., Swenson, N.C., Goyal, S., Madsen, J.R., and Picard, R.W. (2010) *Continuous Monitoring of Electrodermal Activity During Epileptic Seizures Using a Wearable Sensor*. Engineering in Medicine and Biology Society (EMBC), 2010 Annual International Conference of the IEEE, IEEE, pp. 4415–4418.

54 Villar, R., Beltrame, T., and Hughson, R.L. (2015) Validation of the Hexoskin wearable vest during lying, sitting, standing, and walking activities. *Applied Physiology, Nutrition and Metabolism*, **40** (10), 1019–1024.

55 TruPosture (2015) *TruPosture*, http://truposture.com/ (accessed 16 December 2016).

56 Spire (2015) *Spire*, http://spire.io/ (accessed 16 December 2016).

57 Zensorium (2015) *Zensorium*, https://www.zensorium.com/ (accessed 16 December 2016).

58 CBC (2015) *Microsoft's Eyeglasses that Detect Emotions*, http://www.cbc.ca/news/canada/british-columbia/microsoft-is-developing-eye glasses-that-detect-emotions-in-others-1.3062576.

59 moodwings (2015) *Mood Wings*, http://research.microsoft.com/apps/pubs/default.aspx?id=192695 (accessed 16 December 2016).

60 Kientz, J.A., Hayes, G.R., Westeyn, T.L., Starner, T., and Abowd, G.D. (2007) Pervasive computing and autism: assisting caregivers of children with special needs. *IEEE Pervasive Computing*, (1), 28–35.

61 Randell, C., Andersen, I., Moore, H., and Baurley, S. (2005) *Sensor Sleeve: Sensing Affective Gestures.* 9th International Symposium on Wearable Computers–Workshop on On-Body Sensing, pp. 117–123.

62 Pirsiavash, H. and Ramanan, D. (2012) *Detecting Activities of Daily Living in First-Person Camera Views.* Computer Vision and Pattern Recognition (CVPR), 2012 IEEE Conference on, IEEE, pp. 2847–2854.

63 Ermes, M., Parkka, J., Mantyjarvi, J., and Korhonen, I. (2008) Detection of daily activities and sports with wearable sensors in controlled and uncontrolled conditions. *IEEE Transactions on Information Technology in Biomedicine*, **12** (1), 20–26.

64 Kern, N., Schiele, B., and Schmidt, A. (2003) Multi-sensor activity context detection for wearable computing, in *Ambient Intelligence*, Springer, pp. 220–232.

65 Chen, J., Kwong, K., Chang, D., Luk, J., and Bajcsy, R. (2006) *Wearable Sensors for Reliable Fall Detection.* Engineering in Medicine and Biology Society, 2005. IEEE-EMBS 2005. 27th Annual International Conference of the, IEEE, pp. 3551–3554.

66 Amft, O., Junker, H., and Tröster, G. (2005) *Detection of Eating and Drinking Arm Gestures Using Inertial Body-Worn Sensors.* Wearable Computers, 2005. Proceedings. Ninth IEEE International Symposium on, IEEE, pp. 160–163.

67 Nishimura, J. and Kuroda, T. (2008) *Eating Habits Monitoring Using Wireless Wearable In-Ear Microphone.* Wireless Pervasive Computing, 2008. ISWPC 2008. 3rd International Symposium on, IEEE, pp. 130–132.

68 Ashbrook, D. and Starner, T. (2002) *Learning Significant Locations and Predicting User Movement with GPS.* Wearable Computers, 2002.(ISWC 2002). Proceedings of the 6th International Symposium on, IEEE, pp. 101–108.

69 Zhang, D., Chen, C., Zhou, Z., and Li, B. (2012) Identifying logical location via GPS-enabled mobile phone and wearable camera. *International Journal of Pattern Recognition and Artificial Intelligence*, **26** (08), 1260007.

70 Statista (2015) *Number of Smartphone Users Worldwide from 2014 to 2019*, http://www.statista.com/statistics/330695/number-of-smartphone-users-worldwide/ (accessed 16 December 2016).

71 Naqvib, N.Z., Kumar, A., Chauhan, A., and Sahni, K. (2012) Step counting using smartphone-based accelerometer. *International Journal on Computer Science and Engineering*, **4** (5), 675.

72 Nandakumar, R., Gollakota, S., and Watson, N. (2015) *Contactless Sleep Apnea Detection on Smartphones*. Proceedings of the 13th Annual International Conference on Mobile Systems, Applications, and Services, ACM, pp. 45–57.

73 Pamplona, V.F., Mohan, A., Oliveira, M.M., and Raskar, R. (2010) *NETRA: Interactive Display for Estimating Refractive Errors and Focal Range*. ACM Transactions on Graphics (TOG), vol. 29, ACM, p. 77.

74 LiKamWa, R., Liu, Y., Lane, N.D., and Zhong, L. (2013) *MoodScope: Building a Mood Sensor from Smartphone Usage Patterns*. Proceeding of the 11th Annual International Conference on Mobile Systems, Applications, and Services, ACM, pp. 389–402.

75 Lu, H., Frauendorfer, D., Rabbi, M., Mast, M.S., Chittaranjan, G.T., Campbell, A.T., Gatica-Perez, D., and Choudhury, T. (2012) *StressSense: Detecting Stress in Unconstrained Acoustic Environments Using Smartphones*. Proceedings of the 2012 ACM Conference on Ubiquitous Computing, ACM, pp. 351–360.

76 Dybdal, M.L., Agustin, J.S., and Hansen, J.P. (2012) *Gaze Input for Mobile Devices by Dwell and Gestures*. Proceedings of the Symposium on Eye Tracking Research and Applications, ACM, pp. 225–228.

77 Cho, K.S., Choi, I.H., and Kim, Y.G. (2012) Robust facial expression recognition using a smartphone working against illumination variation. *Appl. Math.*, **6** (2S), 403S–408S.

78 Wang, Y.C. and Cheng, K.T. (2011) *Energy-Optimized Mapping of Application to Smartphone Platform-A Case Study of Mobile Face Recognition*. Computer Vision and Pattern Recognition Workshops (CVPRW), 2011 IEEE Computer Society Conference on, IEEE, pp. 84–89.

79 Rachuri, K.K., Musolesi, M., Mascolo, C., Rentfrow, P.J., Longworth, C., and Aucinas, A. (2010) *EmotionSense: a Mobile Phones Based Adaptive Platform for Experimental Social Psychology Research*. Proceedings of the 12th ACM International Conference on Ubiquitous Computing, ACM, pp. 281–290.

80 Khan, A.M., Lee, Y.K., Lee, S., and Kim, T.S. (2010) *Human Activity Recognition via An Accelerometer-Enabled-Smartphone Using Kernel Discriminant Analysis*. Future Information Technology (FutureTech), 2010 5th International Conference on, IEEE, pp. 1–6.

81 Muehlbauer, M., Bahle, G., and Lukowicz, P. (2011) *What Can An Arm Holster Worn Smart Phone Do for Activity Recognition?* Wearable Computers (ISWC), 2011 15th Annual International Symposium on, IEEE, pp. 79–82.

82 Keally, M., Zhou, G., Xing, G., Wu, J., and Pyles, A. (2011) *PBN: Towards Practical Activity Recognition Using Smartphone-Based Body Sensor Networks.* Proceedings of the 9th ACM Conference on Embedded Networked Sensor Systems, ACM, pp. 246–259.

83 Cheng, Y., Chen, K., Zhang, B., Liang, C.J.M., Jiang, X., and Zhao, F. (2012) *Accurate Real-Time Occupant Energy-Footprinting in Commercial Buildings.* Proceedings of the 4th ACM Workshop on Embedded Sensing Systems for Energy-Efficiency in Buildings, ACM, pp. 115–122.

84 Feldman, D., Sugaya, A., Sung, C., and Rus, D. (2013) *iDiary: From GPS Signals to a Text-Searchable Diary.* Proceedings of the 11th ACM Conference on Embedded Networked Sensor Systems, ACM, p.6.

85 Rekimoto, J., Miyaki, T., and Ishizawa, T. (2007) LifeTag: Wifi-Based Continuous Location Logging for Life Pattern Analysis. LoCA, 2007, pp. 35–49.

86 Wang, H., Sen, S., Elgohary, A., Farid, M., Youssef, M., and Choudhury, R.R. (2012) *No Need to War-Drive: Unsupervised Indoor Localization.* Proceedings of the 10th International Conference on Mobile Systems, Applications, and Services, ACM, pp. 197–210.

87 Kadomura, A., Li, C.Y., Chen, Y.C., Tsukada, K., Siio, I., and Chu, H.-H. (2013) *Sensing Fork: Eating Behavior Detection Utensil and Mobile Persuasive Game.* CHI'13 Extended Abstracts on Human Factors in Computing Systems, ACM, pp. 1551–1556.

88 Kim, J.S., Chee, Y.J., Park, J.W., Choi, J.W., and Park, K.S. (2006) A new approach for non-intrusive monitoring of blood pressure on a toilet seat. *Physiological Measurement*, **27** (2), 203–211.

89 Watanabe, K., Watanabe, T., Watanabe, H., Ando, H., Ishikawa, T., and Kobayashi, K. (2005) Noninvasive measurement of heartbeat, respiration, snoring and body movements of a subject in bed via a pneumatic method. *IEEE Transactions on Biomedical Engineering*, **52** (12), 2100–2107.

90 Ark, W.S., Dryer, D.C., and Lu, D.J. (1999) *The Emotion Mouse.* The 8th International Conference on Human-Computer Interaction: Ergonomics and User Interfaces, vol. 1, pp. 818–823.

91 Zimmermann, P., Guttormsen, S., Danuser, B., and Gomez, P. (2003) Affective computing-a rationale for measuring mood with mouse and keyboard. *International Journal of Occupational Safety and Ergonomics*, **9** (4), 539–551.

92 Mota, S. and Picard, R.W. (2003) *Automated Posture Analysis for Detecting Learner's Interest Level.* Computer Vision and Pattern Recognition Workshop, 2003. CVPRW'03. Conference on, vol. 5, IEEE, pp. 49–49.

93 Tapia, E.M., Intille, S.S., and Larson, K. (2004) *Activity Recognition in the Home Using Simple and Ubiquitous Sensors*, Springer.

94 Orr, R.J. and Abowd, G.D. (2000) *The Smart Floor: A Mechanism for Natural User Identification and Tracking*. CHI'00 Extended Abstracts on Human Factors in Computing Systems, ACM, pp. 275–276.

95 Adib, F., Kabelac, Z., Katabi, D., and Miller, R.C. (2014) *3D Tracking via Body Radio Reflections*. Usenix NSDI, vol. 14.

12

Smart Cities and the Symbiotic Relationship between Smart Governance and Citizen Engagement

Tori Okner[1] *and Russell Preston*[2]

[1] *Senior Associate, Principle Group, Boston, MA, USA*
[2] *Design Director, Principle Group, Boston, MA, USA*

CHAPTER MENU

Smart Governance, 344
Case Study – Somerville, Massachusetts, 348
Looking Ahead, 365

Objectives

- To become familiar with smart governance through emerging best practices stakeholder engagement
- To become familiar with the importance of placemaking and human-scaled design with respect to smart governance
- To become familiar with the importance of city planning on the neighborhood scale that connects sincerely with local places
- To become familiar with how to organize an engaging, community-led planning process
- To become familiar with how to make informed decisions about technological solutions that have emerged through community discussions and planning.

This chapter recognizes that cities are at the forefront of innovation and are increasingly competing for scarce resources. It suggests that for smart cities to succeed they must facilitate human connection. This can be achieved through smart governance, circumspect implementation of the tenants of smart cities, along with human-centered planning and design. The symbiotic relationship between citizen engagement and smart governance is explored through a case study on Somerville, Massachusetts. Lessons learned and broad recommendations are then posted for readers' consideration.

Smart Cities: Foundations, Principles and Applications, First Edition.
Edited by Houbing Song, Ravi Srinivasan, Tamim Sookoor, and Sabina Jeschke.
© 2017 John Wiley & Sons, Inc. Published 2017 by John Wiley & Sons, Inc.

12.1 Smart Governance

Today, as the world's population continues to urbanize, demands on cities are ever increasing. Cities must plan for the impacts of climate change, the public health and economic burdens of sprawl development, and the increasing consumer choice to live and work in mixed-use, walkable neighborhoods, and do so while competing for limited resources. The old trope that necessity is the mother of invention has proven true in cities, where, in the absence of national leadership, municipalities are at the forefront of innovation on the issues of the day. From tackling climate change through the C40 to the Urban Food Policy Pact, cities are at the forefront as Benjamin Barber proffered when he imagined what *If Mayors Ruled the World*. Truly, the "optimism and opportunity is coming up from cities...that's where the real action is....it is a global phenomena" [1]. Municipal leaders are striving to do more with less, and while technology can facilitate that, success depends on smart governance (Figures 12.1 and 12.2)

Truly smart governance is more than "public policy keeping pace with technological developments" [2]. Technology alone will not improve our cities. Municipalities interested in embracing the tenants of smart cities – from smart economy, mobility, and environment, to smart people, living, and governance – must effectually engage the local citizenry and leverage technology to do so. Perhaps even more so than the need for research and innovation, the success of the smart cities depends on circumspect implementation.

The application of technology to support the creation of more human-scaled development has immense potential. In the last hundred years, technology has

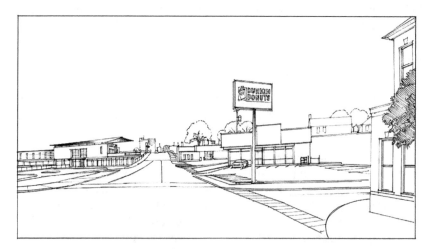

Figure 12.1 Union Square Prospect Street, no build condition. *Source*: Courtesy of David Carrico.

Figure 12.2 Union Square, Prospect Street, proposed. *Source*: Courtesy of David Carrico.

affected the way we organize ourselves and our cities more so than at any other time in history. Municipal water and sanitation systems have decreased the incidence of diseases. The elevator allowed us to reach the sky. The advent of personal automobile and postwar industrialization of the housing market allowed us to supply the American Dream to hundreds of thousands of families. Until only a few decades ago, these last two technologies were perceived as great steps forward; they have since been linked to dozens of social, environmental, and financial catastrophes. The technologies that supported suburban land use also allowed the fundamental pattern for how cities function to be disrupted with unintended consequences. Contemporary use of technology must learn from the past to support the growing challenges cities face today and facilitate sustainable, resilient growth. Smart solutions will be needed to support this shift toward walkable, mixed-use neighborhoods.

Cities that learn to attract growth through the creation of authentic places will be more competitive than those that do not.[1] "The future of thriving and resilient cities is not led by sustaining or innovating around infrastructure and services, but by building the capacity of communities to drive their own shared

1 Places that fully embrace local elements – food, climate, culture – create a more authentic character than those that do not. The generic and ready-made has been a key characteristic supporting the rapid growth of suburbia, and globalization, and in striking reaction to this everyplace, is a "no-place condition." People are increasingly looking for unique experiences that genuinely authentic places provide.

value – to sustain Placemaking" [3].[2] This analytical and objective approach to city building was pioneered by William "Holly" Whyte and popularized in his book *City: Rediscovering the Center*. The not-for-profit Project for Public Spaces has expanded on this work demonstrating that projects focusing on the improvement of social interactions in public space can be the catalyst for investment, change social behaviors, and help to build better connections in a community, oftentimes leading to further improvements. Proper organization of emerging technologies for transportation, communication, development, and finance is the key to this puzzle. It is essential that the technologies support human-scaled details that make these authentic places work. Citizen engagement is crucial to achieving this balance and to understand the unique details of a place.

Early visions of the smart city were built on urban modeling software. Urban dynamics, computer simulations, computer-based urban simulations, predictive city simulators – these efforts to apply system dynamics to urban centers have now largely been dismissed due to their severe constraints [4]. Historically, these constraints have been primarily the limitations of computing power, access to large amounts of current data about a place, and the reliance on incorrect assumptions when designing the modeling. In the early 1970s, these constraints led many cities to fundamentally question the recommendations that these models produced.

Natural metaphors later came into vogue pioneered through the writing of Jane Jacobs, Christopher Alexander, and others. The common element in these foundational works is the idea that cities function as ecosystems with patterns and feedback loops much like those found in nature. This thinking has led to revisions in system dynamics that represent cities now, to more accurately capture the complexity that these ecosystems possess. As technology has advanced and large amounts of real-time data have been opened up for analysis, we are finding that the dynamic modeling of a city's performance is again helping assist community leaders with decision making.

Citizen engagement insures that hyperlocal insights can be leveraged to inform the system and identify leverage points for community improvements and growth. Smart governance is needed to facilitate this fine-grained engagement and to maximize the use of smart cities technology. There is, in short, a symbiotic relationship between citizen engagement and smart governance. To effectively use technology, planners must recognize that cities are made up of a network of individual neighborhoods – neighborhoods that have different problems, opportunities, and constraints. Cities are polycentric and technology can facilitate these local systems, because "by giving lots of touch points

---

2 In this instance, "placemaking," is defined as "a collaborative process by which we can shape our public realm in order to maximize shared value." http://www.pps.org/reference/what_is_placemaking/.

across the city, making the government more visible, you thicken the mesh of democratic engagement" [1]. Smart cities will enliven the nodes, or touch points, but smart governance is necessary to ensure the data are activated; smart governance ensures that data-driven management does not rest on the false certainty of analytics but is informed by citizen engagement [1].

12.1.1 Smart Governance and Smart Cities

Today's smart city is understood to be a place "where information technology is combined with infrastructure, architecture, everyday objects, and even our bodies to address social, economic and environmental problems" [4]. However, smart cities are a means to an end – a more sustainable and vibrant community.

The diffusion of digital technology in the public realm, specifically Internet-based technology, has lagged far behind the private sector. Nevertheless, expectations for "E-government" reflect the heightened expectations for efficiency and transparency enabled by technology in other realms of private life. While traditional government service was based on face-to-face interaction with constituents and paper trails through bureaucracy, there is now less tolerance for inefficiency. Indeed, there is the "expectation of networked delivery" even in governance [5]. There has been a culture shift due to access in technology elsewhere; there is a desire for more immediate access to the information, both for more modern customer service in constituent services and for more open government [4, 6]. Stephen Goldsmith, Professor of Practice at the Ash Center and Director of the Data-Smart City Solutions project, observed that technological tools – from cloud computing to mobile and data mining tools – provide a "breathtaking opportunity" for change if governments embrace them [1].

Where cities are becoming smarter and more sustainable, it is often in response to the demands of their constituents. According to the Ash Center for Democratic Governance and Innovation at Harvard Kennedy School, "people are strengthening their relationships with governments to improve their lives in unprecedented ways...Individuals and groups at the local level are what drive it (civic innovation): they're harnessing access to data and technology to redefine how we interact with where we live" [7].

12.1.2 The Role of Planning & Design

The daily experiences of a city's population represent an often-untapped source of knowledge about how the city could be improved. Tapping into this knowledge base is essential for smart governance. Historically, planning departments have been the primary channel for harnessing this knowledge for the public good. Yet, it is difficult to determine if the facet of the public participating in planning efforts truly represents the issues and desires of the community at large. Smart cities can better utilize information and communication technologies to connect with stakeholders during planning efforts with the objective of bringing broader participations and more input into the creation of these plans. Robust public participation leads to the creation of more authentic plans that are easier to implement. Smart governance can benefit greatly by establishing real-time planning feedback channels within their neighborhoods.

Today, a city needs to be considering how it can improve its streets in 5 years, develop new transportation options in 20 years, and establish the framework for reinvestment over the next 100 years. Municipal planning is essential to envisioning the possibilities. Smart governance empowers citizens to answer the questions and in turn prioritize which technologies are essential to fulfilling a city's short- and long-term goals. Cities that tend to be the most economically sound plan for the future. When a city has a clear physical vision for its future, has organized its capital budget to tie directly to projects and programs to implement that vision, and establishes feedback channels to monitor the progress, it is evident how powerful a systems approach to economic development can be. The technology of smart cities can support the greatest of economic development opportunities and the fundamental purpose of cities, to facilitate the gathering, meeting, and collaboration between humans [4].

Planning cities is never complete. Neighborhoods need to deal with new conditions that emerge every day. Smart cities have the opportunity to deploy and facilitate dynamic communication with their citizens so that planning can occur in real time. Static reports and cumbersome master plans do not help municipalities create healthy communities in the ever-changing economic, environmental, and social world we now live in. Planners need to help communities establish their vision and facilitate ongoing dialogue, react, and learn with each step toward fulfilling that vision.

Once a city establishes dynamic communication channels, planning changes from a linear, government-led process to one that is much more collaborative and iterative. This dialogue leads to a more open design process that can accommodate greater participation from the public in a productive and enjoyable process. The technology being deployed is not complex but when coordinated well can produce a platform that leads to more genuine discussions between neighbors and city officials. This collaboration allows for more sincere plans to be developed that neighbors will be proud to help implement and pleased to live with the results.

12.2 Case Study – Somerville, Massachusetts

12.2.1 Slumerville to Somerville

Lauded as one of the most influential cities in the country and the best-run city in the Commonwealth, Somerville is home to nearly 79,000 in just over

4 square miles [8, 9]. It is one of the most ethnically diverse cities in the United States. Situated a mere 3 miles from Boston, it has emerged at the vanguard of urban leadership, recognized as an All-America City three times [10].

Over the last decade in particular, from the Boston Globe and Boston Magazine to the Huffington Post and The New York Times, article after article debates whether Somerville is the new Brooklyn, the pox of Millennials on the town, the loss of affordable housing, and the emphasis on all things local. Somerville is characterized by the vibrancy and tensions of the postindustrial immigrant community, artist and makers, and young professionals living in the densest city in New England. Recognized as one of the 100 Best Communities for Youth, Somerville is a radically different place today than it was even 10 years ago [11].

Through the 1960s, Somerville was a transit hub and largely industrial, hosting, among other firms: a bleachery, glass works, ice, tube and a safety envelope company, Sears Roebuck and Company warehouse, and a Ford auto assembly plant [12]. A period of deindustrialization and depopulation followed the closing of most of the industrial fixtures in the succeeding decades and the nickname Slumerville took hold. During the same time, a live-work artist development brought new life to a former A&P complex, becoming a model for similar projects around the country and sowing seeds of the creative economy now thriving in Somerville. Tensions of gentrification began with the Massachusetts Bay Transit Authority's expansion of the Red Line from Boston in 1985.

12.2.1.1 Professionalizing City Hall

To some, it seemed evident that development pressure was driving decision making in City Hall; Slumerville was known for corrupt government. Indeed Lester Ralph, Mayor from 1970 to 1978 ran on a campaign to "clean up Somerville." While there had been "a push to professionalize City Hall" beginning with his predecessor, the battle for Assembly Square – a large redevelopment site – revealed a legacy of fiscal opportunism that continued through the millennium.

Somerville is now experiencing two very different approaches to economic development. The first is based on the large-scale, both physical and economic, slow moving, private and public development projects that attract large employers. The second is the small-scale support of a growing community of entrepreneurs, makers, and artists who are building a vibrant creative economy within the network of postindustrial properties found throughout the city. The first is akin to big game hunting, while the second is akin to more economic gardening – both have been successful. The construction of the Assembly Square subway station brought approximately 4500 new jobs to Somerville when Partners Healthcare commenced development of a new corporate headquarters adjacent to the station. Greentown Labs, part of the creative economy centered in Union Square from 2011 brought $97 million in funding to the various start-ups housed within the business incubator space. Ranked the most bike-able community in the Northeast and one of the top ten biking and walking scores in the nation, Somerville understands sustainable transit fuels the local economy [13, 14]. As the Boston region continues to grow, it will be strategic to continue to support these distinct economic development strategies and, even more importantly, make sure that the two strategies work together collectively.

Mayor Curtatone The last decade of change has been shepherd by Mayor Joseph Anthony Curtatone. Curtatone served as an Alderman in the City of Somerville for 8 years before becoming Mayor in 2004, at the age of 38. He was a popular-at-large Alderman, said to bridge Somerville old timers and new progressives.

As Alderman, he was chair of the finance committee. As he explained, "We weren't managing for results...we had a budgetary process where nothing was aligned with outcomes or measures...there were no goals and objectives...there was no accountability. No transparency...we would make decisions on a crisis basis" [15]. The City of Somerville relied on a traditional line item budget until Curtatone led the transition to a performance-based budget.

In the spirit of collaboration with civil society that Curtatone is now known for, he engaged Harvard students to help revamp the city's budgetary process. CommonWealth Magazine, a political observer, explained, "This curious collaboration – all the more striking because of Harvard's rarified reputation and Somerville's sometimes unsavory past – was designed to introduce to the city a new form of financial management known as activity-based budgeting" [16].

Managing for Results As a Mayoral candidate, Curtatone ran on a platform of management reform [17]. He sought out best practices in the public and private sector and aimed to apply them to municipal government [15]. One of the primary goals was transparency, as emphasized by the budgetary reform.

In Somerville, as in other urban centers, "citizen expectations of innovation in public services continue to grow, while budgets shrink" [4]. To improve citizen services while working to bring the city's budget from the red, Curtatone focused on data-driven management, looking to revolutionary programs such as CompStat in New York and CitiStat in Baltimore [17]. Soon after Curtatone entered office, Somerville became the first city in the United States to operate a 311 constituent service center and Connect CTY (also known as a reverse 911) mass notification technology [18].

In *Smart Cities*, Anthony Townsend acknowledges the benefits of implementing data-driven management "as a triage tool for stretching scarce city resources" [4]. However, he warns against seeing it as a holy grail – citing the exposure of police maneuverings to affect CompStat as one cautionary tale.

To prevent such manipulation, Townsend recommends that data need to be "sufficed throughout" City Hall, a notion Curtatone continues to champion.

SomerSTAT Arguably, SomerSTAT, Somerville's data-driven performance management system, served as a transition technology, introducing a smart city lens that is now seen to be "embedded in the culture of (City Hall)" [19].
SomerSTAT is described as an

> an initiative that helps Mayor Curtatone supervise the work of city departments and use timely data to inform decision-making and implement new ideas. Four SomerStat staff study financial, personnel, and operational data to understand what's happening within departments. Then, in regular meetings with City department managers and other key decision-makers, SomerStat helps to identify opportunities for improvement and then track the implementation of those plans. The meetings have become an ongoing conversation among city managers on where the city should be headed. Each meeting allows city managers to better understand how the City can streamline and improve its services to its constituents [17].

While the SomerSTAT team initially collected and analyzed data, they did not see themselves as "masters of the data," but as facilitators leading a collaborative process [20]. Rather than take a "gotcha STAT" approach to data collection, Skye Stewart, current Director of SomerSTAT aims to see the total operations across departments in City Hall, positioning the team, "to identify the pain points across all departments" [20]. Indeed, George Proakis, the Director of Planning for Office of Strategic Planning and Community Development (OSPCD), observed that "STAT has let the Mayor build a very nimble Mayor's office. Previously, there was no interdisciplinary policy staff" [19].

ResiSTAT, a data-driven decision-making program, brings SomerSTAT out into the community both virtually and through regular community meetings. Rather than pushing data out, touting achievements from City Hall, it is designed to strengthen residential ties to City Hall and serve as a two-way process. Data on neighborhood level issues, from waste disposal and snow removal to the city's first report on wellbeing and the municipal budget, are presented online, through a website, listserv, active blog – the "online community forum" – and biannual ward meetings [21, 22]. By "empowering citizens with data" and making Alderman, city staff, and often the Mayor available at meetings, Somerville facilitates active engagement down to the street level. Decision making through ResiSTAT set the stage for the neighbourhood planning that Somerville is now known for, the smart governance practice reflects a culture "pickled in the idea that transparency is king" [23]. It is a prime example of an innovative government program that "will create new ways for citizens to make their voices heard, giving them the ability to provide input into regulations, budgets, and the provision of services" [6].

12.2.2   Planning Somerville

12.2.2.1   SomerVision

When Curtatone entered office, he ushered in an era of governance that strove "to be abnormal." SomerVision, a community-based approach to urban planning and the City's first comprehensive plan, is one of the cornerstones of Curtatone's abnormal approach. It focused on creating shared values and measurable goals until 2030. Lauded with the Comprehensive Planning Award for a municipality over 50,000 population by the American Planning Association Massachusetts Chapter (APA-MA), it was developed over 3 years by a 60-member steering committee and supported by the OSPCD; it is the guiding document for decision making in Somerville [24]. Monthly Steering Committee meetings, a series of community values workshops, showcase events to review the draft plan, a widespread public survey, and discussions through ResiSTAT meetings fostered a strong sense of community ownership. The language of the goals, policies, and actions outlined are the result of the public process, largely unedited by City Hall.

The implementation strategy, including a map and measurable targets on housing, job creation, open space, transformation areas, and transportation, ensure that SomerVision sets measurable goals and a way to evaluate progress. These targets began with back-of-the-envelope calculations that evolved through an analysis of a wide range of variables [25]. Today, SomerSTAT uses SomerVision to evaluate the work of the OSPCD, but perhaps more critically, OSPCD is increasingly trying to use the STAT team as a resource to build out their analytics. The SomerVision process generated tremendous community feedback. In the years since it was published, the City has grappled with how to quantify the results of their robust commitment to public participation; a commitment that continues to deepen and evolve.

12.2.2.2   Somerville by Design

The accountability and transparency Curtatone trumpets extend to a public-engagement model described as "outreach–dialogue–decide–implement." This process is distinct from the "decide–present–defend" model used by planners for decades. OSPCD sees it as a fundamental shift, a "new method for planning (that) acknowledges that the best results usually occur when informed residents collaborate with public officials to establish a vision for their neighborhood's future" [26].

Much like SomerSTAT led to ResiSTAT, bringing data-driven decision making into the community, SomerVision led to Somerville by Design (SBD), urban planning at the neighborhood level. An iterative approach to community-led planning, the SBD process is interactive and relies on a few tried and true strategies: (i) grow a crowd – crowdsource and encourage regular participants to bring new people in; (ii) bring meetings out into public space; (iii) translate – build a diverse crowd; (iv) use feedback loops – review previous decisions made through public process; (v) create interdisciplinary teams – work with consultants with complimentary expertise that balance out the skills on staff; and (vi) get speedy results – upload designs and output in real time to build on momentum from public meetings [27].

A recent process diagram of SBD, created by SomerSTAT and recognized by the Metropolitan Area Planning Council, Boston's regional planning agency, as a leading practice, emphasizes that community process begins without a preordained plan, solutions are crafted through an iterative feedback loop in which problems are identified and best practices and new ideas are explored before arriving at a solution. Decisions are made in coalition.

City Hall departments have been reorganized as a result of SBD, ensuring municipal staff work in teams, not silos, to address the multidisciplinary challenges inherent in sustainable development. The SomerSTAT framework is integrated early on, providing a valuable, data-rich, feedback loop. For smaller cities, such as Somerville, city departments can have a greater impact when they work across departments in multidisciplinary teams that are organized around specific physical boundaries, such as neighborhood by neighborhood. With each new neighborhood that the SBD approach takes on, departmental silos weaken, and interdepartmental teams emerge to address the challenges inherent in the type of sustainable development the community is promoting for their future. The citywide integration of SomerSTAT has provided a valuable, data-rich, framework for this planning and feedback loops.

SBD is orchestrated in collaboration with an impressive breadth of citizen activists. The gear work behind SBD is made transparent through the charrette process to engage stakeholders quickly and meaningfully.[3] Novel ideas are tested through Tactical Urbanism[4] – short-term action to test long-term planning ideas – allowing innovative or misunderstood projects to be vetted by neighborhood stakeholders.

One of the key elements of the SBD planning method is the facilitation of a constructive and critical discussion with stakeholders that informs the ongoing

---

3 A charrette is a multiday planning event that brings together a multidisciplinary team to focus intensely on a subject area. The charrette is held in close proximity to the subject area making attendance by the local stakeholders easy. It includes design studio space as well as pubic meeting space allowing for an iterative discussion to occur between designers and stakeholders so there is complete transparency over the plan creation and giving the professional team a greater degree of understanding for the local constraints, opportunities, and issues. A Somerville by Design charrette is 3–5 days in length.
4 Tactical Urbanism is defined as a city and/or citizen-led approach to neighborhood building using short-term, low-cost, and scalable interventions intended to catalyze long-term change.

improvement of the process. Each neighborhood planning effort is approached with a similar strategy, but the specific engagement tools are calibrated to the unique neighborhood needs. This constant improvement of the process and quick ability to adapt is a key example of how Somerville's smart governance is helping change the way neighborhood planning is done today.

Neighborhood Planning In anticipation of the construction of seven new Green Line subway stations, Somerville launched a series of station area planning efforts in 2012. The first of these plans, Gilman Square, was the testing ground for this approach to planning. The planning process had a number of objectives – most notable to ensure that the local community had strong control over the future of their neighborhood as development pressure increased with the oncoming transit. The Gilman Square Station Area Plan went on to win the Congress for the New Urbanism New England's Urbanism Award for its innovative public-engagement methods as well as it visionary plans for the future station area.

Close interaction with the local community allowed the planning team to collaborate with several long-time property owners. The resulting plan creates a new public square at the intersection of three of the neighborhood's most important streets. Once the station is constructed, it will open directly onto this new Gilman Square resulting in the creation of a remarkable new public space that will include both public and private redevelopment projects, a new park, and prominent civic tower (Figures 12.3 and 12.4). In order to accomplish

Figure 12.3 Gilman Square. *Source*: Courtesy of Russell Preston.

Figure 12.4 Gilman Square, rendering of new public square abutting the proposed Green Line MBTA station. *Source*: Courtesy of David Carrico.

this feat, land trades between public property, both street right of way and existing open space, and private land will need to occur, new streets must be constructed, and the city and the transit agency will need to work in close coordination. This complexity would not be possible without a shared vision established though the SBD process fueling years of implementation ahead.

Prior to the SBD Gilman Square process, no one expected this type of result. No one had considered a new civic space, let alone discussed how one might be built. The SBD method was able to uncover real hidden value for the community. Exploring value creation opportunities that change the public right of way has now become a common task in all neighborhood-planning projects.

In Davis Square, an existing transit-served neighborhood in Somerville, public space is short supply and an important public amenity that the city was looking to improve. While researching previous planning efforts, the Somerville by Design team tackled an unsuccessful, previous plan to convert a small public parking lot into a new plaza (Figure 12.5). In the spirit of Tactical Urbanism, the team decided to resurrect this idea and temporarily convert the space during the course of the multiday design charrette (Figure 12.6). The 12 parked cars that are there most days were removed in exchange for food trucks, live music, and circus acrobats (Figure 12.7). With the installation of comfortable seating and ongoing programming, it was only a matter of minutes before local residents began to use the new public space (Figure 12.8).

Figure 12.5 Davis Square, Cutter Plaza existing conditions. *Source*: Courtesy of Dan Bartman.

Figure 12.6 Davis Square, Cutter Plaza pop-up plaza 1. *Source*: Courtesy of Russell Preston.

Figure 12.7 Davis Square, Cutter Plaza pop-up plaza 2. *Source*: Courtesy of Dan Bartman.

Figure 12.8 Davis Square Cutter Plaza image build out. *Source*: Courtesy of Urban Advantage.

The plaza was not intended to be a spectacle. The team used it to attract the public into the planning process while testing the idea of converting parking to a public plaza permanently. By locating the studio space for the charrette in an empty storefront adjacent to the pop-up plaza, the planning process drew stakeholders who normally do not attend planning events into the creation of a long-term vision for their neighborhood. Through traditional public meetings, plan presentations, and town hall style debate, the previous public review of the proposed plaza was unsuccessful. Tactical Urbanism, coupled with crowd-sourcing tools such as Twitter and Facebook, enabled the team to produce the pop-up plaza in a few weeks and to facilitate a conversation about the space in real time. Together, these tools allowed for a more sincere discussion about what the plaza could do and how it could serve the neighborhood resulting in overwhelming support of the permanent conversion of the parking lot to a pedestrian plaza.

The creative class that resides at, works at, and visits Union Square, another Somerville neighborhood, has made it one of the Boston region's most coveted. It is a center for the arts. It houses one of the country's largest maker spaces, Artisan's Asylum. It will also be one of the first neighborhoods to have a new Green Line subway station constructed. Tensions are high in anticipation of this new transit. In 2014, the city launched its most intensive neighborhood planning effort to date to establish a clear vision for the community in advance of the rail arriving.

In addition to addressing concerns about the potential radical changes to the community's character that the transit access could create, the neighborhood planning was tasked with creating a vision for more than 40 acres of land iden-tified by the city's comprehensive plan as transformation redevelopment. How do you accommodate new transit-oriented development to take advantage of the new subway station and plan for the creation of a regional job center with millions of square feet of new development, all while preserving the authentic character of an innovative and art-oriented community? This question would test the limits of what was possible with the smart approach to planning that is SBD.

In Union Square, the planning team expanded on the typical SBD approach in four phases. The first is coordinating not only the planning team but also key members of the local community. Several events were hosted to crowdsource how to plan with the community; the idea is to plan for the planning together. This effort aims to educate the community on the process, ask how it can be made more sympathetic to local neighborhood conditions, and reiterate that there will be a beginning, middle, and end.

A small group of "community connectors" were organized in Union Square. Their role was to utilize their existing communication networks – through their various blogs, websites, Twitter, Facebook, and email lists – to push the plan-ning team announcements further. This group also included the busybodies that every neighborhood has – people who know everyone, talk to everyone, and are generally the nods in the social fabric of the community. Supporting this naturally occurring communication network in the real world with digital assets allows the planning team to reach wider and deeper into the community than is typically possible. When you answer the "who, what, when, where, and how" questions together with the community, trust is easier to establish and it allows for a more genuine understanding of the community.

The second phase in Union Square consisted of helping the community establish a vision for their future through a series of community workshops. The community braved the worst winter on record to help make a better plan. Dozens of neighbors attended a neighborhood walking tour of Union Square on a Saturday morning at 7 a.m. in January with below-freezing temperatures and 2 ft of snow on the ground (Figures 12.9–12.11). They worked together in a space without heat but with translation services in four languages, conducting visioning and community mapping exercises. Long lauded for their facility with community engagement, mapping has become a much more powerful tool with the use of Google Earth and mobile apps that bring the technology to constituents to manipulate.

The third phase was hosting a multiday design charrette. The key is to bring the designers to the community by setting up a temporary studio where they are able to receive real-time feedback from stakeholders. The charrette concluded with a comprehensive pin-up of all the ideas drawn and developed. Setting up a

Figure 12.9 Union Square, walking tour.

Figure 12.10 Union Square, Prospect Street in January.

Figure 12.11 Union Square, out in the cold.

fully functional design studio in a temporary space that can support the work of dozens of consultants for a week used to be a much more expensive undertaking. In Union Square, the studio took place in a recently vacated post office. By utilizing inexpensive printers, cellular Internet hot spots, and cloud file-sharing services, an old post office was transformed into a fully operational design studio in only a few hours. The technology that allows for multiple consultants, from different firms, to plug into the studio instantly did not exist 5 years ago. Integrating this technology into the process has allowed for resources to be reassigned from administrative cost to higher value programming such as Tactical Urbanism installations or broader stakeholder engagement. This use of project management software and cloud services also allows the city to select a network of consultants to work collaboratively without the need for cumbersome project administration.

The fourth and final stage is to capture all the discussions that folks have following the charrette by hosting a series of plan open houses. These events usually have live music, local food, and a casual atmosphere to encourage discussion. They are designed to be fun events. In Union Square, one of the open houses was held in an underutilized parking lot transformed into a public courtyard with food, seating, and a bike repair kiosk. During the open house, new and refined plans are posted, encouraging conversation both between stakeholders and the planning team and among neighbors. At this stage, it becomes clear which ideas the neighborhood has embraced as their own that, irrespective of the plan document, will take on a life of their own.

These open houses conclude with the publishing of a draft plan, which is open for public comment online, beginning another discussion with the community as detailed elements of the plan are debated through various threads. This phase of the process usually includes another tactical intervention demonstrating several improvements conceived of in the plan. This type of short-term action helps keep what can be a long process for the public fun and interesting while providing the planning team a last point of feedback before publishing the final plan.

The draft of the Union Square Neighborhood Plan received the Comprehensive Planning Award from the American Planning Association's Massachusetts Chapter in 2015 further demonstrating the impact that this innovative approach to neighborhood planning is having on the region. The SBD method would not be possible without the deep support of the city's leadership at the highest level, their commitment to an iterative approach that is not always predictable, and their use of emerging technology to seamlessly connect the project team to the selves and to the community.

Technology in the Somerville by Design Process Proakis says Somerville is continually looking to technology to support a seamless dialogue, where "technology isn't replacing that (community) interaction but augmenting it so it becomes more meaningful." In 2014, Proakis was awarded Planner of the Year by the Massachusetts Chapter for the American Planning Association. Proakis believes that "as a City we do more public communication using technology than anyone else I know" [19].

Technology is used both in the service of community-led dialogue and to facilitate the design process. The municipal team relies upon web-based technologies to interface with one another and the Somerville citizenry, enabling collaboration that, while rooted locally, attracts talent beyond the city limits. Consultants work across platforms, using Asana to manage projects with the City Team and Slack, an instant messaging application, to communicate internally.

At the center of the Somerville by Design, internal process is Dropbox. The cloud-based file-sharing services have allowed the city to establish project folders that team members from multiple offices and, sometimes, spread all over the country to seamlessly share files with simple version control. This virtual server is the bedrock of the method, and without it, this approach would be far more difficult.

A typical planning department's familiarity with graphics software stops at PowerPoint. The use of Adobe's Creative Cloud has been another cloud-based solution that has allowed for the city's planning staff to not only work in the sources files created by the consultant team but also to become key contributors to the creation of deliverables by using InDesign, Illustrator, and Photoshop collaboratively with consultants. The use of this platform is now required by all offices working on SBD projects. Somerville has invested in their staff in a smart way by placing them at the center of creating the multiple deliverables required by this iterative planning process.

In additional to the internal project management and production technology, the integration of a number of externally focused platforms has been needed to further support this sincere approach to placemaking. In order to truly create a transparent planning process, the planning department needed to set up their own communications platforms so that information could be transmitted in real time to the various stakeholders engaged in the multiple planning efforts occurring throughout the city. OSPCD hosts a WordPress Blog and uses the freedom of the blogosphere and an SBD Twitter account as disruptive tools both to push information out in real time and as a tool for engagement. They also developed and managed their own MailChimp email list; most departments and even teams within their own departments have their own listservs [28].

All these tools were activated outside the City's Communications Department and established communication channels. OSPCD staff were not experts in communications, but by integrating this technology into the SBD method they have acquired, sometimes through trial and error, the essential communication skills needed to maximize the impact of these social media platforms.

These tools now facilitate and promote honest dialogue about the future of the city.

Establishing these communications platforms has allowed for the increased use of online surveys and polling to further understand the community's preferences. These platforms have varied from free series such as Survey Monkey to premium ones such as MindMixer. The use of these tools requires that OSPCD juggle sometimes a dozen accounts at a time, with some staff spending up to a quarter of their time on outreach [28].

Furthermore, OSPCD staff seek emerging software to meet their needs. For example, Senior Planner Dan Bartman spent years looking for a web-based platform to gather public feedback through the planning process. Through Internet searches and listserv queries, and after rejecting possibilities that would need to be rebuilt from the ground up and adapted, he discovered MarkUp. With an eye toward advances in the field, Bartman realized the City of Los Angeles was using MarkUp to collect feedback on documents for the Re:CodeLA project. Built by Urban Insight, Re:CodeLA's web design consultant and the planners and programmers behind Planetizen.org, MarkUp was not designed and deployed to be a stand-alone product. OSPCD persuaded Urban Insight to launch OpenComment – a rebranding of MarkUp – with Somerville and Somerville by Design as its first customer and early adopter [29].

Staff emphasized the heightened accountability that follows government forays into social media and open technology. Previously, for instance, remarks made at a conference out of town largely lived and died there. Today, OSPCD and the Mayor's office live-tweet highlights and engage in conversation via Twitter, with experts in the field as well as engaged Somerville residents who use the forum to hold staff members to a new level of accountability [30]. Remarks are picked up and monitored, raising the level of dialogue and breaking down barriers between the community and City Hall. At community meetings, residents expect staff to be up to the minute with information from Open Comment, to access all relevant documents in real time, and for any hard paper designs or surveys to be quickly uploaded for public access. Constituents can live-stream community meetings online, review public documents with ISSUU, participate in open conversation with MindMixer, and submit feedback via Open Comment. However, as Senior Planner Melissa Woods noted, on any given issue, people regularly join the conversation mid-process, explaining they were unaware of the ongoing dialogue.

12.2.2.3 Smart Cities and Planning in Somerville

At present, Somerville is undergoing several future-oriented planning processes, investing in smart governance to sow the seeds of a resilient and sustainable city. From neighborhood, open space, and transit plans, Somerville residents are actively involved in envisioning their future, using technology through the planning process and planning for future technology.

For several years, the City has partnered with Gehl Architects, a firm founded by legendary urbanist Jan Gehl, to foster a human-scale approach to urban life. Curtatone has embraced Gehl, an urban designer known for systematically measuring human behavior in cities, celebrating local research that "has shown that small changes in the build environment can make a huge difference" [31]. Gehl Architects have been studying humans and their use of cities for decades. Most cities have no idea how their parks, plazas, or sidewalks are used. They know more about the flow of automobile traffic at any given intersection than they do about the humans that live, work, or visit those same streets. Somerville has begun to integrate Gehl's systematic method for documenting the use of public space throughout the city in order to establish a baseline for ongoing improvements to the public life of the city.

This process starts by collecting data and by counting people – watching where they walk, what they are doing, and observing how the space functions. These metrics can verify certain assumptions about how the space is functioning and can provide guidance about what improvements can be made to encourage a more vibrancy to unfold. Continued measurement and verification of these human activities will be used to support the ongoing implementation of a larger vision for the area. Somerville has begun to measure what matters.

This focus on the human scale has been made possible through the SBD process. Fine-grained improvements will lead to more walkability, greater reliance on biking as a mode of transit, and great enjoyment of the public realm. These improvements can now be identified and systematically tracked, evidencing short-term gains that may be worth long-term investments in permanent infrastructure. This approach has allowed Somerville, a small city with limited resources, to achieve improvement in public life without the substantial investments often required to upgrade the public infrastructure that has such direct control over the vibrancy of the city's streets and squares.

Observed to be pursuing "an innovation economy agenda," Somerville is employing smart governance and utilizing technology to foster nimble growth [32]. While the Mayor has been pitched on smart cities by IBM, Cisco, Verizon, and a swath of start-ups touting, for example, street lights as cell towers that transmit data and generate municipal revenue, Somerville is "not looking for a global, all encompassing platform" [23]. Wary of both committing to a single technology and burdening constituents with an early adopter tax, the city is discerning in its commitments. The city, in partnership with Code for America, is taking small steps – adding GPS to plows and writing smart technology into union contracts – and bigger steps, partnering with Audi to test self-driving car technology [23]. The partnership with Audi allows Somerville another avenue to explore how to get more out of the city's existing streets. This objective is shared across issue areas; the basic question is, how can the city build on its existing assets, and how can technology facilitate this capitalization? At the same time, Somerville is advocating for innovative technology and services to be brought to scale for small cities [23].

Somerville is moving into a new phase of open data and data management, trying to carve a path that continues to empower staff and residents alike. Many cities launch highly celebrated open data initiatives without a smart governance strategy behind them. Somerville did just the opposite, beginning with SomerSTAT, the city is moving from a well-managed community process to now enlivening its open data platform [30].

While the city has an existing open data portal, the majority is 311 data that is housed by constituent services and static [20]. At present, they are undertaking back-end work to incorporate data from the Assessors, Inspectional Services, Public Safety, and so on. They are moving toward a data dashboard and focused on owning their own data in vendor contracts. They seek out firms that respect that objective, clearly stating the "new expectation that our data is our data all the time" [23]. Moving forward, the Mayor's office aims to develop an application programming interface and transparent code that constituents can tweak [20, 29].

As cities look for ways to become more resilient, officials would be wise to embrace proactive community-led planning and the technology that enables it. The direct public involvement ongoing in Somerville informs City Hall of future opportunities for growth. As Brad Rawson, co-lead on SomerVision and now the Director of Transportation and Infrastructure for Somerville, emphasized, "Community led planning can lead to fiscal stewardship" [25]. Community members guide the plans for improving their neighborhoods; their commitment evidencing a level of community attachment any Somerville resident can attest to [6]. Temporary installations and other Tactical Urbanism approaches test the concepts developed through community-led planning, resulting in a more collaborative public process.

As Somerville hones its approach to community-led planning, the city is increasingly focused on the neighborhood level. This hyperlocal method has resulted in authentic plans for each neighborhood that, when combined, create a richer and more vibrant picture of the city than possible without such a high level of public dialogue. The city is simultaneously committed to integrating the latest technologies into the planning process and planning to ensure that Somerville benefits from smart technology at the appropriate scale.

12.3 Looking Ahead

12.3.1 Lessons from Somerville

Somerville is looking ahead to SomerVision 2.0 – a strategic plan that will be developed through robust community process. Curtatone aspires to guide the city toward the Net Zero and Vision Zero goals while receiving a AAA bond rating. This is the sustainable and resilient future of Somerville. The various goals Curtatone is championing can go hand in hand, but at present there is no organizational policy or strategy in place to ensure that they are aligned. The Mayor's dual commitments to managing for results and for transparency through community process have raised the level of engagement and constituent expectations. To get there, the city will have to further blur the lines across departments, moving beyond OSPCD to build a truly multidisciplinary plan.

Moving forward, City Hall will benefit from a more coordinated approach to technological engagement, especially through social media. At present, the various websites and listservs, while aiming to reach the maximum number of people possible, also ensure that communication is fragmented. Increasing virtual access to public meetings will invite greater participation. Moreover, the use of social media is largely limited to an English language audience. The city's hardworking SomerViva team of language liaisons is working to address these issues. With a new official Somerville City Hall website to be rolled out in 2016, there is potential for more intradepartmental alignment. A mapping feature, for example, will allow community members to self-select their neighborhood and opt-in to hyperlocal notifications.

Constituents routinely asked, "How did you get that number?" While they are eager to have the data readily available, some aspire to a data platform that allows constituents to answer that question themselves. To do so, OSPCD must continue to "STAT the plan" – working in tandem with the SomerSTAT team – and partnering with the community to build capacity. The level of research and analysis possible will grow exponentially once data sets are active on the city's dashboard. The potential for innovation will increase as community members can explore and engage with community assets in a whole new way. It may be possible to set outcomes-based metrics through an open and collaborative process in which community members can see both variables and results.

There is an understandable hesitancy in the public sector's commitment to quantitative targets, to pin down numbers that voters can reference time and again. OSPCD understood that few of their constituents would familiarize themselves with the full plan document; they wanted "headlines" to bring more people into the planning process, with full acceptance of the accountability that follows. Today, more than 5 years after the plan was published, the principle targets have "congealed in people's minds" [25]. The targets in SomerVision have helped build the community's planning skills, but few understand them to be iterative goals. OSPCD is aware that traditional plans are static documents and that "the day a plan is exported from InDesign it is already out of date" [29]. Staff are looking for ways to make plans "living documents", using technology not just to improve access and dialogue but to make planning ever more iterative [29].

By advancing a culture of planning through technology, the legacy of one set of numbers loses its hold as the pace of discussion quickens and language becomes more ephemeral. Curtatone has an ambitious agenda and sees technology as a tool for the curious, posting that his interest in technology is really a quest of the curious that "technology enables curiosity (in the) pursuit of new ideas, to foster ingenuity and innovation" [32]. In his efforts to remake government in the image of a social enterprise, Curtatone encourages people to aim high and then figure out how to reach the stated goal.

Data can smooth the path, allowing for a more informed discourse when filtered through the tacit lens of the community. When the city needs to prioritize the allocation of resources, determining, for instance, how to ensure they are addressing deficiencies in the sewer system, planning for capital projects and greater resiliency, sensors can provide information on weak points in the infrastructure. This hyperlocal information can support a decision to fund repairs in one neighborhood over another, encourage residential storm water management through appropriate plantings, and even win the expansion of existing urban farms and gardens.

Cities flourish where there is a vibrancy to the life humming within, where technology "honors and leverages community assets" rather than molding them [25]. Somerville must continue to foster the authentic places that have instilled such a strong sense of community connection and employ technology to achieve a human-scale, sustainable, and resilient city. Doing so will require systems thinking, drawing on the economy of cities, to maximize the potential as a transit hub and center for innovation. Smart governance will enable smart cities to be in command of their own fate [25].

12.3.2 Recommendations

For cities looking to embrace the technology behind smart cities, smart governance will be critical to successful implementation. Somerville offers a strong case study for a small city, grappling with increasing demands on City Hall under resource-constrained conditions. For those looking to learn from the Somerville experience, the following can pave the way:

1) Plan for the future, the future is now
   Planning for urban growth and health must be an ongoing process. Trading the conventional linear process for one that allows sequential planning efforts to overlap with previous ones is essential.
2) Respect authenticity in placemaking
   Cities are in competition with one another. To foster community connectedness, smart governance must protect and advance authentic places.

3) View cities as dynamic, polycentric systems

The complex web of human life and activity that occurs in cities is what makes them the center for innovation. Smart cities must honor the parts that make the whole.

4) Invite data into all corners of City Hall

Quantitative analysis must be seen as a carrot to achieve results otherwise unimaginable and increase transparency not a stick brought down by management to expose deficiencies.

5) Craft community-led planning processes

Fostering a culture of iterative planning elevates the process with local insight and ensures community buy in.

6) Employ technology to advance engagement

Low-cost technology, when well coordinated, can expand outreach, increase the diversity of perspective by eliminating traditional barriers to participation, invite talent beyond City Hall, and gain trust.

7) Facilitate multidisciplinary teams

Smart cities face wicked challenges that belie traditional government departments. Collaboration is essential to efficiency and success.

8) Work at the neighborhood scale

Neighborhoods are dissimilar. A one-size-fits-all approach is not appropriate; when appropriately applied, technology can illuminate the strengths and weaknesses of each neighborhood, and local insight can interpret the information to improve upon existing assets and shore up necessary deficiencies.

9) Aim high

Setting goals and inviting those within and around City Hall with data to determine those goals are empowering and need not be constraining.

10) Focus on resiliency and sustainability

From fiscal health and public health to educational and environmental policy, smart cities of the future will be increasingly resilient and sustainable (Table 12.1).

Final Thoughts

Somerville, Massachusetts, exemplifies the possibility of municipal leadership and urban innovation. The case study illuminates how smart governance enables best practices in community-led planning and design. Smart cites can integrate these principles with emerging technologies to become more resilient and sustainable.

Table 12.1 Recommendations table.

Recommendation	Rational	Action
Plan for the future, the future is now	Planning for urban growth and health must be an ongoing process	Trade the conventional linear process for one that allows sequential planning efforts to overlap with previous ones is essential
Respect authenticity in placemaking	Cities are in competition with one another; authenticity fosters community and cannot be replicated	Protect and advance authentic places
View cities as dynamic, polycentric systems	The complex web of human life and activity that occurs in cities is what makes them centers for innovation	Honor individual parts of the city while taking a holistic approach
Invite data into all corners of City Hall	Quantitative analysis can achieve results otherwise unimaginable and increase transparency	Use data as a carrot, not a stick
Craft community-led planning processes	Local leadership elevates the process and ensure community buy in	Foster a culture of interactive planning with local insight
Employ technology to advance engagement	Low-cost technology, when well coordinated, can expand outreach, increase the diversity of perspective by eliminating traditional barriers to participation, invite talent beyond City Hall, and gain trust	Integrate technology with low barriers to entry into public sector projects
Facilitate multidisciplinary teams	Smart cities face wicked challenges that belie traditional government departments	Collaborate for the sake of efficiency and to achieve more successful outcomes
Work at the neighborhood scale	Neighborhoods are dissimilar. A one-size-fits-all approach is not appropriate	Use technology to illuminate the strengths and weaknesses of each neighborhood and local insight to interpret the information to improve upon existing assets and shore up necessary deficiencies
Aim high	Ambition and inclusion are empowering	Set goals and invite those within and around City Hall with data to determine those goals
Focus on resiliency and sustainability	From fiscal health and public health to educational and environmental policy, smart cities of the future will be increasingly resilient and sustainable	Apply a resiliency and sustainability lens to every project and make them foundational

Questions

1 How do you define smart governance?

2 What is an authentic place?

3 How can a city's comprehensive plan inform the creation of neighborhood plans?

4 Give an example of a community-led planning process that has informed how the city used data for decision making.

5 Give an example of how data can inform physical planning.

6 How can city government better organize their departments to be more responsive to the concerns of creating authentic places?

7 Name three technologies that can allow city government to get innovative at low cost.

8 How did Somerville overcome its resource limitations to create smart neighborhood plans?

References

1 NewAmerica (2014) Stephen Goldsmith and Susan Crawford at New America, Ash Center for Democratic Governance and Innovation.
2 CSIS. (2014). "Smart Governance: How public policy can keep pace with technological developments." Retrieved from http://csis.org/event/smart-governance-how-public-policy-can-keep-pace-technological-developments (accessed December, 2015).
3 PPS. "What is Placemaking?" Retrieved December, 2015, 2015, from http://www.pps.org/reference/what_is_placemaking/.
4 Townsend, A.M. (2013) *Smart Cities; Big Data, Civic Hackers, and the Quest for the New Utopia New York, New York*, W.W. Norton & Company, Inc.
5 Mechling, J. (2002) Information Age Governance: Just the Start of Something Big? governance.com, in *Democracy in the Information Age* (eds E.C.N.J. Kamarck and S. Joseph), Brookings Institution Press, Washington, DC, pp. 141–160.

6 Goldsmith, S.C. and Susan, C. (2014) *The Responsive City*, Jossey-Bass, San Francisco, CA.

7 Thornton, S. (2015). *"The Smart Chicago Collaborative: A New Model for Civic Innovation"* Retrieved December, 2010, 2015, from https://www.livingcities.org/blog/444-the-smart-chicago-collaborative-a-new-model-for-civic-innovation-in-cities.

8 Keane, T.M.J. (2006) *The Model City*, Boston Globe, Boston, MA, New York Times Company.

9 Schwarz, H. (2014) *The Most Influential Cities in the Country, According to Mayors*, The Washington Post, District of Columbia The Washington Post Company.

10 NCL. (2015). "All-America City Award." Retrieved December, 2015, 2015, from http://www.nationalcivicleague.org/previous-all-america-city-winners/.

11 Byrne, M. (2011) *Somerville among 100 best places for young people*, Boston, MA, New York Times Publishing Group, Boston Globe.

12 SHPC (2013) *A Sampling of Industrial Somerville*, Somerville, MA, Somerville Historic Preservation Commission and the Somerville Bike Committee.

13 Dungca, N. (2014) *Somerville is top biking city around*, Boston Globe, Boston, MA, The New York Times Company.

14 NWS. (2014). "Living in Somerville." Retrieved December, 2015, 2015, from https://www.walkscore.com/MA/Somerville.

15 Goldsmith, S. (2009) *Somerville's Joseph A. Curtatone*. Management Insights, Governing.

16 Preer, R. (2005) *Harvard Students Help Somerville Revamp Its Budgeting Process*, CommonWealth.

17 COS(A) XXX. SomerSTAT. Retrieved from http://www.somervillema.gov/departments/somerstat (accessed December, 2015).

18 COS(B) XXX. "Mayor." Retrieved December, 2015, 2015.

19 Proakis, G. (2015b) *Personal Interview*, Director of Planning. Somerville, MA

20 Stewart, S. (2015). *Personal Interview*, Director of SomerStat. Somerville, MA

21 COS(C). A Report on Wellbeing; The Happiness Project. Retrieved from http://www.somervillema.gov/departments/somerstat/report-on-well--being (accessed December 2015).

22 COS(D) (2015). ResiSTAT http://somervilleresistat.blogspot.com.

23 Hadley, D. (2015). *Personal Interview*, Chief of Staff Somerville, MA

24 APA-MA (2012). APA-MA is pleased to announce winners of the 2012 APA-MA Awards Program which honors outstanding planning projects and planning leadership.

25 Rawson, B. (2015) *Personal Interview*, Director of Transportation and Infrastructure. Somerville, MA.

26 OSPCD 2015. "Our Planning Methodology." Somerville by Design Retrieved December, 2015, from http://www.somervillebydesign.com/about/.

27 Proakis, G. (2015a) *Somerville by Design; Eight Strategies for Public Participation, Lessons from the SBD Charrettes*, OSPCD, Somerville, MA.

28 Woods, M. (2015). *Personal Interview*, Senior Planner. Somerville, MA.

29 Bartman, D. (2015). *Personal Interview*, Senior Planner Somerville, MA

30 Staff, C. o. B. (2015) *Personal Interview*, City Staff. Somerville, MA.

31 COS(E) (2014). City, Congress for NEw Urbanism to Host Screening of "The Human Scale" at Somerville Theatre Jan. 30. Somerville, MA, City of Somerville

32 MIT (2014) Enterprise Forum Cambridge, in *A Tech-Friendly Mayor* (ed. R. Cronk), Massachusetts Institute of Technology, Cambridge, MA.

13

Smart Economic Development

Madhavi Venkatesan

Department of Economics, Bridgewater State University, Bridgewater, MA, USA

CHAPTER MENU
Introduction, 373 Perception of Resource Value, Market Outcomes, and Price, 378 Conscious Consumption and the Sustainability Foundation of Smart Cities, 384

Objectives

- To understand the relationship between smart economic development and sustainability.
- To understand the role of economic and social frameworks in determining economic outcomes.
- To understand the significance of culture, specifically conscious consumption in enabling sustainable outcomes.
- To understand and appreciate the importance of economic education and continuous improvement in fostering long-term smart economic development.

13.1 Introduction

Smart cities are fundamentally sustainable. How sustainability is reached in a city requires holistic assessment and is enabled through technology, specifically with respect to operationalizing efficiency. However, given the present consumerism, fostered economy, perhaps the most significant, powerful, and traction-inducing vehicle for instituting sustainability, is found in enabling conscious consumption. O'Connell [1] views the education of the inhabitant as

Smart Cities: Foundations, Principles and Applications, First Edition.
Edited by Houbing Song, Ravi Srinivasan, Tamim Sookoor, and Sabina Jeschke.

being the catalyst for smart city implementation and finds that education is correlated with government participation. Government participation rates promote establishment of public goods and also ensure that social values promote self-policing of the same. Consistent with this attribution would be that smart city planning, by definition incorporating a sustainability-focused electorate, would embed government-funded infrastructure support in the form of public goods that enable sustainability in reflex activities, such as trash disposal and water use. Ultimately, through both personal decision-making and government facilitation, education promotes a self-reinforcing culture of sustainability.

13.1.1 Significance of Educating for Sustainability

In the United States, consumption contributes to over 65% of gross domestic product (GDP), which since the 1940s has been the international metric for economic progress. Given this linkage and the corresponding focus on GDP growth as a proxy for progress, consumption decisions can have a significant ripple effect throughout a single economy as well as the finite global resource base. Consider, for example, the use of milk cartons. Wax lined, printed paper milk cartons have been created for the transport and preservation of milk from the production to the consumption stage. However, the components of the carton were not developed with waste disposal in mind, rather increasing distribution and sales were the rationale for the carton. As a result, largely related to the focused basis of its creation, the milk carton serves a consumption purpose without consideration to the impact to the environment and potential future human and animal health due to its nonbiodegradable or reusable composition. This illustration on a broader consumption scale provides a simplified perspective to evaluate the underlying values captured in consumption decisions [2, 3]. From this perspective, production for consumption may be expressed as a myopic activity, focused on near-term satiation of a need or want to the exclusion of the evaluation of the impact or ripple effect of the satiation.

The values embedded and communicated within demand and supply determine the manner in which a need or want is attained. To the extent that there is no discussion of the values and behavioral factors assumed and reflected in demand and supply, arguably, implicit values, the values and the subsequent behaviors, become endogenous to the economic system. Therefore, explicit awareness of present behavioral assumptions inclusive of the "unlimited wants" of consumers, profit maximization motivations of producers, and understated resource depletion resulting from externalized costs offer the potential to modify active and embedded behavior.

An understanding of economics specifically oriented toward enabling the development of rational economic agent behavior can raise awareness of the significance of consumption behavior as the activity relates to sustainability, where the defining of sustainability is consistent with the Environmental Protection Agency [4]: "Sustainability creates and maintains the conditions under which humans and nature can exist in productive harmony, that permit fulfilling the social, economic and other requirements of present and future generations." Awareness in turn fosters the development and implementation of conscious and unconscious reinforcement of sustainability, which are the needed elements in driving a culture of sustainability, and in turn provides a foundational element of smart city traction and ultimately success [5].

13.1.2 Economics in Cultural Context

Economics evaluates human behavior relative to wants, needs, and resource allocation within a natural environment. By definition, the parameters of the discipline include other life forms and physical resources needed to maintain both life and environmental regeneration. To the extent that a human culture incorporates nonhuman elements in decision-making, the economic system includes an understanding of the holistic interdependence of living and non-living elements of the planet.

Culture is a significant contributor to what is perceived as valuable and is the determining parameter in the designations that ultimately yield to resource allocation within a society. Given that culture is a learned behavior, culture can either promote or diminish any given society's understanding of the interconnectedness of human and planetary life, thereby determining the extent of the anthropocentric, or human-centered, perspective. The United Nations Educational, Scientific and Cultural Organization (UNESCO) defined culture as a significant component to attaining global sustainability:

Culture shapes the way we see the world. It therefore has the capacity to bring about the change of attitudes needed to ensure peace and sustainable development, which, we know, form the only possible way forward for life on planet Earth. Today, that goal is still a long way off. A global crisis faces humanity at the dawn of the 21st century, marked by increasing poverty in our asymmetrical world, environmental degradation, and shortsightedness in policymaking. Culture is a crucial key to solving this crisis [6].

The inputs and outputs of economic systems are dependent on the value structures of a society, and to the extent that economics explains observable phenomenon and proposes optimal outcomes, the discipline can be both responsible for the maintenance of an economic framework and the catalyst for a change. Economic outcomes in essence mimic the values of the participants in an economic system.

Evaluating the historical cultural progression of human society can promote a stronger understanding of the economic relationship with resource allocation, both intra- and intersociety, and most importantly provide insights with respect to how perceptions of the world are shaped through cultural frameworks at a given point in time. The pace at which cultural attributes evolve may also provide a deeper understanding of why institutional and social frameworks may be inconsistent with the manifestation of contemporary challenges. Viewing economic thought or philosophy over time reveals the dynamic and cultural elements of society, as well as the basis of economic thought that remains in the principles literature in the present period.

13.1.3   Reconciling Economic Theory and Historical Context

The cultural attribution of value is a significant and arguably primary differentiator with respect to the variation in the perspective between societies of the quality of life for both human and nonhuman elements. Examples of surviving written works that provide a foundation or insight with respect to economic activities include Plato's *Republic* and Aristotle's *Politics*. The similarities in economic circumstances as described by the authors are consistent with the phenomenon observable today; however, the evaluation of human behavior as it applied to accumulation of wealth, stratification of society, and the role and impact of gratification were framed within an evaluation and discussion of moral philosophy and ethics, positioning Western economics up to the 18th century within the discipline of moral philosophy and politics. The evolution of the discipline continued through the modern era until the discipline formerly separated from moral and political philosophy through iteration as political economy to its present stand-alone context as economics. The observable mechanics of economic systems were the basis of discussion in conjunction with the human values, whether assumed as innate or culturally inspired. A connection between the qualitative and quantitative aspects of economic outcomes was articulated and addressed as an evolving and dynamic process. From this perspective economics discussions offered both a *normative* and a *positive* perspective, where the former provided opinions and values related to optimization and the latter described observable activity. At the present time, the economics in practice has shed the normative element of the discipline opting for a positive attribution as a means to enhance its standing as a science. In essence the focus on optimization has been to the exclusion of explicit evaluation of prevailing values. Given the significance of embedded values in conscious decision-making, the lack of articulation of values may contribute to the implicit value of outcome-based decision-making that only considers the optimization of the outcome rather than the impact of the outcome to others and future consumption. Perhaps the modification of the defining of the boundaries of present economic thought to be independent of a value-based foundation may in part provide an explanation for the imbalance in sustainable outcome observable today yet credited as being attributable to economic optimization.

The foundation for current economic thought can be found in the writings of Adam Smith [7], Jeremy Bentham [8], David Ricardo [9], and Karl Marx [10] along with many others. However, though all of these authors provided insights related to the human behavior attribution of economics contemporary to their time and though the circumstances described appear contemporary, the context of their writings has often been neglected in lieu of an adoption of an absolute meaning of their opinions. In essence, allowing the commentaries of these authors to embody a universal significance independent of time has arguably enabled the transfer of the theoretical modeling of a society specific to one period to another, independent of the observable change in society.

13.1.4 Significance of Context

The Classical period of economics has become the foundation of the study of the discipline of economics. To a large extent, the economic principles in practice have maintained the theories espoused by the writers and contributors to economic thought contemporary to the period. John Stuart Mill's [11] *Principles of Political Economy* provided a summary of the contributions to economic thought by Adam Smith, David Ricardo, and other significant thought leaders of the 19th century and became a standard text used in the study of economics into the early 20th century. However, of note is that the authors including Mill were relaying behaviors perceived in a society contemporary to their life and questioning aspects of the observed progress of the time including poverty, the role of money, and the potential impact of population growth. Their thoughts were debated discussions and their frameworks were not adopted as immutable facts. Additionally, the issues discussed were similar to those of predecessor Western societies and as evidenced in the moral philosophical discourses of Plato and Aristotle, nearly two millennia earlier. Of significance in the noted persistence of specific economic system fostered social outcomes are the cultural and time specificity of observed similarities. The evaluation of the human condition within a given social and economic framework provides the challenge to economists to both be positive evaluators from the perspective that positive signifies reporting on observable and factual phenomenon and normative participants, where normative requires an expression of value judgment.

Present instruction of economics has eliminated the normative aspects of assessment, reducing economics to mathematical relationships that are addressed in absolute terms rather than in alignment with cultural attributions coincident with their development. Further the seeming lack of attention to values and behavior incorporated within economic assessment has distanced the tangibility of economics, limiting the understanding of the explanatory potential of economics and the application of economics as both a cause and a remedy of unsustainable practices. There is a need to promote and foster an understanding of the role of values in economic outcomes and the sustainability of observed outcomes. In enabling this education, consumers can not only understand the intergenerational impact of their actions, but they can also appreciate and support the requisite sustainability infrastructure and regulations of the smart city.

13.2   Perception of Resource Value, Market Outcomes, and Price

Economics is the social science discipline that evaluates the relationship between human wants and the resources available to satisfy them. In identifying and explaining the relationship between wants and resources, economists use broad generalizations related to human behavior, arguably the most significant of which relates to wants.

Wants are based on the premise that individual economic agents, individuals interacting within the general economy, will always seek to have more of desirable goods and services. Desirable goods include both normal goods, which are goods that an individual will continue to purchase as their income increases, and luxury goods, which are goods that are not needed but are wanted to support an external display or perception of status or wealth. Not all goods are desirable, for example, inferior goods represent a classification of goods and services that will be reduced or eliminated by consumers as their incomes increase.

The behavior of wanting more, sometimes referenced as unlimited wants, is a social value, consistent with consumerism, which is defined as the focused act of consuming goods and services to improve utility, the economic concept that defines the benefit of consumption. Arguably it is not representative of an intrinsic human characteristic but rather a learned behavior. This is an important point. If a behavior is learned, it can be unlearned and a new behavior can emerge, which in turn can produce a different economic outcome.

Resources are broadly defined as including all the inputs in the production of final goods and services that are ultimately tied to the satisfaction of a want. From this perspective, resources could include teak wood trees in the making of furniture, water in the production of soda, and cattle in the production of food. Typically, resources are classified into one of three groupings, which include natural resources, human resources, and capital resources. Trees, water, and cattle are all natural resources. Human labor or entrepreneurship defines human resources, and capital resources consist of man-made objects that can be used to produce goods and services, such as factories and equipment. Regardless of the type of resource, all resources are finite and so by definition can be qualified as scarce [12–14].

Scarcity in economics essentially captures the relationship between wants and the access and availability of resources. For example, one could want a mango and see it hanging high on a tree but not have a ladder to reach it. The good in question is available but it is not accessible. Alternatively, one could stumble on a farmer's market selling mangos only to find that all the mangos on display have been purchased. In this case the mangos are accessible but they are not available. Both of these examples highlight the temporal or time sensitivity of scarcity. In the first example, one could borrow or purchase a ladder but this will take time, and in the second scenario, one can drive or walk to another market, but again, additional time will be required to satisfy the want.

Looking at time in a slightly different manner, a community could require lumber for the construction of new municipal buildings. The lumber required will result in the deforestation of 100 acres. In satisfying the want for lumber today, the community limits access and availability of lumber from the 100 acres over the time period required for the forest to regenerate, creating time-based scarcity.

In a market system, access and availability establishes a perceived scarcity embedded within the supply of a good. Ultimately, the supplier's willingness and ability to sell a specified amount of a good at a prevailing price is assumed to capture the costs of production of the good, implicitly including the scarcity of inputs. As a result it is expected that the higher the degree of perceived scarcity of a resource, the higher its price and, in the case of an input, the resulting price of the final good.

13.2.1 Market Distortions

Market outcomes, price and quantity, are highly dependent on the information that consumers and suppliers have available. Informational asymmetry, where one party has more understanding or knowledge related to a good than another, can create price and quantity outcomes that may not effectively consider scarcity. This results in market inefficiency, a situation where resource use is not efficiently allocated by the market. This is a significant issue and one that consumers are only beginning to understand. For example, abundance is a relative term but it is not inconsistent with scarcity; all resources are scarce. The perception of abundance without the recognition of inherent scarcity of resources can hasten resource depletion.

The production of goods by producers is based on a competitive framework. Additionally, the producer seeks to minimize costs and maximize revenue, to achieve maximum profitability. As a result of the focus on profitability, there is significant incentive for producers to externalize costs of production as a means of cost minimization. Externalizing costs can include pollution discharge, exploitation of regulatory differences between countries, overuse of natural resources, and limited waste disposal and reduction efficiencies. Though in the immediate period this may be beneficial to profitability, it may promote both short-lived unsustainable returns and longer-term environmental and social costs.

Consumers may not be aware of the implicit trade-offs being made as a result of the production of a good. This informational asymmetry can be attributable to many reasons, including a belief that regulatory agencies guarantee safety, to just simply a lack of diligence when assessing goods. For consumers, reliance on market efficiency without an understanding of the embedded incentives of producers can promote negative externalities. In effect the pursuit of satisfying unlimited wants may include effectively delegating environmental and social stewardship to producers whose incentives may not include the evaluation of these parameters. The end result is most readily seen in natural resources, where under pricing due to lack of inclusion of scarcity can lead to extinction or elimination of a resource's availability [15].

In a market-driven economy, such as the United States, the market is credited with efficiently determining the price of an item by implicitly incorporating the costs associated with production. When consumers or producers face low prices for consumption and input purchases, respectively, and the underlying belief is that the price being paid is fully reflective of the cost of the item being purchased, there is less of an incentive for efficient use and higher potential for waste. Price effectively becomes a measure of a resource's worth. When asymmetric or incomplete assessment of scarcity is prevalent, price may not properly indicate the cost of the resource being consumed.

In some areas of the world, forested land has been perceived as abundant, and the resulting price for land has been limited to the perception of present period abundance. The net result of the perception has been excessive global deforestation, resulting in present period-pronounced scarcity in some regions. Decades will be required to promote regrowth of the same lands. Had prices considered the impact of forest harvesting or the price of temporal scarcity, demand would have been lessened. Both consumption and production could have promoted efficient market pricing, leading to sustainable resource use, all from this simple inclusion.

13.2.2 Externalities

Demand and supply yield market outcomes that are assumed to represent an efficient allocation of resources. The price at which the quantity demanded equals the quantity supplied is therefore expected to embody the cost associated with the production and consumption of the good or service. However, production and consumption are not limited to the transactional nature of exchange of the final good at the determined market price. In the process of production and consumption, there are costs that are not factored that impact the well being of the economy at large, and these are referenced as externalities. In essence, externalities arise when an individual or firm engages in activities that influence the well being of others and where no compensation is provided in exchange for the imposition.

Typically externalities are characterized as negative, signifying that the externality yields an adverse outcome. These externalities are referenced as being *negative externalities*. However, there is a potential that a positive outcome could be generated, leading to a positive externality. In the discussion of externalities it often assumed that market participants accept the externalities generated by their actions as acceptable due to their focus on immediate gratification of their needs. For the producer this equates to externalizing the cost of disposal of waste products into waterways and the air where no cost is directly borne to adversely impact profits, but qualitative costs are assessed that may impact the enjoyment and longevity of multiple life forms and generations of human life. For the consumer the externality can be evaluated in the indifference to waste creation at the point of the consumption decision or even the externalities associated with the production of the good or service being purchased. In the case of the former, the cost of disposal of packaging material is typically marginal to zero, relatively negligible, but disposal creates a negative externality in the landfill, incinerator, or recycling plant that could have been avoided with a thoughtful exercise of demand.

At present, the type of internalizing of externalities that has occurred has been limited to quantifying the externality to an overt cost. However, to the extent that the costs may remain unassessed and the market mechanism is not cognizant and focused on the elimination of the externality-based cost, rather than the minimization of overall costs, this process has yielded suboptimal outcomes. For example, assume that a firm produces ambient pollution as a result of incineration of waste. If a governmental regulatory body institutes a fee or cost for pollution, effectively charging the firm for the ability to pollute the air, the producer is able to delegate responsibility for environmental stewardship to the price of pollution. Additionally, depending on the price elasticity of demand for the service offered, the producer may be able to not only transfer the costs now associated with polluting activity to the consumer but also maintain the pollution level. Assuming that the consumer is inelastic, in this example, the negative externality related to internalizing the cost has not changed; instead only the responsibility of pollution has been transferred to a cost, revenue to the regulating body has been generated, and the consumer has suffered erosion in their overall disposable income and purchasing power.

The same type of scenario exists with a permit trading program, where in effect permits are issued for a specific amount of externality emission, allowing economic agents to trade and thereby optimize through again cost minimization. However, the cost minimization is founded on the presumption or delegation of the permit system to fostering socially optimal outcomes, again, relieving the economic agent engaged in the creation of the externality from being directly accountable for qualitative actions. Additionally, the trading of permits assumes that optimal financial outcomes equate to optimal environmental and social outcomes due to the aggregated assessment of pollution. However, to the extent that pollution is not distributed evenly and certain locations may have a disproportionate concentration, the permit systems fail to generate a socially optimal outcome. This may be compounded by the impact of inelasticity, which may allow for the transfer of costs of implementation of the permit program to the economic agents the program was designed to protect.

Externalities are defined as a type of market failure based on the premise that optimal social outcomes result from individual economic agents acting in self-interest. However, if, instead of being a market failure, externalities could be evaluated to assess and develop an optimizing strategy between individual interests and enhanced social outcomes, externalities could be internalized within the market model as a modification of preference. Perhaps externalities only indicate a lack of holistic awareness on the part of the consumer and producer or a cultural bias toward immediate gratification. These characteristics can be potentially modified through education. Optimal and universally acceptable strategies could then be adopted to promote sustainability.

The success of this internalization strategy relies on the development of the educated rational economic agent as a consumer. If consumers are aware of the responsibility inherent in their consumption and are aware of the environmental and social impact of production processes, consumer demand can create the coalescing framework to augment preference to exhibit demand for sustainably produced products. The augmentation in demand does not allow for the opportunity of delegation of responsibility of pollution capacity to a cost or, alternatively, the incorporation within a cost minimization framework; as a result, the change in preference and subsequent modification in demand promote the development of market outcomes that are environmentally and socially optimal from the position of what is supplied. Smart cities inherently require consumer cognizance to promote dynamic continuous improvement consistent with the maintenance of long-term sustainability. However, the focus of the smart city also requires educated consumer inhabitants to ensure the provision of both public and common goods as provided through environmental and government regulation and oversight.

13.2.3 Common Goods

Resources such as air and water have no market price and are considered to be abundant. On the surface, these resources may appear to be unlimited; however, increased population pressures along with externalized costs related to production, such as pollution, have diminished the availability of both potable water and clean air. How could this have occurred?

The lack of price, a market model promoting the focus of profit maximization, and unlimited wants are largely responsible. Consumers have effectively allowed supply to determine demand by not imposing restrictions on how goods can be produced. Producers have focused on short-term profitability in lieu of long-term strategic resource utilization. In the short run both consumers and producers have benefitted, but the cost of consumption and profitability was externalized to other nations, the environment, and future generations. For example, in the 17th century, North American coastal waters were described and recorded as being rich in quantity and diversity of fish; the perception of abundance led over time to overfishing, and presently many varieties are endangered or at the risk of extinction. The cost of fishing included the human and capital costs not the replenishment costs. This yielded an ability to maintain artificially low prices, greater yields for profitability (overfishing), and waste.

An understanding of the perception of scarcity and abundance provides a strong foundation to understanding supply, demand, and market outcomes as these concepts relate to resource allocation and sustainability. To the extent that consumers delegate responsibility for sustainable consumption to producers and producers are focused solely on profit maximization, increased understanding of the responsibility inherent in consumption may provide a catalyst for increasing sustainable production, consumption, and development. As holistic evaluation of consumption is an assumed behavior of the rational economic agent, strengthening the understanding of the role of consumption may be significant in enabling the development of the rational economic agent.

Supply and demand reflect the amount that producers or suppliers of a good or service are willing and able to sell at a particular price and the amount that consumers of a good or service are willing and able to purchase at a particular price, respectively. Though on the surface the concepts of supply and demand appear simple, the characteristics that determine the explicit willingness and ability can be complex. The complications can arise as a result of differences in the preferences, behaviors, cultural values, financial capacity, and resource access and availability to the production process as these relate to suppliers. For consumers or demand, the complications can also be attributed to preferences, behaviors, cultural values, financial capacity, and wealth perception, as well as the perception of value and price, along with access and availability, of other substitute and complementary goods. Where and how the supply and demand interact with each other define a market. A market is comprised of a group of producers (supply) and consumers (demand) for a specific good or service, who collectively, as part of their exchange process, determine the market price or equilibrium price of a good or service.

Price is the natural outcome of the supply and demand relationship. It is indicative of the value of a good based on a consumer's assessment of the costs and benefits of purchasing the good. As consumers become increasingly aware of the environmental and social costs of production, the prevailing price may be corrected either through regulatory imposition of the costs of externalities within the market mechanism or via consumers, who will opt to purchase goods not on price but related to holistic production costs.

It is important to note that the market relationship is dependent on information and understanding of the limits of duty of care. The outcome of the market relationship, price, and quantity can only reflect the embedded preferences and social values depicted in demand and supply. If the market outcome does not meet expectations, the market model is not to blame; rather the prevailing value structure may be the flaw.

Value in this context is related to how resources are valued from the perspective of the quality of care and maintenance we would be willing and able to provide to ensure the protection of the resource. The use of the word "value" is not directly based on market quantification but expresses the hierarchical importance that consumers and producers would attribute to a resource; examples may include the environment, human health, and animal welfare.

Every day consumers make decisions with the collective strength of aggregated individual demand. These decisions influence supply and demand going forward, including the ability of producers to develop new goods and services, as well as resources and technological advances to satisfy both existing demand and projected future demand. Demand is a powerful catalyst in the evolution of market outcomes. However, to a large extent the power of demand is limited both by the fragmentation of consumers due to limited opportunities for coalescing around specific interests and limited consumer understanding of the inherent power of aggregated consumption decisions. From this perspective, understanding how the market functions and the power of consumption in creating sustainable economic outcomes is one aspect of developing into a rational agent.

The values embedded and communicated within demand and supply determine the manner in which a need or want is attained. To the extent that there is no discussion of the values and behavioral factors assumed and reflected in demand and supply, arguably, implicit values, the values and the subsequent behaviors, become endogenous to the economic system. From this perspective, explicit awareness of present behavioral assumptions inclusive of the "unlimited wants" of consumers, profit maximization motivations of producers, and understated resource depletion resulting from externalized costs offer the potential to modify active and embedded behavior.

13.3 Conscious Consumption and the Sustainability Foundation of Smart Cities

The explicit discussion of the embedded assumptions guiding the behavior of the decision-maker is typically not a part of the economic education process. As a result, to the extent that individual economic agents, producers or consumers of a good or service, are bounded by rationality that does not include addressing the impact of externalized or non-quantified costs, the economic discussion does not promote or position the assessment of alternative outcomes. Implicitly and endogenously, the economic discussion establishes and maintains a consumption to production circular flow, focusing on the gratification of consumption and profit taking from production, seemingly eliminating assessment of externalities and holistic dynamics. Returning to the milk carton example provided in Section 13.1, the economic discussion would be limited to the utility gained from consuming the milk and the corresponding profit maximization of the producer. Waste would be regarded as an externality rather than an endogenous aspect of the decision-making process. Additionally, costs are priced into the product through efficient market assumptions. In net consumers would expect that the purchase price is indicative of the holistic cost of the product, and producers would view production costs as being related to market priced inputs, not environmental impacts, during or as part of the life cycle of the good.

As Nelson [16] points out, economics evaluates efficiency with respect to the "use of resources to maximize production and consumption, not by the moral desirability of the physical methods and social institutions used to achieve this end." The factors that are included in an economic evaluation are limited to the tangible quantifiable costs, and costs are overlooked where either a market or regulatory oversight has not provided a monetary justification. From this perspective, the impact of consumption decisions on the environment, economic disparity, or endangerment of other species is not an issue. The market mechanism disenfranchises the consumer from the welfare of those impacted by his/her consumption and promotes the perception that price alone is indicative of the true cost of a good. Nelson notes, "The possibility that consumption should be reduced because the act of consumption is not good for the soul, or is not what actually makes people happy, has no place within the economic value system." The underlying assumption is that consumers are driven to want more. As a result, economic modeling assumes that reduction in consumption in the current period is only addressed through the lens of an increase in consumption in a later period. That the assumption of insatiable want may be taught a learned behavior reinforced through a market model is not even addressed in economics [17].

A general and seemingly applicable assumption is that consumers and producers maximize the benefit related to the opportunity accessible in their particular circumstance. The desire to reach an optimal outcome for a given point in time, as has been noted before, is subjective and specific to how these economic agents view the concept of maximization, which in turn is likely to be highly correlated with cultural values. For example, in indigenous societies there is evidence that a balance between present and future periods along with that of the environmental system, as a whole, was included in decision-making and optimization [18]. In present consumerism-fostered economies, the cultural values are less likely or unlikely to incorporate environmental and social justice parameters proactively. The focus of observable and marketed consumption is immediate gratification. However, as consumer awareness of both the impact of consumption and the power of consumption to modify and catalyze economic outcomes increases, there is growing evidence of a shifting cultural paradigm to one of sustainability.

Markets do fail to produce optimal outcomes. Sometimes this is due to the myopic focus of market participants as in the case of externalities, and in other circumstances it can be attributable to the lack of excludability as in the case of common goods. To some extent cultural values dictate the significance of the adversity related to the creation of externalities or abuse of common goods. The use of market models has been the regulatory mechanism to modify socially nonoptimal outcomes, but through relying on the market mechanism rather than simultaneously including mechanics to promote cultural change, the majority of regulatory interventions to date have had limited to questionable success.

13.3.1 Smart Economic Development

The present global status of recognized environmental degradation, exploitation, and resource depletion tied to understating of holistic of costs and ultimately the pursuit of a narrowly defined consumption-based metric of economic progress, GDP, has promoted an increased multidisciplinary interest in sustainability. By definition the concept of sustainability incorporates the intertemporal allocation of resources through a holistically assessed strategic utilization rate that includes environmental and social justice parameters. From the Brundtland Report,

> Sustainable development is development that meets the needs of the present without compromising the ability of future generations to meet their own needs. It contains within it two key concepts:
>
> the concept of **needs**, in particular the essential needs of the world's poor, to which overriding priority should be given; and
> the idea of **limitations** imposed by the state of technology and social organization on the environment's ability to meet present and future needs.

The defining of sustainability is in close alignment with the objective of the discipline of economics. From Section 13.1, "Economics is the study of human behavior in relation to a resource-constrained world." The ability of economics to add value to sustainability objectives requires the insertion of value parameters or normative thinking in conjunction with the positive or observational stance adopted by the discipline. The catalyst for the value-based practice of economics rests with the ability to promote an understanding of the discipline, establish pervasive rational agent behavior in the economy, and promote attainment of optimal social and environmental outcomes rather than observationally recording the realization of the theory of second best. The latter is embedded in the characterization of market failure and externalities. Establishment of rational agent decision-making and thereby responsible conscious consumption provides the conduit for paradigm shifting from consumerism to sustainability.

A constituency with an understanding of the holistic relationship between consumption and sustainability and having engagement in government [1] are foundational elements in the social policing and infrastructure of smart economic development. This chapter prompts further evaluation in the mechanics of sustainability-focused education and also establishes the view that the measurement of success of smart economic development may not be captured in present, standardly used metrics, namely, GDP, given that the underlying values that support GDP expansion may be inconsistent with the success parameters associated with sustainability.

13.3.2 Next Steps Theory and Practice

Smart cities are sustainable. The defining of sustainability is in close alignment with the objective of the discipline of economics. The ability of economics to add value to sustainability objectives requires the insertion of value parameters or normative thinking in conjunction with the positive or observational stance adopted by the discipline [14]. In comparing economic frameworks, sustainable economics due to its acceptance of the breadth of human attribution is well positioned to both assess and, thereby through evaluation, promote iterative policy implementation to support the inclusion of the paradigm of sustainability as a societal norm.

Rational economics suggests market instruments, while sustainable economics suggest a policy mix. The kind of government programs that are recommended depends on the impact of consumption determinants. For instance, if rational and economic determinants have a great impact on consumer behavior for a certain product, market solutions might fit best and be combined with educational (indirect) instruments. In contrast, if economic determinants have a low impact and if involvement is low, direct instruments should be put in place. So in general, before implementing government programs, it should be investigated which factors influence consumption, and this should include education. This could be done by a mixed methods approach including multivariate data analysis. For this reason, much more research on the specific effects of the government programs and the effects of their combination is necessary.

Further, there is a necessity to operationalize the concept of sustainability and introduce new measuring systems beyond the GDP. Evaluation of the sustainable development of a society and the impact of government policies through the lens of GDP is not sufficient. On the contrary, a focus on the GDP is consistent with the promotion of consumerism [19], which in turn is congruous with resource depletion and externalizing costs. Smart cities will have to be assessed based on nontraditional economic factors, if the objectives of their implementation are to be realized.

Final Thoughts

Smart economic development is dependent upon holistic and routine evaluation of economic and societal frameworks. These frameworks need to be assessed and modified as part of an ongoing continuous improvement process. Fundamentally, what may have been viewed as appropriate action at a point in time may no longer serve the same purpose due to changing environmental, social, and cultural parameters. However, the members of a society have to be both empowered and cognizant of the need for this type of evaluation in order for efficiency and ultimately sustainability to be a realized inter- and intragenerational attribute. From this perspective, the deployment of consumer education programs targeted at defining responsible demand as conscious consumption are a requisite foundation for smart economic development.

Questions

1 What is the role of culture in driving sustainability and smart economic development?

2 Is smart economic development possible without a reassessment of current economic and social frameworks?

3 What role does education play in enabling smart economic development?

4 How does consumption demand impact sustainability?

5 Should smart economic development rely on the same measures of prosperity as a consumerism-fostered economy? Specifically, should capacity, as measured by GDP, be the metric of progress?

References

1 O'Connell, L. (2008) Exploring the social roots of smart growth policy adoption by cities. *Social Science Quarterly*, **89** (5), 1356–1372.
2 Colander, D. (2005) What economists teach and what economists do. *The Journal of Economic Education*, **36** (3), 249–260.

3 Schweitzer, A. (1981) Social values in Economics. *Review of Social Economy*, **39** (3), 257–278.

4 Environmental Protection Agency. (n.d.). *What Is Sustainability?* Retrieved from http://www.epa.gov/sustainability/basicinfo.htm.

5 O'Hara, S.U. (1995) Sustainability: Social and ecological dimensions. *Review of Social Economy*, **53** (4), 529–551.

6 UNESCO (Culture Sector) (2000) *World Culture Report*, United Nations, Paris.

7 Smith, A. (1791) *An Inquiry into the Nature and Causes of the Wealth of Nations*, Basil, Tourneisen and Legrand, p. 22.

8 Bentham, J. (1879) *An Introduction to the Principles of Morals and Legislation*, Clarendon, Oxford, p. 14.

9 Ricardo, D. (1911) *The Principles of Political Economy & Taxation*, J.M. Dent & Sons, London, p. 8.

10 Marx, K. (2009) *Das Kapital*, Regnery Publishing, Inc., Washington, DC Vol. 1.

11 Mill, J.S. (2016) in *Principles of Political Economy with some of their Applications to Social Philosophy*, 7th edn (ed. W.J. Ashley), Longmans, Green and Co., 1909, London.

12 Choi, S. and Ng, A. (2011) Environmental and economic dimensions of sustainability. *Journal of Business Ethics*, **104** (2), 269–282.

13 Czech, B. (2000) Economic growth as the limiting factor for wildlife conservation. *Wildlife Society Bulletin*, **28** (1), 4–15.

14 Venkatesan, M. (2016) *Economic Principles: A Primer, A Foundation in Sustainable Practices*, Kona Publishing, Mathews, NC.

15 Boran, I. (2006) Benefits, intentions, and the principle of fairness. *Canadian Journal of Philosophy*, **36** (1), 95–115.

16 Nelson, R.H. (1995) Sustainability, efficiency, and God: Economic values and the sustainability debate. *Annual Review of Ecology and Systematics*, **26**, 135–154.

17 Knoedler, J.T. and Underwood, D.A. (2003) Teaching the principles of economics: A proposal for a multi-paradigmatic approach. *Journal of Economic Issues*, **37** (3), 697–725.

18 Nerburn, K. (1999) *Wisdom of the Native Americans*, Novato, CA, New World Library, p. 41.

19 Venkatesan, M. (2015). *Values, Behaviors and Economic Outcomes*. Retrieved from https://www.youtube.com/watch?v=ipq5owxlU0g (accessed 19 January 2017).

14

Managing the Cyber Security Life-Cycle of Smart Cities

Mridul S. Barik, Anirban Sengupta, and Chandan Mazumdar

Department of Computer Science & Enginering, Jadavpur University, Kolkata, India

CHAPTER MENU
Introduction, 391
Smart City Services, 393
Smart Services Technologies, 394
Smart Services Security Issues, 396
Management of Cyber Security of Smart Cities, 397
Discussion, 403
Conclusion, 404

Objectives

- To become familiar with smart city services and technologies.
- To become familiar with cyber security and privacy issues associated with different services of smart city.
- To become familiar with the need for management of cyber security and privacy of smart cities.
- To become familiar with different phases of cyber security and privacy management life cycle.

14.1 Introduction

The world is witnessing rapid urbanization, hitherto unseen. Latest UN report [1] says that currently, more than 50/ the world population lives in cities, and this figure is expected to grow up to 70/ population growth, cities have embraced advanced information and communication technology (ICT)-based

Smart Cities: Foundations, Principles and Applications, First Edition.
Edited by Houbing Song, Ravi Srinivasan, Tamim Sookoor, and Sabina Jeschke.
© 2017 John Wiley & Sons, Inc. Published 2017 by John Wiley & Sons, Inc.

solutions to provide smart services to their citizens. This gives rise to the concept of "smart cities" where networked sensors and actuators deployed pan city allow real-time monitoring of the urban environment and also enable timely reaction to events through central control system with minimal or no human intervention. Today, several countries are working toward the development of smart cities that will offer citizen-centric smart services in an efficient and cost-effective manner. There is no universally accepted definition of a smart city, till date. The idea of a smart city may vary, depending on the level of development, willingness to change and reform, resources, and aspirations of the city residents [2]. Most of the smart services are provided through the use of cyber–physical technologies. Traditionally, cyber and physical systems have been independent of each other. Perturbation in either cyber or physical systems is contained within the respective domains. In cyber–physical systems these two domains are tightly integrated, which opens up opportunities for newer types of attacks through exploitation of vulnerabilities of individual domains. Moreover, ensuring the security of resources and privacy of citizens' data is a major concern in any smart city solution. With the growing number of attacks in the cyberspace every day, it is expected that a large attack surface of smart cities will aid cyber criminals to launch large-scale attacks so as to breach security and privacy of resources and data. Hence, planning and implementation of cyber security measures is of utmost importance in the case of smart cities. Historically, it has emerged that any cyber security program consists of both managerial and technical measures. These find ample support in existing cyber security laws and regulations [3, 4], which differ in scope and intensity between nations and business sectors. Interconnected infrastructure in smart cities gives rise to scenarios where an attack on one sector may cause an unrelated, but connected, sector to be compromised as well. Hence, it is extremely important to study and analyze the security requirements of smart cities, keeping in mind both the physical (due to common infrastructure) and logical (intra- or inter-sector) dependencies of different sectors. Based on these requirements, a detailed risk assessment may be undertaken to identify, analyze, and evaluate the security risks within the cyberspace of a smart city. The identified risks will enable the design of smart security measures to help prevent, detect, and/or recover from those risks. However, it should be borne in mind that, like all other security programs, implementation of cyber security for smart cities is not a one-time affair. Security concerns evolve continuously with changes in infrastructure, discovery of new vulnerabilities, emergence of new threats, and enhancements in the regulatory framework. Hence, we propose a life-cycle approach to manage the various phases of cyber security of smart cities. It begins with the establishment of scope and boundaries of the proposed security implementation. This is followed by the identification of security requirements, keeping in mind applicable laws and regulations and specific security concerns and expectations of various stakeholders.

Risk assessment is performed to identify and comprehend the risks that can potentially breach the cyber security of the smart city. The identified risks are then prioritized, and a detailed risk mitigation strategy is designed. This would comprise of identification of security controls and measures that can reduce the risks to an acceptable level. It includes the formulation of cyber security policies and procedures and a selection of appropriate software and hardware security tools. The measures are then implemented, configured, and adopted to mitigate the identified risks. After implementation of security measures, continuous monitoring is undertaken to check whether the implemented controls are able to satisfy the identified security requirements. Such monitoring is usually based on log records and generates several metrics that help to understand the security posture of the smart city. Depending on the monitoring results, corrective actions, if required, are planned and implemented to address specific concerns. Thus, a complete cyber security life cycle for smart cities is proposed in this chapter.

14.2 Smart City Services

Smart cities aim to create a sustainable urban environment by leveraging ICT-based solutions, thereby providing better services to its citizens [5]. The concept of smart services in smart cities is still at an abstract level. A huge amount of research effort is required to clearly understand its principles. Smart services are intended for providing context-aware information to citizens using ICT.

Among different services that define a smart city, this chapter considers only the key services: smart transportation, smart grid, smart water management, and smart healthcare.

Smart transportation is an integral part of any smart city initiative and thrives toward improving citizens' experience while on the move using smart technologies. Smart transport systems combine intelligent transportation systems (ITS) with other vehicular technologies, such as autonomous vehicles, vehicle to vehicle (V2V), and vehicle to infrastructure (V2I), to improve safety, security, efficiency, and environment friendliness of surface transport systems. It also enables provisioning of an array of innovative transport-centric services such as emergency vehicle notification systems, automatic road enforcement, variable speed limits, collision avoidance systems, dynamic traffic light sequence, city traffic monitoring and real-time traveler information, smart parking, advanced driver assistance, highway intelligence, etc.

Smart grid systems deploy networked sensors and meters to gather real-time system data from generation, transmission, and distribution systems. A central analytics engine receives these data through communication networks and sends control commands to on- field intelligent electronic devices, such as programmable logic controllers and supervisory control and data acquisition (SCADA) systems.

Smart water management systems integrate sensors, controls, and analytical components to guarantee that water is efficiently delivered only when and where it is needed. It also implements real- time monitoring of water quality.

Smart health (s-health) [6] is a new concept that envisages provisioning of health services by reaping the benefits of context-aware network and sensing infrastructure of smart cities.

14.3   Smart Services Technologies

Smart city functionalities are enabled by a wide variety of technologies [7] across different service sectors. Some of the important and widely used ones are discussed below.

In smart transportation services an ITS [8, 9] gathers real-time traffic data and uses them to inform automated decisions regarding the function of traffic-related infrastructures. These systems typically include four main components: sensors that gather information on traffic conditions (such as inductive loops for traffic metering), actuators or controllers that make changes to traffic control devices (such as traffic lights), a central computing unit that analyzes data and suggests system adjustments, and communication systems to connect various components. Although traffic communication networks have traditionally used wired links, cities are increasingly adopting various forms of wireless technologies such as Wi-Fi, WiMAX, Bluetooth, GPS, GSM/CDMA, and 3G for such communications. Autonomous vehicle technology enables automobiles to execute safe and efficient commands based on its understanding of the environments in which they operate. Real-time data from variety of embedded sensors are processed by the central onboard computer to make decisions and issue commands to the vehicle's physical control elements such as the steering, acceleration, breaking, and signaling systems. V2V technology uses dedicated short-range communications that allow wireless exchange of data among vehicles traveling in the same vicinity, for significant safety improvements. Vehicles in a V2V network can send and receive data about their location, speed, and relative distance in order to alert drivers regarding potential dangerous situations. For example, when one vehicle brakes suddenly, vehicles behind could get a warning message before they get too close. V2I systems allow physical infrastructures such as traffic signals and ramps to inform vehicles of their presence and to allow vehicles to send information to the infrastructure. Another application of V2I systems would be in situations where the velocities and accelerations of vehicles and inter-vehicle distances would be suggested by the infrastructure on the basis of traffic conditions, with the goal of optimizing overall emissions, fuel consumption, and traffic velocities.

NIST SP 1108 [10] provides a conceptual model where it mandates seven logical domains for smart grids: bulk generation, transmission, distribution, customer, markets, service provider, and operations. A backbone network establishes inter-domain communication. A local area network is used for intra-domain communication. Smart grid communication network is similar to the Internet in terms of complexity and hierarchical structure, but they differ in terms of functional objectives. The basic function of the Internet is to provide data services for users and it adopts a best effort strategy for this. Whereas, power communication networks should ensure reliable, secure, and real-time message delivery and non-real-time monitoring and management. The kind of communication model used in smart grid is either top-down (center to device) or bottom-up (device to center). Peer-to-peer communication model is usually restricted to local area networks of individual domains owing to security concerns. Two widely used protocols in power systems are the distributed networking protocol 3.0 (DNP3) and IEC 61850. Advanced metering infrastructure (AMI) is another technology that brings transparency and efficiency to energy consumption in smart grids. Smart meters measure, store, and transmit energy usage data and voltage data for residences and commercial buildings. Sometimes, AMI systems employ two-way communication technology, often using wireless connections to send and receive data from utilities and system operators.

Future smart water management systems will use different networked and automated technologies in its functional components such as smart water treatment, smart water distribution, smart water storage subsystems, and smart water quality monitoring. Devices such as smart valves and smart pumps will be able to adjust to their environment by means of wireless communication with each other and with a central control system. This allows administrators to maintain system awareness, monitor automatic system functionality, and remotely access and control distribution devices.

Although still at a conceptual phase, technologies in smart healthcare would adopt practices in standard mobile healthcare systems. Additionally it would use cyber–physical infrastructure of other smart service sectors to provide context-sensitive healthcare services [6, 11, 12].

Among other technologies, cloud computing [13, 14] and Internet of Things (IoT) [15, 16] would play significant roles in realizing smart services in future smart cities. Cloud computing would enable robust remote backup functions capable of withstanding local disasters. It is able to provide required information without users being aware of the location of hardware and data.

14.4   Smart Services Security Issues

A smart city gathers data from smart devices and sensors embedded within. It shares those data via a smart communications system that is typically a combination of wired and wireless infrastructure. It then uses smart software engines to make smart decisions for enhanced service performances. Cyber security of smart cities means protection of data, systems, and infrastructure responsible for the city's operations and for the stability and livelihood of its citizens [17].

The very nature of smart city cyber–physical infrastructure opens up a huge attack surface, which is the aggregation of all known and unknown vulnerabilities, taking into consideration existing security controls across all subsystems and networks. The following paragraphs highlight some of the security issues associated with smart city services.

In smart transportation sector, the design of automotive vehicle architectures is still mainly driven by safety and cost factors rather than security. In most of the cases, the current state-of-the-art industry practices for establishing communication between embedded subsystems in automotive cars do not follow standard computer security principles [18]. This renders those autonomous vehicles prone to theft and remote attacks. Moreover, in-vehicle wireless sensor networks, used in applications like tire-pressure monitoring systems (TPMSs) [19], do not employ any cryptographic algorithms for protecting their communications. An in-depth understanding of the vulnerabilities, threats, and attacks is necessary for the development of any defense mechanism [20]. Moreover, solutions based on cryptographic mechanisms should be suitable for resource-constrained embedded systems.

Some of the potential attack scenarios [21, 22] are enumerated below:

- Autonomous vehicles are vulnerable to remotely executed attacks. Attackers able to control one or more autonomous vehicles could cause them to crash.
- Unlike general purpose computing devices, installing system updates or security patches for a car may be expensive and complicated, rendering them vulnerable to zero-day attacks.
- Features like access to Internet while traveling greatly increases a vehicle's attack surface.
- The use of cellular and Bluetooth technology in modern cars also increases the risk of remote attacks.
- Devices in autonomous vehicles that receive external inputs, such as GPS, are vulnerable to signal jamming.
- An attacker could intercept and modify safety-related data as it is communicated over V2V networks, leaving neighboring vehicles blind to actions like sudden braking, acceleration, lane changes, or turning.

Smart grid significantly depends on intelligent and secure communication infrastructures for its operation. Many of the communications technologies currently used in smart grid have known vulnerabilities, which, when exploited, could lead to unreliable system operations, affecting both utilities and consumers [23–25]. The automated meter reading (AMR) technology widely used in smart grid lacks basic security measures to ensure privacy, integrity, and authenticity of data. Moreover, some AMR meters periodically broadcast their energy usage data over insecure wireless links. Attackers can capture this data with modest technical effort and identify unoccupied residences or citizens' routines [26].

SCADA systems are widely used in water distribution systems and sewage treatment plants. These systems are designed primarily to maximize functionality, not security. Attackers with little effort can capture/modify traffic between the central control unit and on-field units such as the pumping stations and valves. Another threat to this system is the theft of sensitive information regarding site maps, details of chemical processes, site security plans, and so on. This renders such SCADA control networks vulnerable to disruption of service, eventually leading to public safety concerns [27]. In the famous Maroochy Shire incidence in Australia in 2000 [28], an attacker penetrated into the SCADA system and caused a large volume of untreated sewage to be released in public places.

Huge amount of citizen information captured in smart city services can endanger the privacy of its citizens [29, 30]. For instance, smart health services might help to mitigate many health-related issues; its ability to gather unprecedented amounts of information could endanger the privacy of citizens. Moreover, from the data collected in a smart city, it would be possible to infer citizens' habits, their social status, and even their religion. All these information are very sensitive, and when they are combined with health information, the result is even more severe.

In a nutshell, protecting the privacy of citizen data and securing the infrastructure is an uphill challenge currently faced by the research community. Thus, it calls for proper management of cyber security in the context of smart cities.

14.5   Management of Cyber Security of Smart Cities

Ensuring the security of resources and privacy of citizens' data is a major concern in any smart city project. The cyberspace is infested with malware and witnesses novel and damaging attacks every day. It is expected that the total connectivity of smart cities will aid cyber criminals to launch large-scale attacks so as to breach security and privacy of resources and data. Hence, design and implementation of cyber security measures is of utmost importance in the case of smart cities.

However, it is almost impossible to achieve absolute security. Cyber security can be visualized as a spectrum that runs from very insecure to very secure.

It is a balancing act that requires the deployment of "proportionate defense." The controls that are implemented should be proportional to the risk. It is important to determine relevant controls by comparing the cost of security with the value of services they are protecting, privacy needs of citizens' data, and efficiency needs of systems.

In other words, implementation of cyber security measures is a management issue, where a fine balance is to be drawn between system efficiency, security expenses, and data protection.

However, cyber security needs of a smart city are not static; they evolve continuously with changes in services and assets, discovery of new vulnerabilities, and emergence of new threats. Moreover, changes in laws, regulations, and contractual obligations also lead to new security requirements. Hence, the process of developing and deploying a proper cyber security program for a smart city is not a one-time affair; rather it is a continual process of analysis, design, implementation, monitoring, and adaptation to changing needs.

Considering the above requirements, we propose a life-cycle approach to manage the various phases of cyber security of smart cities (Figure 14.1). The different phases of this life cycle are detailed in the following subsections.

14.5.1   Scope and Cyber Security Policy Formulation

The purpose of this phase is to identify the scope of implementation of cyber security in a smart city and define its cyber security policy. The scope could be restricted to a single smart service or multiple services, or cover the entire city. It includes exclusions, if any, along with proper justification. Cyber security policy is a set of high-level statements describing the objectives, beliefs, and goals of the city as far as security is concerned but not how these solutions are engineered and implemented. It defines the overall security and risk control objectives that a smart city endorses.

This phase will first enable a city to determine the boundaries and applicability of the security management system to establish its scope. This will be done by considering the following:

- External and internal issues that are relevant to the purpose of the city and that affect its ability to achieve the intended outcome(s) of its cyber security system
- Requirements of interested parties that are relevant to cyber security (e.g., legal and regulatory requirements and contractual obligations)
- Interfaces and dependencies between activities performed by the city and those that are performed by other smart cities.

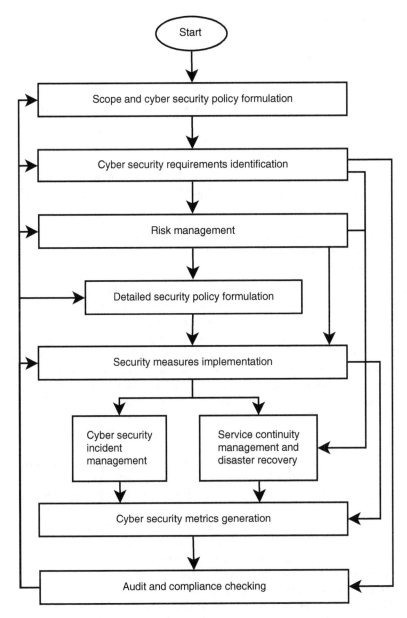

Figure 14.1 Smart city cyber security life cycle.

The other feature of this phase is the development of cyber security policy that

- Is appropriate to the purpose of the city
- Includes cyber security objectives of the city
- Includes a commitment to satisfy applicable cyber security requirements
- Includes a commitment to continual improvement of the cyber security system.

14.5.2 Cyber Security Requirements Identification

This phase will help a city to identify its cyber security requirements vis-a-vis its scope and policy. A city first needs to identify its critical assets. This would include:

- *Primary assets*, comprising of smart services and information assets
- *Supporting assets*, consisting of hardware, software, network, personnel, site, and city's structure.

Details about **smart services** are usually obtained during the *scope and cyber security policy formulation* phase. Other critical assets are identified during this phase. **Site** comprises the information pertaining to physical location of the city. The **city's structure** describes the hierarchy that exists among the administrative functionaries of the city. The security requirements of each of these assets need to be identified. This essentially includes their confidentiality, integrity, and availability requirements. Additionally, legal and contractual obligations may give rise to further cyber security requirements.

14.5.3 Risk Management

This phase will perform service-based risk assessment and treatment. Computation of risk is based on three factors: service value, severity of vulnerabilities that exist within services and related assets, and likelihood of occurrence of threats that can exploit those vulnerabilities.

Service values are computed considering the security requirements of the smart services (identified in the previous phase). Interservice dependencies may be used to adjust the values of these requirements.

Vulnerabilities can be of two types: technical and managerial. Technical vulnerabilities exist within services owing to the presence of weakness in software and hardware assets. Managerial vulnerabilities occur due to lack of (or improper) implementation of policies and procedures. Identification of such vulnerabilities, and computation of their severities, is usually a manual process.

Threats can be categorized as city wide and service specific. While the former applies to all services of a smart city (usually *natural* threats), the latter is applicable to specific services only (usually *environmental* and *human-induced* threats). Past incidents can be used to compute the likelihood of occurrence of threats in a smart city.

The service values, severities of vulnerabilities, and likelihood of occurrence of threats will be combined to derive service-specific risks of a smart city.

A city needs to define a set of risk acceptance criteria pertaining to its services. The computed risk values will be evaluated against these criteria to determine the risks that need treatment. Risks, which are less than the specified risk acceptance threshold, can be accepted, while all other risks need to be treated.

Risks can be treated by avoiding them, applying controls, or transferring them to third parties (insuring, outsourcing, etc.). Treatment of risks requires resources like manpower, infrastructure, time, money, and knowledge/know-how. Security infrastructure includes tools like firewalls, intrusion detection/prevention systems, switches, and anti-malware. Procurement and installation of tools alone will not secure a city. It is important to formulate detailed cyber security policies that will help in the proper utilization and administration of those tools. This is detailed in the next section.

14.5.4 Detailed Security Policy Formulation

This phase will help to generate security manuals, detailed cyber security policies, procedures, guidelines, and documented information (forms and records) templates for implementing security within a smart city. This includes access control policies, acceptable use policies, monitoring policies, and so on. It is also important to maintain records of all cyber security activities. This is achieved by preserving device logs and maintaining specific forms that have been filled by concerned personnel.

14.5.5 Security Measures Implementation

The implementation of security measures includes procurement and installation of identified tools and enforcement of security policies and procedures. It is important to assign responsibilities to specific personnel who will oversee these activities. Moreover, citizens need to be made aware about various security issues and mitigation strategies from time to time. This can be implemented by conducting targeted security awareness and education programs and assessing their effectiveness.

14.5.6 Cyber Security Incident Management

Security policies or controls alone do not guarantee complete protection of a smart city. After controls have been implemented, residual vulnerabilities are likely to remain that can make cyber security ineffective and thus security incidents possible. Further, it is inevitable that new instances of previously unidentified threats will occur. Therefore, it is essential for the city to have a structured and planned approach to:

- Detect, report, and assess cyber security incidents
- Respond to cyber security incidents, including the activation of appropriate controls for the prevention and reduction of and recovery from impacts
- Report security vulnerabilities (weaknesses) that have not yet been exploited and assess and deal with them appropriately
- Learn from cyber security incidents and vulnerabilities, institute preventive controls, and make improvements to the overall approach to cyber security incident management.

It is important for citizens to report cyber security events and incidents (including weaknesses) so that appropriate preventive and corrective actions may be initiated. An event/incident reporting form should be filled up that contains information pertaining to event/incident type, event/incident category, event/incident sub-category, site, date/time of discovery, impact, and so on. After an event/incident has been reported, a senior authority should be required to confirm the same. Finally, after appropriate preventive/corrective actions have been completed, the event/incident will need to be closed with records of implementation details. Appropriate alerts should be generated if an event/incident is not closed within the estimated deadline.

14.5.7 Service Continuity Management and Disaster Recovery

This involves managing the recovery or continuation of smart services in the event of disruption/disaster, and the overall program through training, exercises, and reviews, to ensure that the service continuity plan(s) stays current and up to date. The city should perform impact analysis and risk assessment to determine service continuity needs. Service continuity policy, procedures, and plan may be generated by considering inputs from previous phases as follows:

- *Cyber security requirements identification* – specific cyber security requirements of the services can be derived from this phase.
- *Risk management* – risk values of smart services and assets may be obtained from outputs of this phase; moreover, identified options for risk treatment relevant to continuity management are also obtained.

Based on the above inputs, relevant options for continuity management and disaster recovery, along with benefits and required resources, are identified. The city needs to select a subset of these options (based on cost–benefit analysis, legal, regulatory and contractual obligations, etc.) as its service continuity strategy.

It is important for the administrators to understand the effectiveness and performance of the implemented security processes. This phase will generate cyber security metrics to indicate the current security posture of the city. It would help the administrators take decisions regarding improvements in its security program and consequent investments. Data pertaining to security activities as well as incidents are collected from previous phases. These are analyzed to compute security metrics. These may be presented in the form of a dashboard that displays monthly, quarterly, and annual trends of cyber security metrics of the city.

The purpose of this phase is to measure compliance and identify nonconformance of the implemented security processes against the cyber security requirements of a smart city. This should be performed at regular intervals, and the outcome should be used to plan corrective actions and enhancements. This will help to continually improve the cyber security program of the smart city and provide assurance to all interested parties about the safety of its services.

14.6   Discussion

Maintenance of security of the services of smart city, including the privacy of its citizens, is a critical task. It is not enough to simply assess security requirements and implement appropriate controls during the launch of a smart service. It is important to continuously monitor security processes, identify changes in requirements, and make necessary enhancements. Absence of such continuous security review and implementation could render the services insecure and susceptible to cyber attacks. Consider the case where a smart health system is launched in a city. Before the launch, a comprehensive risk assessment is conducted, and appropriate security measures are implemented. After some time, several citizens begin receiving unwarranted offers of health checkups and medicines at discounted rates from a specific pharmaceutical company. The citizens do not sense any wrongdoing; while some of them reject the offers, others accept them. However, a reassessment of security risks reveals that an insider has connived with the pharmaceutical company and revealed private medical records of citizens. This has allowed the company to launch targeted offers to the public. Detection of such privacy breach, and mitigation of the same, is possible if the proposed life-cycle methodology is followed. It is suggested that there should be a continuous process of implementation of the life-cycle methodology in order to detect and mitigate security risks to smart services. Moreover, reported incidents should also trigger security reevaluation, starting from the appropriate phase of the life cycle (Figure 14.1).

14.7 Conclusion

Adoption of smart city model is inevitable to cope up with the unprecedented urban population growth. Wide deployment of cyber–physical infrastructures, for realizing smart city services, poses as a great threat for city administrators. Managing the cyber security is vital for smooth functioning of smart city operations. In this chapter, we have proposed a life-cycle approach to manage the various phases of cyber security of smart cities. This provides a holistic approach toward managing threats, vulnerabilities, and risks of a city based on perception of security of city administration and the citizens.

Questions

1 Why is cyber security important for smart cities?

2 How can cyber security requirements for smart cities be identified?

3 How does smart transportation work?

4 What are the logical domains for smart grids?

5 List some potential attacks to smart city services.

6 How can adoption of a life-cycle approach help in managing cyber security of smart cities?

7 List the different phases of cyber security life cycle of smart cities.

8 What are the different types of assets of smart cities?

9 What are the various factors that are considered for performing risk assessment?

10 List the various fields of an event/incident reporting form.

References

1 World Urbanization Prospects (2014) *The 2014 Revision, Highlights*, esa.un .org/unpd/wup/highlights/wup2014-highlights.pdf (accessed 20 December 2016).

2 Smart City Mission Statement and Guidelines (2015) *Ministry of Urban Development, Government of India Publication, March 2015.*

3 The Gazette of India (2008) *The Information Technology (Amendment) Act*, Government of India Publication.

4 International Organization for Standardization (2012) *ISO/IEC 27032:2012: Information Technology – Security Techniques – Guidelines for Cybersecurity*, 1st edn, ISO/IEC, Geneva.

5 Bartoli, A., Hernández-Serrano, J., Soriano, M., Dohler, M., Kountouris, A., and Barthel, D. (2011) *Security and Privacy in Your Smart City.*

6 Solanas, A., Patsakis, C., Conti, M., Vlachos, I.S., Ramos, V., Falcone, F., Postolache, O., Perez-Martinez, P.A., Pietro, R.D., Perrea, D.N., and Martinez-Balleste, A. (2014) Smart health: a context-aware health paradigm within smart cities. *IEEE Communications Magazine*, **52** (8), 74–81, doi: 10.1109/MCOM.2014.6871673.

7 Balakrishna, C. (2012) *Enabling Technologies for Smart City Services and Applications.* 2012 6th International Conference on Next Generation Mobile Applications, Services and Technologies, pp. 223–227, doi: 10.1109/NGMAST.2012.51.

8 An, S.H., Lee, B.H., and Shin, D.R. (2011) *A Survey of Intelligent Transportation Systems.* Computational Intelligence, Communication Systems and Networks (CICSyN), 2011 3rd International Conference on, pp. 332–337, doi: 10.1109/CICSyN.2011.76.

9 Kostakos, V., Ojala, T., and Juntunen, T. (2013) Traffic in the smart city: exploring city-wide sensing for traffic control center augmentation. *IEEE Internet Computing*, **17** (6), 22–29, doi: 10.1109/MIC.2013.83.

10 NIST (2010) *Office of the National Coordinator for Smart Grid Interoperability, NIST Framework and Roadmap for Smart Grid Interoperability Standards*, NIST Special Publication 1108, Release 1.0, pp. 1–145.

11 Patsakis, C., Papageorgiou, A., Falcone, F., and Solanas, A. (2015) *s-Health as a Driver Towards Better Emergency Response Systems in Urban Environments.* Medical Measurements and Applications (MeMeA), 2015 IEEE International Symposium on, pp. 214–218, doi: 10.1109/MeMeA.2015.7145201.

12 Liu, S., Li, W., and Liu, K. (2014) *Pragmatic Oriented Data Interoperability for Smart Healthcare Information Systems*. Cluster, Cloud and Grid Computing (CCGrid), 2014 14th IEEE/ACM International Symposium on, pp. 811–818, doi: 10.1109/CCGrid.2014.38.

13 Nowicka, K. (2014) Smart city logistics on cloud computing model. *Procedia – Social and Behavioral Sciences*, **151**, 266–281, doi: 10.1016/j.sbspro.2014.10.025, Green Cities – Green Logistics for Greener Cities, Szczecin, 19-21 May 2014.

14 Khan, Z. and Kiani, S.L. (2012) *A Cloud-Based Architecture for Citizen Services in Smart Cities*. Utility and Cloud Computing (UCC), 2012 IEEE 5th International Conference on, pp. 315–320, doi: 10.1109/UCC.2012.43.

15 Zanella, A., Bui, N., Castellani, A., Vangelista, L., and Zorzi, M. (2014) Internet of things for smart cities. *IEEE Internet of Things Journal*, **1** (1), 22–32, doi: 10.1109/JIOT.2014.2306328.

16 Gaur, A., Scotney, B., Parr, G., and McClean, S. (2015) Smart city architecture and its applications based on iot. *Procedia Computer Science*, **52**, 1089–1094, doi: 10.1016/j.procs.2015.05.122, The 6th International Conference on Ambient Systems, Networks and Technologies (ANT-2015), the 5th International Conference on Sustainable Energy Information Technology (SEIT-2015).

17 Ferraz, F.S. and Ferraz, C.A.G. (2014) *Smart City Security Issues: Depicting Information Security Issues in the Role of An Urban Environment*. Utility and Cloud Computing (UCC), 2014 IEEE/ACM 7th International Conference on, pp. 842–847, doi: 10.1109/UCC.2014.137.

18 Patsakis, C., Dellios, K., and Bouroche, M. (2014) Towards a distributed secure in-vehicle communication architecture for modern vehicles. *Computers & Security*, **40**, 60–74, doi: 10.1016/j.cose.2013.11.003.

19 Rouf, I., Miller, R., Mustafa, H., Taylor, T., Oh, S., Xu, W., Gruteser, M., Trappe, W., and Seskar, I. (2010) *Security and Privacy Vulnerabilities of In-Car Wireless Networks: A Tire Pressure Monitoring System Case Study*. Proceedings of the 19th USENIX Conference on Security, USENIX Association, Berkeley, CA, USA, USENIX Security'10, pp. 21–21, http://dl.acm.org/citation.cfm?id=1929820.1929848 (accessed 20 December 2016).

20 Sagstetter, F., Lukasiewycz, M., Steinhorst, S., Wolf, M., Bouard, A., Harris, W.R., Jha, S., Peyrin, T., Poschmann, A., and Chakraborty, S. (2013) *Security Challenges in Automotive Hardware/Software Architecture Design*. Design, Automation Test in Europe Conference Exhibition (DATE), 2013, pp. 458–463, doi: 10.7873/DATE.2013.102.

21 Humayed, A. and Luo, B. (2015) *Cyber-Physical Security for Smart Cars: Taxonomy of Vulnerabilities, Threats, and Attacks*. Proceedings of the ACM/IEEE 6th International Conference on Cyber-Physical Systems, ACM, New York, NY, USA, ICCPS '15, pp. 252–253, doi: 10.1145/2735960.2735992.

22 McCarthy, C., Harnett, K., and Carter, A. (2014) *Characterization of Potential Security Threats in Modern Automobiles: A Composite Modeling Approach.* Report No. DOT HS 812 074. National Highway Traffic Safety Administration, Washington, DC.

23 Yan, Y., Qian, Y., Sharif, H., and Tipper, D. (2012) A survey on cyber security for smart grid communications. *IEEE Communications Surveys Tutorials*, **14** (4), 998–1010, doi: 10.1109/SURV.2012.010912.00035.

24 Wang, W. and Lu, Z. (2013) Survey cyber security in the smart grid: survey and challenges. *Computer Networks*, **57** (5), 1344–1371, doi: 10.1016/j.comnet.2012.12.017.

25 NIST (2010) *The Smart Grid Interoperability Panel – Cyber Security Working Group, Guidelines for Smart Grid Cyber Security.*

26 Rouf, I., Mustafa, H., Xu, M., Xu, W., Miller, R., and Gruteser, M. (2012) *Neighborhood Watch: Security and Privacy Analysis of Automatic Meter Reading Systems.* Proceedings of the 2012 ACM Conference on Computer and Communications Security, ACM, New York, NY, USA, CCS '12, pp. 462–473, doi: 10.1145/2382196.2382246.

27 Van Leuven, L.J. (2011) Water/wastewater infrastructure security: threats and vulnerabilities, in *Handbook of Water and Wastewater Systems Protection*, Springer New York, pp. 27–46, doi: 10.1007/978-1-4614-0189-6_2.

28 Slay, J. and Miller, M. (2008) Lessons learned from the Maroochy water breach, in *Critical Infrastructure Protection*, Springer, Boston, MA, pp. 73–82, doi: 10.1007/978-0-387-75462-8_6.

29 Martinez-Balleste, A., Perez-Martinez, P.A., and Solanas, A. (2013) The pursuit of citizens' privacy: a privacy-aware smart city is possible. *IEEE Communications Magazine*, **51** (6), 136–141, doi: 10.1109/MCOM.2013.6525606.

30 Perez-Martinez, P.A. and Martinez-Balleste, A.S.A. (2013) *Privacy in Smart Cities – A Case Study of Smart Public Parking.* Proceedings of the 3rd International Conference Pervasive Embedded Computing and Communication Systems, pp. 55–59.

15

Mobility as a Service

Christopher Expósito-Izquierdo, Airam Expósito-Márquez, and Julio Brito-Santana

University of La Laguna, San Cristóbal de La Laguna, Spain

CHAPTER MENU

Introduction, 409
Mobility as a Service, 413
Case Studies on Mobility as a Service, 427
Conclusions and Further Research, 432

Objectives

- Introducing the concept of Mobility as a Service (Maas).
- Discussing how the changes in attitudes and mind of people have encouraged the development of new mobility solutions.
- Becoming familiar with the main information and communications technologies and physical infrastructures associated with the MaaS.
- Presenting the most successful business models arisen in the mobility sector.
- Presenting some of the most innovative projects arisen in the field of transportation over the last years.

15.1 Introduction

Transportation sector has undoubtedly suffered a great revolution over the last decades, mainly thanks to the technological development and innovation. In ancient times, people used to travel long distances by using primitive forms of transportation, such as horse-drawn carriages or galleys, which were uncomfortable, unsafe, and highly time consuming. The widespread deployment of

Smart Cities: Foundations, Principles and Applications, First Edition.
Edited by Houbing Song, Ravi Srinivasan, Tamim Sookoor, and Sabina Jeschke.
© 2017 John Wiley & Sons, Inc. Published 2017 by John Wiley & Sons, Inc.

railways, especially in the mid-19th century, the subsequent appearance of cars, and the consolidation of air and maritime transportation means in the daily life have progressively enabled people to travel longer distances in a cheaper and faster way than at times in the past. However, transportation is an alive sector, and therefore it is essential that it caters for the new mobility requirements and interests of people in the near future [1].

Multiple mobility patterns and expectations with regard to transportation have been already identified in the current society to a greater or lesser degree. A common understanding of these patterns by the diverse stakeholders, such as policy analysts, urban designers, or transportation planners, establishes a solid basis for initiating and evaluating improvement strategies in transportation management systems. By way of an example, leisure travelers in Western society usually seek to use transportation means from an economic outlook, in which low cost is considered as the most relevant decision. This is in contrast with the mobility requirements of workers, students, or business travelers. They usually prefer fast transportation means that enable them to reach their destinations by spending the shortest traveling times. In light of this scenario, the current transportation sector has to cope with many demands and competing factors, such as comfort, cost, availability, quickness, convenience, and safety.

The transportation sector has evolved throughout history in a constant quest to satisfy mobility requirements. Up to now, the successive changes experienced by this sector have turned it into a cornerstone for the development of any economic, social, or political activity of society [2]. The global economy is substantiated by the opportunities derived from the movement of goods within international supply chains, people, and information. In this context, transportation is to a great extent responsible for employment generation and improving economic competitiveness by enabling connectivity between population, agricultural and natural resources, and employment centers. Furthermore, transportation has somewhat encouraged the appearance of new travel behaviors. For instance, the frequency of trips has increased gradually motivated by the wide range of available transportation alternatives at about the same rate as their average traveling time has simultaneously decreased, thanks to the support provided by improvement in terms of efficiency and reliability. The decentralization of urban centers is another indirect consequence arising from possibilities associated with transportation. Indeed, new transportation networks are some of the most sensitive elements behind the spatial structure of metropolitan areas. These are today more compact due to the high population density and control on land use but, at the same time, more accessible through multimodal corridors with high capacity.

Some of the major challenges faced by the transportation sector are brought together in urban areas. Currently, urban population represents more than a half of the global population. It should be nevertheless pointed out that the most striking aspect in this regard is that global population has been doubled in only a few decades. In addition, according to the data provided by the 2015 Revision of World Population Prospects of the United Nations,[1] the continuing growth of the world's population is forecasted to increase by more than 2.5 billion people by 2050. This fact evidences an unrelenting urbanization process with deep impact on the economic and social areas. Furthermore, the urbanization process has also intensified the pressure on urban infrastructures to cope with a broad spectrum of mobility needs. The main reason is that all the urban mobility demands are concentrated on a reduced area in which multiple heterogeneous transportation means must coexist to generate real value. This causes the mobility problems to increase at a pace even faster than the urbanization growth rate. This way, all the future changes in mobility terms must be endorsed by a progressive and adequate development of the existing physical infrastructure network and new efficient systems to enhance service provision.

Moreover, multitude of voices around the world are exerting an increasing pressure on governments, private companies, or international organizations for developing and deploying more sustainable transportation means. The growing concern of the general public, especially in Europe and the United States, about how to reduce the total annual contaminant emissions, noise, or urban air pollution has taken an active role in the political initiatives of international and national governments and their corresponding regional and municipal authorities [3]. The new sustainability-oriented policy actions seek to regulate the new development and innovation in the field of transportation with the aim of counting on a more competitive productive system while safeguarding a high level of environmental protection.

The profound changes in the awareness about sustainability and climate change indicate that mobility solutions in the future must be of course efficient but also in harmony with nature and environment. In this regard, an emerging concept has arisen in recent years: *e-mobility*. It is an element of sustainable mobility that seeks to use electric vehicles in response to the present transportation necessities of citizens. The vehicles are supported by a network grid infrastructure containing energy charging stations, which supplies energy recharging capabilities. Electricity from renewable sources is without doubt the best candidate to deal with the targets in environmental protection matters. The main reason behind its use is the substantial reduction in terms of CO_2 emissions both at local and global levels. Nonetheless, other benefits behind using electricity in mobility are that engines of electric vehicles are at the moment several times more efficient, require lesser maintenance costs, and enable a quieter transportation than those of fossil-fuel-powered vehicles. Interested readers can find a detailed analysis of e-mobility in Europe in the book [4].

---

1 http://esa.un.org/unpd/wpp/Publications/Files/Key_Findings_WPP_2015.pdf

In spite of the emergence and development of modern electric vehicles, countries today have to find new solutions to face the problems derived from the enormous number of private cars on their roads [5]. This is undoubtedly one of the main problems in terms of mobility in urban areas. Indeed, the situation is especially challenging in large metropolitan areas, where the reduced availability of space and high concentration of population give rise to parking space shortage and traffic congestion [6]. Far from changing and from a global level of aggregation, data indicate that the number of private cars is going to be increasing at least over the next years. This is the result of several heterogeneous factors, such as increase in gross domestic product per capita, reduction in production costs, or public transportation inadequacy, among others. However, despite the massive popularity of private cars in developed countries, they are definitely underused assets. This is due to the fact that, in most of cases, they are only used as a limited part of the day and with low levels of occupancy.

New international trends are progressively appearing in practically all economic activities. People are nowadays demanding the availability of assets for its use without the need to buy them. These growing trends have motivated that most private organizations redefine their traditional business models in finding more customer-centric viewpoints. As a result, new service-oriented business models are emerging. The focus on these models is shifting toward creating value through product utilization instead of selling them, as has been the case until now. As broadly discussed in the following section, transportation sector cannot live with its back turned to the changing scenario in mobility matters. The goal in this regard is to provide a more flexible transportation ecosystem that better suits the future mobility challenges by acting as a single point of contact for travelers.

MaaS has recently emerged as a revolutionary change in the conceptualization of future transportation. Broadly speaking, this term comprises a sophisticated conglomerate of heterogeneous transportation means, physical infrastructures, and information and communications technologies (ICTs) working in combination to enable citizens to reach their destinations efficiently. The main objective of these systems is to offer a similar quality of service as privately owned vehicles through a wide range of on-demand mobility alternatives to be selected while preventing any type of commitment to a pre-specified transportation mean. This way, a citizen requiring some point-to-point ride accesses the MaaS system by means of an enabler, such as a smartphone application or the Web, which reports all the existing transportation schedules to reach the desired destination according to different user-defined criteria, such as quickness, cost, or comfort. Consequently, users can enjoy the convenience of private vehicles while being free from parking troubles or maintenance costs, among others.

As aforementioned, the concept of MaaS comprises multitude of technical elements. Each one should be discussed in its own right due to the fact that each has enough individual relevance. This chapter only intends to summarize the functionalities, technologies, and representative projects arisen in the field over the last years. With this goal in mind, the remainder of this chapter describes the changes in attitudes and mind in regard to mobility behind the new generation of *Millennials*, the role of physical transportation infrastructures and ICTs to provide interoperable mobility solutions, the appearance of autonomous and connected vehicles, and the mobility from the perspective of sharing economy. Particularly, Section 15.2 discusses the functional and technical aspects of MaaS systems. Afterwards, Section 15.3 reviews some highlighted projects associated with MaaS. Finally, this chapter ends with some concluding remarks about MaaS.

15.2 Mobility as a Service

Broadly speaking, smart cities are those urban areas that make use of the available information and communications technologies, in short ICT, with the aim of creating public value that in turn improves the quality of life of their citizens [7]. The development of any smart city should be in harmony with the main sustainability pillars: social, economic, and environmental ones. Mobility plays a highlighted role in this context due to the fact that it enables to connect employment centers, natural resources, and residential locations. However, the mobility requirements have been gradually changing over the last years. Citizens are nowadays demanding sustainable mobility solutions that enable them to reach their destinations efficiently instead of remaining committed to any particular transportation mean.

15.2.1 Millennials

The term *Millennials* refers to those generations born in the early 1980s; although there is no unequivocal consensus, most researchers and commentators use these birth years. The precise delineation of the concept varies from one source to another. For instance, Howe and Strauss [8, 9] define the Millennial cohort as that consisting of individuals born between 1982 and 2004. Up to now, other names have been also proposed to denote the Millennials. Some of these are *Generation Y*, *Generation We*, *Global Generation*, *Generation Next*, *Net Generation*, and *Generation 9/11*, among others. Neil Howe and William Strauss predicted that the Millennials will become more like the civic minded *Greatest Generation*, a generation composed of individuals born in the United States between 1901 and 1924 but with a stronger sense of community at both local and global levels [9].

The differentiating characteristic behind the Millennials is that they are tending to do things with the aim of helping their own city or town and also the people who live there. The Millennials have grown up in the presence of computers, the Internet, and multitude of electronic devices such as mobile phones and tablets. Electronic devices make the world increasingly online and socially networked. Additionally, these electronic devices are increasingly aimed at sharing information on the Internet. The Millennials are also able to easily handle new programs, operating systems, and devices. As a general rule, the Millennials may perform computing tasks more quickly and easily than older generations. They have no qualms about sharing their lives online, that is, having a public life on the Internet. The Millennials understand their public life on the Internet, and specifically on social media, as a self-promotion and a promising way to foster connections through online media.

In an increasingly connected and technologically changing world, the transportation sector undoubtedly evolves influenced by what consumers want and expect from it. The Millennials are a generation with a *less driving lifestyle*; instead they prefer to exchange driving for cycling, walking, and using public transportation. Furthermore, they would prefer the latest technological products rather than owning their first car. In fact, they prefer to consume experiences and services rather than having their own possessions. Mimi Sheller, a sociology professor at Drexel University and director of the Mobilities Research and Policy Center said, "the Millennials don't value cars and car ownership, they value technology, they care about what kinds of devices you own."[2] Also, she said, "the percentage of young drivers is inversely related to the availability of the Internet. Why spend an hour driving to work when you could take the bus or train and be online?" This assessment reflects the character and close relationship of the Millennials with technology, the need to be online at anywhere and at anytime, and also the conceptual association with the emerging sharing economy [10]. This fact is also clearly reflected in the analysis of Goldman Sachs, which concludes as follows: "Millennials have been reluctant to buy items such as cars."[3]

Moreover, the aging of society in developed countries presents a risk factor in driving. In the long term, this risk factor implies a greater abandonment of the own vehicle. People are nowadays more likely to consume mobility than buying a private car, and thus individuals are trending toward new service-based mobility models. For instance, the Finnish city of Helsinki has started in this regard to test an innovative idea to make car ownership unnecessary in practice. It is a platform that enables citizens to design route planning composed of different transportation services, from city bikes up to sharing services. Another example is found in the Swedish city of Gothenburg, which has successfully piloted a platform called UbiGo (see Section 15.3.1). Broadly speaking, it is a service that combines public transportation, car rental services, carpooling,

---

2 The New York Times – The End of Car Culture: http://www.nytimes.com/2013/06/30/sunday-review/the-end-of-car-culture.html.
3 Millennials Coming of Age: http://www.goldmansachs.com/our-thinking/pages/millennials.

taxi, and a cycling system in an all-in-one platform. The basic idea behind the previous services is to present the existing transportation options via an interface whose input data are the departure and destination spots as well as user transportation preferences. The output is composed of the possible transportation routes and the payment options. The ease of use and the possibilities given by these services together with the capabilities of the new generations encourage the progressive development of new transportation service models based on relegating the privately-owned vehicles to a secondary position.

15.2.2 Concept of Mobility as a Service

The embattled city growth today poses great challenges that are difficult to solve. Particularly, current cities need to be increasingly *smart* and have efficient strategies to meet strict environmental legal obligations. The growth of cities based on economic, social, and environmental sustainability is one of the major challenges. This is possible by improving the efficiency of cities and proper use of smart solutions. A smart city needs the participation, ideas, and experiences of those stakeholders in the process, public administration, citizens, and private sector. In this regard, the balance of interests among stakeholders is vital to achieve the new smart objectives [7]. The growth of modern cities is beset by the increasing mobility demands of society, which in turn affects the deterioration of the environment. The society requires a high degree of flexibility in terms of mobility and hence the development of new transportation means. This fact encourages the development of sophisticated transportation systems tailored to fulfill the needs of society, in order to ensure an intelligent, safe, and economically efficient mobility. Despite the changing trends, current patterns of mobility have become highly dependent on private vehicles, with consequent impacts on sustainable land use. Transportation is one of the key factors in sustainable development due to its environmental, social, and economic impact. An efficient and flexible transportation system that allows to develop intelligent mobility is necessary for the economic growth and improvement of the quality of life.

Nowadays, the transportation sector is at the beginning of a period of plenty of changes in the way of providing true value, bolstered by innovative products, technologies, and services. Indeed, these elements change user expectations and opportunities every day. Every day the use of new technologies by users of transportation services is more widespread. This allows the development of new services, favoring the transportation user. This evolution of the transportation sector is clearly positive. Nevertheless, the transportation sector has in general provided relatively inflexible services so far, which customers ultimately take or leave. For example, train and bus routes are usually predefined, stops are found at given locations, schedules and rates are fixed, and so on. These predefined services give rise to transportation is not readily adaptable to unforeseen changes of demand and customer expectations. However, this new evolution of the transportation sector is changing this issue. Increasingly, the industry is more focused on the customer, adapting to changing customer needs, and less committed to a specific way of mobility. Customers demand flexibility in mobility services in line with social trends that revolve around part-time work, work at home, social responsibilities, and so on. The mobility services today must be personalized and available on demand in order to meet the customer expectations. The transportation sector should be consequently based on user requirements and their experience.

Over the last few years, the most exciting emerging development to change the model to provide transportation services is the MaaS. Broadly speaking, MaaS is a new paradigm that focuses on providing a single platform for combining all the existing transportation options and provides them to the customers as an integrated and simple solution. The concept was originally introduced in Finland but has gained popularity quickly. Heikkilä [11] defined MaaS as a system in which a comprehensive range of mobility services is provided to customers by mobility operators. Mobile operators are here defined as companies that buy mobility services, such as public transportation, car sharing, and taxi, from service providers, and combine them to give customer service. It is important to emphasize that MaaS is not only a mobile application but is also a change on the way to provide and consume mobility. It breaks the traditional paradigm of owning cars or hiring travel services from point to point. In order to replace owned cars, MaaS must have all possible transportation means. The service provision needs to be contemplated with a number of additional multimodal services. MaaS covers all types of transportation services, does not distinguish between private and public transportation, and unifies and standardizes the type of transportation on the service.

There are two main factors responsible of MaaS: servicizing and sharing economy. *Servicizing* is based on the idea that the user does not want the ownership of a product. Instead, the user prefers to make use of the functionalities or services provided by the product [12]. For this purpose, the provider can get paid for the unit of service provided rather than per unit of product sold. In other words, this means that end users are buying more services, instead of buying products and produce actions themselves. Furthermore, MaaS is related to sharing economy [13], which refers to sharing products and services without possessing the products or resources for oneself. Thanks to this, people can focus on the activity itself rather than worry about the responsibility derived from the ownership of the product or resource. Sharing and servicizing could reduce the need for owning and make efficient use of resources, for example, car sharing, ride sharing, and other sharing transportation alternatives, which currently remain supplied individually. These services are still intended to a specific audience, and generally existing platforms fail when providing favorable conditions for comprehensive provision of mobility services.

The combination of factors such as technological development, servicizing, and sharing economy allows MaaS to provide a transportation model through a common interface for all the existing transportation means. This is done by grouping the services needed by the user in mobility packages. The vision of MaaS is to group services belonging to the transportation sector in a unique, cooperative, and interconnected system that provides mobility services to the end user through the creation of an ecosystem of operators and providers. In order to facilitate understanding of the concept of MaaS, a specific example is defined in the following. It is worth mentioning that the example is fictitious. In fact, with the aim of defining, an example would be necessary to know the preferences and needs of individual users. Figure 15.1 depicts four MaaS packages that are defined to different user segments.

The first package is called business package. This package is designed for people who travel frequently and require high flexibility and availability of transportation. The next package is called family package, designed for families who require unlimited public transportation and use of a car, including taxi service. Urban commuter package is designed for users who focus on the home city area but also need taxis and rental cars occasionally to go beyond the areas served by public transportation. In addition, domestic public transportation is also included for journeys between cities. It is aimed at users who live in central urban areas and do not have a large family unit. Finally the 15-min package is for people who need quick service-providing taxis. Also it includes free transportation in home city areas, for short journeys that do not require a short time.

Business package 800€/month	Family package 1200€/month
– 5 min pickup across the EU – Free taxi in home city – Lease car and road use – Taxi roaming world wide	– Lease car and road use – Shared taxi for family members with 15 min pickup – Home city public transportation for all – Domestic public transportation 2500 km

Urban commuter package 95€/month	15 minutes package 135€/month
– Free public transportation in home city areas – Up to 100 km free taxi – Up to 500 km rental car – Domestic public transportation up to 1500 km	– 15 min from call to pick up by shared taxi – EU wide roaming for shared taxi at 0.5/km – Free public transportation in home city – Domestic public transportation up to

Figure 15.1 Example of MaaS service package.

The benefits of MaaS are multiple and can be viewed from different perspectives. Some of the benefits to all stakeholders are the cost savings as a result of increased service efficiency, the convergence of different transportation means, the end of public versus private transportation, and the new business models in transportation sector. From the standpoint of users, the most obvious benefits in MaaS are custom development of intelligent mobility services that reflect their needs, developing new mobility services, good performance, and accessibility of transportation services custom; improving the user experience; using a platform that combines all transportation means and presented an integrated interface; and providing transportation as a flexible, personalized on-demand service. The public sector, by using information systems, can increase the efficiency of infrastructure, and mobility services provided and distribution of resources based on end-user needs, improving the management of traffic incidents and providing better services with less investment. As a result of improved mobility services, a better environment for new businesses with the public sector as facilitator is generated. The companies may find new markets and business, and new opportunities regarding transportation and infrastructure business. The growth potential of the products and services of information technology and the growth of the technology and transportation market are among the contributions of MaaS, as well as smart transportation connections to all business sectors.

15.2.3 Transportation Infrastructures

Transportation infrastructures containing roads, airports, railways, and seaports are the backbone of any mobility solution. Their main goal is to provide the integrative basis required to ensure the proper functioning of the transportation sector. These infrastructures require huge economic investments both to be built and maintained, but they, in turn, report economic and social benefits, thanks to their potential for new opportunities for interaction in any activity of current society. In fact, they are considered as valuable instruments for strengthening competitive sectors due to their flexible capabilities to connect physically producers, consumers, population, employment centers, natural resources, and so on and consequently creating value for the citizens. In transportation planning, the potential economic impact of a particular improvement in infrastructures is usually assessed by means of market potential indicators. Some outstanding examples in this direction can be consulted in [14, 15].

The current challenges linked to the quick evolution of the global urbanization process and the ever-increasing population mobility needs force cities to assess and invest in the deployment of modern infrastructures that ensure enough provision of a suitable transportation offer to their citizens while, simultaneously, enabling the interchange of freights in multimodal supply chains. Unfortunately, the expansion of the infrastructures is not generally in the same pace as the increase in mobility demands. This scenario is especially dramatic in road transportation, where the projections about the future increase of the number of private cars worldwide evidence a troublesome trend in which this is expected to be doubled in only a few decades. Some of the common barriers found when deploying new transportation infrastructures are the social and environmental concerns of society, the physical limitations associated with overpopulated urban centers in reduced areas, or the shortage of financial sources, among others. In order to overcome the latter fact, many public and private partnership approaches have been widely applied in the transportation sector to afford the required economic investments [16].

Expanding the existing transportation infrastructures has been traditionally seen as the single viable option by urban designers to face with the new mobility requirements. The space limitations and monetary constraints found in practice have given rise to that transportation managers today devise innovative alternatives that take advantage of the potential of new ICTs. A suitable orchestration of all the existing infrastructures and services is required in this regard with the goal of obtaining an optimized and integrated transportation network. Up to now, many cities have already implemented mobility projects based on the vertical integration of their transportation infrastructures and services. However, a horizontal integration and coordination of them must be carried out through technology in order to achieve a significant improvement in future mobility solutions and bring together the benefits of innovation.

Multitude of innovative projects in terms of planning, construction, or maintenance aimed at improving the transportation infrastructures have been implemented incessantly around the world over the last decades. In most of the cases, they have been considered as promising opportunities to reconstruct obsolete infrastructures that lack capacity for current and future mobility volumes. In the following, some highlighted projects are briefly described.

The 95 Express[4] in Miami, United States, comprises a set of toll lanes for road vehicles based upon an interesting *dynamic tolling*. Specifically, their rates vary according to the traffic conditions perceived when entering the lanes. These infrastructures are supported by a complete sensor system to report real-time information about the number of vehicles in the lanes, their speeds, and how close together they are. This information is appropriately exploited to determine the rates of the toll and provide the best conditions possible.

Moreover, the I-4/Selmon Expressway Connector[5] was a complex infrastructure project completed in October 2014 whose main goal was to construct a time-saving link between two of the major transportation corridors in the city of Tampa, United States, and also enable the access of trucks to the Port of

---

4 http://www.95express.com/.
5 http://tbinterstates.com/projects/projectinfo.asp?projectID=175&RoadID=3.

Tampa. One of the distinguishing technical features of the project is found in the elevated nature of the link due to the fact that it had to cross several urban streets. This highway has a sophisticated toll facility based on an all-electronic toll collection system comprising a prepaid toll program and toll-by-plate system.

Another project is the European Railway Traffic Management System.[6] It is an industrial project promoted by the European Union that seeks to provide technology dedicated to control the railway infrastructures of the member states to increase train flows along the routes. Its main objective is to provide a continuous communication-based signaling system with the aim of attaining cross-border rail interoperability through Europe. For this purpose, the project comprises the European Train Control System, which is an automatic train protection system dedicated to control the security of the traffic, and the Group Special Mobile for Railways, whose goal is to enable voice and data communication between railways and trains. The project has the enough potential to maximize the future use of the existing railways by allowing convoys of automated trains to move immediately behind each other.

The last project to introduce here is the Miami Intermodal Center[7] built by the Florida Department of Transportation in the city of Miami, United States. Generally speaking, it was initially conceived as a massive transportation hub that provides seamless intermodal connectivity between all the existing ground transportation means with the primary goal of mitigating the traffic problems around the Miami International Airport (MIA). The project comprised several major elements: the Rental Car Center (RCS, which is dedicated to accommodate all the rental car companies that operate inside the airport); the Miami Central Station (MCS, which provides transportation connections between the transportation options); the MIA Mover – a light-rail people mover system that operates and enables users to move between the MCS and the RCS efficiently; and the redesign of several major corridors to access the MIA.

15.2.4 Information and Communications Technologies

Traditionally, transportation users make use of the information services offered by a punctual transportation service provider for information on available services. These services are often based on the use of ICTs, which provide more efficient and immediate service. For MaaS, managing between transportation service providers and the combination of different transportation services must be transparent. The user must have the perception of attending a unique selling point of transportation services. That is why the digitization of information and ICTs are key factors in the development of MaaS. The aviation sector is a good example of digitization. The user can here choose and combine different

---

6 http://www.ertms.net.
7 http://www.micdot.com.

available travel options. Advances in telecommunications and mobile devices have enabled the development of new ecosystem of applications. The combination of mobile devices and the Internet has changed the way in which information is shared, facilitating improvement and advancement of services and processes. Mobile devices enable users to access and share information instantly and from anywhere. ICTs facilitate the integration between services and users, which share information and use of services. Also, new technologies enable users to take a more dynamic and proactive role as a developer and data producer in any transportation system. The user will no longer be the only consumer in the transportation system. Instead, the whole transportation system will be generated with, to, and by the users. The role of data and information will be crucial. Transportation data, infrastructure data, and physical transportation infrastructures will together compose the essential platform for mobility services.

As in other productive sectors of current society, the availability of relevant information related to the state of the existing transportation infrastructures is a key element in the development of the future mobility services. Without doubts, information is going to be one of the most critical assets in the transportation sector in the decades ahead. The main reason is found in the fact that it enables transportation managers to know how the mobility patterns around the city are and additionally forecast future demands. This can be only achieved by the operational intelligence integration of the traditional isolated information systems and the massive use of advanced business analytics to create a real value.

Digitization and accessibility of data are an essential part of the success of strategies to MaaS. It is the common thread that enables collaboration and integration of the different actors and transportation systems, allowing smarter decisions. In this regard, the strategies of open data by the government are necessary; they are crucial facilitators toward a more open government by improving efficiency in daily operations in transportation, as well as creating opportunities to deliver innovation, services, and business models. The evolution of technology, standards, and best practices ensures efficient handling of large sets of growing data and protection of the confidentiality and privacy of the same. However, there are multitude of technical challenges such as how and where the data should be collected and how to measure the validity, quality, and aging of data. In addition to these challenges, which data are relevant and to whom must be decided.

Recent advances in technologies and techniques in data processing are evolving to handle large volumes of data. And transforming these data into information and knowledge to be exploited by the stakeholders with the aim of making smarter decisions is a crucial issue. Other fundamental aspects in the treatment and storage of information are data privacy, data security, and protection of intellectual property. A compendium of standards, technologies, and work practices that can protect private data must exist. These challenges represent largely successful implementation of technologies for data analysis because errors in these matters would break the trust of users. Efforts on not undermining user confidence are largely required; however, it can be seen how large data are used to improve the quality and speed of decision making and improve the efficiency of public and private entities while enabling better products and services to customers and citizens. Some of the relevant approaches in this regard are the use of simulation for the prediction of mobility demands, optimization in problem solving, and data analysis and modeling. The data allow stakeholders to meet the needs of such individuals and group trends, enabling to personalize services, for example, using real-time geotagged data that allow to route traffic in an optimized way. Surprisingly, in spite of the fact that business analytics is not a new topic for the transportation sector, this has vaguely exploited its capabilities. For instance, executives have generally focused on the number of people using the transportation alternatives so far but without paying attention to the particular requirements, expectations, or desires of the individuals. This evidences the necessity of recognizing the individual citizens in the transportation sector.

15.2.5 Interoperability

In order to provide and integrate mobility solutions, it is necessary to take into account the existing systems and infrastructures, which bring additional technical challenges due to interoperability issues. The interoperability is the key to develop and manage systems and provide competitive solutions in terms of MaaS. The existence of standards is essential for the different suppliers and technologies that can interact with each other. Today, society is experiencing a revolution of the Internet of Things, driven by the emergence of smart devices, wireless sensors, and IP-enabled devices, which encourage the creation and management of *ad hoc* networks of devices and sharing data across devices and systems.

It is common for different suppliers generating devices and systems under their own specifications and communication protocols. The integration of complex systems requires interoperability at three levels. The first is at the technical and syntax level. Generally, it is referred to physical and logical connectivity, infrastructure, and exchange and data structure of messages. The second is a semantic level, concerning the business context and concepts and types of information contained in the messages exchanged. Finally, the organizational level is based on the strategic and tactical objectives of the organization, operational and business procedures, and regulatory elements including economic and legal levels.

The success behind obtaining smart transportation infrastructures and integrated services with the ability to generate true value for cities and make improvements in the quality of life of citizens depends on government agencies and companies involved in the transportation field that collaborate efficiently. The exchange of information between the different actors should be given fluidly. This will be a key factor in the success of improving transportation infrastructure. Generally, the data associated with the infrastructures are collected by different organizations independently and without relationship between them. However, it would be beneficial to receive a global perspective by integrating all available data about the transportation infrastructures. To develop intelligent solutions, these must be supported by the collection of data in real time through infrastructure. These data are exploited in order to offer new services and generally collected through a network of sensors to monitor infrastructure and communicate data through communications networks.

15.2.6 Autonomous Car

In the future, through the use of automated vehicles, also called "autonomous cars," "self-driving cars," or "driverless cars," [17, 18] MaaS will substantially enhance productivity by offering the level of convenience of a private vehicle but without the physical ownership. Broadly speaking, it is an automated car that can drive without a driver by leveraging map and sensor data. Society of Automotive Engineers (SAE) International provides a common taxonomy and definition for automated driving in the standard J3016, identifying six levels of driving automation from driver assist to fully automated driverless vehicles. These levels are descriptives and technical, not legals, and also they do not imply particular order of market introduction. The content levels are level 0 "not automation," level 1 "driver assistance," level 2 "partial automation," level 3 "conditional automation," level 4 "high automation," and level 5 "full automation." The fundamental difference occurs between levels 2 and 3. The driver performs some parts of the driving task at level 2, whereas an automatic driving system can perform the entire driving task at level 3.

Autonomous vehicles contribute to cost savings in transportation services, which could lead to lower production costs of services, and hence utility rates can be lowered even further. This could result in greater use of transportation and, therefore, productivity. Another reason for promoting autonomous cars is safety. Autonomous vehicles promote safe driving by reducing the effect of human error in driving, this being one of the main causes of traffic accidents. It may have more cars on the road, thanks to sensors installed in autonomous cars,thus shortening traffic times. Another interesting aspect is that disability would not be a relevant factor in driving; everyone could drive. In addition, the autonomous vehicle will allow users to spend the traveling time on other activities such as working, relaxing, or accessing entertainment.

There are also some disadvantages concerning the autonomous vehicle. The main one is that the user is concerned with vehicle malfunctions. This may result in a car accident without human action to prevent more serious consequences. Additionally, the autonomous vehicle could have a negative impact on sectors based on professional drivers, especially taxi companies, trucks and buses. Nevertheless, for those drivers to whom driving is a pleasure, they will not understand the concept of autonomous vehicle and are likely to maintain their conventional vehicles. Besides that often it happens with early versions of the technology, at first the adoption of autonomous vehicle will be slow; given the high cost of technology, many people cannot afford it.

15.2.7 Connected Vehicle

Automotive electronics and wireless technologies have evolved rapidly. The combination of both technologies, the convergence of devices, and the changing lifestyles are helping to modify the driving experience beyond the vehicle. Users expect to have the same experiences and connectivity when they are using vehicles as in their homes. The connected vehicle technologies arise to fulfill such demands. Overall, this is an initiative that aims to create wireless networks between vehicles, enabling secure and interoperable communications and exchange information with the environment through which they pass. These communications would occur between vehicles, infrastructure, and any communication device with the ultimate goal of making transportation smarter, safer, and greener, with improved mobility and quality of life of citizens.

One of the advantages that connected vehicle technologies bring is to reduce road accidents and their consequences. For example, if a vehicle is able to identify that a pedestrian is crossing the road or a motorcycle invades the lane, its brakes can react before the driver and in turn alert vehicles that are behind. Also, the connected vehicles will be able to recognize the factors that cause traffic congestion and take countermeasures. Savings in waiting times affect the reduction of fuel consumption. If the vehicles are able to communicate with road signs, other cars, traffic lights, traffic management centers, and so on, automatically the vehicle is able to compute the best route, saving time while driving. Connectivity in vehicles can provide services that free us of the burden. Other advantages are the capacity for communication with the infrastructure of the city to identify free parking spaces, advanced driver assistance systems are electronic systems that provide additional support in certain driving situations, recommendation hosting services along the route, closest re-fuelling services, and so on. Travelers have access to accurate information and travel impact generated by the type of driving.

Cars have evolved to become smarter and to have communication skills with the environment. But the flip side of this is that the new capacity will be exposed to new security threats and privacy of information. The thieves could access services and vehicle information can determine, for example, the positioning of the vehicle, and they could even access cloud services and put personal data at risk. Even connected cars can become vulnerable to malicious software hacks. Hackers could access vehicle networks and cause traffic problems, endangering the safety of the occupants of vehicles in danger.

15.2.8 Sharing Mobility

Sharing economy – or collaborative consumption – refers to a mean to promote sustainable consumption practices through encouraging online peer-to-peer economic activities [10]. The key factor associated with the concept of sharing economy is the shift experienced by the traditional consumption culture, leaving behind the trend of owning assets and acquiring a new attitude in which consumers simply share access to assets and services. Of course, consumers still want to exploit the capabilities of assets, but the fundamental difference found here is that they no longer need to own them as long as these can be easily borrowed at low prices.

Without doubts, the mobility sector is one of the fastest-growing segments closely linked to the sharing economy, thanks to the exponential increase in the number of new business models that has spawned over the last years. Some of the most successful business models are those associated with car sharing, bike sharing, ride sharing, and shared parking. In the following, these are briefly introduced.

Car-sharing services have likely become one of the most disruptive innovations in terms of sustainability related to the transportation market [19]. These are new car-based mobility solutions provided by vehicle manufacturer or car rental firm and dedicated to make available alternative transportation means to citizens struggling with those problems derived from car ownership. In this new scenario, on-demand rides are sold to private individuals, who pay only for the time taken or distance traveled. The core element in car-sharing systems is the temporal and spatial sharing of a fleet of cars by many people over time. In spite of the wide spectrum of existing car-sharing systems, cars in these types of mobility systems are usually available at specific-purpose car bays, which are frequently either strategically scattered around the city or concentrated in public transportation stations. The availability of smartphone and web-based applications enables registered users to book cars online and access them by using personal electronic key cards. Several business models derived from car-sharing systems can be easily identified. The most popular is that proposed by some private companies, which have deployed their own fleets of cars in a given area to be individually rented. Additionally, new alternative business models, such as the peer-to-peer one, have quickly gained stature over the years. Some successful examples of these business models are introduced and discussed in Section 15.3.

Similar to car-sharing services, the bike has been an asset widely used in business models associated with sharing economy. In this case, bike-sharing services provide short-term rentals of a pool of bikes that enable users to do point-to-point trips among a network of stations at closely spaced intervals throughout an urban area. Broadly speaking, every time a user wants to use a bike, he simply goes to his/her nearest docking point, swipes his/her personal card, and accesses the bike. Once the user has reached his/her desired destination, he drops the bike off in his/her nearest docking point to be available for the next trip. In many cities, the use of bike-sharing services is being encouraged from an economical perspective by allowing users free access to the bikes during the first 30 min period. In addition, the bikes are in most of cases equipped with wireless communication technology to enable real-time monitoring of occupancy rates and Global Positioning System to follow their movement. The bike-sharing services have clearly evidenced highlighted operating advantages due to the fact that they help to improve the accessibility and support flexible mobility, reduce gas emissions, enhance public health, or palliate the problems associated with private car ownership [20], among others. Lastly, innovate projects have recently appeared to improve the service provided to the users. This is the case of the project termed Copenhagen Wheel[8], promoted by the city of Copenhagen in Denmark, and the Senseable City Lab of the Massachusetts Institute of Technology. This project has developed a wheel that helps to transform a conventional bike into a hybrid electric one by capturing the energy dissipated while cycling and braking to be used when needed during driving. The wheel also contains an embedded control system, wireless connectivity, and multiple sensors. The latter enable to measure the environmental pollution and share these data to identify the less polluted zones to move around the city.

Moreover, ride-sharing services – also referred to as carpooling services – are mobility solutions in which, in general terms, an arrangement between a driver and one or more individuals, termed riders, who do not belong to the same household and with their own matching itineraries and time schedules enables to share a ride in a personal vehicle. The riders find automated matches for their desired rides through smartphone applications. Specifically, once a rider requests some ride, the applications seek to match the rider mobility requirements with the existing drivers in the system that are offering rides. The vehicle operating expenses derived from the ride in ride-sharing services are usually shared among the driver and the riders. Unlike the previously described car-sharing approaches, the vehicles in these services are privately owned. Furthermore, as pointed out in [21], a distinction should be made on the basis of the role of drivers. From a purist viewpoint, ride-sharing services are those in which a given driver picks up and drops off riders along his/her route. Contrary to this approach, drivers are today acting as traditional taxi drivers in many popular ride-sharing services, such as Uber. See Section 15.3.3 to obtain a description of this service. Finally, it is also worth mentioning

---

8 https://www.superpedestrian.com.

that ride-sharing systems are nowadays considered by urban designers and politicians as a proliferating alternative to get single-occupant cars off the roads and maximize the use of high-occupancy vehicle lanes. Also, they help when alleviating traffic congestion and parking shortage and mitigating the problems derived from gas emissions.

Finally, the shared parking initiatives are aimed at exploiting the existence of empty private parking spaces in a collaborative way by multiple citizens. It is well known that parking spaces are conventionally underused assets, especially in city centers. This means that they are only used for a reduced portion of the time, and they, in addition, follow some predictable utilization patterns. The main goal behind shared parking initiatives in this regard to make an economical profit from an empty space in private facilities. This way, every time the owner of a private parking space is not using its parking space, this can be put as accessible to be used by another citizen in return for remuneration. The most extended business model in this type of service is described as follows. An owner publishes in a management application its particular schedule of availability of the parking space with additional information about location, dimensions, pricing rates, or access mode. The citizens demanding some parking space can check the list of existing ones close to him/her to be rented through a smartphone application and select and pay for that which fits with their needs.

15.3 Case Studies on Mobility as a Service

This section is wholly devoted to present and analyze some relevant projects associated with the concept of *MaaS* introduced in Section 15.2. It is not intended to be an exhaustive review of the field but rather a brief overview of some trends and promising initiatives arisen over the last years.

15.3.1 UbiGo

UbiGo[9] is the name of an innovative transportation broker service that spanned over the whole city of Gothenburg, the second largest city in Sweden, in 2013. It has arisen as a service piloted in the Go:Smart project[10], which seeks to deal with some of the societal objectives of Sweden as regards urban mobility. In particular, it promotes a more sustainable transportation of people in urban areas with the aim of mitigating the excessive pressure exerted by private car ownership on the environment and physical infrastructures and its impact on the quality of life of the citizens while also opening new green business opportunities.

---

9 http://www.ubigo.se/las-mer/about-english.

10 http://www.mistraurbanfutures.org/en/project/gosmart

On the technical side, UbiGo is supported through a new technological solution based upon a smartphone application and intended to become an all-in-one urban mobility interface, mainly thanks to cooperative relationships among different public, private, and research partners. Its overarching goal is to provide citizens everyday travels by accessing a wide variety of alternative transportation means to private cars. In fact, the available mobility service combines, for instance, public transportation, car-sharing, rental car service, taxi, and a bicycle system. Nevertheless, these service choices are transparently provided as a single offer to the users via service packages.

Essentially, UbiGo was conceived as a MaaS operator that comprises a revolutionary business and subscription model. The proposed business model of the commercial operator behind UbiGo is based on flexible monthly user subscriptions in which additional trips can be progressively purchased on demand. Specifically, a householder – representing himself/herself and/or a group of members of his/her family – is allowed to define their own mobility requirements and their credit limits to consume transportation services through a joint account. The subscribed users are allowed to make bookings on their desired trips at any moment. Additionally, the use of sustainable transportation means in UbiGo is promoted by providing a reward in the form of virtual credit to their users, which can be used to pay other services such as cultural and artistic activities.

It is worth mentioning that UbiGo was initially assessed by a provider-driven field operational test [22] involving 70 paying households in Gothenburg during a 6-month period (from November 2013 until April 2014). All of them were offered a mobile phone subscription. Some of the early insights drawn from the field operational test about the main practical incentives for users in terms of new mobility services such as convenience and economic advantage, and the added value perceived from its use are discussed in [23].

Finally, UbiGo is additionally the name of the company that today plans to relaunch a commercial version of the service during 2016 in Sweden. With this goal in mind, a recent agreement of cooperation between the company and Ericsson has been announced. Specifically, Ericsson will supply the MaaS-IT platform and UbiGo will provide the business concept. UbiGo is also in a position to support potential MaaS operators in other cities to implement a mobility service faster and cheaper.

15.3.2 car2Go

car2go[11] is a flexible free-floating one-way car-sharing provider created by the multinational automotive corporation Daimler AG and established initially in the city of Ulm, Germany, in April 2009. Up to now, it has progressively

---

11 https://www.car2go.com.

expanded its services globally in multitude of cities around the world, such as Austin, United States, Vancouver (Canada), Madrid (Spain), and Rome (Italy), among others. However, it is also worth pointing out that the poor market penetration in cities such as London, United Kingdom, or South Bay, United States, has caused the service to be no longer available at present.

car2go offers citizens the opportunity of accessing a large fleet of compact gasoline- and electric-powered two-seater vehicles distributed around their city to take one-way trips. Unlike traditional car-sharing systems based on docking at predefined stations [24], the main distinguishing feature of the service provided by car2go is its high flexibility of use, which is achieved through trips that are completely discretionary. This means that the trips can start and finish at any point within the designated operating area of the city due to the fact that the vehicles have no fixed positioning, which makes them easily accessible. This prevents users from returning the vehicles to their pickup locations after use. In fact, users can leave the operating area of car2go freely, but they have to go back to it with the aim of finishing their trips.

Every time a user requires a one-way trip, one of the vehicles distributed in the city can be taken freely. The current locations of the available vehicles close to him/her are also given over a web or smartphone application, which act as technological enablers of the service offered by car2go. Thanks to the Global Positioning System technology incorporated into all the vehicles, their positions can be checked in real time. Furthermore, the user can book his/her desired vehicle at most 30 min before the trip starts. This way, the vehicle is ready to be used when the user arrives. The access to a vehicle is granted by swiping a personal radio-frequency identification card in front of the reader on the windshield. Once inside the vehicle, the user is prompted for a personal password and required to answer a few questions about its current state of cleanliness via the navigational touch screen before starting the trip. Lastly, the user has only to leave the vehicle in a parking spot at his/her destination and lock it to be available for the next trip. It is also worth mentioning that car2go has recently launched a mobile-based access service as an alternative to personal identification cards. It is a smartphone application feature that allows users to begin their trips via their smartphones. A recent overview of the main car-sharing programs worldwide and an interesting discussion about them is provided in [25].

The business model of car2go is based on subscription without any permanent membership fee. Registered users only pay for their particular renting duration of the vehicles, which means that no fixed charges exist. Also, neither minimum nor maximum time limits are required when renting the vehicles. Finally, it is worth pointing out that car2go tries to deal with the problems derived from the usual parking shortage by allowing free street parking. In addition, there is evidence that car2go has a positive environmental impact by reducing the CO_2 emissions [26].

15.3.3 Uber

Uber[12] is a technology company founded in San Francisco, United States, created in 2010, and with over 500 employees and growing. This company offers a free software platform available for mobile devices that allows users, through a tracking system, to order a private car to travel from one location to another. The platform is capable of allowing the user to track his/her vehicle and receive a text message confirming when the driver is arriving. In this regard, the driver is able to inform the user through the platform that is coming to his/her location. Furthermore, the driver does not know the phone number of the user. However, the driver can contact the user if this cannot be found in the departure location. In addition, the company does not own any vehicle. Instead, it works under a system of cooperation with drivers and private units. Thus, Uber seeks in summary to match passengers with drivers.

When using Uber, no cash is exchanged. Instead, the charge of the trip is done through an electronic payment credit card the user entered when registering in the application. Upon finishing the ride, this is charged electronically, and the receipt with the details of the trip is sent to the user by email. Then, the user can evaluate the service of the driver and vice versa and finally check the route taken through a map. As discussed, Uber is a company that does not require a specific-purpose license to its drivers. There is no regulation for which a driver may not discriminate against a passenger. Uber sets the taxi fares. Generally, there is a premium rate during peak and flat rates for the remaining hours. These rates are based on distance, car type, and time. Regarding earnings, 80% of the rates are for the driver and the remaining 20% for the company. Despite the loss of 20% of profits, drivers of Uber as a rule earn more than the traditional taxi services.

Statistics and numbers support the success of Uber as a ride-sharing company. Uber is available in 60 countries and over 300 cities worldwide with 1 million rides daily, made by more than 8 million users. In some major US cities such as Los Angeles and San Francisco, in the first quarter of 2015, users preferred the Uber service than the traditional taxi, with greater use of the Uber service. There are many reasons why customers prefer to book Uber versus taxis. The main reasons are to have a fixed price before booking, tracking drivers on the map, comfort without cash payment, and one-tap rides.

Uber is one of the major companies that have been a strong competition to the taxi industry and independent drivers. Different taxi companies, independent drivers, and even governments have taken legal action against Uber. These legal actions are based on unfair competition from Uber to taxis because the company does not pay taxes or license fees, and that drivers are not trained and endanger passengers. Uber is currently involved in hundreds of lawsuits and even banned in some countries and cities. However, governments and taxi

---

12 https://www.uber.com.

companies have not been able to completely stop the operations of Uber, mainly because its operations are conducted over the Internet, which carries underdeveloped legal implications.

Finally, it is worth mentioning that there are other alternatives to Uber in the market. For example, Lyft[13] is a technology company founded in California in June 2012. Its mobile phone platform facilitates peer-to-peer ride sharing by connecting passengers who need a ride with drivers who have a car.

15.3.4 RideScout

RideScout[14] is a mobile application that aggregates information from all possible options of public, private, and social transportation services. RideScout gives the possibility to seek, in its system of aggregation of public and private transportation, nearby buses, subways, taxis, trains, sedan services, bike-share services, carpooling, and car-sharing programs. The user just types his/her origin and destination, and all transportation service options will be listed through a simple interface. It also includes a strong social component that allows creating user groups, Google, Twitter, and Facebook integration, in order to allow planning trips in groups.

Some of the extra features are (i) the possibility to sort different transportations by time and cost, and (ii) users have the possibility to book, pay, and qualify the journey within the application. An interesting feature of the application is its gamification, which allows the user to know the environmental impact based on the different transportation means used in each journey. However, the features discussed are the most basic the user would expect from an application of this type. There are other more advanced features applied in certain places of deployment of the application. For example, in Washington DC, United States, and with the help of different partners, RideScout is able to report free spaces in the nearest Capital Bikeshare stations and also warns the user when approaching the stops and it is time to request to stop and when it is time to go to the stop to take transportation service.

Importantly, as an aggregator of transportation services, from conception of the project idea, RideScout has done a great number of relationships with municipalities, universities, large organizations, and key transportation industry players. The mobile app can add information of metro, other public transportation networks, Sidecar, car2Go, taxi-hailing app Hailo, and Capital Bikeshare. Unfortunately, Uber does not authorize the inclusion of its information, as it maintains a policy of not sharing information with other sites. RideScout has created and consolidated alliances with different transportation industry players to provide MaaS as understood from the company, and this strategy of seeking partners should continue in the future. Rachel

---

13 https://www.lyft.com.
14 http://ridescout.com.

Charlesworth, RideScout's spokeswoman, said that Uber was continually seeking new partners and alliances to expand its services. However, initially there were difficulties to convince the different actors related to transportation being added to RideScout and free inclusion in an active point-of-sale marketplace. The inclusion of information of new transportation agents is one of the key success factors of RideScout.

At first RideScout was launched in Austin, United States, with an iPhone version for the mobile application, accumulating 6000 downloads from March to November 2013. Proven success with the launch of Austin, the development team decided to develop a version for Android and launch the application in Washington DC and San Francisco. RideScout expanded from a few cities to 69 cities within just a few months in the summer of 2014. Nowadays, RideScout works in many cities in the United States and enjoys an excellent host. According to estimates by the developer, in 2018 more than 4 million users will use RideScout to find and compare transportation options.

15.4 Conclusions and Further Research

Over the last decades, multiple transportation means have competed actively for gaining the highest possible market share within the mobility sector. However, they are nowadays facing a great challenge: adapting themselves to the new mobility requirements and expectations of current society.

People are today exerting a high pressure on transportation industry stakeholders with the aim of having new transportation means that enable them to reach their destinations efficiently. This situation has become especially problematic in urban areas, where the limited land availability increases the parking space shortage and traffic congestion. Additionally, urban areas must cope with the growing concern of the general public about how to reduce emissions of toxic and harmful substances linked to motorized transportation that contribute largely to global warming. Furthermore, despite the efforts carried out by public transportation transit through economic incentives or improvement in terms of reliability and quality of service to minimize traffic congestion, environmental emissions, and mitigate problems associated with urban sprawl, among others, it has usually encountered many difficulties for its massive deployment. As a result, the private car is still today the most attractive transportation mean for most of people due to its instrumental benefits and symbolic and affective aspects.

The technological development and innovation in the transportation sector is transforming the conceptualization of mobility at a dizzying rate. Particularly, the emphasis is shifting toward the capability to reach destinations efficiently and under complementary heterogeneous criteria such as convenience, comfort, or cost. Nowadays, the mobility is acquiring a role of just-in-time service, where the individual requirements must be satisfied on demand by the transportation providers. In this scenario, *MaaS* has emerged as a multimodal and seamless form to manage the current urban mobility while preventing any type of commitment to a pre-specified transportation mean. Its main goal is to provide a solid transportation solution by enabling access to a comprehensive range of sustainable mobility services in such a way that the users can define their particular mobility patterns and satisfy their own necessities when and how they prefer. Some immediate consequences of new business models developed under the paradigm of Mobility as a ServiceMaaS are energy savings, increase of customer satisfaction, and overcoming the limitations on infrastructure expansion in urban areas. Also, the adoption of ICTs in these mobility solutions gives rise to new intelligent and connected transportation alternatives that encourage the improvement in the quality and efficiency of the existing mobility services. Particularly, these provide real-time information to the users with the aim of minimizing their waiting times, making onward connections faster, and reducing their ecological footprints.

Acknowledgments

This work has been partially funded by the Spanish Ministry of Economy and Competitiveness (projects TIN2012-32608 and TIN2015-70226-R). Airam Expósito-Márquez would like to thank the Canary Government for the financial support he receives through his/her post-graduate grant.

Final Thoughts

The concept of MaaS is introduced in this chapter. The main functionalities, technologies, and representative projects arisen in the field of transportation over the last years are analyzed. All of them have encouraged a shift toward reaching the destinations efficiently at the expense of any type of commitment to a pre-specified transportation mean.

Questions

1   How did the mobility patterns change over the last few years?

2   What are the main components of an efficient mobility solution?

3   How do new transportation means save money and time?

4   What are the pros and cons of using sharing mobility solutions?

5   What are the relevant projects associated with Mobility as a Service?

References

1 Votolato, G. (2007) *Transport Design: A Travel History*, Reaktion Books.
2 Agbelie, B.R. (2014) An empirical analysis of three econometric frameworks for evaluating economic impacts of transportation infrastructure expenditures across countries. *Transport Policy*, **35**, 304–310.
3 Gärling, T. and Steg, L. (2007) *Threats from Car Traffic to the Quality of Urban Life: Problems, Causes, and Solutions*, Elsevier.
4 Leal-Filho, W. and Kotter, R. (eds) (2015) *E-Mobility in Europe. Trends and Good Practice*, Springer International Publishing.
5 Kent, J.L. (2014) Driving to save time or saving time to drive? The enduring appeal of the private car. *Transportation Research Part A: Policy and Practice*, **65**, 103–115.
6 Li, Z. and Hensher, D.A. (2012) Congestion charging and car use: a review of stated preference and opinion studies and market monitoring evidence. *Transport Policy*, **20**, 47–61.
7 Dameri, R.P. and Rosenthal-Sabroux, C. (2014) Smart city and value creation, in *Smart City*, Progress in IS (eds R.P. Dameri and C. Rosenthal-Sabroux), Springer International Publishing, pp. 1–12.
8 Howe, N. and Strauss, W. (1991) *Generations: The History of America's Future, 1584 to 2069*, Quill, New York.
9 Howe, N. and Strauss, W. (2000) *Millennials Rising: The Next Great Generation*, Vintage, New York.
10 Heinrichs, H. (2013) Sharing economy: a potential new pathway to sustainability. *GAIA*, **22** (4), 228–231.
11 Heikkilä, S. (2014) Mobility as a service – a proposal for action for the public administration, Case Helsinki. Master's thesis. Aalto University.
12 Toffel, M.W. (2008) *Contracting for servicizing*, Harvard Business School Technology & Operations Mgt. Unit Research Paper, (08-063).
13 Hamari, J., Sjöklint, M., and Ukkonen, A. (2015) The sharing economy: why people participate in collaborative consumption. *Journal of the Association for Information Science and Technology*, **67** (9), 2047–2059.
14 López, E., Monzón, A., Ortega, E., and Mancebo-Quintana, S. (2009) Assessment of cross-border spillover effects of national transport infrastructure plans: an accessibility approach. *Transport Reviews*, **29** (4), 515–536.
15 Condeço-Melhorado, A., Gutiérrez, J., and García-Palomares, J.C. (2011) Spatial impacts of road pricing: accessibility, regional spillovers and territorial cohesion. *Transportation Research Part A: Policy and Practice*, **45** (3), 185–203.
16 Delmon, J. (2009) *Private Sector Investment in Infrastructure Project Finance – PPP Projects and Risks*, Wolters Kluwer, Austin, TX.

17 Jo, K., Kim, J., Kim, D., Jang, C., and Sunwoo, M. (2014) Development of autonomous car −Part I: distributed system architecture and development process. *IEEE Transactions on Industrial Electronics*, **61** (12), 7131–7140.

18 Jo, K., Kim, J., Kim, D., Jang, C., and Sunwoo, M. (2015) Development of autonomous car −Part II: a case study on the implementation of an autonomous driving system based on distributed architecture. *IEEE Transactions on Industrial Electronics*, **62** (8), 5119–5132.

19 Shaheen, S.A. and Cohen, A.P. (2008) *Worldwide carsharing growth: an international comparison*. Institute of Transportation Studies, working paper series, Institute of Transportation Studies, UC Davis.

20 Ricci, M. (2015) Bike sharing: a review of evidence on impacts and processes of implementation and operation. *Research in Transportation Business & Management*, **15**, 28–38. Managing the Business of Cycling.

21 Stiglic, M., Agatz, N., Savelsbergh, M., and Gradisar, M. (2015) The benefits of meeting points in ride-sharing systems. *Transportation Research Part B: Methodological*, **82**, 36–53.

22 Leminen, S., Westerlund, M., and Nyström, A.G. (2012) Living labs as open-innovation networks. *Technology Innovation Management Review*, **2** (9), 6–11.

23 Sochor, J., Strömberg, H., and Karlsson, I. (2015) The added value of a new, innovative travel service: insights from the UbiGo field operational test in Gothenburg, Sweden, in *Internet of Things. IoT Infrastructures, Lecture Notes of the Institute for Computer Sciences, Social Informatics and Telecommunications Engineering*, vol. **151** (eds R. Giaffreda, D. Cagavnova, Y. Li, R. Riggio, and A. Voisard), Springer International Publishing, pp. 169–175.

24 Shaheen, S.A. and Cohen, A.P. (2013) Carsharing and personal vehicle services: worldwide market developments and emerging trends. *International Journal of Sustainable Transportation*, 7 (1), 5–34.

25 Shaheen, S.A., Chan, N.D., and Micheaux, H. (2015) One-way Carsharing's evolution and operator perspectives from the Americas. *Transportation*, **42** (3), 519–536.

26 Firnkorn, J. and Müller, M. (2011) What will be the environmental effects of new free-floating car-sharing systems? The case of car2go in Ulm. *Ecological Economics*, **70** (8), 1519–1528.

16

Clustering and Fuzzy Reasoning as Data Mining Methods for the Development of Retrofit Strategies for Building Stocks

Philipp Geyer[1] and Arno Schlueter[2]

[1] *Department of Architecture, Faculty of Engineering Science, Katholieke Universiteit Leuven, Belgium*
[2] *Institute of Technology in Architecture, ETH Zurich, Switzerland*

CHAPTER MENU
Introduction, 438
Method, 440
Application Case, 442
Data Sources and Preprocessing, 443
Clustering, 448
Fuzzy Reasoning, 456
Mixed Fuzzy Reasoning and Clustering, 459
Postprocessing: Interpretation and Strategy Identification, 459
Comparison and Discussion of Methods, 464
Conclusion, 467

Objectives

- To familiarize the reader with data mining methods for the development of effective retrofit strategies for a building stock, including energy efficiency measures (EEMs) and automated network identification (ANI) for smart energy networks
- To identify the benefits and methodological differences between sparse information approaches, that is, the type–age classification, and novel approaches based on information available from building catalogs and databases, measurements, as well as data mining methods in smart city contexts
- To introduce readers to hierarchical agglomerative clustering and fuzzy reasoning as intuitive data mining methods for strategy development for the building stock
- To provide a guide for the application of such methods to the readers' own situation

Smart Cities: Foundations, Principles and Applications, First Edition.
Edited by Houbing Song, Ravi Srinivasan, Tamim Sookoor, and Sabina Jeschke.
© 2017 John Wiley & Sons, Inc. Published 2017 by John Wiley & Sons, Inc.

16.1 Introduction

A successful energy transition to a low-carbon built environment requires the full retrofit of the building stock. However, the current retrofit rates of 1–2% per year are far too low to reduce emissions in time [1]. The investments required in retrofitting are high but are currently not sufficiently focused on energy efficiency and the reduction of greenhouse gas emissions. For this reason, specific retrofit strategies and programs must be developed. This task incurs a problem in that it needs to define those energy efficiency measures (EEMs) that lead to the greatest reductions of emissions, or other environmental impacts, at the lowest cost. Given a building stock varying from hundreds to ten thousands of buildings in a city, it is obvious that planners cannot examine every building to determine the potential of EEMs. In this situation, classification schemes or groups are typically used to characterize the building stock and to develop retrofit measures for each group. A conventional scheme for this purpose is type–age classifications [2]. Given the situation that more data is becoming available through the application of smart city technologies, new techniques can be developed.

Such technologies, when applied to smart cities and regions, provide much more information on buildings and their environment using sensor technologies, operational monitoring, and building stock databases. These data offer new approaches for the development of smart energy retrofit strategies and smart supply systems at an urban level that have the potential to outperform traditional methods. The utilization of this information will play an important role in future planning support systems and decision-making, as has been highlighted in recent publications [3]. There are different approaches that address energy performance simulation and analysis at the urban level [4], [5]. Furthermore, Quan *et al.* [6] integrate energy performance calculations with geographic information systems (GIS) environments and compared the results to monitoring data. However, these implementations often stay at the energy analysis level and are not taken to the level of strategy development.

16.1.3.1 Clustering

Therefore, in this chapter, we propose data mining methods to develop retrofit strategies for building stocks. In contrast to other approaches, the primary criterion for strategy development is the cost efficiency of the EEM. The approach focuses on strategy development and decision-making. The objective is to find buildings that react similarly to a variety of EEMs to develop a strategy that aims not only at individual buildings but also at groups of buildings. For instance, previous applications of clustering typically either focus on building parameters or on energy consumption, such as in the case of Santamouris *et al.* [7], who applied clustering to a database of 320 schools in Greece and built groups based on energy consumption with climatic normalization. Gaitani *et al.* [8] identified the typical building properties and parameters of the schools by k-means clustering. Jones *et al.* [9] clustered a building stock by building properties, such as heated ground floor area, facade, and window-to-wall ratio. However, this use of data mining does not provide direct information for the planning of energy retrofits.

16.1.3.2 Conditions of Data Mining in Building Stock

The application of data mining methods to energy retrofits must consider the specific application context of building stock retrofits. First, the building stocks consist of sets of spatially distributed objects (buildings in a city or community) with certain characteristics that determine their performance. Second, they interact in a systemic way, which means that they are linked by their energy systems (district energy systems, smart grids, etc.) and economic systems (subsidy schemes, budgets, and market). Given these systemic interdependencies, selected EEMs relate not only to one building but also to the entire building stock. Thus, retrofit measures need to be developed at the building stock level in a smart way. This means that it is not only necessary to consider the effect at the building level but also the effect at the level of energy systems and economic systems.

Applied data mining methods identify groups of buildings for exactly defined purposes, that is, to find buildings that react similarly to retrofitting measures. This allows for the development of intelligent systemic strategies instead of isolated approaches to individual buildings. Such intelligent strategies exploit the similarities and complementarities of buildings as well as their spatial relationships. As a result, building retrofit strategies can be defined for buildings that respond similarly to certain measures. Moreover, those buildings that are well suited to the creation of efficient district networks can be identified, further pointing to the possibility of creating new energy networks or the extension of existing ones.

16.1.3.3 Contents of the Chapter

After the introduction, a state-of-the-art review and a description of the case study used throughout the chapter are presented. Then, we describe the data preprocessing steps necessary to establish a consistent database. The main part describes the two key methods, namely, algorithmic clustering and fuzzy reasoning, and their application. Both methods are applied using spatial and non-spatial data as well as combined data. Next, we highlight methods of postprocessing that enable the derivation of strategies from the results, as well as to visualize them for planners and decision-makers. Finally, the two methods are compared with respect to their performance, support of certain search objectives, and robustness to incomplete and nonheterogeneous databases. Specific details of the application of the individual methods have been published previously [10, 11].

Clustering and fuzzy reasoning methods provide the classification needed to exploit similarities and spatial proximity. This allows for strategies for groups, subsystems, automated network identification (ANI), and so on. The main difference between the two methods that will be highlighted is that hierarchical agglomerative and partitioning clustering are pre-assessment methods, whereas fuzzy reasoning is a post-assessment method.

16.2   Method

Conventional building stock management relies on sparse data—a hangover from the era prior to the application of computers to the management of the built environment. The most frequently used method is the type–age classification, like the one that was developed by the Institut Bauen und Umwelt (IBU) in Germany [12], which was recently extended Europe-wide [2]. These methods rely on the types and ages of buildings as criteria whereby they are assigned to groups. From these groups, one representative reference building is typically selected, and EEMs are developed for that building. The resulting strategy is then applied to the entire group, as shown in Figure 16.1. However, the effect of EEM on the different buildings in such a group based on type–age data is not very homogeneous. Therefore, they do not form a very good basis for recommending EEMs [10].

The data that can be acquired using smart city technologies provide options for improving building stock management. The starting point, shown in Figure 16.2, is data generated by sensors, smart meters, surveys, simulations, and so on; this information is stored in databases and Geographic Information Systems (GIS). With the development of information modeling at the building level (Building Information Modeling, or BIM, [13], [14]) including its standards (such as Industry Foundation Classes, IFC, [15]) and at the urban level (GIS) including its standards (such as CityGML, [16]), well-structured approaches to the management of these data are currently evolving. However, to establish a building stock database, preprocessing of the data is required to ensure consistency. The resulting comprehensive database provides an important foundation for the application of data mining techniques.

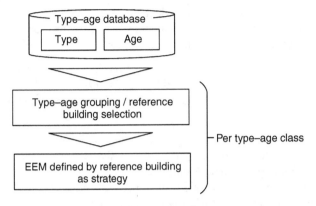

Figure 16.1 Conventional type–age classification for deriving groups for building stock management.

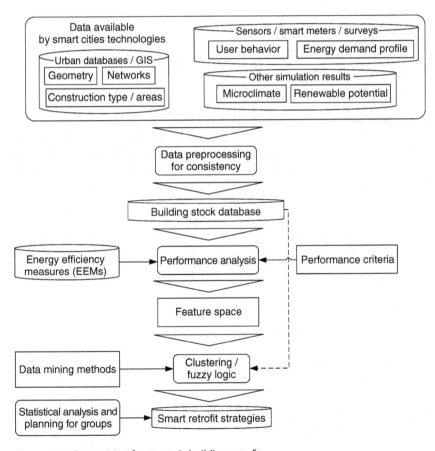

Figure 16.2 Data mining for strategic building retrofit.

The basis of data mining is the definition of a feature space defined by the aspects or dimensions to be used by clustering or fuzzy reasoning to classify the data. There are two ways of constructing a feature space:

1) The parameters found in the database are used as features, like those shown by Santamouris *et al.* [7], Gaitani *et al.* [8], and Jones *et al.* [9].
2) Data analysis is used to construct the feature space to which data mining is applied.

The second method is the one that we propose to facilitate performance-based decision-making for building stock management. Based on a consistent database of the building stock, performance analysis is executed. First, a set of EEMs is needed, and second, the definition of performance criteria are required (Fig. 16.2). Feature selection depends on the individual case. The best EEMs for feature construction are measures that are feasible and that significantly affect building performance. In Section 16.3, we demonstrate a feature selection as part of the case study.

The feature space results from the application of a the performance analysis of EEMs for each building in the database. According to its response to each EEM, each building is positioned accordingly in the feature space. The units of the feature space are defined by the units of the performance criteria.

The constructed feature space forms a basis for the application of data mining methods. In this chapter, we focus on clustering and fuzzy logic. However, the constructed feature space also allows the application of other methods. The aim of the proposed methods is the derivation of groups within the feature space. Furthermore, we use data mining methods to find buildings with certain properties, for example, old heating systems. For this purpose, data from the building stock database is used directly, as indicated by the dashed arrow in Figure 16.2. This gives rise to a hybrid approach.

For those groups identified by data mining, for which the number is usually low, it is a relatively easy task to develop retrofitting strategies. In postprocessing, the application of statistical analysis can help to determine the characteristics of a group and to derive an appropriate EEM. Finally, assigning an annual financial budget leads to a time-based transformation scenario.

16.3   Application Case

We apply the methods described in this work and explore the results using the case study of Zernez, a Swiss community, which was studied as part of the interdisciplinary research project "Zernez Energia 2020." The objective of the research project was to transform the building stock of the community, consisting of 309 buildings of different types, towards zero-emissions operation

(Figure 16.13). To develop a transformation strategy, this objective was approached from the perspectives of urban transformation, local and regional energy infrastructure, material flows, and building retrofit. As part of the project, a comprehensive building database of all the buildings in the community was established, featuring over 50 parameters per building, including age, building envelope surfaces and materials, heating and electricity consumption, type and age of building systems, occupation, and geospatial data.

A sampling of these parameters, including building geometry, heated floor area, current type of heating system, roof orientation, and available roof surface was used for the performance analysis of EEMs. As a set of feasible and successful EEMs previously applied to buildings in the community, the following singular and combined measures were chosen:

- Increase in insulation of the envelope (walls and roof) to reduce heat losses.
- Exchange of heating system by, for example, replacing oil heating with a heat pump or by connecting to a district heating system.
- Add renewable energy generation, such as photovoltaics and/or solar thermal.
- Combinations of the previously mentioned strategies, such as an increase in the amount of insulation combined with the replacement of the heating system and/or addition of photovoltaics/solar thermal generation.

As part of the overall objectives of the research project, the final energy demand and CO_2 emissions, and the cost efficiency of the EEMs were chosen as performance criteria. Clustering and fuzzy logic were utilized for the evaluation, identification, and aggregation of buildings featuring EEMs with similar cost efficiencies [10], [11] to develop a custom retrofit strategy for the entire building stock and thus yield quicker CO_2 mitigation at a less cost, compared with conventional strategies, as shown in Section 16.8.3.

16.4 Data Sources and Preprocessing

The aim of the data preparation is the creation of a consistent and uniform database. Consistent means that measurement errors are removed; uniform means that the data are available for all instances with the same unit and with the same reference.

16.4.1 Data Sources and Modeling

An increasing amount of building stock data is available that can serve as a basis for the methods introduced in this work. This includes data on building type, location, age, construction, date of retrofit, and energy consumption. Access to such data is greatly dependent on the country and local policies. However, an increasing number of public and private entities have started to make these data available to researchers and practitioners. Such publicly available data can be expanded by using survey methods and measurements. These data can then be used to build customized bottom-up models that represent the energy behavior of buildings, which can then be used directly for energy simulation or the calibration of simulation models.

16.4.1.1 Public Databases and Datasets

Publicly available databases provide an increasing amount of data for assessment and modeling. The US Government, for example, has recently provided access to energy data as a measure to accelerate energy efficiency investment and innovation [17].

Meanwhile in Switzerland, the Federal Register of Buildings and Housing, provided by the Federal Statistical Office, provides data such as location, footprint, age, and type of heating system for residential buildings in a georeferenced database [18]. This can be enhanced by obtaining georeferenced, geometric data such as that available from the Federal Office of Topography. The Swiss Buildings Database presents every building in 3D including the exact roof geometries, which can be used to estimate solar generation potential [19].

In Asia, Singapore has released the average household energy consumption for Housing and Development Board (HDB) apartments, which make up 80% of its residential housing stock [20]. To foster innovation in the improvement of energy efficiency and energy conservation in the residential sector, Singapore has released over 12 million electricity consumption records with a 30-min resolution covering a period of 8 months and 8 million gas consumption records [21].

The spatiotemporal resolution and aggregation of data of such datasets can, however, be heterogeneous. This constrains their usability to, for example, applications where high-resolution temporal and/or spatial data are necessary.

16.4.1.2 Protocols and Surveys

Gathering building data by surveying and protocoling is one of the most commonly utilized techniques, especially in the context of energy audits [22]. The core objective is to obtain data to determine the actual energy use, together with the impacting parameters such as conversion efficiencies, characteristics of the building envelope, and occupant behavior. Such protocols may include energy bills and records that are normalized to exclude external influences [22]. Gathering all of the relevant building data, however, can be cumbersome and dependent on owners or operators documenting energy consumption data, while executing measures, and changes to the building envelope and service systems.

To assess not only one but many buildings, surveying techniques can be used to obtain data from the building owners and users themselves. A large-scale example is the US Residential Energy Consumption Survey executed on a representative sample of over 12,000 housing units throughout the United States [23].

On a smaller scale, for a research project aiming at developing a zero emissions retrofit strategy for the Swiss village of Zernez, all of the building owners in the community were surveyed. The results constituted a comprehensive database of 50 parameters per building, resulting in over 15,000 data points that described the entire building stock in the village. These data were georeferenced and stored in a GIS database for further modeling and analysis, resulting in an optimized retrofit strategy for the village's entire building stock [24]. The quality of these data, however, is dependent on the owners' or inhabitants' knowledge and records of energy consumption, retrofit measures, and changes. As nonexperts are not too familiar with the energy data and building properties, there is a certain risk of there being faulty data entries.

16.4.1.3 Sensor Measurements and Bottom-Up Modeling

Measurements of building energy consumption, interior and exterior environmental states, and building service systems have become common place as a result of decades of research and implementation of energy characterization, as in the case of energy ratings. Unfortunately traditional approaches to measurements, however, often involve significant effort and cost, while being intrusive to the inhabitants [25]. Recently, a new generation of low-cost, minimally intrusive measurement approaches have been devised. These are based on the widespread use of sensors to capture temperatures, energy flows, and occupant behavior. Such approaches use wireless sensor networks (WSNs) to connect multiple sensor nodes and directly route the data to servers for postprocessing [26]. Low-cost sensor networks are often based on affordable and flexible hardware and open-source protocols such as Arduino or Raspberry Pi [27, 28]. Other even simpler approaches use data loggers that are provided to the inhabitants to measure environmental properties such as room temperature. A single data logger can provide helpful data to categorize buildings according to their performance [25].

Measured data has been used since the 1970s to establish building-specific, bottom-up energy models using parameter estimation [29]. Some of the key objectives are to keep the number of required parameters small [30, 31] and to facilitate the long-term extrapolation of energy performance from short-term measurements [32, 33]. Furthermore, aspects of occupancy can be factored into the calculations [34]. Presenting an example that combines a cost-efficient WSN with bottom-up modeling, Nagy *et al.* [35] use multisensor data to generate a simplified steady-state model for approximating the heating energy demand. This model can then be used to investigate the effect of retrofit measures to identify optimal strategies for a specific building, offering significant improvements over conventional techniques.

16.4.2 Construction of Feature Space

Using a consistent and uniform database, the effect of measures can be determined. As this step must be performed for all buildings and all measures, it should be performed automatically and should not cause an excessively high computational load. Furthermore, the method for constructing the feature space depends on the data available in the database. Here, energy analysis serves as a representative example in this chapter; other types of analysis employed for sustainable building design might be applied in a similar way. Furthermore, we assume that, for all buildings, the use, the building geometry, the age of the building, and its heating system are known. Subject to additionally available data, further information can be derived:

1) If the heating and electrical energy consumption, the heating system type, and the construction properties are available, then the CO_2 emissions and further environmental impact can be calculated, reflecting the actual user behavior. The effect of retrofit measures can be derived relative to the current energy consumption and the changes in construction.
2) If only construction properties are available, then the effect on energy consumption and emissions can be assessed using statistical user profiles and assumptions for building operation. Either a pseudo-static calculation or dynamic simulation can be used to determine the energy consumption and respective CO_2 emissions.

If no additional data are available, the approach is bound to the same granularity of information as the type–age classification and will lead to similar results.

16.4.2.1 Metamodeling

To perform an energy analysis and further analysis tasks, metamodeling is a useful method. This method involves the use of surrogate models based on mathematical approaches, such as the response surface method (RSM) and support vector regression (SVR), or on approaches based on artificial intelligence, such as artificial neural networks (ANNs), replacing either the simulation results or monitored data. This includes interpolation capabilities to avoid the need for simulation or measurements for further intermediate data. More importantly, it speeds up the analysis. Using simple methods, such as enhanced RSM, we can obtain very good results for representing the energy demand [47]. Such an approach can be coupled with building stock management [10].

16.4.2.2 The Feature Space

The results of a performance analysis of the retrofit measures for each building provide the source for describing the feature space. This procedure results in the matrix **F** describing the feature space and constituting the effects e of m measures on n buildings:

$$\mathbf{F} = e_{n,m} \tag{16.1}$$

For the effect e, the units of the performance indicator are selected in advance. In different examinations, we deemed that it would be useful to use cost efficiency as performance indicator. We define cost effectiveness as the achieved emission reduction per invested economic means. However, depending on the task, the use of other performance indicators is also possible. As we discuss later, it is possible to use multiple performance indicators with different units. This produces a feature matrix with multiple effects with respect to measures:

$$\mathbf{F} = e_{n,k} \text{ with } k = p \cdot m \tag{16.2}$$

where p is the number of performance indicators. In this situation, the application of scaling and normalization is helpful. While scaling needs to be adopted to each individual case, normalization to a range between 0 and 1 can be achieved simply:

$$\mathbf{F} = \frac{e_{n,m} - e_{\min,m}}{e_{\max,m} - e_{\min,m}} \tag{16.3}$$

Furthermore, the application of compression can be beneficial because this moves outliers closer to the main groups and, in clustering and visualization, avoids groups with a very low number of instances. An example of compressions is

$$e_{\text{compr}} = \begin{cases} e_{n,m} & \text{if } \tilde{e} <= 1 \\ c_m \tilde{e}^{0.1} & \text{if } \tilde{e} > 1 \end{cases} \quad \text{with } \tilde{e} = \frac{e_{n,m}}{c_m} \tag{16.4}$$

where c_m is the compression boundary that can be set as a percentile, for example, to 90% so that the highest 10% per measure are compressed. An alternative to compression is the use of logarithmic functions.

16.5   Clustering

In this section, we address the use of clustering to identify similar instances in the feature space. With a defined feature space, an advantage of clustering is that no further knowledge is required and similarity means similar reactions to EEMs in the first step, as described in Section 16.5.2. Section 16.5.3 makes an extension for spatial relations as is required for energy networks. These spatial relations must not be confused with the feature space. While the feature space only describes the similarity in terms of virtually defined performance dimensions, spatial relationships denote the distance between buildings according to their geographic location.

16.5.1   Nonspatial Clustering

Algorithmic clustering is an appropriate means of identifing similarly reacting groups. The two main methods are hierarchical agglomerative clustering (HAC) and partitioning clustering [36], [37], [38]. These methods have two different function principles that operate as follows: hierarchical clustering assigns every instance to one cluster and merges then step-by-step the closest clusters; partitioning clustering works with a predefined number of cluster centers, assigning each instance to the closest center; algorithms are then used to minimize the distance between centers and cluster members by reassigning members. The results of both methods are similar: instances close to each other are related to groups, as outlined in Figure 16.3. Because the feature space dimensions depend on the effects of the EEM, the clustering identifies groups of buildings that react similarly to EEMs.

From a user's point of view, two differences between the clustering methods are important. First, partitioning clustering requires defining the number of clusters in advance, whereas hierarchical clustering goes through every possible number of clusters. Thus, the latter scans all cluster numbers and allows the selection of one number without a starting constraint. Usually, dendrograms, like that shown for Zernez in Figure 16.4, serve this purpose.

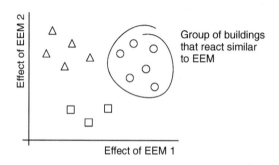

Group of buildings that react similar to EEM

Figure 16.3 **Principle of clustering according to the reaction to EEM.**

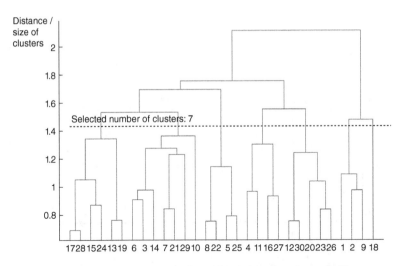

Figure 16.4 Example dendrogram for hierarchical clustering. *Source:* [10]).

Second, the computational performance of the methods scales differently. Typically, hierarchical clustering has a computational complexity of $O(N^2)$ [39], which means that the computation time is proportional to the squared number of buildings. For the Zernez case with 300 buildings, the computation time of clustering was insignificant (<1 s on a standard office computer). However, the quadratic behavior leads to a significant increase of computation time in case of, e.g., 30,000 buildings, which is a typical number for a city or a region. In contrast, partitioning clustering algorithms behave better. For instance, the partitioning clustering algorithm k-means perform better with a complexity of $O(NKd)$ with N objects, K clusters, and d dimensions [39], which means a linear increase in the calculation time with the number of buildings.

The result of the clustering application assigns each building to one cluster; no duplicate association is allowed. The use of fuzzy cluster association that allows multiple membership of buildings in clusters is not considered for strategy development. Figure 16.5 shows, as scatterplot matrix, the clustering result for the building stock of Zernez within nine EEMs or combinations of these forming a nine-dimensional feature space based on cost efficiency in terms of the reduction g CO_2 eq. per invested Swiss Franc. To perform this clustering, nine representative EEMs and combinations were selected, a feature space was constructed as described above, and HAC was applied. The number of clusters was selected by the dendrogram shown in Figure 16.4, and two clusters with outliers were subsequently merged with other clusters. The scatterplot matrix allows us to identify scatterplots in that there is a clear separation between the clusters, as shown in Figure 16.6.

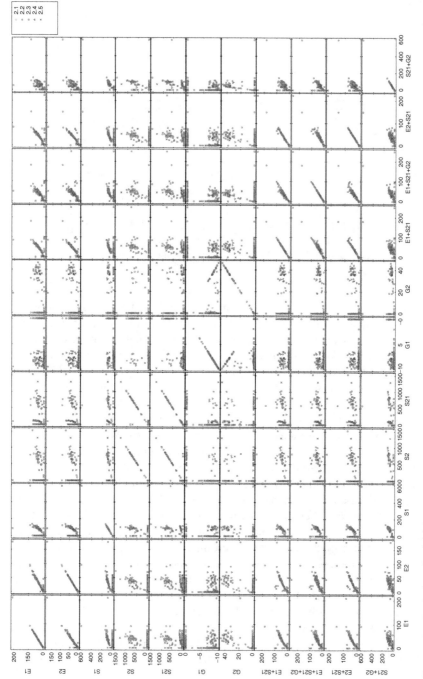

Figure 16.5 Typical result of applying clustering to feature space based on cost efficiency in terms of reduction in g CO_2 eq. per invested Swiss Franc.

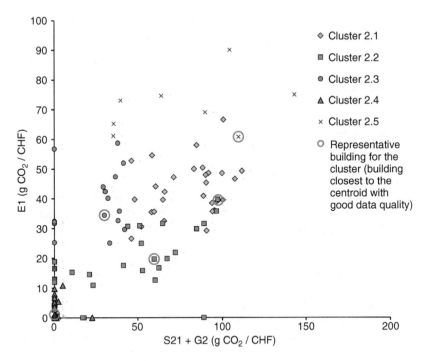

Figure 16.6 One selected scatterplot from the matrix in Figure 16.5 that shows the instances and their allocations to clusters in detail.

16.5.2 Spatial Clustering

Energy transformations for attaining sustainable buildings cannot stop at the building level. Retrofit strategies also need to consider the urban level including energy transport between buildings in networks. Network costs for piping, cables, and further network infrastructure, as well as transport losses, strongly rely on the geometry of the network. For this reason, it is necessary to consider the spatial distribution of buildings when clustering is applied for strategies that involve energy networks. This is particularly true for small networks, such as thermal microgrids, which we have described in detail for application to Zernez [11].

The basic idea of spatial clustering is to find buildings that are located geographically close together and which have complementary characteristics, as outlined in Figure 16.7. For example, a building with residual heat can be coupled with other buildings with heat demand by a thermal network. This type of clustering leads to automated network identification (ANI) for energy transport.

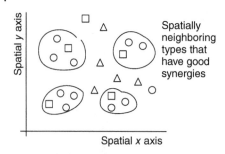

Spatially neighboring types that have good synergies

Figure 16.7 Spatial clustering identifies geographically close buildings with complementary characteristics, such as buildings with residual heat and those with more heat demand.

16.5.2.1 Managing Nonuniform Effect Dimensions

Spatial clustering involves a mixed feature space composed of different units. In this situation, scaling and normalization are needed, as described in Section 16.4.2. Additionally, a normalization function f_n might serve to level out differences in the feature space dimension:

$$f_n(x) = \frac{x - x_{min}}{x_{max} - x_{min}} \tag{16.5}$$

Furthermore, weighting by a weight factor w_d per dimension d might assign a priority to some of the feature dimensions to achieve results that are well-machted to the purpose. The effect of weighting is shown in the next subsection for the case of spatial and nonspatial features. A further solution avoiding the mixing of these two feature types is a two-step approach that is discussed in the subsection after the next one.

16.5.2.2 Results of One-Step Spatial Clustering

One-step clustering assigns all features, both the spatial and nonspatial ones, to a single feature space. To this feature space, the method applies one of the available clustering methods. In consequence, this feature space consists of very different dimensions and units. The experimental clustering shown in Figure 16.8 includes the locations of buildings linked to the unit of meters, the reductions of CO_2 emissions in kg eq. CO_2 per year and square meter, and the ages of the building heating systems in years. Given this situation, first, normalization followed by weighting, as defined by Eq. 16.5, are used to derive a uniform feature space. Furthermore, compression based on a sigmoid function served to avoid outlier clusters with only one or a few buildings:

$$f_c(x) = \frac{1}{1 + e^{a(0.5-x)}} \text{ with } a = 5 \tag{16.6}$$

The effect of the compression is shown in Figure 16.8b. The transformation moves the outer regions of the feature space in particular to the middle, thus reduces the number of outliers.

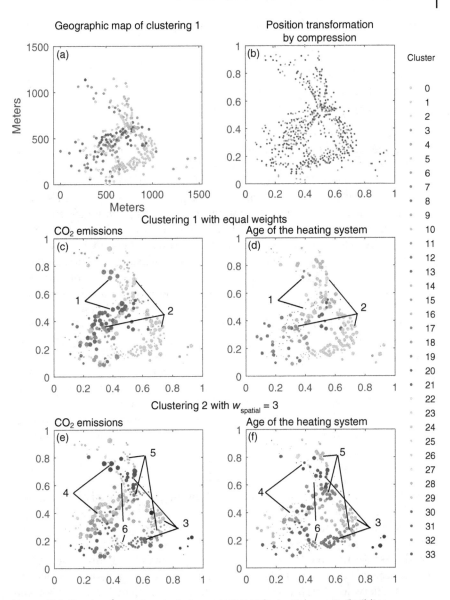

Figure 16.8 Results of one-step clustering experiment for spatial aspects (building locations) and nonspatial aspects (CO_2 emissions and age of the heating system) based on Zernez data. Panels (a), (c), and (d) show an equally weighted clustering (Clustering 1). Panels (e) and (f) show results with spatial aspects weighted three times higher than nonspatial ones (Clustering 2). Panel (b) shows the applied compression (blue dots) in relation to the original position (gray dots).

These data with normalization and compression give the following feature space:

$$F = \begin{bmatrix} w_1 \ f_c(f_n(e_{\text{location } x})) \\ w_2 \ f_c(f_n(e_{\text{location } y})) \\ w_3 \ f_n(e_{CO_2 \text{ emissions}}) \\ w_4 \ f_n(e_{\text{Heating system age}}) \end{bmatrix} \tag{16.7}$$

Figure 16.8a–d shows the results using a uniform weighting (all $w = 1$), such that all the features have the same priority. In the post-processing procedure, 50 clusters are filtered to exclude all those clusters with only one building, giving a final number of 37 clusters. The procedure first identifies those clusters of buildings having high emissions for which the age of the heating system is low or unknown (Cluster 1). Second, it identifies those buildings with high emissions and an old heating system (Cluster 2). However, for these clusters, the spatial connectivity is too low to warrant the establishment of energy networks. For this reason, the weighting of the spatial feature dimensions was increased by a factor of three ($w_1 = 3$ and $w_2 = 3$). This considers the importance of the spatial aspect for the realization of energy networks and leads to much more compact clusters, as shown in Figure 16.8e and f. These clusters slow three main types: First, clusters of buildings with high emissions and old heating systems (Cluster 3), which are the first priority clusters for thermal microgrids, as we argued in Schlueter *et al.* [11]. Second, some clusters have high emissions, but the age of the heating system is either low or unknown (Cluster 4); some clusters have old heating systems and low emissions (Cluster 5). These two clusters have second-order priority, which makes them candidates to be merged with first-order clusters only if they are close to them. Finally, there are clusters with low emissions and new heating systems or systems of an unknown age that are not of interest (Cluster 6).

16.5.2.3 Two-Step Spatial Clustering Results

The one-step procedure is subject to the drawback of mixing all of the features that have different meanings for retrofitting, as illustrated by the example data. For this reason, it makes sense to use a specific clustering procedure consisting of several steps. In the shown case, two steps are the best choice: (i) identify candidates for thermal energy networks using a nonspatial clustering and (ii) from these candidates, identify those buildings that are spatially close enough to form a network. This procedure first tackles the nonspatial identification and then the spatial grouping.

In the example, the first clustering procedure identifies, in the database, those buildings with high emissions or old heating systems. The feature space for this purpose is

$$F = \begin{bmatrix} f_n(e_{CO_2 \text{ emissions}}) \\ f_n(e_{\text{Heating system age}}) \end{bmatrix} \tag{16.8}$$

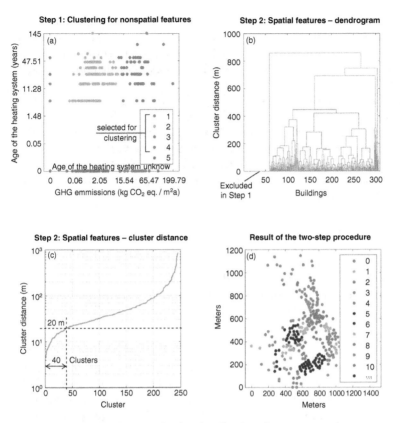

Figure 16.9 Two-step clustering for the identification of energy networks.

Figure 16.9a shows the clustering results for the first step. There are three clusters with old heating systems and low (Cluster 1), average (Cluster 2), and high (Cluster 3) emissions, all of which are candidates for networking. Furthermore, for many of the buildings, the age of the heating system is not known. Those that have high emissions (Cluster 4) are included; those with low emissions (Cluster 5) are excluded.

Given these candidates, the second step involves pure spatial clustering. The dendrogram in Figure 16.9b shows the complete spatial clustering process carried out by a HAC algorithm. The distance measure, which has a major impact on the result, as described in Geyer *et al.* [10], is the shortest distance:

$$d(r, s) = \min(\text{dist}(x_{r,i}, x_{s,j})) \text{ with } i \in (1, \dots, n_r), j \in (1, \dots, n_s) \qquad (16.9)$$

Figure 16.8c, which is a plot of distances between the clusters, serves to convert the technological distance criterion to a number of clusters. Furthermore, from previous observation [11], we know that a distance of less than 20 m between buildings leads to there being a high probability of the application of economic thermal microgrids. According to the cumulative sum in the dendrogram (orange line), this makes 40 clusters. After generating these 40 clusters, a further simple filtering algorithm serves to exclude those clusters with only one building.

This procedure results in the 28 groups shown in Figure 16.9d. Cluster 0 includes all those buildings excluded by Step 1 and by the one-building filter in Step 2. The clustering exhibits spatially compact groups of buildings that are appropriate for energy networks, although some clusters or some buildings in clusters exhibit larger distances. However, as the information is intended to assist with planning, we would expect human judgment to be involved anyway, which would eliminate these outliers.

16.6 Fuzzy Reasoning

Fuzzy reasoning can serve a similar data mining purpose in energy retrofit strategies. While clustering starts with a very low amount of information in the feature space, fuzzy reasoning requires more definitions before application and thus is not as neutral as clustering.

16.6.1 Nonspatial Fuzzy Reasoning

The basis for fuzzy reasoning, as described by Harris [40], is membership functions that define whether an instance is part of a fuzzy set. The core of fuzzy reasoning is the use of sets that allow a partial membership between zero and one. This allows intuitive filtering that is close to natural language definitions. For instance, in the case of a retrofit, if buildings with high emissions are the criterion, the fuzzy function can be set up such that buildings with moderate emissions are partly included in this set. Thus, they can also be considered if they match the other criteria very well.

16.6.1.1 Ramp Membership Functions

Easily handled functions that are used for membership definition are ramp-shaped functions, as shown in Figure 16.10. Such membership functions μ are defined by the lower and upper limits x_l and x_u, as follows:

$$\mu(x, x_l, x_u) = \begin{cases} x \leq x_l & 0 \\ x > x_l \wedge x < x_u & \dfrac{x - x_l}{x_u - x_l} \\ x \geq x_u & 1 \end{cases} \tag{16.10}$$

In Figure 16.11, the first two functions show examples of such membership functions for identifying buildings for retrofitting. The first function filters all

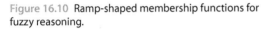

Figure 16.10 Ramp-shaped membership functions for fuzzy reasoning.

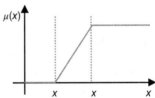

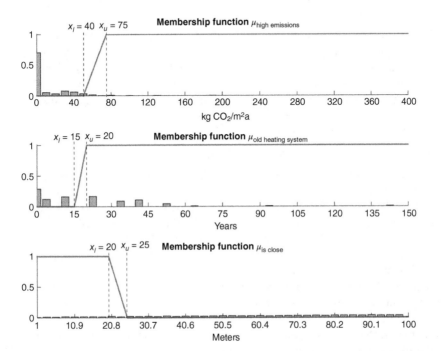

Figure 16.11 Membership functions for retrofitting strategies in Zernez.

the buildings that have high CO_2 emissions, as defined by the ramp from 40 to 75 g CO_2 eq. emissions per square meter and year. The second function includes all those buildings with an old heating system, as indicated by the ramp from 15 to 20 years. Note that the first function relies on the performance analysis according to the scheme described in Section 16.2, while the second function relies directly on one category from the database. This shows how mixing of the information sources is possible.

For retrofitting, the intersection of the two fuzzy sets is an group of buildings of particular interest in that they produce high emissions and have an old heating system. This set is defined by

$$\mu_{\text{heating system exchange candidates}} = \mu_{\text{much emissions}} \cap \mu_{\text{old heating system}} \qquad (16.11)$$

By applying defuzzification, as described by Harris [40], a group of buildings is identified for which the updating of the heating system would be particularly relevant. This is equivalent to the groups identified in the previous section by clustering methods without spatialization. However, in contrast to clustering, the following criteria for the retrofitting strategy must be developed beforehand so that the emergence of unexpected solutions is less likely to occur.

16.6.2 Spatial Fuzzy Reasoning

While these fuzzy functions deal only with data for a single building, micro-grids involve more buildings and their spatial configuration. For this purpose, the evaluation is extended using spatial information with fuzzy functions. A membership function for the proximity of two buildings is set up. This spatial extension of the membership function uses a Euclidian distance function d:

$$d(B_i, B_j) = \min(\sqrt{(x_{j,m} - x_{i,n})^2 + (y_{j,m} - y_{i,n})^2})$$

$$\text{with indices } m = 1..m_{max}; n = 1..n_{max}, \tag{16.12}$$

where x and y are the corner coordinates for the shortest distance between two buildings B_i and B_j. An additional algorithm determines the closest corners to use for this purpose. The pairwise comparison of all buildings forms the distance matrix D with a size n^2 for n buildings; for the case of 306 buildings, this is approximately 93'000. The application of the membership function $\mu_{\text{is close}}$, as shown in Figure 16.11, to D identifies all the pairs of buildings that are sufficiently close to form a microgrid.

After evaluating the data using the three membership functions, we used a fuzzy logic function to identify candidates for thermal microgrids. This function identifies those building pairs that have high emissions or an old heating system and that are also sufficiently close to form a microgrid. For economic reasons, we use an "or" conjunction between $\mu_{\text{much emissions}}$ and $\mu_{\text{old heating system}}$, as it also makes sense to connect a building with an old heating system even if it does not produce high emissions. The fuzzy logic function used to attain this is:

$$\mu_{\text{micro grid affine}} = (\mu_{\text{much emissions}} \vee \mu_{\text{old heating system}}) \wedge \mu_{\text{is close}} \tag{16.13}$$

The resulting membership function $\mu_{\text{micro grid affine}}$ delivers a microgrid matrix M with building pairs that could be used to configure a microgrid. A further algorithm identifies clusters, which are the proposed microgrids, from this pair matrix by agglomerating the connected buildings in the matrix. This method was applied to the dataset of the example. The results in Figure 16.12 show very compact clusters. However, due to the strict criteria defined by the membership functions, considerably fewer clusters are available relative to HAC. These clusters include only buildings that match all the criteria. For further details, see Schlueter *et al.* [11].

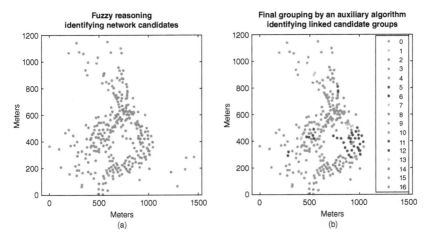

Figure 16.12 Groups of buildings identified using a spatialized fuzzy reasoning approach.

16.7 Mixed Fuzzy Reasoning and Clustering

Another means of performing non-spatial and spatial grouping involves combining both methods. For non-spatial filtering, when the criteria are defined, fuzzy reasoning is an appropriate method. The following membership function serves this purpose:

$$\mu_{\text{micro grid candidates}} = \mu_{\text{much emissions}} \vee \mu_{\text{old heating system}} \tag{16.14}$$

The resulting candidates are shown in Figure 16.13a; these data exclude any geographical information and thus deviate slightly from the results shown in Figure 16.12a. In a next step, the HAC algorithm is applied only to these candidates, together with filter to exclude those clusters with only one building. Figure 16.13b shows the 21 clusters identified in this way. The resulting groups are very similar to those of pure fuzzy reasoning described in the previous section. Slight deviations occur because of the different spatial evaluations. While fuzzy reasoning adopts a fixed cutoff criterion for a distance that has been defined in advance, HAC scans different distances during the clustering procedure.

16.8 Postprocessing: Interpretation and Strategy Identification

16.8.1 Data Visualization

The results of the clustering and reasoning process must be visualized before they can be interpreted by domain experts. In fact, making sense of the results

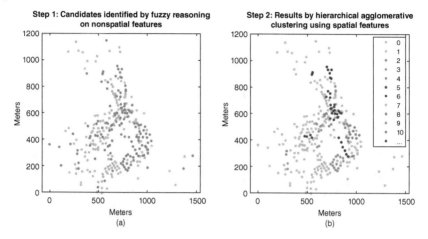

Figure 16.13 Mixed identification procedure using fuzzy reasoning to filter the building stock, followed by hierarchical agglomerative clustering for spatial group identification. (a) Building candidates identified by fuzzy reasoning. (b) Network candidates identified by hierarchical agglomerative clustering.

is crucial, as it represents the human factor in an otherwise unsupervised, automated process [41]. Clustering results are commonly visualized using histograms, parallel coordinate plots, and scatterplots. On a scatterplot, each result candidate is plotted onto a graph with up to five dimensions, using the x-, y-, and z-axes as well as dot shapes and colors. The distribution and spatial vicinity of the candidates is used to determine clusters, outliers, and trends [42]. Additionally, especially when using interactive graphs, filtering, zoning, and zooming can be used to select specific groups of candidates across multiple dimensions.

Due to the limitations of visually displaying more than five dimensions, a scatterplot matrix is used to visualize higher dimensions and each dimension versus every other dimension is visualized [42]. A scatterplot matrix of n variables therefore has n^2 panels, each one containing a scatterplot of two variables [43].

Figure 16.5, for example, uses a scatterplot matrix of nine variables to visualize the results of clustering for the optimal building retrofitting of the Zernez community. Scatterplot matrices, however, tend to become overwhelming when a large number of variables are displayed. This can be mitigated by choosing specific arrangements, such as selecting the most relevant or placing the most relevant in a prominent way, for example, organizing the data so that the most interesting variables appear close to the diagonal [43].

As the building stock is inherently spatial, clustering information can also be visualized using geospatial mapping techniques. By coloring building shapes according to their cluster affiliation in, for example, a GIS environment, additional information on the spatial characteristics of cluster members and the cluster boundaries of clusters otherwise not be apparent in the numerical data can be provided. Another mapping technique for displaying georeferenced data is the use of heat maps. A heat map uses color coding to display the result values for a selected dimension (Figure 16.14). This can be combined with additional geometric features such as scaling of the z-axis to display a second dimension. For example, Fonseca and Schlueter [44], used heat maps to visualize the clustering results of the energy simulation data for a neighborhood. Recently, powerful visualization engines such as CartoDB [45] have been made available to effectively visualize geo-spatiotemporal data in many different ways.

16.8.2 Data Mining Postprocessing

After clusters or groups have been identified, they can be used for further analysis to verify the results and further enhance knowledge discovery. Cluster postprocessing may entail modifying the results according to prior knowledge, manual corrections, filtering, cluster merging, or splitting [46]. Further in-depth analysis may also be executed on the cluster either as a whole or among the cluster candidates. As an example of postprocessing performed on the entire cluster, we use the identified clusters to develop an optimal timeline of retrofits as part of a zero emissions strategy a community that is outlined in the next subsection [24]. Having identified clusters of candidates that respond similarly to a set of specific measures, high-impact clusters can be processed faster to have a higher effect in less time. Furthermore, as the optimal measures for the candidates are similar, retrofits can be executed for multiple buildings, thus saving cost and effort. It should be noted, however, that these results contain uncertainties due to the nature of the data collection and assumptions in the process, as mentioned above. To develop an individual retrofit design, these results and the underlying parameters must be verified and possibly even adopted by an expert.

For an example of postprocessing among the cluster candidates, we evaluated the cost efficiency of thermal microgrids among candidates identified by fuzzy logic [11]. The identified candidates that currently use fossil fuels for heat generation systems, which have reached the end of life, and that are within a certain distance from each other, qualify to be connected to a local thermal microgrid. Depending on the characteristics of the candidates, such a microgrid is more cost efficient than individual solutions. To assess the cost efficiency, the generation and distribution losses and costs are estimated for the microgrids, which are then ranked according to their cost-efficiency.

Figure 16.14 Heat maps for the case study: (a) heating energy consumption, (b) association with different clusters of response to retrofit measures.

16.8.3 Transformation Strategy Development

The result of data mining, as performed before, is the creation of groups of buildings either for the application of similar EEMs or as candidates for forming networks; in postprocessing, cost-efficient measure combinations and network configurations are developed. These results provide information for transformation strategies, overarching policy making, and financing programs. Proposed measures for individual buildings that belong to the identified groups however must be validated by experts using detailed performance analysis. In this section, we transform the recommended measures to time-based transformation strategies that describe the level of emissions reduction that can be attained at a given point in time with a given yearly budget.

16.8.4 Strategy Development

To achieve a quick transformation, a smart strategy exploits the derived data by first addressing those groups exhibiting the best cost efficiency. Example data from the Zernez case [10] in Table 16.1 show such a sequence of clusters starting with the most effective types. Such a sequence forms the basis for a transformation strategy. Given the number and sizes of the buildings, the total volume and cost of retrofits per cluster can be determined (Table 16.1, rightmost column). Further assigning an annual budget for retrofitting allows us to determine how the annual CO_2 emissions decrease, as shown in Figure 16.15. The comparison of the "clustering" strategy with the "business as usual" and "doubled investments" strategies demonstrates that strategic transformation based on clustering is able to achieve a far better effect, leading to double the emissions reduction (80%) relative to spending more investments on retrofitting in the "doubled investments" scenario (40%). This benefit relies on the fact that the data mining approach makes it possible to start with those measures offering the highest cost efficiency. This involves starting from the low hanging fruits in order to achieve emission reductions quickly, turning investments into effect most effectively.

16.8.5 Policy Development and Retrofit Programs

These transformation strategies can be translated into policies. A statistical analysis of the parameters within the groups shows typical parameter patterns. For instance, Clusters 2.1 and 2.2 in the Zernez data mainly use oil heating systems and have well-insulated façades, which make them appropriate for the very cost-effective use of heat pumps driven by local low-emission hydropower. With this information, in addition to approaching individual building owners,

Table 16.1 Developed strategies with average cost efficiency per cluster.

Cluster	Strategy developed for the cluster	Cost efficiency in g CO_2/CHF	Total cost in CHF
2.1	EEM S21: Exchange of the heating system by district heating if available, else by a heat pump	258	3.0 Mio.
2.2	EEM S21: Exchange of the heating system by district heating if available, else by a heat pump	205	1.4 Mio.
2.5	EEM E2+S21: Exchange of the heating system by district heating if available, else by a heat pump and insulation of the building envelope	77	4.8 Mio.
2.3	EEM E2+S21: Exchange of the heating system by district heating if available, else by a heat pump and insulation of the building envelope	40	6.4 Mio.
2.4	None	0	0

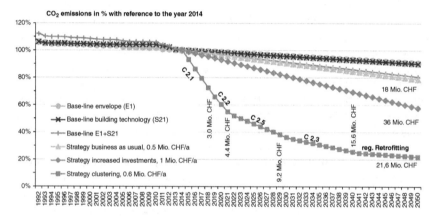

Figure 16.15 Transformation strategies for the Zernez case study.

it is possible to set up very effective policies and implement them through retrofit programs.

16.9 Comparison and Discussion of Methods

Algorithmic clustering and fuzzy reasoning are used for a similar purpose: to identify buildings that react similarly to EEMs and are located close together. However, the potential and performance of these data mining methods are different, which is discussed in this section.

16.9.1 Information

First, the information to be defined prior to the application of the methods differs:

- Algorithmic clustering does not require criteria definition before the data mining procedure. The only required information is the number of clusters in the case of partitioning clustering. All the criteria, rules, and strategy development follows clustering by analyzing the characteristics of the instances in the clusters.
- In contrast, fuzzy reasoning, requires the definition of criteria in the form of membership functions and optionally the definition of a reasoning logic governing how to combine the membership functions and the resulting fuzzy sets.

This has a significant effect on the characteristics of the performed search. While algorithmic clustering performs a wide, unconstrained search that demands subsequent interpretation, fuzzy reasoning requires only some interpretation to define the criteria beforehand but then exactly searches and groups the building stock according to those criteria. Therefore, clustering better suits heuristic discovery, while fuzzy reasoning supports the matching of exact criteria, for example, for applying a specific technology.

16.9.2 Scaling

Furthermore, if applied to large building stocks, as in the case of a smart city, scaling is of high importance. The methods behave differently in terms of scaling:

- Algorithmic clustering scales, as mentioned above, either linear (partitioning clustering) or quadratic (hierarchical clustering) depending on the number of buildings. Both methods can be used in nonspatial and spatial contexts.
- Fuzzy reasoning, as described above, scales linearly with the number of buildings in a nonspatial context. The spatial context is expected to exhibit a scaling behavior close to that of a quadratic scaling due to the pairwise comparison in the distance matrix.

The scaling behavior exhibits an advantage over methods that use prior information, such as partitioning clustering, as well as nonspatial fuzzy reasoning, which suggests that prior information reduces the computational load of grouping. However, at the same time, the search area is limited by prior information and its restricting effect.

In terms of spatialization, partitioning clustering applied to geographic building data appears to offer the best scaling behavior. However, this combination requires the assumption of a number of clusters beforehand, which usually leads to a trial-and-error approach. In contrast, the definition of closeness by a fuzzy function and the use of HAC do not require such assumptions. As shown in Figures 16.12 and 16.13, spatial fuzzy reasoning and HAC lead to similar results. The main difference is that the fuzzy approach allows the direct definition of a distance criterion (membership function "is close"), whereas HAC requires indirect control by deriving the number of clusters from the dendrogram. Therefore, for situations where the ideal distance is known, such as in the case of thermal microgrids, we recommend the application of spatial fuzzy reasoning.

Fuzzy reasoning has a further advantage in that fuzzification is very intuitive and automatically includes scaling and normalization by its membership functions. It is close to using natural language formulation and translates criteria directly for evaluation. Criteria can easily be combined by logic formulations. Moreover, the method does not require extensive postprocessing in terms of analyzing the groups for certain characteristics as the method already defines the characteristics beforehand.

16.9.3 Robustness

Furthermore, data quality has a different impact on the results obtained with the two methods. A dataset for a building stock usually includes outliers and missing or faulty data. The methods react differently to such faults:

- Algorithmic clustering is quite sensitive to data faults. Outliers and missing data replaced by default values often lead to sparsely populated pseudo clusters that later need to be removed manually in postprocessing. Compression overcomes this issue to some extent.
- Fuzzy logic is more robust in terms of such faults. Outliers do not lead to pseudo groups as predefined membership functions have the same effect as fixed boundaries and thus fixed clusters. This avoids a frequent problem encountered in automated grouping.

To conclude this comparison, we recommend different applications of data mining in building stock management:

- **A Broad Scan of Strategic Groups with no Prior Assumptions:** For this task, HAC is the best option. The cluster size and distance in dendrograms quickly provide information about the separation of the clusters to determine a good number of clusters. The poor scaling behavior of the method is the only drawback. Spatial properties can be included as further dimensions in the feature space to make this process more of a press-one-button approach if the feature space is defined. A two-step approach provides more details about nonspatial and spatial characteristics of the clusters.
- **Criteria for Clustering are Known:** In this case, fuzzy logic should be used due to its ability to clearly set up and combine criteria. A typical situation is a technology or a retrofit measure to be examined for its potential.

Spatialization is possible using the described approach of a distance matrix and a membership function describing closeness.

- **Application to Large Building Stocks:** In this case, the better scaling behavior of partitioning clustering should be exploited. Some trials of this clustering method should be performed. The number of clusters needs to be estimated from the characteristics of the feature space, and the results need to be checked for appropriate cluster separation. Spatialization for partitioning clustering is possible but difficult; in most spatial contexts, it is hard to estimate the number of clusters. If estimating the number of clusters and partitioning clustering are not successful, then we recommend to proceed to the fuzzy reasoning approach and estimating the membership functions.

Finally, extension and combination of the described methods with other related approaches is possible. For instance, uncertainty and continuous evaluation of the performance play an important role in developing building stock retrofit strategies as they concern long-term periods. Approaches considering uncertainty, like those proposed by Hopfe [48], Heo *et al.* [49], and Manfren *et al.* [50], can easily be integrated with fuzzy reasoning. Membership functions reflect such uncertainty models very well. To reduce the degree of uncertainty, recent approaches, such as those by Hong *et al.* [51] and Heo *et al.* [49], use monitoring data to calibrate the models. Such calibration can be integrated into the fuzzy logic method and the clustering method. Furthermore, changes in the external conditions, such as the economic climate, might require frequent evaluation, as in the case of the NPV-based multistep approach proposed by Kumbaroğlu and Madlener (2012). The fast automated evaluation of both, algorithmic clustering and fuzzy reasoning support such multistep evaluations.

16.10   Conclusion

Data mining methods, such as clustering and fuzzy logic, have the potential to support the planning and strategic development of building stock retrofitting to address the current demand for energy-efficient and sustainable buildings. Although they use different mechanisms and perform slightly differently, both methods can serve the same purpose: structuring large sets of building stock data for energy retrofitting. This allows to focus investments on those building groups and strategies that have the optimum effect for a given investment budget.

Upcoming smart cities and regions will generate large volumes of available data. This will place more importance on data mining and complementary methods of intelligent computation. The exploitation of data-intensive smart city technologies requires intelligent computer support to gain strategic information and support the intelligent transformation of building stock. The presented approach is expected to show a significantly better adaptation of the proposed measures and strategies compared to conventional classification approaches, such as the type–age classification. While the type–age approach relies on many uncertain assumptions in terms of similarity, typical utilization, construction, and performance of the buildings in a given group, data mining provides a means of exploiting the available data, to identify similarities, and to define well-performing strategies. Therefore, we recommend the use and further development of such methods, given data are available, to improve the efficiency and impact of building retrofit. We expect that the new methods will be of increasing importance for planning and strategic development of an energy-efficient and sustainable built environment.

Final Thoughts

In this chapter, we have discussed the application of hierarchical agglomerative clustering and fuzzy reasoning as data mining methods for the building stock management and strategic planning. We adopted both methods to the development of retrofit strategies and to the characteristics of the available data. This includes the necessary steps of pre- and postprocessing. Furthermore, we have given recommendations for the application of the adopted methods to different situations in building stock management. Finally, we would like to emphasize that traditional sparse information approaches, such as type–age classification, do not fully exploit the potential of smart cities and their huge amount of available data and that this asset requires the development and application of intelligent computation methods to allow us to fully benefit from the available data.

Questions

1 Why are data mining methods necessary for planning in the context of smart cities?

2 What is the methodological difference between hierarchical agglomerative clustering and fuzzy reasoning, and what effect does this difference have?

3 What information is required before applying the methods?

4 Which pre- and postprocessing steps are required when using data mining?

5 What criteria are required for selecting hierarchical agglomerative clustering or fuzzy reasoning?

6 How is a clustering result transformed into a transformation strategy?

Acknowledgments

We thank our collaborators in the "Zernez Energia 2020" research project for the exchange of data related to cost, energy conversion, and building energy calculations. This research was partly funded by the Swiss Commission for Technology and Innovation (CTI).

References

1 Roberts, S. (2008) Altering existing buildings in the UK. *Energy Policy*, **36** (12), 4482–4486. doi: 10.1016/j.enpol.2008.09.023

2 IWU (Institut Wohnen und Umwelt) (2012) *Typology Approach for Building Stock Energy Assessment*. Main Results of the TABULA project, www .building-typology.eu (accessed Dec 2014).

3 Geertman, S., Jr, J.F., Goodspeed, R., and Stillwell, J. (2015) *Planning Support Systems and Smart Cities*, Lecture Notes in Geoinformation.

4 Robinson, D., Haldi, F., Kämpf, J., Leroux, P., Perez, D., Rasheed, A., and Wilke, U. (2009) CitySim. Comprehensive Micro-Simulation of Resource Flows for Sustainable Urban Planning. In Wurtz, E. (Ed.). Building Simulation 2009. Proceedings of BS2009, 11th Conference of IBPSA, International Building Performance Simulation Association, Scotland, 1083–1090.

5 Reinhart, C., Dogan, T., Jakubiec, J.A., Rakha, T., Sang, A. (2013) Umi–an urban simulation environment for building energy use, daylighting and walkability. In Wurtz, E. (Ed.). Building simulation 2013. Proceedings of BS2013, 13th conference of IBPSA (International Building Performance Association), France, 476–483.

6 Quan, S.J., Li, Q., Augenbroe, G., and Brown, J. (2015) Urban Data and Building Energy Modeling. A GIS-Based Urban Building Energy Modeling System Using the Urban-EPC Engine, in *Planning Support Systems and Smart Cities*, Springer International Publishing, pp. 447–469. doi: 10.1007/978-3-319-18368-8_24

7 Santamouris, M., Mihalakakou, G., Patargias, P., and Gaitani, N. (2007) Using intelligent clustering techniques to classify the energy performance of school buildings. *Energy and Buildings*, **39** (1), 45–51. doi: 10.1016/J.enbuild.2006.04.018

8 Gaitani, N., Lehmann, C., Santamouris, M., and Mihalakakou, G. (2010) Using principal component and cluster analysis in the heating evaluation of the school building sector. *Applied Energy*, **87** (6), 2079–2086. doi: 10.1016/j.apenergy.2009.12.007

9 Jones, P., Lannon, S., Williams, J. (2001) *Modelling building energy use at urban*. Proceedings of the 7th International IBPSA conference, 175–180.

10 Geyer, P., Schlüter, A., and Cisar, S. (2016) Application of clustering for the development of retrofit strategies for large building stocks. *Advanced Engineering Informatics*. doi: 10.1016/j.aei.2016.02.001

11 Kohler, N., Hassler, U., and Paschen, H. (1999) *Stoffströme und Kosten in den Bereichen Bauen und Wohnen – Konzept Nachhaltigkeit*, Springer, Berlin Heidelberg. doi: 10.1007/978-3-642-58503-6_1

12 Eastman, C. (2011) *BIM handbook – a guide to building information modeling for owners, managers, designers, engineers, and contractors*, Wiley, Hoboken, NJ.

13 Borrmann, A., König, M., Koch, C., and Beetz, J. (2015) *Building Information Modeling – Technologische Grundlagen und industrielle Praxis*, Springer.

14 buildingSMART (2013) *Industry Foundation Classes IFC2x4*, http://www .buildingsmart-tech.org/specifications/ifc-releases/ifc4-release/ (accessed Apr 2013).

15 Kolbe, T.H. and Zlatanova, S. (2009) Representing and Exchanging 3D City Models with CityGML, in *3D Geo-Information Sciences* (eds J. Lee and S. Zlatanova), Springer, Berlin Heidelberg, pp. 15–31. doi: 10.1007/978-3-540-87395-2_2

16 The White House, Office of the Press Secretary (2016) *Fact Sheet. Cities, Utilities, and Businesses Commit to Unlocking Access to Energy Data for Building Owners and Improving Energy Efficiency*. Whitehouse.gov. https:// www.whitehouse.gov/the-press-office/2016/01/29/fact-sheet-cities-utilities-and-businesses-commit-unlocking-access. (accessed January 2016).

17 Bundesamt für Statistik (BFS) (2016) Eidg. Gebäude- und Wohnungsregister. https://www.housing-stat.ch/ (accessed February 2016).

18 SWISSTOPO (2016) *swissBUILDINGS3D 2.0*. http://www.swisstopo.admin .ch/internet/swisstopo/en/home/products/landscape/swissBUILDINGS3D_ V2.html (accessed February 2016).

19 Singapore Energy Market Authority (EMA) (2016) Energy Statistics. https:// www.ema.gov.sg/statistics.aspx (accessed February 19).

20 Urban Prototyping (UP) Singapore (2016) *E3 Hackathon. Urban Prototyping (UP) Singapore*. http://www.upsingapore.com/events/e3-energy-efficiency-everyone-hackathon/. (accessed February 2016).

21 Santamouris, M. (2010) *Energy Performance of Residential Buildings. A Practical Guide for Energy Rating and Efficiency*, Taylor & Francis.

22 U.S. Energy Information Administration (EIA) (2016) *Residential Energy Consumption Survey (RECS)*. http://www.eia.gov/consumption/residential/ reports/2009/methodology-end-use.cfm. (accessed February 2016).

23 Geyer, P., Schlueter, A., and Cisar, S. (2014) *A Performance-Based Clustering Model for Retrofit Management of Building Stocks*. In The 21st International Workshop on Intelligent Computing in Engineering, Cardiff, UK.

24 Papafragkou, A., Ghosh, S., James, P.A.B. *et al.* (2014) A Simple, Scalable and Low-Cost Method to Generate Thermal Diagnostics of a Domestic Building. *Applied Energy*, **134**, 519–530. doi: 10.1016/j.apenergy.2014.08.045

25 Frei, M., Nagy, Z., and Schlueter, A. (2015) Towards Data-Driven Building Retrofit In Proceedings of CISBAT 2015. Lausanne, Switzerland.

26 Arduino: *What Is an Arduino? - Learn.sparkfun.com.* 2016. https://learn.sparkfun.com/tutorials/what-is-an-arduino, (accessed Apr 2016).

27 Raspberry Pi (2016) *Teach, Learn, and Make with Raspberry Pi.* https://www.raspberrypi.org/ (accessed Apr 2016)

28 Rabl, A. (1988) Parameter Estimation in Buildings. Methods for Dynamic Analysis of Measured Energy Use. *Journal of Solar Energy Engineering*, **110** (1), 52–66. doi: 10.1115/1.3268237

29 Sonderegger, R. (1978) *Dynamic Models of House Heating Based on Equivalent Thermal Parameters*, Princeton Univ, NJ.

30 Sonderegger, R. (2010) *Diagnostic Tests Determining the Thermal Response of a House*, Lawrence Berkeley National Laboratoryhttp://escholarship.org/uc/item/4x96z8c5 (accessed November 2015).

31 Subbarao, K. and Anderson, J.V. (1983) A Graphical Method for Passive Building Energy Analysis. *Journal of Solar Energy Engineering*, **105** (2), 134–41. doi: 10.1115/1.3266356

32 Subbarao, K. and Flowers, L. (1985) *Balanced Measurement/Calculation-Based Approach to Building Energy Analysis*. SERI/TP-253-2381; CONF-840819-15, Solar Energy Research Inst, Golden, CO (USA) http://www.osti.gov/scitech/biblio/6179169.

33 Rabl, A. and Rialhe, A. (1992) Energy Signature Models for Commercial Buildings. Test with Measured Data and Interpretation. *Energy and Buildings*, **19** (2), 143–54. doi: 10.1016/0378-7788(92)90008-5

34 Nagy, Z., Rossi, D., Hersberger, C. *et al.* (2014) Balancing Envelope and Heating System Parameters for Zero Emissions Retrofit Using Building Sensor Data. *Applied Energy*, **131** (October), 56–66. doi: 10.1016/j.apenergy.2014.06.024

35 Xu, R. and Wunsch, D.C. (2009) *Clustering – IEEE series on computational intelligence*, Wiley, Oxford.

36 Abonyi, J. and Feil, B. (2007) Classical fuzzy cluster analysis, in *Cluster Analysis for Data Mining and System Identification* (eds J. Abonyi and B. Feil), Birkhäuser, Basel, pp. 1–45. doi: 10.1007/978-3-7643-7988-9_1

37 Hansen, P. and Jaumard, B. (1997) Cluster analysis and mathematical programming. *Mathematical Programming*, **79** (1), 191–215. doi: 10.1007/BF02614317

38 Xu, R. and Wunsch, D.I. (2005) Survey of clustering algorithms, Neural Networks. *IEEE Transactions on*, **16** (3), 645–678.

39 Harris, J. (2006) *Fuzzy logic applications in engineering science*, Springer, Dordrecht.

40 Chen, K. and Ling, L. (2004) VISTA. Validating and Refining Clusters Via Visualization. *Information Visualization*, **3** (4), 257–70. doi: 10.1057/palgrave.ivs.9500076

41 Fayyad, U.M., Wierse, A., and Grinstein, G.G. (2002) *Information Visualization in Data Mining and Knowledge Discovery*, Morgan Kaufmann.

42 Hurley, C.B. (2004) Clustering Visualizations of Multidimensional Data. *Journal of Computational and Graphical Statistics*, **13** (4), 788–806. doi: 10.1198/106186004X12425

43 Fonseca, J.A. and Schlueter, A. (2015) Integrated Model for Characterization of Spatiotemporal Building Energy Consumption Patterns in Neighborhoods and City Districts. *Applied Energy*, **142**, 247–265.

44 *CartoDB* (2016) *CartoDB*. https://cartodb.com/ (accessed February 2016).

45 Rubel, O., Weber, G.H., Huang, M.Y. *et al.* (2010) Integrating Data Clustering and Visualization for the Analysis of 3D Gene Expression Data. *IEEE/ACM Transactions on Computational Biology and Bioinformatics*, **7** (1), 64–79. doi: 10.1109/TCBB.2008.49

46 Geyer, P. and Schlüter, A. (2014) Automated metamodel generation for Design Space Exploration and decision-making – A novel method supporting performance-oriented building design and retrofitting. *Applied Energy*, **119** (0), 537–556. doi: 10.1016/j.apenergy.2013.12.064

47 Hopfe, CJ. (2009) Uncertainty and sensitivity analysis in building performance simulation for decision support and design optimization, PhD diss., Eindhoven University.

48 Heo, Y., Augenbroe, G., and Choudhary, R. (2013) Quantitative risk management for energy retrofit projects. *Journal of Building Performance Simulation*, **6** (4), pp. 257–268, http://dx.doi.org/10.1080/19401493.2012.706388.

49 Manfren, M., Aste, N., and Moshksar, R. (2013) Calibration and uncertainty analysis for computer models – A meta-model based approach for integrated building energy simulation. *Applied Energy*, **103** (0), pp. 627–641, http://dx.doi.org/10.1016/j.apenergy.2012.10.031.

50 Hong, T., Yang, L., Hill, D., and Feng, W. (2014) Data and analytics to inform energy retrofit of high performance buildings. *Applied Energy*, **126**, pp. 90–106, http://dx.doi.org/10.1016/j.apenergy.2014.03.052.

51 Kumbaroğlu, G. and Madlener, R. (2012) Evaluation of economically optimal retrofit investment options for energy savings in buildings. *Energy and Buildings*, **49**, pp. 327–334, http://dx.doi.org/10.1016/j.enbuild.2012.02.022.

17

A Framework to Achieve Large Scale Energy Savings for Building Stocks through Targeted Occupancy Interventions[1]

Aslihan Karatas[1], Allisandra Stoiko[2], and Carol C. Menassa[1]

[1] Department of Civil and Environmental Engineering, University of Michigan, Ann Arbor, MI, USA
[2] Department of Chemical Engineering, University of Michigan, Ann Arbor, MI, USA

CHAPTER MENU
Introduction, 474
Objectives, 475
Review of Occupancy-Focused Energy Efficiency Interventions, 476
Role of Occupants' Characteristics in Building Energy Use, 481
A Conceptual Framework for Delivering Targeted Occupancy-Focused Interventions, 483
Case Study Example, 490
Discussion, 493
Conclusions and Policy Implications, 494

Objectives

- To become familiar with occupants' energy use characteristics and their major impact on building energy use.
- To introduce a multilevel framework for occupancy energy use intervention strategies in buildings.
- To become familiar with the MOA methodology to collect occupancy energy use characteristics data.
- To introduce a link between occupants' energy use characteristics and energy policy tools.

---

[1] Contents of this chapter originally appeared in the following article: Karatas, A., Stoiko, A., and Menassa, C. (2016) A framework for selecting occupancy-focused energy interventions in buildings. Building Research and Information, Special Issue: Building governance and climate change: regulation and related policies, Taylor and Francis. 44(5–6), 535–551. This chapter is reprinted with permission of the publisher (Taylor & Francis Ltd, http://www.tandfonline.com).

Smart Cities: Foundations, Principles and Applications, First Edition.
Edited by Houbing Song, Ravi Srinivasan, Tamim Sookoor, and Sabina Jeschke.
© 2017 John Wiley & Sons, Inc. Published 2017 by John Wiley & Sons, Inc.

17.1 Introduction

In the United States, the existing building sector accounts for 40% of the total energy consumption by the built environment [1]. UNEP [2] reported that the existing building sector represents an excellent opportunity to achieve large-scale energy use reductions in a cost-effective manner through efficiency and conservation strategies. The objective is to alleviate economic, environmental, and social problems associated with diminishing natural resources and global warming. To maintain a large-scale energy use reduction in the existing building sector, researchers have emphasized the need for decision-makers (e.g., policy-makers) to carefully analyze the effect of individual occupant behaviors and their determinants on the energy consumption of buildings to ensure that they can design effective energy policy tools [3, 4].

Energy policy tools aimed to achieve large-scale energy savings for a stock of buildings (e.g., university campus, community, city) should integrate occupancy-focused interventions that implement knowledge-based solutions, persuasion, reward/penalty systems, and technological solutions [5, 6]. To achieve this, energy policy tools can be designed based on one of two basic ways that intend to (i) directly affect individual behaviors through inducement tools and regulation tools (i.e., incentives and/or sanctions) and (ii) affect the environment in which individual behaviors are manifested through knowledge tools that indirectly influence behavior (i.e., knowledge based and/or persuasion) [7]. Regulatory tools, often referred to as government command and control systems, are intended to change behavior by forcing people to obey the law or other regulations without providing a promise of a positive incentive. Inducement tools aim to motivate an individual through the promise of a reward or penalty to behave in a certain way without the level of government coercion inherent in regulations. On the other hand, knowledge tools are intended to change individual behavior based on the provision of information. These methods strive to achieve a desired outcome through continuous monitoring of how individual or group behaviors change [7].

Despite the fact that policy analysts often suggest knowledge tools as a starting point for policy intervention because of their noncoercive nature [7], energy policy tools are mostly designed as either regulatory tools (e.g., occupancy-focused interventions that implement technological solutions such as HVAC scheduling and resets according to outside conditions) [8–11] or inducement tools (e.g., occupancy-focused interventions that implement reward/penalty systems such as varying energy costs with on- and off-peak consumption) [10, 12, 13].

The aforementioned approach for designing energy policy tools often leads to three main challenges for decision-makers (e.g., policy makers) when applied in the context of energy reduction in the built environment. First, complying energy policy tools with the regulations and incentives mostly resulted in inefficiencies due to higher costs of implementation of appliance standards, building energy codes, financial incentives, and public sector energy leadership programs [5, 7, 14, 15].

Second, energy policy makers often ignore the significant impact of occupants on the energy use in buildings by assuming that all occupants have the standard behavior pattern (e.g., a fixed set-point room temperature preference) [16–19]. Azar and Menassa [17] carried out a comprehensive analysis using energy simulation to study the impact of nine occupancy-related actions (e.g., after-hours equipment and light use) on energy use in commercial buildings of different sizes and located in 10 different weather climates across the United States. The results from these individual impacts were found to be as high as 30%. The combined effects of some of the occupancy-related actions resulted in an increase in energy use in the building in excess of 50%. Other studies such as Moezzi et al. [20], Sanchez et al. [21], and Webber et al. [22] support these results and emphasize that significant energy reductions can only be achieved if building occupants are engaged in the process.

Third, Hand [7] highlighted that lack of information or capacity is mostly the primary barrier, which can be overcome by knowledge tools that relay the appropriate information to the occupants. The main difference in this case is that it is unnecessary to incentivize or sanction the target in order to elicit the desired behavior. Therefore, when designing policy measures aimed at reducing energy use in a large stock of buildings, it is important to identify the diverse energy use characteristics of occupants that significantly contribute to environmental problems and the factors (e.g., occupancy-focused interventions) that make sustainable behavioral patterns attractive [4].

To address and overcome the aforementioned challenges in the development of large-scale energy policy tools, this study focuses on the development of a conceptual framework that proposes multilevel building energy use intervention strategies. This is achieved by examining the fundamental differences among occupancy-focused intervention strategies and identifying which interventions are more effective based on the occupants' characteristics and energy use profiles. To accomplish this, the framework presented in this study adopted MOA approach from the consumer and social marketing field to establish an analogy that enables occupancy-focused intervention strategies, in this case considered as advertisements, to encourage the building occupants to adopt the desired energy use characteristics. This framework assists decision-makers to propose cost-effective large-scale energy policy tools to deliver energy efficiency occupancy-focused interventions and to evaluate the benefits and effectiveness of different energy policy tools before the actual implementation on a large stock of buildings.

17.2 Objectives

The aim of this chapter is to address the identified research gap in the development of energy policy tools (i.e., their higher costs for building occupants and managers and ignoring the impact of occupants' impact on buildings energy consumption) and present a conceptual framework for assisting policy makers in designing effective energy policy tools. This framework proposes a multi-level building energy use intervention strategy that is capable of systematically evaluating the effects of occupancy-focused interventions on the occupants' behavior and also designing energy policy tools tailored to various behavioral and energy use characteristics of occupants. To accomplish this, the proposed framework is developed to answer the following questions:

1) How can occupants' energy use profile be measured and classified before and after the implementation of energy policy tools? The answer to this question helps determine the energy use profiles (i.e., occupants' situational characteristics pre- and postexposure to any intervention) to evaluate the benefits and effectiveness of different energy policy tools before actual implementation on a large stock of buildings (e.g., in a residential community, city, or campus environment).

2) What type of building energy use intervention strategies can be selected by decision-makers to achieve the required energy reductions at lower costs based on the occupants' energy use characteristics? This supports decision-makers in formulating effective energy policy tools to achieve cost-effective occupancy-focused intervention programs that increase the attractiveness of pro-environmental behavior and encourage a more sustainable behavior pattern.

3) How can the proposed framework be implemented in real-life examples to achieve large-scale energy reductions by delivering interventions to the occupants through the most effective energy policy tools? The results of a case study of a real building demonstrate the capabilities of the proposed framework in evaluating the intervention strategies and designing effective energy policy tools based on the occupants' energy use characteristics.

17.3   Review of Occupancy-Focused Energy Efficiency Interventions

To engage occupants in reducing energy use in a stock of buildings (e.g., university campus, community, city), energy policy tools should integrate occupancy-focused interventions that implement knowledge-based solutions, persuasion, reward/penalty systems, and technological solutions [5, 6]. Existing occupancy-focused interventions are investigated based on a comprehensive literature review related to knowledge translation theory [23–29], social marketing theory [30–32], and law and public policy [33, 34]. Accordingly, a detailed description of each intervention is provided below.

17.3.1 Knowledge-Based Interventions

Knowledge-based interventions involve the presentation of informative messages to invoke voluntary behavior change. It attempts to raise awareness about the benefits of changing a behavior without presenting an explicit reward or penalty. Through outlets like posters, articles, videos, and brochures, researchers have attempted to convey information to influence consumers in contexts such as encouraging vegetable consumption, stair use, and sunscreen use [35–37]. Sahota et al. [37] found that students increased vegetable consumption by 0.3 servings per day after 1 year of healthy living education and class-wide discussions. Armstrong et al. [35] determined that sunscreen use increased to 1.6 days/week after adults viewed a video about the sun's effect and the importance of sunscreen and read a pamphlet with similar information. This method has generally experienced moderately positive results, with modest improvements in the desired behavior.

In energy use context, several studies used knowledge-based interventions to reduce energy use in buildings and were divided into four main categories. The first category used information distribution outlets (e.g., posters, videos, brochures) to influence occupants [38–42]. For example, Hayes and Cone [39] created a poster that described methods to reduce electricity consumption and gave out consumption statistics. Results showed that the distribution of energy consumption facts and reduction guidelines does not appear to effectively influence occupants. Zografakis et al. [43] used schoolwork and a project, including field visits and other hands-on approaches, to educate students on energy-efficient behavior. Both students and their parents consequently behaved in more energy-efficient ways, including turning off lights and closing the windows when the heater is on, but some benchmarks experienced no significant changes. While information distribution is a widely tested form of education, it has provided mixed results on its own.

The second category considered interactive programs with more personalized approaches to information delivery [44, 45]. McMakin et al. [45] created site-specific video programs to motivate military families to conserve energy. Combined with other awareness materials, the program resulted in a 3% reduction in gas use and 7% reduction in energy use after 1 year. However, Geller [44] found that motivated individuals attending a 3-h workshop on energy conservation did not tend to apply skills either immediately or 6 weeks later. This implies that the short-term delivery of information may not be effective, and an extended program may yield better results.

The third category considered feedback methodology that is based on comparing current energy use with historical use [46, 47]. External feedback provides consumers with personalized evaluation and a means to monitor progress. These studies generally concluded that even if feedback method is an effective knowledge-based method for building energy reduction, it may become less effective in changing behaviors over an extended period of time (e.g., 2 years).

Finally, the last category focused on peer comparison of monthly or quarterly energy use [48–50]. Allcott [48] determined that energy use decreased approximately 2% with these reports and began to regress between quarterly reports. Ayres et al. [51] discussed studies that experienced energy savings of 1.2–2.1%. One of the studies, however, showed no statistical difference between monthly or quarterly reports. Ayres et al. [51] concluded that peer comparison is more effective in decreasing energy use than energy consumption information alone. Furthermore, participants' energy use dropped 16% after the first peer comparison period and 32% after the second peer comparison, both relative to baseline levels. These studies generally concluded that peer comparison is a successful means to reduce energy use.

17.3.2 Persuasion Interventions

Persuasion involves providing rewards to encourage a favorable behavior. Researchers studied persuasion interventions to incite voluntary changes through incentives like money and fast food [52, 53]. Flora and Flora [54] studied students who participated in the "Book It!" program for free pizza or had financial incentives from parents if the students read. They determined that these forms of persuasion increased the amount of reading and may have helped increase enjoyment of reading and literacy. John et al. [53] provided financial incentives for weight loss in obese veterans and determined that incentives provided greater weight loss. However, follow-ups after a maintenance period revealed no significant net weight loss between financial incentive and control groups. These studies show that persuasion appears to be an effective strategy to encourage people to adopt a desired behavior, but the behavior may be reduced if a persuasion method is later removed. In the context of energy use, several studies have been conducted to study the effect of incentives for reducing energy use. These are divided into two main categories.

The first category considered monetary incentives, both small and large, to decrease energy use [55–60], as well as their respective long-term effects. Katzev and Johnson [58] provided occupants with $3.00–$10.00 for reducing energy use during 2-week periods. They measured no significant difference in energy use between control groups during the study or follow-up. Winett et al. [61] provided households with variably high ($0.30 per 1% reduction) and low ($0.013 per 1% reduction) rewards for decreased kilowatt-hours of electricity compared to the average. Initially, both high- and low-reward groups saved energy, with the high-reward group decreasing 16% more than the low-reward group. After receiving the reward, however, there was a larger increase in energy use by the high-reward group than the low-reward group. This larger relapse indicates that high-reward groups may become less intrinsically motivated than low-reward groups. While monetary incentives appear to be an effective persuasion method to reduce energy use, low-to-medium monetary rewards could lead to more sustained motivation for the occupants.

The other category of studies considered pledging campaigns as incentives to encourage sustainable energy conservation behavior [52, 60, 62]. These studies found that pledging to an action, such as by signing a promise, resulted in higher commitment. Schick and Goodwin [63] reported that in a voluntary "pledge to save," participants decreased energy use by a factor of three when compared with non-pledgers. Boyce and Geller [62] observed that signing a pledge to give thank-you cards to others for pro-environmental behavior resulted in significantly increased activity. These promise signers gave 3.0 cards per week compared with 0.5 cards per week. This indicates that pledging is an effective persuasion method to encourage occupants to reduce energy use.

17.3.3   Penalty Interventions

Penalty interventions consist of negative consequences, often legal, that discourage an unfavorable behavior. It provides a set of rules that result in sanctions and penalties for noncompliance. For example, Farchi et al. [64] examined a point system used to discourage risky driving behavior. They studied Italy's highly publicized new driving point system, where drivers began with 20 points and lost points for various traffic violations, ultimately having their licenses revoked. Results were significant where the emergency department visits decreased by 12% over 3 years. Additionally, several studies have focused on the influence of penalties in the context of energy conservation.

For example, multiple studies considered dynamic building control by altering energy costs with on- and off-peak consumption [12, 13, 65]. These studies observed that increased price of on-peak electricity reduced the consumption of energy during that period. Heberlein and Warriner [65] determined that consumers with higher price ratios (8:1) used less on-peak electricity than those with lower price ratios (2:1). Therefore, it appears that increasing energy cost during on-peak periods may be very effective in reducing energy use particularly by increasing the difference in ratio between on- and off-peak costs.

17.3.4 Technology Interventions

Technology interventions consist of tools and systems that automatically solve problems without continued human influence. It is often dictated by laws and other regulations that require nonvoluntary changes for compliance. Research suggests that careful building design, including incorporation of building automation systems (BASs), can dramatically reduce energy use [8, 10, 11, 66–68]. Mathews et al. [68] determined that the most lucrative BAS strategies include heating and air conditioning scheduling according to outside conditions, air bypass control on cooling coils, and reset and setback control. By optimizing these systems, predicted annual energy savings were 66%, which would result in a 30% reduction in a building's total energy consumption. Ruzelli [10] reported actual total savings of 10–15% in buildings featuring a general BAS.

Existing buildings have many options to decrease energy use by retrofitting and replacing electrical components [69, 70]. Harvey [70] observed an average energy savings of 68% in buildings changing from noncondensing to condensing boilers. Furthermore, Environmental and Energy Study Institute [69] asserts that compact fluorescent light bulbs use 66% less energy, while modern energy-efficient appliances use up to 40% less energy than their non-energy-efficient equivalents.

17.3.5 Building Energy Use Interventions in Energy Policy Design

To achieve large-scale energy reductions in a stock of buildings, aforementioned interventions (i.e., knowledge based, persuasion, penalty, technology) should be delivered efficiently and cost-effectively through energy policy tools (i.e., knowledge, inducement, and regulatory tools). From a policy making perspective, relaying the appropriate information to the target is mostly the major barrier to obtain the desired behavior [7]. Therefore, knowledge-based interventions can be delivered through knowledge energy policy tools that enable or encourage voluntary behavior change of the occupants with minimum economic and environmental costs, as shown in Figure 17.1. Persuasion interventions can also be delivered through knowledge energy policy tools to change the occupants' energy use behavior based on the given information and in the desired manner. This will also provide voluntary behavior change for the occupants to produce and maintain energy use reduction in buildings over time.

If extreme energy use patterns are observed among the building occupants, lower-level interventions such as knowledge based and persuasion might not be sufficient to effectively reduce the energy use. Therefore, there is a need to supplement knowledge-based methods with interventions from higher levels of interventions (i.e., penalty, technology), which result in higher economic and environmental costs, as shown in Figure 17.1. Penalty interventions can be delivered through inducement energy policy tools, since they mainly aim

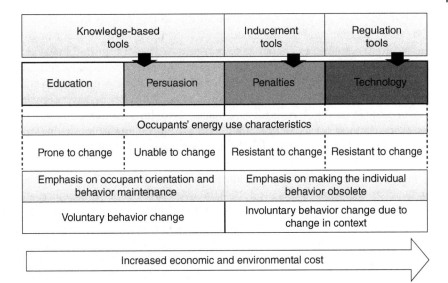

Knowledge-based tools		Inducement tools	Regulation tools
Education	Persuasion	Penalties	Technology

Occupants' energy use characteristics			
Prone to change	Unable to change	Resistant to change	Resistant to change
Emphasis on occupant orientation and behavior maintenance		Emphasis on making the individual behavior obsolete	
Voluntary behavior change		Involuntary behavior change due to change in context	

Increased economic and environmental cost

Figure 17.1 Building energy use intervention strategies.

to motivate individuals through the promise of a reward or penalty to change their behavior in the desired manner. Moreover, technology interventions can be delivered through regulatory energy policy tools, since they are intended to change individuals' behavior through legislative command and control systems [7] that render the individual behavior obsolete. In both cases, the focus is mostly on involuntary (forced) behavior change of the occupants, at increased economic costs.

17.4   Role of Occupants' Characteristics in Building Energy Use

Policies aimed that to engage occupants in energy reduction, strategies should simultaneously investigate possible determinants of their energy use characteristics (e.g., knowledge, attitudes) and the effectiveness of interventions [71, 72]. To identify the role of occupants' characteristics in building energy use and their impact on the interventions, the proposed framework in this study attempts to measure three main metrics of occupancy characteristics: motivation, opportunity, and ability. These measures are often used in consumer and social marketing field to determine product popularity and potential for sale and accordingly design marketing strategies. In this case, the analogy assumes intervention strategies as advertisements enticing the building occupants to adopt certain energy use characteristics.

Several studies from consumer and social marketing studied MOA model to identify the consumers' information processing level and accordingly main influential factors in encouraging consumers to adopt desired behavior [73–79]. Maclnnis et al. [76] proposed an MOA model to explain certain outcomes influenced by the extent of brand information processing from advertisements. Based on this model, consumers' MOA levels have a great impact on the level of processing brand information during advertisements or after exposure to ads. Hastak et al. [80] emphasized the important mediation role that MOA plays in determining the communication effectiveness of ads in consumer research.

Consumers loyal to a particular product usually have high MOA levels that help facilitate adoption of this product. Moreover, Bigné et al. [81] applied an MOA model to explore the key drivers of online airline ticket purchases intentions and also to identify which perceived channel benefits are more effective for consumers when using the Internet to purchase airline tickets. Results from this study highlighted that the MOA model is a useful methodology to predict consumers' ticket purchasing intentions that explains 55% of variations in adopting a desired behavior.

In consumer and social marketing field context, motivation is defined as a goal-directed arousal to engage consumers in the desired behavior to process brand information in the advertisement [72, 82–84], opportunity as executional factors (e.g., exposure time to ads) that are not in the control of consumers to enable desired actions [76, 79, 81, 85], and ability as consumers' perception of their capacity to access the brand information and interpret this information to create new knowledge structures [74, 76, 79, 81, 86].

In light of the aforementioned multidisciplinary approach combining social sciences and marketing, this study investigated an analogy between MOA characteristics of people to process brand information in their environment and MOA levels where occupancy-focused intervention strategies can be regarded as advertisements enticing the building occupants to adopt certain energy use characteristics. By using this analogy, multilevel intervention strategies and their related energy policy tools can be developed to reduce energy use in buildings based on identified MOA characteristics of building occupants. In energy policy designing context, the MOA level of occupants can be measured to identify the role of occupants' characteristics in building energy use and accordingly design the occupancy-focused intervention strategies and energy policy tools to be effective. A detailed explanation for the adopted methodology of occupants' energy use MOA characteristics for this presented framework is provided below:

The motivation (M) level of an occupant refers to a particular occupant's perceived personal relevance in terms of needs, goals, values, and the level of involvement with the information (e.g., external stimuli) presented in the energy intervention strategy. A strong link indicates a high M level and potential for the occupant to consider adopting the intervention strategy.

The opportunity (*O*) level of an occupant represents an important precondition for both motivation (*M*) and ability (*A*) and is directly related to the immediate environment of the occupants and how that affects the availability, accessibility, and time allocated for comprehension of the energy use knowledge.

The ability (*A*) level of an occupant measures a given occupant's proficiencies in interpreting energy use knowledge. This ability is largely dependent on the occupant's prior knowledge about energy use and conservation acquired through experience (e.g., asking occupants to turn light off before leaving their offices).

17.5   A Conceptual Framework for Delivering Targeted Occupancy-Focused Interventions

When designing energy policy measures to reduce building energy use, it is important to identify occupants' behaviors that significantly contribute to environmental problems and then identify the factors that make sustainable behavioral patterns attractive [4]. Therefore, the framework in this study proposes a multilevel intervention strategy targeted toward the diverse human characteristics to sustain energy use reduction in large building stocks over time. To achieve this, the presented framework for designing effective energy policy tools includes five main stages (see Figure 17.2): (i) measuring the occupants' preexposure MOA level; (ii) clustering occupants' preexposure MOA level and identifying their energy use profiles (i.e., occupants' situational characteristics prior to any intervention); (iii) measuring occupants' postintervention exposure MOA levels; (iv) clustering postexposure MOA levels and identifying energy use profiles (i.e., occupants' situational characteristics after any intervention) to determine the effectiveness of intervention strategies; and (v) identifying the link between occupants' MOA levels and the multilevel energy efficiency intervention strategies and accordingly energy policy tool types (i.e., knowledge tools, inducement tools, and regulation tools).

17.5.1   Measuring the Impact of Occupancy Characteristics on Building Energy Use

This stage of framework development focuses on measuring the occupants' energy use profile to evaluate the impact of occupancy characteristics on building energy use. Therefore, a set of metrics were identified based on MOA energy use characteristics of occupants and organized in Tables 17.1–17.3.

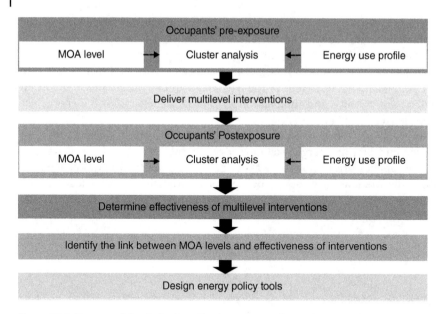

Figure 17.2 Framework for designing effective energy policy tools.

Table 17.1 Metrics for measuring occupancy motivation level for energy conservation.

Metrics of construct	Preexposure measures	Postexposure measures
Measures of motivation (M) Self-related knowledge (internal stimuli) Needs Goals Values Energy use knowledge (external stimuli) Level of energy use Impact and consequences	Assess *self-awareness* about the importance of energy use knowledge Measure *desire to receive* energy use knowledge Detect *norms of avoiding* energy use knowledge (e.g., not interested in attending workshops or receiving e-mails)	Assess *self-awareness* about the importance of energy use knowledge Measure *desire to process (continue to receive)* energy use knowledge Detect *norms of avoiding* energy use knowledge (e.g., still not interested in attending workshops or receiving e-mails)

The motivation (M) level of an occupant can be measured as the perceived link between an occupant's needs, goals, and values (self-related knowledge) and the level of involvement with the energy use information (e.g., level, impact, and consequences), as shown in Table 17.1. M is a key factor if lower levels of intervention (e.g., knowledge based) and accordingly knowledge energy tools

Table 17.2 Metrics for measuring occupancy opportunity level for energy conservation.

Metrics of construct	Preexposure measures	Postexposure measures
Measures of opportunity (*O*) Availability Amount of information Information format Modality Rate of exposure to information	*Determine availability:* Availability of energy conservation control system (e.g., indoor lighting control) Office condition level (e.g., physical characteristics of offices) *Determine number of times:* Attend awareness seminars Read information on general advertisement boards (self-reported) Read e-mails (ask for response with a blank e-mail) Discuss with peers (self-reported)	*Measure information recall by recording:* Number of arguments about impact of energy use on global environment Number of arguments about impact of energy use on building footprint Number of arguments related to the benefits of conservation methods

Table 17.3 Metrics for measuring occupancy ability level for energy conservation.

Metrics of construct	Preexposure measures	Postexposure measures
Measures of ability (*A*) Energy use prior knowledge Impact Consequences Conservation strategies	Measure *extent* conservation strategies are used (e.g., estimate number of times consciously turn lights off at end of day) Measure *subjective knowledge* of energy use relative to average person Perceived effectiveness of intervention strategy to reduce impacts/consequences Measure *actual knowledge* (i.e., factual information): Terminology Possible impacts/consequences Criteria to evaluate impacts/consequences	Measure *improvement in actual knowledge* (i.e., factual information): Terminology Possible impacts/ consequences Criteria for evaluating impacts/consequences Factors related to conservation strategies Perceived effectiveness of conservation strategies to reduce impacts/ consequences

are to be successful. High M level of an occupant can be identified by the occupants' desire to adjust room temperature (e.g., setting the office heating point to a lower temperature during unoccupied hours) and lighting system (e.g., turning off the office lights when not in use) and adopt energy conservation strategies (e.g., bringing an extra jacket instead of setting heating point to higher degrees). Lack of motivation implies that occupants will resist any change and will require higher levels of intervention (e.g., penalties) and energy policy tools (i.e., inducement and regulatory tools) to improve their M level and get involved in energy conservation.

The opportunity (O) level of occupants can be measured by factors such as the amount of information, the format of the information (e.g., organized by energy reduction target like heating set point and the impact of changing that on energy use or by measures to avoid changing the heating set point like wearing layers), the modality (e.g., information presented in workshops gives occupants short period of time to process vs. having the information available on bulletin boards affording occupants' continuous opportunities to attend to and comprehend), and how often the information available is (e.g., weekly e-mails) (see Table 17.2).

The ability (A) level of occupants can be measured by evaluating their prior knowledge about energy use, its impact, and consequences as well as knowledge about possible conservation strategies (see Table 17.3). The probability that occupants will process energy reduction information from intervention strategies is directly related to this preexisting knowledge. The A level of occupants can also be measured by perceived energy consumption knowledge level and actual level of knowledge on energy consumption facts. One set of studies has shown that people need to have sufficient ability (e.g., self-efficacy) before they can actively care enough to take environmentally responsible actions that benefit others [87].

The identified metrics presented in Tables 17.1–17.3 enable decision-makers to measure occupants' preexposure MOA level and energy use profiles (i.e., occupants' situational characteristics prior to any intervention) and postexposure MOA levels (i.e., occupants' situational characteristics after any intervention) to determine the effectiveness of intervention level (e.g., knowledge based, persuasion, or combination) to use for a given energy reduction strategy (e.g., encouraging occupants to turn office lights off when not in use).

17.5.2 Clustering Occupants' MOA Levels and Energy Use Profiles

In this stage of the framework development, occupants' preexisting MOA levels are classified into main target clusters relative to their potential to process energy use characteristics. Azar and Menassa [16] identified the different occupants' energy use characteristics as "high energy consumers" that represent occupants that overconsume energy, "medium energy consumers" that represent occupants making minimal efforts toward energy savings, and "low energy

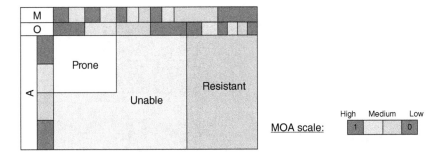

Figure 17.3 MOA levels of occupants.

consumers" that represent occupants that use energy efficiently. Based on these identified characteristics, this study classified the occupants' MOA levels into three categories (see Figure 17.3): "prone to change" (i.e., occupants who are willing to adopt energy reduction strategies immediately), "unable to change" (i.e., occupants who are willing to adopt energy reduction strategies immediately but do not have the necessary knowledge and tools to do that), and "resistant to change" (i.e., occupants who are unwilling to adopt energy reduction strategies regardless of whether they have the necessary knowledge and tools or not). As shown in Figure 17.3, each MOA level of an occupant is measured using the "MOA scale" that ranges from 1 (represented in green color as high motivation level) to 0 (represented in red color as low motivation level).

After identifying occupants' preexposure energy use characteristics, the effectiveness of intervention strategies is determined by measuring the occupants' postexposure MOA level. The actual energy use data before and after the application of the intervention to a group of occupants can be compared to develop effectiveness distributions for the four levels of intervention strategies relative to the overall preexisting MOA level (i.e., prone, unable, or resistant) in the building. Occupants in a certain energy use profile of the building stock are expected to adopt and implement the intervention strategy based on the effectiveness level of interventions. The effectiveness of each strategy is expected to be highest when it corresponds to a certain MOA level (e.g., persuasion is most effective when overall MOA level is in the medium range).

17.5.3 Identifying Multilevel Building Energy Use Intervention Strategies

This stage of the framework development focuses on linking the occupants' MOA levels to the multilevel energy efficiency intervention strategies, and accordingly energy policy tool types (i.e., knowledge tools, inducement tools, and regulation tools), as shown in Table 17.4. First, the occupant MOA levels (e.g., "prone to react" that refers to the occupant who has self-interest and consistent with societal goals and is willing to adopt energy reduction strategies without additional reinforcement) are presented in Table 17.4. Then,

Table 17.4 Characteristics of intervention strategies from building energy conservation perspective.

Occupants' MOA characteristics	Energy policy tools	Intervention	Benefits and solutions presented to target	Expected benefits/costs	Occupant reaction	Time to achieve benefits
MOA = Prone Strong self-interest and consistent with societal goals Merely uninformed occupant No additional reinforcement necessary	Knowledge tools	*Knowledge based*: attempts to teach and create awareness about benefits of a particular behavior *Objective*: influence knowledge, attitudes, and beliefs	*Suggests* benefits but does not deliver them explicitly *Does not add new choices* (uses existing choices) *Target is required to initiate quest/solution to achieve benefit*	No explicit reward/penalty	Un-coerced free choice behavior *Voluntary compliance*	Promise of future potential payback Unable to reinforce directly
MOA = Unable to slightly resistant Strong self-interest but insufficiently consistent with societal goals and reinforcement in self-interest		*Persuasion*: offers reinforcing incentives/consequences *Objective*: invite voluntary exchange	*Offers benefits they want* *Adds choices* with comparative advantage and favorable cost–benefit relationships *Target receives solutions* through well-known distribution channels	Positive reward/punishment delivered when exchange transaction is completed	Un-coerced free choice behavior *Voluntary exchange* (self-monitoring – self-sanctioning)	Direct and timely exchange for desired behavior Direct reinforcement Expects free market exchange

MOA = Resistant Existing self-interest cannot be overcome with additional rewards through exchange	Inducement tools	*Penalties:* prescribes a body of rules of action/conduct *Objective:* coerce and threaten to achieve nonvoluntary compliance	*Forces benefits by providing external motivation in the absence of internal one Adds proffered choices Target bound by legal force*	Sanctions, penalties, and legal consequences for noncompliance	Coerced behavior *Nonvoluntary compliance*	Direct and timely exchange for desired behavior Direct reinforcement
	Regulation tools	*Technology:* control behavior change and referred to governmental authority and legitimacy *Objective:* coerce and threaten to achieve nonvoluntary compliance		Law or other costly regulations without requiring a promise of a positive incentive		

these MOA levels were linked to energy policy tools and their related four energy efficiency intervention strategies (i.e., knowledge based, persuasion, penalties, and technology). Subsequently, these intervention strategies were classified as the benefits and solutions presented to target (e.g., adding choices with comparative advantage and favorable cost–benefit relationships), the expected benefits/costs (e.g., sanctions, penalties, and legal consequences to behave in a certain way without the level of government coercion inherent in regulations for noncompliant energy use behavior), the expected occupant reactions of the target to each intervention (e.g., un-coerced choice behavior that refers to the change of energy use behavior voluntarily), and the time to achieve expected benefits (e.g., direct and timely exchange for desired energy use behavior that refers to direct reinforcement by the government command and control systems), as shown in Table 17.4.

Accordingly, Table 17.4 provides a guideline for policy makers to design cost-effective and efficient energy policy tools to deliver multilevel building energy use intervention strategies based on identified energy use charac-teristics of occupants in large building stocks. For example, if the majority of occupants in a large building stock are identified as "prone" to conserve energy, then decision-makers should focus on designing knowledge energy policy tools to deliver knowledge-based interventions for the occupants. These interventions are based on teaching and creating awareness about benefits of energy reduction strategies that only provide suggestions for the occupants and comply with voluntarily behavior change to adopt energy reduction strategies.

17.6   Case Study Example

Data was collected from employees of an energy efficiency consulting com-pany that occupies a single floor in an 85,000 ft^2 building located in Madison, Wisconsin. The offices are equipped with an intelligent BAS with a centralized monitoring and control of the indoor environment to maintain the operational performance of the facility and comfort of building occupants. This BAS sys-tem provides occupants with different levels of occupancy control over build-ing environment. For example, unlike the occupants in single offices, not all surveyed occupants had the capability to control their environment through individual thermostats. In both cases, the opportunity level of the occupants is expected to vary.

To identify occupants' MOA level preexposure to occupancy-focused interventions, an online survey was distributed to the occupants in the case study building. This survey includes 39 questions and focuses on evaluating occupants' control level on energy systems (e.g., indoor lighting control), office environment conditions, energy conservation motivation level, and energy conservation knowledge level. Previous studies (e.g., medical field, ticket purchasing website) found out that motivation is directly associated with most behaviors [77, 81]. However, opportunity and ability affect behaviors only when motivation is present and therefore moderate the impact of motivation on behaviors. Thus, the survey is developed by stating a set of research hypotheses and their relevant measures to investigate occupants' energy use characteristics through assessing their MOA level on adopting energy-saving behaviors. These hypotheses are designed based on the extended context of MOA levels of occupants in energy use characteristics and incorporated with a set of measures that is identified based on a comprehensive literature review.

The company surveyed had a total of 39 employees. Of those employees, 19 responded to the online survey with a response rate of 49%. Occupants' energy use characteristics and preexposure to occupancy-focused interventions were identified using k-means clustering analysis performed in MATLAB (2014). Framework implementation was accomplished in three steps: (i) all survey questions are categorized to measure M, O, or A; (ii) M, O, and A for each respondent are determined as the average of the responses to the corresponding questions; (iii) k-means was used to cluster the occupants into three main clusters. Each of the resulting clusters had the number of occupants shown in Figure 17.4a with the centroids (i.e., the average M, O, and A of the respondents in each cluster, respectively) as follows: for resistant to react ($M = 0.12$, $O = 0.33$, $A = 0.27$), for the unable to react ($M = 0.73$, $O = 0.20$, $A = 0.87$), and for prone to react ($M = 0.89$, $O = 0.80$, $A = 0.84$).

These results can be interpreted as follows:

i) 13 occupants (i.e., 68% of total respondents) with prone-to-react behavior are those with flexible behavior characteristics and are willing to adopt energy reduction strategies without additional reinforcement. These occupants have very high MOA levels when measured on a scale of 0–1. This result is expected given the company profile that the respondents work for. For example, an occupant in this group has self-motivation to adjust the plug loads when not in use to conserve energy, is satisfied with the thermal comfort and lighting quality levels at his/her office environment, and has a high level of perceived self-knowledge capacity on energy conservation.

ii) Three occupants (i.e., 16% of total respondents) with unable-to-react behavior have self-interest to adopt energy reduction strategies voluntarily but do not have the necessary ability and/or occupancy control level to do that (e.g., no/low control on thermostat settings in his/her office room). This is reflected in the low average O level and very high M and A levels that occupants in this category had. Even if occupants in this group have similar M and A levels to occupants with prone-to-react behavior, their lower O level makes them unable to react to change their energy conservation behavior. For example, an occupant in this group is not able

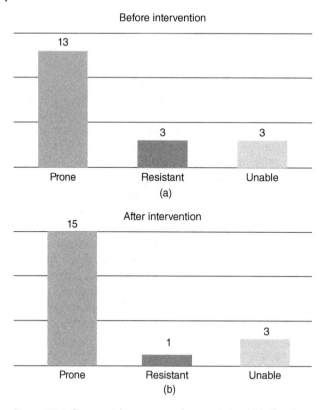

Figure 17.4 Occupants' energy use characteristics: (a) before intervention, (b) after intervention.

to turn off the lights when not in use due to the lack of control on lighting, even if he/she has self-motivation to adjust lights when not in use and a high level of knowledge capacity on energy conservation strategies about lights. In this particular case, there are not much behavioral interventions that can be done to change the profile of these respondents to prone to react.

iii) Three occupants (i.e., 16% of total respondents) with resistant-to-react behavior are reluctant to adopt energy reduction strategies. Occupants in this category had very low M level with moderate-to-low O and A levels. For example, occupants in this category are identified as unwilling to turn off their computers when not in use while exhibiting strong knowledge about the methods to reduce plug loads in their office space.

Due to the higher percentage of prone-to-react occupants in this case study, the conceptual framework presented here proposed that energy savings can be achieved effectively by delivering knowledge-based interventions

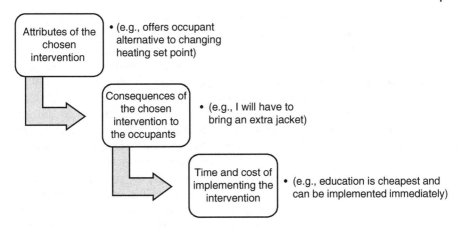

Figure 17.5 Criteria for evaluating intervention strategies.

to the occupants. Accordingly, the proposed intervention strategy for this case study is evaluated using three criteria: attributes (e.g., offers occupant alternative to changing heating set point), consequences (e.g., I will have to bring an extra jacket, I will be rewarded for turning lights off with a gift card to my favorite store), and time and cost of implementing the intervention (e.g., knowledge-based intervention is the cheapest as it does not require an exchange of rewards or penalties for achieving required change and can be implemented immediately), as shown in Figure 17.5.

Moreover, occupants' behavior change after delivering the education interventions is also investigated to highlight the impact of intervention on the occupants' behavior. Azar and Menassa [88] used simulation analysis and estimated the occupancy conversion rate from extreme (i.e., resistant) to low (i.e., prone) users at 10% when discrete interventions are used (e.g., educational campaigns). This will result in an increase in the number of prone-to-react occupants. Accordingly, this study assumed 10% occupancy energy use profile conversion rate to analyze the energy use behavior change after exposure to education intervention. The results from this analysis are presented in Figure 17.4b. In this case, two resistant-to-react occupants are converted to prone-to-react occupants by increasing their motivation level through education interventions. Additional interventions (e.g., rewards like gift cards) can be used to entice the remaining resistant occupant to adopt more conservative energy use behavior.

17.7 Discussion

In energy policy making, common approaches, such as assuming that all occupants have the standard behavior pattern (e.g., a fixed set-point room temperature) and presuming that people are rational agents making considered decisions based on available information and resources, have resulted in unintended consequences of individual behaviors and diminished the impacts and effectiveness of interventions [3, 89]. To overcome these devastating consequences, it is important to systematically evaluate the effects of interventions and their impacts on the occupants' behavior in a stock of buildings [71, 72]. Therefore, this study proposed a conceptual framework that provides a linkage between energy policy tools and their related multilevel building energy use intervention strategies to systematically evaluate and change the environmental behavior of occupants and accordingly achieve large-scale energy reductions. Using this framework, occupancy-related actions (e.g., after-hours equipment use; occupied and unoccupied hours; cooling and heating temperature set points) can be examined, and well-tuned interventions to change the relevant occupants' behaviors can be provided efficiently and cost effectively.

Moreover, the set of criteria presented in Figure 17.5 can be used for all four intervention strategies (i.e., education, persuasion, penalties, and technology) to guide decision-makers in the selection of intervention strategies by providing them with research-based evidence of their effectiveness to achieve the required energy reductions given the occupants' MOA characteristics. Accordingly, decision-makers can determine (i) how occupants' energy use profile in a stock of buildings can be measured and clustered pre- and postimplementation of energy policy tools and (ii) how to design cost-effective energy policy tools to deliver multilevel energy efficiency occupancy-focused interventions that promote actions for improving sustainable behavior pattern. The conceptual framework presented in this study demonstrates a scalable, adaptable, and expandable approach able to support decision-makers (e.g., energy planners, local administrators) in identifying the most effective energy policies and strategies.

17.8 Conclusions and Policy Implications

The objectives of this study are to present a framework that defines the relationship between occupants' energy use characteristics and the effectiveness of occupancy-focused intervention strategies. This approach will lead to the design of cost-effective and efficient energy policy tools for large-scale energy savings in building stocks (e.g., university campus, community, and city). To achieve this, a comprehensive literature review was conducted on intervention strategies that aim to engage occupants in reducing energy use in buildings. Subsequently, an analogy was investigated between MOA characteristics of people to process brand information in their environment and MOA levels where occupancy-focused intervention strategies can be regarded as advertisements enticing the building occupants to adopt certain energy use characteristics. Then, a conceptual framework was presented that proposes a multilevel intervention strategy tailored to various occupants' energy use characteristics to maintain energy use reduction in buildings. The presented framework for designing effective energy policy tools was developed in three steps: (i) quantitatively measuring occupants' preexposure MOA level and energy use profiles (i.e., occupants' situational characteristics prior to any intervention) and postexposure MOA levels to determine the effectiveness of intervention level (e.g., knowledge based, persuasion, or combination) to use for a given energy reduction strategy (e.g., encouraging occupants to turn office lights off when not in use); (ii) clustering occupants' pre- and postexposure MOA levels and energy use profiles; and (iii) linking occupants' MOA levels to the multilevel energy efficiency intervention strategies and accordingly energy policy tool types (i.e., knowledge tools, inducement tools, and regulation tools). This framework will assist researchers and policy makers in (i) systematically evaluating the effects of occupancy-focused interventions on occupants' behavior and (ii) designing and implementing energy policy tools targeted to diverse human energy use characteristics to deliver occupancy-focused intervention programs efficiently and cost effectively.

This framework will also help in encouraging pro-environmental behavior for occupants in a stock of buildings and providing a more sustainable behavior pattern.

Currently the authors are testing the implementation of the proposed framework in real case study stock of buildings. Data from the occupants of actual buildings and their energy consumption are being collected from several buildings at the University of Michigan campus. Then, the effectiveness of intervention strategies on different occupants' characteristics will be investigated to validate the proposed framework on a large-scale building stock. It is also noteworthy to mention that the proposed framework in this chapter mainly focuses on the collective impact of individuals on energy use of large building stocks. Authors' future study will also consider the effect of the interactions between occupants by understanding their underlying social network topology (SNT), which defines the relationships between occupants and the strength of their influence on each other based on the frequency of interaction and interdependence. Understanding SNT will help decision-makers to analyze how energy knowledge diffuses between occupants and how this diffusion contributes to enhance the impact of the chosen intervention strategy on improving the postexposure MOA levels and the building's energy consumption.

Questions

1 Why is it important to consider occupants' energy use characteristics when considering energy reduction interventions in buildings?

2 Name the different levels of occupancy-focused interventions in buildings.

3 What is the origin of the motivation, opportunity, and ability model?

4 What are the motivation metrics in the context of building occupants?

5 What are the opportunity metrics in the context of building occupants?

6 What are the ability metrics in the context of building occupants?

Acknowledgment

The authors would like to acknowledge the financial support for this research received from the US National Science Foundation (NSF) CBET 1407908 and 1349921. Any opinions and findings in this chapter are those of the authors and do not necessarily represent those of NSF.

References

1 EIA (2014) *How Much Energy Is Consumed in Residential and Commercial Buildings in the United States?*

2 UNEP (2014) *Sustainable Buildings and Climate Initiative Promoting Policies and Practices for Sustainability*, UNEP.

3 Schweiker, M. and Shukuya, M. (2010) Comparative effects of building envelope improvements and occupant behavioural changes on the exergy consumption for heating and cooling. *Energy Policy*, **38**, 2976–2986.

4 Von Borgstede, C., Andersson, M., and Johnsson, F. (2013) Public attitudes to climate change and carbon mitigation—Implications for energy-associated behaviours. *Energy Policy*, **57**, 182–193.

5 Azar, E. and Menassa, C.C. (2014) A comprehensive framework to quantify energy savings potential from improved operations of commercial building stocks. *Energy Policy*, **67**, 459–472.

6 Carrico, A.R. and Riemer, M. (2011) Motivating energy conservation in the workplace: An evaluation of the use of group-level feedback and peer education. *Journal of Environmental Psychology*, **31**, 1–13.

7 Hand, L.C. (2012) *Public Policy Design and Assumptions About Human Behavior*, Western Political Science Association's Annual Conference.

8 Akridge, J.M. (1998) *High-albedo roof coatings--Impact on energy consumption (No. CONF-980123--)*, American Society of Heating, Refrigerating and Air-Conditioning Engineers, Inc., Atlanta, GA.

9 Jollands, N., Waide, P., Ellis, M. *et al.* (2010) The 25 IEA energy efficiency policy recommendations to the G8 Gleneagles Plan of Action. *Energy Policy*, **38**, 6409–6418.

10 Ruzelli, A. (2010) *Proceedings of the 2nd ACM Workshop on Embedded Sensing Systems for Energy-Efficiency in Building*. ACM, Zurich, Switzerland, p. 93.

11 Wong, N.H., Cheong, D., Yan, H. *et al.* (2003) The effects of rooftop garden on energy consumption of a commercial building in Singapore. *Energy and Buildings*, **35**, 353–364.

12 Braun, J.E. (1990) Reducing energy costs and peak electrical demand through optimal control of building thermal storage. *ASHRAE Transactions*, **96**, 876–888.

13 Kouveletsou, M., Sakkas, N., Garvin, S. *et al.* (2012) Simulating energy use and energy pricing in buildings: The case of electricity. *Energy and Buildings*, **54**, 96–104.

14 Levine, M., Urge-Vorsatz, D., Blok, K., Geng, L., Harvey, D., Lang, S., Levermore, G., Mehlwana, A.M., Miragedis, S., and Novikova, A. (2007) *Residential and Commercial Buildings*. Climate change 2007; Mitigation. Contribution of Working Group III to the Fourth Assessment Report of the IPCC. Cambridge University Press, Cambridge, United Kingdom and New York, NY, USA.

15 Lopes, M.A.R., Antunes, C.H., and Martins, N. (2012) Energy behaviours as promoters of energy efficiency: A 21st century review. *Renewable and Sustainable Energy Reviews*, **16**, 4095–4104.

16 Azar, E. and Menassa, C. (2011a) An agent-based approach to model the effect of occupants' energy use characteristics in commercial buildings. *Computing in Civil Engineering, American Society of Civil Engineers (ASCE)*, Miami, FL, 536–543.

17 Azar, E. and Menassa, C.C. (2011b) Agent-based modeling of occupants and their impact on energy use in commercial buildings. *Journal of Computing in Civil Engineering*, **26**, 506–518.

18 Mahdavi, A. and Pröglhöf, C. (2009). *Toward Empirically-Based Models of People's Presence and Actions in Buildings*, Proceedings of building simulation, pp. 537–544.

19 Tanimoto, J., Hagishima, A. (2009) *Total Utility Demand Prediction Based on Probabilistically Generated behavioral Schedules of Actual Inhabitants*.

20 Moezzi, M., Iyer, M., Lutzenhiser, L., and Woods, J. (2009) *Behavioral assumptions in energy efficiency potential studies*, California Institute for Energy and Environment (CIEE), Oakland, Calif.

21 Sanchez, M., Webber, C., Brown, R. *et al.* (2007) *Space heaters, computers, cell phone chargers: How plugged in are commercial buildings?* Lawrence Berkeley National Laboratory.

22 Webber, C.A., Roberson, J.A., McWhinney, M.C. *et al.* (2006) After-hours power status of office equipment in the USA. *Energy*, **31**, 2823–2838.

23 Backer, T.E. (1991) Knowledge utilization The third wave. *Science Communication*, **12**, 225–240.

24 Bzdel, L., Wither, C., and Graham, P. (2004) *Knowledge Utilization Resource Guide.* Knowledge Utilization and Policy Implementation.

25 Estabrooks, C.A., Thompson, D.S., Lovely, J.J.E., and Hofmeyer, A. (2006) A guide to knowledge translation theory. *Journal of Continuing Education in the Health Professions*, **26**, 25–36.

26 Kaufman, L. and Rousseeuw, P.J. (2009) *Finding groups in data: an introduction to cluster analysis*, John Wiley & Sons.

27 Norcross, J.C., Koocher, G.P., and Garofalo, A. (2006) Discredited psychological treatments and tests: A Delphi poll. *Professional Psychology: Research and Practice*, **37**, 515.

28 Spring, B., Walker, B., Brownson, R. *et al.* (2008) *Definition and competencies for evidence-based behavioral practice (EBBP),* Counsel for Training in Evidence-Based Behavioral Practice, Evanston, IL.

29 Wensing, M., Bosch, M., and Grol, R. (2010) Developing and selecting interventions for translating knowledge to action. *Canadian Medical Association Journal*, **182**, E85–E88.

30 Grier, S. and Bryant, C.A. (2005) Social marketing in public health. *Annual Review of Public Health*, **26**, 319–339.

31 Kotler, P., Roberto, N., and Lee, N. (2002) *Social marketing: Improving the quality of life*, Sage Publications, Thousand Oaks, Calif.

32 Peattie, K. and Peattie, S. (2009) Social marketing: A pathway to consumption reduction? *Journal of Business Research*, **62**, 260–268.

33 Hedlund, J. (2000) Risky business: safety regulations, risk compensation, and individual behavior. *Injury Prevention*, **6**, 82–89.

34 Houston, D.J. and Richardson, L.E. (2007) Risk compensation or risk reduction? Seatbelts, state laws, and traffic fatalities. *Social Science Quarterly*, **88**, 913–936.

35 Armstrong, A.W., Idriss, N.Z., and Kim, R.H. (2011) Effects of video-based, online education on behavioral and knowledge outcomes in sunscreen use: a randomized controlled trial. *Patient Education and Counseling*, **83**, 273–277.

36 Blamey, A., Mutrie, N., and Tom, A. (1995) Health promotion by encouraged use of stairs. *BMJ*, **311**, 289–290.

37 Sahota, P., Rudolf, M.C., Dixey, R. *et al.* (2001) Randomised controlled trial of primary school based intervention to reduce risk factors for obesity. *BMJ*, **323**, 1029.

38 Agha-Hossein, M., Tetlow, R., Hadi, M. *et al.* (2014) Providing persuasive feedback through interactive posters to motivate energy-saving behaviours. *Intelligent Buildings International Journal*, **7**, 1–20.

39 Hayes, S.C. and Cone, J.D. (1977) Reducing residential electrical energy use: Payments, information, and feedback. *Journal of Applied Behavior Analysis*, **10**, 425–435.

40 Hutton, R.B. and McNeill, D.L. (1981) The value of incentives in stimulating energy conservation. *Journal of Consumer Research*, **8**, 291–298.

41 Marans, R.W. and Edelstein, J.Y. (2010) The human dimension of energy conservation and sustainability: a case study of the University of Michigan's energy conservation program. *International Journal of Sustainability in Higher Education*, **11**, 6–18.

42 Midden, C.J., Meter, J.F., Weenig, M.H., and Zieverink, H.J. (1983) Using feedback, reinforcement and information to reduce energy consumption in households: A field-experiment. *Journal of Economic Psychology*, **3**, 65–86.

43 Zografakis, N., Menegaki, A.N., and Tsagarakis, K.P. (2008) Effective education for energy efficiency. *Energy Policy*, **36**, 3226–3232.

44 Geller, E.S. (1981) Evaluating energy conservation programs: Is verbal report enough? *Journal of Consumer Research*, **8**, 331–335.

45 McMakin, A.H., Malone, E.L., and Lundgren, R.E. (2002) Motivating residents to conserve energy without financial incentives. *Environment and Behavior*, **34**, 848–863.

46 Dolan, P. and Metcalfe, R. (2013) *Neighbors, Knowledge, and Nuggets: Two Natural Field Experiments on the Role of Incentives on Energy Conservation*, Centre for Economic Performance, LSE.

47 Van Houwelingen, J.H. and Van Raaij, W.F. (1989) The effect of goal-setting and daily electronic feedback on in-home energy use. *Journal of Consumer Research*, **16**, 98–105.

48 Allcott, H. (2011) Social norms and energy conservation. *Journal of Public Economics*, **95**, 1082–1095.

49 Ayres, I., Raseman, S., and Shih, A. (2012) Evidence from two large field experiments that peer comparison feedback can reduce residential energy usage. *Journal of Law, Economics, and Organization*, **29**, 992–1022.

50 Siero, F.W., Bakker, A.B., Dekker, G.B., and Van Den Burg, M.T. (1996) Changing organizational energy consumption behaviour through comparative feedback. *Journal of Environmental Psychology*, **16**, 235–246.

51 Ayres, I., Raseman, S. and Shih, A. (2009) *Evidence from Two Large Field Experiments that Peer Comparison Feedback can Reduce Residential Energy Usage*. 5th Annual Conference on Empirical Legal Studies Paper. Available at: http://ssrn.com/abstract=1434950. Accessed on: May 05, 2016.

52 Burn, S.M. and Oskamp, S. (1986) Increasing community recycling with persuasive communication and public commitment. *Journal of Applied Social Psychology*, **16**, 29–41.

53 John, L.K., Loewenstein, G., Troxel, A.B. *et al.* (2011) Financial incentives for extended weight loss: a randomized, controlled trial. *Journal of General Internal Medicine*, **26**, 621–626.

54 Flora, S.R. and Flora, D.B. (2012) Effects of extrinsic reinforcement for reading during childhood on reported reading habits of college students. *The Psychological Record*, **49**, 1.

55 Alahmad, M.A., Wheeler, P.G., Schwer, A. *et al.* (2012) A comparative study of three feedback devices for residential real-time energy monitoring. *IEEE Transactions on Industrial Electronics*, **59**, 2002–2013.

56 Darby, S. (2006) The effectiveness of feedback on energy consumption. *A Review for DEFRA of the Literature on Metering, Billing and direct Displays*, **486**, 2006.

57 Handgraaf, M.J., Van Lidth de Jeude, M.A., and Appelt, K.C. (2013) Public praise vs. private pay: Effects of rewards on energy conservation in the workplace. *Ecological Economics*, **86**, 86–92.

58 Katzev, R.D. and Johnson, T.R. (1984) Comparing the effects of monetary incentives and foot-in-the-door strategies in promoting residential electricity conservation. *Journal of Applied Social Psychology*, **14**, 12–27.

59 McClelland, L. and Cook, S.W. (1980) Promoting energy conservation in master-metered apartments through group financial incentives. *Journal of Applied Social Psychology*, **10**, 20–31.

60 Whitsett, D.D., Justus, H.C., Steiner, E., Duffy, K. (2013) *Persistence of Energy Efficiency Behaviors over Time: Evidence from a Community-Based Program.*

61 Winett, R.A., Kagel, J.H., Battalio, R.C., and Winkler, R.C. (1978) Effects of monetary rebates, feedback, and information on residential electricity conservation. *Journal of Applied Psychology*, **63**, 73.

62 Boyce, T.E. and Geller, E.S. (2001) Encouraging college students to support pro-environment behavior effects of direct versus indirect rewards. *Environment and Behavior*, **33**, 107–125.

63 Schick, S. and Goodwin, S. (2011) *Residential Behavior Based Energy Efficiency Program Profiles*, Bonneville Power Administration.

64 Farchi, S., Chini, F., Rossi, P.G. *et al.* (2007) Evaluation of the health effects of the new driving penalty point system in the Lazio Region, Italy, 2001–4. *Injury Prevention*, **13**, 60–64.

65 Heberlein, T.A. and Warriner, G.K. (1983) The influence of price and attitude on shifting residential electricity consumption from on-to off-peak periods. *Journal of Economic Psychology*, **4**, 107–130.

66 Erickson, V.L., Achleitner, S., and Cerpa, A.E. (2013) *POEM: Power-Efficient Occupancy-Based Energy Management System*, Proceedings of the 12th international conference on Information processing in sensor networks. ACM, pp. 203–216.

67 Erickson, V.L. and Cerpa, A.E. (2010) *Occupancy Based Demand Response HVAC Control Strategy*, Proceedings of the 2nd ACM Workshop on Embedded Sensing Systems for Energy-Efficiency in Building. ACM, pp. 7–12.

68 Mathews, E., Botha, C., Arndt, D., and Malan, A. (2001) HVAC control strategies to enhance comfort and minimise energy usage. *Energy and Buildings*, **33**, 853–863.

69 Environmental and Energy Study Institute (2006) *Energy-Efficient Buildings: Using Whole Building Design to Reduce Energy Consumption in Homes and Offices*.

70 Harvey, L.D. (2009) Reducing energy use in the buildings sector: measures, costs, and examples. *Energy Efficiency*, **2**, 139–163.

71 Abrahamse, W., Steg, L., Vlek, C., and Rothengatter, T. (2005) A review of intervention studies aimed at household energy conservation. *Journal of Environmental Psychology*, **25**, 273–291.

72 Steg, L. and Vlek, C. (2009) Encouraging pro-environmental behaviour: An integrative review and research agenda. *Journal of Environmental Psychology*, **29**, 309–317.

73 Buurma, H. (2001) Public policy marketing: Marketing exchange in the public sector. *European Journal of Marketing*, **35**, 1287–1302.

74 Celsi, R.L. and Olson, J.C. (1988) The role of involvement in attention and comprehension processes. *Journal of Consumer Research*, **15**, 210–224.

75 Machleit, K.A., Madden, T.J., and Allen, C.T. (1990) Measuring and modeling brand interest as an alternative ad effect with familiar brands. *Advances in Consumer Research*, **17**, 223–230.

76 MacInnis, D.J., Moorman, C., and Jaworski, B.J. (1991) Enhancing and measuring consumers' motivation, opportunity, and ability to process brand information from ads. *Journal of Marketing*, **55**, 32–53.

77 Moorman, C. (1990) The effects of stimulus and consumer characteristics on the utilization of nutrition information. *Journal of Consumer Research*, **17**, 362–374.

78 Polonsky, M.J., Binney, W., and Hall, J. (2004) Developing better public policy to motivate responsible environmental behavior–an examination of managers' attitudes and perceptions towards controlling introduced species. *Journal of Nonprofit & Public Sector Marketing*, **12**, 93–107.

79 Rothschild, M.L. (1999) Carrots, sticks, and promises: a conceptual framework for the management of public health and social issue behaviors. *Journal of Marketing*, **63**, 24–37.

80 Hastak, M., Mazis, M.B., and Morris, L.A. (2001) The role of consumer surveys in public policy decision making. *Journal of Public Policy & Marketing*, **20**, 170–185.

81 Bigné, E., Hernández, B., Ruiz, C., and Andreu, L. (2010) How motivation, opportunity and ability can drive online airline ticket purchases. *Journal of Air Transport Management*, **16**, 346–349.

82 Govindaraju, R., Hadining, A.F., and Chandra, D.R. (2013) *Physicians' Adoption of Electronic Medical Records: Model Development Using Ability–Motivation–Opportunity Framework, Information and Communication Technology*, Springer, pp. 41–49.

83 Richins, M.L. and Bloch, P.H. (1986) After the new wears off: the temporal context of product involvement. *Journal of Consumer Research*, **13**, 280–285.

84 Zaichkowsky, J.L. (1985) Measuring the involvement construct. *Journal of Consumer Research*, **12**, 341–352.

85 Hallahan, K. (2001) Enhancing motivation, ability, and opportunity to process public relations messages. *Public Relations Review*, **26**, 463–480.

86 Parra-Lopez, E., Gutierrez-Tano, D., Diaz-Armas, R.J., and Bulchand-Gidumal, J. (2012) *Travellers 2.0: motivation, opportunity and ability to use social media* Social Media in Travel, Tourism and Hospitality, Ashgate, pp. 171–185.

87 Geller, E.S. (1995) Actively caring for the environment - an integration of behaviorism and humanism. *Environment and Behavior*, **27** (2), 184–195.

88 Azar, E. and Menassa, C.C. (2015) Evaluating the impact of extreme energy use behavior on occupancy interventions in commercial buildings. *Energy and Buildings*, **97**, 205–218.

89 Prendergrast, J., Foley, B., Menne, V., and Isaac, A.K. (2008) *CreATures of HABiT?, The art of behaviour change*, The Social Market Foundation, London.

18

Sustainability in Smart Cities: Balancing Social, Economic, Environmental, and Institutional Aspects of Urban Life

Ali Komeily[1] and Ravi Srinivasan[2]

[1] *School of Construction Management, University of Florida, Gainesville, FL, USA*
[2] *M.E. Rinker, Sr. School of Building Construction Management, University of Florida, Gainesville, FL, USA*

CHAPTER MENU

Introduction, 503
Sustainability Assessment in Our Cities, 506
Sustainability in Smart Cities, 508
Achieving Balanced Sustainability, 512

Objectives

- To become familiar with the concept of sustainable development especially in the context of urban life
- To become familiar with the current status of urban sustainability assessment and their shortcomings
- To become familiar with definitions of smart cities vis-à-vis sustainable city milestones
- To become familiar with direct and indirect contributions of smart cities in regard to sustainability goals in cities

18.1 Introduction

In about past three centuries, we have observed two important trends in humans' life. The first trend is associated with the population growth; in 1750, at the start of the Industrial Revolution, the world's population was about 700 million. By 1900, this figure stood at 1.5 billion, a double increase in 150 years. The population had doubled again by 1960, and this time it only took 60

Smart Cities: Foundations, Principles and Applications, First Edition.
Edited by Houbing Song, Ravi Srinivasan, Tamim Sookoor, and Sabina Jeschke.
© 2017 John Wiley & Sons, Inc. Published 2017 by John Wiley & Sons, Inc.

years to reach 3 billion, despite the effects of World Wars I and II. The trend continued to accelerate, with the population doubling again in 39 years to 6 billion by 1999 and adding another billion in just one decade to reach today's total of more than 7 billion.

The second trend is associated with the urban population; in 1800 only 3% of the world population was urbanized; this figure increased to 30% and 47% in years 1950 and 2000. In April 2008, the world passed the 50% urbanization mark. In addition to the increasing trend in urban population, there is an increase in metropolitan size as in 1950 there was only one city worldwide with a population of over 10 million; in 2015 that figure stood at 30. The United Nations Population Fund (2007) reported that by 2030 it expects that 7 out of 10 people will live in urban areas. The urban areas of the world are expected to absorb all the population growth expected over the next four decades while at the same time drawing in some of the rural population [1].

Unarguably, these trends pose an immense challenge for cities around the world and have stirred up discussions on how to plan, prepare, mitigate possible risks, and create opportunities. The discussions frequently involve stakeholders from different sectors and levels of society. Among others, sustainable development is one of the hotly debated topics. Sustainable development achieved its first culmination by the publication of Brundtland (1987) report "Our common future," which coined the definition of sustainable development as to meet "the needs of the present without compromising the ability of future generations to meet their own needs" [2].

Three decades after the Brundtland report and being widely embraced by public and private sectors, understandings and interpretations of sustainable development have evolved. Despite some obscurity and lack of clear definition, which is not necessarily a negative point as it creates the room for further discussion, there is a broad consensus on the fact that the concept of sustainable development must encompass "environmental," "social," and "economic" dimensions. These dimensions are directly intermingled with each other, making it meaningless to pursue one while ignoring others. Whereas environmental sustainability relates to making decisions with the intent of protecting the natural environment, social sustainability is about actively supporting the capacity of current and future generations to create healthy and livable community by promoting equity, diversity, livability, democracy, and so on. Economic sustainability refers to using resources wisely, efficiently, and responsibly for long-term benefits (Figure 18.1).

Additionally, "institutional" sustainability, as a new dimension of sustainable development, is gaining momentum. Institutional aspect of sustainability, which is sometimes referred to as political or governmental aspect, relates to the policies, governing principles, structures, regulations, and role of citizens in cities (Figure 18.2).

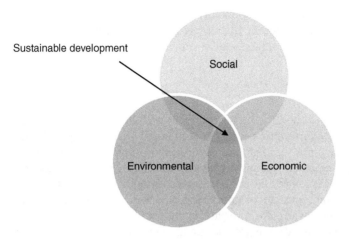

Figure 18.1 The commonly used depiction of sustainable development based on the three pillars of sustainability.

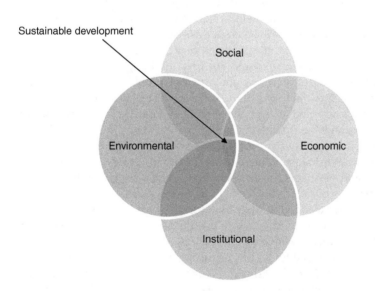

Figure 18.2 The proposed graphic for all dimensions of sustainable development, which includes social, economic, environmental, and institutional.

This chapter is organized as follows: Section 18.1 introduces the concept of sustainability and its dimensions. Section 18.2 focuses on the assessment of sustainability, its current practices, and the existing limitations and shortcomings. Section 18.3 reviews the definitions of smart cities, provides a complete list of such definitions from perspective of various stakeholders, and discusses how the concept of sustainability has been adopted and integrated in such definitions. Finally Section 18.4 discusses how smart cities can help in improving limitations in sustainability assessment.

18.2   Sustainability Assessment in Our Cities

The fast-paced urbanization has caused cities to account for more than 75% of the global energy consumption and 80% of the total greenhouse gas (GHG) emissions; great deal of overall global resource consumption happens in the cities. The role of cities in sustainable development has gained prominence, and the concept of sustainable cities has received significant political momentum worldwide as the central focus for driving worldwide sustainability. Cities are now at the forefront of efforts for achieving goals such as social justice, eliminating poverty, reduction in greenhouse gas emissions, better air quality, conservation of energy, economic vitality, and reduction of unemployment.

Although the desire to monitor and quantify sustainability in a meaningful way is well founded and widespread, the subject of sustainability and its assessment, afflicted by widespread information gaps, vagueness, ill definition, and uncertainties, has curiously lagged in this regard [3]. Without any doubts sustainability problems are interdisciplinary, very complex, and to some degree subjective. But the complexity should not be a disabling factor for pursuing rigorous analyses. In fact, such are the types of problems that data-driven solutions have much to offer. One of the key benefits of smart cities systems is their ability to harness and analyze the streaming data. The resulting analysis can help understand the underlying processes and then be used to change or influence policies and initiatives aimed at specific groups of citizens – or even the individual urbanite. While the ownership of data generated through smart devices and sensors is still an open issue – and a key one in terms of its monetization by corporations, its security, and ethical use – there is also clear potential for real-time and long-term monitoring of specific types of data [4]. This is where the process of urban sustainability assessment can benefit from the unique opportunity provided by smart cities.

Establishing reliable methods for measuring sustainability is currently a major issue, which acts as the driving force in the discussion on sustainable development. Developing tools that can reliably measure sustainability is a prerequisite for identifying non-sustainable processes, informing design-makers of the quality of products, monitoring impacts on aspects of human life [5], and moving the decision-making process toward more rigorous, quantitative, and empirical foundations. The motivations for measuring sustainability are multiple: policy and decision-making, environmental management, advocacy, participation, consensus building, research, and analysis.

At the high level, there are three different target groups whose attitudes toward clarity of sustainability assessment differ: scientists, decision-makers, and citizens [6]. Whereas scientists are interested primarily in statistically useable and possibly nonaggregated data, the decision-makers require aggregated data, as well as supplementary data related to established goals and criteria. Individual users and citizens prefer aggregation of data to one value (i.e., an index) [7].

The political will to assess and monitor the sustainability of cities resulted in a series of sustainability assessment tools and rating systems. Buildings were among the first elements of built environment that went under scrutiny, and a series of assessment tools, mostly known as green building rating systems, were developed around the world. Over time, both planners and policymakers increasingly came to understand the importance of neighborhoods as the building blocks of cities [8] and the nearest social, economic, environmental, and institutional level to citizens in which sustainability can be meaningfully assessed. More than 20 different neighborhood sustainability assessment tools and rating systems were created in a span of 10 years by different organizations (governmental and nongovernmental) around the world. In the meantime, sustainability of cities as a whole was also a subject for studies, and several indices and rating methodologies were created. Figure 18.3 shows a schematic view of the levels of sustainability assessment, and Appendix 1 lists some of the building- and neighborhood-level sustainability assessment tools and rating systems.

These tools have contributed to increasing the environmental awareness among the actors involved. They also help sharing criteria and objectives of sustainability among professionals, for which these tools are a method and framework of reference for evaluating their projects. Moreover, they guide toward better practices, if not the best practice, and can facilitate legal and political agenda in some cases [9] and improve the market demand and supply [10]. In this sense, it is clear that these tools are increasingly promoting sustainable design and practices. However, there are serious criticisms to the way these tools perform sustainability assessment, which questions their ability to effectively perform a *balanced sustainability assessment* in the era of data-driven solutions [9]. The balance in the assessment is meant from various aspects, as shown in Figure 18.4, which are [11]:

- Procedural balance: Focus on stronger citizens' role in the process of development and assessment

Figure 18.3 Multiple levels of sustainability assessment in built environment.

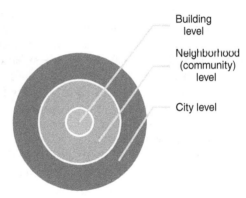

Building level

Neighborhood (community) level

City level

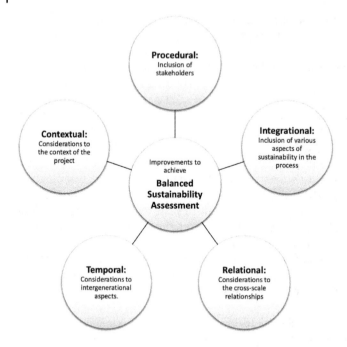

Figure 18.4 Different aspects of balanced assessment of sustainability. *Source:* Komeily and Srinivasan (2016) [11]. Reproduced with permission of Procedia Engineering.

- Contextual balance: Focus on the contextual specificities of each region
- Integrational balance: Focus on balanced assessment of sustainability by considering its four aspects
- Relational balance: Focus on cross-scale relationship between neighborhoods, buildings, and infrastructures
- Temporal balance: Focus on intergenerational aspect of sustainability

18.3 Sustainability in Smart Cities

Since the 1990s, a great deal of research has been focused on smart cities in the form of academic, industrial, and governmental publications. By reviewing the literature it becomes evident that smart city is still a fuzzy concept, which lacks definitional clarity. The initial definitional focus of the concept of smart cities was on the significance of ICT with regard to modern infrastructures within cities. Later, some experts criticized such definitions as being too technically oriented and suggested to add a strong governance-oriented approach that emphasizes the role of social capital and governance into this conceptual framework.

Currently, the term "smart cities" is frequently used with reference to almost any technology-driven urban initiative, encompassing a broad range of aspects of urban life such as quality of life and welfare, sustainability, social cohesion, economic growth, and governance targeted for building new cities or restructuring existing ones. The mainstream discourses present it as "a semantic synthesis and a conventional wisdom evoking efficiency, accountability, self-reliance" [12]. A key set of its critiques has focused around its definitional boundaries, by essentially asking what is it that comprises a "smart" city? And what – if any – should be the performance targets, characteristics, and indicators through which the claims of smart cities can be assessed and evaluated? [13] However, the broad unbounded nature of these definitions makes operationalization of the smart cities concept ambiguous and difficult [14]. These obscurities can represent an important obstacle to policymakers, by making it difficult to recognize smart cities, measure the "smartness" performance of a city, and put in place appropriate policies to incentivize the development of smart cities.

To demonstrate the diversity of urban life aspects, which are suggested to be included in smart cities concept, by experts from academia, industry, and governmental agencies, a word cloud was applied to the key words of smart cities definitions, which we identified from 60 publications. The result is shown in Figure 18.5.

Figure 18.5 The result of word cloud on 60 definitions of smart cities collected from a diverse set of publications including academia, corporate, and government perspectives.

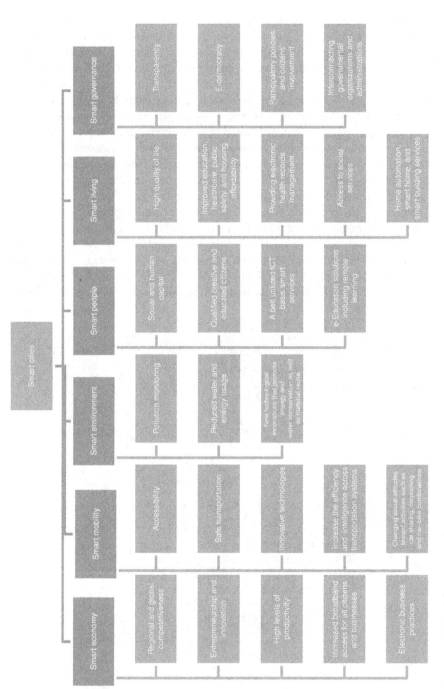

Figure 18.6 Typical dimensions of smart cities.

As Figure 18.5 demonstrates, sustainability is one of the issues frequently mentioned in smart cities definitions, especially in the more recent publications. A group of select definitions of smart cities that emphasize on importance of sustainability issues are included in Appendix 2.

Since the past decade, several initiatives have been started by the governments or the industry leaders. Examples of such efforts are European Digital Cities, InfoCities, INTELCITY roadmap, EUROCITIES (funded by European Digital Agenda), Songdo International Business District (funded by South Korea government) and Smarter Planet project (IBM) [15].

Literature shows that typically six main dimensions for smart cities are defined. As shown in Figure 18.6, these dimensions are smart economy, smart mobility, smart environment, smart people, smart living, and smart governance. However, sustainability is defined only in terms of environmental practices; we argue that this is in fact an incorrect approach to the definition of urban sustainability. As discussed earlier sustainable development pursues goals toward four main dimensions covering broad aspects of urban life. Hence, as Figure 18.7 demonstrates, the dimensions of smart cities are directly or indirectly aligned with sustainability goals and impact them accordingly.

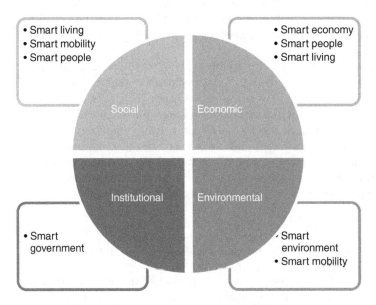

Figure 18.7 Smart cities dimensions versus sustainability dimensions.

18.4   Achieving Balanced Sustainability

As discussed earlier the sustainability assessment suffers from imbalance. In this section we focus on how smart cities can help in improving the sustainability of the cities and enhance the balance in its assessment.

18.4.1   Improving Procedural Balance

In current practice a limited number of stakeholders are involved in indicators selection, policy, and decision-making process. Including the parties that are affected by the evaluation can help unraveling human relationships within the community and facilitate entwining objective and subjective factors [16]. Previous studies have indicated that current NSA tools are mostly expert led, that is, top-heavy, and have failed to include all stakeholders, which might have different priorities and concerns both in criteria selection and their weightages. Especially, inclusion of citizens' opinion in assessment process is another field for improvement in current NSA tools; citizen-based systems are more successful in measuring community activity, individual happiness, satisfaction with local area, and perception of community spirit [17, 18]. Smart cities are a facilitator for such inclusion and creating balance.

Smarter cities start from the human capital side rather than blindly believing that ICT can automatically create a smart city [13, 19]. Therefore, the label "smart city" should refer to the capacity of clever people to generate clever solutions to urban problems.

Smart cities initiatives allow members (whether individual citizen or community as a whole) of the city to participate in the governance and management of the city and become active users [20]. This is known to be called e-democracy. E-democracy facilitates greater participation in governments and enhances effective governance using the smart cities ICT platform. Leveraging on the capabilities of smart cities (i.e., the Internet and mobile technology), e-democracy has the potential of creating new forms of engagement, deliberation, and collaboration in the political process to make democratic processes more inclusive and transparent [21–23].

18.4.2   Improving Contextual and Temporal Balance

By definition, sustainability calls for applying an integrative approach by taking into account various factors, their relationships, and interdependencies. As [24] noted "context is the most influential element of the assessment, and it must be intended in a large, comprehensive way, by disaggregating physical aspects – like geography, climate, etc. – and non-physical aspects – like legislation, local habits, etc., all in one culture – of a place." Local adaptability, sensitivity to the context, development type, and regional priorities are necessary for achieving an "asset-based" and "considerate to local values"

Table 18.1 Some of the defining factors in neighborhood context, for sustainability assessment.

Defining factors in neighborhood context

Physical	Operational	Socioeconomic	Environmental	Institutional
Block shapes and sizes	Transit stops, routes, and frequency	Local culture	Climate	Regulations and codes
Street design and pattern	Bike lanes	Demographics	Land form and topography	Policies and guidelines
Connectivity	Accessibility	Safety and crime rates	Water bodies	Special districts and designations
Building types and uses		Landmarks and historic features	Wildlife	Land ownership, easements
Surface material		Walkability		
		Neighborhood assets		
		Land use and diversity		
		Shopping and service areas		

assessment. Current tools and rating systems have considered regional context in a limited fashion; for example, LEED-ND assigns up to 4 out of 110 points to the regional priorities, and the remaining points are awarded to project regardless of their specific circumstance.

A proposed list of variables that contribute to the context of an urban area is shown in Table 18.1. It is clear for urban context, being dependent upon so many variables, that there could be a variety of outcomes, each specific for the urban area under study.

A smart city provides a strong platform for better contextual data collection and consequently improved understanding of differences in urban areas and action priority identification. In the following section, it will be shown how some of the suggested physical factors could be used for identifying different urban contexts. Though they have a limited scope, the following examples can show how urban context could be different from various perspectives.

18.4.2.1 City Blocks as a Contextual Variable

Physical features of cities are among the most influential factors in the way residents perceive their city. In quantifying the physical features of a city, different variables might be used, for example, street design, number of intersections and cul-de-sacs, area of parcels, and so on. One of the interesting variables that can carry useful information about overall street design and structure, in both macro- and microlevels, is a city block. As suggested by Louf and Barthelemy

[25], in contrast to street networks, city blocks can be defined without ambiguity as they are legally defined as the smallest area of private properties delimited by public rights-of-way; hence, by using blocks, it is easier to extract the information related to the visual aspects of the street network and its configurations. Blocks are indeed simple geometrical objects (polygons) whose properties are easily measured [25]. For the purpose of this chapter, blocks are analyzed by two factors: (i) surface area and (ii) shape. The shape of a block is measured using shape factor Φ, which is defined as the ratio between the area of the block and the area of the circumscribed circle c (Eq. 18.1):

$$\Phi = \frac{A}{A_c} \tag{18.1}$$

A Macrolevel Analysis: A Case of Miami, Orlando, and Tampa   Street design of cities has a direct consequence on the overall urban context; to show such differences in cities, Miami, Tampa, and Orlando, which are among main cities in Florida, were selected for a comparative analysis. According to the previously mentioned legal definition of a city block, block shapefiles for the three cities were generated using an ArcGIS toolbox developed by the authors. Then the area and shape factor for each block were calculated. Figure 18.8 compares the kernel density estimation (KDE) of Φ for Miami and Orlando and Miami and Tampa. The peak for Miami is located around $\Phi = 0.50$, which is close to Φ of a square ($\sim$0.64). For city blocks in Orlando, an almost bimodal distribution is visible, peaking around $\Phi = 0.10$ and $\Phi = 0.5$. The distribution of blocks in Tampa looks similar to that of Miami by peaking around $\Phi = 0.45$ to 0.60. From these values it could be concluded that the street patterns in Miami and Tampa have a more grid-like structure (resulting in higher number of blocks with shape factor close to the square's), while in Orlando blocks demonstrate heterogeneity in shapes, suggesting significant curvilinear or irregular street structures throughout the city. Such differences directly change the contextual characteristics of cities on a macroscale and impact factors such as accessibility, connectivity, and walkability of neighborhoods.

A Microlevel Analysis: Deeper Look at Miami versus Tampa   In macrolevel analysis similar overall block shape distributions were observed between Miami and Tampa. In order to gain a deeper understanding of these cities vis-à-vis each other, the blocks were taken a step forward by dividing them based on their surface areas into three logarithmic bins.

The conditional probability distribution $P(\Phi|A) \, P(A)$ of shapes, for a given bin, was drawn and is shown in Figure 18.9.

- Bin 1: block surface area [1000, 10,000)
- Bin 2: block surface area [10,000, 100,000)
- Bin 3: block surface area [100,000, 1,000,000)

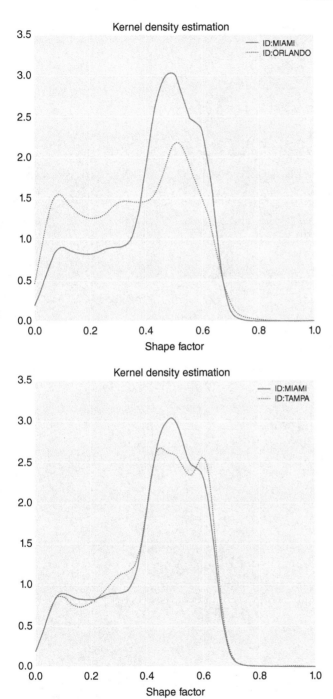

Figure 18.8 *x*-Axis represents Φ, and *y*-axis represents the density. Green solid plots relate to KDE of Miami. Red dotted plot relates to KDE of Orlando and Tampa.

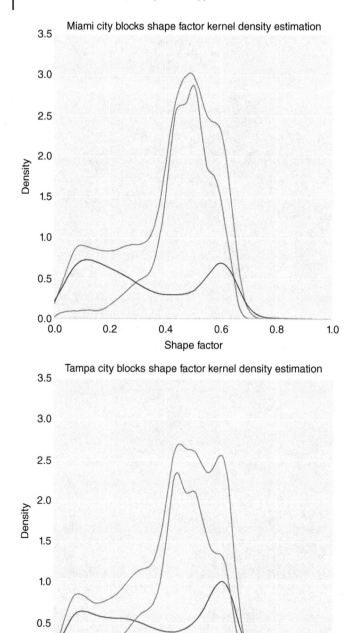

Figure 18.9 *x*-Axis represents Φ, and *y*-axis represents the density. For both cities, green plot relates to overall KDE, blue plot relates to KDE of blocks located in bin 1, red plot relates to KDE of block located in bin 2, and yellow plot relates to KDE of block located in bin 3.

This deeper look into the two cities' block shapes for any given area bins reveals some of the differences in their street designs and structures. Clearly, Miami has a higher overall density peak. In Miami, 63% of blocks fall in bin 2 and Φ for that bin peak is around 0.50; in Tampa 55% of blocks fall in bin 2 and the peak is around $\Phi = 0.40$. This shows a more prevalent gridiron street structure in Miami compared with Tampa. In contrast, Tampa's smaller blocks' (located in bin 1) shape factors globally maximize around $\Phi = 0.6$, while this is not true for Miami.

Comparison of Neighborhoods in Gainesville Given the transitional aspect of the urban-built environment, within a city, a variety of physical characteristics could be observed. To demonstrate this, the City of Gainesville, FL, is used as a case study. A series of physical measures related to every parcel in the city are collected, as shown in Table 18.2. Given the variability of urban forms and lack of a clear standard for defying urban form, an unsupervised machine learning technique is employed to identify the patterns in urban area.

There are more than 106,000 parcels in the City of Gainesville. For every parcel, the measurements in Table 18.3 are calculated. This would create a matrix with dimensions of 106000 * 6; hence k-means clustering is used for its ability to handle large datasets.

Given Gainesville is a small town, it was assumed that there would be three primary urban contexts in terms of physical aspects, namely, (i) downtown area, (ii) middle ring suburbs, and (iii) sprawls. Figure 18.11 shows these contextual patterns after the clustering data are converted into spatial data and mapped in GIS.

Table 18.2 Measures used for clustering Gainesville, FL, based on it urban form.

Variable	Measure	Definition
Int_Count	Intersection count	Total number of intersections in the buffer area around the parcel
CDS_Count	Cul-de-sac count	Total number of cul-de-sacs in the buffer area around the parcel
Resd_Count	Residential count	Total number of single-family lots in the buffer area around the parcel
Rsd_Median	Residential size	Median size of residential properties in the buffer area around the parcel
Str_Length	Street length	Total length of streets in the buffer area around the parcel
Green_PCT	Percentage undeveloped area	Through land cover image analysis, the percentage of undeveloped area is calculated in the buffer area around the parcel

Table 18.3 The first four parcels with their corresponding measurements.

FID	Int_ Count	CDS_ Count	Resd_ Count	Rsd_Median	Str_Length	Green_ PCT
0	0	0	7	1,179.10	0	1
1	0	0	3	15,000.76	0	1
2	66	22	6	37,704.42	225,752.13	0.90
3	93	32	8	37,913.68	340,782.36	0.94
4	98	35	11	37,898.44	453,934.94	0.94

Clearly, by defining the context on a set of comprehensive physical, operational, socioeconomic, environmental, and institutional indicators, different patterns from a given urban area emerge. Such differences call for smart priority allocation.

Another important factor in sustainability is that it cannot be limited to a certain time dimension. By definition, sustainability is about the present and future generations. Consequently, it is important for a successful assessment to adopt an intergenerational approach to achieve temporal balance. Changes in demographics, climate, resources, and economy are part of every community's life, and any assessment tool must be able to consider these changes. Hence it is necessary to have a lifetime approach toward projects. Current tools and rating systems have not paid enough attention to the dimension of time in their assessment. Some tools have broken down their certification process in separate stages from design to development and completion stage, but their assessment abruptly ends as the project becomes operational. A main reason for this can be attributed to the prominent role of project developer in using NSA and its voluntary nature. However, in smart cities there are mechanisms that can precisely resolve the issues of context specificities and changes over different time frames.

One of the most important factors affecting the context of a city is called urban form (or urban morphology). There are many studies that show the impact of urban form on social, economic, and environmental indicators [26–31]. Since urban form could be quantifiably measured and monitored, a key enhancement of this approach is its ability to distinguish different neighborhoods and regions even within one city, as a city can have different contextual characteristics (i.e., central business districts, urban, suburban areas). Additionally, over time by changes in urban form, the contextual characteristics can be modified accordingly as shown in Figure 18.10 (Figure 18.11).

18.4.3 Improving Integrational Balance

Several studies have demonstrated that the current assessment approaches focus more on the environmental aspect, thus raising concern over adoption

Figure 18.10 Iterative process of context
identification over time.

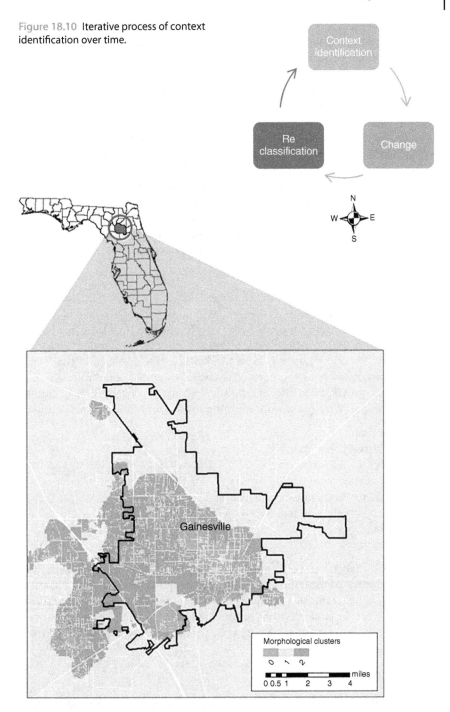

Figure 18.11 Result of *k*-means clustering, with three clusters on more than 160,000 parcels
located in Gainesville, FL.

of a physical-/material-based approach to sustainability. Although the criteria to reduce environmental impact of the project are necessary in contributing to overall sustainability, it is essential to consider all aspects of sustainability in an equitable manner. A sustainable neighborhood should promote healthy social life and relationships as well as bolstering local economy, production, and economic power, that is, from household level to community and city level. Criteria such as affordable housing, safe and inclusive community, integral policies, and local economy and job are still not adequately considered; achieving intragenerational equity necessitates addressing socioeconomic criteria.

This unbalanced focus has been attributed to lack of equal knowledge on how to measure social, economic, and institutional sustainability (compared with environmental sustainability) and limited knowledge on conceptualization of both sustainability [32] and sustainable assessment [33].

Smart cities platform can help resolving the issue with limited knowledge and information especially in social and institutional aspects. In the following we discuss these two aspects in details.

18.4.3.1 Institutional and Governing Aspect

The ICT-based governance is known as smart governance and it widely represents a collection of technologies, people, policies, practices, resources, social norms, and information that interact to support city governing activities. The importance of institutional aspect in smart cities is so crucial that some experts have called it as the core of smart cities initiatives. Although little literature on smart cities addresses issues related to governance, Chourabi *et al.* [20] prepared a table of important factors in governance of smart cities:

- Collaboration
- Leadership and champion
- Participation and partnership
- Communication
- Data exchange and open data platforms
- Service and application integration
- Accountability
- Transparency

The e-governance can help transparency in governing agency, reduce corruption, improve participation of citizens and experts, decrease bureaucratic processes, hasten response time, and increase citizens' satisfaction. Additionally, since great deals of services are provided within an ICT-based platform, tracking and assessment of governing agencies could be significantly improved.

18.4.4 Current Developments: Sustainability Information Modeling Platforms

In this section, the necessary improvements for urban sustainability were discussed in order to meet the demands and at the same time exploit the technological capabilities of the 21st century. An important aspect of any platform acting as a solution for discussed urban sustainability challenges is its ability to integrate all aspects of sustainability. Such a platform could then be used to be integrated into city dashboards as a monitoring and predication system for cities' processes and policymaking. In this regard, there has been progress. One of the platforms aiming to bring different aspects of urban sustainability is Dynamic Sustainability Information Modeling (Dynamic-SIM) platform. Currently under development, Dynamic-SIM offers a sound solution to the problems related to operating tools in silos by being integrative and collaborative using an extensible environment spanning seamlessly across a variety of scales from city to communities and campuses and then to building scale. This ability to cut across scales, considering site location and its contextual factors as discussed in [34], is what allows for this platform to achieve its ultimate goal: to model, simulate, and visualize the linkages and interplay of all elements of the urban fabric of smart cities. The Dynamic-SIM collects and stores historical data for learning and pattern recognition. Additionally, it is able to be one of the receiving ends for data collected from IoT in order to perform accurate analyses. Dynamic-SIM platform has the ability to (i) track materials and energy flows and recommend optimal use of renewable and nonrenewable sources with minimal environmental impact, (ii) produce auditable environmental reports and sustainability strategies toward integrative environmentalism of smart cities, (iii) characterize upstream and downstream life cycle impacts of campuses and neighborhoods, (iv) identify best and worst performers and identify and implement efficiency measures using reduced order models, (v) "walk into" building zones to investigate abnormal energy usages and/or system faults, (vi) conduct "what if" scenarios and data analytics of building energy components, and (vii) perform energy efficiency and environmental accounting procedures from within the platform.

The Dynamic-SIM platform, besides enabling modeling, simulating, and visualizing sustainability performance of smart cities, will provide a roadmap toward integrative sustainability assessment of smart cities. Using the integrated environmental accounting structure, the sustainable performance of smart cities can be characterized via a net balance sheet.

Final Thoughts

In this chapter we discussed the concept of urban sustainability, its assessment, and limitations in current practices of sustainability assessment. We also overviewed the definitions of smart cities, the prominence of sustainability vis-à-vis smart cities concept. And finally we discussed how smartness in our cities can move our cities toward sustainable development goals and improve the way sustainability assessment is performed. With this in mind we can define a smart city as a city where ICT is used to achieve a balanced and intergenerational sustainability in urban life.

Questions

1 What is the definition of sustainable development and why does it matter?

2 Name and describe three common limitations in sustainability assessment tools that are currently used in practice.

3 How can smart technologies used in a city improve our understanding of social dimension of urban life?

4 What is a smart community and why is it called optimal scale for sustainability assessment?

5 Why should a successful sustainability assessment consider intergenerational issues?

Appendix 1

	Rating system	Country/region
Neighborhood level	LEED-ND	United States
	Enterprise Green Community	United States
	Green Land Development	United States
	BREEAM Communities	United Kingdom
	One Planet Communities	United Kingdom
	CASBEE-UD	Japan
	EarthCraft Communities (ECC)	United States – Greater Atlanta
	DGNB for Urban Development	Germany
	Green Star – Communities	Australia
	GSAS for Districts	Qatar
	Green Mark for Districts	Singapore
	GBI Township	Malaysia
	Neighborhood Sustainability Framework	New Zealand
	HQE2R	France
	ECOCITY	EU
	Green Townships	India

(*continued*)

	Rating system	Country/region
	Aqua for Neighborhood	Brazil
	Pearl Community for Estidama	UAE
	BEAM Plus Neighborhood	Hong Kong
	EnviroDevelopment	Australia
	BERDE for Clustered Development	Philippines
Building level	BREEAM	United Kingdom
	CASBEE	Japan
	CEPAS	Hong Kong
	ENERGY STAR	United States
	EPLabel	United Kingdom
	Estidama Pearl Rating System	UAE
	Green Globes	United States
	HQE	France
	LEED	United States
	Living Building Challenge	United States
	NABERS	Australia
	SB Tool	International
	SPiRiT	United States
	Three Star System	China

Appendix 2

Source	Definitions/features	Sustainability				Country	Category
		SOC	ECO	ENV	INS		
Toppeta (2010)	An important goal of smart cities must be improvement in sustainability and livability	✓				Italy	Academic
González et al. (2011)	A smart city is a public administrative service or authority that delivers (or aims to deliver) a set of new generation services and infrastructure, based on information and communication technologies. Services provided by smart cities should be easy to use, efficient, responsive, open, and sustainable for the environment		✓	✓	✓	France*	Academic

(continued)

Source	Definitions/features	Sustainability				Country	Category
		SOC	ECO	ENV	INS		
Caragliu *et al.*	A city is smart when investments in human and social capital and traditional (transport) and modern (information and communication technology (ICT)) communication infrastructure fuel sustainable economic growth and a high quality of life, with a wise management of natural resources, through participatory governance	✓	✓	✓	✓	Italy	Academic
Alcatel Lucent (2011)	The smart city concept is really a framework for a specific vision of modern urban development. It recognizes the growing importance of ICTs as drivers of economic competitiveness, environmental sustainability, and livability. By leveraging ICT as a core element of their development, the smart cities of the future will foster economic growth, improve the lifestyle of citizens, create opportunities for urban development and renewal, support eco-sustainability initiatives, improve the political and representative process, and provide access to advanced financial services	✓	✓	✓	✓	France	Industry
Gartner (2011)	A smart city is based on intelligent exchanges of information that flow between its many different subsystems. This flow of information is analyzed and translated into citizen and commercial services. The city will act on this information flow to make its wider ecosystem more resource efficient and sustainable. The information exchange is based on a smart governance operating framework designed to make cities sustainable			✓	✓	NA	Industry

(continued)

Source	Definitions/features	Sustainability				Country	Category
		SOC	ECO	ENV	INS		
Batty *et al.* (2012)	A city, in which ICT is merged with traditional infrastructures, coordinated and integrated using new digital technologies. These technologies establish the functions of the city and also provide ways in which citizen groups, governments, businesses, and various agencies who have an interest in generating more efficient and equitable systems can interact in augmenting their understanding of the city and also providing essential engagement in the design and planning process				✓	United Kingdom	Academic
Bakıcı *et al.* (2012)	Smart city as a high-tech intensive and advanced city that connects people, information, and city elements using new technologies in order to create a sustainable greener city, a competitive and innovative commerce, and an increased life quality	✓	✓	✓		Spain	Academic
Azkuna (2012)	A city can be considered as "smart" when its investment in human and social capital and in communications infrastructure actively promote sustainable economic development and a high quality of life, including the wise management of natural resources through participatory government	✓	✓	✓	✓	Spain	Government
Schaffers *et al.* (2012a)	A smart city is referred to as the safe, secure, environmentally green, and efficient urban center of the future with advanced infrastructures such as sensors, electronics, and networks to stimulate sustainable economic growth and a high quality of life. *A smart city as a high-tech intensive and advanced city that connects people, information, and city elements using new	✓	✓	✓	✓	Italy	Industry

(continued)

Source	Definitions/features	Sustainability				Country	Category
		SOC	ECO	ENV	INS		
	technologies in order to create a sustainable greener city, a competitive and innovative commerce, and an increase in the quality of life with a straightforward administration and maintenance system of the city						
Barrionuevo *et al.* (2012)	Being a smart city means using all available technology and resources in an intelligent and coordinated manner to develop urban centers that are at once integrated, habitable, and sustainable						
European Commission (2014)	Smart cities combine diverse technologies to reduce their environmental impact and offer citizens better lives. This is not, however, simply a technical challenge. Organizational change in governments – and indeed society at large – is just as essential. Making a city smart is therefore a very multidisciplinary challenge, bringing together city officials, innovative suppliers, national and EU policymakers, academics, and civil society	✓		✓	✓	EU	Government
European Commission (2013)	Smart cities should be regarded as systems of people interacting with and using flows of energy, materials, services, and finance to catalyze sustainable economic development, resilience, and high quality of life; these flows and interactions become smart through making strategic use of information and communication infrastructure and services in a process of transparent urban planning and management that is responsive to the social and economic needs of society	✓	✓	✓	✓	EU	Government/

(*continued*)

Source	Definitions/features	Sustainability				Country	Category
		SOC	ECO	ENV	INS		
Homeier (2013)	A "city" can be defined smart when systematic information and communication technologies and resource-saving technologies are used to work toward a post fossil society, reduce resource consumption, and enhance permanently citizens' quality of life and the competitiveness of local economy – thus improving the city's sustainability. The following areas are at least taken into account: energy, mobility, urban planning, and governance. An elementary characteristic of a smart city is the integration and cross-linking of these areas in order to implement the targeted ecological and social aspects of urban society and a participatory approach	✓	✓	✓	✓	Austria	Government/
IBM (2013)	A smarter city uses technology to transform its core systems and optimize finite resources. Because cities grapple on a daily basis with the interaction of water, transportation, energy, public safety, and many other systems, a vision of smarter cities is a vital component of larger plans. At the highest levels of maturity, a smarter city is a knowledge-based system that provides real-time insights to stakeholders, as well as enabling decision-makers to proactively manage the city's subsystems. Effective information management is at the heart of this capability, and integration and analytics are the key enablers			✓		Japan	Industry
Siemens (2014)	A smart city is made up of three main parameters to make sure that there is an overall development of energy, healthcare, buildings, transport, and water management in a city:	✓	✓	✓		Germany	Corporate

(*continued*)

Source	Definitions/features	Sustainability				Country	Category
		SOC	ECO	ENV	INS		
	• Environmental care – with right technologies, cities will become more environmentally friendly • Competitiveness – with the right technologies, cities will help their local authorities and businesses to cut costs. • Quality of life – with the right technologies, cities will increase the quality of life for their residents						
Oracle (2014)	One manifestation of the Oracle iGovernment vision is Oracle's Solutions for Smart Cities, which will address the ever-increasing need to provide businesses and citizens with transparent, efficient, and intelligent engagement with their local authority/ administration – through any channel – for any purpose, from information requests and government program enrolment to incident reporting or scheduling inspections to complete online start-up of a local business. Development, implementation, and refinement of such a multichannel single point-of-contact platform to all government organizations lays the foundation for a range of additional capabilities from business recruitment and retention to self-selecting and interest- and knowledge-based communities among citizens to improve management of civil contingencies and emergency disaster planning				✓	United States	Corporate
Fujitsu (2014)	It takes more to build a smart city than simply using ICT to link and manage social infrastructure. Providing new values and services that residents truly need is also essential. Generating the					Japan	Corporate

(continued)

Source	Definitions/features	Sustainability				Country	Category
		SOC	ECO	ENV	INS		
	knowledge to arrive at solutions by continuing to closely examine local issues, while putting this information into the equation when analyzing the enormous amount of data from smartphones, various sensors, meters, and other devices, is a crucial task. Achieving it requires that Fujitsu put ICT to work to establish a sustainable social value cycle and create new innovations						
Hitachi (2014)	A smart city is typically defined as "an environmentally conscious city that uses information technology (IT) to utilize energy and other resources efficiently." In Hitachi's vision, a smart city is one that seeks to satisfy the desires and values of its residents, with the use of advanced IT to improve energy efficiency and concern for the global environment as prerequisites, and in so doing maintains a "well-balanced relationship between people and the Earth"	✓		✓		Japan	Industry
Schneider Electric (2014)	A smart city is an efficient city, a liveable city, and economically, socially, and environmentally sustainable city. This vision can be realized today using innovative operational and IT and leveraging meaningful and reliable real-time data generated by citizens and city infrastructure. Five steps to make a city smart: 1. Vision: setting the goal and the roadmap to get there 2. Solutions: bringing in the technology to improve the efficiency of the urban systems	✓	✓	✓	✓	United Kingdom	Industry

(continued)

Source	Definitions/features	Sustainability				Country	Category
		SOC	ECO	ENV	INS		
	3. Integration: combining information and operations for overall city efficiency						
	4. Innovation: building each city's specific business model						
	5. Collaboration: driving collaboration between global players and local stakeholders						
Schneider Electric India (2015)	The most effective definition of a smart city is a community that is efficient, liveable, and sustainable, and these three elements go hand in hand. Traditionally, water, gas, electricity, transportation, emergency response, buildings, hospitals, and public services systems of a city are separate and operate in silos independent of each other. A truly efficient city requires not only that the performance of each system is optimized but also that these systems are managed in an integrated way to better prioritize investment and maximize value. An efficient city also starts a community on the path to become competitive for talent, investment, and jobs by becoming more liveable. A city must work to become a pleasant place to live, work, and play. It must appeal to residents, commuters, and visitors alike. It must be socially inclusive, creating opportunities for all of its residents. It must provide innovative, meaningful services to its constituents. Liveability plays a critical role in building the talent pool and the housing market and in providing cultural events, which can bring memorable experiences, international attention, and investment to the community.	✓		✓		India	Corporate

(*continued*)

Source	Definitions/features	Sustainability				Country	Category
		SOC	ECO	ENV	INS		
	A sustainable community is one that reduces the environmental consequences of urban life and is often an output of efforts to make the city more efficient and liveable. Cities are the largest contributors of carbon emissions; the highways, public spaces, and buildings we rely on to live, work, and play emit the bulk of each city's emissions. Implementing efficient, cleaner, and sustainable operations in all of these areas is critical to minimizing a city's environmental footprint. Cities must also look at other methods of achieving sustainability, including resource efficiency, regenerating aging districts, ensuring robustness of systems, and incorporating design and planning in harmony with their natural ecosystem, as opposed to simply living in them						

References

1 Martine, G. and Marshall, A. (2007) State of World Population 2007: unleashing the potential of urban growth, in *State of World Population 2007: Unleashing the Potential of Urban Growth*, UNFPA.

2 Brundtland, G., Khalid, M., Agnelli, S., Al-Athel, S., Chidzero, B., Fadika, L., and Singh, M. (1987) Our Common Future (\'Brundtland report\').

3 Babcicky, P. (2012) Rethinking the Foundations of Sustainability Measurement: The Limitations of the Environmental Sustainability Index (ESI). *Social Indicators Research*, **113** (1), 133–157. doi: 10.1007/s11205-012-0086-9

4 Caprotti, F. (2015) Building the smart city: Moving beyond the critiques. Retrieved from https://ugecviewpoints.wordpress.com/2015/03/24/building-the-smart-city-moving-beyond-the-critiques-part-1/ (accessed December 19, 2015)

5 Afgan, N. (2004) Sustainability assessment of hydrogen energy systems. *International Journal of Hydrogen Energy*, **29** (13), 1327–1342. doi: 10.1016/j.ijhydene.2004.01.005

6 Braat, L. (1991) The predictive meaning of sustainability indicators, in *In search of indicators of sustainable development*, Springer, Netherlands, pp. 57–70.

7 Krajnc, D. and Glavič, P. (2005) How to compare companies on relevant dimensions of sustainability. *Ecological Economics*, **55** (4), 551–563. doi: 10.1016/j.ecolecon.2004.12.011

8 Searfoss, L. (2011) Local Perspectives on HUD's Neighborhood Stabilization Program. *Report for National Community Stabilization Trust.*

9 Komeily, A. and Srinivasan, R.S. (2015) A need for balanced approach to neighborhood sustainability assessments: A critical review and analysis. *Sustainable Cities and Society*, **18**, 32–43.

10 Reed, R., Wilkinson, S., Bilos, A., and Schulte, K.W. (2011) *A Comparison of International Sustainable Building Tools–An Update.* In The 17th Annual Pacific Rim Real Estate Society Conference, Gold Coast (pp. 16–19).

11 Komeily, A. and Srinivasan, R.S. (2016) What is Neighborhood Context and Why does it Matter in Sustainability Assessment? *Procedia Engineering*, **145**, 876–883.

12 Santangelo, M. and Vanolo, A. (2014) *The Smart City: Critical Perspectives.* Retrieved from https://www.jiscmail.ac.uk/cgi-bin/webadmin?A2=crit-geog-forum;9f84a112.1405 (accessed December 19, 2015)

13 Hollands, R.G. (2008) Will the real smart city please stand up? Intelligent, progressive or entrepreneurial? *City*, **12** (3), 303–320.

14 Certomà, C. and Rizzi, F. (2015) *Smart Cities for Smart Citizens: Enabling Urban Transitions Through Crowdsourcing.* Retrieved from https://ugecviewpoints.wordpress.com/2015/08/19/smart-cities-for-smart-citizens-enabling-urban-transitions-through-crowdsourcing/ (accessed December 19, 2015).

15 Cardone, G., Foschini, L., Bellavista, P. *et al.* (2013) Fostering participation in smart cities: A geo-social crowdsensing platform. *Communications Magazine, IEEE*, **51** (6), 112–119.

16 Scerri, A. and James, P. (2010) Accounting for sustainability: Combining qualitative and quantitative research in developing 'indicators' of sustainability. *International Journal of Social Research Methodology*, **13** (1), 41–53.

17 Hardi, P. and Zdan, T. (1997) *Assessing Sustainable Development: Principles in Practice*, The International Institute for Sustainable Development, Winnipeg.

18 Morse, S. and Fraser, E.D. (2005) Making 'dirty' nations look clean? The nation state and the problem of selecting and weighting indices as tools for measuring progress towards sustainability. *Geoforum*, **36** (5), 625–640.

19 Shapiro, J.M. (2006) Smart cities: quality of life, productivity, and the growth effects of human capital. *The review of economics and statistics*, **88** (2), 324–335.

20 Chourabi, H., Nam, T., Walker, S., Gil-Garcia, J. R., Mellouli, S., Nahon, K., and Scholl, H.J. (2012) *Understanding Smart Cities: An Integrative Framework*. System Science (HICSS), 2012 45th Hawaii International Conference (pp. 2289–2297). IEEE.

21 Coleman, S. and Gotze, J. (2001) *Bowling Together: Online Public Engagement in Policy Deliberation*, Hansard Society, London, pp. 39–50.

22 Ayo, C.K., Adebiyi, A.A., and Fatudimu, I.T. (2008) E-Democracy: A requirement for a successful E-Voting and E-Government implementation in Nigeria. *International Journal of Natural and Applied Sciences*, **4** (3), 310–318.

23 Shirazi, F., Ngwenyama, O., and Morawczynski, O. (2010) ICT expansion and the digital divide in democratic freedoms: An analysis of the impact of ICT expansion, education and ICT filtering on democracy. *Telematics and Informatics*, **27** (1), 21–31.

24 Conte, E. and Monno, V. (2012) Beyond the building-centric approach: A vision for an integrated evaluation of sustainable buildings. *Environmental Impact Assessment Review*, **34**, 31–40.

25 Louf, R. and Barthelemy, M. (2014) A typology of street patterns. *Journal of the Royal Society Interface*, **11** (101), 20140924.

26 Banister, D., Watson, S., and Wood, C. (1997) Sustainable cities: Transport, energy, and urban form. *Environment and Planning B: Planning and Design*, **24** (1), 125–143. doi: 10.1068/b240125

27 Bramley, G. and Power, S. (2009) Urban form and social sustainability: The role of density and housing type. *Environment and Planning B: Planning and Design*, **36** (1), 30–48. doi: 10.1068/b33129

28 Camagni, R., Gibelli, M.C., and Rigamonti, P. (2002) Urban mobility and urban form: the social and environmental costs of different patterns of urban expansion. *Ecological Economics*, **40** (2), 199–216. doi: 10.1016/S0921-8009(01)00254-3

29 Dempsey, N., Bramley, G., Power, S., and Brown, C. (2011) The social dimension of sustainable development: Defining urban social sustainability. *Sustainable Development*, **19**, 289–300.

30 Marshall, J.D. (2008) Energy-efficient urban form. *Environmental Science and Technology*, **9** (42), 3133–3137.

31 Muñiz, I. and Galindo, A. (2005) Urban form and the ecological footprint of commuting. The case of Barcelona. *Ecological Economics*, **55** (4), 499–514. doi: 10.1016/j.ecolecon.2004.12.008

32 Pope, J., Annandale, D., and Morrison-Saunders, A. (2004) Conceptualising sustainability assessment. *Environmental Impact Assessment Review*, **24** (6), 595–616.

33 Lawrence, D.P. (1997) The need for EIA theory-building. *Environmental Impact Assessment Review*, **17** (2), 79–107.

34 Komeily, A. and Srinivasan, R. (2015) *GIS-Based Decision Support System for Smart Project Location*. In 2015 International Workshop on Computing in Civil Engineering.

19

Toward Resilience of the Electric Grid

Jiankang Wang

Department of Electrical and Computer Engineering, Ohio State University, Columbus, OH, USA

CHAPTER MENU

Electric Grids in Smart Cities, 536
Threats to Electric Grids, 544
Electric Grid Response under Threats, 553
Defense against Threats to Electric Grids, 558

> *"Expect the best, plan for the worst, and prepare to be surprised."*
>
> - Denis Waitley

Objectives

- To understand the structure and basic operating principles of the electric grid, its connection to smart cities and interdependence with other critical infrastructures.
- To understand possible threats to electric grids from natural disasters and malicious intents. As some of these threats have manifested themselves through historical events, others were recently enabled by advancement of smart cities' cyber infrastructures.
- To understand the cyber and physical causal chains in the electric grid from initial threats to final consequences, and to show existing metrics that enable us to quantify these consequences and probabilities of their occurrences.
- To discuss the responding mechanisms of the electric grid and other sections in smart cities, which prevent, intercept and mitigate these threats and make the electric grid resilient.

Smart Cities: Foundations, Principles and Applications, First Edition.
Edited by Houbing Song, Ravi Srinivasan, Tamim Sookoor, and Sabina Jeschke.
© 2017 John Wiley & Sons, Inc. Published 2017 by John Wiley & Sons, Inc.

19.1   Electric Grids in Smart Cities

Electricity provides lifeblood to urban operations. Urban infrastructures are operated on electricity for food preservation, water treatment, heat and light, phone service, Internet, factories, hospitals, and emergency response. Disruption of electricity will degrade or cease all these services and normal function of critical municipal departments, such as fire stations and police stations. In smart cities, which host many highly integrated and interdependent layers, prolonged disruption would seriously affect every infrastructure. An outage of more than 10 days will result in 80% of economic activity ceasing [1]. For power delivery systems in cities, continuously supplying electricity of satisfactory power quality remains the top priority.

Today's power systems are described by the industry as *"reliable,* but not *resilient* [2]."* With sufficient redundancies, they can easily handle failures of small numbers of components due to probabilistic events. For example, you may not notice any lights flickering if a district transformer were taken offline due to an accidental short circuit, because its load was transferred to the backup margin of other transformers in the service area. However, reliability to handle random disturbance does not naturally lead to resilient response to specific events. While a power system's reliability can be improved by adding redundancies, its resilience requires awareness of the existence of system vulnerabilities and strategic planning of the system's configuration.

Threats that cannot be handled by redundancies are mainly of two kinds: weather hazards and malicious attacks. Unlike random events, they are much less controllable and usually engender high impact. Extreme weather hits power systems in a wide range and could physically damage multiple critical grid assets in a fairly short duration, making backup transfer impossible. For example, in 2012, the Northeastern United States was struck by Hurricane Sandy. Over 100,000 primary electrical wires were destroyed, several substation transformers exploded, and numerous substations were flooded. This led to the disconnection of approximately 7 million people [3].

Compared with extreme weather, malicious attacks to power delivery systems could be more dangerous, because they usually involve malicious intent. They may aim at disrupting power to a service area or target a critical grid asset. Redundancies designed to absorb random disturbances of small magnitudes are incapable of defending against these intentional actions of high intensity. Moreover, attacks are usually launched to the most vulnerable parts of power delivery systems. In 2013, attackers near San Jose disabled 17 substation transformers by shooting through a chain-link fence. The bullet holes caused the transformers to leak thousands of gallons of oil and ultimately overheat. The backup capacity of neighborhood transformers barely supported the system through the attack; only because it occurred in early morning when the load was low, the backup capacity became sufficient [4].

More sophisticated malicious attacks take place on the cyber level of power systems. As the power delivery system is evolving, an increasing number of sensors applied to the grid allow for both improved flexibility and increasing automation. However, such an increase in automation also increases the number of on-ramps for cyber attacks. The latest publicized cyber attack on industrial control systems was the Stuxnet worm. The objective was to corrupt a specific type of programmable logic controller (PLC) by rewriting parts of the code and turning it into the attacker's agent. With modifications, it could become a serious threat to power grids [2]. In smart cities, this increased integration with the communication infrastructure can leave the grid vulnerable as layer upon layer of connectedness results in an increasing amount of trust in suppliers.

Above reliability, a power system resilient to these threats is sought after for smart cities. As written in the US Presidential Policy Directive 21, "resilience is the ability to prepare for and adapt to changing conditions and withstand and recover rapidly from disruptions." To build a resilient grid, the following questions need to be addressed: (i) What can go wrong? (*possible threats*); (ii) What are the consequences? (*grid response and causal chain*); (iii) How likely are the consequences to occur? (*failing conditions*); and (iv) How is a normal status restored? (*resilient response*). In this chapter, we wish to answer these questions by presenting both real examples and technical principles of power systems' response to cyber and physical threats.

The rest of this section briefly introduces the background of power systems and terminologies. Section 19.2 presents real threats experienced by power systems in history, together with modeling methods for the threats. Section 19.3 unfolds the causal chains of the past major events to power systems, evaluates the consequences of possible attacks, and presents vulnerability assessment methods. Methods to improve resilience of power systems are discussed in Section 19.4. We assume that readers have basic knowledge of electrical engineering but no prior knowledge in power systems.

19.1.1 Structure of Power Systems

19.1.1.1 Vertical Structure

Modern power systems can be dissected into three layers: (i) a physical power grid (denoted as the grid), including generators, power delivery equipment, and lines, loads, and protective devices; (ii) a control layer, including central and local decision-making units; and (iii) a monitoring layer, including meters, actuators, and hardware to store and transmit measurements (see Figure 19.1). In the context of cyber–physical systems, the power grid is physical, whereas the monitoring and control layers are its cyber subsystems. The monitoring layer collects, transmits, processes, and analyzes the measurement over various service areas and at various rates. The control layer determines the current system status based on processed measurements and makes decisions accordingly.

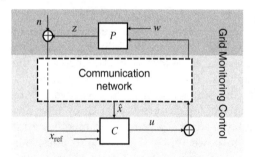

Figure 19.1 Vertical structure of power systems.

w System disturbance z Measurement
n Measurement noise $\hat{x}$ Estimated state
u Operation command x_{ref} Desired state

These decisions are passed onto intelligent electronic devices (IEDs), which finally trigger mechanical actions on the (physical) grid.

The interaction of this layered structure may be illustrated by an example of loss minimization in a local distribution grid. On the monitoring layer, nodal meters measure voltage and current values and transmit them through the supervisory control and data acquisition (SCADA) system [5]. On the control layer, the processed voltage and current values are input to the loss minimization algorithm in the distribution management system (DSM) software. Finally, on the grid, the voltage regulators receive control commands and change their tap positions accordingly. Following the voltage adjustment, power flows on the grid are redistributed, rendering minimum power delivery losses.

The physical grid consists of four sections: the generation section where primary energy is converted into electric power (electricity), transmission and distribution sections that transport this power, and the load section that uses power. Cities rarely host the generation and transmission sections. Large generating units are usually located outside densely populated areas. Urban electric grids take in power from their upstream transmission networks, transform it into lower voltages, and deliver it to commercial centers, small enterprises, residential users, factories, and municipal departments; they mainly consist of the distribution and load sections. Because of their function, we refer to the part of power systems in cities, including both its physical and cyber subsystems, as *power delivery systems*.

19.1.1.2 Transmission

The transmission system (above 138 kV) transports power from generating units to all the power delivery systems within its operation region. Events striking power systems at the transmission level, if not mitigated in time, may roll on to cascading blackouts to downstream power delivery systems. The transmission network is composed of power lines and stations/substations. In cities where real estate is valuable and transmission towers cannot be afforded, transmission lines sometimes adopt insulated cables buried underground.

19.1.1.3 Station and Substation

The connection between power delivery systems and transmission systems occurs at distribution substations. Stations and substations are common targets of both physical and cyber attacks. Stations and substations house transformers, switchgear, measurement instrumentation, and communication equipment. High-voltage (HV) transformers carry 60–70% of the nation's electricity. They are designed and manufactured to custom specifications for a specific network application. If an HV transformer is damaged, replacement could take from 5 to 20 months and cost from $2 million for a 230 kV unit to $7.5 million for a 765 kV unit [6]. The situation becomes more threatening as the power system will be operated under high stress and in a more vulnerable condition until replacement is completed.

Cyber attacks may happen at substations to switchgear and measurement instrumentation. Switchgear includes circuit breakers and other types of switches used to disconnect parts of the transmission network for system protection or maintenance. With the promotion of power system automation, more of these switchgear become digital and remotely controllable [7]. Cyber attacks that falsify control command can de-energize an important circuit and cause large-scale disturbances. Measurement instrumentation collects voltage, current, and power data for monitoring control and metering purposes. Communication equipment transmits these data to control centers and also allows switchgear to be controlled remotely. Cyber attacks to any of these devices will sabotage timely remedial actions when a system goes through abnormal status. Examples of these attacks are given in Section 19.2.

19.1.1.4 Distribution

Power delivery systems consist of distribution networks, loads, and their control and monitoring layers. Primary distribution networks operate from 2 kV to 35 kV, delivering electricity from substations, which are connected to transmission level. Transformers on primary distribution networks further step down the voltage to 120 V/480 V. The wires that operate at this low voltage level and are directly connected to end users are secondary distribution networks [8]. Power networks that are operated between 35 kV and the transmission voltage level (138 kV) are referred to as the sub-transmission level[1].

In cities, distribution networks are usually formed in mesh topologies, whereas in suburban and rural areas radial topologies are more dominant.

---

1 Some utilities and operators refer to networks that operate between 4 kV and 110 kV as primary distribution level. The voltage level classification varies by operators, electric administrative regions, and countries.

Compared with radial topologies, mesh networks are more expensive but embody higher reliability and operation flexibility. In a contingency, they permit a load to be served through an alternative path when there is a problem in the original path. On the other hand, they may complicate the deployment of protection schemes due to the presence of multiple power flow paths. Cyber attackers can take advantage of this by hacking into the distribution management systems (DMS) and cause electricity disruption.

19.1.2 Operation of Power Systems

Power systems are operated through a combination of automated control and actions that require direct human intervention. The objective of real-time operation is to ensure that the system stays stable and protected while meeting end-user power requirements. System stability requires a precise balance between power generation and consumption at all times. This requirement can be explained by the *swing equation*

$$M\ddot{\delta} + D\dot{\delta} + P_D = P_G$$

When the total power generation P_G does not meet the total demand P_D, the generation's power angle δ changes with time and results in frequency deviation. The frequency deviation also depends on the system inertia M and damping D, which can be improved through transmission technologies, such as flow control devices and power system stabilizers [9]. The equation expresses the energy balance dynamics within power systems. Consider the entire US power system as one generator whose rotor rotates at 60 Hz, the standard frequency in North American power systems. (In actuality, the rotation speed is likely to be different and varied for many generators. Here we "group" all these generators to rotate at one speed for explanatory convenience.) If there is more consumption than generation ($P_D > P_G$), the rotor will release kinetic energy by rotating slower to make up the energy supply needed. Likewise, if there is more generation than consumption ($P_D < P_G$), the rotor will store the extra generation as kinetic energy and rotate faster. This behavior is known as "inertial response." The rotor's position gives the power angle δ, and its movement dynamics are described by the frequency $\dot{\delta}$ and the frequency's deviation $\ddot{\delta}$.

Sudden change of load or loss of generating units will lead to unbalance. Physical and cyber attacks at generators and power transmission lines can easily create such unbalance. For example, cyber attacks can maneuver multiple generators simultaneously. If the balance is not regained in a short time, the system power angles become unstable – its frequency can exceed allowable bounds. An unstable frequency can lead to unstable voltages and result in equipment damage as well as blackouts. This phenomenon of transient instability does not originate from power delivery systems, which do not have large generating units and only absorb power but not export power to elsewhere. It happens at the transmission level and propagates to power delivery systems downstream.

19.1.2.1 Control

The principle of control is to maintain the balance between generation and load, ensuring that the frequency and voltage of the power system stay within the established system standard. The grid response under disturbance depends on its control mechanisms.

Governor control is the initial corrective action to any change of load or generation. The governor is a device that controls the mechanical power driving the generator via the valve limiting the amount of primary energy, such as steam, water, or gas flowing to the turbine. The governor acts only in response to locally measured changes in the generator's output frequency, which is related to the rotor's rotation speed. The governor responds within seconds, and the rotational speed of the rotor (thus the output frequency of the generator) stays constant after the governor's response [9]. The resulting frequency will be lower than the initial frequency, 60 Hz, if the load dropped or higher if the load increased.

Governor control is implemented to regain the local unbalance between the electric load and mechanical output of a generator. Automatic generation control (AGC) is used to bring all the generators within a control region back to 60 Hz. AGC can refer to two related concepts: (i) a device installed on generators and (ii) the control scheme that is implemented through these devices. A control area only has a select number of generators installed with AGC, which are of large capacities or at vital locations in the transmission network topology. After governor control, the control center in a region measures the difference between actual and scheduled net power flows with other control areas. This difference is called the area control error (ACE) [10]. A positive ACE means that the total amount of generation within the control area is more than the scheduled amount, because of a previous load drop and a governor's response. Receiving this ACE, the control center automatically sends signals to generators equipped with AGC to decrease the generator's output. Attacks at control systems have been studied for contaminating ACE signals or manipulating AGC on generators directly. Due to the natural influence of AGC-equipped generators, malicious manipulation will cause severe consequences on a large scale. Examples of these attacking models are presented in Section 19.2.

Beyond a certain level of power imbalance, generation reserves will be resorted to. The generation reserves in a control region are usually made of additional generating units that are on standby or those that can ramp up their output on request [10]. However, when all other options for balancing power have been exhausted, the system operator must proactively reduce the load, generally referred to as load shedding. Load shedding can be accomplished in the form of temporary power interruption to contracted users, voltage reductions, or rotating blackouts. Sometimes when rapid response is demanded, load shedding is automatically implemented through emergency control and protection schemes.

19.1.2.2 Scheduling

Scheduling ensures the long-term balance between generation and load. It determines which generating units should operate and at what power level, and it is accomplished on a predetermined fixed time interval. The objective is stated via economic means, to minimize cost, and subject to generation and transmission constraints. Scheduling consists of economic dispatch and unit commitment, each covering two overlapping time ranges.

Economic dispatch minimizes the overall production cost of electricity, considering the power delivery losses, by optimally allocating projected demand to generating units that are online. Computers at control centers run optimization algorithms, typically every 5 or 10 min, to determine the dispatch for the next hour and send these decisions to all the generators. Unit commitment decides when generators should start up and shut down, as it is not practical to keep all the generators online all the time. It is typically done 1 day ahead and covers dispatch for periods ranging from 1 to 7 days [10, 11].

Due to their financial meaning, the scheduling process is likely to be targeted by attackers motivated by increasing their own benefits or harming competitors' benefits. The consequence, nevertheless, is more than monetary loss and can lead to efficiency reduction, even frequency instability. Since the malicious intent may come from someone who actually owns the physical generating assets, the attack path is not confined to the cyber layers of power systems and can be difficult to defend against.

19.1.2.3 Protection

Another important aspect of the operation of power systems is protection. Protection devices are applied to specific equipment, including motor loads, generators, transmission lines, and other power delivery equipment. They serve to protect both grid assets and humans in contact with them. The coordinated operation of a group of protective devices on electric grids is defined as a protection scheme[2]. Protection schemes are designed to benefit an entire power system rather than only protecting individual devices. Protection is achieved using sensing equipment (including current/voltage transformers and digital sensors), de-energizing equipment (such as circuit breakers and reclosers), and relays, which are the decision-making units.

When a fault is detected on the electric grid, the corresponding protection scheme selects the range of circuits to be de-energized. This range may be as limited as the faulted device itself or as great as a full section of the distribution network. The selection should minimize the loss of load and also depends on the system operation status, fault type, and location [12]. By removing the

---

2 In some context, protection schemes are referred to decision-making logics embedded in protection devices.

circuit, the electric grid benefits from eliminating the stress and preventing further equipment damage and any associated expensive and lengthy repairs. Protective action must be fast (usually in fractions of a second) and accurate.

Power systems, in order for the protection schemes to work, are designed with sufficient redundancies, so that they can withstand the removal of one or several elements without unduly stressing the overall physical grid. As discussed in Sections 19.3 and 19.4, the inherent strength of the physical grid is the best defense against catastrophic failures. However, if the system is already stressed for some reason, such as equipment outages, heavier than normal loads, or extreme weather, this corrective action may exacerbate the situation and result in wide-area outages.

By targeting protection schemes, cyber attacks can cause power failure in a fast and direct way. One fear is that if attackers can hack into relays, they could de-energize circuits by tripping circuit breakers. Another possibility is to suppress protective actions under real faults or to launch joint cyber–physical attacks, with the physical attack to cause faults. Studies on both of these attack models are presented in Section 19.2.

19.1.2.4 Distribution Automation

Compared with transmission networks, distribution networks are usually planned with less expensive technologies, mainly because their scale of impact is limited to local areas. During recent years, distribution automation (DA) has been promoted in urban power delivery systems. In densely populated urban areas, like New York City, cities that concentrate highly functional government departments, like Washington, D.C., or areas that encompass manufacturers and harbor information entities, like Silicon Valley, electric operation cost is nontrivial, and failure of power delivery systems will lead to substantial economic losses.

IEEE defines a DA system as one that enables an electric utility to remotely monitor, coordinate, and operate distribution components in a real-time mode from remote locations. The principal objective of DA can be summarized as energy conservation and self-healing under faults [13]. Two prominent examples are automated fault detection, isolation, and restoration (FDIR) and optimization of system voltage and power flows. In an automated distribution network, voltage values are collected and power flows are monitored in real time. Once a fault is detected, circuit breakers and sectionalizers are controlled remotely to de-energize the faulted circuit and reroute the power to restore the loads. With regard to energy conservation, power delivery losses can be minimized with voltage control devices such as switched capacitor banks, load tap changers (LTCs) on the substation transformers, or voltage regulators installed along distribution feeders.

On the one hand, DA reduces the risk of power failures under probabilistic events, like tree intrusion into power lines, and disastrous weather. On the other hand, its communications, IT infrastructure, and sensors create vulnerabilities to cyber attacks. As renewable energy sources, such as roof photovoltaic (PV), are aggressively promoted under the smart cities context, urban electric grids can no longer be considered as passive power sinks. If manipulated by adversaries, DA systems can even drive flow back to its upstream transmission level. Correlated attacks at the distribution level can cause catastrophic consequences.

19.2   Threats to Electric Grids

Power systems are designed with redundancies to handle random disturbances of small magnitudes, such as tree intrusion to power feeders or automobiles striking power poles. However, the design cannot afford infinite redundancies to withstand or quickly recover from intentional destruction of critical components or damage inflicted simultaneously on multiple components. Threats that cannot be handled by redundancies mainly come from weather hazards and malicious attacks. This section presents historical examples of threats to power systems and their threat models.

19.2.1   Threats to the Physical Grid

The vulnerability of modern power systems is increased by the fact that the power industry is moving from vertically integrated structure to deregulated entities, which are responsible for their own economic benefits in a competitive electricity market environment [14]. Under the vertical structure, utilities have the obligation of providing reliable and satisfactory service. The recent restructuring of the power industry put forward the cost pressure from consumers and regulators, deferring investment to strengthen and upgrade the grid. Consequently, many parts of transmission networks are heavily stressed, which would directly affect the downstream power delivery systems in any adverse occasion.

19.2.1.1   Threats from Weather Hazards

In the 2013 report from the National Research Council (NRC) on power delivery systems, weather hazards are identified as the main threat to the national grid and the original cause of 60% cascading blackouts worldwide in the past 50 years, since the 1965 blackout [3].Disastrous weather events directly destroy or damage power delivery components simultaneously and on a great scale. Stations and substations are less affected. The major damage is concentrated on the transmission and distribution networks, as transmission lines and distribution feeders are sparsely located over a distance. Strong wind and heavy ice can easily bring down transmission lines and distribution feeders, interrupting the power delivery passages. Storms and earthquakes can take out a transmission tower, which disables all circuits amounted to it, manifesting domino effect. In the winter of 1998, an ice storm struck Quebec and the Northeastern United States; 2.3 million customers lost power. In 1999, Taiwan was affected by transmission tower collapse due to earthquake [3].

Storms take an annual toll on many distribution networks. Fortunately, utilities are usually prepared for such emergencies and often pool their resources to aid each other in restoring service. The single-contingency criterion $(N - 1)$ and multiple-contingency criterion $(N - k)$ network planning is used in reliability planning for this purpose [8]. DA technologies that enable faster reconfiguration of a network are often deployed in cities, which demand high reliability, to reduce the weather-related losses. If a weather event is widespread and long lasting, neighboring utilities might lose their backup, and this method becomes ineffective. For example, in 1990, Hurricane Sandy caused Long Island to lose all ties to Connecticut and New Jersey, and New York City lost all ties to New Jersey. The flooding resulted in over 2 million customer outages. "While utilities may typically prepare for an N-1 or N-2 event...It is an N-90 contingency," as noted by the New York Independent System Operator (NYISO) [3].

Extremely hot weather can overheat the transmission lines and distribution feeders. This heat itself, though causing extra power delivery losses, is unlikely to damage the lines and feeders. However, overheated lines and feeders will sag, which lower the clearance to ground and result in short circuit by touching a tree branch or other ground objects. Such incidents are more often observed in Asian countries at the distribution level, where the ground clearance of distribution feeders is lower [15].

Among all weather disasters, solar storms are rare in occurrence (happening around every 11 years) but could entail great impact on transmission and distribution networks, transformers, and relays. Its interaction with the Earth's magnetic field can produce auroral currents, or electrojects, that follow generally circular paths around the two magnetic poles at altitudes of 100 km or more. In March 1989, the entire Hydro-Quebec power system, including the major metropolitan areas of Montreal and Quebec City, was blacked out due to a gigantic solar storm. Several large power transformers also failed elsewhere in Canada and the United States, and hundreds of relays misoperated. Power systems in the northern hemisphere suffered voltage depressions and unusual swings in real and reactive power flows [16].

19.2.1.2 Threats from Malicious Attacks

Malicious attacks on the physical grid of power systems, though generally more fearsome, entail much less cost compared with disastrous weather and are unlikely to cause damage on a large scale. In general, there are three types of malicious attacks: small-scale vandalism by small groups with limited technical sophistication, well-planned terrorist activities, and strategic war activities.

Bulk transmission substations have unique security concerns to terrorist attacks in that they are relatively soft targets; they are vulnerable to standoff attacks as well as penetration attacks by adversaries compromising the substations' perimeter fences. HV transformers are difficult to replace, which could take as long as 15 months and cost as much as US$7.5 million. At the occurrence of transformer attacks, power systems may go under transient instability, and damage to HV transformers can separate power delivery systems from generation for long periods. Such attacks were launched on a frequent basis in the United Kingdom, by the Irish Republican Army, and in Iraq, by insurgent groups [6].

During wartime, electric grids were once important targets. In particular, air planners tended to become enamored with the vulnerability of electric grids to air strikes [17]. Historically, there have been four basic strategies behind attacks on national electric systems: to cause a decline in civilian morale, to inflict cost on the political leaders to induce a change, to hamper military operation, and to hinder war production [17]. However, hardly any war attacks on electric grids were successful in history. On the one hand, the military is relatively unaffected by a loss of power, as they are a high-priority user and can be self-supported. As power technologies evolve, many military bases are innovated into a microgrid [18]. On the other hand, the induced civilian damage was usually politically counterproductive. Therefore, threat models of war attacks on electric grids are less considered.

19.2.1.3 Models for Threats on the Physical Grid

Two approaches are common in modeling threats on electric grids: high-level probabilistic models and high-risk multiple contingencies. High-level probabilistic models do not model power system physics, that is, they allow neglecting time-dependent dynamics and grid configuration changes under threats. The initial overall system stress is represented by upper and lower bounds on a range of random initial component loadings. The model is parameterized by initial disturbances from weather [19]. High-risk multiple contingencies use a deterministic approach. Failures of components are selected from the *worst case* rather than the most likely case or every likely case. They are used to evaluate a power system's performance under the loss of a group of its most critical components, from which a plan is concluded to reinforce the power grids.

Both methods are commonly used in modeling damage from attacks and weather to the grid. Engineers and researchers have their own preference in choosing one method over the other, depending on applications. For instance, in power system planning, high-risk multiple contingencies are used by utilities, while the high-level probabilistic method is more preferred in estimation of cascading failures among layers of civil infrastructures.

19.2.2.1   How Secure Is the Cyber Layer of Power Systems?

The monitoring layer and control layer of a power system consist of sensors and automated and manual controls, all of which are tied together through communication networks. Early communication networks used in power systems were carefully isolated and tightly controlled. Historically, utilities typically own and operate at least parts of their own telecommunication systems, which often consist of a fiber-optic or microwave backbone connecting major substations, with spurs to smaller sites [20]. However, economic pressure from deregulation of the power industry pushed utilities to make greater use of commercially available communications and other equipment that was not originally designed with security in mind. Traditional external entities including suppliers, consumers, regulators, and even competitors now must have access to segments of the network [20].

From a security perspective, such interconnections with office and electronic business systems through other layers of communication have created vulnerabilities. Intrusions intending to bring down the power system can follow the same passage of a utility's daily operation. For example, remote access to a substation network from a corporate office or location is not uncommon for control and maintenance purposes. Dial-up, virtual private network (VPN), and wireless are available between remote access points and the substation local area network (LAN) [2, 21]. These access points are potential sources of cyber vulnerabilities.

Figure 19.2 illustrates the intrusion points in a power system communication network. Intrusions can originate from remote access points within a control center or a substation. If an intrusion originates from communication networks, it needs to pass the firewalls within the control center or substation of the target. As an alternative, intrusions can contaminate user interfaces of the control center or substation directly and gain access to other facilities within the target. Activities that compromise these remote access points could steal critical information (such as grid configurations) and cause electricity disruption by opening circuit breakers. In some cases, these malicious activities could result in physical damage of key grid components, as was demonstrated in the experiment of generation self-destruction by a cyber attack, launched by the Department of Homeland Security, in March 2007 [22].

19.2.2.2   Classification of Cyber Attacks

In EPRI's 2000 survey, utilities ranked bypassing controls, integrity violation, and authorization violation [23] as their top 3 perceived threats among all attacking methods in Table 19.1. The surveyed attacking methods, however, pause at the level of communication networks and do not reflect the final

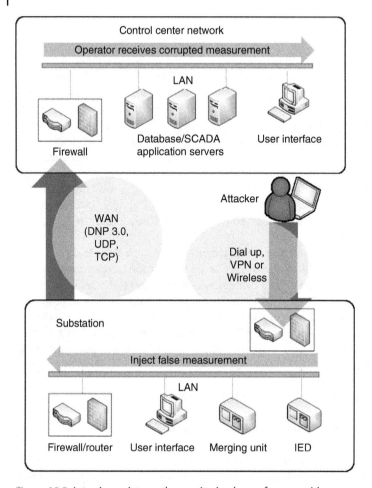

Figure 19.2 Intrusion points on the monitoring layer of power grids.

consequences inflicted on power grids. Here we present a comprehensive classification of cyber attacks, which describes their attacking paths on the vertical structure of power systems: control attack, monitoring attack, control–monitoring (CM) attack, and monitoring–control (MC) attack. Figure 19.3 illustrates their attacking strategies. The rest of this section explains these four classes of cyber attacks in detail.

19.2.2.3 Attacks on the Control Layer

The attacking classes on the control layer include a control attack and CM attack, illustrated in Figure 19.3b. In control attacks, hackers may hijack IEDs and falsify decisions. Knowledge of the original decision u is not necessary. This class of attacks is essentially the same as physical attacks, in the sense that the

Table 19.1 Attacks on communication networks.

Attack	Definition
Authorization violation	Accessing the system without proper access rights
Bypassing controls	Exploiting system flaws or weaknesses by an authorized user in order to acquire unauthorized privileges
Denial of services	Deliberate impeding of legitimate access to information
Eavesdropping	Acquiring information flows, by listening to radio or wireline transmissions or by analyzing traffic on a LAN
Illegitimate use	Knowingly or unknowingly intruding on system resources
Indiscretion	Indiscriminately opening information files and so on
Information leakage	(Users) unintentionally providing information to a disguised third party
Integrity violation	Modifying or destructing messages and the computer infrastructure with no authorization
Intercept/alter	Intercepting and altering information flows, usually by accessing databases and modifying data
Masquerade	Posing as an authorized user on a network and attaining full rights of an authorized user, often enabled by having other users' passwords
Replay	Using information previously captured without necessarily knowing what it means
Repudiation	Impeding users' access to the entities from which they undertook some action such as sending a message or receiving information
Spoof	Deceiving users and applications to performing functions on behalf of an attacker, by falsifying the legitimacy of the attacking software/service

grid's configuration is altered by control signals in a way similar to mechanical operation. For instance, the action of opening a circuit breaker through control logic and that of physically opening it on-site results in the same physical consequence on the grid.

In 2008, hackers turned out lights in multiple cities in an unnamed country, as revealed by a senior CIA analyst. The hackers demanded extortion payments before totally disrupting the power supply of the entire country. To date, the identities of the hackers remain unknown, but the intrusion was confirmed as via the Internet [2]. In the same year, a penetration test was performed in the United States by a team of security experts. With simple rootkits, the red team broke into the security system of an unnamed utility within the first few minutes of testing [24]. More worrisome control attacks might happen on the transmission level, where losing any key components such as transmission lines or a generating unit could result in instability of system frequency and voltage.

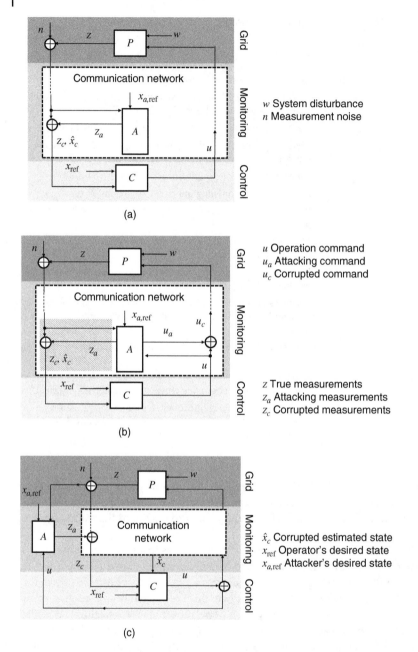

Figure 19.3 Four classes of cyber attacks on power systems. Control–monitoring attack is illustrated by (b). General control attack is illustrated also by the same diagram of (b) if taken out the shaded area. (a) Monitoring attack. (b) Control–monitoring (CM) attack. (c) Monitoring–control (MC) attack.

Compared with physical attacks, the adverse consequences of attacks on the control layer can be magnified in two ways: (i) more sophisticated manipulation of the IEDs is possible, and (ii) a large number of IEDs can be manipulated simultaneously. For example, Ref. [25] shows coordinated control of a group of generators in a manner that destabilizes other machines in the system and forces them to be disconnected from the system. Reference [26] shows control attacks on the distribution level. The attacker destabilizes voltages in an inverter-based power distribution grid by falsifying reference signals of the voltage droop controllers.

If the control attack is coupled with an attack that contaminates measurements, system operators would not be able to detect the malicious activities for a long time. After a certain point, the system may not be restorable, and wide-area outages could occur. This class of attacks targets control functions and are accessorized/facilitated by injecting false measurements on the monitoring layer. We call these CM attacks. In network security, similar attacking strategies are referred to as covert attacks [27]. The attacking strategy is illustrated in Figure 19.3b. By injecting attacking measurements $\mathbf{z_a}$, the attacker corrupts system measurements of real power, magnitudes of voltage, and current. The corrupted measurements $\mathbf{z_c}$ result in incorrect estimation of power system state variables (voltage magnitudes and phase angles) $\hat{\mathbf{x}}_c$. The corrupted measurements and estimated states disguise the malicious control activities $\mathbf{u}_c$. The operation status of the electric grid is driven from the operator-desired state $\mathbf{x}_{ref}$ to the attacker's desired state $\mathbf{x}_{a,ref}$ without being detected.

19.2.2.4 Attacks on the Monitoring Layer

Hovering over the control layer and the grid, the monitoring layer provides the power system's status in real time and inspects the condition of critical assets. With developed metering and communication technologies, the automation level at all layers of power grids can be greatly advanced. However, connecting a large number of devices over a wide area raises the concern of attacks on the monitoring layer. Moreover, attacks on the monitoring layer can manipulate information at higher resolutions and of nuanced values over a wider area [5, 28–30]. This causes great difficulties in detection and defense against these attacks; yet the adverse consequences resulting from these attacks are complex, uncertain, and hard to predict.

Attacks on the monitoring layer include monitoring attacks and MC attacks. Monitoring attacks aim at sabotaging and disturbing the information collection process in power system operation. MC attacks contaminate measurements to manipulate decision-making process. Their attacking strategies are illustrated in Figure 19.3a and c.

The effectiveness of monitoring attacks largely depends on its launch timing. Under critical system conditions, if measurements are not correct or delayed in transmission, operators will not be able to take proper control actions, resulting in severe consequences. For monitoring attacks on real-time measurements, one problem is how to inject false measurements that can bypass the detection of state estimators. The two conditions of injecting false measurements are that (i) they must be within the error tolerance of a state estimator and (ii) they must align with data structures of actual measurements, which are implied from the network topology of the target electric grid [31]. If they satisfy these two conditions, attacking measurements can be constructed to manipulate the state estimation in arbitrary ways.

In monitoring attacks, an attacker's goal is modeled as causing the maximum erroneous deviation from actual measurements in its pth norm; false measurements z_a are constructed to satisfy the mathematical form: $\max|z - z_a|_p$. The attacking models are extended by considering the constraints on an attacker's resources, for example, meter numbers and accessible locations [5]. Consequences of monitoring attacks include implying incorrect power system topologies and erroneous power system states, misleading state estimators to reject the actual measurements of un-attacked meters (so-called data framing, borrowing the note of framing innocent ones as criminals) [32, 33]. However, there is clear indication of what physical damage monitoring attacks will inflict on the grid.

Unlike monitoring attacks, MC attacks require great knowledge of system operation strategies and configurations. MC attacks not only make operators (or automated control programs) see what the adversary wants them to see but also fool operators to take actions that are unwarranted, so that the operators themselves damage the infrastructure, not the attacker. By injecting false measurements z_a, MC attackers *manipulate* control command u in order to accomplish the attacking goal $x_{a,ref}$ on the physical grid.

MC attacks are more difficult to detect, mainly for two reasons. First, they are accomplished via longer attack paths, which are initiated from the monitoring layer and completed on the physical grid. Second, existing mechanisms against malicious actions, which inspect control decisions' inconsistency with system states, are incapable of detecting MC attacks. Under MC attacks, control decisions are driven by corrupted measurements. Thus, control output and system states are always consistent. Reference [34] demonstrates using minimax formulation the reconstruction of attacking measurement vectors based on the consequences caused by an MC attack on the physical grid. Reference [35] exemplifies an MC attack aimed at local voltage instability by rerouting the voltage measurements in a power distribution grid.

MC attacks demand the most sophistication of attackers. Nevertheless, they could be the most harmful cyber attacks to power systems among the four classes, because of their potential consequences, difficulty of detection, and impact scale.

19.3   Electric Grid Response under Threats

The evolution of modern power systems was motivated by a series of black-outs and system failures. Compared with the power system that we had in 1895, when the power station at Niagara Falls was completed, today's electric grid[3] is highly interconnected and much more complex. Control and protection technologies have been under continuous innovation to ensure the normal operation of the electric grid against all types of threats. Nevertheless, the electric grid is not under a bulletproof shell. The ultimate consequence of a power system's failure to handle a threat can degrade power quality, electricity interruption, and even blackout. These consequences could be amplified or translated to other meanings in the context of urban operation.

This section looks into the formation of adverse consequences on the electric grid under threats. We found that this topic is even more important in some sense than merely mentioning power systems' control and protection schemes against certain threats; while the latter tells the existing capability of today's power systems, the formation of the adverse consequences reveals the vulnerabilities of the power system and direction of future grid reinforcement. Moreover, considering the threats from malicious attacks, this topic is critical in grid forensics. Knowing the formation of adverse consequences can bring insights into possible attack paths, which could span both the cyber and physical layers of power systems, and into the attacker's initial move.

In fact, any adverse consequences could have many causal chains that lead to them. Fortunately, there are only limited physical phenomena that result in undesirable performance on the electric grid. This section starts with introducing these phenomena, followed by a few real examples of failure mechanisms. Mathematical models for grid response and assessment methods for grid vulnerabilities are presented at the end of this section.

19.3.1   (Unwanted) Physical Phenomena on Power Grids

19.3.1.1   Circuit Faults
A circuit fault is formed by an undesired topological change inside power delivery equipment or on power transmission and distribution networks. The basic topological changes are short circuit, by adding an undesired electrical path, and open circuit, by eliminating a functional (but not necessarily active) electrical path.

Faults can occur within generators, motors, or transformers, on cables, or even within protective devices, such as a circuit breaker. Sometimes they can

---

3 A power system consists of three layers: the monitoring layer, the control layer, and the physical grid. The electric grid specifically refers to the physical grid. These two terminologies are defined in Section 19.1 under the vertical structure of power systems.

directly result in permanent damage of these grid assets. Fortunately, these faults are not common, as most of the grid assets are under regular inspection and encapsulated. Even though they have not happened in reality yet, cyber attacks could cause a fault on a generator, motor, or transformer by operating them above thermal ratings over a long period. This was demonstrated in the 2007 Aurora Generator Test by the Idaho National Laboratory [22].

If a fault occurs on transmission and distribution lines, they are line faults. Line faults are most common because lines are exposed to the environment. Bad weather and physical attacks can easily cause line faults, both short circuit and open circuit. Short-circuit faults are often triggered by lightning strikes. Extremely hot weather can cause line sags and make tree contact, and wind and ice loading may cause insulator strings to fail mechanically, both resulting in short-circuit faults. Storms can topple towers and poles, leading to open-circuit faults. Physical attacks often cause open-circuit faults. In Columbia, the Fuerzas Armadas Revolucionarias de Colombia (FARC)carried out hundreds of attacks on a monthly basis against transmission and distribution facilities [22].

Not only do line faults greatly reduce the power delivery capabilities of the transmission and distribution networks, but they are also accompanied by many other severe consequences. At the fault point, fire and even explosion may happen. There may be destructive mechanical forces due to high currents. The high currents in the faulted system may overheat equipment, reducing their useful life. In addition, unsymmetrical faults (if the faults do not occur at all three phases equally) will cause unbalanced operation of the electric grid; open-circuit faults on transmission lines that are carrying high currents may cause transient (frequency) instability, which entails an adverse impact over a great scale on the power grid.

19.3.1.2 Frequency Instability

Frequency instability of a power system depends on the system configuration and operating mode. Since power systems rely on synchronous machines to generate electrical power, a necessary condition for stable and nominal system frequency is that all synchronous machines remain "in step." Physically, this means all generators' rotors rotate at the same pace. When the configuration and operating mode (e.g., electricity demand or generation capability) are subjected to a change, this pace is disturbed. As introduced in Section 19.1, the system frequency starts to increase if the electricity generation amount exceeds the demand, and vice versa. The disturbance is said to result in frequency instability if some generators lose synchronism with the rest of the system. In other words, their rotors run at a higher or lower speed than the system synchronous speed. Unstable system frequency will result in large fluctuation in the generators' power output, current, and voltage.

Based on the causes, the magnitude of the disturbances over a duration can be large or small. Continuous small variations in loads and generation are small disturbances. With the control and scheduling methods (introduced in Section 19.1), modern power systems can remain in stable mode under small disturbance. However, cyber and physical attacks could temporarily disable the control and monitoring functions of power systems. This instability that follows can be of two forms: (i) rotor oscillations of increasing amplitudes and (ii) a steady increase in rotors' rotating pace of a group of generators [36]. These instability phenomena after small disturbance are *small-signal instability*.

If a disturbance is of large magnitude and happens over a short duration, it might result in *transient instability*. Loss of running generating units and active transmission lines because of malicious attacks and weather hazards and sharp increase in demand due to extreme hot or cold weathers are all transient disturbances. A system's ability to remain stable depends on both the initial system operation state and severity of the disturbance. Unstable grid response involves large excursion of generator rotor angles [36]. Modern power systems are designed to operate as stable for a selected set of contingencies. However, this selection is hard to make exhaustive and currently incapable of handling attacks that aim at the vulnerable control and protection of the system.

19.3.1.3 Voltage Instability

A power system is said to be voltage stable if after any disturbance it is able to restore the voltage to a steady state and to have this voltage of an acceptable magnitude. The main condition of voltage instability is that the power system cannot meet the demand for reactive power, which may be caused by generators' excitation limit, transmission lines' high inductance, load characteristics, or even the voltage control devices on transmission and distribution networks that are supposed to support voltage [37].

Another phenomenon often mentioned with voltage instability is voltage collapse. They are two different concepts. Voltage instability is a local phenomenon but may have a widespread impact. Voltage collapse is more complex and is the result of a sequence of events accompanying voltage instability, leading to a low-voltage profile in a significant part of the electric grid. Voltage collapse is difficult to simulate and does not have fixed patterns. Despite its severe adverse impact, it is unlikely to be manufactured by malicious attackers.

Voltage instability is sometimes associated with frequency instability. The gradual loss of synchronism of generators would result in low voltage at the intermediate points in the network [37]. Studies have shown that malicious attacks become a great concern of system voltage, even if the grid was evaluated as voltage stable under regular disturbances. These attacks could happen at both the transmission and distribution levels. In particular, in the urban electric grid that is microgrid and integrated with renewable energies, high reactive power demand can be manipulated to lead to unstable voltage of loads across the city [26].

Subject to threats, final damages on the electric grid may be minimized and limited to a local area or propagate to wide areas. To study failing mechanisms on electric grids, we summarize the patterns of unfolding sequences of cascading blackouts in history:

1) Initially the electric grid is operated under stress. The cause can be unusually hot or cold weather (e.g., in the Western US blackout of 1996, the very hot weather led to sharp load increases and great amount of power export on the interties from the Pacific Northwest to California [38]) and planned equipment maintenance (in the 2003 Sweden/Danish blackout, two 400-kV lines and HVDC links were out of service due to maintenance [39]).

2) Generating units and/or transmission lines are lost, which will further stress the grid. In the Western US blackout of 1996, it was a 500-kV line outage. In the North American blackout of 2003, a generator was tripped, and so was a 345-kV line. In the 2003 Italy blackout, a few intertie lines were tripped between Italy and Switzerland [39–41].

3) Inappropriate control and protective actions are taken, because of operators' mistake, settings of the algorithms, or inherent flaws of protective device design.

   a) Information sharing plays a key role here and may greatly affect the operator's decision at critical moments. In the 1996 Western US blackout, the simulation in the Western System Coordinating Council (WSCC) database showed stable frequency response, whereas the actual grid response was oscillatory unstable. The incorrect information partially led to the delayed operation intervention [42].

   b) As the grid becomes more interconnected, the operating conditions may vary at a wider range. Settings of algorithms, in particular for protective devices, need to adapt to this widened range of system operating modes. In the 1965 Northeast blackout, the primary cause was the setting of remote breaker-failure admittance relays mis-tripped. The settings were based on the load level 9 years earlier, which was lower than the normal load carried on that November 9 blackout [42, 43].

   c) Distance relays performing Zone-3 tripping is another well-recognized cause. When the grid is operated under high current and low voltage, the impedance seen by the backup relay is less, which leads to the backup relay's operation. This Zone-3 operation will further stress the grid, overload other lines, and lead to other dynamic issues [19].

4) The subsequent events usually involve frequency instability and possibly voltage instability. Emergency protection schemes can stop the events from cascading by splitting the systems according to plan and by shedding loads to maintain supply and demand within each island. If this action is not taken, the grid may end up with widespread voltage collapse and uncontrolled system splitting (as a few groups of generators, and each group is in synchronism but not in synchronism with other groups) [44].

It is worrisome that this pattern of cascading blackout might be reproduced by adversaries. Theoretically, attackers with sufficient resources can coordinately launch attacks on the grid as trigger events while hacking into the monitoring and control layer, sabotaging the restoration of the power system. If attacks were launched under bad weather, the consequence can be worsened. Thus, the methods used to prevent cascading failures resulting from random events are also invaluable to prevent malicious intentions.

19.3.3 Modeling and Vulnerability Assessment Methods of Grid Response

Dynamic response of the electric grid is highly complex, involving phenomena of different timescales and at all grid components. For real-time operation purposes, grid response is simulated for a few phenomena separately; in offline studies of past events, which allow more time and sophisticated modeling, the interactions of these phenomena may be considered in more comprehensive models. The following tools are often used in practice to study different phenomena in grid response [9, 14, 19, 45].

- Static security assessment (SA) – SA use fast approximate power flow methods, sometimes DC power flow, to screen a preselected list of contingencies. The identified harmful contingencies are further analyzed with full power flow.
- Transient security assessment (TSA) – TSA examines transient frequency instability at post-contingency conditions with time resolutions of 0.01–10 s. The transient contingencies are preselected loss of grid components and various faults.
- Voltage security assessment (VSA) – VSA examines voltage instability at post-contingency conditions by solving the steady-state power flow. VSA analyzes the response of AGC, tap changers, and generators' excitation systems in the order of seconds.
- Small-signal analysis (SSA) – SSA examines small-signal instability of the power system by using linearized system models.
- Available transfer capacity (ATC) analysis – ATC analysis combines the functions of SA, TSA, VSA, and SSA to compute the available transfer capacity between specified points of receipt and points of delivery.

The vulnerability of power systems can be assessed for the entire grid or assessed at the most vulnerable grid component. The vulnerability of the entire electric grid depends on the grid's loading conditions. For the capacity distribution of grid components, inherently, the grid's capability of handling the loads is determined by its topology. Based on this fact, a category of vulnerability assessment uses network graph theory to model the failure mechanism. Overloaded components are removed, and the further stressed grid is subject to an accelerated rate of overloading and failure. References [46–48] use the small-world network concept in social science to count the remote connections in a power network. The loss of those remote connections will distribute power to other lines, further decreasing the transfer capacity of the grid. References [47, 49] use the scale-free network concept, similar to the small-world concept, but instead count vital buses in a power network. Reference [50] uses the betweenness centrality theory to assess the load on nodes or links. As power flows through the lines of least electric impedance, a line is more loaded by its electric character and location on the network. By removing the overloaded lines and nodes, this method reveals the propagation of failures in a cascading blackout.

The vulnerable grid components are identified under two conditions: (i) they tend to fully or partially lose normal function when subjected to threats, and (ii) the degraded or ceased functions are important to maintain the system normal operation. Sometimes the second factor may be determinant; a grid asset is considered vulnerable because of its criticalness, even if it is not intuitively vulnerable. HV transformers are a good example [22]. For other grid assets, which possess similar functions, methods that consider their roles in grid response are used to identify the vulnerable/critical assets. In [51–53], a bi-level optimization problem is posed to identify the vulnerable transmission lines of a power grid as

$$\max_{l \in L} \min_{p \in D} \sum_i c_i p_i$$
$$\text{s.t. } g(p, l) \leq b$$

The narrative of the formulation is that attackers will choose the lines ($l \in L$) that maximize the system operation cost ($\sum_i c_i p_i$). The grid response is modeled by the inner problem as the operator sheds loads to maintain supply and demand balance ($g(p, l) \leq b$) after the attack while minimizing the load-shedding cost ($\sum_i c_i p_i$). The vulnerable lines are identified under the worst scenario, in which attackers deduce the grid response in advance and make moves accordingly.

19.4 Defense against Threats to Electric Grids

Defending against threats to electric grids includes three stages: detecting the existence of the threats, identifying the types and sources of the threats, and mitigating the negative impacts from the threats. The focus of technology development in each stage may depend on the nature of threats. For weather hazards, fault allocation methods that identify the damaged circuits and restoration technologies carry more interest. This aligns with our intuition: as many parts of the electric grid may be struck by a storm, the priority is to isolate the problematic circuits and restore electricity to critical services. In contrast, for malicious attacks on the cyber layers of power systems, detection methods are indispensable in order to proceed to the other two stages.

This section gives an overview of solutions to the threats faced by today's power systems. These solutions fall into three categories: technical recommendations, technological schemes (such as remedial control and protection schemes), and mathematical models of defense methods in each defense stage. More emphasis is put on the last category. These mathematical models either provide systematic frameworks of implementing the recommendations or bring insights into the development of different technologies. By presenting these methods at different layers of power systems, we hope to construct for readers a directory of research and future development in electric grid defense.

Finally, we have to point out the incompleteness of this section. Despite our best attempt at describing defense solutions from the standpoint of power systems, in the context of smart cities, the defense is perhaps taken from a higher level, monitoring the threats to multiple layers of infrastructures and coordinating the efforts from all the layers to mitigate the negative impact. We hope to exploit this topic in further studies and look forward to any feedback from readers.

19.4.1 Recommendations for Grid Resilience Enhancement

Modern power systems are reliable in responding to random disturbances of small magnitudes. The threats that cannot be handled by redundancies are primarily weather hazards and malicious attacks, which further fall into physical attacks on grid assets and cyber attacks that disturb monitoring and control functions of power systems.

Recommendations based on weather hazards include facility repair, service restoration, and consequence management. To ensure fast repair, practices and technologies that improve system-wide instrumentation and the ability of near real-time state estimation should be developed on both the transmission and distribution levels. Techniques that enable fast fault localization and isolation are particularly important [54, 55]. Without these measures, system operators would select a restoration plan under disadvantageous observations. On the distribution level where the cities' loads are connected, DA and load-shedding management technologies should be available [13]. Utilities and municipal service entities should work with local private and public sectors to identify critical customers and plan a series of technical and organizational arrangements [1]. During the times of system stress, noncritical loads may be shed or reduced through brown rotations in order to relieve the loading stress of power delivery equipment and supply–demand balance [56, 57]. After faults are isolated and disturbances are cleared, restoration of service should be prioritized for critical customers and sectors by reconfiguring the network, local storages, and distributed generators [13].

Recommendations based on physical threats from malicious attacks mainly focus on system hardening and crime surveillance. The attack targets include transmission and distribution lines, substations, control centers, communication devices, and other power delivery equipment [22]. Targeting generation facilities is less likely, because power plants are usually protected by external architectures and trained personnel [3].

Transmission and distribution lines are widely distributed; it is very difficult to completely protect all key components. Standards (including EOP-003-1 and TPL-001-0 to TPL–004-0) and techniques need to be enforced that identify lines and sets of most impact on post-contingency cost [19]. In the meantime, regional operators should improve electronic surveillance that uses sensors and monitoring equipment, along with information-processing equipment, to allow rapid identification of and response to multisite attacks [3, 22]. Substations and control centers are the choke points of power delivery. There is a need to develop and dedicate specific security equipment to these facilities, such as cameras, sensors, intrusion detection devices, access controls, improved lighting and perimeter security fencing, and buffer zone security [3, 22]. In addition, to keep the operator aware of the power grid's status, physical security and energy independence of communication infrastructure must be ensured [57, 58].

Recommendations based on cyber threats from malicious attacks are made to both utilities and upstream vendors of communication and remote-controllable devices. Utilities should consider isolating the critical systems from judicious interconnections; wireless links should not be used to implement critical functions. As control systems need data from other systems and vice versa, interchange over public networks is sometimes not avoidable. However, such interconnections representing security risks should be designed with care. Firewalls with strong authentication and integrity checks on network protocols, patch management, and network traffic monitoring should be implemented. On the substation level, communication to relays and other monitoring equipment should be conducted under strict requirements of security, such as minimizing connectivity and ensuring strict authentication.

SCADA vendors and vendors of IEDs with communication functions should establish security practices and provide security services throughout a product's life cycle [20]. Vendor support is needed to remediate the unnecessary exposure and vulnerabilities caused by excessive services and unpatched systems [59].

19.4.2 Core Technologies for Power System Resilience

19.4.2.1 Emergency Control and Protection

When the electric grid is under high stress and failure starts to propagate, emergency schemes that coordinate the control and protection efforts to deal with this situation are then deployed to preserve system stability, maintain overall system connectivity, and avoid serious equipment damage. These emergency schemes are usually referred to as SPS (Special Protection System) or RAS (Remedial Action Scheme) [60, 61]. A good example in the United States is the SPS installed in the Florida Power and Light system [62]. Many studies and practice have been devoted to the development of SPS, but so far there is not a rigorous standard defining its functionalities and operation. Technologies that are commonly regarded as essential to SPS include, but are not limited to, adaptive relaying that adjusts tripping settings based on the stress level of the grid, automatic reactive power compensation using microprocessor controls, phasor measurement units (PMU) that enable near real-time synchronized monitoring over wide areas, and digital control and protection at power plants that block unit tripping during stressed grid operation [63]. Typical key operations in SPS are as follows:

1) When great imbalance is observed between generation and loads and reserves are exhausted, underfrequency load shedding and undervoltage load shedding are deployed for noncritical loads. These load-shedding relays are installed at planned locations, and their action will not affect service to critical loads. The tripping range of frequency and voltage are usually preset. Sometimes, remote control tripping is also allowed.

2) If load shedding does not clear the imbalance, the system may be separated into predefined islands. This intentional separation is distinct from the islands that are formed by coherent swings among generation groups in frequency instability. Out-of-step relays are installed at limited locations to control the propagation of failures.

3) Out-of-step relays are also installed at generating units to prevent permanent equipment damage due to frequency excursion. When the system is assessed under high stress, these relays are blocked to prevent equipment loss that exacerbates the situation.

4) When the power system cannot meet reactive power demand, static VAR compensator (SVC) and static compensator (STATCOM) are coordinated to prevent voltage instability.

5) After faults are isolated and disturbances are cleared, selected gas turbines are reconnected and controlled to restored service. This black-start procedure must be coordinated among generating units and with the load recovery level.

19.4.2.2 Restoration from the Distribution Level

Restoration from the distribution level is often associated with two concepts: distributed generation (DG) and microgrids. Compared with conventional power plants, DG are connected at the distribution level, have much smaller capacity (from a few kilowatts to megawatts), and often use renewable energy as the primary source [64]. They are identified as valuable resources to maintain service to critical loads, such as hospitals, fire stations, and police stations [57]. Light-emitting diode (LED) traffic and street lights paired with distributed energy resource (DER) can operate when the bulk power supply is not accessible [65]. Recent studies show that DERs that are installed with battery storage can be deployed in black start and achieve faster and more efficient restoration [66].

Microgrids are distribution networks consisting of DG, energy storage, and loads. They normally operate connected to the main grid but can break off and run in an autonomous mode in emergent situations [64]. One successful example of microgrid in grid's resilience is the Sendai Microgrid in Sendai City, Japan. In the Tohoku Region Pacific Coast Earthquake, 2011, the inland areas of the Tohoku and Kanto regions suffered from damage to lifeline utilities due to the shutting down of power plants and collapse of infrastructure. Electric power was partially restored after 3 days of the earthquake. Nevertheless, the Sendai Microgrid was able to supply power to loads continuously within its service area. The autonomous operation started by battery supplies, followed by gas engine restoration [67].

Microgrids have clear boundaries; they are connected to the main grid at a point of common coupling that maintains voltage at the same level of the main grid. During the islanding operation, a switch can separate the microgrid from the main grid automatically or manually. In the autonomous mode, the generation and storage are coordinated, and some loads are shed, in order to maintain the power balance within the microgrid [68]. After the main grid is cleared from the threats, by regulating DG, a microgrid is synchronized to the same frequency as the main grid and reconnected to it. A microgrid comes in a variety of designs and sizes. It can be a single facility like the Santa Rita Jail microgrid in Dublin, California, or a distribution network that powers a large area like in Fort Collins, Colorado.

19.4.3 Development of Defense Methods against Threats

19.4.3.1 Defense on the Physical Grid

To defend against physical threats, the electric grid needs to be planned for long-term infrastructure reinforcement and to prepare an operation for contingent conditions. Defense methods on electric grid planning focus on identifying critical grid assets. If ceasing or degrading its function will induce great impact on the rest of the system, then the asset is a critical asset. This definition is similar to vulnerable grid components.

Some grid assets are critical by nature, for example, substation transformers and HV transformers, while others may depend on its operating condition and location. For widely distributed transmission lines, it is financially impossible to harden or monitor every individual line. In [51–53, 69], a bi-level optimization problem is posed to identify the critical transmission lines of a power grid as

$$\max_{l \in L} \min_{p \in D} \sum_i c_i p_i$$

s.t. $g(p, l) \leq b$

The transmission lines selected by the attacker induce the maximum impact post-contingency. In the attack, the attacker possesses asymmetrical information, knowing the contingent rescheduling policy that the operator will initiate after the chosen transmission lines are taken off. While the operator tries to minimize the attack's impact, the attacker maximizes this impact by "foreseeing" the operator's strategy. This model is also referred as the offender–defender model, where the power system operator happens to play the defender's role.

The transmission lines are identified by whose removal will most significantly reduce the power transfer capability (whereas these transmission lines themselves are not necessarily of the greatest volt-amp capacity). The small-world network concept was used to describe the electric grid's response on a high level. Based on the topology of a transmission network, we can calculate the clustering coefficient C and characteristic length path L as

$$C = \frac{1}{N} \sum_i \frac{2E_i}{k_i(k_i - 1)} \text{ and } L = \frac{1}{N(N - 1)} \sum_{i \neq j} d_{ij}$$

where N is the number of nodes in the network, E_i is the number of edges in the subgraph associated with node i, k_i is the number of adjacent nodes of i, and d_{ij} is the shortest path length between nodes i and j. The transmission lines are selected if L is increased without these lines.

In operation defense, a tri-level model is proposed in [53] as an extension of [52, 69]. The objective of the formulation is

$$\min_{p \in D} \max_{l \in L} \min_{p \in D} \sum_i c_i p_i$$

which reschedules against the attacker's selection in the bi-level optimization model. This model is commonly referred as the defender–attacker–defender model.

The interaction between the attacker and defender can go on for virtually infinite iterations, and the challenge is presented as finding the solution to this problem. In the simple case as presented earlier, where the attacker and operator have exactly the same objective of opposite optimizing directions, the problem is a zero-sum game. A Nash equilibrium can be found, at which the defender can no longer reduce the negative impact and the attacker can no longer increase it [70, 71]. However, in reality, the operator is unlikely to game with the attacker; she/he has her/his own priority of operating the grid. The solution to the problem can be complex.

19.4.3.2 Defense on the Control Layer

In Section 19.2, cyber threats from malicious attacks are classified as monitoring attacks, control attacks, MC attacks, and CM attacks. To protect the power system against these threats, defense methods on the control layer are represented by terminating susceptive control actions, hardening communication infrastructure, active signaling, and inference control.

If malicious intentions have been successfully detected and identified, the operator can simply terminate the control functions on the target. For example, Internet-based load altering attacks are studied in [72]. These attacks attempt to control and change loads that are accessible through the Internet in order to damage the grid through circuit overflow or disturbing the balance between demand and supply. Once such attacks are detected, immediate actions are suggested to shut down automated load scheduling or curtail the targeted loads.

Hardening communication infrastructure resembles the idea of reinforcing critical grid assets. The communication links to be hardened are identified by estimating the value of investment (alternatively, by estimating the cost without hardening the object). The mathematical models usually take the form of a bi-level mixed-integer problem, with an objective

$$\min_{\alpha_i} \max_{\alpha_j \neq \alpha_i} C(\alpha_1, \ldots, \alpha_N)$$

where C is the negative impact from the attack and α_i is the object to protect. The condition $\alpha_j \neq \alpha_i$ can be included in the constraints; defining the attack is ineffective to the hardened objects. In [72], this method is applied to identify the minimum load groups to protect, so that Internet-based load attacks result in the least negative impact. In [73], attackers attempt to obtain economic benefits by controlling generators' output in electricity markets. Using the same method, the most influential generators are identified.

In CM attacks, measurements are corrupted to cover the malicious control actions. Therefore, the malicious control is difficult to detect directly. Reference [74] studies a general cyber–physical system. The attacker launches a replay attack by replacing the actual measurements with the recorded measurements. The attacked system correspondingly performs the control of recorded action. In this way, the attacker manipulates the system control, which may be harmful to the actual conditions. The proposed solution for the operator is to inject a control signal Δu_k on top of the optimal control u_k^*. If the system is under attack, the asymptotic expectation of X^2 detector will have a higher value due to the Δu_k. Thus, the attack can be detected. The weakness of active signaling is sacrificing control performance by deviating from the u_k^*.

In MC attacks, measurements are corrupted to manipulate power system operation. The defense challenge arises for the control functions that rely on real-time measurement input. Even if an MC attack is detected, it is difficult to identify and restore the contaminated data set in real time. In this case, inference control is applied. In [75], ACE is contaminated in the secondary frequency control. The untrusted measurements are discharged and the control center is "flying blind." The proposed defense method predicts the AGC performance based on load forecast and obtains a new ACE. For different possible load scenarios, the mean of the new ACE is used to issue generator correction during an MC attack.

19.4.3.3 Defense on the Monitoring Layer

The main function of the monitoring layer is to collect data for decision support on the control layer and to inspect the performance of the physical grid. Defense methods on the monitoring layer include the following categories:

(i) Assessing system vulnerability to threats based on operating status. In order to be successful, attacks on measurements, including monitoring attacks, MC attacks, and CM attacks, must not trigger the alarm of state estimators. Therefore, these attacks are either unobservable or cause very small deviations that are within the tolerance of the state estimator detector [76]. In [77], security indices, for unobservable sparse attack α_k and for small magnitude attack β_k, are created by solving

$$\alpha_k := \min_c \|Hc\|_0 \text{ and } \beta_k := \min_c \|Hc\|_1 \text{s.t. } 1 = \sum_i H_{ki} c_i$$

where $\|Hc\|_0$ denotes the number of nonzero elements in the attacking vector $a = Hc$, c is an arbitrary vector, and H is the Jacobian matrix derived from power flow equations. $\|Hc\|_1$ is the 1-norm of Hc.

Control attack is studied in [28], which considers a general linear time-invariant (LTI) cyber–physical system. The work proposes the condition of static detectability and dynamic detectability of malicious control action $u_K(t)$. These conditions and security indices depend on the system topology and operating conditions. They provide a security assessment of the entire system.

(ii) Detecting anomalies in measurements and control actions. In [75], a statistical method and a temporal method are proposed to detect anomalies in ACE signal. In the statistical method, ACE's value is checked against a probability distribution summarized from its past values. To prevent an attacker from gradually changing the ACE signal, the temporal method compares the difference between the sum of expected and actual ACE over a time period of interest.

Reference [78] proposes two anomaly detectors and uses a Bayesian methodology for integrating the output of these detectors. The control actions are first to check against an invariant induction detector and then with the artificial ant approach. The invariant induction detector may not cover all relationships in the data, while the artificial ant approach sometimes tends to cause a false positive. Integrating the outputs from the two detectors through a Bayesian framework complements the weakness of these two detectors and provides a satisfactory detection rate.

References [28, 79] develop centralized and distributed detection methods for control attacks in power systems. The detection filters are developed as a residual signal $r(t)$, which equals zero if and only if the system is not under attack. The distributed detection filter reduces the requirement of continuous communication of measurements to the control center. However, both detection filters are not able to detect attacks that hide in the transient dynamics or that are aligned with the noise statistics.

(iii) Identifying and localizing compromised components in power systems. Identification and localization are similar concepts in the sense that they pick out the corrupted data from the normal ones, while identification usually finds the corrupted data based on a certain set of predefined attacking patterns. These patterns may contain more information that may reveal the attacker's identity or appear in temporal sequences that reveal the attacker's next move.

Reference [79] proposes a centralized procedure consisting of designing a residual filter to determine whether a predefined set coincides with the attack set. The output of the attack identification filter for the attack set K will be a residual signal r_K. The attack set K is identifiable if and only if $r_K(t) = 0$ for all t. Reference [76] proposes a detector based on generalized likelihood ratio test (GLRT) and shows that the detector is asymptotically optimal, in the sense that it offers the fastest decay rate of mis-detection probability. The GLRT is further reduced to a non-convex optimization problem. Given the attacker has access only to a small number of meters (i.e., the attacking set is sparse), the problem is solved by exhaustively searching through all sparsity patterns.

(iv) Restoring the contaminated measurements. The restoration may be applied directly to the measurements or to the system states, which are estimated from the state. Reference [80] proposes a decoder and the maximum number of compromised meters that the decoder can tolerate and still unambiguously recover the correct state.

Detection and measurement restoration are also solved as a joint problem. Reference [81] defines an estimation performance measure. The state is estimated as the recovered data by minimizing the performance measure subject to constraints on the tolerable levels of detection errors.

(v) Identifying meters to be secured. The selection of meters is related to the identifiability and detectability condition of the attacks. In problem formulation, it is equivalent that the measurement is ensured to be accurate if its meter is secured.

For unobservable measurement attacks, the attack sets are usually interacting and conforming [5]. Therefore, by securing some meters, the completeness of the interacting and conforming attack set is broken. The measurement residuals will exceed the detection threshold, and the attack becomes detectable to the state estimator [31]. An example of a load redistribution attack is presented in [34]. A combinatorial optimization identification algorithm is proposed to check the meter sets, which once secured will give a measurement residual greater than the bad data detector's threshold. Another example of topology attack is studied in [33], and meters to be secured are identified through a network graph approach. It proves that any topology attack is not detectable if protecting all injection meters on a vertex set B and flow meters on a spanning tree $\mathcal{T}$, given that $(B, \mathcal{T})$ covers the whole graph.

The effect of meter protection is also quantified by the cost of attack. Reference [82] selects meter sets to protect by maximizing the minimum attack cost and the average attack cost, where the attack cost is transformed from the detectability of the attack.

In [79], the number of meters to be secured is equalized to identifiability of a monitoring attack. The attack set K is identifiable if and only if a Kron-reduced LTI system is without corrupted measurement.

In addition to securing measurements, the system state value can be secured directly. Reference [83] proposes a method to place PMU. With known accurate states, the corrupted estimation can be easily identified. This method, although attractive, is subject to high cost of the PMU.

Final Thoughts

Electric grids play critical roles in urban operation, providing electricity to citizens and supporting other critical infrastructures. This chapter discusses four important questions for building a resilient electric grid in smart cities: (i) What can go wrong? (possible threats); (ii) What are the consequences? (grid response and causal chain); (iii) How likely are the consequences to occur? (failing conditions); and (iv) How is a normal status restored? (resilient response). We answer these questions by presenting both real examples and technical principles of power systems' response to cyber and physical threats.

In particular, we distinguish the concepts of reliability (which could be enhanced by optimally increasing the redundancy of power delivery equipment of the electric grid) and resilience (which requires more sophisticated and intelligent analysis and operation in real-time). We show that many favorable features of smart cities, such as advanced cyber-infrastructure and interdependent urban operations, could not only enable more resilient grid operation against some threats (e.g., natural disasters) but also significantly increase the risk and impact of cyber-threats. The electric grid's configuration and its operation strategies should be carefully designed by considering all these factors and to be integrated with other urban functions.

Questions

1 What are the differences between reliability, robustness and resilience for the electric grid? Alternatively, think of if we can use reliability metrics (e.g., SAIFI and SAIDI) to evaluate the robustness and resilience of electric grid? Why or why not?

2 What is the cyber-physical structure of the electric grid? Can you depict a diagram and point out possible attack paths/ vulnerabilities on it?

3 Use the sections to describe an electric grid. What section/s cover the geographical range of cities?

4 Cyber-attacks can be defended against on both the cyber-space and physical grid. Name a few defense methods.

5 What metrics can we use to evaluate the consequences of threats to the electric grid? What metrics can we use to evaluate the resilience of the grid?

References

1 Stanford University (2001) *Critical infrastructure assurance guidelines for municipal governments planning for electric power disruptions.*
2 Liu, C.C., Stefanov, A., Hong, J., and Panciatici, P. (2012) Intruders in the grid. *IEEE Power and Energy Magazine*, **10** (1), 58–66.
3 Cooke, D. (2013) *The Resilience of the electric power delivery system in response to terrorism and natural disasters.*
4 Halper, E. and Lifsher, M. (2014) *Attack on electric grid raises alarm.*
5 Liu, Y., Ning, P., and Reiter, M.K. (2011) False data injection attacks against state estimation in electric power grids. *ACM Transactions on Information and System Security (TISSEC)*, **14** (1), 1–33.
6 Parfomak, P.W. (2014) Physical Security of the US Power Grid: High-Voltage Transformer Substations. Tech. Rep., Congressional Research Service, https://fas.org/sgp/crs/homesec/R43604.pdf (accessed 24 January 2017).
7 Northcote-Green, J. and Wilson, R.G. (2006) *Control and Automation of Electrical Power Distribution Systems*, vol. **28**, CRC Press.

8 Willis, H.L. (2004) *Power Distribution Planning Reference Book*, CRC Press.

9 Kundur, P., Balu, N.J., and Lauby, M.G. (1994) *Power System Stability and Control*, vol. **7**, McGraw-hill, New York.

10 Gómez-Expósito, A., Conejo, A.J., and Cañizares, C. (2008) *Electric Energy Systems: Analysis and Operation*, CRC Press.

11 Morales, J.M., Conejo, A.J., Madsen, H., Pinson, P., and Zugno, M. (2013) *Integrating Renewables in ELECTRICITY MARKETS: OPERATIONAL PROBLEMS*, vol. **205**, Springer Science & Business Media.

12 Anderson, P.M. (1998) *Power System Protection*, John Wiley & Sons, Inc.

13 Momoh, J.A. (2007) *Electric Power Distribution, Automation, Protection, and Control*, CRC Press.

14 Denny, F.I. and Dismukes, D.E. (2002) *Power System Operations and Electricity Markets*, CRC Press.

15 Seifi, H. and Sepasian, M.S. (2011) *Electric Power System Planning: Issues, Algorithms and Solutions*, Springer Science & Business Media.

16 Kappernman, J. and Albertson, V.D. (1990) Bracing for the geomagnetic storms. *IEEE Spectrum*, **27** (3), 27–33.

17 Griffith, T.E. Jr. (2009) *Microgrids and Active Distribution Networks, IET Renewable Energy*, The Institution of Engineering and Technology.

18 IET Renewable Energy Series (2009) *Microgrids and Active Distribution Networks*, The Institution of Engineering and Technology.

19 Baldick, R., Chowdhury, B., Dobson, I., Dong, Z., Gou, B., Hawkins, D., Huang, H., Joung, M., Kirschen, D., Li, F. *et al.* (2008) *Initial Review of Methods for Cascading Failure Analysis in Electric Power Transmission Systems IEEE PES CAMS Task Force on Understanding, Prediction, Mitigation and Restoration of Cascading Failures.* Power and Energy Society General Meeting-Conversion and Delivery of Electrical Energy in the 21st Century, 2008 IEEE, IEEE, pp. 1–8.

20 Ericsson, G.N. (2010) Cyber security and power system communication essential parts of a smart grid infrastructure. *IEEE Transactions on Power Delivery*, **25** (3), 1501–1507.

21 Hong, J., Liu, C.C., and Govindarasu, M. (2014) Integrated anomaly detection for cyber security of the substations. *IEEE Transactions on Smart Grid*, **5** (4), 1643–1653.

22 Morgan, M.G., Amin, M., Badolato, E., Ball, W., Nae, A., and Gellings, C. (2007) *Terrorism and the electric power delivery system.*

23 EPRI (ed.) *Communication Security Assessment for the United States Electric Utility Infrastructure.*

24 Kabay, M. (2010) *Attacks on Power Systems: Hackers, Malware.*

25 DeMarco, C.L., Sariashkar, J., and Alvarado, F. (1996) *The Potential for Malicious Control in a Competitive Power Systems Environment.* Control Applications, 1996, Proceedings of the 1996 IEEE International Conference on, IEEE, pp. 462–467.

26 Teixeira, A., Paridari, K., Sandberg, H., and Johansson, K.H. (2015) *Voltage Control for Interconnected Microgrids Under Adversarial Actions*. Emerging Technologies & Factory Automation (ETFA), 2015 IEEE 20th Conference on, IEEE, pp. 1–8.

27 Wang, W. and Lu, Z. (2013) Cyber security in the smart grid: survey and challenges. *Computer Networks*, **57** (5), 1344–1371.

28 Pasqualetti, F., Dörfler, F., and Bullo, F. (2011) *Cyber-Physical Attacks in Power Networks: Models, Fundamental Limitations and Monitor Design*. Decision and Control and European Control Conference (CDC-ECC), 2011 50th IEEE Conference on, IEEE, pp. 2195–2201.

29 Ten, C.W., Govindarasu, M., and Liu, C.C. (2007) *Cybersecurity for Electric Power Control and Automation Systems*. Systems, Man and Cybernetics, 2007. ISIC. IEEE International Conference on, IEEE, pp. 29–34.

30 Duan, D., Yang, L., and Scharf, L.L. (2011) *Phasor State Estimation from PMU Measurements with Bad Data*. Computational Advances in Multi-Sensor Adaptive Processing (CAMSAP), 2011 4th IEEE International Workshop on, IEEE, pp. 121–124.

31 Abur, A. and Exposito, A.G. (2004) *Power System State Estimation: Theory and Implementation*, CRC Press.

32 Kim, J., Tong, L., and Thomas, R.J. (2014) Data framing attack on state estimation. *IEEE Journal on Selected Areas in Communications*, **32** (7), 1460–1470.

33 Kim, J. and Tong, L. (2013) On topology attack of a smart grid: undetectable attacks and countermeasures. *IEEE Journal on Selected Areas in Communications*, **31** (7), 1294–1305.

34 Yuan, Y., Li, Z., and Ren, K. (2011) Modeling load redistribution attacks in power systems. *IEEE Transactions on Smart Grid*, **2** (2), 382–390.

35 Teixeira, A., Dán, G., Sandberg, H., Berthier, R., Bobba, R.B., and Valdes, A. (2014) *Security of Smart Distribution Grids: Data Integrity Attacks on Integrated Volt/VAR Control and Countermeasures*. American Control Conference (ACC), 2014, IEEE, pp. 4372–4378.

36 Kundur, P., Paserba, J., Ajjarapu, V., Andersson, G., Bose, A., Canizares, C., Hatziargyriou, N., Hill, D., Stankovic, A., Taylor, C. *et al.* (2004) Definition and classification of power system stability IEEE/CIGRE joint task force on stability terms and definitions. *IEEE Transactions on Power Systems*, **19** (3), 1387–1401.

37 Van Cutsem, T. and Vournas, C. (1998) *Voltage Stability of Electric Power Systems*, vol. **441**, Springer Science & Business Media.

38 De La Ree, J., Liu, Y., Mili, L., Phadke, A.G., and Dasilva, L. (2005) Catastrophic failures in power systems: causes, analyses, and countermeasures. *Proceedings of the IEEE*, **93** (5), 956–964.

39 Andersson, G., Donalek, P., Farmer, R., Hatziargyriou, N., Kamwa, I., Kundur, P., Martins, N., Paserba, J., Pourbeik, P., Sanchez-Gasca, J. *et al.*

(2005) Causes of the 2003 major grid blackouts in north America and Europe, and recommended means to improve system dynamic performance. *IEEE Transactions on Power Systems*, **20** (4), 1922–1928.

40 Pourbeik, P., Kundur, P.S., and Taylor, C.W. (2006) The anatomy of a power grid blackout. *IEEE Power and Energy Magazine*, **4** (5), 22–29.

41 Sun, K. and Han, Z.X. (2005) *Analysis and Comparison on Several Kinds of Models of Cascading Failure in Power System*. Transmission and Distribution Conference and Exhibition: Asia and Pacific, 2005 IEEE/PES, IEEE, pp. 1–7.

42 Taylor, C.W. and Erickson, D.C. (1997) Recording and analyzing the July 2 cascading outage [Western USA Power System]. *IEEE Computer Applications in Power*, **10** (1), 26–30.

43 Yamashita, K., Li, J., Zhang, P., and Liu, C.C. (2009) *Analysis and Control of Major Blackout Events*. Power Systems Conference and Exposition, 2009. PSCE'09. IEEE/PES, IEEE, pp. 1–4.

44 Liu, C.C. and Li, J. (2008) *Patterns of Cascaded Events in Blackouts*. 2008 IEEE Power and Energy Society General Meeting-Conversion and Delivery of Electrical Energy in the 21st Century.

45 Häger, U., Rehtanz, C., and Voropai, N. (2014) *Monitoring, Control and Protection of Interconnected Power Systems*, vol. **36**, Springer.

46 Surdutovich, G., Cortez, C., Vitilina, R., and da Silva, J.P. (2002) *Dynamics of "Small World" Networks and Vulnerability of the Electric Power Grid*. VIII Symposium of Specialists in Electric Operational and Expansion Planning, Brazil.

47 Wang, X.F. and Chen, G. (2003) Complex networks: small-world, scale-free and beyond. *IEEE Circuits and Systems Magazine*, **3** (1), 6–20.

48 Zongxiang, L., Zhongwei, M., and Shuangxi, Z. (2004) *Cascading Failure Analysis of Bulk Power System Using Small-World Network Model*. Probabilistic Methods Applied to Power Systems, 2004 International Conference on, IEEE, pp. 635–640.

49 Liu, Y. and Gu, X. (2007) Skeleton-network reconfiguration based on topological characteristics of scale-free networks and discrete particle swarm optimization. *IEEE Transactions on Power Systems*, **22** (3), 1267–1274.

50 Hines, P. and Blumsack, S. (2008) *A Centrality Measure for Electrical Networks*. Hawaii International Conference on System Sciences, Proceedings of the 41st Annual, IEEE, pp. 185.

51 Salmeron, J., Wood, K., and Baldick, R. (2004) Analysis of electric grid security under terrorist threat. *IEEE Transactions on Power Systems*, **19** (2), 905–912.

52 Salmeron, J., Wood, K., and Baldick, R. (2004) Optimizing Electric Grid Design Under Asymmetric Threat (II). Tech. Rep., Naval Postgraduate School Monterey CA Dept of Operations Research, http://www.dtic.mil/dtic/tr/fulltext/u2/a423274.pdf (accessed 24 January 2017).

53 Salmeron, J., Wood, K., and Baldick, R. (2009) Worst-case interdiction analysis of large-scale electric power grids. *IEEE Transactions on Power Systems*, **24** (1), 96–104.

54 Seethalekshmi, K., Singh, S., and Srivastava, S. (2008) *Wide-Area Protection and Control: Present Status and Key Challenges*. 15th National Power Systems Conference, Bombay, India, pp. 169–175.

55 Phadke, A.G. and Thorp, J.S. (2009) *Computer Relaying for Power Systems*, John Wiley & Sons, Inc.

56 Little, R.G. (2003) *Toward More Robust Infrastructure: Observations on Improving the Resilience and Reliability of Critical Systems*. System Sciences, 2003. Proceedings of the 36th Annual Hawaii International Conference on, IEEE, pp. 9–pp.

57 Amin, M. (2000) National infrastructures as complex interactive networks, in *Automation, Control, and Complexity: An Integrated Approach* (eds T. Samad and J. Wayrauch), John Wiley & Sons, Inc., New York, pp. 263–286.

58 Wind (2015) *Building a Smarter Smart Grid: Counteracting Cyber-Threats in Energy Distribution*, http://www.windriver.com/whitepapers/building-a-smarter-smart-grid/Energy-Security_white-paper.pdf (accessed 24 January 2017).

59 INL (2011) *Vulnerability Analysis of Energy Delivery Control Systems*, Idaho National Laboratory, https://energy.gov/sites/prod/files/Vulnerability%20Analysis%20of%20Energy%20Delivery%20Control%20Systems%202011.pdf (accessed 24 January 2017).

60 Liu, C.C. (2004) *Strategic Power Infrastructure Defense (SPID)*. Power Engineering Society General Meeting, 2004. IEEE, IEEE, pp. 3–Vol.

61 Horowitz, S.H. and Phadke, A.G. (2008) *Power System Relaying*, vol. **22**, John Wiley & Sons, Ltd.

62 Rehtanz, C. and Bertsch, J. (2002) *Wide Area Measurement and Protection System for Emergency Voltage Stability Control*. Power Engineering Society Winter Meeting, 2002. IEEE, vol. 2, IEEE, pp. 842–847.

63 Phadke, A.G. and Thorp, J.S. (2008) *Synchronized Phasor Measurements and their Applications*, Springer Science & Business Media.

64 Marnay, C., Rubio, F.J., and Siddiqui, A.S. (2000) *Shape of the Microgrid*, Lawrence Berkeley National Laboratory.

65 Castro, M., Jara, A.J., and Skarmeta, A.F. (2013) *Smart Lighting Solutions for Smart Cities*. Advanced Information Networking and Applications Workshops (WAINA), 2013 27th International Conference on, IEEE, pp. 1374–1379.

66 Lopes, J.P., Moreira, C., and Resende, F. (2005) Control strategies for microgrids black start and islanded operation. *International Journal of Distributed Energy Resources*, **1** (3), 241–261.

67 Marnay, C., Aki, H., Hirose, K., Kwasinski, A., Ogura, S., and Shinji, T. (2015) Japan's pivot to resilience: how two microgrids fared after the 2011 earthquake. *IEEE Power and Energy Magazine*, **13** (3), 44–57.

68 Guerrero, J.M., Vasquez, J.C., Matas, J., Vicuna, D., García, L., and Castilla, M. (2011) Hierarchical control of droop-controlled AC and DC microgrids a general approach toward standardization. *IEEE Transactions on Industrial Electronics*, **58** (1), 158–172.

69 Alvarez, R.E. (2004) Interdicting electrical power grids. PhD thesis. Monterey, California. Naval Postgraduate School.

70 Wood, R.K. (1993) Deterministic network interdiction. *Mathematical and Computer Modelling*, **17** (2), 1–18.

71 Roy, S., Ellis, C., Shiva, S., Dasgupta, D., Shandilya, V., and Wu, Q. (2010) *A Survey of Game Theory As Applied to Network Security*. System Sciences (HICSS), 2010 43rd Hawaii International Conference on, IEEE, pp. 1–10.

72 Mohsenian-Rad, A.H. and Leon-Garcia, A. (2011) Distributed internet-based load altering attacks against smart power grids. *IEEE Transactions on Smart Grid*, **2** (4), 667–674.

73 Negrete-Pincetic, M., Yoshida, F., and Gross, G. (2009) *Towards Quantifying the Impacts of Cyber Attacks in the Competitive Electricity Market Environment*. PowerTech, 2009 IEEE Bucharest, IEEE, pp. 1–8.

74 Mo, Y. and Sinopoli, B. (2009) *Secure Control Against Replay Attacks*. Communication, Control, and Computing, 2009. Allerton 2009. 47th Annual Allerton Conference on, IEEE, pp. 911–918.

75 Sridhar, S. and Govindarasu, M. (2014) Model-based attack detection and mitigation for automatic generation control. *IEEE Transactions on Smart Grid*, **5** (2), 580–591.

76 Kosut, O., Jia, L., Thomas, R.J., and Tong, L. (2010) *Malicious Data Attacks on Smart Grid State Estimation: Attack Strategies and Countermeasures*. Smart Grid Communications (SmartGridComm), 2010 1st IEEE International Conference on, IEEE, pp. 220–225.

77 Sandberg, H., Teixeira, A., and Johansson, K.H. (2010) *On Security Indices for State Estimators in Power Networks*. 1st Workshop on Secure Control Systems (SCS), Stockholm, 2010.

78 Jin, X., Bigham, J., Rodaway, J., Gamez, D., and Phillips, C. (2006) *Anomaly Detection in Electricity Cyber Infrastructures*. Proceedings of the International Workshop on Complex Networks and Infrastructure Protection, CNIP.

79 Pasqualetti, F., Dorfler, F., and Bullo, F. (2013) Attack detection and identification in cyber-physical systems. *IEEE Transactions on Automatic Control*, **58** (11), 2715–2729.

80 Fawzi, H., Tabuada, P., and Diggavi, S. (2011) *Secure State-Estimation for Dynamical Systems Under Active Adversaries*. Communication, Control, and Computing (Allerton), 2011 49th Annual Allerton Conference on, IEEE, pp. 337–344.

81 Tajer, A. (2014) *Energy grid state estimation under random and structured bad data*. Sensor Array and Multichannel Signal Processing Workshop (SAM), 2014 IEEE 8th.

82 Dán, G. and Sandberg, H. (2010) *Stealth Attacks and Protection Schemes for State Estimators in Power Systems*. Smart Grid Communications (SmartGridComm), 2010 1st IEEE International Conference on, IEEE, pp. 214–219.

83 Giani, A., Bitar, E., Garcia, M., McQueen, M., Khargonekar, P., and Poolla, K. (2013) Smart grid data integrity attacks. *IEEE Transactions on Smart Grid*, **4** (3), 1244–1253.

20

Smart Energy and Grid: Novel Approaches for the Efficient Generation, Storage, and Usage of Energy in the Smart Home and the Smart Grid Linkup

Julian Praß[1], Johannes Weber[1], Sebastian Staub[2], Johannes Bürner[1], Ralf Böhm[1], Thomas Braun[1], Moritz Hein[3], Markus Michl[1], Michael Beck[4], and Jörg Franke[1]

[1] Institute for Factory Automation and Production Systems (FAPS), Friedrich-Alexander University of Erlangen-Nürnberg, Germany
[2] Chair of Energy Process Engineering, Friedrich-Alexander University of Erlangen-Nürnberg, Germany
[3] Chair of Measurement and Control Systems, University of Bayreuth,, Germany
[4] Institute of Separation Science and Technology, Friedrich-Alexander University of Erlangen-Nürnberg, Germany

---

CHAPTER MENU

Introduction, 576
Generation of Energy, 576
Storage of Energy, 581
Smart Usage of Energy, 587
Summary, 600

Objectives

- To become familiar with novel approaches focusing on smart generation, storage, and usage of energy
- To become familiar with small wind turbines and combined heat and power plants in order to generate energy in future smart homes
- To become familiar with smart thermal storage heating systems and a possibility of connecting the smart homes to the smart grid
- To become familiar with energy-efficient radiation heating, comfort-oriented heating, smart waste heat usage, and a novel concept of ventilation with heat recovery that uses a friction ventilator as central unit

Smart Cities: Foundations, Principles and Applications, First Edition.
Edited by Houbing Song, Ravi Srinivasan, Tamim Sookoor, and Sabina Jeschke.
© 2017 John Wiley & Sons, Inc. Published 2017 by John Wiley & Sons, Inc.

20.1 Introduction

In accordance with the Kyoto Protocol and the Directive 2009/28/EC of the European Union (EU) that aim at reducing the greenhouse gas emissions and increasing the amount of renewable energy generation, future smart homes will not only have to be more efficient, with regard to the energy consumption, but will also have to contribute to the generation, storage, and distribution of energy.

In this context various applications along the energy chain in the home and urban environment that build upon each other will be discussed.

20.2 Generation of Energy

The starting point will be the generation of renewable energy on the level of individual homes and housing estates using aerodynamically and aeroacoustically optimized small wind turbines as well as combined heat and power (CHP) micro plants using organic Rankine cycles (ORCs) to complement solar energy. In the field of small wind turbines, the three goals – noise reduction in order to fulfill user requirements, high energy output, and small plant costs – have to be satisfied. Generation of electric and thermal energy by CHP micro plants incorporating ORCs and micro-fluidized bed furnaces makes this technology efficient on a small scale and therefore available for single-family homes.

20.2.1 Aerodynamically and Aeroacoustically Optimized Small Wind Turbines

Wind energy is gaining importance in the last decades due to the emerging awareness of the need for environmentally sustainable power generation. The reason for this perception rests on the understanding of the limitation of fossil fuel reserves and of the negative effect of greenhouse gases on the environment, which are released by burning fuels for electricity production [1]. Therefore, the main focus nowadays is put on the development of renewable energy, which leads to the installation of more decentralized energy production devices such as roof-mounted photovoltaic modules or small wind turbines in urban and suburban areas as shown in Figure 20.1. While solar energy and large, horizontal wind turbines (>100 kW) are very popular in Germany, small wind turbines (<100 kW) and especially micro wind turbines (<5 kW) are facing some difficulties, such as [2]:

- High investment costs
- Too little energy generation
- Noise level
- Visual impact on buildings
- Difficult performance in turbulent and complex wind situations
- Less durability

Figure 20.1 Office building with roof-mounted vertical axis wind turbines [3].

To overcome these problems, current research deals with methods to increase the aerodynamic power performance while lowering the sound emissions, by developing new blade designs, reducing investment costs, and integrating small wind turbines in urban environments. Especially the installations in urban areas are difficult due to very complex wind situations including fast-changing wind speed with altitude and different turbulence intensity, which has a huge impact on the power generation and the durability of such turbines. Also a lower noise level is an important factor for social acceptance of such turbines. In Germany there is a strict law of Noise Emission Control, named "Technical Instructions on Noise Abatement - TA Lärm." This law defines binding noise emission values for different zones in urban areas. The maximum level during the day in general residential areas is fixed as 55 dB (A) while it is only 40 dB (A) at night [4].

In general, small wind turbines can be divided according to their axis in horizontal and vertical wind turbines. Due to their high technical maturity, the horizontal axis wind turbines are dominating the market, while the vertical design is less prominent. Integrating small wind turbines in urban environment leads to different requirements, which can be possibly satisfied by the vertical ones. Current investigations in the area of small wind turbines have focused on vertical axis wind turbines (VAWTs), which in comparison with horizontal axis wind turbines are characterized by the following advantages [2]:

- Insensitivity to wind direction, which avoids the need of a yaw system
- Applicability in presence of turbulent streams
- Lower noise level [5]

The most widely known design of VAWTs is called Darrieus turbine, which was developed by the French engineer Georges Darrieus in the early 20th century [2]. This type is driven by lift force, which is the most efficient way to convert wind energy into mechanical energy. Contrary to that, a typical drag-type device is the Savonius rotor, which was invented by the Finnish architect Sigurd Savonius. Due to the low power coefficient of maximally 30%, the Savonius rotor is not used for commercial energy generation [2]. Within the last years, experimental and numerical studies were performed in the field of VAWTs: M. Mohamed investigated the aerodynamic performance of the straight Darrieus turbine (H-rotor) with different airfoil shapes in order to maximize the aerodynamic power output [6]. Furthermore, a big focus is placed on dynamic stall, which strongly limits the aerodynamic behavior of the turbine, for example, by the studies of Ferreira [7]. Mertens investigated the performance of an H-Darrieus in the skewed flow on a roof [8]. Weber has put his research focus on the aeroacoustics of different Darrieus wind turbines [9].

In order to get a better physical understanding of the aerodynamics and aeroacoustics of the H-Darrieus turbine, a complementary approach consisting of experimental measurements, computational fluid dynamics (CFD) simulations, and computational aeroacoustic (CAA) simulations were performed. After gathering experimental data in an anechoic wind tunnel, a transient flow simulation was carried out using a commercial CFD software. Following that, an in-house code was used for acoustic post-processing. The code is called sound prediction by surface integration (SPySI) and is based on the porous Ffowcs Williams–Hawkings method [10]. Such tools for noise prediction can be very useful in wind turbine design in order to optimize the system with respect to noise and aerodynamic performance. The mentioned experimental measurements in an anechoic wind tunnel took place at the University of Erlangen-Nürnberg. The advantages of carrying out measurements in the wind tunnel are:

- Controlled and reproducible flow conditions
- Acoustic and aerodynamic measurements
- Clean and turbulent inflow conditions

The experimental measurements were validated with the help of CFD simulations and showed very good agreement.

20.2.2 Combined Heat and Power Micro Plants Using Organic Rankine Cycles

The worldwide growing energy demand necessitates a more sensitive and responsible handling of available energy resources. One possible method for reaching a more efficient energy management is CHP production. This technology merges the commonly split generation of heat and electricity. The overall efficiency can reach up to 90% and exceeds the efficiency of split generation, often significantly [11].

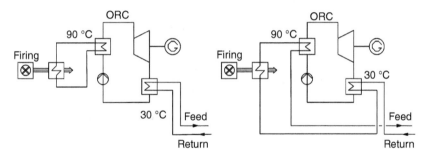

Figure 20.2 Two possible setups for linking an ORC plant with a heating system.

CHP is quite popular for distributed power generation in cities and small towns. Common solutions require large heating grids, which have to be built and maintained to distribute the heat. Furthermore the heat transport in such grids suffers high losses between the CHP plant and the consumer. In order to overcome these disadvantages, "micro scale" CHP covers just the demand of single consumers and can be installed at single-family homes. To get an idea of the needed size of such a system, the energy consumption in one year of a German household can be approximated to about 12,000 kWh of heat [12] and 3500 kW h of electrical energy [13].

One technical approach to realize such an autonomous energy supply is the installation of an ORC to existing heating systems as shown in Figure 20.2. The principle of the ORC process is quite similar to common steam power cycles. A working fluid gets pumped into a boiler where it gets evaporated. After that it passes through an expander and condenses afterward in a heat exchanger. In contrast to the "classical" Rankine cycle, the working fluid is an organic liquid instead of water. The advantage of organic media is that they have a lower boiling point than water. This makes it possible to convert low-temperature heat into electricity. The use of ORC for recovering waste heat has been approved for many years. Typically they are realized from $50 \, kW_{el}$ to $2 \, MW_{el}$ [14] with an electrical efficiency between 6% and 17% [11]. Smaller systems cannot be realized due to economic reasons. The investment costs are too high and are not covered by the electrical power production.

As shown in Figure 20.2, there are at least two different options for placing the ORC into the heating system. In the first one the boiler connects to the evaporator of the ORC only. The waste heat of the ORC process is then used to cover the heat demand. In the second system the waste heat after the evaporator is additionally used to feed the heating system. That causes a higher temperature for the feed and a lower temperature for the return to the boiler. One difference between these two possible setups is that in the first version, two different and separated cycles are necessary. The second setup can be realized just with one.

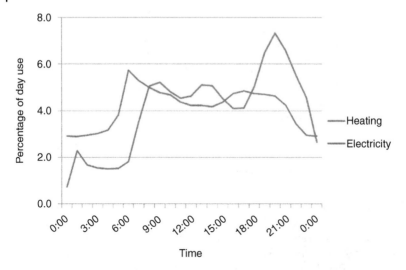

Figure 20.3 Qualitative chart of the heat and electricity demand of a single house during one day.

In the first case the efficiency of the ORC is optimized but the firing efficiency is limited. The opposite effects are realized in the second case. Due to the lower return temperature to the boiler, the efficiency of the firing rises. On the other hand the electrical efficiency of the ORC process shrinks.

The German start-up company ORCAN Energy AG is developing small- and microscale ORCs. Their approach is to take only mass-produced and cheap articles from, for example, the automotive production as parts for the ORC plant. The resulting system is not the most technically efficient, but due to significantly lower investment costs, it is more economically feasible. This concept even allows an ORC plant with a power output of 1 kW. This size, with a resulting electrical efficiency of 5–10%, is perfectly suitable for use in a CHP microscale application.

The challenge is now the integration of this CHP system into the dynamic environment of a household. Figure 20.3 illustrates qualitatively the fluctuating heat and power need over one day during the winter season. It shows the continual gap and time offset between the heat and power requests. With the application of a CHP system, the production of these two sizes will be linked and there would be always a mismatch. Just in combination with a smart control system and storages for heat or power or both, a CHP supply of single- and multifamily homes can be realized. By the installation of a smart controlling system, the fluctuations of the electrical power demand can get smoothed. This reduces the needed size of energy storages. The cheapest and best proven-in-use option for buffering energy in the suggested application would be a hot-water storage tank. These tanks are already quite common in most existing installations with solid fuel firings, solar thermal installations, and heat pump systems. With the installation of such a buffer, the running time of the firing system gets nearly independent from the heat demand of the costumers. This allows the CHP setup to react first of all on the electrical power requests. In times with a low heat but high electricity demand, the CHP system can start working. The power request is going to be covered and the excess heat is placed in the storage tank. In times with a low electrical power demand, for example, at night or early in the morning, the stored heat can be used to heat up the living space.

The realization and demonstration of such a smart, flexible, and efficient system is the task of ongoing projects at the EHome Center. An interdisciplinary team of engineers and computer scientists are working on the technical feasibility of a CHP system applicable for single- and multifamily homes.

20.3 Storage of Energy

In this section a novel approach for storing a surplus of renewable electric energy in local energy networks by using thermal storage heating systems in homes enhanced by modern smart grid communication technology and control algorithms will be presented. User-centered energy monitoring accompanying this approach will also be analyzed since it plays a major role for creating user acceptance in this context. Furthermore the intelligent distribution of electric energy between the smart home and the smart grid will be discussed. New approaches are needed in order to address the migration of homes from energy consumers to ProStumers – entities that can serve as producer, storage, and consumer of energy.

20.3.1 Thermal Storage Heating Systems

In response to Fukushima's nuclear disaster in March 2011, the German government defined an energy concept with several climate and energy objectives. The main contents of this law are the reduction of greenhouse gases and the increase of electricity production from renewable energy sources, as well as the reduction of primary energy consumption. The implementation of this law requires fundamental changes in the current method of energy supply, as all German nuclear power plants are about to be shut down in the near and medium future. Furthermore, old coal-fired plants will also gradually be switched off in order to reduce the usage of coal and lignite as energy sources. The resulting energy supply gap is to be bridged by an increased use of renewable energies. According to the climate and energy objectives of the Federal Ministry for Economic Affairs and Energy [15], these goals are to be achieved through wind turbines and photovoltaic systems.

An energy supply that relies on renewable sources has the disadvantage that it is highly dependent on external and uncontrollable factors, such as weather conditions; photovoltaic systems can only generate electrical energy when enough solar radiation is available and wind turbines heavily depend on sufficient wind speeds. This results in a potentially high fluctuation of energy supply, imposing many challenges on energy suppliers in order to maintain the current grid infrastructure operational as it does not provide sufficient storage capacity to compensate for temporal variance. This is why energy generation and energy consumption have to match as closely as possible to prevent the power grid from collapsing. Nowadays it is already possible that an oversupply of renewable energy occurs during low demand-side load periods. By reaching the energy goals and thus increasing the amount of renewably generated energy, this problem will gradually increase until 2050. In case of periods with oversupply, ways to obviate this collapse scenario are doing demand-side management or using energy storage. Because of the scarceness of processes that are suitable for demand-side management, for example, load shifting with the help of compressed air systems [16], different ways of storing energy are topics of current research. More than 99% of the overall installed storage is provided by pump storage power plants. In Germany, for example, 6.6 GW of this type of power plant are available, which produced 4 TW h of electrical energy in 2014 [17]. Other technologies such as battery packs or flywheel storages are too expensive to be economically feasible [18]. Within residences, several options of energy storage are feasible. In Germany, besides heat pumps and hot-water tanks, about 1.4 million night storage heaters are available, which contain a potential for energy storage of about 20 TW h each year, which is a five times higher potential than all German pump storage power plants have. This potential of already existing storage heaters could be used until affordable alternative storage systems are developed.

The approach of using this technique for load shifting is not new at all. In the 1960s the principle of using electric energy storages for generating additional loads was implemented to increase the efficiency of poorly controlled power plants. As the ramp-up times, especially of nuclear and coal-fired plants, are too long for short-term shutdowns and restarts, it is more economical for power plant operators to run their facilities continuously loading storage heaters with the surplus of electrical energy during the night. Today there is a similar problem with renewable sources where generation capacity is not as easily plannable and predictable. An approach to address this problem is to load the existing thermal storages during periods of energy surpluses produced by photovoltaic, wind turbines, or other renewable resources.

In order to use this kind of storage, a new way of communication is needed. These days, the communication between the power supplier and single households is based on ripple control. One of the disadvantages of this type of communication is its unidirectionality. For demand-side management information exchange in both directions is necessary since the energy supplier has to send commands for loading and therefore needs to receive information about the

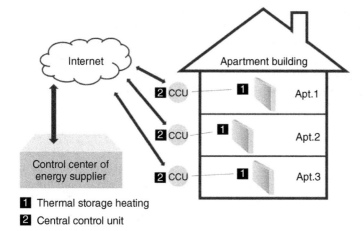

Figure 20.4 Schematic representation of the cross-linked system components and their communication paths.

capacity of storage in the coverage area of interest. For this request, a technology other than ripple control is required. One possibility that allows for bidirectional communication is a connection via the Internet. Figure 20.4 shows a schematic representation of the cross-linked system components and their communication paths. The energy supplier is connected via the Internet to a central control unit (CCU) in each household with storage heaters. Via power line communication, all storage devices are connected with the CCU. A master control unit located at the energy supplier requests data from a weather forecast service, allowing for a reliable prediction of the energy generation from renewable sources like wind power and photovoltaic. In this way the energy supplier is able to generate an individual charging curve for each household for approximately one day ahead of time. This schedule can contain arbitrary charging cycles, where each cycle equals a 15 min interval. This cycle time has

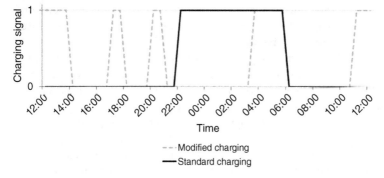

Figure 20.5 Comparison of standard and modified charging of thermal storages.

been found to be a reasonable compromise with respect to load shifting and also reduces the required amount of data to be exchanged between each electrical storage heater and the master control unit of the energy supplier. By using this kind of modified charging control, storage heaters strongly depend on the weather conditions. As shown in Figure 20.5, the time periods of loading are very flexible unlike previous approaches. With this configuration load shifting can be achieved.

20.3.2 Connecting Smart Homes to the Smart Grid

In order to tackle the transition of the electric power supply toward a renewable-based generation plant system, it is necessary to fully exploit locally available energy sources and to generate production surpluses in rural areas for the supply of urban agglomerations and industrial centers. By this requirement the current top-down approach of electrical energy supply is reversed, and multidirectional energy flows occur as a result. A particular challenge is to control the large number of decentralized generation plants, which often exhibit low generation capacities and unsteady behavior of supply.

A new approach is needed in order to address the migration of homes from energy consumers to ProStumers – entities that can serve as producers, storages, and consumers of energy. For this purpose it is necessary that all ProStumers autonomously exchange information with adjacent plants, storages, and consumers via an exchange platform – taking up the cyber – physical system paradigm [16]. The idea is to create a swarm containing all existing plants that independently regulates its generation and consumption behavior according to the prevailing supply and demand for electric power. As shown in Figure 20.6, all devices in the electric energy system are interconnected and include microcontrollers that have interfaces to all measuring instruments and control units. Thus the microcontrollers are capable of detecting the present state of electricity production or consumption, available storage capacity, and further information on, for example, status or health of the plants. Furthermore information on current price limits, optimal plant operations, and temporal restrictions are consigned in the data memory of the microcontroller. Based on all available information, the microcontrollers offer and request amounts of electrical energy on an exchange platform associated with the local segment of the electrical network. The service installed on the platform allows the plants to negotiate the exchange of electric power based on defined criteria such as price or available storage capacities. The structure of the information system is analogous to the network topology of the grid, and nodes of the electrical network – the transformers, which are also the links between the network segments – serve as balance nodes of the information network. Amounts of electrical energy that could not be received or delivered among the plants connected to the platform service are offered or requested on the platform

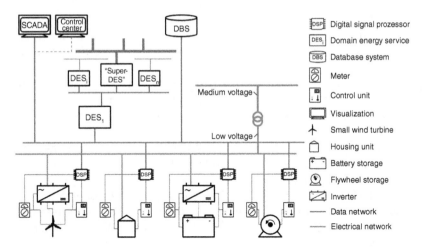

Figure 20.6 Structure of an agent-based energy grid.

service at the next higher hierarchical level. A history of relevant system data is stored in the database system and can be viewed in the supervisory control and data acquisition (SCADA) system, which also supports exception and fault handling [19].

The central element of this approach is the desire to balance the energy flows between producers, consumers, and storages on the lowest possible level of the distribution system. A higher degree of self-sufficiency of the lower levels of the system leads to lower power flows on higher network levels. Thus transmission losses can be reduced, and, provided that the subnetworks on lower levels are equipped with blackout recovery and island operation capability, impacts of major disturbances can be diminished. Distribution network operators and local utilities at the level of distribution networks aiming to implement an agent-based system of electrical power supply are faced with a number of challenges.

Firstly, it is desirable that within the observed network segment, the existing generation plants, storages, and consumers are matched with respect to their power, their capacity, and their patterns of production or consumption. Otherwise a high degree of self-sufficiency over a longer period of time is not possible, or demand-side management measures of significant degree have to take place in situations where amounts of energy cannot be obtained from higher network levels at an affordable price. This ultimately leads to a loss of comfort and in consequence to a lower customer satisfaction. Secondly, an investment in construction and operation of an information technology infrastructure, which connects all resources to the network, is required. In addition to all producers and storage devices, consumers have to be connected to the data network via smart metering or smart home gateways or comparable other interfaces, which detect the power consumption and manage the behavior of the various devices. These gateways at the interface between the home network and the electric power grid represent the consumption of the whole entity toward the grid and ensure the supply of all domestic appliances. Therefore the home appliances provide information about their load-time curve and forecast data as well as possibilities and restrictions of individual demand-side management to the gateway. The gateway derives the expected power consumption from this information and communicates it to the adjacent domain energy service (DES), to which it is connected. Unplannable, fluctuating generation plants in the households, for example, photovoltaic plants and small wind turbines, as well as spontaneous electricity needs of residents, complicate the load forecast. Larger stand-alone generating plants and industrial customers, which are supplied at higher voltage levels, show volatile production and consumption load profiles, too. Consequently a dynamic system behavior is required and short-term as well as long-term energy storage may be deemed necessary. For the connection of renewable generation plants and industrial customers to an intelligent electrical power system, also regulatory issues and process-related requirements from production have to be taken into account. In Germany, regenerative power plants under the renewable energy act (*Erneuerbare-Energien-Gesetz* (*EEG*)) may preferably feed energy into the grid. This primacy is to ensure within an intelligent, agent-based, self-managing system of electrical power supply. To avoid costly production downtime and for a continuous operation of critical production processes such as sensitive glass melts, industrial customers are dependent on a reliable supply. Therefore, security of supply will be the subject of special supply contracts between the industrial customers and the operators of smart grids. The margin for control and regulation measures of the intelligent system is further reduced by these requirements, and sources of instability are being created.

For a stable grid operation with fluctuating feed and spontaneous development of consumption, a high frequency of coordination between the different systems is necessary. Since transactions for the exchange of energy consist of multiple steps and immediate reaction to unexpected events is vital, a powerful information infrastructure, which is also open to the expansion of new facilities, has to be installed. Existing approaches for the negotiation of the exchange of amounts of energy based on prices, for example, the negotiation algorithm in the DEZENT project of TU Dortmund University, consider operation intervals of 500 ms and below [20].

In contrast, decentralized distribution structures in rural areas without broadband connection exist, where only individual generation plants are equipped with communication modules operating on the global standard of mobile (GSM) communication. The simultaneous upgrade of all systems in a distribution area to intelligent ProStumers as well as providing a data network with sufficient bandwidth is associated with large financial expenditure and is hard to afford for many network operators.

Nevertheless first providers appear on the market who connect generating plants and storage devices to a smart grid service. Utilities have to spend approximately 5.000 Euros per unit for the equipment. For a distribution system including several hundreds of small plants and storage, devices incur costs of several hundreds of thousands to millions of euros. An economic advantage of upgrading the equipment to intelligent ProStumers arises as soon as the promotion and privileging of producers by the renewable energies act (EEG) is omitted. From this point, it is for the plant operators or the cyber–physically equipped, communication-capable plants themselves to find a buyer for the amount of generated energy. A distribution network operator who has established a suitable information infrastructure can offer plant operators a market platform as a service. In this scenario, plant operators contribute by single or recurring payments to the upgrading of their equipment.

20.4   Smart Usage of Energy

Concerning the energy consumption of smart homes, a multitude of solutions are presented that have the potential to significantly decrease the energy consumption of residential buildings. Among these are energy-efficient radiation heating systems powered by renewable electric energy. In this case plasma technologies are used to spray the heating system on virtually any surface and therefore providing heat selectively where needed. Moreover various approaches for energy-efficient heating systems providing a maximum of comfort are presented, such as concepts that use sensor-based presence detection in order to control the temperature level of radiators intelligently and the combination of planar carbon–fiber-based infrared (IR) radiant heating elements with a conventional hot-water system equipped with a heat pump.

Furthermore approaches for decreasing or using the waste heat from refrigerators' and freezers' condensers in order to enhance the energy efficiency of household appliances will be demonstrated. Various system concepts such as using the condenser waste heat to support the preparation of the domestic hot-water supply or the usage of heat pipes and cold storages in order to provide cold from the environment for these applications will be evaluated. Additionally an innovative compact and silent ventilation system with heat recovery that increases the living comfort and saves energy will be explained.

20.4.1   Energy-Efficient Radiation Heating

In order to reduce the energy consumption of buildings, it is important to notice that most of the energy is used to fulfill the residents' heating requirements.

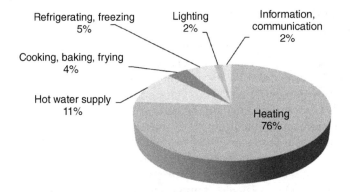

Figure 20.7 Energy consumption per household [21]. *Source:* Bumann (2015) [21]. Reproduced with permission of Energieverbrauch im haushalt. URL http://www .richtigbauen.de/info/en/pics1/en49.h1.gif.

Figure 20.7 shows the energy consumption per household, broken down to consumer groups. A novel approach to realize a noticeable reduction of this amount is given by panel heating systems that operate with renewable energy in order to achieve the political environmental goals. In this research topic the use of a panel heating system, which is produced by an innovative cold atmospheric plasma beam technology, is presented.

Current heating systems show several disadvantages. Usually separate rooms are required for fuel tanks and central-heating systems. To connect the central-heating system with the individual radiators, a pipe system of high investment and maintenance costs is essential. The radiators operate on the principle of convection heating, which leads to energy losses in case of opened windows and doors. On top of that natural convection due to density differences between warm and cold air leads to internal air movement and thus blows up dust, which negatively influences residents, especially people who suffer from allergies. Accordingly a solution to reduce the energy costs and to increase the physical comfort of living is required.

The feeling of comfort depends on many factors such as room air temperature, surface temperature of space-enclosing walls, and floor temperature [22]. The comfort chart indicates the correlation between air room temperature and surface temperature [23]. This chart shows that most people feel comfortable with a surface temperature of about 22 °C and a room air temperature of only 15 °C, which is a state of conditions that can be created using radiant heating. This type of heat transfer cannot be described by the laws of thermodynamics but only by means of quantum mechanics [24]. In reality both effects, convection and radiation, occur simultaneously, but with panel heating the radiation component predominates. Therefore, the energy loss during ventilation periods can be drastically reduced since the energy is not stored in the air. Many studies have already analyzed this phenomenon. The results confirm the fact that radiation heating is capable to support residents in saving energy [25].

With this basic knowledge and research results, it is mandatory to investigate possibilities to produce panel heating systems in an economic way in order to profit from the potentials of radiation heating. In current research the use of an additive method to build up an electrical conductive structure has been evaluated. The method is characterized by a high flexibility in application since it can be applied in a single process step without preparation and follow-up treatment. Almost any surface can be coated, for instance, plasterboards, wood panels, and laminate. By using the cold active atmospheric plasmadust® technology, tin and copper circuit paths can be generated on three-dimensional surfaces. Using microscale metal powder that is melted in the plasma flame by ignition of nitrogen gas, paths can be printed by a thermomechanical interconnection between substrate and surface. Figure 20.8 shows the process of the plasmadust method schematically.

Many studies have been carried out in order to identify the influence of different parameters of the process on the resulting path [26, 27]. It turned out that particularly the distance between nozzle and substrate, the speed of coating, and the numbers of crossings have the most significant impact on the coating quality. The resistance of the conductor track has been determined in several tests that confirmed the suitability of the technology for radiation heating. Moreover, the geometry, that is, the profile of the conductor, has been measured with a laser microscope. Furthermore, different current intensities have been applied to the components causing resistance heating, which can be detected

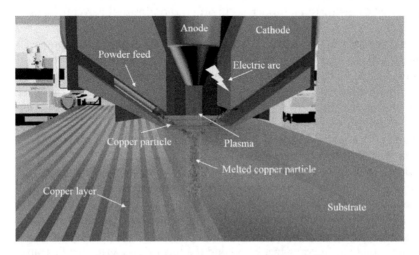

Figure 20.8 Schematic representation of the process plasmadust®.

Figure 20.9 Different specimens coated with an additive structured conductor.

with a thermal camera. In addition investigations to show the adhesion strength of several substrates have been carried out [27].

The aforementioned investigations showed that the best outcome can be generated by a distance of 15 mm between nozzle and substrate. The optimal feed rate for the plasma technology is 100 mm/s. Experiments also show the best results with three or a maximum of four crossings. It has to be noted that the overall efficiency of the process decreases with increasing number of crossings due to longer process times. The average height resulting from ideal process parameters is about 27.16 μm, and the conductor track width that can be seen in Figure 20.9 is approximately 6.51 mm at a sample length of 57.00 mm. When current is applied on the ends of the specimen to get a temperature of 30°C, a power of 0.11 W is needed. An output of 1.57 W is required to achieve a temperature of 60°C. For safety reasons the temperature should not be higher than 60°C. Future research deals with the simulation of the production process as well as the material properties in order to optimize the overall heating efficiency.

Some possible applications of the room heating system are shown in Figure 20.10. The walls are covered with an additive structured metallization to provide a comfortable warmth. Figure 20.10 also demonstrates that heat is only provided where it is really necessary in order to save energy. There are many alternative applications, beyond the heating of rooms, which could be realized with the plasmadust technology such as the heating of wind turbines to increase the economics and to protect people from ice shedding. Many people desire different temperatures at their place of work. By installing individual heating system at desks, an individual temperature can be set for each place of work. Furthermore the problem of steaming up mirrors may be tackled by using this technology.

A large amount of residential property in Germany and other countries comprises older buildings with a hot-water central-heating system using radiators. Owing to the poor building insulation and the line losses caused by the transport of the hot water to the radiators, the energy efficiency is relatively bad.

Figure 20.10 Various applications for additive manufactured heating structures.

Typically, the occupants keep the room temperature constant all day no matter whether the room is occupied or not. They only change the desired value for the radiator thermostats when they open the windows. This behavior leads to increased energy consumption because the heat losses rise with the room temperature. The renovation of the building insulation or the substitution of the radiators by a panel heating system in combination with a new central-heating installation would improve the efficiency. Unfortunately, in case of an apartment building, the one-off amount to be invested would be high, and it would not be possible to reconstruct the apartments successively.

Normally, high supply temperature needed by the radiators rules out heat pumps for the renovation of existing older buildings, whereas such pumps are often used in the new-building sector. This may be remedied by the combination of planar carbon–fiber-based IR radiant heating elements with a conventional hot-water system equipped with a heat pump. A lower supply temperature of the radiators increases the seasonal coefficient of performance (SCOP) of modern heat pumps, but then the radiators cannot keep the room temperature at a comfortable level. To maintain the thermal comfort, the IR radiant heating elements are added as another source of heat. Modern planar heating elements have a thickness of 0.4 mm and can be easily installed, preferably at the ceiling of a room, at low cost. These elements heat up the enveloping surface of the room, and even an air temperature decreased by some degrees Celsius leads to a proper level of thermal comfort. This combination of heating systems reduces the energy consumption of the building. Nevertheless, the coefficient of performance (COP) of the heat pumps decreases especially in winter when its operation is needed most. During the heating period, the low outside temperature combined with the line losses could lead to a COP of one. Consequentially, the combination of the two technologies has to be examined regarding the best trade-off between the operating conditions of the two.

In contrast to conventional heating systems, the proposed system allows for large rooms to be heated but partially. This requires appropriate measurement systems. Through the radiators and the IR heating elements, the control unit keeps the room temperature at a basic level. While doing so, it optimizes the use of both systems under energy-saving and economic aspects. Only the occupied zones are heated additionally by the IR heating elements to improve the local comfort level. Therefore, the room temperature has to be measured at several relevant places, and, in addition to that, persons in the room have to be localized. Especially in the living environment, the localization sensor system must respect the occupants' privacy and anonymity. Taking this into consideration, various possible solutions for indoor localization have been examined and compared regarding their capability to detect moving and motionless persons, as shown in Table 20.1.

The final controlled variable in such a heating system is the thermal comfort level of the occupants. This cannot be measured directly but must be inferred from measurable physical variables. The heating system is likely to comprise several feedback control loops serving to keep these variables in certain ranges to ensure the thermal comfort (in other words, the measurable variables are intermediate controlled variables).

To quantify the thermal comfort level from the measured values, it is advantageous to calculate a single variable from the measured ones. The predictive mean vote (PMV) according to DIN EN ISO 7730 [28] is the commonly used measure of thermal comfort. Some quantities needed to calculate the PMV are difficult to measure so newer models have been developed, which rely on a

Table 20.1 Qualitative comparison of different sensors to detect persons in rooms.

Sensor type	Detect moving	Detect motionless	Anonymity & acceptance	Availability	Costs
Video camera	+	+	−	+	0
Radar	+	+	−	+	0
Passive (pyroelectric) IR sensor	+	−	+	+	+
Thermopile array (low resolution)	+	+	+	+	−
CO_2 sensor	−	−	+	+	0
Time-of-flight sensor	+	+	+	−	0

reduced set of quantities. According to the adaptive comfort standard (ACS), thermal comfort in a naturally ventilated building depends on the mean outdoor air temperature [29]. The suggested value to control is the operative temperature. Unfortunately, the ACS model is valid only at an outdoor temperature above 10 °C. Below, the optimal indoor operative temperature T_{op} should be in the range of 18.4 °C $\leq T_{op,ACS} \leq$ 23.4 °C, which differs slightly from the proposed range of 20 °C $\leq T_{op,ISO} \leq$ 24 °C by DIN EN ISO 7730 [28]. In both ranges 90% of the persons should feel comfortable.

We have implemented a sensor system that, following the findings of [30], is able to measure the operative temperature directly. Based on this measurement system, an intelligent control system will be implemented, which responds to the occupants' presence and behavior while maintaining their privacy and thermal comfort with the goal to reduce the overall energy consumption for heating.

20.4.2 Comfort-Orientated Heating in Smart Homes – Overview and Field Study

Concerning the energy consumption of households, heating energy plays the major role. In the EU, roughly 60–80% of the total energy used by households is required for providing a comfortable room temperature [31]. So if smart homes shall contribute to energy efficiency and conservation of natural resources, this field surely has to be addressed.

This potential has been recognized at first in the research sector, for example, [32], but in the last years, more and more smart heating control systems addressing various needs have become available in the market as shown in Figure 20.11. A closer look at the technical features of the solutions shows that they can be divided into different categories determined by their target group. On the one hand there are systems that focus on adjusting the temperature in a single room by controlling individual radiators. On the other hand there are solutions that directly control the central-heating boilers or pursue a hybrid approach. Another important differentiating feature is the way the systems get their reference value for the temperature. Here, there exist systems that apply a self-learning approach using sensor-based presence detection as well as ones that can be programmed directly on the device or via apps.

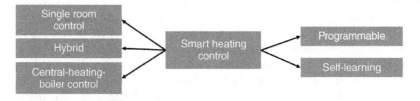

Figure 20.11 Classification of smart heating control systems.

While the technical aim of the various solution approaches always is to provide the best possible energy efficiency at undiminished living comfort, the reason for their diversity stems from the different living environments and manifold user preferences that have to be taken into account. The challenge is that these individual factors vary from household to household. Systems that influence the central-heating boiler can only be applied by homeowners but are not suited for the millions of residents in the usual multifamily rental apartment buildings. The latter represent the target market for individual room controls that can be easily retrofitted on radiators in individual rooms. The other significant factors are system costs and user preferences concerning the user interface. Simple technical solution approaches allow the stand-alone use of smart thermostatic radiator valves and are available at cheap prices. They require manual programming of comfort temperatures for different times of the day or week on a small onboard display by the user. Such systems are well suited for the price-conscious user who is willing to undertake the effort of manually programming individual thermostatic radiator valves. Users looking for a more comfortable way of controlling their smart home and willing to spend more money to find their needs fulfilled in system concepts that link the radiator valves to smart home control stations, which in turn allow a web- or app-based graphical configuration of heating curves. Users not willing to dive into the world of smart home graphical user interfaces and not able – for example, elderly people – or willing to manually configure the thermostatic radiator valves on small onboard displays are addressed by intelligent self-adjusting systems powered by learning algorithms. Additionally energetically self-sustaining systems using energy-harvesting principles are also available, reducing user effort even further by eliminating the need for a regular change of batteries.

In order to evaluate the energy-saving potential of smart heating systems, an exemplary system was installed in a multifamily rental apartment residence. The en:key system [33] was used in the scope of this context since it is a state-of-the-art technology addressing many core prerequisites for smart heating systems like reducing the necessary physical interaction with the heating control as well as autonomously adapting to the occupancy of rooms and household characteristics [34]. The main result from this field study is that energy savings of more than 25% can be achieved on average per household, as shown in Figure 20.12 [35]. In this context it has to be pointed out that the maximum energy-saving potential based on the results of the study seems even bigger when taking into account that the intelligent individual room control system was installed in all rooms, but not in all of them the system was able to realize savings. There are some rooms that, due to their comfort requirements or their utilization cycles, are not well suited for the system, which leads to an increased energy consumption for these rooms. The sleeping room is usually one of the cooler rooms in an apartment with little heating requirements. A system with automated occupancy detection recognizes persons in that

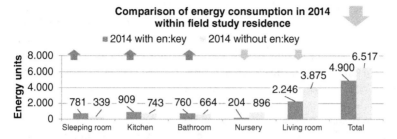

Figure 20.12 Performance of smart home heating systems with intelligent individual room control – field study results.

room and heats it up while going to bed in the evening or dressing in the morning, leading to heating periods not intended by the user. In this context the kitchen and bathroom are other critical rooms. The kitchen on the one hand gets heated up by cooking, so less heating would be required in advance, whereas the bathroom only requires punctual higher temperatures, for example, when leaving the shower or heating up a towel. Based on the system's learning algorithm, this leads to longer periods of elevated temperature than necessary in consequence increasing the energy consumption. Exceptions like these cannot be covered by the current version of the intelligent individual room control system since it has no knowledge about the rooms it is situated in [35]. But there are also rooms that fit the learning algorithm well like nurseries or living rooms. Taking into account that especially the latter one is a large, warm, and highly frequented room in the typical household, this is where the intelligent individual room control system shows its true benefits, achieving energy savings of more than 30% [35]. By replacing the system by another technical approach better suited for the requirements of bathrooms, sleeping rooms, and kitchens or modifying the learning algorithm to consider the requirements of these rooms, the full potential of smart heating applications could be accessed.

20.4.3 Smart Waste Heat Usage and Recovery from Refrigerators and Freezers

In the residential sector refrigeration and freezing appliances are responsible for about 17% of the electrical energy demand [36]. In general there are many different processes to achieve low temperatures, for example, sorption-based systems such as absorption chillers or chemical heat pumps, sterling gas chillers, or vapor injection chillers. Nevertheless, the electrically driven compression chiller is by far the most widely used technology for supplying cooling, especially in the domestic living area. The counterclockwise cycle process, which is implemented in most modern household refrigerators and freezers, works with the aim to transfer heat from a lower to a higher temperature level. As a result energy can be delivered to the environment, for example, the kitchen. Most attempts to improve the energy efficiency of household refrigerators and freezers focus on optimizing the thermal insulation, utilization of high efficient compressors, or enhanced heat transfer in the evaporator unit and the condenser unit [37]. Furthermore, the research concerning new refrigerants is focused on increasing the energetic efficiency of the thermodynamic cycle process and decreasing the global warming potential [38, 39]. A new approach to increase the energy efficiency of domestic refrigerators and freezers is to optimize the whole energy system by coupling different energy sources and sinks. This procedure is state of the art in the industrial sector and known as heat integration [40]. In the domestic sector several approaches for increasing the energy efficiency and for recovering the waste heat from refrigerators and freezers in order to reduce the primary energy demand of the whole household are possible [41].

The first concept idea uses the waste heat of refrigeration processes for the DHW preparation. As mentioned before heat is produced during the thermodynamic cycle process as a result of condensing the refrigerant in the condenser. The amount of heat produced depends on the temperature levels, more precisely on the pressure difference between evaporator and condenser. With the efficiency of the cycle process, which is called coefficient of performance, it is possible to calculate the heat output of the condenser as a function of the compressor power input. Standard A^{+++} refrigerators and freezers operate with an annual electric energy demand of 115 and 172 kW h, respectively, according to the energy label regulation for household appliances published by the European Commission in 2009 [42]. Using these values it is possible to calculate an annual waste heat of 540 kW h for the refrigerator and 568 kW h for the freezer. The heat can then be transferred to the cold fresh water of the DHW system. The waste heat temperature level is hereby limited by the condenser temperature, which is typically in the range of 30–40 °C for a household refrigerator. Therefore, the concept can be used mainly for preheating, as the required hot-water temperature is preferably 60 °C. Moreover it is necessary to include a heat storage to overcome the occurring time difference between heat sink and source. One possible simple solution for a heat storage is a hot-water storage tank. Besides the refrigeration applications, there are other home appliances, like dish washers or washing machines, that can be seen as potential waste heat sources.

Another concept evaluates the possibility to achieve higher energy efficiency for cooling processes by using the environment as an infinite thermal heat sink. With the help of heat pipes as efficient heat exchangers and the integration of a cold storage, it is possible to cover a part of the energy demand by passive cooling. Thus, the overall energy demand for the cooling unit can be significantly reduced. The performance of the energy storage loading process and therefore the efficiency of the concept depend strongly on the outside temperature, as well as on the design of the heat pipes. Using a common refrigerator with a cooling temperature of approximately 5 °C, this concept can be applied for outside temperatures below 0 °C. For freezing applications a direct passive cooling approach is not reasonable because of the necessary low-temperature levels of −18 °C in the freezer compartment. Therefore, it is proposed to combine the condenser of the freezer with a suitable cold storage. With the cold process working with the cold storage, the condenser temperature of the chiller is decreased, and the temperature shift between evaporator and condenser is smaller. A reduced temperature difference results in a higher COP and therefore in a lower energy demand for the compression chiller. The storage can be charged with heat pipes, whenever the outside temperature is below the storage temperature.

The last approach is based on a possible form of load management for photovoltaic energy, by using the surplus electricity to load a low-temperature thermal storage with a compression chiller. This cold storage can be used to cover the required cooling capacity. The concept requires the operation of the refrigeration process in a way that fits to the generation profile of the photovoltaic system. A meaningful implementation of the concept is only possible with the help of a suitable cold storage, for example, a phase change material. The operational behavior and the controlling of the storage system considering several boundary conditions, such as the radiation profile and the cooling demand, are important for the system efficiency.

20.4.4 Ventilation with Heat Recovery

Due to improved building insulation in order to reduce heating requirements, the natural exchange of air between buildings and their environment diminished significantly. As a result CO_2, humidity and volatile organic compounds (VOCs) accumulate inside inhabited buildings and thus lower the overall air quality. Conventional airing by opening windows or doors in order to allow for natural exchange of air is not the best possible solution to restore an enjoyable air quality. Fresh cold air from the outside has to be reheated after airing, which contradicts the aim of reducing heating requirements. Furthermore, pollution such as dust, pollen, or noise can be transported inside the building during aeration. In order to address this problem, the market offers ventilation systems with heat recovery that mostly feature similar operating principles: thermal energy gets transferred from warm air to cold air via a crossflow heat exchanger. Therefore, both required air flows are created by axial or radial fans. As a result of the fans with rotating blades, noise is generated with a characteristic frequency, the so-called blade pass frequency. Furthermore the conventional heat exchangers lead to a remarkable pressure drop inside the ducts that has to be compensated by the fans. As a consequence, the overall electrical energy consumption of these systems is high, which, in combination with the investment costs, leads to long payback periods for the costumers.

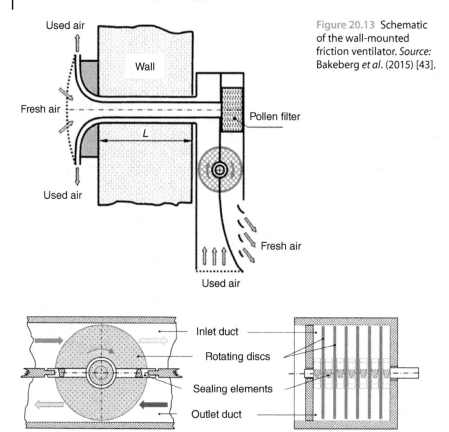

Figure 20.13 Schematic of the wall-mounted friction ventilator. *Source:* Bakeberg *et al.* (2015) [43].

Figure 20.14 Operating principle of the friction ventilator [44].

A new concept for a ventilation system with regenerative heat recovery that combines the main elements of comparable conventional systems, that is, the two fans and the heat exchanger, in one single compact friction ventilator has been developed in order to reduce production costs, energy consumption, and noise emissions. As shown in Figures 20.13 and 20.14, the bladeless friction ventilator consists of round discs with smooth surfaces, mounted within a fixed distance from one another on a shaft, which is installed between two separate ducts.

The energy transfer from the driven rotor to the fluid is achieved by momentum exchange due to the relative velocity between the fluid and the corotating discs in contrast to lift forces occurring from rotating blades as in conventional ventilators. This momentum exchange is often referred to as fluid friction in the literature, which is why this type of turbomachine is named *friction ventilator*. Sealing elements between the rotating discs separate the inlet duct from the outlet duct. As mentioned before the rotating discs also act as heat exchanger. This is achieved by the fact that the discs periodically contact the fluid of both ducts so that they can extract thermal energy from the warmer air in the outlet duct and transfer it to the colder fresh air inside the inlet duct. As a result there is no need for a high pressure drop in the heat exchanger unit, which leads to a reduced energy consumption. Furthermore the combination of the three main elements of conventional ventilation systems with heat recovery in one friction ventilator reduces the production costs and thus the overall investment costs for the costumer. Another advantage is the lack of blades on the rotor that would inevitably lead to noise at the blade pass frequency. As the bladeless rotor transports fluid in a continuous way, it does not produce periodical pressure fluctuations and thus avoids one of the sources of characteristic ventilator noise.

The general suitability of friction ventilators to transport fluid has already been known and investigated for many years. Tesla [45] patented the concept of smooth discs mounted on a rotating shaft to serve as an efficient, mechanically simple, and economically competitive pump. Rice [46] carried out intensive research on this kind of turbomachinery and confirmed the possibility of low noise operation and reasonable efficiency. Experimental investigations of a friction ventilator with the purpose to serve as a ventilation system with heat recovery on a test bench that allows for temperature differences in the ducts as well as precise pressure difference, volume flow, and temperature measurements as shown in Figure 20.15 have been carried out and documented by Renz *et al.* [44]. The analysis of a wide range of different rotors and operating points showed heat recovery rates up to 40% that were almost unaffected by the distance between the discs and thus demonstrated the general suitability of the concept as a ventilation system with heat recovery. The measured characteristic curves showed an almost linear behavior with different gradients depending on the disc distance. Numerical investigations have been carried out by Praß [47] that show good agreement with the measurements.

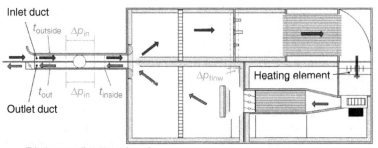

Figure 20.15 Schematic of the test bench used to investigate the friction ventilator [44].

Current and future research deals with still existing challenges of the system including the investigation of the optimum operation point, considering hydro-dynamic efficiency, low noise, and efficient heat recovery. As demonstrated by Renz *et al.* [44], the turbomachine shows an operating characteristic depending on the Reynolds number, that is, the ratio of inertial to viscous forces acting on the fluid. The improved flow control through the turbomachinery from suction to pressure side as well as the flow around the sealing elements is subject to current research. Especially the possibility of shaping the sealing channels and the ducts in such a way that the resulting pressure profile inside the ducts reduces the overflow of the sealing elements is examined in more detail. Furthermore the shape and materials of the discs are important since the frequency of temperature change of the discs is directly dependent on the rotational speed of the shaft. This leads to a short residence time of the discs in each of the ducts requiring high thermal conductivity of the material in order to enable fast heating and cooling of the discs inside outlet and inlet duct, respectively. By further improvement of heat recovery rates and hydrodynamic efficiency, the friction ventilator can be used as a silent, low-cost, mechanically simple, and low-maintenance ventilation system with heat recovery that exhibits short payback periods and is thus attractive for costumers.

20.5   Summary

In this chapter novel approaches for the efficient generation, storage, and usage of energy in the smart home environment and their link to the smart grid have been presented. Innovative concepts and their feasibility for these purposes have been demonstrated. These concepts included small wind turbines and CHP micro plants for energy generation. Furthermore, thermal heating systems that serve as energy storage and a promising concept to connect smart homes to the smart grid have been introduced. Finally, radiation heating and comfort-orientated heating systems as well as the possibility of waste heat usage and an innovative concept of ventilation with heat recovery have been explained.

Final Thoughts

In this chapter, we discussed a variety of innovative concepts that contribute to energy-efficient future smart homes. The aim was to demonstrate that the possibilities of energy generation, storage, and usage are not limited nowadays to technologies such as photovoltaics, battery storage systems, and smart metering. This chapter showed that current research goes far beyond optimizing well-known technologies. This approach is crucial in order to minimize energy waste and to maximize energy efficiency of future smart homes.

Questions

1 Which possibilities exist in order to generate energy in smart homes complementary to photovoltaics?

2 What are the challenges of small and micro turbines in urban areas?

3 Why does it make sense to use night storage heaters for load shifting?

4 What are ProStumers?

5 What are the advantages of circuit paths produced via plasmadust method?

6 What are advantages of a friction ventilator used as ventilation system with heat recovery compared with conventional ventilation system with heat recovery?

References

1 Manwell, J.F. (2009) *Wind Energy Explained: Theory, Design and Application*, 2nd edn, John Wiley & Sons, Ltd.
2 Paraschivoiu, I. (2002) *Wind Turbine Design with Emphasis on Darrieus Concept*, Presses Internationales Polytechnique, Montreal.
3 Bilfinger Envi Con GmbH (2016) *Envi Con and Plant Engineering GMBH*, http://www.envi-con.bilfinger.com/ (accessed 21 December 2016).
4 UM Welt Pakt Bayern (1998) *Vollzitat: Sechste Allgemeine Verwaltungsvorschrift zum Bundes-Immissionsschutzgesetz (Technische Anleitung zum Schutz gegen Lärm - TA Lärm) vom 26.08.1998, GMBl. vom 28.08.1998*, S. 503, http://www.izu.bayern.de/recht/detail_rahmen.php?pid=1104010100119 (accessed 24 January 2017).
5 Iida, A., Mizuno, A., and Fukudome, K. (2004) *Numerical Simulation of Aerodynamic Noise Radiated from Vertical Axis Wind Turbines*.
6 Mohamed, M.H. (2012) *Performance Investigation of H-Rotor Darrieus Turbine with New Airfoil Shapes*, pp. 522–530.
7 Ferreira, C.S. (2009) The near wake of the VAWT 2D and 3D views of the VAWT aerodynamics. PhD thesis. TU Delft.
8 Mertens, S. (2006) Wind energy in the built environment. PhD thesis. TU Delft.

9 Weber, J., Becker, S., Scheit, C., Grabinger, J., and Kaltenbacher, M. (2015) Aeroacoustics of Darrieus wind turbine. *International Journal of Aeroacoustics*, **14** (5-6), 883–902.

10 Farassat, F. (1996) *Introduction to Generalized Functions with Applications in Aerodynamics and Aeroacoustics.*

11 Karl, J. (2006) *Dezentrale Energiesysteme: Neue Technologien im liberalisierten Energiemarkt*, 2nd edn, Oldenbourg, München [u.a.].

12 Destatis Statistisches Bundesamt (2015) *Gesamtwirtschaft & umwelt - material- & energieflüsse - energieverbrauch - statistisches bundesamt (Destatis): Wärmebedarf*, https://www.destatis.de/DE/ZahlenFakten/GesamtwirtschaftUmwelt/Umwelt/UmweltoekonomischeGesamtrechnungen/MaterialEnergiefluesse/Tabellen/EnergieRaumwaerme.html (accessed 21 December 2016).

13 Destatis Statistisches Bundesamt (2015) *Gesamtwirtschaft & umwelt - material- & energieflüsse - energieverbrauch - statistisches bundesamt (Destatis): Stromverbrauch*, https://www.destatis.de/DE/ZahlenFakten/GesamtwirtschaftUmwelt/Umwelt/UmweltoekonomischeGesamtrechnungen/MaterialEnergiefluesse/Tabellen/StromverbrauchHaushalte.html (accessed 21 December 2016).

14 Schuster, A., Karellas, S., Kakaras, E., and Spliethoff, H. (2009) Energetic and economic investigation of organic Rankine cycle applications. *Applied Thermal Engineering*, **29** (8-9), 1809–1817, doi: 10.1016/j.applthermaleng.2008.08.016.

15 BMWi.de BETA (2010) *Bundesministerium für Wirtschaft und Technologie, (BMWi) Öffentlichkeitsarbeit - Energiekonzept für eine umweltschonende, zuverlässige und bezahlbare Energieversorgung, 11019 Berlin*, https://www.bmwi.de/BMWi/Redaktion/PDF/E/energiekonzept-2010,property=pdf,bereich=bmwi2012,sprache=de,rwb=true.pdf (accessed 24 January 2017).

16 Böhm, R., Bürner, J., and Franke, J. (2015) Smart factory meets smart grid: cyber physical compressed air systems enable demand side management in industrial environments. *Applied Mechanics and Materials*, **805**, 25–31.

17 Quaschning, V. (2015) *Regenerative Energiesysteme - Technologie, Berechnung, Simulation*, Carl Hanser Verlag.

18 Sterner, M. and Stadler, I. (2014) *Energiespeicher - Bedarf, Technologie, Integration*, Springer.

19 Häger, U., Rehtanz, C., and Voropai, N. (2014) *Monitoring, Control and Protection of Interconnected Power Systems*, Springer, Berlin, Heidelberg.

20 Lehnhoff, S. (2010) *Dezentrales vernetztes Energiemanagement - Ein Ansatz auf Basis eines verteilten adaptiven Realzeit-Multiagentensystems*, 1st edn, Springer, Berlin Heidelberg, New York.

21 Bumann, M.G. (2015) *Energieverbrauch im haushalt*, http://www.richtigbauen.de/info/en/pics1/en49.h1.gif (accessed 21 December 2016).

22 Mayer, E. (1989) Physik der thermischen Behaglichkeit. *Physik in unserer Zeit*, **20**, 97–103.

23 Kircheis, K. and Kircheis, R. (2009) *Der Klang der Räume - Die Kraft lebendiger Räume*, FQL Publishing.

24 Baehr, H.D. and Stephan, K. (2013) *Wärme- und Stoffübertragung*, vol. **8**, Springer.

25 Meier, C. (1999), *Die Tragödie der Strahlung*, http://www.richtigbauen.de/info/phy/pm/pm07.htm (accessed 21 December 2016).

26 Franke, J. (2013) *Räumlich elektronische Baugruppen (3D-MID)*, Carl Hanser Verlag.

27 Schramm, R. and Franke, J. (2013) *Electrical Functionalization of Thermoplastic Materials by Cold Active Atmospheric Plasma Technology*. Electronics Packaging Technology Conference (EPTC 2013), 2013 IEEE 15th.

28 DIN EN ISO 7730:2006-05 (2006) *Ergonomie der thermischen Umgebung - Analytische Bestimmung und Interpretation der thermischen Behaglichkeit durch Berechnung des PMV- und des PPD-Indexes und Kriterien der lokalen thermischen Behaglichkeit.*

29 de Dear, R.J. and Brager, G.S. (2002) Thermal comfort in naturally ventilated buildings: revisions to {ASHRAE} standard 55. *Energy and Buildings*, **34** (6), 549–561, doi: 10.1016/S0378-7788(02)00005-1. special Issue on Thermal Comfort Standards.

30 Simone, A., Babiak, J., Bullo, M., Landkilde, G., and Olesen, B.W. (2007) Operative temperature control of radiant surface heating and cooling systems. in *Proceedings of Clima 2007 Wellbeing Indoors* (eds O. Seppänen and J. Säteri), FINVAC, Helsinki, Finland, https://www.irb.fraunhofer.de/CIBlibrary/search-quick-result-list.jsp?A&idSuche=CIB+DC6897 and http://www.irbnet.de/daten/iconda/CIB6897.pdf (accessed 24 January 2017).

31 Eichhammer, W. and Lapillonne, B. (2015) *Synthesis: Energy Efficiency Trends and Policies in the EU*. http://www.odyssee-mure.eu/publications/br/synthesis-energy-efficiency-trends-policies.pdf (accessed 21 December 2016).

32 Lu, J., Sookoor, T., Vijay, S., Gao, G., Holben, B., Stankovic, J., Field, E., and Whitehouse, K. (2010) *The Smart Thermostat: Using Occupancy Sensors to Save Energy in Homes.*

33 Kieback&Peter (2015) *enkey*, http://www.enkey.de (accessed 21 December 2016).

34 Fabi, V., Camisassi, V., Bellifemine, F., Bella, V., and Corgnati, S.P. (2015) *Benefits and Disadvantages of Existing Smart Heating Control Systems: A Critical Survey*, http://iet.jrc.ec.europa.eu/energyefficiency/conference/eedal2015 (accessed 21 December 2016).

35 Michl, M., Kettschau, A., and Schneider, C. (2015) *Potentialanalyse selbst-lernender Einzelraumtemperaturregelungen zum energieeffizienten Heizen*, pp. 436–449.

36 EnergieAgentur.NRW (2015) *Erhebung wo-bleibt-der-strom*, http://www .energieagentur.nrw.de/presse/singles-verbrauchen-strom-anders-15327.asp (accessed 21 December 2016).

37 Gholap, A. and Khan, J. (2007) Design and multi-objective optimization of heat exchangers for refrigerators. *Applied Energy*, **84** (12), 1226–1239, doi: 10.1016/j.apenergy.2007.02.014.

38 Lee, Y. and Su, C. (2002) Experimental studies of isobutane (R600a) as the refrigerant in domestic refrigeration system. *Applied Thermal Engineering*, **22** (5), 507–519, doi: 10.1016/S1359-4311(01)00106-5.

39 Rasti, M., Hatamipour, M., Aghamiri, S., and Tavakoli, M. (2012) Enhancement of domestic refrigerator's energy efficiency index using a hydrocarbon mixture refrigerant. *Measurement*, **45** (7), 1807–1813, doi: 10.1016/j.measurement.2012.04.002.

40 Kemp, I.C. (2007) *Pinch Analysis and Process Integration - A User Guide on Process Integration for the Efficient Use of Energy*, 2nd edn, Butterworth-Heinemann, Oxford.

41 Beck, M., Müller, K., and Arlt, W. (2015) Energetische Evaluierung von Kältespeichern und Abwärmenutzung für Kühl- und Gefrieranwendungen. *Chemie Ingenieur Technik*, **87** (7), 957–965, doi: 10.1002/cite.201400106.

42 Verordnung (EG) Nr. 643/2009 (2009) *Der Kommission - Anforderungen an die umweltgerechte Gestaltung von Haushaltskühlgeräten*, Amtsblatt der Europäischen Union.

43 Bakeberg, C., Becker, S., Beede, T., Pauer, R., and Schlücker, E. (2015) *Device for pumping two flows*. WO2014EP00197 20140125.

44 Renz, A., Becher, M., Becker, S., and Pauer, R. (2015) *Regenerative heat exchanger with flow friction ventilator*.

45 Tesla, N. (1913) *Fluid propulsion*. USA Patent 1.061.142.

46 Rice, W. (2003) Tesla turbomachinery, in *Handbook of Turbomachinery*, 2nd edn, Chapater 14 (eds E. Logan and R. Roy), Marcel Dekker, New York, pp. 861–874.

47 Praß, J. (2015) *Untersuchung einer dezentralen Raumbelüftungsanlage mit Wärmerückgewinnung mittels zweifach teilbeaufschlagtem, mehrflutigem Querstrom-Reibungsventilator*, pp. 420–429.

21

Building Cyber-Physical Systems – A Smart Building Use Case

Jupiter Bakakeu, Franziska Schäfer, Jochen Bauer, Markus Michl, and Jörg Franke

Institute for Factory Automation and Production Systems (FAPS), Friedrich-Alexander University of Erlangen-Nürnberg, Germany

---

CHAPTER MENU

Foundations – From Automation to Smart Homes, 606
From Today's Technologically Augmented Houses to Tomorrow's Smart Homes, 608
Smart Home: A Cyber-Physical Ecosystem, 612
Connecting Smart Homes and Smart Cities, 629
Conclusion and Future Research Focus, 631

Objectives

- To become familiar with the concepts of Building automation, home automation, smart homes, smart living, ambient assisted living (AAL) and smart cities.
- To become familiar with the concepts of cyber-physical system (CPS), internet-of-things (IoT), self-organization and multi-agent systems (MAS) in a smart-home environment.
- To have a good understanding of the challenges of building smart home and smart building.
- To become familiar with the main use cases and scenarios of smart homes and smart buildings.
- To become familiar with the state-of-the-art smart homes solutions and their limitations.
- To provide an example of a smart home implementation based on self-organization principles.

In recent years, the paradigm of CPS finds its way in many sectors of society. In the field of production technology, self-organizing decentralized systems

Smart Cities: Foundations, Principles and Applications, First Edition.
Edited by Houbing Song, Ravi Srinivasan, Tamim Sookoor, and Sabina Jeschke.

based on CPS are the central part of visions for automated production called Industry 4.0 in Germany [1] or Industrial Internet in the United States [2]. Other areas such as energy, mobility, and health also lean on these concepts in order to reinvent their technological backbone. The application of the CPS in the energy domain is addressed in more detail in Chapter Chapter 23 where an energy system based on micro plants, storage systems, and consumers controlled and interconnected by smart grid components is described. Last but not least, the field of private living is also influenced by the developments in the domains of CPS, Internet of Things (IoT) and Internet of services [3] where there is a clear trend to intelligent and interconnected distributed systems. The situation in this area is presented in detail in this chapter. Starting from a classification of smart home, the state-of-the-art and unfolding solution approaches are explained, and a model architecture is developed.

21.1 Foundations – From Automation to Smart Homes

This section classifies the term smart home, presents its origins, and points out connections to other relevant concepts influencing it. First of all, the term automation denotes the transfer of work from men to machines that are able to carry out the designated tasks independently [4]. Men are only involved in the previous planning and realization process [4]. It started out in the field of production on a quest for increasing efficiency. Nowadays, it makes its way in all fields of living such as mobility, energy, and health. Also, the building sector is influenced by these ideas leading to the term building automation.

Building automation denotes the interconnection of sensors and actors within a building by communication networks with the goal of automatically carrying out operational sequences [5]. This leads to the automatic monitoring and control of building functions. Typical scenarios in this context are heating control or security mechanisms. In general, the necessary planning and installation tasks are carried out by specialized personnel. Typical target applications in the context of building automation are industrial or public buildings as well as offices or malls.

Home automation is a specialization of building automation and denotes the automation of private residences. Therefore, it targets significantly smaller installations than building automation but with mostly the same goals. In accordance with it, it also addresses the automation of building functions by using an interconnected network of sensors and actors [6]. But in contrast to building automation, it also focuses on providing a user-friendly user interface. Furthermore, in addition to system solutions that are planned and installed by designated specialists, there are systems that can be installed by end users. Usually

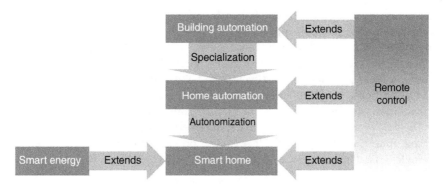

Figure 21.1 Classification of the smart home notion.

the automated building functions comprise use cases from the fields of energy, entertainment, health, security, and comfort.

The term smart home finally extends the notion of home automation by aiming at an intelligent connection of sensors and actors (Figure 21.1). Although it is sometimes used as a synonym for home automation, it can be clearly distinguished from it by focusing on the aspect of intelligent control [7, 8]. In this context, the system not only carries out previously defined operational sequences by simple switching operations based on a predefined pattern, but it is also able to react spontaneously according to individual user needs using its inherent capabilities of self-control.

In the light of the rise of renewable energies for power supply that are organized in a decentralized smart grid, smart homes are more and more influenced by the concept of smart energy. Smart energy addresses the intelligent connection of smart homes to the smart grid by smart meters in order to provide load shifting and storage capacities and therefore allow for a stable and efficient electrical energy supply [9].

Finally, remote control extends the functionality of building automation and home automation as well as smart homes by means of remote control systems and therefore enhancing user comfort and extending control possibilities. Remote control in this context is considered as a separately addressed extension of smart home functionality since a lot of literature sees it as an optional part [10]. To further clarify, the means of interacting with an automation system remotely in this context envisions a connection via data networks. The usage of a radio control device does not fall in this category.

The classification of automation systems in the context of private living shows the clear need to integrate automated sensor–actor interactions in building automation in order to make smart building use cases available. This requires communication technologies that are coined for the demands of various home applications that cannot be provided be a single communication standard. This fact has been realized by the various players in the smart home environment and is addressed by interoperability efforts channeled within different consortiums (AllSeen Alliance, OSGI Alliance, etc.). But having in mind the definition of a smart home as an intelligent system environment powered by self-control capabilities, the question arises how this self-control can be set in motion.

The following sections explain in detail how such a smart home installation based upon the principles of self-organization can be developed using technologies based on semantic communication.

21.2 From Today's Technologically Augmented Houses to Tomorrow's Smart Homes

A considerable amount of research has been carried out toward making long-standing smart home visions technically feasible. The technologically augmented homes made possible by these works are starting to become a reality, yet living in and interacting with such homes or buildings has introduced significant complexity for the user while offering a limited amount of benefits. As smart home technologies are slowly and increasingly adopted, the knowledge we gain from its use suggests a need to revisit the opportunities and challenges they pose in order to appropriately define the vision of smart homes and smart buildings installation with self-organization properties.

21.2.1 Smart Home: Past, Present, and Future

Smart homes were an underexplored field of research two decades ago, but with the recent advancements in communication technologies, various efforts ranging from scientific approaches in academia to research and development in industry has been carried out in order to gain additional understanding of the new challenges that arise from the use of the emerging technologically augmented homes. An excellent synthesis of these challenges for smart homes and buildings was presented in 2001 by Edwards and Grinter [11]. In their early vision, Edwards and Grinter presented seven challenges that arise from the ways in which we expect smart homes to be deployed and inhabited. These challenges address technical questions of interoperability, decentralized control, and reliability. They discuss the social concerns about the adoption of domestic technologies and the implications of such technologies. They also present the design issues of the smart home vision and the human–machine interaction challenges that arise from such environments.

To date, the limited amount of research into smart home has been primarily focused on the technical possibilities. In 2008, a survey on smart homes made by Chan *et al.* [12] mentions a significant progress in certain areas such as remote management and elderly assistance enabled by the increasing availability of personal computers; the rapid development of miniature, autonomous, and wireless sensors; and the massive adoption of cellular phones with embedded sensors such as accelerometers and gyroscopes. They also include a discussion on wearable and implantable devices and assistive robots. In their vision, the main challenges were still in user needs, acceptability and satisfaction as well as the reliability and efficiency of sensory systems and data processing software. Standardization of information and communication systems and legal and ethical issues was also discussed.

As people's expectations of what technology can do for them are changing, the vision of what a smart home entails is continuously evolving as well. In 2009, another survey carried out by Chan *et al.* [13] included wearable, implantable, and microcapsule devices as smart home features and stated a user-centered design as one of the important issues. Future perspectives on smart homes as part of a home-based healthcare network were also discussed.

As the vision of the intelligent house evolves, the need to align the smart home functionality with the user behaviors becomes more and more evident as Grinter predicted. Location awareness also becomes more and more relevant. Further study was conducted on location awareness in an intelligent environment. Hightower and Borriello summarized the location detection techniques and discussed several taxonomies of the location detection system [14]. Their survey categorized the properties of location detection systems according to physical position, symbolic location, and absolute and relative measurements. Issues related to accuracy, precision, measurement scale, and cost were also evaluated to compare different location systems. Manley *et al.* categorized location detection systems based on the mobility of objects and tagging capability [15]. They also discussed the possibilities presented by location awareness in a smart environment to solve the design issues of smart homes.

Context awareness is an important prerequisite to create an intelligent environment in smart homes. Robles and Kim discussed context-aware tools for smart home implementation [16]. The review presents a brief overview of rule-based smart home architecture, aware community systems, networked robots, and context-aware gateways. A comparison between various smart home protocols, for example, X10, ZigBee, and Wi-Fi, is also provided.

Security issues have been also investigated in the last decade. Pishva and Takeda presented a taxonomy of security threats in smart homes [17]. They discussed different types of attacks ranging from user impersonation to firmware alteration and suggested some prevention methodologies. A summarized threat-likelihood level is also presented, which categorized attack possibility according to appliance type and attack category.

To summarize, the vision of smart homes that cleverly support their inhabitants through technology has already been addressed to some extent and can feasibly be addressed from a technical point of view. But these new technologies have also introduced new challenges. For example, there is an increasing difficulty of maintaining and securing home networks due to the invisibility of connections introduced by wireless networks and an increasing complexity of installations due to a larger quantity and diversity of devices. Mennicken *et al.* [18] summarize these challenges as follows:

- *Meaningful Technologies*: The major advances of smart home technologies will lie in supporting the goals and values of inhabitants achieving their peace of mind and increasing their feeling of being connected with their homes.
- *Privacy and Security*: The increased connectedness of our homes raises questions about what data are being collected, whether they are transferred outside of the domestic environment, and how they are being accessed.
- *The Increasing Complexity of the Domestic Spaces*: Besides the larger diversity and quantity of devices in a smart home environment, supporting multiple users with conflicting behaviors becomes more and more difficult.
- *Human–Home Collaboration*: Multimodal and context-aware interaction between smart homes and their occupants remain a nontrivial task.

To address these issues, future trends tend to rely on the use of available technologies such as IoT to implement the smart home vision.

IoT is one of the main communication developments in recent years. It makes our everyday objects (e.g., health sensors, industrial equipment, vehicles, clothes) connected to the Internet and to each other. According to [19], the basic concept behind IoT is the pervasive presence around us of various wireless technologies such as radio-frequency identification (RFID) tags, sensors, actuators, and mobile phones, in which computing and communication systems are seamlessly embedded. Through unique addressing schemes, these objects interact with each other, and cooperate to reach common goals. In fact, this interconnection allows the objects surrounding us to share data, to interact, and to act autonomously on behalf of their users. This prospect opens new doors toward a future, where the real and virtual world merge seamlessly through the massive deployment of embedded devices [20].

From the perspective of a private use, smart home is one of the most interesting applications of IoT. For instance, energy management could be improved through the control of home equipment such as air conditioners, refrigerators, and washing machines. Another illustration of IoT applications in the smart home sphere relies on social networking paradigms. Indeed, an interesting development would be using a Twitter-like concept. In this concept, various objects in the house can periodically tweet the readings, which can be easily followed from anywhere [21]. From the perspective of business use, environmental monitoring can be achieved by keeping track of the number of occupants and by managing the utilities within a building. In addition, from the perspective of utility services, smart grids, AAL, and health monitoring are few of the most interesting applications.

Several challenges stand between the conceptual idea of IoT applied to smart homes and the full deployment of its applications into our daily life. In fact, the successful deployment of smart home is closely related to the establishment of a standard architecture, which should cover IoT characteristics and support future extensions, the same way current Internet architecture achieved during the past 40 years. A well-defined, scalable, and secure architecture is required to bring the smart home concept closer to reality.

In the literature, several architectures have been proposed [22–25]. Nevertheless, each of them brings a share of drawbacks and fails covering all IoT characteristics. These characteristics can be summarized as follows:

- *Distributivity*: IoT will likely evolve in a highly distributed environment. In fact, data might be gathered from different sources and processed by several entities in a distributed manner.
- *Interoperability*: Devices from different vendors will have to cooperate in order to achieve common goals. In addition, systems and protocols will have to be designed in a way that allows objects (devices) from different manufacturers and from different domains to exchange data and work in an interoperable way.
- *Scalability*: A wide range of objects are expected to be part of the smart home infrastructure. Thus, systems and applications that run on top of them will have to manage an unprecedent amount of generated data.
- *Resources Scarcity*: Both power and computation resources will be highly scarce since devices are more and more miniaturized and the power wall is almost reached.
- *Security*: User's feelings of helplessness and being under some unknown external control could seriously hinder IoT's deployment in the smart home environment.

Furthermore, the control of heterogeneous smart devices embedded in an unpredictable physical world is challenging due to the dynamically changing environment, the conflicting behavior of household devices, and the incompatible user preferences in a multiuser environment. To challenge these points, CPS can be one of the potential approaches. A CPS in [26–28] is defined as the integration of computation with physical processes. In these mechatronic systems, embedded computers and networks monitor and control the physical processes. Usually feedback control loops are used leading to an interlocking of the physical and the virtual world. Applications of CPS in a home environment arguably have the potential to help its successful deployment. However, computing and networking technologies today may unnecessarily impede the design of CPS-based home environments. For example, the lack of temporal semantics and adequate concurrency models in computing and today's "best effort" networking technologies makes predictable and reliable real-time performance difficult. Software component technologies including object-oriented design and service-oriented architectures are built on abstractions that match software better than physical systems. Although embedded systems have always been held to a higher reliability and predictability standard than general-purpose computing, the physical world is however not entirely predictable. CPS-based home environment will not operate in a controlled environment and must therefore be robust to unexpected conditions and adaptable to subsystem failures [29].

As presented earlier, smart home environments can also be viewed as real-time CPS with a large diversity in system architecture and components. In addition to the challenges in developing traditional distributed real-time systems (discovery, synchronization, concurrency, etc.), smart home environments should provide solutions for interoperability between devices by relying on semantic communication, context and location awareness, abstractions for sensors/actuators and event representation, and mechanisms for service compositions [29].

21.3 Smart Home: A Cyber-Physical Ecosystem

The vision of smart homes comprises a clever support for the inhabitants' needs and goals through a nearly invisible presence of technology. In the development of new products and to realize this vision, nature can serve as a role model. If you think, for example, of an ant colony, you can observe self-organization in its perfection. Every single ant senses certain signals of their colleagues and knows intuitively how to act for achieving a common global goal of the whole colony. The collective performance of various ants leads to emergence effects seeming impossible for a single individuum without this kind of networking. Ants are just one part of the big community of living organisms. The conjunction of this community with the nonliving components including their interaction as a system is called ecosystem [30]. The transfer of this role model to our technical world leads to the term "digital ecosystem." This kind of ecosystem is a distributed, adaptive, and open socio-technical system, which includes the aspects self-organization, scalability, and sustainability [31]. As collaboration of heterogeneous entities, a smart home or smart building can be classified as such a system.

Technology is already part of our social community and will be an integral part of our future homes and buildings supporting an adaptive, intelligent communication and certain decision-making processes. These homes or buildings are a dynamic complex of living beings, intelligent technical devices, and nonliving components interacting as a functional unit. This complex interrelationship has the ability of self-regulation to a certain degree. Human–machine interaction plays an important role in this construct as social action and interaction influence technology and vice versa. In this chapter, we start with the user's and developer's perspective in the form of scenarios and associated use cases. The realization process of these scenarios is addressed in the interoperability chapter as communication basis, followed by the application of agent-based techniques and decentral control aspects to form the socio-technical ecosystem smart home or smart building. Later, we present AAL as a special field of application. Finally, we describe the role of the smart home in the context of smart cities.

21.3.1 Use Cases and Smart Home Scenarios

An intelligent home has to be as individual as its residents. Therefore, the user is a very important element of the socio-technical ecosystem. Scenarios and use cases of different complexities can help defining the user's (possible) functional needs to exploit the full potential to his benefit, also resulting in a higher acceptance. Every scenario has a functional goal, narrated as interactions of user roles in a real-world context. The description can comprise different levels of complexity, that is, single system versus a whole system of systems or a single apartment versus a whole building.

The applications and resulting scenarios are numerous. A categorization of the functional possibilities of the smart home or building can help to get an overview of potential fields of application. Generally, there is no common perspective on how to categorize a smart home or building. If we focus on the smart home, different approaches are presented, for example, by Strese *et al.* [32] and Gentner and Wagner [33]. A representation of an extended version of Glasberg's perspective is shown in Figure 21.2. His point of view as basis seems to be the most comprehensive description of the connection between devices, functions, and applications in a smart home. Therefore, it can be perfectly used for scenario orientation. The center of Glasberg's "circle" describes the technical fundamentals including communication and intelligent devices. The outer circle is one of the most interesting parts for the categorization of smart home applications and scenarios. It is divided into the application field's energy, security, health, entertainment, and comfort including example applications. We extended the original version by adding the field "energy" out of an analysis of the cited approaches. Furthermore, we aggregated Glasberg's field "entertainment" and "work & communication" to the term "entertainment" as work is based on communication techniques that can be used for both – private and business – applications. The examples can be considered as sources of

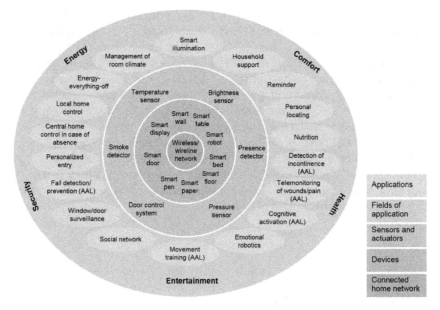

Figure 21.2 Adapted version of Glasberg's smart home representation [34] including fields of application, functional possibilities, and certain components.

ideas with invitation to enlargement. A special, somehow combined, field of application is the field of ambient assistant living (AAL), which includes general life support aspects as well as support for long-term care. This special issue with focus on CPS is addressed in detail in Section 21.3.5.

Out of these ideas specific scenarios can be depicted to represent a user's need in a real-world context with a set of specific assumptions and outcomes. Scenarios can also be defined as description of a very specific and representative human–machine interaction often derived from a generic scenario called use case. Use cases represent an interaction between system and external actor, comprising a main behavior, alternative behavior(s), and exceptions [35]. They define functionalities and targets of a system without exactly telling how this target will be technically achieved.

As already mentioned, a full presentation of every possible use case of smart buildings or even smart homes would go beyond the scope of this book chapter – and actually does not exist. But for some fields of application, for example, energy, there exists a valuable collection of use cases in [36], covering energy use cases for different building typologies such as smart homes, residential buildings, office buildings, data centers, and hotel buildings. The presented use cases are divided in high- and low-level use cases and depicted as UML diagram.

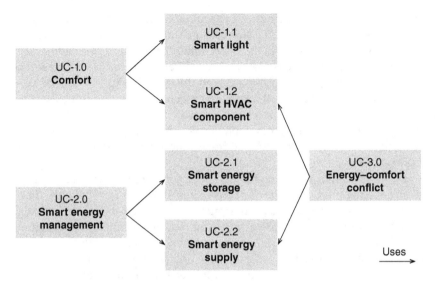

Figure 21.3 For the development of our architecture, we started with two comfort and smart energy use cases resulting in a conflict use case.

For the scope of our research described in detail in Section 21.3.4, we started with simple comfort use cases (UC-1.0, UC-1.1, UC-1.2) comprising smart light and smart heating/ventilation/air conditioning (HVAC) and smart energy use cases (UC-2.0, UC-2.1, UC-2.2), resulting in an energy-conflict use case (UC-3.0), as shown in Figure 21.3.

One possible example scenario for UC-1.2 could be a "Return-Home" scenario of Mrs Robinson. After a day at work, Mrs Robinson returns home at 6 p.m. It is already dark outside. She enters the living room, and the light automatically turns on. As she finishes work every day at the same time, the smart thermostats have already warmed up the room to her comfort temperature. The other climate values, CO2 content and humidity, are also on a perfect level. After preparing a meal in the kitchen, she spends the whole evening in the living room. Because of her continuous presence the CO2 content value increases. She does not like cold draught, and the value does not extend to a critical limit. So the window opens automatically to regulate the room climate after Mrs Robinson goes to bed.

One basic requirement to realize the presented use cases including the aforementioned scenario is the interoperability among the necessary technical devices. This aspect is addressed in the following section.

21.3.2 Interoperability for CPS-based Smart Home Environments

The classification of the different smart homes and smart buildings use cases shows a clear need to integrate automated sensor–actor interactions in private homes and commercial buildings. This requires communication technologies that are coined for the demands of various home applications. This fact has been realized by the various players in the smart home environment and is addressed by interoperability efforts channeled within different consortiums. However, due to the rapid progress of technology and the rise of a large number of heterogeneous devices, a variety of independent communication protocols were created, establishing a complex scenario to ensure interoperability between the devices. As a result, even the development of simple applications in this heterogeneous ecosystem of devices requires advanced programming skills and a considerable amount of time.

Smart homes and smart buildings are equipped with various devices and subsystems that generate heterogeneous data with respect to their functionalities and operation. Such scenario transforms the home/building environment into a data intensive entity that requires complex coordination and management leading to significant implementation problems. The first problem resides in the fact that the functionalities of smart homes and smart buildings depend by definition on higher heterogeneity of subsystems built with different specifications and protocols. Managing these different subsystems contributes toward rapid growth of residential gateways, which makes the integration task harder and time consuming. Another problem occurs due to the differences in operating systems, programming language, and hardware for those subsystems. The devices and subsystems come from different vendors and are developed in isolation and independently without considering the requirements for interoperation.

According to the IEEE Standard Computer Dictionary [37, 38], interoperability is defined as the ability of two or more systems or components to exchange information and to use the information that has been exchanged. For smart home and smart building systems, interoperability is related to the information exchange between two or more devices and the use of this information to perform a joint task execution in a federated manner without the need of external participation.

Different levels of interoperability can be found in the literature. From a technical point of view, we can distinguish four levels [39–41]. The hierarchy is illustrated in Figure 21.4:

- *No Interoperability* describes stand-alone systems.
- *Technical or Networks Interoperability* is the case when a communication protocol exists for exchanging data between participating systems allowing the exchanges of bits and bytes. Examples of common network interoperability standards are transport control protocol (TCP), user datagram protocol (UDP), and file transfer protocol (FTP).
- *Syntactic Interoperability* is the case when a common structure exists for exchanged information, which includes structured data and information.

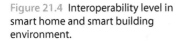

Figure 21.4 Interoperability level in smart home and smart building environment.

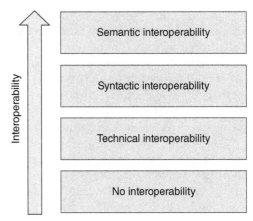

Mechanisms to understand the data structure in messages exchanged between two entities are provided. Protocols such as simple object access protocol (SOAP) and mechanisms such as asynchronous publish/subscribe count as examples.

- *Semantic Interoperability* is the case when the meaning of data is shared based on the common use or predefined ontology.

Many solutions are proposed by consortiums and industries as well as researchers in tackling the interoperability issues in smart homes and smart buildings. The first approach used by researchers was to implement smart gateways (consisting of hardware and software modules) as protocol converters between defined home networks that exist in smart home environment. An example is the interface between UPnP and LonWorks as implemented by Chemishkian and Lund [42]. Later, researchers built integration platforms, which are able to communicate over a variety of different protocols and function as central servers for the device communication [43–45]. These integration platforms are able to convert messages from different formats to others and communicate over different physical layers using a number of protocols. However, most of these middleware-based solutions require a central infrastructure to manage the communication between the involved devices.

Having the definition of a smart home as an intelligent system environment powered by self-organization capabilities in mind, one solution seems to be getting recognized as an ideal approach for interoperability. This solution relies on the use of semantic communication protocols such as OPC UA as backbone.

Based on a client/server architecture, OPC UA defines a platform-neutral middleware, which enables data acquisition and information modeling. It ensures a secure and reliable communication between different subsystems [46]. It also specifies the exchange of real-time information between subsystems from different manufacturers. OPC UA servers include an information model that allows users to organize data and their semantics in a structured manner. The information model, which constitutes the address spaces of OPC UA servers, supports some object-oriented features such as objects and classes (object and object type), composition, and inheritance. The information model also supports concepts such as references, which identify how nodes are related to each other and generally reflect an operation between the two, such as "Node A contains Node B" [47].

With the OPC UA middleware, it is possible to build an information model on top of each device of a smart home or a smart building environment. This information model supports the device identification and describes the data structure, the functionality, and the services provided. Invoking a service from another device is done by means of method call. Through subscriptions, devices can monitor the state and data of others devices or subsystems including smart sensors and therefore initiate actions based on these changes without the need of an external system. The possibilities opened OPC UA, which best suits our vision of a CPS-based smart home and smart building powered by self-organization capabilities.

To summarize, we presented in this section the importance of interoperability between the devices forming smart homes and smart buildings environments and concluded that OPC UA is best suited for this requirement since it allows us to define an ontology at each device level. In the next section, we present the problem of decentralized coordination and cooperation and the application of agent-based theory as a possible solution.

21.3.3 Decentralized Coordination and Cooperation: Applying Agent-Based Theory to Adaptive Architectural Environments

Due to the dynamically changing environment, the conflicting behavior of household devices and the incompatible user preferences in a multiuser environment, the vision of smart buildings with heterogeneous devices embedded in an unpredictable physical world is challenging. In this infrastructure, devices from different application domains are interconnected and are represented by their own intelligence that communicates information on operating status and needs, collects information, and responds in ways that most benefit its owners. Adaptivity in this context implies that the smart devices are able to sense and respond to changes in external circumstances, adjust internal functions, and independently adopt actions meeting current needs.

The realization of such an architecture requires a high-performance infrastructure capable of providing fast and intelligently optimized local responses coordinated with a higher level global goal in order to prevent or contain rapidly evolving adverse events. Centralized systems are too slow for this purpose. A distributed architecture of self-controlled and self-organized devices can enable local decision making that meets higher level goals. Constant interactions and transactions of smart devices therefore move the smart building environment beyond central control to a collaborative network of almost biological complexity (see Section 21.3.1). In these cases, smart devices provide their functionality as a service and are able to consume online services in order to better address their internal goals.

Over the last decade, a number of modern techniques and methods to deal with the complexity of designing a system with the ability to effectively and intelligently manage and coordinate the operation of multiple independent systems and subsystems in buildings have been proposed [48–50]. Among these modern and advanced techniques, designs based on the multi-agent paradigms stand out mainly as a result of its distributed, scalability, and autonomous properties [51, 52].

21.3.3.1 Multi-Agent Systems (MAS)

The concept of intelligent MAS has been used extensively in the field of artificial intelligence (AI) and is closely related to the subject of distributed problem solving [49, 53]. An agent refers to an entity that functions continuously and autonomously in an environment in which other processes take place and other agents exist to achieve both its individual goal and the collective objective of system [54, 55]. Such agents are described by the following properties [56]:

- *The Ability to Perceive, Communicate, and Interact with the Environment.*
- *Autonomy* refers to the principle that agents can operate on their own without the need for human guidance. Agents follow a specified goal defined by objectives.
- *Cooperation with Other Agents* refers to the social abilities of agents, for example, the ability to interact with other agents and possibly humans via some communication protocols.
- *Learning*: integrated feedback loops in the behaviors of agents enable them to be rational.

One of the main advantages of using agent-based control paradigm is its ability to provide independent, loosely coupled entities that encapsulate some specific functionality and interact with each other to solve tasks. MAS can be employed to solve the problems which are complex, difficult, or impossible for an individual agent to solve. MAS are suitable for the domains involving interaction between different people/organizations with different goals and proprietary information.

Unlike other design paradigms, agents can distinguish between the level of task completion or problem solving from the level of control [55]. This is achieved by sharing a minimum amount of information between entities and asynchronous operation via message exchanges [56]. An ideal rational agent is also able to choose the actions that would maximize its expected performance with little or no human influence based on its own perception of the environment as well as its built-in and acquired knowledge.

Four distinct types of agents can be distinguished in the literature:

- *Simple Reflex Agents* based on condition–action rules. They are stateless devices that do not have memory of past world states.
- *Agents with Memory* that have an internal memory used to keep track of past state of the world.
- *Agents with Goals* that consider future state of the world and choose accordingly their actions based on the current world state and the desirable goals.
- *Utility-Based Agents* that allow decision making by comparing options between conflicting goals and choose between likelihood of success and importance of goal.

21.3.3.2 Potentials of MAS for Smart Building Applications

The MAS paradigm provides a flexible and reusable platform, which permits easy addition and removal of agents in the system [57]. In addition, MAS provides a decentralized solution for solving the problems of control, management, and coordination of the heterogonous systems that make up smart homes and smart buildings. As a result, a distributed and improved manageability of building operation as well as an improved coordination of occupant's interaction with various components can be achieved, which leads to a global improvement of both energy and comfort performance of smart homes and smart buildings.

As described earlier in this chapter, user comfort and energy efficiency are the main goals of smart buildings. Balancing between energy efficiency and occupant's comfort is a challenging task and can only be achieved at room level where the occupants' needs and behavior can be precisely determined. The task is even harder at higher integration levels since the monolithic structure of current building control and management systems allows limited information exchange between various building subsystems. Ali and Kim [58] defined user comfort in a building using three parameters: thermal comfort, visual comfort, and air quality. Although the notion of comfort in buildings has evolved, many solutions to efficiently increase the comfort of the building's occupants have been implemented [59, 60]. However, most of these solutions give a passive role to the building occupant in the operation and determination of the comfort conditions or are unsustainable or economically unviable. It has been shown most recently that providing building users with more control possibilities could make occupants more tolerant to a wider range of comfort conditions and therefore could lead to an improvement of the energy consumption [61]. On the other hand, allowing users to manage their comfort conditions can potentially result in conflicts with the building managers whose goal is the optimal operation cost of the smart building. The application of MAS as demonstrated by the authors in [62] can be used to efficiently manage these interactions and thus achieve optimal building performance.

With the recent advancements in information and sensor technologies, it is now possible to obtain fine-grained information on user presence and behavior. By exploiting this information at each level of the smart environment, that is, workspace level, room level up to the building level, the building's energy demand for comfort can be more precisely determined. The scalability, decentralized, and cooperative properties of agents also give it more advantages over alternative solutions in matching the building operations with the user behaviors and handle these interdependences as well as interactions between systems, subsystems, and users thus contributing toward improved system efficiency and energy performance of buildings [63]. In the next section, we describe the role of semantic agents for CPS-based smart building to provide an insight of how the aforementioned context-sensible information can be generated and used in a smart building environment.

21.3.3.3 Applying MAS to Smart Building Environment

The use of concepts from artificial intelligence in combination with MAS has been demonstrated by a number of authors to improve both energy and comfort performances of buildings. Many of these solutions use utility-based agents [64, 65], simple reflex agent [66, 67], and agents with memory [68, 69]. The MAS software used in these applications can be divided into three conceptual levels. The lowest levels are perceptual and physical systems that provide real-time information of what is happening in the building. The next level provides a uniform agent-based interface to everything installed in the building such as device drivers so that the agent interfaces eliminate any network problem occurring from different parts of the system. The top levels can be defined as application or service layers that prepare particular functionality or provide particular services for each room of the building.

Most of these approaches rely on a central multi-agent platform such as JADE [70], which is the central point of information processing where the agent system is running. These solutions do not suit the vision of a CPS-based smart building where physical devices and its counterpart agents must be tightly coupled together to form a unique entity. A CPS is an entity, where every physical component is connected, respectively, to a semantic agent as cyber component, comprising a local ontology, a semantic service, and the agent itself. This vision can only be realized with the development of decentral agent frameworks and architectures.

Our proposed architecture unifies the physical level and the agent level in a single middleware software used by every device. Each agent in our MAS is assigned to a particular task and offers this task as a service to other agents. Collaboration among agents is mediated by asynchronous messages. From a conceptual point of view, our MAS is composed of four layers as illustrated in Figure 21.5.

Physical Layer or Device Abstraction Layer The physical layer is intended to hide different communication protocols and formats of data provided by the underlying hardware component, in order to offer the upper layer entity abstraction layer a unified data format. Such a hardware component could, for example, be a temperature sensor that can provide temperature values in different formats using different protocols (I^2C, Bluetooth, etc.). A common denominator interface is also added on the top of this layer to provide a universal and uniform interface.

Entities Abstraction Layer with Local Ontology This layer provides a one-to-one mapping between the physical entities and a proxy that instantiates a local ontology model. This layer is responsible for the interpretation of the raw data coming from lower layer and the maintenance of the local ontology. It maps changes in sensor values into events and forwards them with contextual information. On the one hand, it provides an interface to the higher layers to get actual sensor readings and information about the current state of the device. On the other hand, it forwards control orders coming from upper layers to the underlying actuators.

Agent or Control Layer This layer implements the actual intelligent agent of the smart device. It is responsible for the interagent communication, the control of the underlying hardware component, and the selection and computation of the device actions. The implemented agent is an agent with a memory whose goal is to maximize the effects of the tasks the device is assigned to, for example, heating a room in case of a thermostat.

Service Layer This layer hosts the services and other applications that are tightly coupled to the infrastructure. Examples of services and applications in this layer are local light control (quick reaction to turn on light when presence is detected) and real-time energy management at the device level.

With regard to Figure 21.4, a common ontology is needed in the entities abstraction layer in Figure 21.5. There are lots of separately developed ontologies with a different focus in the field of smart home such as context awareness [71] or energy efficiency [72, 73]. But a common ontology development including different important stakeholders such as manufacturer, users, ontology specialists, and others is still missing and is virtually important.

In this section, we presented the important role of agent-based architecture for the realization of CPS in a smart building environment. After a short introduction to MAS and the potentials for agent-based control in future buildings and smart homes, we described a possible semantic agent framework as part

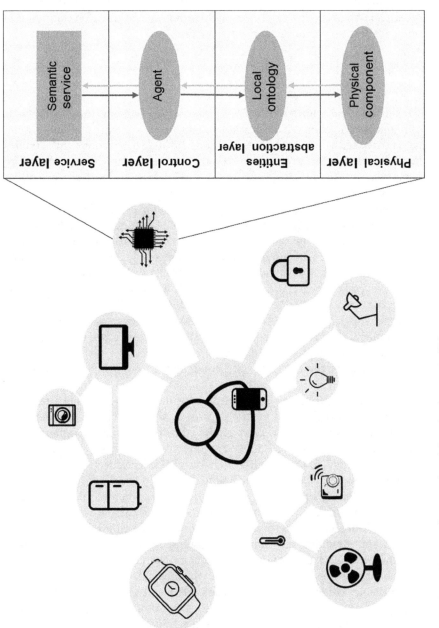

Figure 21.5 Semantic agent framework and the role of agents in CPS.

of a context-aware CPS. The realization of future interacting CPS as a part of an intelligent building is on its way, although there is still a wide field for future research in realizing decentral agent architectures and a common communication base on the level of semantic interoperability.

21.3.4 Application: A Decentralized Control in Private Homes Based on the Paradigms of the Industry 4.0

To illustrate the concept of a CPS-based smart building installation with self-configuration and self-organization abilities, we implemented smart home use cases based on the paradigms of "Industry 4.0." Descriptions of the implemented and proposed scenarios as well an overview of the system architecture are provided. In addition, we also analyze and discuss the performance of the implemented prototype.

The concept of "Industry 4.0" provides a vision for the application of CPS in industrial automation [74]. As stated in [75], the key requirements for the factory of the future include smart objects that are technical devices with a decentralized intelligence surrounded by an *ad hoc* universal networking that insures communication ability between the smart objects. Self-configuration, adaptability and agility, the use of Internet standards for cable-based and/or wireless communication (e.g., TCP/IP, Ethernet, Bluetooth, Wi-Fi), vertical integration in the network, and advanced human–machine interface are also some of its key requirements. This vision of a decentralized architecture based on CPS and the IoT has shown impressive results in the industry sector [76–79] and has the potential to solve the challenges presented by the long-standing smart home vision.

Based on the promising results from the industry sector, the presented project aims to replace the central control of smart home environments by intelligent sensors and actuators equipped with semantic communication technologies such as the OPC unified architecture (OPC UA) presented by the OPC Foundation. and thus builds a technical residential ecosystem that can support demographically sensitive living.

OPC UA defines a platform-neutral middleware, which enables data acquisition, information modeling, and secure and reliable communication between the plant floor and the enterprise [80]. The use of this well-established technology in a smart building environment as a standardized interoperable communication framework enables the involved components/devices to semantically describe themselves and hence makes their information and resources mutually available. Using a multi-agent-based intelligent management system, for example, *ad hoc* cooperation for joint task execution can be realized in order to achieve benefits for building owners and tenants.

As sensor and actuator networks mature, they become a core utility of smart homes and enable the running of many CPS applications. Integrating these components into a private home makes smart home installations highly complex. A modern technical equipment that is originally designed to increase the habitant comfort and energy efficiency becomes nontransparent and uncontrollable, which reduces their acceptance and thus also the market penetration. By relying on a decentralized organizational approach based on self-organizing, autonomous components, the acceptance of new technologies can be achieved by making the installation, operation, maintenance, and care of the devices equipped with increasingly complex functions more intuitive. In this case, the interoperability of the system components must be ensured by a middleware based on a semantic communication. The required flexibility must also support *ad hoc* connection in order to make the living environment dynamic and robust to disturbances. As a side effect, residents are no longer committed to specific manufacturer product since the exchange of devices is facilitated.

Being motivated by the advantages that such an architecture proposes, we created a similar paradigm for smart homes. We implemented a smart home scenario in which the conflict between comfort maximization and energy efficiency is illustrated. As depicted in Figure 21.6, the scenario includes only the living/dining room of a smart apartment testbed, which is equipped with actuators (controllable windows, lights, and thermostats) and smart sensors (light sensors, temperature sensors, proximity sensors, and air quality sensors) at fixed position in the room. We assume that the apartment can be owned and inhabited by an elderly person who might be physically challenged in various ways. Our smart home scenario, therefore, aims to maximize the comfort of the habitant by providing localized lighting, heat, and air quality and at the same time optimize the overall energy consumption. With these configurations, special conflict situations can occur, for example, when the thermostat is on and the window is opened.

The proposed architecture relies on the client/server architecture of the OPC UA middleware as a lightweight platform and language-independent method to exchange information between the components. Each sensor and actor has an OPC server to provide information about itself and an OPC client to gather information about other components and the environment. The basis of the provided information is a previously created OPC UA model (see Figure 21.7 that encodes information about the topology of the system, the position of the components in the room, the capacity and capability of each component, its behavior, and its internal state.

Once a component is connected to the network, it registers itself to the discovery server. It then sends a discovery request to discover other components in the neighborhood. Later, the actor or sensor browses the information model of each discovered component to gather information about their location, capacity, and provided services and selects those who are in the same room and compatible for possible cooperation. An example to illustrate this

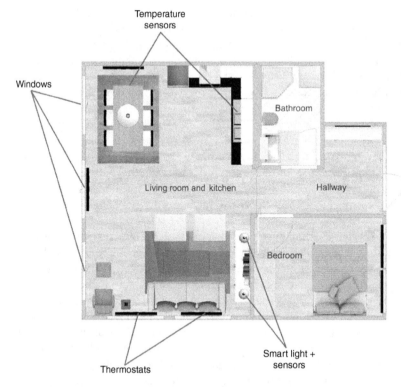

Figure 21.6 Overview of the implemented smart home environment. The living room is equipped with smart sensors (temperature, light, air) and smart actuators (window, light, thermostats, ventilators) that communicate and negotiate their actions in order to efficiently support the individual resident's needs.

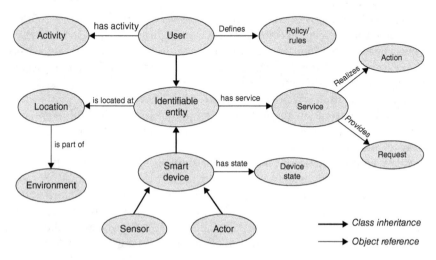

Figure 21.7 Excerpt of the implemented information model.

can be thermostats and temperature sensors as depicted in Figure 21.8. If the counterpart component is eligible for cooperation, the first component requests an authorization to open a session on the second component through an OPC UA function call. In the case of a positive response, the component opens a session on the second component and makes a subscription to the desired information, which could be, for example, real-time sensor data. Each time a sensor's value changes, the actor decides individually and locally of the action it takes by following predefined rules and policies.

The experimental results show that the stability of the system depends mostly on the policies and rules that its embedded agent follows. It is, therefore, essential to consider the complete system and simulate the system as a MAS during the design of the information model of each component. Nevertheless, it can also be stated that many concepts of "Industry 4.0" are perfectly applicable to the private home sector.

In this section, we presented an application of agent-based theory for a decentralized control of smart devices in private homes. The following section presents some applications of smart homes environments to support AAL.

21.3.5 Application: Ambient Assisted Living (AAL)

As mentioned before, the term smart home describes a residence equipped with technology. With the help of this technology, a common set of goals is achieved to create benefits for the resident. Examples of such goals are the management of energy consumption or an increase in the availability of home-based healthcare for elderly or disabled people. Obviously, end users' benefits will be maximized if all the available sensors, actuators, devices, and services are able to communicate and consider use cases from common major smart home domains (i.e., home automation, energy management, health management). Moreover, if existing use cases are enriched by web services, additional user benefits will be achieved. The architecture related to this web service approach is known as a service platform [81], (Figure 21.9). According to this mindset, a smart home will be connected to a smart city, and the end user will be able to enjoy all the benefits of a connected life. Hereon, we focus on the health-related use cases, also referred to as AAL.

Common AAL use cases include fall detection and health data monitoring combined with automated calls to relatives or service centers. The typical connections between the cyber and the physical worlds currently hardly exist. In the future, more interaction will likely occur, if the context awareness improves. Examples include fall detection in combination with automated door opening and personalized light management. People who are suffering from dementia benefit from light profiles that reflect normal sunlight. Other people like the flexible light management systems that are currently available, that is, Philips Hue. With the help of person detection and identification, smart

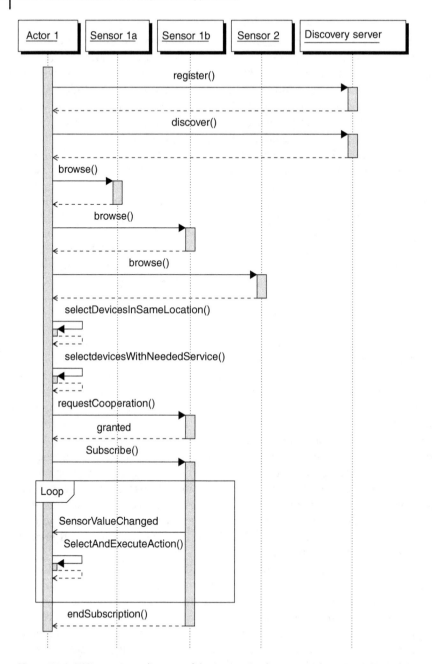

Figure 21.8 UML sequence diagram of the interaction between a thermostat (Actor 1 in room 1) and surrounding smart sensors (Sensor 1a and 1b in room 1 and sensor 2 in room 2). After discovering the devices in the neighborhood, the thermostat selects the sensors of interest based on the location and provided services. Afterward, the thermostat issues a negotiation to cooperate with the sensor. If the sensor agrees to the cooperation, the thermostat opens a secure session and subscribes to the smart sensor's events. The smart actor selects and performs its actions based on the sensor events received.

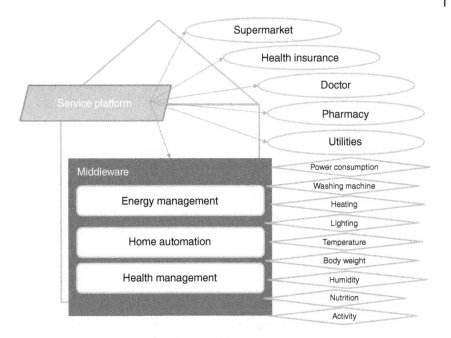

Figure 21.9 Representation of a smart home as a service platform on the Internet of Things (IoT). External web services are called through available APIs. The communication between the sensors and actuators of different vendors occurs by using middleware. Furthermore, the service platform offers a human–machine interface (HMI).

homes can control the lighting according to the current user's preferences. Furthermore, with the help of available web services, smart homes are able to request data from wearable devices (smart watches, activity trackers, camera-enriched glasses). In sum, existing AAL solutions (see Figure 21.2) could be enriched using automated procedures, which will lead to changes in the physical world.

In general, the market situation for existing smart home and AAL solutions is difficult due to a lack of domain- and vendor-independent interoperability. The connected life approach, on the one hand, maximizes users' benefits, so it is advantageous for improving the current market situation. On the other hand, the requirements for interoperable, secure, and easy-to-install solutions are increasing as well, especially for AAL-related scenarios, because there are often stringent country-specific law regulations related to medical data exchanges.

21.4 Connecting Smart Homes and Smart Cities

Driven by the advances in hardware technologies, smart homes, smart buildings, and larger ensembles such as airports, hospitals, or university campuses are already equipped with a multitude of mobile terminals, embedded devices as well as connected sensors and actuators. The opportunities provided by these smart environments will play a significant role in the consideration of future smart cities. In such a setting, smart buildings are expected to play a crucial role in coping with the challenges of urbanization and demographic changes, for example, regarding sustainability, energy distribution, mobility, health, or public safety/security.

Furthermore, smart cities envisage efficiencies in urban infrastructure and utilities, smart communication and data networks, as well as comfortable environments for work and daily life. All of these will enable occupants to be smart and productive users. Smart buildings are a critical enabler of such comfort and productivity. Thus, developing smart buildings would go a long way in creating smart cities and enhancing the overall quality of life.

A clear tendency to cross-linking individual system components with one another or with other systems has been observed in the development of building technologies in the last 20 years. In such a networked system, smart buildings, in which heating/cooling systems, lighting, and other energy consumers can be controlled centrally (e.g., even anywhere via smartphone or tablet), are particularly promising. In these cases, smart buildings are equipped with intelligent devices that take into account factors such as the anticipated electricity price, the weather forecast, and standard values for a good indoor climate and reduce its energy demand in a targeted manner during peak load phases while maintaining the proper level of comfort for building occupants. These smart buildings are able to manage their energy demand more precisely and link up with other buildings to form micro grids. This stabilizes primary grids, compensates for supply fluctuations, and reduces overall energy demand and cost. The smart buildings of the future will also use many types of temporary energy storage units, such as the electric vehicles, thermal storage units, water tanks, and mechanical units as well as flywheels.

People spend 80–90/% of their lives in buildings, making buildings an integral part of their ecosystem. With the advent of new technologies, the role buildings play is being redefined from a static environment to a more dynamic and interactive space that impacts the lifestyles, well-being, and productivity of their occupants [1]. As the population gets older, technical assistance systems that can support and facilitate daily activities in a nonintrusive way are needed. This results in a strong interlocking of technical and social systems, which increases the use of AAL technologies in everyday life. The acquisition of vital and environmental data is done using wearable and spatially distributed sensor systems. The technologies used in the AAL environment can be implemented modularly and networked together and with other systems to allow an adaptation to individual needs and to the individual environment. Through an integrated view of the data, an optimal assistance can be provided at the individual and at the city level. To achieve this, an interdisciplinary approach involving various medical, technological, sociological, and economic areas is required.

Building information modeling (BIM) is commonly understood as 3D "virtual building models." In particular, for large public infrastructure projects, the term BIM comes with different depths of structured IT models to be used during the entire life cycle of a building from the concept/design phase through the construction, operation, management, and decommissioning. BIM and the environment data captured by smart buildings give a strong insight into the health of city's infrastructures. By exposing its insight as services, smart buildings and smart homes will enable a more efficient use of urban infrastructure and utilities.

21.5   Conclusion and Future Research Focus

This chapter explored the concept of smart home and smart building based on IoT and CPS technologies. By reviewing the challenges of the smart home vision, we highlighted that CPS-based smart environments are best suited to support the integration of heterogeneous devices with different application domains and architectures. By exploring typical smart home use cases and scenarios, we showed that the interoperability between the components of the smart building and a decentralized coordination, in addition to the controls of these components, are the main requirements for such systems. We then proposed an approach based on OPC UA that supports self-description by using a common ontology and self-organization through the use of MAS theory. The presented implementation example also showed that most concepts of CPS applied in the industry sector can be transposed to the smart homes and smart buildings domain.

Having shown the promising potentials of CPS-based smart home and smart buildings through the implemented application, future works should focus on addressing some of its limitations, which include the development of a common integrated ontology, which should ease the interoperability and the integration of the heterogeneous smart devices forming smart homes and smart buildings environments. From a practical point of view, it is evident that building such CPS-based environments will be hampered due to the difficulties associated with the information exchange requirements. Therefore, it is essential to build effective information on aggregation models that cut across all levels of the building abstraction.

We believe that the future of smart cities will consider smart homes and smart buildings as a key component to effectively cope with the challenges of urbanization and demographic changes, especially those regarding sustainability, energy distribution, mobility, health, or public safety/security. Smart homes and smart buildings will be connected to various smart city service providers, which will enable an automation and an optimization of these services. Therefore, a generalization and a validation of our approach in other smart environments and IoT application domain within the smart city context should be studied first. In this context, real-time requirements are more critical than the home domain and justify an early adoption and extension of CPS design methodologies.

Final Thoughts

This chapter explored the concept of Smart Home and Smart Building based on IoT and CPS technologies. By reviewing the challenges of the Smart Home vision, we highlighted that CPS-based smart environments are best suited to support the integration of heterogeneous devices with different application domains and architectures. By exploring typical smart home use cases and scenarios, we showed that the interoperability between the components of the smart building and a decentralized coordination, in addition to an intelligent control of these components, are the main requirements for such systems. We then proposed an approach based on OPC UA that supports self-description by using a common ontology and self-organization through the use of MAS theory. The presented implementation example also showed that most concepts of CPS applied in the industry sector can be transposed to the smart homes and smart buildings domain.

Questions

1   What is the definition of building automation, home automation, smart home and what are the main challenges of building smart homes and smart buildings?

2   Give a classification of smart home uses cases and scenarios?

3   Name and describe two common limitations of currently used in practice smart home architectures.

4   What are the advantages of CPS based smart homes architectures?

5   What is the definition of interoperability and why does it matters in a smart environment?

6   Describe one or two AAL use cases?

References

1 Kagermann, H. and Wahlster, W.H.J. (2013) *Deutschlands Zukunft als Produktionsstandort sichern – Umsetzungsempfehlungen für das Zukunftsprojekt Industrie 4.0*, München [u.a.], https://www.bmbf.de/files/ Umsetzungsempfehlungen_Industrie4_0.pdf (accessed 20 December 2016).

2 Hardy, Q. (2015) *Consortium Wants Standards for 'Internet of Things'*, http://bits.blogs.nytimes.com/2014/03/27/consortium-wants-standards-for-internet-of-things/?_php=true&_type=blogs&_php=true&_type=blogs&_r=5 (accessed 20 December 2016).

3 Michl, M., Schäfer, F., Bauer, J., and Franke, J. (2015) *Von der Industrie 4.0 zur Wohnung 4.0 - Einsatzpotenziale verteilter Systemansätze im industriellen und privaten Umfeld*, 01-02/2015.

4 Heinrich, B. and Linke, P.G.M. (2015) *Grundlagen Automatisierung – Sensorik, Regelung, Steuerung*, Springer Fachmedien, Wiesbaden.

5 Kranz, H.R. (2009) *VDI Richtlinie 3814 - Part I: Building Automation and Control Systems (BACS) - System Basics*, VDI - Verein deutscher Ingenieure, Düsseldorf.

6 Pinkert, K., Adam, M., Behle, J.M., Runge, M., Brucke, M., and Witte, M. (2009) *Leitfaden zur Heimvernetzung*, BITKOM, Berlin.

7 Deloitte (2015) *Licht ins dunkel – erfolgsfaktoren für das smart home*, http://www2.deloitte.com/content/dam/Deloitte/de/Documents/technology-media-telecommunications/TMT-Studie_Smart%20Home.pdf (accessed 20 December 2016).

8 Gentner, A., Wagner, G., and Esser, R. (2015) *Smart Home – Zukunftschancen verschiedener Industrien*, München [u.a.], https://www.bmbf.de/files/ Umsetzungsempfehlungen_Industrie4_0.pdf (accessed 20 December 2016).

9 Paradiso, J., Dutta, P., Gellersen, H., and Schooler, E. (2011) Smart energy systems. *Pervasive Computing*, **10**, 11–12.

10 Esser, R., Schidlack, M., and Wagner, G. (2015) *Vor dem boom – marktaussichten für smart home*, https://www.bitkom.org/Publikationen/ 2014/Studien/Marktaussichten-fuer-Smart-Home/141023_Marktaussichten_ SmartHome.pdf (accessed 20 December 2016).

11 Edwards, W. and Grinter, R. (2001) At home with ubiquitous computing: seven challenges, in *Ubicomp 2001: Ubiquitous Computing*, Lecture Notes in Computer Science, vol. **2201** (eds G. Abowd, B. Brumitt, and S. Shafer), Springer Berlin Heidelberg, pp. 256–272, doi: 10.1007/3-540-45427-6_22.

12 Chan, M., Estève, D., Escriba, C., and Campo, E. (2008) A review of smart homes–present state and future challenges. *Computer Methods and Programs in Biomedicine*, **91** (1), 55–81, doi: 10.1016/j.cmpb.2008.02.001.

13 Chan, M., Campo, E., Estève, D., and Fourniols, J.Y. (2009) Smart homes – current features and future perspectives. *Maturitas*, **64** (2), 90–97, doi: 10.1016/j.maturitas.2009.07.014.

14 Hightower, J. and Borriello, G. (2001) Location systems for ubiquitous computing. *Computer*, **34** (8), 57–66, doi: 10.1109/2.940014.

15 Manley, E.D., Nahas, H., and Deogun, J. (2006) *Localization and Tracking in Sensor Systems*. Sensor Networks, Ubiquitous, and Trustworthy Computing, 2006. IEEE International Conference on, vol. 2, pp. 237–242, doi: 10.1109/SUTC.2006.83.

16 Robles, R.J. and Tai-hoon, K. (2010) *Review: Context aware tools for smart home development*.

17 Pishva, D. and Takeda, K. (2008) Product-based security model for smart home appliances. *IEEE Aerospace and Electronic Systems Magazine*, **23** (10), 32–41, doi: 10.1109/MAES.2008.4665323.

18 Mennicken, S., Vermeulen, J., and Huang, E.M. (2014) *From Today's Augmented Houses to Tomorrow's Smart Homes: New Directions for Home Automation Research*. Proceedings of the 2014 ACM International Joint Conference on Pervasive and Ubiquitous Computing, ACM, New York, NY, USA, UbiComp '14, pp. 105–115, doi: 10.1145/2632048.2636076.

19 Atzori, L., Iera, A., and Morabito, G. (2010) The internet of things: a survey. *Computer Networks*, **54** (15), 2787–2805, doi: 10.1016/j.comnet.2010.05.010.

20 Abdmeziem, M., Tandjaoui, D., and Romdhani, I. (2016) Architecting the internet of things: state of the art, in *Robots and Sensor Clouds, Studies in Systems, Decision and Control*, vol. **36** (eds A. Koubaa and E. Shakshuki), Springer International Publishing, pp. 55–75, doi: 10.1007/978-3-319-22168-7_3.

21 Gubbi, J., Buyya, R., Marusic, S., and Palaniswami, M. (2013) Internet of things (IoT): a vision, architectural elements, and future directions. *Future Generation Computer Systems*, **29** (7), 1645–1660, doi: 10.1016/j.future.2013.01.010.

22 Mitsugi, J., Sato, Y., Ozawa, M., and Suzuki, S. (2014) *An Integrated Device and Service Discovery with UPNP and ONS to Facilitate the Composition of Smart Home Applications*. Internet of Things (WF-IoT), 2014 IEEE World Forum on, pp. 400–404, doi: 10.1109/WF-IoT.2014.6803199.

23 Viani, F., Robol, F., Polo, A., Rocca, P., Oliveri, G., and Massa, A. (2013) Wireless architectures for heterogeneous sensing in smart home applications: concepts and real implementation. *Proceedings of the IEEE*, **101** (11), 2381–2396, doi: 10.1109/JPROC.2013.2266858.

24 Souza, A.M. and Amazonas, J.R. (2013) *A Novel Smart Home Application Using An Internet of Things Middleware*. Smart Objects, Systems and Technologies (SmartSysTech), Proceedings of 2013 European Conference on, pp. 1–7.

25 Han, D.M. and Lim, J.H. (2010) Design and implementation of smart home energy management systems based on ZigBee. *IEEE Transactions on Consumer Electronics*, **56** (3), 1417–1425, doi: 10.1109/TCE.2010.5606278.

26 Wan, K., Man, K.L., and Hughes, D. (2010) Specification, analyzing challenges and approaches for cyber-physical systems (CPS). *Engineering Letters*, http://www.engineeringletters.com/issues_v18/issue_3/EL_18_3_14.pdf (accessed 20 December 2016).

27 Lee, E. (2010) *CPS foundations*. Design Automation Conference (DAC), 2010 47th ACM/IEEE, pp. 737–742.

28 Lee, E.A. (2008) *Cyber Physical Systems: Design Challenges*. International Symposium on Object/Component/Service-Oriented Real-Time Distributed Computing (ISORC), invited Paper, http://chess.eecs.berkeley.edu/pubs/427 .html (accessed 20 December 2016).

29 Singh, J., Hussain, O., Chang, E., and Dillon, T. (2012) *Event Handling for Distributed Real-Time Cyber-Physical Systems*. Object/Component/ Service-Oriented Real-Time Distributed Computing (ISORC), 2012 IEEE 15th International Symposium on, pp. 23–30, doi: 10.1109/ISORC.2012.12.

30 Tansley, A.G. (1935) The use and abuse of vegetational concepts and terms. *Ecology*, **16** (3), 284–307, http://www.jstor.org/stable/1930070 (accessed 20 December 2016).

31 Briscoe, G. and De Wilde, P. (2006) Digital ecosystems: evolving service-orientated architectures, in *1st International Conference on Bio Inspired Models of Network, Information and Computing Systems*, (eds T. Suda and C. Tschudin), IEEE Press, p. 17, doi: 10.1145/1315843.1315864.

32 Strese, H., Seidel, U., and Knape, T. (2010) *Smart home in deutschland: untersuchung im Rahmen der wissenschaftlichen begleitung smart home in deutschland: untersuchung im Rahmen der wissenschaftlichen begleitungzum programm next generation media (NGM) des bundesministeriums für wirtsch, aft und technologie*, http://www.vdivde-it.de/publikationen/ studien/smart-home-in-deutschland-untersuchung-im-rahmen-der- wissenschaftlichen-begleitung-zum-programm-next-generation-media-ngm- des-bundesministeriums-fuer-wirtschaft-und-technologie/at_download/pdf (accessed 20 December 2016).

33 Gentner, A. and Wagner, G. (2013) *Licht ins Dunkel: Erfolgsfaktoren für das Smart Home*, https://www2.deloitte.com/content/dam/Deloitte/de/ Documents/technologymedia (accessed 20 December 2016).

34 Glasberg, R. and Feldner, N. (2009) *Lleitfaden zur heimvernetzung*, https://www.bitkom.org/Publikationen/2008/Leitfaden/BITKOM- Studie-Konsumentennutzen-und-persoenlicher-Komfort/Studie- Konsumentennutzen.pdf (accessed 20 December 2016).

35 Alexander, I. and Maiden, N. (2004) Scenarios, stories, and use cases: the modern basis for system development. *Computing and Control Engineering*, **15** (5), 24–29, doi: 10.1049/cce:20040505.

36 Schitkow, I., Burón, J.L., Socorro, R., Dherbécourt, Y., Bonneau, D., Privat, G., and Haro, H. (2011) *Smart Buildings Scenario Definition: Deliverable d 4.1 of the Eu-Project Future Internet of Smart*

Energy (FINSENY), http://www.fi-ppp-finseny.eu/wp-content/uploads/2012/05/D4.1_Smart-Buildings-scenario-definition_v1.1.pdf (accessed 20 December 2016).

37 IEEE Std 610. (1991) *IEEE Standard Computer Dictionary: A Compilation of IEEE Standard Computer Glossaries*, IEEE, pp. 1–217, doi: 10.1109/IEEESTD.1991.106963.

38 C4ISR Architecture Working Group and others (1998) Levels of information systems interoperability (LISI). *US DoD.*

39 Young, P., Chaki, N., Berzins, V., and Luqi (2003) *Evaluation of Middleware Architectures in Achieving System Interoperability.* Rapid Systems Prototyping, 2003. Proceedings. 14th IEEE International Workshop on, pp. 108–116, doi: 10.1109/IWRSP.2003.1207037.

40 Athanasopoulos, G., Tsalgatidou, A., and Pantazoglou, M. (2006) *Interoperability Among Heterogeneous Services.* Services Computing, 2006. SCC '06. IEEE International Conference on, pp. 174–181, doi: 10.1109/SCC.2006.59.

41 Ge, Y., Qiu, X., and Huang, K. (2010) *Conceptual Interoperability Model of NCW Simulation.* Information Theory and Information Security (ICITIS), 2010 IEEE International Conference on, pp. 911–914, doi: 10.1109/ICITIS.2010.5689791.

42 Chemishkian, S. and Lund, J. (2004) *Experimental Bridge lonworks/sup /spl reg///upnp/spl trade/1.0.* Consumer Communications and Networking Conference, 2004. CCNC 2004. 1st IEEE, pp. 400–405, doi: 10.1109/CCNC.2004.1286895.

43 Järvinen, H. and Vuorimaa, P. (2014) *Interoperability for Web Services Based Smart Home Control Systems.* Proceedings of the 10th International Conference on Web Information Systems and Technologies, pp. 93–103, doi: 10.5220/0004948500930103.

44 Yang, C., Yuan, B., Tian, Y., Feng, Z., and Mao, W. (2014) *A Smart Home Architecture Based on Resource Name Service.* Computational Science and Engineering (CSE), 2014 IEEE 17th International Conference on, pp. 1915–1920, doi: 10.1109/CSE.2014.351.

45 Albuquerque, H.J.O. and de Aquino Junior, G.S. (2014) *A Proxy-Based Solution for Interoperability of Smart Home Protocols.* Complex, Intelligent and Software Intensive Systems (CISIS), 2014 8th International Conference on, pp. 287–293, doi: 10.1109/CISIS.2014.96.

46 Matthias, D., Leitner, S.-H., and Mahnke, W. (2009) *OPC Unified Architecture.*

47 IEC 62541-3:2010). (2011) *OPC Unified Architecture – Part 3: Address Space Model; English version en 62541-3:2010*, pp. 1–217.

48 Shaikh, P.H., Nor, N.B.M., Nallagownden, P., Elamvazuthi, I., and Ibrahim, T. (2014) A review on optimized control systems for building energy and comfort management of smart sustainable buildings. *Renewable and Sustainable Energy Reviews*, **34**, 409–429, doi: 10.1016/j.rser.2014.03.027.

49 Dounis, A. and Caraiscos, C. (2009) Advanced control systems engineering for energy and comfort management in a building environment–review. *Renewable and Sustainable Energy Reviews*, **13** (6–7), 1246–1261, doi: 10.1016/j.rser.2008.09.015.

50 Sharples, S., Callaghan, V., and Clarke, G. (1999) A multi-agent architecture for intelligent building sensing and control. *Sensor Review*, **19** (2), 135–140, doi: 10.1108/02602289910266278.

51 Reinisch, C. and Kastner, W. (2011) *Agent Based Control in the Smart Home*. IECON 2011 – 37th Annual Conference on IEEE Industrial Electronics Society, pp. 334–339, doi: 10.1109/IECON.2011.6119275.

52 Mokhtar, M., Stables, M., Liu, X., and Howe, J. (2013) Intelligent multi-agent system for building heat distribution control with combined gas boilers and ground source heat pump. *Energy and Buildings*, **62**, 615–626, doi: 10.1016/j.enbuild.2013.03.045.

53 Wang, Z., Wang, L., Dounis, A.I., and Yang, R. (2012) Multi-agent control system with information fusion based comfort model for smart buildings. *Applied Energy*, **99**, 247–254, doi: 10.1016/j.apenergy.2012.05.020.

54 Shoham, Y. (1994) Agent oriented programming: an overview of the framework and summary of recent research, in *Knowledge Representation and Reasoning Under Uncertainty, Lecture Notes in Computer Science*, vol. **808** (eds M. Masuch and L. Pólos), Springer, Berlin Heidelberg, pp. 123–129, doi: 10.1007/3-540-58095-6_9.

55 Wooldridge, M. and Jennings, N.R. (1995) Intelligent agents: theory and practice. *Knowledge Engineering Review*, **10**, 115–152.

56 El Fallah Seghrouchni, A., Florea, A., and Olaru, A. (2010) Multi-agent systems: a paradigm to design ambient intelligent applications, in *Intelligent Distributed Computing IV, Studies in Computational Intelligence*, vol. **315** (eds M. Essaaidi, M. Malgeri, and C. Badica), Springer, Berlin Heidelberg, pp. 3–9, doi: 10.1007/978-3-642-15211-5_1.

57 Kwak, J.Y., Varakantham, P., Maheswaran, R., Chang, Y.H., Tambe, M., Becerik-Gerber, B., and Wood, W. (2014) TESLA: an extended study of an energy-saving agent that leverages schedule flexibility. *Autonomous Agents and Multi-Agent Systems*, **28** (4), 605–636, doi: 10.1007/s10458-013-9234-0.

58 Ali, S. and Kim, D.H. (2013) Energy conservation and comfort management in building environment. *International Journal of innovative Computing, Information and Control*, **9** (6), 2229–2244.

59 Cole, R.J., Robinson, J., Brown, Z., and O'shea, M. (2008) Re-contextualizing the notion of comfort. *Building Research & Information*, **36** (4), 323–336, doi: 10.1080/09613210802076328.

60 Ackerly, K. and Brager, G. (2013) Window signalling systems: control strategies and occupant behaviour. *Building Research & Information*, **41** (3), 342–360, doi: 10.1080/09613218.2013.772044.

61 Nicol, J. and Humphreys, M. (2002) Adaptive thermal comfort and sustainable thermal standards for buildings. *Energy and Buildings*, **34** (6), 563–572, doi: 10.1016/S0378-7788(02)00006-3.

62 Liu, K., Duangsuwan, J., and Huang, Z. (2002) Intelligent agents enabling negotiated control of pervasive environments. *Energy and Buildings*, **38** (6), 99–122, doi: 10.1016/S0378-7788(02)00006-3, special Issue on Thermal Comfort Standards.

63 Labeodan, T., Aduda, K., Boxem, G., and Zeiler, W. (2015) On the application of multi-agent systems in buildings for improved building operations, performance and smart grid interaction – a survey. *Renewable and Sustainable Energy Reviews*, **50**, 1405–1414, doi: 10.1016/j.rser.2015.05.081.

64 Brous, J. (2007) *Pacific Northwest Grid Wise Testbed Demonstration Projects*, http://www.pnl.gov/main/publications/external/technical_reports/PNNL-17079.pdf (accessed 20 December 2016).

65 Warmer, C., Hommelberg, M., Roossien, B., Kok, J., and Turkstra, J. (2007) *A Field Test Using Agents for Coordination of Residential Micro-CHP*. Intelligent Systems Applications to Power Systems, 2007. ISAP 2007. International Conference on, pp. 1–4, doi: 10.1109/ISAP.2007.4441634.

66 Yu, D.Y., Ferranti, E., and Hadeli, H. (2013) *An Intelligent Building that Listens to Your Needs*. Proceedings of the 28th Annual ACM Symposium on Applied Computing, ACM, New York, NY, USA, SAC '13, pp. 58–63, doi: 10.1145/2480362.2480376.

67 Abras, S., Pesty, S., Ploix, S., and Jacomino, M. (2010) Advantages of MAS for the resolution of a power management problem in smart homes, in *Advances in Practical Applications of Agents and Multiagent Systems*, *Advances in Intelligent and Soft Computing*, vol. **70** (eds Y. Demazeau, F. Dignum, J. Corchado, and J. Pérez), Springer, Berlin Heidelberg, pp. 269–278, doi: 10.1007/978-3-642-12384-9_32.

68 Rutishauser, U., Joller, J., and Douglas, R. (2005) Control and learning of ambience by an intelligent building. *IEEE Transactions on Systems, Man and Cybernetics, Part A: Systems and Humans*, **35** (1), 121–132, doi: 10.1109/TSMCA.2004.838459.

69 Davidson, E., McArthur, S., Yuen, C., and Larsson, M. (2008) *AuRA-NMS: Towards the Delivery of Smarter Distribution Networks Through the Application of Multi-Agent Systems Technology*. Power and Energy Society General Meeting – Conversion and Delivery of Electrical Energy in the 21st Century, 2008 IEEE, pp. 1–6, doi: 10.1109/PES.2008.4596672.

70 Cavone, D., De Carolis, B., Ferilli, S.P., and Novielli, N. (2011) *An agent-based approach for adapting the behavior of a smart home environment*.

71 Gu, T., HangWang, X., Keng Pung, H., and Qing Zhang, D. (2004) *An ontology-based context model in intelligent environments.* http://www-public .it-sudparis.eu/~zhang_da/pub/Ontology-2004-2.pdf.

72 Fensel, A., Tomic, S., Kumar, V., Stefanovic, M., Aleshin, S.V., and Novikov, D.O. (2013) *Sesame-s: Semantic smart home system for energy efficiency.* Informatik-Spektrum, **36** (1).

73 Reinisch, C., Kofler, M., Iglesias, F., and Kastner, W. (2011) *Think home energy efficiency in future smart homes.* EURASIP Journal on Embedded Systems, (1), 104 617, doi: 10.1155/2011/104617.

74 Schlick, J. (2012) *Cyber-Physical Systems in Factory Automation – Towards the 4th Industrial Revolution.* Factory Communication Systems (WFCS), 2012 9th IEEE International Workshop on, pp. 55, doi: 10.1109/WFCS.2012.6242540.

75 Detlef Zühlke, W.W. and Mittelbach, K. (2012) *SmartFactoryKL,* http:// smartfactory.dfki.uni-kl.de/en/content/pressekonferenz2 (accessed 20 December 2016).

76 Perez, F., Irisarri, E., Orive, D., Marcos, M., and Estevez, E. (2015) *A CPPS Architecture Approach for Industry 4.0.* Emerging Technologies Factory Automation (ETFA), 2015 IEEE 20th Conference on, pp. 1–4, doi: 10.1109/ETFA.2015.7301606.

77 Shellshear, E., Berlin, R., and Carlson, J. (2015) Maximizing smart factory systems by incrementally updating point clouds. *IEEE Computer Graphics and Applications,* **35** (2), 62–69, doi: 10.1109/MCG.2015.38.

78 Vogel-Heuser, B., Weber, J., and Folmer, J. (2015) *Evaluating Reconfiguration Abilities of Automated Production Systems in Industrie 4.0 with metrics.* Emerging Technologies Factory Automation (ETFA), 2015 IEEE 20th Conference on, pp. 1–6, doi: 10.1109/ETFA.2015.7301441.

79 Flatt, H., Koch, N., Rocker, C., Gunter, A., and Jasperneite, J. (2015) *A Context-Aware Assistance System for Maintenance Applications in Smart Factories Based on Augmented Reality and Indoor Localization.* Emerging Technologies Factory Automation (ETFA), 2015 IEEE 20th Conference on, pp. 1–4, doi: 10.1109/ETFA.2015.7301586.

80 Damm, M., Leitner, S.H., and Mahnke, W. (2009) *OPC Unified Architecture,* Springer, Berlin Heidelberg, p. 1, ISBN: 978-3-540-68899-0.

81 Weidner, R., Redlich, T., and Wulfsberg, J.P. (2015) *Technische Unterstützungssysteme.*

22

Climate Resilience and the Design of Smart Buildings

Saranya Gunasingh[1], Nora Wang[2], Doug Ahl[3], and Scott Schuetter[3]

[1] Seventhwave, Chicago, IL, USA
[2] Pacific Northwest National Laboratory, Washington, DC, USA
[3] Seventhwave, Madison, WI, USA

CHAPTER MENU
Climate Change and Future Buildings and Cities, 642
Carbon Inventory and Current Goals, 644
Incorporating Predicted Climate Variability in Building Design, 646
Case Studies, 648
Implications for Future Cities and Net-Zero Buildings, 662

Objectives

- To understand the large ecological footprint of cities that stems from high energy use in buildings and to become familiar with existing policies and programs that address the issue.
- To provide a brief overview of several voluntary programs initiated by city and national governments and professional organizations to mitigate GHG emissions. The shortcomings of such programs, which are a hindrance to achieve emission reduction goals and are examined here.
- The chapter familiarizes the concept of climate-resilient building design to building designers and owners and presents the need to be proactive with current new construction and building retrofits.
- Using climate data of the past for future buildings leads to misrepresentation of projected future energy use. The research work discussed in this chapter identifies a method to address this uncertainty, so that both building owners and city infrastructure can be adequately equipped for climate induced energy use variations.

Smart Cities: Foundations, Principles and Applications, First Edition.
Edited by Houbing Song, Ravi Srinivasan, Tamim Sookoor, and Sabina Jeschke.
© 2017 John Wiley & Sons, Inc. Published 2017 by John Wiley & Sons, Inc.

22.1   Climate Change and Future Buildings and Cities

We design, construct, and operate buildings in a world affected by climate change. Climate change has caused an average global temperature increase of 1.5 °F over the last century and is predicted to cause an additional increase of 0.5–8.6 °F in the next century [1]. Climate change is often portrayed as a problem of the future, but in fact the science of climate change has been around for many decades now. The first Intergovernmental Panel on Climate Change (IPCC) was convened in 1988 with the goal of understanding the environmental, social, and economic impacts of climate change and to gather the latest scientific information on the subject (IPCC) [2]. Since then, the IPCC has published five Assessment Reports that provide the latest scientific knowledge of climate changes, observed outcomes, and predicted future uncertainties and risks. According to the Fifth Assessment Report, the global surface temperature is projected to rise over the twenty-first century under all assessed emission scenarios [3].

The impacts of climate change go beyond temperature increase. According to the IPCC's Fifth Assessment Report [3], many aspects of climate change and associated impacts will continue for centuries, and it is very likely that

- Heat waves will occur with a higher frequency and longer duration.
- Occasional cold winter extremes will continue to occur.
- Mean precipitation will decrease in dry regions.
- Sea level will rise in more than about 95% of the ocean's total area.
- Food security will be threatened and renewable surface water and ground water resources will decline in most dry subtropical regions.

An important attribute of smart cities of the future will be resilient infrastructure that can adapt to the environmental changes with proactive design and planning strategies. Architects and engineers have a professional responsibility to design the built environment for low environmental impact while also being resilient to the effects of climate change. This chapter presents energy analysis methods for design professionals and policymakers that can be applied to current industry practices, so that future buildings and cities are best equipped to manage the impacts from the changing climate.

As a fundamental concept, resilience is gaining ground in future building design and city development. It has broad implications. How can a building remain operable without access to electrical and other utilities (passive survivability)? How can a building protect human life during an extreme event and quickly recover? Climate-resilient buildings will be designed to maintain comfortable indoor conditions during extreme heat or cold events and to serve as utility bridges or generators when public infrastructure is damaged [4]. Resilience has also become an important strategy of city development, especially for coastal cities that have suffered from hurricanes. For example,

New York City Mayor de Blasio created the Office of Recovery and Resiliency in March 2014 to implement the strategies laid out in *PlaNYC: A Stronger, More Resilient New York* [5]. The Stevens Institute of Technology has been exploring engineering resilience and using digital technology to overcome the limits of legacy infrastructure in New York City. The Rockefeller Foundation is supporting a "100 Resilient Cities" program [6] to help cities develop resilience plans and install resilience officers in city staff. The resilience plans cover areas such as infrastructure, water, and natural disaster preparedness.

While future buildings and cities are affected by the changing environment, their inhabitants continue contributing to global climate change. Climate scientists agree that this increase in global temperature is likely caused by greenhouse gas (GHG) emissions from burning fossil fuels. A recent report published by the International Energy Agency [7] shows that buildings are responsible for about one-third of global primary energy consumption and about one-third of total direct and indirect energy-related GHG emissions. With no effective actions to improve building energy efficiency, energy demand in buildings will rise globally by 50% by 2050 based on the projected growth [7].

Limiting the rise in temperature to 3.6 °F above pre-industry levels has been discussed as a global goal. Representatives from 196 nations adopted climate agreement at COP21 Talks (The 2015 United Nations Climate Change Conference) in Paris [8]. The agreement acknowledges that the threat of climate change is urgent and potentially irreversible and reinforces the 3.6 °F target. In addition, the agreement recognizes that limiting the temperature increase to 2.7 °F would significantly reduce the risks and impacts of climate change. More than 30 countries, including the United States, have committed to the Paris climate change agreement of 2015 [8]. The Paris climate agreement takes effect in 2020 and is designed to prevent the global mean temperature from rising no more than 5.6 °F. In addition to the Paris agreement, state and local governments around the world are developing climate action plans to assist in mitigating the socioeconomic, public health, infrastructure, and energy impacts from the changing climate. Although the outcomes of these efforts in controlling global warming are difficult to quantify, building designs in the future will need to be adaptable to the future climate conditions.

One important aspect of developing smart buildings and cities is to reduce GHG emissions through a combination of technology and policy actions. To achieve the 3.6 °F goal, building GHG emissions need to be reduced *to a quarter* of the current level by 2050 [7]. To reach the aggressive goals of reducing GHG emissions by 75%, we need to explore new building paradigms. As Albert Einstein observed, we can't solve problems by using the same kind of thinking we used when we created them. For example, the US Department of Energy (DOE) and Pacific Northwest National Laboratory initiated the Buildings of the Future Scoping Study, which envisions what US mainstream commercial and residential buildings could become in 100 years [9]. With climate change as the future context, the study suggested that solutions to a sustainable and resilient built environment might be more effective at the community or urban level. For instance, not every building can achieve a zero-net-energy[1] goal due to the limitations of their geographic locations, available roof areas, energy demand associated with building functions, and economic constraints. A zero-net-energy campus, neighborhood, or community may be more feasible, although this requires supporting infrastructure and transaction platforms [29].

22.2 Carbon Inventory and Current Goals

The Global Infrastructure Basel (GIB) reports that we have yet to build 75% of the world's infrastructure that will exist by 2050 [11]. Most of these young cities will be located in developing nations, placing more stress on already fragile ecosystems and limited natural resources. While this rapid rate of urbanization and extraordinary stress on limited urban resources is a cause for concern, it presents an opportunity to rethink urban design. This projected rapid urbanization is a onetime opportunity to reinvent city design and implement new urban planning principles and smart city strategies to curb anthropogenic heat generation.

To set achievable goals for emissions reductions and to develop apt methods to reach these goals, policymakers should have the knowledge of existing emissions estimates. This process of identifying all emissions sources and estimating emissions is referred to as "greenhouse gas inventorying." The Global Protocol for Community-Scale Greenhouse Gas Emission Inventories (GPC) (WRI, C40, ICLEI) [12] lays out a comprehensive accounting framework and reporting method of inventorying GHG emissions. Cities are the source of 70% GHG emissions worldwide. This framework addresses many gaps and inconsistencies that currently exist in accurately estimating and reporting the carbon footprint of cities. Much like an energy code that offers a stringent framework for building design and retrofit, this resource serves as a common global standard for accounting and reporting principles, calculating emissions, setting goals, and monitoring performance.

Emissions from the process of generating, delivering, and consuming energy are classified as stationary energy sources, and this has been identified as the largest category of emissions in cities. This group comprises residential and commercial buildings as well as industrial and agricultural processes involving energy generation and consumption. For buildings and other stationary emissions sources, emissions are calculated by multiplying fuel consumption by the

---

1 Zero-net-energy buildings are defined as energy-efficient buildings where, on a source energy basis, the actual annual delivered energy is less than or equal to the on-site renewable exported energy [10].

emissions factor for the fuel type. Fuel consumption is calculated by the energy use intensity (EUI) multiplied by the building area. One of the four suggested methods for calculating fuel consumption is from energy modeling. Energy models can estimate the total energy use buildings at source and site, broken down by fuel type and end energy uses.

The architectural society has developed a series of programs and initiatives (e.g., AIA+2030, 2030 Districts, 2030 Palette) to meet incremental fossil fuel reduction goals over the years toward the ultimate goal of zero emissions. The series of Architecture 2030 challenges [13] ask the global architecture and building community to achieve carbon-neutral status (defined as a building that is designed and constructed to require a greatly reduced quantity of energy to operate, meeting the balance of its energy needs from sources that do not produce GHG emissions and therefore result in zero-net emissions) for all new buildings and major renovations by 2030 by implementing efficiency and sustainable strategies, generating on-site renewable power, and/or importing clean energy. In 2014, the International Union of Architects (UIA) adopted the 2050 Imperatives, a road map to achieve a series of total annual GHG emissions reduction goals for the buildings sector and phase out carbon among existing buildings by 2050 [14].

The C40 Cities Climate Leadership Group (C40) is a network of more than 80 megacities around the world that have committed to addressing climate change [15]. There are two means by which cities target carbon emissions reduction: reduce energy demand and/or switch to clean energy. Fifty-seven percent of the C40 cities are implementing energy-efficient building technology, using performance rating methods, and reducing industrial GHG emissions to realize reductions in energy demand. Fifty percent of C40 cities are also switching to low carbon fuels and/or focusing on clean energy generation, as on-site renewable energy generation technology becomes more affordable.

The top 10 most implemented actions reported by C40 cities include [16]:

- Adding insulation to buildings
- Performing energy audits and providing advice
- Certifying energy performance through a rating system
- Benchmarking energy use
- Improving heating and cooling efficiency
- Installing smart meters
- Installing efficient lighting systems
- Purchasing green electricity
- Implementing building energy management systems
- Submetering

The report notes that three of the top five action items – audits, certification, and benchmarking – are directed at measuring building performance rather than optimizing building design and operation.

Space conditioning is a climate-dependent energy end use and it alone accounts for 38% of total building energy use. However, out of the top 10 action items commonly reported by cities, only two measures can be identified as optimization strategies for high performance buildings – adding insulation and improving heating and cooling equipment efficiency. The eight other measures focus on measurement, energy performance metrics, and certifications [16].

Reducing energy use offers an optimum path to build resilience to changing climate by lowering grid dependency for normal building operation. Improving the energy performance of new and existing buildings should be one of the focus areas of smart city developments. This effort should first begin with the design community – architects and engineers should be aware of the impact of regional climate on building energy performance. Understanding the interaction between climate and energy consumption is essential because energy conservation measures (ECMs) will result in widely varying energy performance outcomes, depending on the climatic profile of the region. Designers should have the knowledge and the tools to incorporate the most climate adaptive and resilient features in their designs.

The impact of climate change on future building operation is generally not factored into design and technology selection decisions by cities. High performance building design should be based on customized local strategies rather than one-size-fits-all approach, but has not been viewed in this light by policymakers. This oversight presents a serious limitation in current industry practices, since there are no policies or guidelines for architects and engineers to incorporate the most adaptive design upgrades in their building projects. We illustrate this limitation through three case studies in the following sections.

22.3 Incorporating Predicted Climate Variability in Building Design

The American Institute of Architects (AIA) 2030 Commitment requires new construction projects to reach carbon neutrality by 2030. The challenge requires architectural firms to design buildings to be 70% more efficient than baseline energy use, progressively reducing energy consumption by 10% for every 5 years and reaching carbon neutrality by 2030. The AIA 2014 Progress Report acknowledges that energy simulation, used early in design and continued over to design completion, has been the primary technology for buildings and architects to meet 2030 goals. One-quarter of all projects with design influenced by energy modeling met their energy targets, and another 26% of projects came in close to reaching their targets. In comparison, nearly 80% of non-modeled projects failed to meet energy reduction targets by 20% or more. With the benefits of energy simulation well established, the AIA is now working toward introducing and increasing simulation use in architectural design [17]. The work discussed in this chapter is based on understanding the right approach to energy modeling to estimate the current and predicted future energy use of buildings and therefore emissions from the built environment.

Energy simulation is a commonly used method for estimating building energy consumption, peak demands, and energy cost. The process uses sophisticated software, which can be used to analyze whole building energy models. Whole building energy simulation typically involves the process of creating two models with the same geometry, weather data, and operational schedules – a baseline, which represents a code-compliant building, and a proposed model, which captures design and energy efficiency upgrades. The difference in annual energy consumption between the two models is the estimated annual energy savings for the proposed building compared with the baseline. Simulations are typically done on hourly time steps for annual energy use estimates but can also be computed for shorter time steps depending on the level of detail and accuracy that is required.

Energy models are indispensable tools for optimization of high performance building design. In other words, energy models serve as testbeds for building designers to test multiple design parameters – such as building massing, orientation, materials, and different HVAC system choices – and obtain energy and cost implications for each design iteration. Access to energy cost savings estimates during the design process is very advantageous to building owners since return on investment (ROI) numbers are often the primary drivers for up-front investment decisions on incorporating high performance building technology. Accuracy of energy models is critical since both investment decisions and annual operating costs are often budgeted based on cost estimates from energy models. Model accuracy depends greatly on the many design and building use pattern inputs, one of which is the regional weather file.

The simulation software uses regional climate data such as temperature, humidity, solar radiation, and wind speed from weather files either from private sources or from weather stations. In the United States, the most commonly used weather file formats in building energy models are Typical Meteorological Year 2 (TMY2) and Typical Meteorological Year 3 (TMY3), which represent hourly weather data averages from 1961 to 1990 and 1976 to 2005, respectively [18]. TMY2 dataset contains readings from 239 locations, while TMY3's data points cover 1020 locations [19].

As building performance is closely tied to climate, higher temperatures and increased climate variability will impact building energy consumption and demand. However, a limitation of current energy analysis practices is the uncertainty resulting from not factoring in future climate predictions. New building construction and major retrofits that are currently being designed will have an expected median lifespan of 70–75 years in the future [20]. From various climate model predictions, it is expected that the future climate will vary vastly from current patterns in terms of both averages and minima and maxima.

Mainstream building simulation practices do not incorporate future climate predictions in building energy analysis. To overcome this limitation, Seventh-wave developed a method of using building energy models coupled with future climate data in an effort to improve estimates of energy consumption over the building's lifespan. In the following text, we outline a method for estimating climate change impacts on building energy performance. This method can be used to identify design strategies that offer the most resilience to high climate variability and can be easily incorporated into current industry practices. Evaluating future energy impacts due to climate variability allows us to take steps to increase the resilience of buildings and communities.

This method was developed to study the potential climate change impacts on energy consumption of buildings at NASA's John C. Stennis Space Center (SSC) in southern Mississippi and later applied to two other case studies in Chicago, IL, and Fort Collins, CO. With three distinct studies to identify strategies for reducing climate impacts on building energy use in three different geographic locations, we suggest that this method can be replicated to any building type and geography. Evaluating future energy impacts due to climate variability allows us to take precautionary steps to increase the resilience of buildings and communities by reducing energy usage and GHG emissions as well as providing credible energy cost budgets.

Our objective was to apply this method for analyzing climate change impacts on building energy performance to identify strategies for designing climate-resilient energy-efficient commercial buildings. We used building simulation modeling to first understand how climate change may impact energy performance and peak demand. We then quantified the sensitivity of specific energy efficiency strategies to future climate data. We focused on those strategies that help building owners and developers meet their business needs. This work extends and complements several previous studies by showing how building energy modeling is used in the design process to quantify the effect of energy efficiency strategies.

22.4   Case Studies

22.4.1   Modeling Methodology

The general method we used, whether applied to existing buildings or new construction, followed the same basic steps. Our three case studies consisted of one existing set of buildings, the SSC (case 1, ASHRAE climate zone 2A) and two new construction multifamily buildings in Chicago (case 2, ASHRAE climate zone 5A) and Fort Collins (case 3, ASHRAE climate zone 5B) [21].

To gather data for simulation models, we acquired design documents such as building drawings and specifications for new construction buildings. For existing buildings, we collected the same documentation where available and additionally performed level 1 building audits to fill in any information gaps and obtain a better understanding of the facilities' use type and operation schedule. With existing buildings, we also had the advantage of access to monthly historical energy consumption data.

We built energy models in DOE2 [22] using eQUEST as a graphical user interface. The first case for which we developed our models was for a subset of existing buildings at the SSC. NASA's John C. SSC in southern Mississippi is a campus with 142 buildings encompassing a variety of usage types, such as offices, laboratories, and testing facility, and ranging in size from 53 to 700,000 ft². We modeled 39 buildings that accounted for 85% of SSC's total source energy use. A key observation of the energy use profile of the facility was that it consumes double the national average of electricity use due to the industrial and research activities. Natural gas consumption accounted for only 7.5% of the total source energy use.

Building geometry (footprint, number of floors) for the SSC models was based on satellite imagery and building square footage provided by facility staff. Interior zoning was predominately set to perimeter-core with specific zoning only occurring for areas with loads significantly different from the building as a whole (i.e., warehouse adjacent to an office). Windows were modeled as approximated window to wall ratios taken from site photos. Because of the age of many of the buildings, we could not determine the precise assembly properties for roofs, walls, and windows. For these cases, we assumed the roof had R-10 insulation. We assumed the walls to be 12″ medium-weight concrete with minimal insulation, and the windows to be single paned with clear glazing. For the handful of newer buildings, we assumed code-required minimum values of insulation and window properties from ASHRAE 90.1-2004. Occupancy density was provided by facility staff. The buildings were predominately occupied between 6:00 a.m. and 6:00 p.m. as corroborated by facility staff. We preliminarily set lighting to code-required values from ASHRAE 90.1-2004 for the building's predominant use type (i.e., 1.0 W/ft² for buildings that were mostly office). No daylighting controls were reported for the modeled buildings. We initially set miscellaneous loads to default values outlined in COMNET's Commercial Buildings Energy Modeling Guidelines and Procedures [23] for a given building's predominant use type. Infiltration flow rates were approximated according to guidelines published by Pacific Northwest National Laboratory [24]. We modeled HVAC system types according to input from facility staff. The majority of primary HVAC systems for the existing buildings were variable air volume with hot water reheat. Cooling was provided by water-cooled chillers, while heating was provided by atmospheric boilers. The efficiencies for the HVAC equipment were

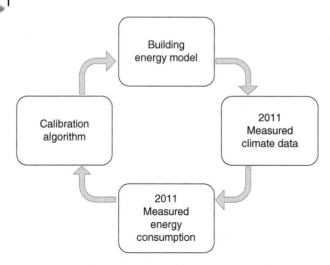

Figure 22.1 Model calibration process.

preliminarily set to code-required minimum values as outlined in ASHRAE 90.1-2004. No demand control ventilation controls were found in the modeled buildings, and only one instance of energy recovery ventilation was found.

These models were calibrated to reflect the buildings' actual annual source energy consumption by comparing the initial modeling results with the actual monitored energy usage for calendar year 2011. We used actual meteorological year (AMY 2011) weather data in our models corresponding to the same period as the measured energy usage (Figure 22.1). Discrepancies between the modeled and actual performance were assumed to be the result of uncertainty in model inputs such as envelope properties, lighting power, plug load equipment power, infiltration flow rates, outdoor airflow rates, and HVAC equipment efficiencies.

We then used the Nelder–Mead simplex optimization algorithm [25] to calibrate each of the models to actual monthly energy use data. The algorithm searches for the energy model input parameter set that minimizes an objective function comparing modeled energy use with actual energy use. Our choice for objective function follows ASHRAE Research Project 1051 and Guideline 14 for energy model calibration and evaluation. We used goodness of fit (GOF) as our objective function, which is based on the coefficient of variation of the root mean squared error between modeled and measured monthly energy consumption, weighted by the annual cost of each fuel type. The convergence criterion for the objective function was set to 15% for each model (i.e., GOF < 15 percent for each building model). We inspected all calibrated model parameters to ensure values fell within acceptable ranges based on our understanding of the building and our engineering experience. Quality checks were also performed on model results. Cooling load, economizer operation, and reheat controls were each rigorously explored to determine proper performance.

After the calibration algorithm had been applied to each building energy model, we had a set of models that represented energy use for the existing buildings under current climate conditions. Once the calibration was finalized, we input typical meteorological year (TMY) weather data representing years 1997–2012 into our models such that our existing models represent buildings operating under the current climate scenario.

The modeling approach was similar for the new construction buildings (cases 2 and 3) except that models were developed using the design documents to reflect an ASHRAE 90.1-2010 Appendix G baseline and they were not calibrated to reflect actual energy consumption. The next step was to select the future climate scenarios for analysis.

22.4.2 Analysis of Climate Scenarios and Impacts

We screened 11 future climate model datasets provided by the North American Regional Climate Change Assessment Program (NARCCAP) ([26], updated 2014). Each of these datasets contained projected climate data on a 30-mile (50 km) grid encompassing all of North America. We selected the grid point closest to the particular building under analysis. Each dataset has data across a variety of climate variables on 3-h intervals for the years 2041–2070. Due to the computationally intensive nature of simulating all future climate scenarios, we instead qualitatively chose datasets representing low impact (Weather Research and Forecasting + Community Climate System model, WRFG + CCSM) and high impact (Canadian Regional Climate model + Community Climate System model, CRCM + CCSM) scenarios, as shown in Figure 22.2. We then selected an extreme year with hottest average summer temperature and a median year with median average annual temperature from each future dataset to use with the SSC building models. Our analysis represents four scenarios: future middle-average annual (WRFG + CCSM 2052), future middle-maximum summer (WRFG + CCSM 2069), future high-average annual (WRFG + CCSM 2052), and future high-maximum summer (CCRM + CCSM 2069). In this manner, our results would bracket the potential range of climate impacts without taking an inordinate amount of computational time.

We had planned to use the full set of climate variables in our energy models. However, some climate variables, such as dew point temperature, wind speed and direction, cloud fraction, and so on, produce only secondary effects on building energy consumption. In order to minimize the impact of those variables, we used NARCCAP data pertaining only to dry-bulb temperature, wet-bulb temperature, atmospheric pressure, and corresponding atmospheric

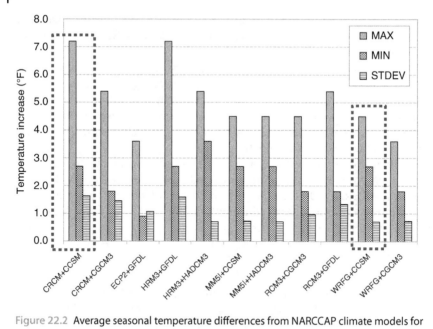

Figure 22.2 Average seasonal temperature differences from NARCCAP climate models for Mississippi region.

variables that could be calculated directly from these primary variables (e.g., enthalpy). All other variables were held constant between current and future scenarios. The NARCCAP datasets contain raw data on future climate variables and cannot be used in energy models. We processed the data to a format that is readable by simulation software (eQUEST) and verified data consistency for all variables to fall within an expected range [27]. We used the future climate data in our models to estimate the impact of climate change on annual electric consumption, annual natural gas consumption, peak electricity demand, peak cooling demand, peak heating demand, and annual utility cost (Figure 22.3).

Once we quantified the expected range of impacts, we developed mitigation strategies to offset them. For all three cases, we analyzed a range of standard ECMs affecting building envelope, lighting, and HVAC systems. The impact of each ECM was individually quantified and ranked based on its effectiveness at saving energy for the specific building type in the specific future climate.

22.4.3 Results

22.4.3.1 Case 1: Southern Mississippi Space Center

We developed and calibrated 32 models representing buildings consuming over 80% of SSC's annual energy consumption under current climate conditions. Modeled total energy use was within 5.5% and 2.1% of measured annual data for electricity and natural gas, respectively. The coefficient of determination

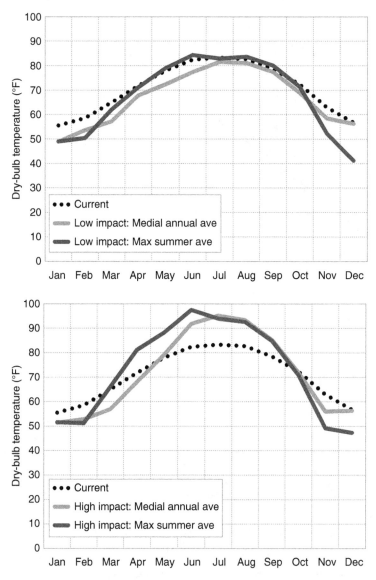

Figure 22.3 Low and high impact climate data.

between measured and modeled energy use improved noticeably from uncalibrated models (0.86) to calibrated models (0.98), as shown in Figure 22.4.

Our two climate change scenarios (low and high impact) were selected as the year with the average annual dry-bulb temperature closest to the median of all years from the future climate model datasets with the smallest and biggest

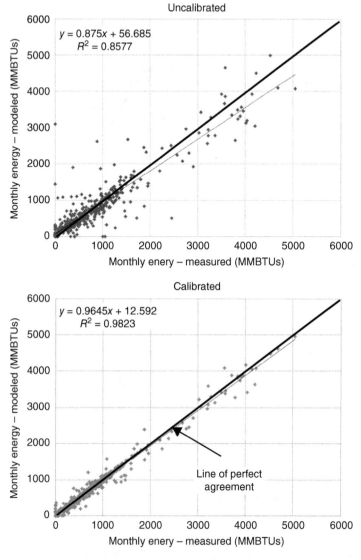

Figure 22.4 Monthly energy use – uncalibrated versus calibrated models.

impacts, respectively (Table 22.1). Our analysis for the SSC site shows a general cooling trend for the low impact scenario that does not align with the warming trend of the larger region. The low impact future climate scenario indicates average annual temperatures that are 4 °F (2 °C) lower compared with the current climate and a maximum annual temperature that is 7 °F (4 °C) higher. The high impact future climate scenario shows no change in annual average

Table 22.1 Energy use – measured versus modeled.

	2011 kW h	2011 therms
Measured	79,576,860	1,227,706
Modeled	83,950,556	1,253,386
%	5.2%	2.0%

Table 22.2 Dry-bulb temperature summary for TMY and future climate scenarios at SSC.

	TMY	Low impact	High impact
Average Tdb (°F)	71	67	71
Maximum Tdb (°F)	102	109	121
Minimum Tdb (°F)	26	25	26
Heating degree days (base 65 °F)	1248	1842	1859
Cooling degree days (base 65 °F)	3322	2498	4269

temperature but an increase of 19 °F (11 °C) for the maximum annual temperature. Additionally, both scenarios project colder winters and a corresponding increase in heating degree days.

Table 22.2 compares temperature metrics between climate scenarios: TMY (represents the present climate 1997–2012), low impact (future), and high impact (future).

Using these different climate scenarios, we quantified the impact of climate change on SSC building performance. Figure 22.5 illustrates this impact, with each bar graph representing the range of impacts bracketed by the low and high impact future climate scenarios.

Table 22.3 reports estimated variation in future energy use of SSC campus compared with current climate conditions. Total site energy consumption increased over current climate conditions for each climate scenario we examined. Our models showed an increase of between 4.3% and 11.3% in annual electricity consumption for the low and high impact future scenarios, respectively. Interestingly, they project that peak cooling demand will decrease 4.7% under the low impact scenario due to lower projected summer humidity levels (even though peak dry-bulb temperatures are expected to increase). The low impact peak electric demand decreases 2.4%, following the reduction in peak cooling.

Conversely, the high impact scenario projects a cooling peak demand increase of 36.8% and an electrical peak demand increase of 19%. This reflects the significantly higher dry- and wet-bulb temperatures projected under this

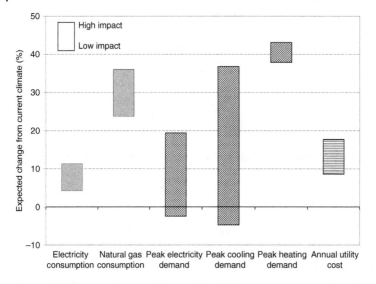

Figure 22.5 Expected range of climate change impacts on energy use and energy cost.

Table 22.3 Future electricity, gas, and peal electric demand variation from current conditions.

Scenario	Annual electricity consumption (percentage increase relative to current)	Annual natural gas consumption (percentage increase relative to current)	Peal electric demand (percentage increase relative to current)
Current	–	–	–
Future middle – avg annual	4.3%	23.8%	−2.4%
Future middle – max summer	3.3%	16.4%	−3.7%
Future high – avg annual	11.3%	36.0%	19.4%
Future high – max summer	11.7%	23.9%	17.2%

scenario. Total gas consumption increased 24% and 36% for the low and high impact scenarios, respectively, with a corresponding increase in peak heating demand of 38% and 43%. This follows the generally lower and more variable wintertime temperatures projected under both climate scenarios. The total projected annual energy cost is expected to increase 9% and 18% for the low and high impact scenarios, respectively. For a facility of this size, this translates to around $1 million dollars.

We then simulated a number of typical ECMs and ranked their effectiveness at mitigating climate change impacts into two groups: primary strategies were the most effective, while secondary strategies had a smaller but sizable mitigation effect. Table 22.3 outlines the most effective climate change mitigation strategies at SSC.

The three primary strategies include improving roof insulation, upgrading the water-cooled chillers, and installing ventilation energy recovery wheels. Additional roof insulation indirectly reduces the cooling and heating loads at SSC during the more extreme summers and winters by reducing the amount of energy used by the heating and cooling equipment.

Upgrading to more efficient chillers directly reduces the amount of cooling energy needed to offset the increased need for cooling during hotter summers. The energy recovery ventilation will recover energy from the exhaust airstream, reducing the wasted energy already used to condition the hotter or colder outside air.

We also identified four secondary strategies, as shown in Table 22.4. The first three strategies – increasing wall insulation, installing high performance windows, and sealing air leaks – indirectly reduce energy use by isolating the conditioned indoor environment from the outdoor climate. The fourth strategy – upgrading to condensing boilers – directly reduces the amount of heating energy needed to offset the increased need for heating during the

Table 22.4 Climate change adaptation strategies at SSC.

Primary strategies	Description
Roof insulation	Add additional roof insulation, minimum R-20
Cooling equipment	Upgrade to high efficiency centrifugal chillers; minimum 0.639 kW/ton, 0.45 kW/ton-IPLV
Energy recovery ventilation	Install enthalpy wheel energy recovery systems on exhaust with bypass and modulation control; 70%+ latent effectiveness, ~0.7″ ΔP
Secondary strategies	
Wall insulation	Add additional wall insulation, 2″ continuous insulation
High performance windows	Replace existing windows with low conductivity glass and thermally broken frames; maximum assembly U-value of 0.35
Tighter envelope	Install continuous air–vapor barrier using spray on air barrier or spray foam to seal all roof penetrations (piping, ductwork, electrical) at both the top and the deck level
Heating equipment	Upgrade to condensing gas-fired boilers; 90%+ thermal efficiency

colder winters. A number of additional strategies were analyzed, but found not to be particularly effective at offsetting the impact of climate change. These were predominantly strategies that affected internal loads, such as more efficient lighting.

22.4.3.2   Case 2: Chicago Multifamily Building

We explored the energy impacts of future climate variability in a 428,000 square foot multifamily development, the Reserve at Glenview, located in the suburbs of Chicago, Illinois. We applied the method developed for the SSC study to understand the energy use, quantify the impacts of climate change on building performance, and understand the resilience to climate change from each design upgrade. However, since this was a new construction project in the early design stages compared with the existing facility in case 1, we did not have actual building energy consumption data to use for calibration. The model therefore reflected a theoretical baseline in line with the applicable building energy code.

For this case, we selected the two future climate datasets (low and high) based on the year with the lowest and highest average annual dry-bulb temperature. The low impact climate data for the Chicago site shows average annual temperatures that are 1 °F (0.5 °C) lower compared with the current climate and a maximum annual temperature that is 1 °F (0.5 °C) higher. The high impact future climate scenario shows average annual temperatures that are 4 °F (2.2 °C) higher compared with the current climate and a maximum annual temperature that is 21 °F (11.7 °C) higher.

Table 22.5 compares temperature metrics between climate scenarios: TMY (represents the present climate 1997–2012), low impact (future), and high impact (future).

Using these different climate scenarios, we quantified the impact of climate change on the Chicago multifamily building's performance. Figure 22.6 illustrates this impact, with each bar graph representing the range of impacts bracketed by the low and high impact future climate scenarios.

Total site energy consumption increased over current climate conditions for each climate scenario we examined. For the project's baseline, our models

Table 22.5 Dry-bulb temperature summary for TMY and future climate scenarios in Chicago.

	TMY	Low impact	High impact
Average Tdb (°F)	50	49	54
Maximum Tdb (°F)	96	97	117
Minimum Tdb (°F)	−9	0	−4
Heating degree days (base 65 °F)	6586	6734	5825
Cooling degree days (base 65 °F)	3099	2993	4342

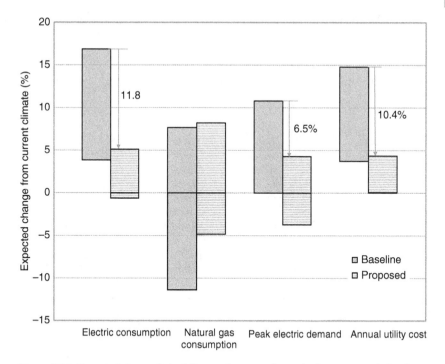

Figure 22.6 Expected change in building performance for each climate scenario for the Chicago multifamily building. Proposed includes 12 possible energy conservation measure upgrades over baseline code.

showed an increase of between 4% and 17% in annual electricity consumption for the low and high impact future scenarios, respectively. The low impact peak electric demand decreases 1% following a reduction in peak cooling. Conversely, the high impact scenario projects an electrical peak demand increase of 11%. This reflects the significantly higher dry-bulb and wet-bulb temperatures projected under this scenario. Total gas consumption decreased 11% and increased 8% for the low and high impact scenarios, respectively. This follows the generally more variable wintertime temperatures projected under future climate scenarios. The total projected annual energy cost is expected to increase 4% and 15% for the low and high impact scenarios, respectively. For a facility of this scale, this translates to a cost increase of $0.04–$0.13 per square foot or $17,000–$56,000 in additional annual energy costs.

We then simulated the actual ECM pursued for this project – partition insulation, wall insulation, high performance windows, efficient interior lighting, efficient exterior and garage lighting, demand control ventilation, efficient fans, garage, ventilation controls, efficient water heating, efficient HVAC, infiltration reduction, ERV type, and effectiveness – and quantified their effectiveness at

Table 22.6 Climate change adaptation strategies for the Chicago multifamily building.

Strategies	Description
High performance windows	Specify windows with low conductivity glass and thermally broken frames; maximum assembly U-value of 0.29
Ventilation controls	Install CO_2 sensors to reduce ventilation during low occupancy periods. Variable speed makeup air unit (MAU) fans
HVAC efficiency	Upgrade split system, rooftop unit (RTU), and MAU cooling efficiencies, upgrade to condensing furnaces; 90%+ thermal efficiency
Fan power reduction	Implement strategies to reduce fan power. In RTUs and MAUs, implement premium efficiency direct drive, use high efficiency filters, oversize any ductwork, and limit AHU face velocity to 350 fpm or less. In split systems, use electrically commuted fan motors
Tighter envelope	Install continuous air–vapor barrier using spray on air barrier or spray foam to seal all roof penetrations (piping, ductwork, electrical) at both the top and the deck level
Energy recovery ventilation	Install energy recovery devices to recover heat from exhaust air

mitigating climate change impacts. Table 22.6 outlines the most effective climate change mitigation strategies for the Chicago multifamily building.

The primary strategies include improving windows, using demand control ventilation, upgrading the cooling and heating efficiencies, reducing fan power, reducing infiltration, and employing energy recovery ventilation. The combined effect of these ECMs is illustrated in Figure 22.6. The ECMs combine to reduce the impact of climate change on annual electric consumption by 12%, peak electric consumption by 6.5%, and annual utility cost by 10%. They had a negligible impact on natural gas consumption.

22.4.3.3   Case 3: Fort Collins Multifamily Building

In order to help us understand how geographic location might influence our results, we repeated the analysis done for Chicago using climate data from Fort Collins, Colorado. We chose Fort Collins because of the community's progressive policies toward sustainability practices, and we were asked by a local power producer to examine these potential effects.

We used the CCSM/CGSM3 model data from NARCCAP [26] to represent both current and future conditions for building energy modeling and comparison purposes. We chose the CCSM model data because summary data suggested a high impact (high temperature change) for the region. We did not analyze other scenarios.

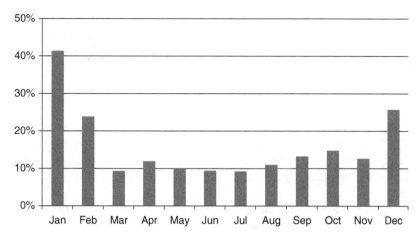

Figure 22.7 Modeled 2050 average monthly air temperature change from current climate.

Data compiled from one regional weather station indicated that average temperature at that station had increased at a rate of 0.15 °F (0.08 °C) per year and 0.13 °F (0.07 °C) per year for the past 30 years for the months of July and January, respectively. The future climate model data for model year 2050 shows an average monthly increase in winter temperatures from 20 to 40% over a current modeled climate year, whereas average summer temperature change is almost 10% (Figure 22.7).

We used the building energy model from the Chicago multifamily case study to examine Fort Collins climate impacts. We simulated the same 12 ECMs under both climate scenarios with a baseline case simulated without any energy efficiency upgrades.

Baseline annual electricity consumption for the building increased 2%, peak demand increased 6%, and natural gas consumption decreased 18%. Peak cooling demand increased almost 30%.

Table 22.7 shows that almost 80% of the energy savings under current climate conditions came from efficient fans (37%), efficient lighting (16%), infiltration

Table 22.7 Comparison of top four energy conservation strategies and their expected energy savings under current and future climate scenario.

Current	Future
Efficient fans, 37%	Efficient fans, 39%
Efficient lighting, 16%	Efficient lighting, 20%
Infiltration reduction, 15%	Demand control ventilation, 11%
Demand control ventilation, 11%	High performance windows, doors, 8%

reduction (15%), and demand control ventilation (11%). In the future climate scenario, almost 80% of the energy savings comes from efficient fans (39%), efficient lighting (20%), demand control ventilation (11%), and high performance windows and doors (8%).

22.4.4 Limitations of the Study and Future Work

The study has established a method to incorporate future climate variability in building design, so variations in energy consumption and cost over the building's lifespan can be accounted for. Many challenges encountered during the study need further investigation to further refine the accuracy of projected future energy use.

Although 11 different NARCCAP climate models are currently available, we have selected only the high and low impact scenarios. Each of these datasets is further scoped to 2 years with median temperatures and hottest summers. This process of selecting four annual weather datasets made the simulation study feasible within our scope. Dry-bulb temperature, wet-bulb temperature, and atmospheric pressure are the primary variables incorporated in our energy models from future climate models. Other climate variables are held constant between current and future climate files.

Future work on this topic will include occupant thermal comfort studies and life cycle analysis from nonlinear energy savings potential resulting from a changing climate profile.

22.5 Implications for Future Cities and Net-Zero Buildings

The results of these three cases are consistent with several previous studies that show the expected change in building energy consumption should follow the same trend as the expected change in temperature. However, building designers, energy managers, and program planners need more detailed information in order to make sound decisions related to addressing climate resiliency in buildings. We attempted to quantify the effects of climate change on the energy efficiency choices that need to be made when constructing a new commercial building or retrofitting an existing one. Building owners and design teams often do not implement the full range of choices available due to budget constraints, payback expectations, and design and scheduling concerns. In our cases, we showed how simple rank ordering of energy efficiency measures using simulation modeling and climate change data would likely affect how energy efficiency is implemented or even perceived by a building owner. For example, in the Fort Collins case, where infiltration reduction accounted for almost 15% of the expected energy savings under a current climate, it only accounted for 4% of the expected savings under a future climate.

Determining the reasons for this unexpected change was beyond the scope of our work, but it represents an area of new research needed to understand how simulation modeling with future climate data can be used to help develop design criteria for constructing and operating climate-resilient buildings. Another study provides a useful starting point in this regard. They suggest modeling with the 50th percentile of climate change data and examining "hard" mitigation measures to reduce expected climate risk and using "soft" approaches for greater than expected changes [28].

While we apply efficient building technology to new construction and existing buildings, it also contributes to lessening future energy impacts from climate change. By using the approaches of previous studies with the one we developed for this study, we can target adaptation strategies that are most effective for a given region and building type. While results of this study are specific to southern Mississippi, Chicago, and Fort Collins, the method could be replicated with other building types and geographic locations. For large organizations that spend millions of dollars on utility bills, changing climate patterns represent a potentially significant risk to facility investments. This study infers that standard energy efficiency approaches are effective strategies to adapt to projected energy use changes and imparts knowledge to industry practitioners on ways to identify and implement climate mitigation resiliency strategies in building design.

Buildings currently rely largely on centralized networks for energy. A completely centralized utility network makes cities and buildings vulnerable to power outages, extreme weather, and other catastrophic events [4]. A centralized supply model completely separates generation from consumption. While making the production end more efficient, this model also adds energy consumption to the delivery systems. Decentralized networks (such as rooftop solar and local storage) for generation and delivery of energy and other building services can provide complementary solutions. A successfully decentralized energy network requires reliable capability to handle peak demand.

Zero-net-energy building development is a major component of a decentralized energy network. Energy analysis that factors in the future climate predictions has a great implication for the development of future zero-net-energy buildings and communities. Zero-net-energy buildings have been advancing from research to reality, although the market is still very small, and such buildings still require a connection to an electric grid that provides power on demand [4]. According to the most current status update from the New Buildings Institute, the number of building projects in the United States targeting or achieving the zero net goal increased from 60 in 2012 to 160 in 2014 [29]. Placed in context, the United States has 5.57 million commercial buildings [30] and is projected to add between 1.7 and 2.4 billion square feet of new additions by 2040 [31].

In California, Executive Order B-18-12 [32] requires that "all new State buildings and major renovations beginning design after 2025 be constructed as Zero Net Energy facilities with an interim target for 50% of new facilities beginning design after 2020 to be Zero Net Energy." The US DOE aims to create integrated solutions for marketable zero-net-energy homes by the year 2020 and commercial buildings by the year 2025 [33]. The ASHRAE Vision 2020 Ad Hoc Committee envisions that the building community will produce market-viable zero-net-energy buildings by the year 2030 [34].

In summary, the changing climate patterns and increased extreme weather conditions may make today's working system unreliable in the not-so-distant future. Therefore, climate resilience plays a critical role in the design and development of tomorrow's smart buildings and cities.

Final Thoughts

The built environment is one of the largest sources of GHG emissions. Reducing energy use in buildings has the twofold effect of decreasing energy dependence while simultaneously contributing to lower GHG emissions. This chapter discusses various efforts at curbing emissions and processes by which the mitigation targets can be realized.

We also discuss the impact of future climate variation on building energy use and identify a method to incorporate climate resilience planning as part of the design process. This method is demonstrated with a case study of NASA's SSC and later replicated in two other geographic locations. Resilience planning is a critical element of smart city planning, since cities will experience a vastly different energy use profiles compared with current energy usage patterns.

Questions

1 What are the widely adopted emission reduction initiatives and have the programs made measurable progress toward reaching their goals?

2 What are steps that local and city governments should execute so they can track GHG emissions and keep emission reduction commitments?

3 How does the built environment contribute to GHG emissions and what are the ways to address emissions reduction at the building level?

4   How can building designers identify the most adaptive strategies to be incorporated in design to offer the most resilience to both future average and extreme weather variations?

5   When resilience planning has not been considered in building design, what adverse outcomes would likely be dealt with over the building's lifespan?

References

1 Environmental Protection Agency (2014) *Climate Change: Basic Information*. Retrieved from United States Environmental Protection Agency: http://www3.epa.gov/climatechange/basics/ (accessed 21 December 2016).

2 IPCC (n.d.) *Organization*. Retrieved from Intergovernmental Panel on Climate Change: http://www.ipcc.ch/organization/organization.shtml (accessed 21 December 2016)

3 IPCC (2014) *Climate Change 2014: Synthesis Report. Contribution of Working Groups I, II and III to the Fifth Assessment Report of the Intergovernmental Panel on Climate Change* [Core Writing Team, R.K. Pachauri and L.A. Meyer (eds.)]. Geneva, Switzerland: IPCC.

4 Wang, N., et al. (2016). *Past and Current Visions for Buildings: Energy, Sustainability and Beyond.* Manuscript submitted for publication.

5 PlanNYC (2013) *A Stronger, More Resilient New York. The City of New York.* Retrieved from http://www.nyc.gov/html/sirr/html/report/report.shtml (accessed 21 December 2016)

6 100 Resilient Cities (2015) Retrieved from http://www.100resilientcities .org/#/-_/ (accessed 21 December 2016)

7 International Energy Agency (2013) *Transition to Sustainable Buildings: Strategies and Opportunities to 2050,* International Energy Agency, Paris.

8 COP21 (2015) *United Nations Conference on Climate Change.* Retrieved from http://www.cop21.gouv.fr/en (accessed 21 December 2016)

9 Wang, N., and Goins, J., (2015) *Buildings of the Future Scoping Study: A Framework for Vision Development.* Pacific Northwest National Laboratory. Retrieved from http://www.pnnl.gov/main/publications/external/technical_reports/PNNL-24126.pdf (accessed 21 December 2016)

10 DOE (n.d.) *Reference: A Common Definition for Zero Energy Buildings.* Retrieved from http://energy.gov/sites/prod/files/2015/09/f26/A%20Common%20Definition%20for%20Zero%20Energy%20Buildings.pdf (accessed 21 December 2016).

11 GIB (2014) *Removing the Bottleneck for Infrastructure Investments, Resilient Cities,* Global Infrastructure Basel.

12 WRI, C40, ICLEI (n.d.) *The Global Protocol for Community-Scale Greenhouse Gas Emission – An Accounting and Reporting Standard for Cities.*

13 Architecture 2030 (2015) Retrieved from Architecture 2030: http://architecture2030.org/2030_challenges/2030-challenge/ (accessed 21 December 2016).

14 Architecture 2030 (2014) *Roadmap to 2030. Submission to the Ad Hoc Working Group on the Durban Platform for Enhanced Action. Version: June 4, 2014 (Amended).* Retrieved from http://architecture2030.org/files/roadmap_web.pdf (accessed 21 December 2016).

15 C40 Cities (n.d.) *About C40.* Retrieved from C40 Cities: http://www.c40.org/about (accessed 21 December).

16 C40 Cities Climate Leadership Group (2014) *Climate Action in Megacities 2.0 - C40 Cities Baseline and Opportunities.*

17 American Institute of Architects (2014) *AIA 2030 Commitment - 2014 Progress Report.*

18 US DOE (2015) *EnergyPlus Energy Simulation Software* Retrieved from US Department of Energy – Energy Efficiency and Renewable Energy: http://apps1.eere.energy.gov/buildings/energyplus/ (accessed 21 December 2016).

19 RReDC (2015) *National Solar Radiation Database.* Retrieved from Renewable Resource Data Center, National Renewable Energy Laboratory: http://rredc.nrel.gov/solar/old_data/nsrdb/1991-2005/tmy3/ (accessed 21 December 2016)

20 US DOE (2012) *Commercial Sector Characteristics.* Retrieved from Buildings Energy Data Book: http://buildingsdatabook.eren.doe.gov/TableView.aspx?table=3.2.7 (accessed 21 December 2016).

21 Schuetter, S. *et al.* (2014) Future climate impacts on building design. *ASHRAE Journal*, **56** (9), 36–44.

22 DOE2 (2013) *DOE-2 based Building Energy Use and Cost Analysis Software.* Retrieved from www.doe2.com (accessed 21 December 2016).

23 COMNET (2013) *Commercial Energy Services Network MGP Manual: Modeling guidelines & Procedures.* Retrieved from www.comnet.org/mgp-manual accessed 12/2013.

24 Gowri, K., Winiarski, D., and Jarnagin, R. (2009) *Infiltration Modeling Guidelines for Commercial Building Energy Analysis PNNL-18898*, PNNL.

25 Nelder, J., and Mead, R., (1965) A simplex method for function minimization. *The Computer Journal*, **7**, 308–313.

26 Mearns, L., Mearns, L.O., W.J. Gutowski, R. Jones, L.Y. Leung, S. McGinnis, A.M.B. Nunes, and Y. Qian (2007, updated 2014) *The North American Regional Climate Change Assessment Program dataset, National Center for Atmospheric Research Earth System Grid data portal, Boulder, CO. Data downloaded 2015-01-15.*

27 Ahl, D. *et al.* (2013) *Impact of Climate Variability on the Energy Use and Economics of NASA FAcilities*, Seventhwave.

28 Coley, D. *et al.* (2012) A comparison of structural and behavioural adaptations to future proofing buildings against higher temperatures. *Buildings and Environment*, **55**, 159–166.

29 New Buildings Institute (2014) *Getting to Zero Status Update: A look at the projects, policies and programs driving zero net energy performance in commercial buildings.* Portland, Oregon. Retrieved from http://newbuildings .org/sites/default/files/2014_Getting_to_Zero_Update.pdf (accessed 21 December 2016)

30 CBECS (2012) *Commercial Building Energy Consumption Survey (CBECS).* Retrieved from U.S. Energy Information Administration: http://www.eia .gov/consumption/commercial/data/2012/ (accessed 21 December 2016).

31 Energy Information Administration (2015) *Annual Energy Outlook.* Retrieved from http://www.eia.gov/forecasts/aeo/data/browser/#/?id=5-AEO2015.

32 Executive Order B-18-12 (2012) *Executive Order B-18-12.* Retrieved from https://www.gov.ca.gov/news.php?id=17508 (accessed 21 December 2016).

33 Office of Energy Efficiency & Renewable Energy (2016) *Multi-Year Program Plan.* Retrieved from Energy.Gov - Office of Energy Efficiency & Renewable Energy: http://energy.gov/eere/buildings/downloads/multi-year-program-plan (accessed 21 December 2016)

34 ASHRAE (2008) *ASHRAE Vision 2020: Producing Net Zero Energy Buildings.* Atlanta, Georgia. Retrieved from file:///C:/Users/sgunasingh/Downloads/20080226_ashraevision2020.pdf

23

Smart Audio Sensing-Based HVAC Monitoring

Shahriar Nirjon[1], Ravi Srinivasan[2], and Tamim Sookoor[3]

[1] Department of Computer Science, University of North Carolina at Chapel Hill, Chapel Hill, NC, USA
[2] M.E. Rinker, Sr. School of Building Construction Management, University of Florida, Gainesville, FL, USA
[3] Senior Professional Staff, The Johns Hopkins University Applied Physics Laboratory, Laurel, MD, USA

CHAPTER MENU

Introduction, 669
Background, 671
The Design of SASEM, 675
Experimental Results, 685

Objectives

- To become familiar with HVAC faults and their acoustic characteristics.
- To become familiar with distributed *acoustic sensing system*.
- To become familiar with cloud-based remote sensing and monitoring system.

23.1 Introduction

Centralized HVAC systems are the primary means to control the indoor climate and maintain occupants' comfort in over 88% of the commercial buildings in the United States [1]. They are also one of the most expensive systems in commercial buildings – in terms of both installation/replacement cost and energy consumption. Failure of an HVAC system is therefore detrimental to our well-being as well as to the finances. A notable consequence of faulty HVAC systems is the "sick building syndrome" – which leads to respiratory problems

Smart Cities: Foundations, Principles and Applications, First Edition.
Edited by Houbing Song, Ravi Srinivasan, Tamim Sookoor, and Sabina Jeschke.
© 2017 John Wiley & Sons, Inc. Published 2017 by John Wiley & Sons, Inc.

attributed to poor ventilation, low or high humidity, and unfiltered airborne particles and chemical pollutants in buildings. With adults spending over 40% of their average daytime at workplaces, proactive prognosis of HVAC system performance cannot be overlooked.

Most HVAC failures are fixable but are extremely costly [2]. A number of HVAC problems, if not repaired early, lead to costlier repairs, or even the need to replace the system entirely. For instance, if the blower motors are left running at a compromised state, they strain other heating and cooling components. Likewise, continued running of a faulty condenser fan stresses the system and causes compressor failures [3]. Identifying and repairing such problems early can save building owners a considerable amount of money. However, due to the lack of an effective, low-cost, and continuous assessment and prognosis mechanism for detecting underperforming HVAC units, it is extremely difficult to determine whether a repair, retrofit, or permanent retirement of an HVAC system is warranted. For a systematic prognosis and life-cycle management of centralized HVAC systems, what we need is a robust, inexpensive, and easily deployable system, so that impending failures can be detected early. Such a system saves money and helps us breathe healthy. It also reduces negative environmental impacts of HVAC systems as it helps in decreasing the number of retired units that find their ways to the landfills, which, in turn, reduces escaped contaminants that have been shown to deplete the ozone layer [4].

A key to repairing HVAC systems before a total failure is early identification of problems. Similar to many other mechanical systems, noise is a key indicator of impending failures in HVAC systems. For instance, squealing or screeching could indicate a bad belt or motor bearing problem [21]. InspectAPedia [22], the free encyclopedia of building and environmental inspection, testing, diagnosis, and repair, has a detailed classification of HVAC noises. We hypothesize that – by employing a smart, low-cost distributed acoustic sensor system that uses machine learning algorithms to learn and classify these noises in real time – we are able to detect faulty HVAC components, predict HVAC failures, and help building owners predictively maintain their HVAC systems in a cost-effective manner.

We propose a Smart Audio SEnsing-based Maintenance (SASEM) system that has a single unifying intellectual focus, that is, enabling predictive maintenance of building equipment by autonomously monitoring and analyzing their acoustic emissions. Using audio signatures to predict equipment failure requires more than simply connecting a microphone to a digital signal processor; it requires the development of novel hardware and software that are low cost, low maintenance, and easy to deploy and take into consideration the variations in noises produced by different equipment, acoustically hostile building environments, false positives and negatives during classification, and privacy issues. We propose novel hardware and middleware for audio data gathering using wireless acoustic sensor networks and cloud computing and effective machine learning-based classifiers to identify acoustic characteristics of building equipment. The three specific aims of the system are as follows:

- Building a Low-Cost Sensing Platform: To build a low-cost acoustic sensing platform for audio data collection and onboard processing. The platform will be built using off-the-shelf hardware components and is expected to cost around $100–$200 per typical air handling room of an HVAC system, which is less than $0.10 per square foot in commercial buildings.
- Acoustic Modeling of HVAC Systems: To address the fundamental challenge of uniquely identifying acoustic characteristics of centralized HVAC systems. We will devise algorithms to automatically learn acoustic models of different HVAC system components. A higher-level model is used to further learn the operational states of an HVAC system. These models are used to monitor deviations of an HVAC system from its regular operation and to identify and localize faults.
- Decision Support System: To develop an interactive decision support system that provides real-time visualization of an HVAC system's health status and help assess the cost and benefits of repairing versus an early retirement of the system based on inputs such as the detected failures, different primary energy types, future energy price trends, and trends in monetary and nonmonetary benefits.

In order to test, evaluate, and demonstrate the capabilities of the integrated HVAC life-cycle management system, a live testbed, referred to as the "living laboratory," will be implemented. The testbed has HVAC systems from four buildings at the University of Florida (UFL) and Oak Ridge National Laboratory (ORNL) campuses.

23.2 Background

23.2.1 HVAC Failure Detection

HVAC maintenance may be reactive, preventive, or predictive. Reactive approaches lead to increased economic loss owing to unexpected equipment downtimes. The goal of preventive and predictive maintenance approaches is to avoid equipment failure and extend their useful lifetime through constant monitoring. While preventive maintenance is mostly periodic in nature, predictive maintenance is based on the actual state of the system. A number of approaches for monitoring system states have been proposed in the literature, which includes using data collected from one or more components [5, 6]; neural network-based maintenance [7, 8]; vibration, IR cameras, and ultrasonic analysis [9]; probabilistic fault prediction subject to a gradual deterioration process [10]; and Kalman filter-based estimations [11]. Studies have indicated that predictive methods are more effective than preventative ones in terms of cost, time, and safety. Currently, a standard technique called *Fault Detection and Diagnosis* is used for predictive maintenance of HVAC systems in commercial buildings. This technique uses both model- and data-based methods to decide maintenance schedules and relies heavily on sensor data such as airflow rate, air temperature, air pressure, and electricity usage [12–20]. These sensors are typically installed in new buildings as they are built. For existing buildings, it is extremely challenging and cost prohibitive to install these sensors.

23.2.2 HVAC Failures and Acoustics

Acoustic sensing methods rely on the rich information provided by sound, where small shifts and changes in its spectro-temporal characteristic reliably indicate differences in the behavior, performance, or content of a system. Examples include acoustic pulse reflectometry [23, 24] and acoustic emission analysis [25]. Soundscape capture and analysis [26] has been used to learn about the diversity of sound sources including sounds generated by the environment [27] and biological organisms [28]. Over the past few decades, microphones and microphone arrays have been implemented in order to monitor the propagation of sound in urban environments and to detect acoustic events [29, 30]. More recently, the same principles have been used in monitoring factory machinery and the maintenance and detection of faults in engines using their noise signatures [31, 32]. Because of the reliability of sound behavior, sound emission analysis serves as a dependable alternative to other sensing modes for predictive maintenance. Typical faults in the air handling room of an HVAC system are discussed in [73], and we summarize them in Table 23.1.

The table shows four major types of HVAC components, devices for each category, typical faults for each type of device, and the potential for using acoustic sensors in detecting a fault. All actuator faults and some of the equipment faults are detectable from their sounds. These are labeled "YES" in the last column to indicate so. These devices produce mechanical noises that change as they wear out. On the other hand, sensors and controllers are digital, and because they do not produce any sounds, their failures are not directly detectable. However, many of these faults are interrelated. For example, failure of a temperature sensor will have an effect on the fan, on heating and cooling coils, and, to some extent, on the duct. Hence, these failures can be inferred by learning the causal relationship between different components. These are labeled as "No/causal" in the table. Finally, some of the devices make noises that are too generic that human hearing is not capable of distinguishing them. However, by carefully training machine learning classifiers, these sounds can be distinguished from one another. In the next section, we provide some results from our preliminary experimentations, which show that these sounds are classifiable with acoustic sensors.

Table 23.1 Faults in HVAC systems. OA, RA, EA, SA, and MA stand for outside, return, exhaust, supply, and mixed air.

Category	Device	Typical faults	Acoustic detection
Equipment	Fan	Pressure drop is increased	Subject to test
		Overall failures of supply and return fans	Yes
		Decrease in the motor efficiency	Yes
		Belt slippage	Yes
	Duct	Air leakage	Subject to test
	Heating coil	Fouling (fin and tube) leads to reduced capacity	Subject to test
	Cooling coil	Fouling (fin and tube) leads to reduced capacity	Subject to test
	Preheating coil	Fouling and reduced capacity	Subject to test
Actuator	OA, RA, and EA dampers	A damper is stuck or in a faulty position	Yes
		Air leakage at fully open and closed positions	Yes
	Heating cooling coil, preheating coil valve	A valve is stuck, broken, or in wrong position	Yes
		Leakage occurs at fully open and closed valve	Yes
Sensor	SA, MA, OA, RA temp	Failures of a sensor are offset, discrete, or drift	No/causal
	MA, OA, RA humidity	Failures of a sensor are offset, discrete, or drift	No/causal
	OA, SA, RA flow rate	Failures of a sensor are offset, discrete, or drift	No/causal
	SA and zone pressure	Failures of a sensor are offset, discrete, or drift	No/causal
Controller	Motor modulation	Unstable response	Yes
	Heating/cooling valve	Unstable response	No/causal
	Flow difference	The system sticks at a fixed speed	Subject to test
	Static pressure	Unstable response	Subject to test
	Zone temperature	Unstable response	No/causal

Based on literature review, our preliminary investigation, and ongoing deployment efforts, we are certain that acoustic sensing is an efficient, reliable, and cost-effective approach for building system fault detection and diagnosis due to its ease of data acquisition as well as some strong correlations between system working conditions and its operation sound and noise. The hypothesis of audio sensing-based fault detection and diagnosis is that a change in sound and noise reveals the operational condition of HVACs, and faults can be detected by tracking these sounds. However, not all building system faults may be directly detected through audio sensing as some of them may not impact the operation sound and noise, that is, sensor and controller faults. Our basic three-step strategy for acoustic-based HVAC fault detection is as follows:

- Identify: Faults that can be identified by operation sound- and noise-related building system faults need to be identified. For air handling units (AHUs), there are four major types of faults that can be classified into two groups – the mechanical faults (i.e., equipment and actuator faults) and the digital faults (i.e., sensor and controller faults). The mechanical faults are caused by the failure of a certain device or component; hence they lead to the change in HVAC sound and noise, which can be identified directly. The digital faults are mainly caused by the failure of the sensor and controller; they may impact the control and energy efficiency of the system, but do not cause the change of operation sound and noise. These faults need to be identified indirectly from their causal effects on other noise-producing components.
- Detect: Once targeted system faults are identified, the next step is to detect them. Detection is based on the hypothesis that the operation sound and noise remain stable or in a certain pattern when there is no fault. Hence we can detect system faults by detecting whether the stable status of system sound and noise is broken or whether there is occurrence of abnormal or rare audio pattern. Faults can also be sensed using thermal imaging and vibration analysis as a cross-check to the acoustically fault diagnosis.
- Diagnose: The last step is to diagnose system faults. This process is conducted based on the hypothesis that different types of system faults should cause different change in system operation sound and noise. Hence, different system faults can be detected by tracking and comparing their corresponding operation sound and noise. In this step, an important process is to intelligently classify the different types of system faults. In general, this process is reached through machine learning method, of which a learning model that is built by learning the relation between audio data and different types of faults will be used to classify different types of faults. To establish this learning model, we may need both audio data and the ground truth data to train and test the model. The ground truth will be derived from the analysis of building energy management data, which requires involvement of expertise to manually diagnose different system faults.

23.3   The Design of SASEM

Our goal is to transform the way predictive maintenance of building HVAC systems is conducted. Toward achieving this, the design for SASEM embarks on three specific aims as shown in Figure 23.1. First, an embedded sensor platform, system architecture, and APIs to support multiple types of sensor nodes to gather acoustic characteristics of HVAC systems will be designed, developed, and tested. Second, acoustic signal characterization and classification algorithms will be developed to create sound libraries of centralized HVAC systems, which will be used to detect and localize faulty components of an HVAC system. Third, a strategic optimization model will be developed as a decision support system to enhance environmental sustainability of centralized HVAC systems. System admins will be able to access this via an online visualization platform. Adequate validation will be performed using a "living laboratory" where embedded devices are deployed and integrated with the visualization platform. Acoustic data of centralized HVAC systems will be gathered in real time and processed for environmental decision making from four buildings – the three are in the UFL campus and the fourth is the Flexible Research Platform situated in ORNL campus.

23.3.1   Building a Low-Cost Sensing Platform

Wireless sensor networks (WSNs) that gather acoustic signals have been used in applications ranging from locating specific animal calls [33] and classifying frog vocalization [34] to shooter localization [35] and speech/music discrimination [36]. The use of WSNs for smart building management has been gaining momentum in sensing environmental variables such as temperature and humidity. Such data have been used for the purposes of energy disaggregation and monitoring, building system control, and fault detection [12–20]. While a number of sensing modalities have been used in the existing literature, the use of audio sensors in building management has remained underexplored. A notable exception is TinyEARS [37], where audio sensors are used to detect the on/off status of appliances for energy disaggregation analysis.

In one of our previous works [38], a smart thermostat has been developed to retrofit existing HVAC systems in homes in order to increase their energy efficiency. When a house is unoccupied, the system sets the HVAC system to a more energy efficient setting. In a follow-on work [39], the authors developed RoomZoner that extended the concept of occupancy-driven HVAC control to reduce the energy wasted in unoccupied rooms of occupied houses. This system was evaluated in a home by retrofitting its existing HVAC system with wirelessly controllable air vents and thermostats that were controlled based on inputs from motion sensors (for occupancy detection) and temperature sensors. In both projects, the energy consumed by the HVAC system was measured using a real-time in-home energy management system called The

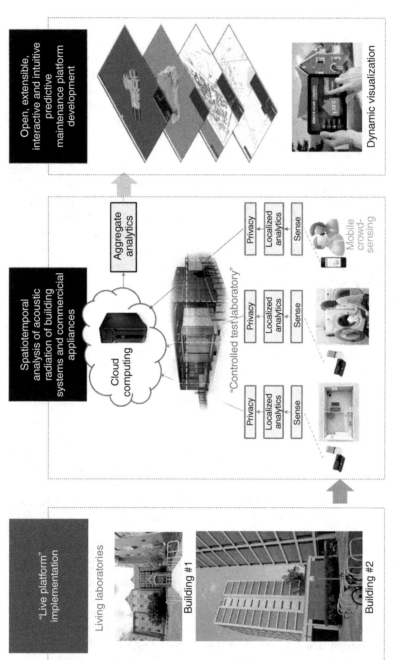

Figure 23.1 Overview of the proposed research showing the interactions among the three specific aims of this project.

Figure 23.2 Sensing platform consisting of a USB condenser microphone, a BeagleBone Black SoC, and USB wireless accessory.

Energy Detective [40]. In instrumenting homes to evaluate these HVAC control systems, a considerable amount of knowledge in deploying a "living laboratory" was gained [41].

Robust and efficient integration of recent developments in WSNs and cloud computing technologies for the purpose of acoustic anomaly detection of HVAC systems has not been previously attempted. A critical challenge is developing a low-cost hardware platform with a sufficiently small form factor that can capture the necessary audio signatures at the required fidelity while filtering out privacy-compromising sounds.

To overcome the challenges and to develop a novel technology for sound-scaping centralized HVAC systems, we design an embedded audio monitoring device that includes a microphone, a microcontroller, and a Wi-Fi module. We developed a similar device consisting of a condenser microphone, a BeagleBone Black embedded device, and a USB Wi-Fi device (Figure 23.2) for detecting an occupant's identity and mood. These devices are capable of continuously recording audio data at an appropriate sampling rate as well as performing real-time onboard data processing and classification. The embedded devices will be placed at different locations throughout the building as part of a network and will coordinate with one another to process audio data and transmit events to a base station. In addition to the on-device processing to filter out speech and other privacy-compromising information, this project will leverage cloud-based services to perform the more sophisticated aggregate analytics, house large datasets, and train classification models.

Our approach leverages the convergence of low-power and ubiquitous mobile computing devices with powerful cloud computing services [42–46]. In addition to the embedded devices, we opportunistically integrate other emplaced microphones and mobile devices such as smartphones and tablets. Android apps for these devices will be created that filter, tag, and analyze the data. The architecture will be flexible enough to automatically identify the set of devices, collect the desired data, and produce configurations that improve the sensing and data processing on these devices. When a problem occurs, occupants and admins will be able to report the ground truth such as a system failure or noise levels. To encourage occupants, we will provide incentives. As part of this architecture, we will also design an audio sensing middleware and an API to remove the complexity from the application developers to interface with a variety of physical sensors and enable local analytics and data processing on heterogeneous devices (ARM Linux, Android, etc.).

Unique research challenges arise from mobile crowd sensing paradigm ranging from participatory and opportunistic data collection, proper incentive mechanisms, transient network communication, and big data processing. These challenges include the following:

- Privacy and Security: Because of the involvement of human participants, several issues regarding the privacy and security of data, for example, sensitive information such as human voice and location being captured, need to be addressed. Since every individual has a different perception of privacy, developing privacy techniques that address variation in individual preferences is needed.
- Quality of Data and Trust: Data gathered by malfunctioning or malicious sensor nodes need to be addressed in order to make sure that [47] user data are not revealed to untrustworthy third parties, which affects the privacy and security of cloud-assisted applications.

23.3.2 Acoustic Modeling of HVAC Systems

Recently, there has been an explosion of audio recognition applications. Some examples include speaker identification [48], speech recognition [49], emotion and stress detection [50, 51], conversation and human behavior inference [52, 53], music recognition [54], and music search and discovery [55]. There are other types of applications that fall into the non-voice, nonmusic category, such as a cough detector [56], heartbeat counter [57], and logical location inference [58, 59]. These examples are limited in number and acoustic sensing is often one of multiple sensing modalities. There are existing works that consider multiple types of acoustic events: SoundSense [60] considers speech, music, and ambient sound and provides a mechanism to label clusters of ambient sounds to extend the set of classes, Jigsaw [18] considers speech and sounds related to activities of daily living, and TinyEARS attempts at acoustic filtering inside a building focusing on energy use [37].

Previously, we built a general-purpose acoustic event detection platform called Auditeur [61]. Compared with prior work in the literature, Auditeur was positioned as a developer platform rather than just a system detecting a set of sounds. Auditeur provided APIs to developers to enable their applications to register for and get notified of a wide variety of acoustic events such as speech, music, sounds of vehicles, etc. Auditeur achieved this ability of classifying general-purpose sounds by utilizing tagged soundlets: a collection of crowd-contributed, short-duration audio clips recorded on a smartphone along with a list of user-given tags and automatically generated contexts. The cloud hosted the collection of tagged soundlets and provided a set of services, which were used by the smartphones, to upload new soundlets and to obtain a detailed classification plan to recognize sounds specified by the list of tags as parameters.

Although audio recognition science is mature, there is a serious lack of work in understanding the anomalous acoustic signals generated by building systems. There are several problems that make this challenging. First, because of complexities of real-world operation, the system must be resilient to ambient noise, attenuation, and reverberation in the received audio signal. Second, because of the wide range of equipment being monitored, flexible and adaptive online learning must be employed. Third, because audio monitoring devices are placed in living spaces with people, strategies for addressing privacy concerns are paramount. Lastly, a major challenge is variations in make and model of HVAC systems, particularly as products from different manufacturers, and even different models from the same manufacturer, may produce different sounds. Thus, audio signatures should be robust to these variations.

A key element to our approach begins with building a "sound library" of tagged audio clips from various appliances and generating acoustic models for detecting anomalous acoustic events. The sound library contains clips of normal and anomalous audio signals collected from HVAC systems. Models for recognizing normal operation, anomalies, and events are generated in the cloud and pushed to the device. The particular benefits of onboard classification are that the application can continue to monitor audio signals without maintained connectivity and raw audio signals do not have to leave the premises for processing. When an equipment is annotated dysfunctional, the tagged event with associated audio features can be used to train the system to be able to predict when the next event occurs.

To recognize different operational states of an equipment, we investigate classification as well as contextual anomaly detection algorithms. Most existing anomaly detection algorithms consider only acute changes to the signal to recognize an anomaly called a point anomaly [62]. However, this might not be suitable in our scenarios where the environment is highly dynamic, new noise-emitting devices may be installed into the environment, and the operating conditions in the home could change. An acoustic event is defined as a short-duration sound (1–3 s) that has significantly higher acoustic energy

Figure 23.3 Acoustic event detection pipeline.

content than the background noise. These types of events are identifiable and their type is recognizable using a typical five-stage acoustic processing pipeline (Figure 23.3).

The process of continuous acoustic event detection starts with a preprocessing stage, which captures audio from the microphone, converts the byte stream into a stream of fixed size frames (typically 32–64 ms), and applies some standard acoustic processing activities such as filtering, windowing, and noise compensation as needed by the application context. Each frame is then passed through a frame-level feature extraction stage. In this stage a number of time and frequency domain acoustic features are extracted. The exact number and type of acoustic features that can characterize a specific acoustic event depends on the application scenario. However, a few acoustic features are commonly seen to characterize a wide variety of sounds, for example, a feature called the mel-frequency cepstral coefficients (MFCCs) is used in voice recognition, music identification, and machine acoustics.

After the feature extraction stage, frames are classified using a frame-level classifier, which acts as an admission controller and decides whether or not to process a frame any further. This step is employed to avoid unnecessary processing of uninteresting frames (e.g., in continuous acoustic event detection, most of the time a microphone is listening to the background noise as nothing is happening in the surrounding environment). If a frame is admitted to the window-level feature extraction stage, a fixed number of consecutive frames are gathered to form a window, and a number of statistics (e.g., mean and variance) are computed per feature per window. Finally, a window-level classifier classifies each window.

Acoustic samples for some types of HVAC faults may be obtained from sources such as HVAC manufacturers and repairers. A number of HVAC faults can also be artificially introduced into the system by injecting controlled randomness into HVAC controllers. However, availability of acoustic samples for all types, or even for a decent set of faults, cannot be assumed for practical reasons. On the other hand, collecting acoustic data by ourselves while waiting for a component to fail may take an indefinite amount of time. Hence, we devise a strategy to deal with this issue that employs unsupervised machine learning algorithms to learn the "regular pattern" in acoustic time series data, and discover, if the system has deviated from its regular operation at any point in time. This is a continuous learning and classification task that is susceptible to high false-positive rates at the early stage of learning. To mitigate this, we employ a human-in-the-loop approach, where a knowledgeable human operator will be notified with the location of the deviated and seemingly faulty component, who will make the final call. In the case of an actual fault or a false alarm, the acoustic models will be retrained to include this new knowledge, as labeled by a human. The PIs have successfully used such a combination of supervised and unsupervised learning approach in one of their works on human activity discovery and recognition [74]. The same principle will be applied in HVAC context to discover and detect faults with an improved precision.

An HVAC system is composed of various units – each contributing to the aggregate sound to which our proposed sound acquisition system is listening. In order to define and classify the health status of each component of the HVAC system, it is essential to create an acoustic model for each of the individual component as opposed to a single global model of the entire system. A global acoustic model of the entire system can only tell whether or not the whole system is functioning perfectly. Although a finer level of inference is theoretically possible by using a multi-output inference model, this is not practical since to train such a system we would have to keep just one component of the system on while shutting down the rest of the system – which is not possible in an HVAC system. Because of this, we propose an ensemble approach, where a number of embedded audio sensing devices work in concert – each providing its own individual view of the system, and a central node (a machine in the cloud) aggregating the individual decisions to identify the component of the system that is problematic.

As an illustration, consider the scenario of Figure 23.4, where three independent embedded acoustic sensing systems are monitoring an HVAC system. Some high-level components of the HVAC system, for example, the heat

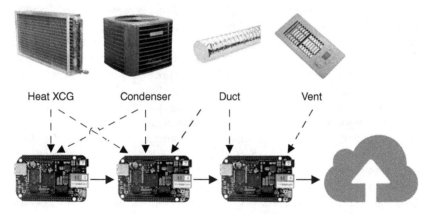

Figure 23.4 Distributed and ensemble approach toward detection of faulty components in HVAC systems.

exchanger, condenser, duct, and vent, are shown inside smaller boxes. Each one of these devices only learns and detects "abnormality" of the system from its own perspective. However, due to the placement of the devices with respect to the system components, when a particular component fails, opinions of some of the sensors must be valued higher than others. For example, the leftmost sensor system strongly hears sounds from the heat exchanger and the condenser unit (shown with arrows), whereas acoustic signals from the duct and the vents are weaker when they reach the leftmost sensor (no arrows). Now, if this sensor ever detects an anomalous behavior of the entire system, it is more likely that the faulty component is either the heat exchanger or the condenser or both, rather than the duct or the vent. It is possible to accurately locate the faulty component since all such individual decisions, as well as the weights of each decision (calculated a priori), are known to the cloud.

Because our embedded device will be installed inside a building, it will be in the presence of the residents. Therefore, the system needs to be able to filter out speech from the microphone input or be able to do classification on-device instead of allowing the signal with speech data to leave the premises. A possible solution to this is to either do the classification entirely on the embedded device by pushing the models down from the cloud or to be able to automatically subtract, or scramble, human voice from the signal, which will be decided as we progress with the project. A possible approach for this project would be to employ a modified granular synthesis technique [63] to blur the spectrum of the voice, thus making it unrecognizable. This technique would maintain the quality, texture, and characteristics of the audio, without compromising the privacy of individuals.

23.3.3 Decision Support System

The Building Information Model (BIM) allows digital representations of buildings and its mechanical systems. The US General Services Administration released the "BIM Guide for Energy Performance" as a method to strengthen the reliability, consistency, and usability of predicted building energy use and energy cost results [64–67]. With BIM's widespread adoption by architects and engineers including portfolio managers, a "dynamic" approach to BIM could enable not only visually tracking impending failures and equipment decay, its environmental cost, and the overall health of centralized HVAC systems but also response to state changes for immediate feedback and action.

The power of BIM can be harnessed to provide HVAC status information. In a previous project, we developed a Dynamic-BIM (dBIM) platform in collaboration with the Building Technologies Research and Integration Center, ORNL [68]. This platform has been further extended [69–72] to integrate 3D heat transfer analysis and enable high-performance parallel computing. Through a grant from the US Department of Energy–International and the

South African National Energy Development Institute (SANEDI), we are expanding the work to include environmental impacts, for example, impact on health and ecosystem services, as well as emissions in the scalable dBIM platform. Most recently, we received the Global Innovation Initiative grant from the US Department of State to develop novel approaches to employing green infrastructure to enhance urban sustainability.

Two major research tasks in designing and developing a decision support system for centralized HVACs are (i) devising an optimization model to help decision makers choose the optimal timeframe for retiring centralized HVAC systems with a short-term and long-term decision horizon and (ii) developing an integrated platform for domain modeling, simulation, and visualization that enables interoperability with newer and legacy systems to seamlessly transfer BIM for representing buildings and their systems.

We design a strategic optimization model that will help assess the cost and benefits of repairing versus an early retirement of an HVAC system, based on inputs such as the detected failures, the operational performance (both short term and long term), different primary energy types, future energy price trends, and trends in monetary and nonmonetary benefits.

Using a defensible set of input assumptions, grounded in the existing scientific and economic literatures, the avoided environmental damages will be translated into economic benefits. Uncertainties around the monetization of environmental impacts and related assumptions will be updated in line with the advancements of the science and economics of climate impacts. Building owners and property managers will be able to visualize and evaluate the benefits likely to result from permanently retiring underperforming systems.

Several monetary and nonmonetary decision factors will be analyzed during the decision making. These are summarized in Table 23.2.

We design a visualization platform that enables importation of BIM overlaid with prognosis and health information of HVAC systems. This is essential to provide advanced warning of failures to owners and property managers through a user-friendly visual interface. Figure 23.5 shows how data from buildings is processed, analyzed, and visualized and how the system provides real-time feedback to building owners and project managers, who can act on the information, completing the loop. The platform integrates domain models, enables simulations, and provides visualizations to aid decision making by the stakeholders. The integrated decision platform is extensible and scalable – spanning seamlessly across a variety of scales – from building to neighborhood to city scale. This ability to cut across scales will allow this platform to achieve its ultimate goal, that is, to model, simulate, and visualize the linkages and interplay among all elements of the urban fabric of smart cities. Besides using sensors and information and communication technologies' advancements, the proposed integrated visualization platform will allow users to "walk" into building zones to investigate abnormal energy usages and/or

Table 23.2 Monetary and nonmonetary decision factors.

Monetary factors	Nonmonetary factors
• Direct cost of electricity saved from future operational energy, repairs, and maintenance if the equipment is retired at the optimal time period (unit: $). Directly affects the institution and its economic well-being • Social cost of carbon from the avoided emissions of CO_2 and NO_x that are expected to result through early retirement of underperforming equipment (unit: $ per metric ton). Affects the society we live and proudly call "our home – the Earth"	• Energy saved (units: kW h; MMTCO2e) • Stratospheric ozone benefits (unit: MMTCO2e); by avoiding the release of CFC and HCFC refrigerant and foam-blowing agent, we can effectively reduce stratospheric ozone destruction through recovery and reclaiming HFC, HCFC, and greenhouse gases • Climate benefits (unit: MMTCO2e), owing to the emissions avoided from energy saved, stratospheric ozone benefits, durable goods recycling, and hazardous substances properly treated

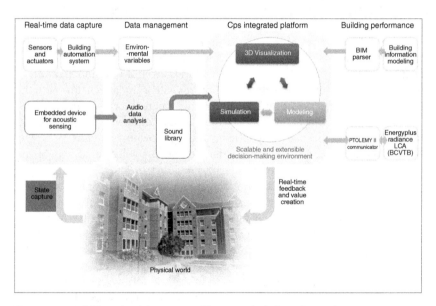

Figure 23.5 Full life cycle of the proposed integrated platform for predictive maintenance.

system faults, conduct "what-if" scenarios and data analytics of building energy components, and perform energy efficiency and environmental accounting procedures from within the platform.

There are limitations in the monetization of social cost of carbon owing to uncertainty, speculation, and lack of data on future greenhouse gas emissions, effects of past and future emissions on the climate, and the translation of these environmental impacts into economic damages. However, using a defensible set of assumptions, we use the social cost of carbon to estimate social benefits of reducing CO_2 emissions using the proposed system for early retirement of centralized HVAC systems. Furthermore, as the decision platform allows stakeholders to perform actions such as "semantic tagging," there is always a possibility of the occurrence of undesirable data clutter that may hinder the viewer's understanding of the content. To overcome this problem, a hierarchical approach to data exploration will be permitted. As the user becomes familiar with data analytics and intuitive interface mechanisms, the potential of data clutter can be eliminated entirely. That said, the visualization platform would take a minimalist (minimal user interface (UI) elements) and intuitive UI approach for effective usability.

23.4   Experimental Results

23.4.1   Spectral Analysis of HVAC Sounds

We have conducted a set of experiments to understand the variability of acoustic signatures in HVAC systems. A portable, dual-track audio recorder (Zoom H1 handy recorder) was used for recording audio samples of HVAC equipment located in three AHU rooms situated in three buildings in the UFL campus – Harn Museum, Phillips Center, and Southwest Recreation Center. Figure 23.6 shows these three test environments. The audio samples were recorded at 44.1 kHz in WAV format with 16 bits per sample signal encoding, and the duration of each clip was 60 s. While all three AHU rooms comprised AHUs of similar types, the hot water boilers were of different types in the Southwest Recreation Center.

Table 23.3 shows the spectrogram plots of 11 audio clips recorded near 5 different AHUs – a pump, a fan, and 3 boilers inside these three buildings. Two of the clips also contain human voices in the background. The horizontal axis of each of the spectrograms denotes time (up to 60 s). The vertical axis denotes the frequency range of -30 dB (lighter shades) to -150 dB (darker shades). We observe that different units have different spectral characteristics – which can be leveraged to identify a unit. For example, a pump, a fan, and a boiler operate in the frequency ranges of -53.4 ± 13.6, -62.3 ± 7.4, and -28.3 ± 5.24, respectively. In addition to that, some of these units have special acoustic signatures identifiable by detecting a special acoustic event, for example, the pump in H2 occasionally makes a sound that excites all frequencies. The presence of voice is identifiable by observing the change in human voice frequency range in the spectrograms – for example, the same fan in H3 and H4 shows slightly different spectral plots due to the presence of voice. Therefore, by using spectral subtraction, voice can be removed from the audio signals to preserve privacy.

Figure 23.6 AHU rooms: (a) Harn Museum, (b) Phillips Center, and (c) Southwest Recreation Center.

The location also has an effect on the acoustic characteristics, for example, boilers in Phillips Center (P1 and P2) are similar in spectral characteristics and boilers in SW Rec Center (R1 and R2) are also similar to each other, but there is a significant difference between boilers from these two places. Hence, we hypothesize that by modeling the spectral characteristics of each unit, we will be able to identify the unit's type and location, remove background noise and human speech, and detect expected/unexpected events.

Although our preliminary investigation suggests that the planned research is promising, there are at least two caveats. Firstly, in a real deployment, there will be cases where multiple units will be making sounds at the same time, and we need to identify them separately. This is, however, a variant of the classic cocktail party problem, which is solvable when there are a sufficient number of acoustic sensing devices. Secondly, in this preliminary study, we collected audio samples from close proximity to the unit. This may not be always possible in an actual deployment. Hence, placing a sensor at a suitable place will be a challenging problem to solve. To handle this, we plan to employ a feedback mechanism during the sensor deployment phase, which will guide us in

Table 23.3 Spectrogram plots of different AHU units within the three buildings.

Location	Equipment	Spectrogram plot (x axis: time 60 s, y axis: Freq. [−30 dB, −150 dB])
H1: Harn Museum	Between AHUs	
H2: Harn Museum	Pump	
H3: Harn Museum	Fan	
H4: Harn Museum	Fan + voice	
P1: Phillips Center	Boiler	
P2: Phillips Center	Boiler	
R1: SW Rec Center	Boiler	
R2: SW Rec Center	Boiler	

choosing an optimal position for a sensor. Finally, in a networked acoustic sensing environment, there are always chances of collisions in wireless transmissions, which may introduce unpredictable delays in detecting faulty units. This will be taken care of by synchronizing all wireless transmissions from all of our sensing units that are within the range of one another.

23.4.2 Longer-Term Deployment

Inspired by the preliminary results of spectral analysis, we conduct a longer-term experiment where we monitor the AHU room of Harn Museum in real time using a smartphone network. We use smartphones to quickly set up a three-node acoustic sensor network that captures acoustic signals and seven other types of onboard sensor readings (accelerometer, gyroscope, humidity, light, temperature, magnetometer, and pressure) from three AHUs of the HVAC. This setup is not low cost; in fact, each mobile device (Samsung Galaxy S4) costs around $500 in the consumer market. But for a quick deployment and to initiate the data collection as early as possible, this was our best option. A custom smartphone application has been developed to capture audio signals at 44.1 kHz and other sensors at 100 ms interval. After capturing, we perform basic data cleaning operations, including detection and removal of human voice, and ship them to a secure server maintained by the

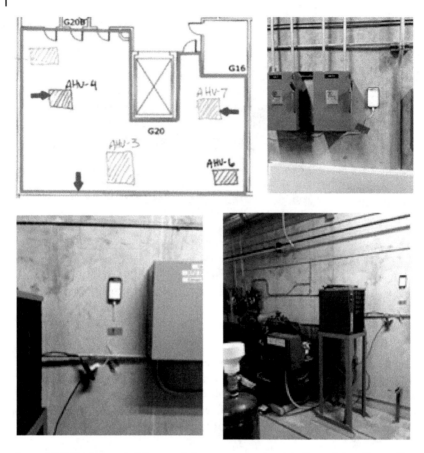

Figure 23.7 Deployment of the smartphone sensor network in three air handling unit rooms.

UFL. From the server, these data are periodically downloaded and analyzed in MATLAB (Figure 23.7).

As discussed earlier, we employ an unsupervised learning approach to model and encode the regular pattern in the acoustic time series data and to discover if a running HVAC system has deviated from its regular pattern of operation at any point in time. At first, the audio stream is converted to a stream of 50-ms frames and passed through a frame-level feature extraction stage where MFCCs are computed for each frame. Then k-means algorithm is used to cluster similar audio frames. Cluster assignment is used as an encoding for each frame. This step maps acoustic frames to k clusters. Finally, we compute the transition probabilities between each pair of clusters. Once the transition probabilities are in steady state, a sequence of unlikely transitions would mean that the HVAC's behavior is unusual with respect to the currently learned model. Note that

	C1	C2	C3
C1	0.076	0.017	0.013
C2	0.017	0.759	0.004
C3	0.013	0.004	0.097

(a) Transition matrix 1

	C1	C2	C3
C1	0.066	0.01	0.008
C2	0.01	0.814	0.002
C3	0.008	0.002	0.08

(b) Transition matrix 2

Figure 23.8 The similarity between transition matrices computed on two different time spans are noteworthy. This phenomenon provides an empirical validity of our acoustic modeling strategy. (a) Transition matrix 1. (b) Transition matrix 2.

modeling the normal HVAC is a continuous learning task, which is susceptible to high false-positive rates at the early stage. To mitigate this, we employ a human-in-the-loop approach.

In Figure 23.8, two transition matrices computed on two completely disjoint and independent time spans (524,280 frames each) are shown. The similarity between corresponding cell values tells us that our proposed acoustic-based HVAC state modeling strategy is fairly stable. With more data and larger number of clusters, these two matrices will converge to steady-state values, which can be used to discover potential faults in HVAC systems.

Final Thoughts

Through SASEM, we aim to develop and mature the science of using acoustic signals for system assessment prognosis of centralized HVAC systems. Our next step is to build an energy-efficient low-cost sensing platform comprising a network of embedded devices. Moreover, we will develop a decision support system with an optimization model and a visualization platform. To elaborate, the optimization model will help decision makers choose the optimal timeframe for retiring centralized HVAC systems with short-term and long-term decision horizons. The visualization platform will allow acoustic-based system assessment and prognosis through simulation and learning and enable interactive user functionalities for data analysis and control.

Questions

1 What are the types of HVAC faults that are detectable by acoustic sensing?

2 What are the three key strategical considerations for acoustic-based HVAC fault detection?

3 Discuss the benefits of predictive HVAC maintenance.

4 Why an unsupervised learning method is necessary for acoustic-based HVAC monitoring?

Acknowledgement

The authors would like to acknowledge the financial support for this research received from the US National Science Foundation (NSF) EAGER #1619955. Any opinions and findings in this chapter are those of the authors and do not necessarily represent those of NSF.

References

1 US CENSUS Bureau. Available: http://www.census.gov/ (accessed January 2015).

2 Turpin, J.R. (2009) *To Repair or Replace: That Is the Key Question*. Retrieved April 24, 2015, from The Air Conditioning, Heating and Refrigeration NEWS: http://www.achrnews.com/articles/110056-to-repair-or-replace-that-is-the-key-question

3 Novell Custom HVAC (2014) *Common Repairs*. Retrieved April 24, 2015, from Novell Custom Heating & Air Conditioning Co.: http://www.novellcustom.com/common-repairs/

4 Miller, E. (2015) *What To Do With Your Old HVAC Unit*. Retrieved April 24, 2015, from Snyder Heating and Air Conditioning: http://blog.snyderac.com/blog/what-to-do-with-your-old-hvac-unit

5 Hosek, M. Krishnasamy, J., and Prochuzka, J. (2006) Intelligent condition-monitoring and fault diagnostic system for predictive maintenance. US Patent No.: US 7882394 B2.

6 Watt, J. (1994) Predictive Maintenance Programs (PMP's) in Small HVAC Applications: Analysis of Available Products and Technology. Report prepared for MEEN 662.

7 Fu, C., Ye, L., Liu, Y. *et al.* (2004) Predictive maintenance in intelligent-control-maintenance system for hydroelectric generating unit. *IEEE Transactions on Energy Conversion*, **19**, 179–186.

8 Lin, C. and Tseng, H. (2005) A neural network application for reliability modelling and condition-based predictive modeling. *International Journal of Advanced Manufacturing Technology*, **25**, 174–179.

9 Liggan, P. and Lyons, D. (2011) Applying predictive maintenance techniques to utility systems. *Official Magazine of ISPE*, **31** (6), 1–7.

10 Zhao, Z., Wang, F., Jia, M., and Wang, S. (2010) Predictive maintenance policy based on process data. *Chemometrics and Intelligent Laboratory Systems*, **103**, 137–143.

11 Yang, S.K. (2002) An experiment of the state estimation for predictive maintenance using Kalman filter on a DC motor. *Reliability Engineering and System Safety*, **75**, 103–111.

12 Schein, J., Bushby, S.T., Castro, N.S., and House, J.M. (2006) A rule-based fault detection method for air handling units. *Energy and Buildings*, **38** (12), 1485–1492.

13 Sun, B. *et al.* (2014) Building Energy doctors: An SPC and Kalman filter-based method for system-level fault detection in HVAC systems. *IEEE Transactions on Automation Science and Engineering*, **11** (1), 215–229.

14 Dong, B., Zheng, O.'N., and Li, Z. (2014) A BIM-enabled information infrastructure for building energy fault detection and diagnostics. *Automation in Construction*, **44**, 197–211.

15 Capozzoli, A., Lauro, F., and Khan, I. (2015) Fault detection analysis using data mining techniques for a cluster of smart office buildings. *Expert Systems with Applications*, **42** (9), 4324–4338.

16 Du, Z. *et al.* (2014) Fault detection and diagnosis for buildings and HVAC systems using combined neural networks and subtractive clustering analysis. *Building and Environment*, **73**, 1–11.

17 Li, S. and Wen, J. (2014) A model-based fault detection and diagnostic methodology based on PCA method and wavelet transform. *Energy and Buildings*, **68**, 63–71.

18 Li, Y. *et al.* (2014) Experimental study on electrical signatures of common faults for packaged DX rooftop units. *Energy and Buildings*, 77, 401–415.

19 Yan, K. *et al.* (2014) ARX model based fault detection and diagnosis for chillers using support vector machines. *Energy and Buildings*, **81**, 287–295.

20 Zhao, X., Yang, M., and Li, H. (2014) Field implementation and evaluation of a decoupling-based fault detection and diagnostic method for chillers. *Energy and Buildings*, **72**, 419–430.

21 Baird, B.L. (2013) *Your HVAC system: 5 Sounds You Don't Want To Hear*. Retrieved April 24, 2015, from Angie's List: http://www.angieslist.com/articles/your-hvac-system-5-sounds-you-dont-want-hear.htm

22 InspectAPedia (2015) *HVAC Noise Descriptions & Recordings*. Retrieved April 24, 2015, from Free Encyclopedia of Building & Environmental Inspection, Testing, Diagnosis, Repair: http://inspectapedia.com/noise_diagnosis/HVAC_Noise_Descriptions.php

23 Ware, J.A. and Aki, K. (1969) Continuous and discrete inverse-scattering problems in a stratified elastic medium. I. Plane waves at normal incidence. *The Journal of the Acoustical Society of America*, **45** (4), 911–921.

24 Sharp, D.B. and Campbell, D.M. (1997) Leak detection in pipes using acoustic pulse reflectometry. *Acta Acustica United With Acustica*, **83** (3), 560–566.

25 Vahaviolos, S.J. (ed.) (1999) *Acoustic Emission: Standards and Technology Update* Vol. 1353, ASTM International.

26 Schafer, R.M. (1977) *The Tuning Of The World*, Alfred A. Knopf.

27 Kull, R.C. (2006) Natural and urban soundscapes: The need for a multi-disciplinary approach. *Acta Acustica United With Acustica*, **92** (6), 898–902.

28 Krause, B. (1987) Bioacoustics: Habitat ambience & ecological balance. *Whole Earth Review*, **57**, 14–15.

29 Mydlarz, C. *et al.* (2014) *The Design of Urban Sound Monitoring Devices* Audio Engineering Society Convention 137, Audio Engineering Society.

30 Kang, J. (2006) *Urban Sound Environment*, Taylor & Francis, New York, p. 304.

31 Czech, P. (2013) Diagnosing a Car engine fuel injectors' damage, in *TST* (ed. J. Mikulski), Springer, pp. 243–250.

32 Adaileh, W.M. (2013) Engine fault diagnosis using acoustic signals. *Applied Mechanics and Materials*, **295**, 2013–2020.

33 Hu, W., Bulusu, N., Chou, C.T. *et al.* (2009) Design and evaluation of a hybrid sensor network for cane toad monitoring. *ACM Transactions on Sensor Networks (TOSN)*, **5** (1), Article 4.

34 Wang, H., Elson, J., Girod, L., Estrin, D., and Yao, K. (2003) *Target Classification and Localization in Habitat Monitoring*. In Proc. of Acoustics, Speech, and Signal Processing Conf.

35 Simon, G., *et al.* (2004) *Sensor Network-Based Countersniper System*. Proceedings of the 2nd international conference on Embedded networked sensor systems. ACM.

36 Saunders, J. (1996) *Real-time Discrimination Of Broadcast Speech/Music*. *Acoustics, Speech, and Signal Processing*, IEEE International Conference on. Vol. 2. IEEE.

37 Guvensan, M.A., Taysi, Z.C., and Melodia, T. (2013) Energy monitoring in residential spaces with audio sensor nodes: TinyEARS. *Ad Hoc Networks*, **11**, 1539–1555.

38 Lu, J., et al. (2010) *The Smart Thermostat: Using Occupancy Sensors to Save Energy in Homes*. Proceedings of the 8th ACM Conference on Embedded Networked Sensor Systems. ACM.

39 Sookoor, T. and Whitehouse, K. (2013) *Roomzoner: Occupancy-Based Room-Level Zoning of a Centralized HVAC System*. Proceedings of the ACM/IEEE 4th International Conference on Cyber-Physical Systems. ACM.

40 Energy, Inc. (n.d.) *The Energy Detective*. Retrieved from The Energy Detective Electricity Monitor: http://www.theenergydetective.com/ (April 26, 2015)

41 Hnat, T. W., *et al.* (2011) *The Hitchhiker's Guide to Successful Residential Sensing Deployments.* Proceedings of the 9th ACM Conference on Embedded Networked Sensor Systems. ACM.

42 Dinh, H.T., Lee, C., Niyato, D., and Wang, P. (2013) A survey of mobile cloud computing: Architecture, applications, and approaches. *Wireless Communications and Mobile Computing*, **13**, 1587–1611.

43 Kumar, K. and Lu, Y.H. (2010) Cloud computing for mobile users: Can offloading computation save energy? *Computer*, **43** (4), 51–56.

44 Fernando, N., Loke, S.W., and Rahayu, W. (2013) Mobile cloud computing: A survey. *Future Generation Computer Systems*, **29** (1), 84–106.

45 Huang, D. (2011) Mobile cloud computing. *IEEE COMSOC Multimedia Communications Technical Committee (MMTC) E-Letter*, **6** (10), 27–31.

46 Guan, L., Ke, X., Song, M., and Song, J. (2011) *A Survey of Research on Mobile Cloud Computing.* In Proceedings of the 2011 10th IEEE/ACIS International Conference on Computer and Information Science (pp. 387–392). IEEE Computer Society.

47 Pournajaf, L., Xiong, L., Garcia-Ulloa, D.A., and Sunderam, V. (2014) *A Survey on Privacy in Mobile Crowd Sensing Task Management.* Technical Report TR-2014-002, Department of Mathematics and Computer Science, Emory University.

48 Lu, H., Bernheim Brush, A.J., Priyantha, B., Karlson, A. K., and Liu, J. (2011) *Speaker Sense: Energy Efficient Unobtrusive Speaker Identification on Mobile Phones*, Proceedings of the 9th international conference on Pervasive computing, June 12-15, 2011, San Francisco, USA.

49 Walker, W., Lamere, P., Kwok, P. *et al.* (2004) *Sphinx-4: A Flexible Open Source Framework For Speech Recognition*, Sun Microsystems, Inc., Mountain View, CA.

50 Lu, H., Frauendorfer, D., Rabbi, M., Mast, M.S., Chittaranjan, G.T., Campbell, A.T., Gatica-Perez, D., and Choudhury, T. (2012) *StressSense: Detecting Stress in Unconstrained Acoustic Environments Using Smartphones*, Proceedings of the 2012 ACM Conference on Ubiquitous Computing, September 05-08, 2012, Pittsburgh, Pennsylvania.

51 Rachuri, K.K., Musolesi, M., Mascolo, C., Rentfrow, P.J., Longworth, C., and Aucinas, A. (2010) *Emotion Sense: A Mobile Phones Based Adaptive Platform For Experimental Social Psychology Research*, Proceedings of the 12th ACM international conference on Ubiquitous computing, September 26-29, 2010, Copenhagen, Denmark

52 Miluzzo, E., Cornelius, C.T., Ramaswamy, A., Choudhury, T., Liu, Z., and Campbell, A.T. (2010) *Darwin Phones: The Evolution of Sensing and Inference on Mobile Phones*, Proceedings of the 8th international conference on Mobile systems, applications, and services, June 15-18, 2010, San Francisco, California, USA.

53 Miluzzo, E., Lane, N.D., Fodor, K., Peterson, R., Lu, H., Musolesi, M., Eisenman, S.B., Zheng, X., and Campbell, A.T. (2008) *Sensing Meets Mobile Social Networks: The Design, Implementation And Evaluation Of The Cenceme Application*, Proceedings of the 6th ACM conference on Embedded network sensor systems, November 05-07, 2008, Raleigh, NC, USA

54 Wang, A.. *An Industrial-Strength Audio Search Algorithm*. In ISMIR '03.

55 Sound Hound. Retrieved from http://www.soundhound.com/ (accessed 20 April 2015).

56 Larson, E.C., Lee, T.J., Liu, S., Rosenfeld, M., and Patel, S.N. (2011) *Accurate And Privacy Preserving Cough Sensing Using A Low-Cost Microphone*, Proceedings of the 13th international conference on Ubiquitous computing, September 17-21, 2011, Beijing, China

57 Nirjon, S., Dickerson, R.F., Li, Q., Asare, P., Stankovic, J.A., Hong, D., Zhang, B., Jiang, X., Shen, G., and Zhao, F. (2012) *Musical Heart: A Hearty Way Of Listening To Music*, Proceedings of the 10th ACM Conference on Embedded Network Sensor Systems, November 06-09, 2012, Toronto, Ontario, Canada.

58 Azizyan, M., Constandache, I., and Choudhury, R.R. (2009) *SurroundSense: Mobile Phone Localization Via Ambience Fingerprinting*, Proceedings of the 15th annual international conference on Mobile computing and networking, September 20-25, 2009, Beijing, China

59 Chon, Y., Lane, N.D., Li, F., Cha, H., and Zhao, F. (2012) *Automatically Characterizing Places With Opportunistic Crowdsensing Using Smartphones*, Proceedings of the 2012 ACM Conference on Ubiquitous Computing, September 05-08, 2012, Pittsburgh, Pennsylvania.

60 Lu, H., Pan, W., Lane, N.D., Choudhury, T., and Campbell, A.T. (2009) *SoundSense: Scalable Sound Sensing for People-Centric Applications on Mobile Phones*, Proceedings of the 7th international conference on Mobile systems, applications, and services, June 22-25, 2009, Kraków, Poland

61 Lu, H., Yang, J., Liu, Z., Lane, N.D., Choudhury, T., and Campbell, A.T. (2010) *The Jigsaw Continuous Sensing Engine For Mobile Phone Applications*, Proceedings of the 8th ACM Conference on Embedded Networked Sensor Systems, November 03-05, 2010, Zürich, Switzerland

62 Chandola, V., Banerjee, A., and Kumar, V. (2009) Anomaly detection: A survey. *ACM Computing Surveys*, **41** (3), Article 15 (July 2009), 58 pages.

63 Roads, C. (1988) Introduction to granular synthesis. *Computer Music Journal*, **12** (2), 11–13.

64 GSA (2009) *BIM Guide Series: 02 – GSA BIM Guide for Spatial Program Validation*, US General Services Administration.

65 Senate Properties (2007) *BIM Requirements, Volume 3 – Architectural Design*, Senate Properties, Finland.

66 Statsbygg (2011) *BIM Manual version 1.2, Norwegian Ministry of Government Administration, Reform and Church Affairs*, Statsbygg.

67 ASHRAE BIM Guide (2009) *American Society for Heating*, Refrigeration and Air-Conditioning Engineers, Atlanta, GA.

68 Srinivasan, R.S., Kibert, C.J., Fishwick, P., Ezzell, Z., Thakur, S., Ahmed, I., and Lakshmanan, J. (2012) *Preliminary Researches in Dynamic-BIM (D-BIM) Workbench Development*, In Proceedings of Winter Simulation Conference, Berlin, Germany, 2012.

69 Srinivasan, R.S., Kibert, C., Fishwick, P., Thakur, S., Lakshmanan, J., Ezzell, Z., Parmar, M., and Ahmed, I. (2013) *Dynamic-BIM (D-BIM) Workbench for Integrated Building Performance Assessments*, In Proceedings of the Advances in Building Sciences Conference, Madras, India, 2013.

70 Srinivasan, R.S., Thakur, S., Parmar, M., and Ahmed, I. (2014) *Toward a 3D Heat Transfer Analysis in Dynamic-BIM Workbench*. In Proceedings of iiSBE Net Zero Built Environment Symposium held in Gainesville, FL, 6-7 March, 2014

71 Srinivasan, R.S., Thakur, S., Parmar, M., and Ahmed, I. (2014) *Towards the Implementation of a 3D Heat Transfer Analysis in Dynamic-BIM Workbench*. In Proceedings of 45th Winter Simulation Conference to be held in Savannah, GA, 7-10 December, 2014

72 Agdas, D. and Srinivasan, R.S. (2014) *Parallel Computing in Building Energy Simulation*. In Proceedings of 45th Winter Simulation Conference to be held in Savannah, GA, 7-10 December, 2014.

73 Yu Y, Woradechjumroen D, Yu D. A review of fault detection and diagnosis methodologies on air-handling units. *Energy and Buildings*, 2014, **82**, 550–562.

74 Nirjon, S., Greenwood, C., Torres, C., Zhou, S., Stankovic, J.A., Yoon, H.J., Ra, H.K., Basaran, C., Park, T., and Son, S.H. (2014) Kintense: A robust, accurate, real-time and evolving system for detecting aggressive actions from streaming 3D skeleton data. In Pervasive Computing and Communications (PerCom).

24

Smart Lighting
Jie Lian

Charles L. Brown, Department of Electrical and Computer Engineering, University of Virginia, Charlottesville, VA, USA

CHAPTER MENU

Introduction, 697
Background, 698
Smart Lighting Applications, 699
Visible Light Communication (Smart Lighting Communication) System, 701
Conclusion and Outlook, 718

Objectives

- To become familiar with the concept of smart lighting
- To become familiar with the development of smart lighting
- To become familiar with the applications of smart lighting

24.1 Introduction

Due to the fast growth of the global population and the development of the modern cities, about 50% of the global population lives in the cities. In addition, this ratio will increase continuously to 70% by 2020 [1]. Such large urban population is a new challenge for city management and city planners. Therefore, intelligent buildings and environment, especially in regard to efficient energy management, should be designed and applied properly.

Considering energy consumption, electrical energy is widely used in lighting, industrial production, communication, broadcasting, and so on. Lighting occupies about 20% of electrical energy consumption around the world. In addition, around 6% of greenhouse gases is generated for lighting [2]. Additionally,

Smart Cities: Foundations, Principles and Applications, First Edition.
Edited by Houbing Song, Ravi Srinivasan, Tamim Sookoor, and Sabina Jeschke.
© 2017 John Wiley & Sons, Inc. Published 2017 by John Wiley & Sons, Inc.

municipalities and other local governments are prioritizing on lighting to save energy [3].

The 288-city survey emphasizes how to use smart lighting systems in smart cities for energy saving. Many cities have paid attention on smart lighting systems in indoor and outdoor environments [4]. From the survey, about 30% of the cities that were surveyed voted "LEDs/energy-efficient lighting" as their top priority for the next a few years. In addition, light-emitting diode (LED) is voted to be the most promising solution for reducing greenhouse gases [3].

Regarding lighting technologies, LEDs are widely used as lighting sources for outdoor and indoor illumination due to high power efficiency and long lifetime. Smart lighting system using LEDs as light sources and data transmitters can become a significant component for smart city and a professional solution for illumination and data transmission with low power consumption. This solution is being powered with the embedded Internet protocol connectivity, making a feasible end-to-end wireless connections between mobile terminals (smartphones, tablets, etc.) and LED lights [1].

For indoor applications, smart lighting systems play an important role. In particular workplaces, specific lighting conditions and illumination are essential for users' visual comfort and working efficiency. In other words, eye safety and illumination requirements need to be considered [5]. Moreover, 20% of the total energy of office buildings in the United States are used by indoor illumination [6]. This means efficient lighting control can significantly improve the energy efficiency of buildings.

24.2 Background

The development of smart lighting technique consists of two stages [7]. The first stage is that the conventional bulbs are replaced by the new technology, LEDs. In addition, LEDs are used as dominant light source for public illumination. In this stage, the researches are focused on improving the performance of LEDs. Higher efficiency, lower cost, and longer lifetime are the characteristics of LEDs that we pursue. The second stage of development of smart lighting is the lighting control algorithm. Adaptive lighting control solutions are the aims for this stage to make smart lighting system have higher performance efficiency, satisfy users' illumination requirements, and provide wireless network connections.

For the first stage, the semiconductor materials, structure, and fabrication can affect the performance of LEDs. Some researchers develop a high-efficiency InGaN quantum-well LEDs with microspheres [8]. With the help of this special structure and material, the efficiency enhancement of 1.7 times is achieved. In order to improve the quantum efficiency of LEDs, another research group proposed a method to combine high flexibility of polymer films with nitride nanowires [9]. The output light power has an enhancement of 38%

Table 24.1 Power saving measurement collected during 6 days.

GROUP	DAY 1	DAY 2	DAY 3	DAY 4	DAY 5	DAY 6
Box office 1	44%	33%	59%	85%	15%	25%
Box office 2	37%	52%	37%	68%	10%	22%
Meeting room	12%	33%	33%	65%	12%	33%
Showroom	72%	90%	52%	72%	44%	52%

Total average power saving 43%

Source: Magno et al. (2015) [12].

to the conventional GaN-based LEDs by using a hybrid structure of straight nanorods [10].

In the second stage, the smart lighting systems are illumination systems that use feedback measurements from a network of various sensors. Considering the quality of output light and power consumption, some researchers propose an optimization framework for control of non-square smart lighting systems with saturation constraints [11]. In [11], two solutions are proposed: one is based on the interior point algorithm, and the other one is based on Newton–Raphson method with projection. With the help of wireless sensor network, smart lighting systems can be controlled efficiently [12]. After experiments, the total power consumption is reduced by more than 40%. Results of the successive 6 days are shown in Table 24.1.

24.3 Smart Lighting Applications

Smart lighting technique is one of the most significant branches in smart cities, which can be used in many aspects such as dimmable outdoor environment lighting, vehicle safety, indoor illumination and communications, smart lighting positioning, and so on.

Outdoor smart lighting systems consist of public lighting, wireless sensor network, Internet, cloud computing, and other techniques. The diagram of a typical outdoor smart lighting system is shown in Figure 24.1. With the help of sensor networks, the outdoor public lighting can be dimmed. To save energy, the system maintains lighting at a low level until vehicle or pedestrian motion is detected [13]. When the area is occupied, the street light can be dimmed to a higher level to satisfy the illumination requirements and keep safety. To save energy and prolong lamps lifetime, some researchers propose an optimization algorithm to reduce the energy consumption and increase lamp life [14]. From the results in [14], the total energy consumption is decreased by 20%, and the average lamp life is increased by 100% in 12 working hours. Similar to the smart

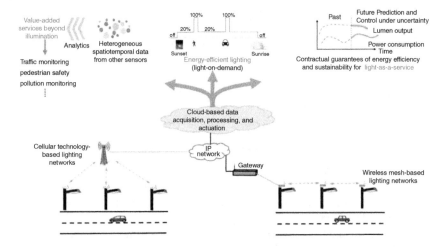

Figure 24.1 Outdoor smart lighting infrastructure for data-driven applications. *Source*: Murthy *et al.* (2015) [3].

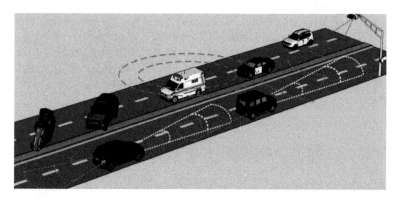

Figure 24.2 Illuminations of smart lighting in an outdoor configuration between several vehicles. *Source*: Cailean *et al.* (2015) [17].

lighting system for the street illumination, parking lot can also use the smart lighting control technique to save energy [15]. The dimming and brightening depend on whether people or vehicles are detected in the monitored area.

The smart lighting can also be applied in vehicles for vehicle safety. The smart automotive lighting can provide illumination and signaling, reliable communications, and so on [16]. An example of smart lighting for vehicles in a highway configuration is shown in Figure 24.2. This vehicle smart lighting system can detect potential risks in advance and provide early warnings to the driver. Therefore, the probability of traffic accidents can be reduced [16].

Indoor and outdoor positioning is another important application for smart lighting technique. Since the lighting system might be installed everywhere,

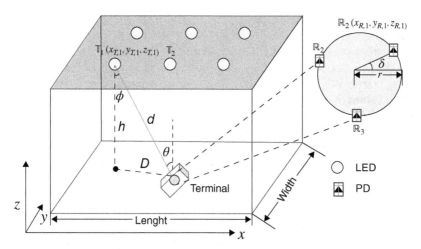

Figure 24.3 Indoor positioning using smart lighting technique. *Source*: Xu *et al.* (2016) [20].

the smart lighting positioning must be a trend of development. An example of indoor positioning using smart lighting is shown in Figure 24.3. For outdoor smart lighting positioning system, bad weather conditions such as rain and fog can easily affect the performance of positioning systems. However, with the help of image sensor, multiple photodetectors, and some algorithms, the outdoor smart lighting positioning is still a promising technique [18]. Indoor positioning systems can be installed on the base of the existing lighting systems. The average positioning error in the smaller environment is 3.78 cm [19].

The most popular application for smart lighting is smart lighting communications. This system uses visible light as communication media and specially designed LEDs work as data transmitters. In the following section, smart lighting communication is introduced in detail. In most cases, the smart lighting communication is known by visible light communication (VLC).

24.4 Visible Light Communication (Smart Lighting Communication) System

With the rapid development of the technology, high data rate transmission will be demanded in our daily lives. Considering the spectrum of radio frequency (RF) communications is so congested, and the data transmission rate of RF communications cannot satisfy the huge demand of the big data transmission, VLC has emerged as a new possible technology for the next generation communications [21]. In VLC systems, white LEDs are used as transmitters for communications and can become the dominant indoor communication method due to its many advantages compared with RF communications. The VLC systems are built as a dual system (illumination and data transmission) and have higher privacy than RF communication systems. LEDs are efficient light sources, and they have long life expectancy [22]. Because of the advantages over RF communications, VLC can become the dominant indoor communication method [23–25].

VLC can also be applied not only in indoor communications but also in outdoor short-range communications, such as vehicle-to-vehicle communications. As the number of vehicles increases every year, an urgent action is needed to prevent and reduce the traffic accidents as well as improve road safety [26]. Since the vehicles' headlights and taillights are usually composed of LEDs, the vehicles can communicate with each other using VLC. Then, an intelligent transportation system (ITS) can be built to improve the road safety and traffic flow based on VLC network.

In addition, the VLC technique can be used for many applications, such as smart lighting, mobile connectivity, healthcare, underwater communications, location-based services, and so on. Therefore, the applications in VLC have a great potential to increase in the next decades. These applications using VLC technique can change the pattern of people's life. According to the latest market research report [27], the VLC market is expected to grow from USD 327.8 million in 2015 to USD 8502.1 million by 2020, at a compound annual growth rate (CAGR) of 91.8% between 2015 and 2020.

24.4.1 System Description

24.4.1.1 Transmitter and Receiver Model

The quality of signals received at the users depends on the transmitters, the receiver, and the channel models. In this section, we describe the transmitter and receiver models.

Multiple-LED lamp model in Figure 24.4 is proved to have a better coverage of illumination and better communication performance [25]. For this model, there are multiple LEDs with different inclination angles for each lamp, and each LED can be controlled separately.

The receivers' model we used is proposed in [28]. For this model, there are V photodetectors with different inclination angles for each receiver; the structure of the multi-detector model is shown in [29]. On this basis, the received signal for each user is a combination of the signals from each detector. To get the optimal signal-to-interference-plus-noise ratio (SINR), we propose a combined optical power allocation and received signal linear combination scheme (Figure 24.5).

24.4.1.2 Indoor VLC Channel Model

In this section, we analyze the indoor VLC channel model and show the derivation of the channel model with the 1-LED and 25-LED lamp models.

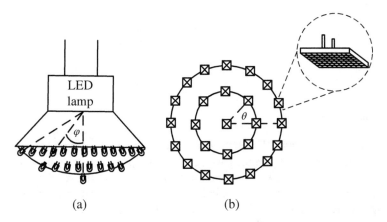

Figure 24.4 The structure of the multiple-LED array lamp: (a) side view and (b) bottom view.

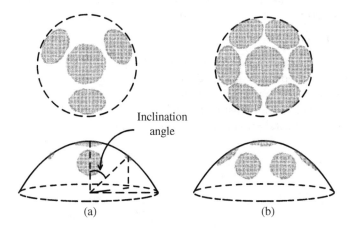

Figure 24.5 Multi-detector model structure, (a) four-detector model, top and side view, (b) seven-detector model, top and side view (similar to [28]).

For indoor VLC systems, white LEDs work as transmitters and photodetectors work as receivers. Since the visible light is incoherent, intensity modulation and direct detection (IM–DD) are employed in VLC systems (Figure 24.6). After the receiver, the received signal can be represented as

$$Y(t) = rX(t) \otimes h(t) + N(t) \tag{24.1}$$

where r represents the responsivity, $\otimes$ is convolution, $X(t)$ is the transmitted optical intensity, and $N(t)$ is the additive noise. $h(t)$ is the indoor channel impulse response, which, using ray tracing, can be modeled as

$$h(t) = \sum_{n=1}^{N} a_n \delta(t - t_n) \tag{24.2}$$

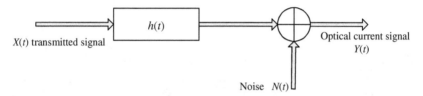

Figure 24.6 Basic indoor VLC channel model.

where a_n and t_n represent the path gain and transmission time delay, respectively. N is the number of multipath components. The channel gains and transmission time delays in (24.2) depend on the light paths to the receiver.

Because of the principles of optics, the light rays from the transmitter can be classified into two types. They are the line of sight (LOS) rays and diffused rays, as shown in Figure 24.7. These two types cause the multipath effect in indoor VLC system. Thus, the indoor VLC channel transfer function can be approximated by Ghassemlooy and Popoola [30]

$$H(f) = H_{\text{LOS}}(f) + H_{\text{diff}}(f), \tag{24.3}$$

where $H_{\text{LOS}}(f)$ is the contribution due to the LOS, which is basically independent of the modulation frequency, and it depends on the distance between transmitter and receiver and on their orientation with respect to the LOS. $H_{\text{diff}}(f)$ is the diffused part, the intensity of which is less than the LOS part. The impulse response of the indoor channel can also be represented as

$$h(t) = \sum_{k=0}^{\infty} h^{(k)}(t)$$

$$= \underbrace{h^{(0)}(t)}_{\text{LOS part}} + \underbrace{\sum_{k=1}^{\infty} h^{(k)}(t)}_{\text{Diffused part}} \tag{24.4}$$

where k is the number of reflections.

The intensity of the LOS rays and diffused rays follows the Lambertian law. The Lambertian radiant intensity model can be defined as [31]

$$R_0(\phi) = \begin{cases} \frac{m+1}{2\pi} \cos^m(\phi) & \text{for } \phi \in [-\pi/2, \pi/2] \\ 0 & \text{for } |\phi| \geq \pi/2 \end{cases} \tag{24.5}$$

where m is the Lambertian mode of the light source and ϕ is the radiation angle for transmitters as shown in Figure 24.8. The maximum radiated power is reached when $\phi = 0$. The Lambertian mode m is related to LED's semiangle $\Phi_{1/2}$ by

$$m = \frac{\ln 2}{\ln(\cos \Phi_{1/2})}. \tag{24.6}$$

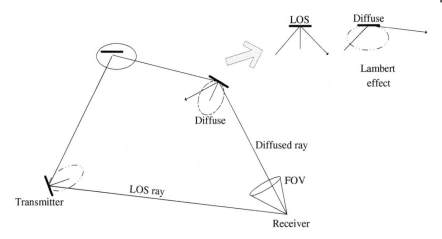

Figure 24.7 Light rays classification.

The detector effective area can be modeled as a function of the incident angle, ψ, as [31]

$$A_{\text{eff}}(\psi) = \begin{cases} A_r \cos \psi & -\pi/2 \leq \psi \leq \pi/2 \\ 0 & |\psi| > \pi/2 \end{cases}, \tag{24.7}$$

where A_r is the area of the detector as shown in Figure 24.8. We assume that the detector cannot be active beyond the field of view (FOV) angle Ψ_c. Therefore,

Figure 24.8 LOS light rays model.

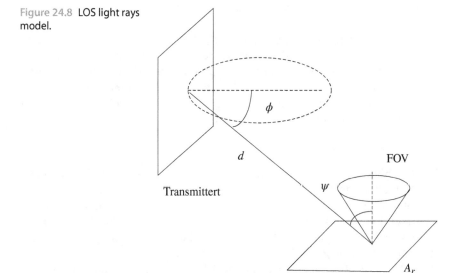

the LOS link can be described as

$$H_{LOS} = \begin{cases} \frac{A_r(m+1)}{2\pi d^2}\cos^m(\phi)\cos(\psi) & -\Psi_c \leq \psi \leq \Psi_c \\ 0 & \text{elsewhere} \end{cases} \tag{24.8}$$

Thus, the impulse response of LOS part can be described as

$$h^{(0)}(t) = h_{LOS}(t) = \frac{A_r(m+1)}{2\pi d^2}\cos^m(\phi)\cos(\psi)\delta\left(t - \frac{d}{c}\right) \tag{24.9}$$

The diffused part is

$$h^{(k)}(t) = A_r L_0 L_1 L_2 \dots L_k \gamma^k \delta\left(t - \frac{d_0 + d_2 + \dots + d_k}{c}\right) \tag{24.10}$$

where $L_0, L_1, \dots, L_k$ represent the link attenuations, γ is the reflection coefficient, and A_r is the area of the detector:

$$L_0 = \frac{(m+1)\cos^m(\phi_0)\cos(\psi_0)}{2\pi d_0^2}$$

$$L_1 = \frac{\cos^m(\phi_1)\cos(\psi_1)}{\pi d_1^2} \tag{24.11}$$

$$\vdots$$

$$L_k = \frac{\cos^m(\phi_k)\cos(\psi_k)}{\pi d_k^2}$$

where d_0 is the distance of the LOS link, d_k represents the distance of the kth bounce link, ϕ is the radiation angle, ψ represents the incident angle, and d and c represent distance between transmitter and receiver and light speed, respectively [32].

The impulse response of the indoor channel can be obtained by ray tracing algorithm, and an example of the impulse response is shown in Figure 24.9. In this figure, we can find the impulse response composed of the LOS and diffused components. The very long tail represents the diffused components. This long tail can introduce intersymbol interference (ISI) when symbol rate is high. The ISI would affect the communication performance.

24.4.2 VLC MIMO Technology

In an RF multiple input and multiple output (MIMO) system, the signal can be sent by multiple antennas from transmitters and received by multiple antennas at receivers. Similarly, for VLC systems, the MIMO technique can also be used. Figure 24.10 shows a configuration diagram of a typical indoor VLC MIMO system. In this figure four LED lamps are used for room lighting as well as for transmitting independent data streams simultaneously [33].

In a VLC MIMO system, if the number of transmitters is Q and the number of receivers is K, this MIMO system channel can be represented as a $Q \times K$

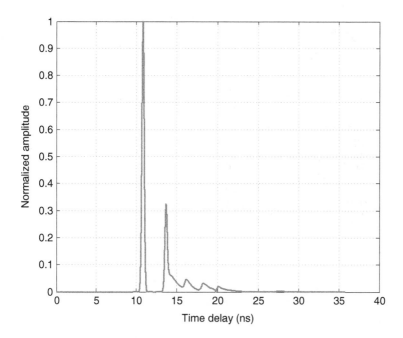

Figure 24.9 Normalized impulse response of multi-LED lamp model with semiangle= 30° at (1.25, 0.625, 0).

matrix. In this section, we assume that there is no ISI; thus the MIMO channel matrix can be described as

$$\mathbf{H} = \begin{pmatrix} h_{11} & h_{12} & \cdots & h_{1K} \\ h_{21} & h_{22} & \cdots & h_{2K} \\ \vdots & \vdots & \ddots & \vdots \\ h_{Q1} & h_{Q2} & \cdots & h_{QK} \end{pmatrix} \tag{24.12}$$

where h_{ij} represents the channel attenuation from the ith transmitter to the jth receiver. For the indoor VLC system, h_{ij} has been described in [25]:

$$h_{ij} = \frac{A_r \cos\langle \vec{r}_{ij}, \vec{n}_j \rangle}{2\pi d_{ij}^2} (m+1)\cos^m \langle \vec{r}_{ij}, \vec{l}_i \rangle \tag{24.13}$$

Using the MIMO technique, the received signal is given by

$$\mathbf{y} = \mathbf{H}^{\mathrm{T}}\mathbf{x} + \mathbf{n}, \tag{24.14}$$

where $\mathbf{y} = (y_1, y_2, \ldots y_K)^{\mathrm{T}}$, y_k represents the signal received by user k, $\mathbf{x} = (x_1, x_2, \ldots x_Q)^{\mathrm{T}}$, x_k is the transmitted signal from transmitter k, $\mathbf{n} = (n_1, n_2, \ldots n_K)^{\mathrm{T}}$, and n_k is the additive noise at receiver k. The noise we model here is the thermal noise and background noise (shot noise is assumed to be small).

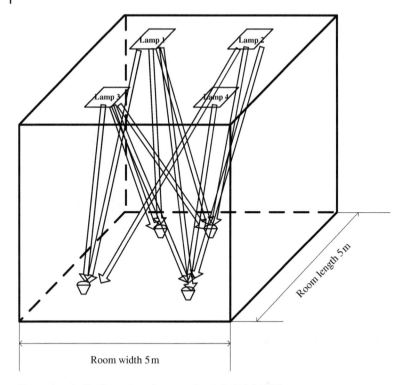

Figure 24.10 Configuration diagram of a typical VLC MISO system.

24.4.2.1 Modulation Schemes in VLC Systems

The conventional modulation schemes adopted in RF communications cannot be readily applied in VLC directly, because the visible light is incoherent. IM–DD are used, so nonnegative signals are transmitted.

24.4.3 On–Off Keying (OOK)

OOK is the simplest technique that can be used in VLC systems. In OOK, the intensity of an optical source is directly modulated by the information sequences, which is usually binary. For a sequence, a bit "one" can be represented by an optical pulse, we call it "on." For the "on," the entire bit duration is occupied. On the contrary, a bit "zero," we call it "off," can be represented as a blank duration.

For OOK, both the non-return-to-zero (NRZ) and return-to-zero (RZ) schemes can be applied. In the NRZ, the whole bit duration is occupied by the pulse when transmitting "1." But in the RZ scheme, only partial duration of bit can be occupied.

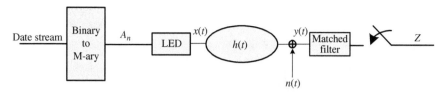

Figure 24.11 Block diagram of *M*-PAM schemes.

24.4.3.1 *M*-ary Pulse Amplitude Modulation (*M*-PAM)

M-ary Pulse Amplitude Modulation (M-PAM) can offer higher bandwidth efficiency than OOK, since more bits can be transmitted using one pulse in M-PAM. In M-PAM, a pulse is sent in each symbol duration, where the pulse amplitude takes on one of the M possible levels. The bandwidth efficiency is defined as

$$\eta = \frac{R_b}{B} = \log_2 M, \tag{24.15}$$

where R_b is bit rate and B represents the bandwidth. The block diagram of *M*-PAM scheme is shown in Figure 24.11. In this block diagram, $n(t)$ represents the noise, and $y(t)$ is the received signal at the receiver. The channel impulse response $h(t)$ is described in Section 24.4.1. After the matched filter and sampling, we can get the signal ready for demodulation and decision.

An example for 4-PAM modulation is given in Figure 24.12 to help us understand the principle of the *M*-PAM scheme. The data stream ready for transmission is "000111100111" in Figure 24.12. For 4-PAM, we divide the stream into groups containing 2 bits. Here, the stream can be divided into "00", "01", "11", "10", "01", and "11". Converting these binary numbers into 4-ary ones, we get 0, 1, 3, 2, 1, and 3 for each group. For 4-PAM, the transmitters just need to send the corresponding power levels.

Figure 24.12 4-PAM modulation.

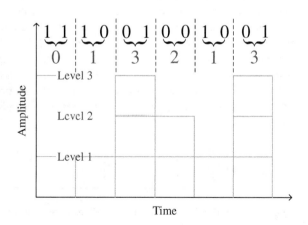

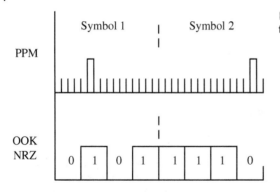

Figure 24.13 Time waveforms for PPM and OOK.

24.4.3.2 Pulse Position Modulation (PPM)

PPM is an orthogonal modulation technique that has a higher power efficiency than OOK. However, to achieve a same bit rate with OOK, more bandwidth is required when using PPM. Thus, for the band-limited indoor channel, the PPM scheme may introduce some ISI. The comparison of waveforms between OOK and PPM is shown in Figure 24.13.

24.4.4 Multiuser VLC Systems

In this section, some conventional multiple access schemes for VLC systems are introduced. Studying from RF communication systems, a cellular structure based on MIMO technique is introduced.

24.4.4.1 Multiple Access Schemes

Time Division Multiple Access (TDMA) Time division multiple access (TDMA) is a conventional multiple access scheme, which is used in the digital 2G cellular system such as global system for mobile (GSM) communications in RF. The users in TDMA transmit signals in rapid succession, one after the other, each using its own time slot as shown in Figure 24.14. For VLC systems, the TDMA can be used directly. The data can be modulated using OOK, PAM, or PPM. The advantages for TDMA are very obvious. The structures of transmitter and receiver can be designed simply. There is almost no multiple access interference (MAI) if synchronization between different users can be guaranteed.

Figure 24.14 A TDMA stream divided into different time slots for different users.

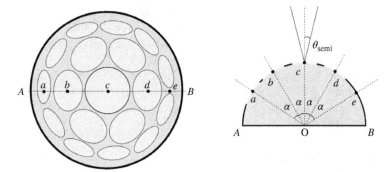

Figure 24.15 The shape of an 18-element angle diversity transmitter. *Source*: Chen and Haas (2015) [34]. Reproduced with permission of IEEE.

Space Division Multiple Access (SDMA) Space division multiple access (SDMA) is a multiple access scheme that uses the space information of users to separate them. In RF communications, this scheme is usually used in satellite communication systems. In SDMA, an antenna array is used as the transmitter. From the antenna array, multiple narrow beams are generated according to the users' positions. In this way, the multiple users using SDMA can share the same time slot. In VLC system, SDMA technique can also be applied [34]. In this paper, LED arrays with narrow beamwidth were used as transmitters as shown in Figure 24.15. By turning on different transmitter elements, the angle diversity transmitter can generate narrow light beams of different directions.

Optical Code Division Multiple Access (OCDMA) In our indoor VLC system, a code division multiple access (CDMA) technique is employed to provide multiple access for simultaneous users. Because of the intensity modulation used in VLC, the CDMA code is different in optical communication than in RF communication. To implement the CDMA technique in optical communications, choosing the proper optical CDMA (OCDMA) code is a significant step.

Nowadays, OCDMA is receiving increasing attention due to its enhanced information security. In an OCDMA system, different users share a common communication medium; thus multiple access is achieved by assigning OCDMA codes to different users [35]. Therefore, the transmitted signal for user i can be described as $d_i\mathbf{c}_i$, where d_i is the intended data for user i and $\mathbf{c}_i = \{c_i[1], \ldots, c_i[L]\}$ represents the length-L OCDMA code for user i.

An important type of OCDMA code is the optical orthogonal code (OOC) that was proposed for IM–DD OCDMA systems [36]. OCDMA codes must satisfy the following conditions [37]:

1) The peak autocorrelation function of the code should be maximized.
2) The cross-correlation between any codes should be minimized.
3) The side lobes of the autocorrelation function of the code should be minimized.

Conditions (1) and (2) ensure that the MAI is minimized, and condition (3) ensures the synchronization process at the receiver. In this chapter, since we consider only the downlink, the codes for all users are transmitted synchronously.

An OOC is usually represented by $(L, w, \lambda_a, \lambda_c)$, where L is the code length, w is the code weight, λ_a is the upper bound on the autocorrelation value for a nonzero shift, and λ_c is an upper bound on cross-correlation values. The conditions for OOC are [35]

$$R_{c_i c_j}[m] = \sum_{l=1}^{L} c_i[l]c_j[l+m] \le \lambda_c \quad \forall m \tag{24.16}$$

where c_i is the ith code word and

$$R_{c_i c_i}[m] = \sum_{l=1}^{L} c_i[l]c_i[l+m] \le \lambda_a \quad \forall m \ne 0 \tag{24.17}$$

There is a special case of $\lambda_a = \lambda_c = \lambda$ that the OOC is represented by (L, w, λ) [35]. Table 24.2 gives us examples for code word indexes (the positions of the "ones" in the code word). For example, the index $\{1, 2, 4\}$ with length 7 represents the code word in Figure 24.16.

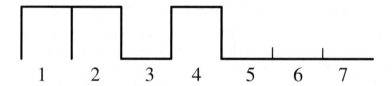

Figure 24.16 OOC $\{1, 2, 4\}$ with length of 7.

Table 24.2 OOC $(L, 3, 1)$ sequence indexes for various length.

L	Sequence index, when $L \le 49$, $\lambda_a = \lambda_c = 1$
7	$\{1, 2, 4\}$
13	$\{1, 2, 5\}\{1, 3, 8\}$
19	$\{1, 2, 6\}\{1, 3, 9\}\{1, 4, 11\}$
25	$\{1, 2, 7\}\{1, 3, 10\}\{1, 4, 12\}\{1, 5, 14\}$
31	$\{1, 2, 8\}\{1, 3, 12\}\{1, 4, 16\}\{1, 5, 15\}\{1, 6, 14\}$
37	$\{1, 2, 12\}\{1, 3, 10\}\{1, 4, 18\}\{1, 5, 13\}\{1, 6, 19\}\{1, 7, 13\}$
43	$\{1, 2, 20\}\{1, 3, 23\}\{1, 4, 16\}\{1, 5, 14\}\{1, 6, 17\}\{1, 7, 15\}\{1, 8, 19\}$

Source: Ghafouri-Shiraz and Karbassian (2012) [37].

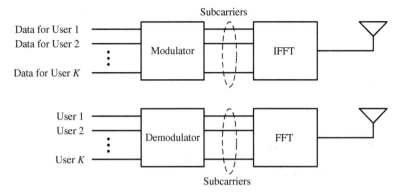

Figure 24.17 Block diagram of OFDMA.

Orthogonal Frequency-Division Multiple Access (OFDMA) Orthogonal frequency-division multiple access (OFDMA) is a novel multiple access scheme based on OFDM technique, which is used in 4G LTE. Multiple access in OFDMA is achieved by assigning subsets of subcarriers to individual users, which is a multiuser version of OFDM. The basic principle of OFDMA is shown in Figure 24.17.

Conventional OFDM generates complex-valued bipolar signals, which need to be modified in order to become suitable for VLC. This operation effectively maps the information symbols onto subcarriers in different frequency bands. A real OFDM signal can be obtained but reduces the system bandwidth by half. This approach has been widely accepted in the literature for the generation of a real OFDM signal. The resulting waveform, however, is still bipolar in nature (it has a positive and negative part). A number of approaches have been proposed for the creation of a unipolar signal. For example, a DC bias can be added to the original bipolar signal. This scheme is known as DC-biased optical OFDM (DCO-OFDM) [38]. Another approach is asymmetrically clipped optical OFDM (ACO-OFDM) [39]. In this scheme, only the odd-indexed subcarriers in the OFDM frame are modulated with information.

24.4.4.2 Cellular Structure for VLC Systems

In VLC systems, a tiny cellular network, attocell network, can provide wireless access for users in an indoor environment. Different from the hexagonal cell in RF communication network, we propose a circular cell as shown in Figure 24.18. The radius of the cell is artificially defined for users in the indoor environment, such that only if the user is located in the access area can it be served by this lamp. To make sure that all area in an indoor environment can be covered, there may be some overlap area between cells as shown in Figure 24.19. If there are some users located in the overlapped area, they can be served by both lamps.

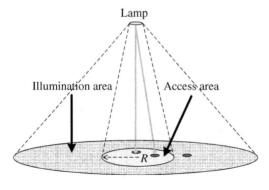

Figure 24.18 Circular cell for indoor VLC systems.

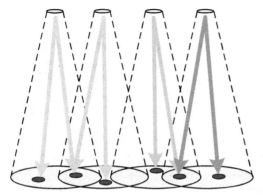

Figure 24.19 Multicell configuration in VLC systems.

Centralized Power Allocation Algorithm for Multicell Indoor VLC Systems A centralized power allocation joint optimization (PAJO) algorithm is introduced in this section [25]. According to this algorithm, we maximize the minimum SINR of all the users and design an MMSE receiver for each user. A block diagram of the PAJO algorithm is shown in Figure 24.20. For PAJO, the transmitted signal from each LED depends on d_1 to d_K as shown in Figure 24.20. d_K is the intended data for user K. The power allocation for each LED is different, which depends on the channel feedback from the users. Since the location of each user is different, the channels of the different users are different. Therefore, we need to allocate the transmitted power to compensate the channel loss for different users.

Distributed Power Allocation Algorithm for Multicell Indoor VLC Systems A distributed power allocation scheme, weighted distributed power allocation joint optimization (WD-PAJO) method is introduced in this section [40].

For WD-PAJO, the indoor area can be classified into a single-covered area, double-covered area, triple-covered area, and so on, as shown in Figure 24.21a. In this algorithm each LED lamp works independently and just serves the users that are located in its access area. Thus, there is only one lamp per optimization

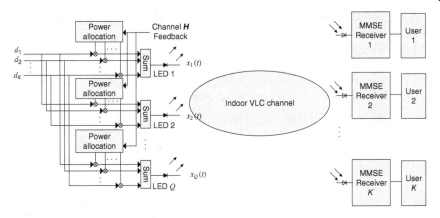

Figure 24.20 Block diagram of the proposed PAJO algorithm to support *K* users simultaneously.

A: Single covered area
B: Double covered area
C: Triple covered area
R: Radius of the access area

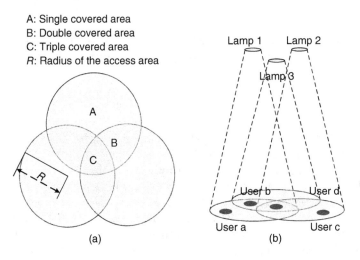

Figure 24.21 WD-PAJO principles: (a) WD-PAJO geometry structure with multiple users and (b) covered area classification.

thread. To eliminate the MAI, each lamp does power allocation optimization by maximizing the minimum weighted SINR for all the users in its own access area. The SINR is weighted by the number of lamps transmitting to the user so that users served by many lamps are not unduly advantaged. For lamp *i*, the algorithm computes

$$\mathbf{P}_i = \arg\max_{\mathbf{P}} \min_{k \in A_i} \gamma_k \cdot \text{SINR}_k, \tag{24.18}$$

where $\mathbf{P}_i$ is the power allocation matrix for the *i*th lamp, A_i is the access area of lamp *i*, and γ_k is the number of lamps serving user *k*.

Figure 24.21b shows an example of how WD-PAJO works. In this example, lamp 1 serves users a, b, and c. Lamps 2 and 3 serve users b and c and users c and d, respectively. Unlike the PD-PAJO, all the three lamps must work independently. Since the three lamps do not communicate with each other, they can do the power allocation optimization in parallel. For this example the objective functions for the three lamps can be represented as

$$\text{Lamp1} : \mathbf{P}_1 = \arg\max_{\mathbf{P}} \min\{\text{SINR}_a, 2\ \text{SINR}_b, 3\ \text{SINR}_c\}$$

$$\text{Lamp2} : \mathbf{P}_2 = \arg\max_{\mathbf{P}} \min\{2\ \text{SINR}_b, 3\ \text{SINR}_c\} \qquad (24.19)$$

$$\text{Lamp3} : \mathbf{P}_3 = \arg\max_{\mathbf{P}} \min\{3\ \text{SINR}_c, \text{SINR}_d\}$$

In general, the WD-PAJO algorithm can be accomplished via the same steps as the PD-PAJO. The only difference is that for WD-PAJO there is only one lamp in each thread, so no communication between lamps is needed.

24.4.5 Practical Considerations

In this section, some practical considerations such as ISI, nonlinearity of LEDs, illumination requirements, and dimming control in indoor VLC systems are analyzed and discussed.

24.4.5.1 Intersymbol Interference (ISI)

In a practical indoor VLC system, because of reflections of walls and furniture, the indoor channel is dispersive. In addition, the bandwidth of LEDs is bandlimited (usually 3 dB bandwidth is 100 MHz); therefore, when the transmitting symbol rate is higher than the overall bandwidth of the channel (LED and indoor channel), the signal must be distorted and the ISI is introduced.

There are two approaches to diminish the ISI. First approach is to reduce the symbol rate. To make sure a relatively high bit rate is achieved, OFDM technique and multilevel modulation schemes such as M-PAM can be used. However, either OFDM or multilevel modulation scheme may introduce nonlinear distortion because of nonlinearity of LEDs. Therefore, the second approach, equalization, becomes more and more popular. Some researchers have proposed methods to design an equalization circuit in [41, 42] to improve the transmitted data rate, which can achieve up to 340 Mb/s using OOK modulation scheme with BER at 2×10^{-3} [43]. There are also some related works to VLC system equalization concentrating on the ISI elimination, which is caused by the dispersive indoor channel. The zero forcing algorithm is generally applied to mitigate the effects of ISI in infrared wireless systems in [44]. Artificial neural network (ANN)-based equalizer is a feasible way to build an equalizer at the users' port [45]. In this paper, with the help of the ANN equalizer, the bit rate can achieve 170 Mb/s using OOK. Some researchers proposed frequency domain equalization algorithms in VLC, which can reduce

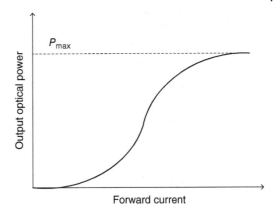

Figure 24.22 Nonlinear transfer function of a LED.

the ISI and improve the data rate [46] by designing an equalizer in frequency domain. Linear equalizer (LE) and decision feedback equalizer (DFE) can also be applied in VLC systems. For LE and DFE, training sequences are needed. The coefficients of the LE and DFE can be trained. From the results in [47], both the LE and DFE can reduce the ISI.

24.4.5.2 Nonlinearity of LEDs

LEDs are commonly used light sources for lighting systems, and they also serve as transmitters in indoor VLC systems. The optical power from LEDs is driven by an input electrical signal that carries information. Due to the structure of LEDs and the principles of generating light, the relation between the output optical power and the input current can be modeled as a nonlinear function as shown in Figure 24.22. This nonlinearity of LEDs introduces a nonlinear distortion on transmitted optical signals.

24.4.5.3 Illumination Requirements and Dimming Control

Considering the total radiation power limit and the illumination requirements of the users, the constraints can be represented as

$$
\begin{cases}
\sum_{k=1}^{K} P_{qk} \leq P_o \\
P_{qk} \geq 0 \\
|\sum_{q=1}^{Q} \sum_{j=1}^{K} h_{qk} P_{qj} \eta - P_{rk}| \leq \Delta \\
k = 1, 2, \ldots, K
\end{cases}
\tag{24.20}
$$

where P_{rk} is the required received power for illumination of user k and Δ is the illumination tolerance. This constraint is to make sure that the illumination level at these users will not be too dark or too bright. η represents the CDMA code weight to CDMA code length ratio, which decides the illumination level.

To illuminate the room uniformly in space, we define virtual users that must have a fixed illumination but no data transmission. These E users are distributed

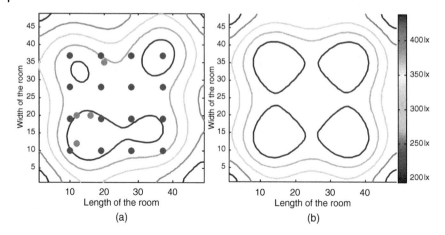

Figure 24.23 Illumination distribution comparison of (a) data transmission case and (b) no data transmission case. The small dots represent the 4 virtual users and 16 real users, with 10% tolerance, and the 25-LED model. .From Jie Lian

uniformly in the room. For these virtual users that require just illumination, we use the constraint

$$| \sum_{q=1}^{Q} \sum_{j=1}^{E} h_{qe} P_{qj} \eta - P_{re} | \leq \Delta \qquad e = 1, 2, \dots E \qquad (24.21)$$

where h_{qe} is the channel between the qth transmitter to the eth virtual user, P_{re} is the required received power illumination for user e, and E is the number of virtual users that only need illumination.

Figure 24.23a shows a contour plot of the illumination distribution for 4 users with both data transmission and illumination requirements, plus 16 virtual users with illumination requirements only. Figure 24.23b shows the illumination distribution without data transmission requirements. Comparing these two figures, the illumination distribution in (a) is still smooth and flat. That is to say, setting illumination constraints prevents the lighting system from creating too dark and too bright spots in the room, and the illumination requirements at all the user locations are satisfied.

24.5 Conclusion and Outlook

This chapter introduces the smart lighting system and its applications. The smart lighting system is an energy-efficient system that can be widely used in indoor and outdoor public illumination. Using LEDs as light sources and data transmitters, VLC system can support high rate wireless connections. In this chapter, we analyze the brief principles of VLC system. For multiuser VLC system, we propose centralized and distributed power allocation algorithms using MIMO technique. From the theoretical analysis and numerical results, we conclude that the smart lighting system can support high rate wireless access and specific illumination requirements with low energy consumption and high power efficiency.

Although some practical problems need to be considered, smart light VLC can become a practical augmentation technology with high security. Increasing interest from academia and industry shows that smart light VLC system can be successfully commercialized in the coming years [48].

Final Thoughts

In this chapter, we discussed the concept of smart lighting, its development, and its applications. The smart lighting technique is one of the most significant branches in smart cities. Smart lighting can be applied in outdoor public lighting systems, vehicle safety lighting, positioning systems, and smart lighting communication systems. In this chapter, we emphatically introduced indoor smart lighting communication principles, power allocation algorithms, and practice considerations. After analysis, we can define smart lighting system as a combination system, which includes illumination, safety alarm, positioning, and communication. From the experiment results, the smart lighting system is an environmental friendly system that can save energy.

Questions

1   What is the definition of smart lighting?

2   Name and describe four common applications for smart lighting technique.

3   Name four multiple access modulation schemes and describe the advantages for each scheme.

4   What are the limitations of indoor smart lighting communication systems?

References

1 Castro, M., Jara, A.J., and Skarmeta, A.F.G. (2013) *Smart Lighting Solutions for Smart Cities.* 2013 27th International Conference on Advanced Information Networking and Applications Workshops (WAINA), pp. 1374–1379.
2 IEA (2006) *Lights Labours Lost.* OECD/IEA.
3 Murthy, A., Han, D., Jiang, D., and Oliveira, T. (2015) *Lighting-Enabled Smart City Applications and Ecosystems Based on the IoT.* 2015 IEEE 2nd World Forum on Internet of Things (WF-IoT), pp. 757–763.

4 T.M.L. Partnership (2014) *Energy Efficiency and Technologies in America's Cities, A 288-City Survey*, http://usmayors.org/pressreleases/uploads/2014/0122-report-energyefficiency.pdf (accessed 21 December 2016).

5 Baniya, R., Maksimainen, M., Sierla, S., Pang, C., Yang, C.-W., and Vyatkin, V. (2014) *Smart Indoor Lighting Control: Power, Illuminance, and Colour Quality*. 2014 IEEE 23rd International Symposium on Industrial Electronics (ISIE), pp. 1745–1750.

6 Prez-Lombard, L., Ortiz, J., and Pout, C. (2008) A review on buildings energy consumption information. *Energy and Buildings*, **40**, 394–398.

7 Karlicek, R.F. (2012) *Smart Lighting - More Than Illumination*. Communications and Photonics Conference (ACP), 2012 Asia, pp. 1–2.

8 Zhao, P., Jiao, X., and Zhao, H. (2012) *Analysis of Light Extraction Efficiency Enhancement for Ingan Quantum Wells Light-Emitting Diodes with Microspheres*. Energytech, 2012 IEEE, pp. 1–6.

9 Tchernycheva, M., Dai, X., Messanvi, A., Zhang, H., Neplokh, V., Lavenus, P., Guan, N., Julien, F.H., Rigutti, L., Babichev, A., Eymery, J., and Durand, C. (2015) *Nitride Nanowire Light Emitting Diodes*. 2015 17th International Conference on Transparent Optical Networks (ICTON), pp. 1–4.

10 Huang, J.K., Liu, C.Y., Chen, T.P., Huang, H.W., Lai, F.I., Lee, P.T., Lin, C.H., Chang, C.Y., Kao, T.S., and Kuo, H.C. (2015) Enhanced light extraction efficiency of GaN-based hybrid nanorods light-emitting diodes. *IEEE Journal of Selected Topics in Quantum Electronics*, **21** (4), 354–360.

11 Afshari, S. and Mishra, S. (2015) *An Optimization Framework for Control of Non-Square Smart Lighting Systems with Saturation Constraints*. 2015 American Control Conference (ACC), pp. 1665–1670.

12 Magno, M., Polonelli, T., Benini, L., and Popovici, E. (2015) A low cost, highly scalable wireless sensor network solution to achieve smart led light control for green buildings. *IEEE Sensors Journal*, **15** (5), 2963–2973.

13 Zotos, N., Pallis, E., Stergiopoulos, C., Anastasopoulos, K., Zotos, N., Bogdos, G., and Skianis, C. (2012) *Case Study of a Dimmable Outdoor Lighting System with Intelligent Management and Remote Control*. Telecommunications and Multimedia (TEMU), 2012 International Conference on, pp. 43–48.

14 Mahoor, M., Najafaabadi, T.A., and Salmasi, F.R. (2014) *A Smart Street Lighting Control System for Optimization of Energy Consumption and Lamp Life*. 2014 22nd Iranian Conference on Electrical Engineering (ICEE), pp. 1290–1294.

15 Zhang, Z., Mistry, A., Yin, W., and Venetianer, P.L. (2012) *Embedded Smart Sensor for Outdoor Parking Lot Lighting Control*. 2012 IEEE Computer Society Conference on Computer Vision and Pattern Recognition Workshops, pp. 54–59.

16 Yu, S.H., Shih, O., Tsai, H.M., Wisitpongphan, N., and Roberts, R.D. (2013) Smart automotive lighting for vehicle safety. *IEEE Communications Magazine*, **51** (12), 50–59.

17 Cailean, A.M., Cagneau, B., Chassagne, L., Dimian, M., and Popa, V. (2015) Novel receiver sensor for visible light communications in automotive applications. *IEEE Sensors Journal*, **15** (8), 4632–4639.

18 Do, T.H. and Yoo, M. (2015) *Potentialities and Challenges of VLC Based Outdoor Positioning*. 2015 International Conference on Information Networking (ICOIN), pp. 474–477.

19 Nakazawa, Y., Makino, H., Nishimori, K., Wakatsuki, D., and Komagata, H. (2014) *Led-Tracking and ID-Estimation for Indoor Positioning Using Visible Light Communication*. Indoor Positioning and Indoor Navigation (IPIN), 2014 International Conference on, pp. 87–94.

20 Xu, W., Wang, J., Shen, H., Zhang, H., and You, X. (2016) Indoor positioning for multiphotodiode device using visible-light communications. *IEEE Photonics Journal*, **8** (1), 1–11.

21 Noshad, M. and Brandt-Pearce, M. (2014) Application of expurgated PPM to indoor visible light communications Part I: single-user systems. *Journal of Lightwave Technology*, **32** (5), 875–882.

22 Komine, T. and Nakagawa, M. (2004) Adaptive detector arrays for optical communication receivers. *IEEE Transactions on Consumer Electronics*, **50** (1), 100–107.

23 Mehmood, R.M.R., Elgala, H., and Haas, H. (2010) *Indoor MIMO Optical Wireless Communication Using Spatial Modulation*. 2010 IEEE International Conference on Communications (ICC), pp. 1–5.

24 Ding, J., Wang, K., and Xu, Z. (2014) *Wireless Infrared Communications*. 2014 International Symposium on Communication Systems, Networks Digital Signal Processing, pp. 1159–1164.

25 Lian, J., Noshad, M., and Brandt-Pearce, M. (2014) *Multiuser MISO Indoor Visible Light Communications*. 2014 Asilomar Conference on Signals, Systems and Computers, (Invited paper), pp. 1729–1733.

26 Luo, P., Ghassemlooy, Z., Minh, H.L., Bentley, E., Burton, A., and Tang, X. (2014) *Fundamental Analysis of a Car to Car Visible Light Communication System*. 2014 Communication Systems, Networks Digital Signal Processing (CSNDSP), pp. 1011–1016.

27 prweb *Visible Light Communication (VLC) Market & Li-Fi Technology Market worth $6,138.02 Million by 2018 - New Report by MarketsandMarkets*, http://www.prweb.com/releases/visible-light/communication-market/prweb11241854.htm (accessed 21 December 2016).

28 Chen, Z., Tsonev, D., and Haas, H. (2014) *Improving SINR in Indoor Cellular Visible Light Communication Networks*. 2014 IEEE International Conference on Communication, pp. 3383–3388.

29 Lian, J. and Brandt-Pearce, M. (2015) *Multiuser Multidetector Indoor Visible Light Communication System*. Opto-Electronics and Communications Conference, pp. 1–3.

30 Ghassemlooy, Z. and Popoola, R.W. (2013) *Optical Wireless Communications: System and Channel Modeling with Matlab*, CRC Press, Boca Raton, FL.

31 Kahn, J. and Barry, J. (1997) Wireless infrared communications. *Proceedings of the IEEE*, **85** (2), 265–298.

32 Lee, K. and Park, H. (2011) *Channel Model and Modulation Schemes for Visible Light Communications*. 2011 IEEE 54th International Midwest Symposium on Circuits and Systems (MWSCAS), pp. 1–4.

33 Zeng, L., O'Brien, D., Minh, H., and Faulkner, G. (2009) High data rate multiple input single output (MISO) optical wireless communications using white LED lighting. *IEEE Journal on Selected Areas in Communications*, **27** (9), 1654–1662.

34 Chen, Z. and Haas, H. (2015) *Space Division Multiple Access in Visible Light Communications*. 2015 IEEE International Conference on Communication (ICC), pp. 5115–5119.

35 Zheng, Z., Chen, T., Liu, L., and Hu, W. (2015) *Experimental Demonstration of Femtocell Visible Light Communication System Employing Code Division Multiple Access*. Optical Fiber Communication Conference (OFC), pp. 1–3.

36 Noshad, M. and Brandt-Pearce, M. (2014) Application of expurgated PPM to indoor visible light communications Part II: access networks. *Journal of Lightwave Technology*, **32** (5), 883–890.

37 Ghafouri-Shiraz, H. and Karbassian, M.M. (2012) *Optical CDMA Network Principle, Analysis and Applications*, Wiley Press.

38 Dissanayake, S.D. and Armstrong, J. (2013) Comparison of ACO-OFDM, DCO-OFDM and ADO-OFDM in IM/DD systems. *Journal of Lightwave Technology*, **31** (7), 1063–1072.

39 Dissanayake, S.D. and Armstrong, J. (2006) Power efficient optical OFDM. *Electronics Letters*, **42** (6), 370–372.

40 Lian, J. and Brandt-Pearce, M. (2015) *Distributed Power Allocation for Indoor Visible Light Communications*. 2015 IEEE Global Communication Conference (GLOBECOM), http://www.ece.virginia.edu/~optcom/ Publications/GLOBECOM2015.pdf (accessed 21 December 2016).

41 Huang, X., Shi, J., Li, J., Wang, Y., and Chi, N. (2015) A Gb/s VLC transmission using hardware preequalization circuit. *IEEE Photonics Technology Letters*, **27** (18), 1915–1918.

42 Wang, Y., Tao, L., Wang, Y., and Chi, N. (2014) High speed WDM VLC system based on multi-band CAP64 with weighted pre-equalization and modified CMMA based post-equalization. *IEEE Communications Letters*, **18** (10), 1719–1722.

43 Li, H., Chen, X., Huang, B., Tang, D., and Chen, H. (2014) High bandwidth visible light communications based on a post-equalization circuit. *IEEE Photonics Technology Letters*, **26** (2), 119–122.

44 Lee, D. and Kahn, J. (1998) *Coding and Equalization for PPM on Wireless Infrared Channels*. 1998 IEEE Global Communications Conference (GLOBECOM), pp. 201–206.

45 Haigh, P., Ghassemlooy, Z., Rajbhandari, S., Papakonstantinou, I., and Popoola, W. (2014) Visible light communications: 170 Mb/s using an artificial neural network equalizer in a low bandwidth white light configuration. *Journal of Lightwave Technology*, **32** (9), 1807–1813.

46 Grobe, L. and Langer, K.-D. (2013) *Block-Based PAM with Frequency Domain Equalization in Visible Light Communications*. IEEE Globecom Workshops (GC Wkshps), pp. 1070–1075.

47 Komine, T., Lee, J., Haruyama, S., and Nakagawa, M. (2009) Adaptive equalization system for visible light wireless communication utilizing multiple white LED lighting equipment. *IEEE Transactions on Wireless Communications*, **8** (6), 2892–2900.

48 Pathak, P.H., Feng, X., Hu, P., and Mohapatra, P. (2015) Visible light communication, networking, and sensing: a survey, potential and challenges. *IEEE Communications Surveys Tutorials*, **17** (4), 2047–2077.

25

Large Scale Air-Quality Monitoring in Smart and Sustainable Cities

Xiaofan Jiang

Department of Electrical Engineering, Columbia University, New York, NY, USA

CHAPTER MENU

Introduction, 726
Current Approaches to Air Quality Monitoring and Their Limitations, 729
Overview of a Cloud-based Air Quality Monitoring System, 731
Cloud-Connected Air Quality Monitors, 733
Cloud-Side System Design and Considerations, 736
Data Analytics in the Cloud, 739
Applications and APIs, 748

Objectives

- To become familiar with the concept of air quality, metrics for evaluating air quality, and adverse health effects of high levels of particulate matter in the context of urban cities.
- To become familiar with the current status of air quality monitoring technologies and their shortcomings.
- To become familiar with the process of designing an end-to-end system for air quality monitoring, incorporating sensors, embedded systems, wireless communication, cloud, and data analytics.
- To become familiar with large-scale physical data analytics through a combination of signal processing and machine learning techniques.
- To become familiar with the development of open platforms for supporting web services and mobile applications.

Smart Cities: Foundations, Principles and Applications, First Edition.
Edited by Houbing Song, Ravi Srinivasan, Tamim Sookoor, and Sabina Jeschke.
© 2017 John Wiley & Sons, Inc. Published 2017 by John Wiley & Sons, Inc.

25.1   Introduction

A clean and sustainable environment is an integral part of future smart cities. In particular, air is one of the most important shared resources on our planet. Unfortunately, the quality of air has deteriorated significantly over the past years for many cities.

According to a 2015 report by BusinessInsider [1], Delhi, India, has the worst air quality in the world, with average particulate matter levels that qualified as "very unhealthy" by the World Health Organization (WHO). Beijing, China, has also made headlines in recent years for encountering some of the worst air quality measurements recorded, particularly during winters (Figure 25.1). A recent study conducted by Peking University and Greenpeace shows that "Over a quarter of a million people in some of China's major cities could have their lives cut short because of high levels of air pollution" [2]. On the other side of the world, air quality has also become a serious concern for metropolitan cities in the United States. In a 2015 "State of the Air" study conducted by the American Lung Association [3], major cities including Los Angeles, Pittsburgh, New York, and Salt Lake City are among the worst in the United States; the same study finds that "nearly 138.5 million people – almost 44 percent of the nation – live where pollution levels are too often dangerous to breathe."

Among the various dimensions of air quality, particulate matter with diameters less than 2.5 μm (PM2.5) has gained a lot of attention recently because of its significant impact on our respiratory system. Medical studies have shown that PM2.5 can be easily absorbed by the lungs, and high concentrations of PM2.5 can lead to respiratory diseases [4] or even blood diseases [5]. We focus on PM2.5 due to the recent widespread attention. However, the same systems, models, and techniques introduced in this chapter can be easily applied to other dimensions of air pollution, such as ozone, nitrogen dioxide, and sulfur dioxide.

(a)   (b)

Figure 25.1 Air quality in Delhi, India, is ranked the worst in the world by BusinessInsider using information from WHO (a). Beijing, China, has also gained worldwide attention recently for its dangerous levels of PM2.5 (b).

AQI category	Index values	Previous breakpoints (1999 AQI) ($\mu g/m^3$, 24-h average)	Revised breakpoints ($\mu g/m^3$, 24-h average)
Good	0–50	0.0–15.0	0.0–12.0
Moderate	51–100	>15.0–40	12.1–35.4
Unhealthy for sensitive groups	101–150	>40–65	35.5–55.4
Unhealthy	151–200	>65–150	55.5–150.4
Very unhealthy	201–300	>150–250	150.5–250.4
Hazardous	301–400	>250–350	250.5–350.4
	401–500	>350–500	350.5–500

Figure 25.2 Air Quality Index published by EPA for PM2.5 [6]. *Source:* Agency, E.P., The national ambient air quality standards for particle pollution. URL http://www3.epa.gov/airquality/particlepollution/2012/decfsstandards.pdf.

PM2.5 is typically measured in terms of its concentration in units of $\mu g/m^3$. For example, a PM2.5 of $10\,\mu g/m^3$ means the total weight of particles with diameter less than 2.5 μm is $10\,\mu g/m^3$. To help people better understand the health impact of different concentrations of PM2.5, the US Environmental Protection Agency (EPA) has published an Air Quality Index (AQI) for PM2.5 [6], shown in Figure 25.2. In this chart, PM2.5 concentrations above 100 are considered unhealthy. In comparison, the annual average PM2.5 level in Delhi was $153\,\mu g/m^3$, and Beijing hit $755\,\mu g/m^3$ on a particularly bad day during the winter of 2013 [7].

Based on evidences of airborne particulate matter and its "adverse health effects at exposures that are currently experienced by urban populations in both developed and developing countries," the WHO published a set of guidelines for PM2.5 [8], as shown in Figure 25.3. For annual mean concentrations, the guideline is $10\,\mu g/m^3$, while 24-h concentrations is set at $25\,\mu g/m^3$. It is clear that many cities around the world still have a long way to go to hit these targets.

Meanwhile, people are looking for better ways to monitor the quality of air in their immediate environment in order to take appropriate actions such as wearing masks or staying at home. While there are many smartphone applications that report publicly available air quality data at the city or district level, they cannot tell the actual air quality people breathe in, which is much more relevant and valuable. This is particularly important since we spend most of our time inside enclosed spaces such as homes and offices, where the air quality may deviate significantly from the outside.

Conventional methods for monitoring air quality are either large and expensive or inaccurate. In recent years, we start to see inexpensive monitors based on dust sensors. But they typically rely on static calibration curve, resulting in

	PM10 (µg/m³)	PM2.5 (µg/m³)	Basis for the selected level
Interim target-1 (IT-1)	150	75	Based on published risk coefficients from multicenter studies and meta-analyses (about 5% increase of short-term mortality over the AQG value)
Interim target-2 (IT-2)	100	50	Based on published risk coefficients from multicenter studies and meta-analyses (about 2.5% increase of short-term mortality over the AQG value)
Interim target-3 (IT-3)*	75	37.5	Based on published risk coefficients from multicenter studies and meta-analyses (about 1.2% increase in short-term mortality over the AQG value)
Air quality guideline (AQG)	**50**	**25**	Based on the relationship between 24-h and annual PM levels

Figure 25.3 WHO air quality guidelines and interim targets for particulate matter: 24-h concentrations. *Source:* WHO (2006) [8]. Air Quality Guidelines: Global Update 2005. Particulate Matter, Ozone, Nitrogen Dioxide and Sulfur Dioxide, World Health Organization.

large errors. Because of the intrinsic accuracy–cost trade-off of PM sensors, it is challenging to be both accurate and affordable.

In this chapter, we present a novel cloud-based approach to this problem. This approach can be easily extended to other types of environmental monitoring in smart cities. Instead of using accurate but expensive stand-alone sensors, we create a client–cloud system consisting of custom-designed sensor front ends and a real-time data analytics engine in the cloud. In addition to using data from custom sensors, we collect a fusion of data, such as meteorology data and location data, which are used by calibration and inference algorithms in the analytics engine.

In *Part I* of this chapter, we guide readers through the design of two sensor front ends – a stationary air quality sensor that connects to the cloud via Ethernet and GPRS and a portable sensor that connects to the smartphone via Bluetooth 4.0.

In *Part II*, we go into details on the systems design and data analytics in the cloud. We introduce a combination of data analysis and machine learning techniques for signal conditioning, sensor calibration, and inference. An artificial neural network (ANN) is trained and used for online sensor calibration; Gaussian process is used to further improve the accuracy of sensors and infer the value at locations where sensors are not available. By building a model of particulate matter using "big sensor data," crowd-sourcing data from sensing frond ends and by offloading analytics to the cloud, this approach is able to achieve good accuracies at much lower cost than any previous solutions.

In *Part III*, we describe the design of APIs, interfaces, and web services for the third-party developers to create applications on top of this systems. In addition, we present a couple of applications created on top of our platform, including a smartphone app that relays raw dust sensor readings, plus the smartphone's own sensor readings such as GPS and IMU, to cloud via 3G or Wi-Fi.

25.2   Current Approaches to Air Quality Monitoring and Their Limitations

There are several approaches to measuring PM2.5. Some existing products measure PM2.5 directly using various types of sensing technologies, while some other approaches indirectly estimate PM2.5 using computational methods, each with their advantages and disadvantages.

Satellite remote sensing of surface air quality has been studied intensively in past decades [9]; this method can help people obtain a general idea of the air quality of the surface. However, this category of methods is easily influenced by clouds and would be sensitive to other factors, such as humidity, temperature, and location. In addition, the results inferred from satellite images only reflect the air quality of the atmosphere rather than the ground air quality that people care more about.

To get a clear image of air quality, most countries nowadays deploy air quality stations. These official air quality stations usually use TEOM (tapered element oscillatory microbalance) [10], to measure the air quality by weighting PM2.5 concentrations accumulated on a filter over several hours (Figure 25.4a). This type of devices is large and expensive (on the order of 300K dollars). A more affordable class of monitors are based on light-scattering method and cost between 300 and 10K dollars. For example, Dylos [11] (Figure 25.4b) is a popular portable particle counter that costs around 300 dollars, while the TSI-3330 [12] is more accurate but costs around 5000 dollars. While 300 dollars is acceptable for some individuals, it is still too expensive for most people or for dense deployments. Over the past year or two, we start to see a number of dust-sensor-based devices aimed at particulate monitoring. These devices typically cost around 100–200 dollars. However, by empirically evaluating these devices, we conclude that they rely on static calibration curves, together with poorly designed air-flow mechanisms, resulting in completely unusable readings under many conditions (with error as large as 300%).

To overcome some of the cost limitations to direct monitoring in large areas, various methods are used to estimate air quality. One such approach is based on simple interpolation using reports from nearby official air quality monitor stations. This method is usually employed by websites trying to visualize AQIs across large areas. As air quality varies in locations nonlinearly, the inference accuracy is quite low.

(a)   (b)

Figure 25.4 (a) Thermo Scientific 1405-DF TEOM Continuous Dichotomous Ambient Air Monitor; (b) Dylos DC1100 Air Quality Monitor.

Other approaches include classical dispersion models, such as Gaussian plume models, operational street canyon models, and computational fluid dynamics. These models are in most cases a function of meteorology, street geometry, receptor locations, traffic volumes, and emission factors (e.g., g/km per single vehicle), based on a number of empirical assumptions and parameters that might not be applicable to all urban environments; however, these parameters are difficult to obtain precisely, and the result is not very accurate either [13].

Recently, big data reflecting city dynamics have become widely available, and a group of researchers seek to infer the air quality using machine learning and data mining techniques. In the "U-Air" [14], the authors infer air quality based on AQIs reported by a few public air quality stations and meteorological data, taxi trajectories, road networks, and POIs (point of interests). This approach shows some early promises of using "big" data while revealing some current limitations. For example, since their approach estimates AQI using a feature set based on historical data, their model cannot respond quickly enough to changes in PM2.5 concentration, which often changes on an hourly basis, leading to large errors at times. There are also crowd-sourcing or participatory sensing approaches [15, 16] to solve the air quality monitoring problem.

In this chapter, we approach this problem using a combination of direct monitoring, modeling and inference, and crowd-sourcing.

25.3 Overview of a Cloud-based Air Quality Monitoring System

To address the numerous challenges in city-scale air quality monitoring that is both accurate and affordable, we introduce a novel client–cloud system, called AirCloud [17]. AirCloud uses a heterogeneous set of data sources as inputs. These data are stored and analyzed by the *air quality analytics engine* in the cloud, to provide accurate device calibration and fine-granularity estimation based on GPS location. AirCloud further provides a number of *web services and APIs* for third-party application developers.

25.3.1 Data Sources

There are various types of data sources that can help us improve the accuracies of air quality estimation. For example, we can collect public weather data to learn the relationships between weather information such as temperature, humidity, pressure, wind speed, and air quality. In order to train the calibration and the inference model, we can construct a system with several classes of air quality data sources – the public air quality stations use very expensive instruments and their data are most accurate; Thermo [18] is less expensive but almost as accurate as the public stations; our system can also use less accurate air quality sensors such as Dylos. Since Thermo is very accurate, we use it to train our estimation models and as the ground truth to evaluate the performance of the online calibration model. But to get a more dense coverage of PM2.5 spatially, we need to design and implement our own PM2.5 instruments – *AQM* and *miniAQM* – as described in more detail in Section 25.4. It is much less expensive, and we can deploy them densely in cities. Combined with crowd-sourcing and our calibration and inference algorithms, AirCloud can provide accurate, calibrated results.

25.3.2 Data Representation and Storage

To describe multiple kinds of data sources in a uniform format and to store and query this data efficiently, we need an unified way to represent data. Here, we use sMAP [19], a sensor data protocol originated from UC Berkeley, as the data representation and storage system. It is designed as the standard specification for physical sensor data. sMAP provides an efficient database for time-series data, plus a powerful query language. We need to perform the calibration and

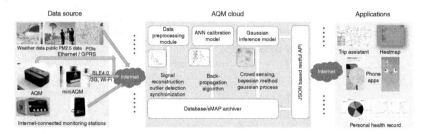

Figure 25.5 Architecture of AirCloud system.

inference online; however, we also have to ensure response speed and security. According to these specific needs, we add a back-end asynchronous service and authentication module to the system.

25.3.3 Air Quality Analytics Engine

Air quality analytics engine is the data processing and mining module, which contains the offline training model and online calibration and inference models; it takes the raw sensor data and computes the accurate calibration and inference results.

It can be observed that PM2.5 is heavily influenced by meteorology factors [14]. We can leverage this knowledge to exploit the dependencies between the sensor error and these meteorology factors. We collected 5 months of data, which contain AQM sensor data, Thermo ground truth data, and meteorology data, to train an ANN model. From this model, AirCloud learns the nonlinear relationship between AQM sensor readings and Thermo readings.

As shown in Figure 25.5, the system first uses the data preprocessing module to reconstruct the raw sensor data. Next, using the offline model we have trained using historical data, AirCloud computes a calibrated PM2.5 reading using real-time AQM data and weather data. Lastly, an online inference model based on Gaussian process is used to infer and predict PM2.5 concentrations at locations where direct sensors are not available or to further calibrate AQM sensor. We describe the details in Section 25.6.

25.3.4 APIs and Applications

AirCloud provides several web services and RESTful APIs for developers. For example, GPS-AIR interface can help developers get accurate PM2.5 estimation value if the GPS location data is valid; device-calibration interface can give back the device calibration result; data-driver interface can help developers easily add, get, or search data from our database. We describe some of these applications, including smartphone apps in more detail in Section 25.7.

25.4   Cloud-Connected Air Quality Monitors

PM2.5 concentrations vary significantly over space, especially for metropolitan cities where pollution sources are multifaceted. In addition, as we can observe from official PM2.5 monitoring station data, PM2.5 concentration changes at an hourly rate. As a result, direct monitoring is necessary.

However, none of the existing monitors satisfy our needs; they are either overly expensive or not accurate enough. To solve this problem, we design and build our own Internet-connected PM2.5 monitors, AQM and miniAQM, as shown in Figure 25.6. The AQM costs about $60 at 10K quantity. We take a novel approach of using inexpensive sensors at the front end but rely on the analytic algorithm in the cloud to improve the accuracies.

There are three kinds of data sources in all:

- Public data: weather data, public PM2.5 data, and POIs, which can be obtained using web crawler.
- Existing monitors: we use Thermo and Dylos connected with laptop as our highly accurate data sources.
- AQM and miniAQM: these are our custom-designed low-cost monitoring devices.

We describe AQM and miniAQM in detail in the following sections.

25.4.1   Sensor Selection

Because of own cost constraints, sensors suitable for our front ends are typically FIR-based light-reflecting sensors, commonly called dust sensor, which

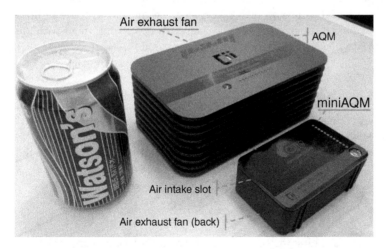

Figure 25.6 Cloud-connected air quality monitors – AQM and miniAQM.

are used extensively inside air purifiers as a rough indicator of air quality. Several choices are available, such as SHINYEI and SHARP. We installed two sensors inside AQM, on opposite sides and facing opposite directions. This design choice enables AQM to self-calibrate and increases the speed of converging to a usable value. This dust sensor arrangement also enables us to automatically detect errors and sensor failures during long-term deployment. For mini-AQMs, only one sensor is used to save space since portability is a priority for miniAQM.

25.4.2 Mechanical Design

While designing AQM and miniAQM hardware, we need to pay close attention to the mechanical structures since they play a significant role in the data accuracies of this type of sensors. In particular, the air flow determines how long and how fast the particles pass through the "detection window" of the sensor and needs to be carefully controlled. Ideally, an air pump should be used to maintain constant air flow; but air pumps are too expensive. Instead, we experimented with many different designs, both passive and active air flow. One early design completely exposes the sensors. This worked well most of the time but does not cope well with high-speed wind or if large particle become dominant. We then experimented on enclosure holes and strain screen in an attempt to decrease the effect of wind. But we found that this design often leads to a self-circulating local environment and will not exchange air with the outside. The final design is using a fan running at a low fixed speed to emulate the air pump, as shown in Figure 25.6. We install two sensors inside AQM on opposite sides and facing opposite directions. In the dust sensor (SHINYEI PPD42NJ), air is self-aspirated with the current of air generation mechanism with a built-in heater, so if we place the fan around AQM device, the fan will affect the air flow. By placing the fan on top of the device, we can ensure exchange of air with the outside to prevent a self-circulating local environment.

25.4.3 Data Communication

There are different communication approaches to connect AQM and miniAQM with cloud server. AQM, shown on the bottom of Figure 25.7, is the stationary version that connects to the cloud via Ethernet or GPRS. miniAQM, shown on the top of Figure 25.7, is the portable version, which connects to the smartphone over Bluetooth 3.0 or 4.0, and to the cloud via the mobile phone's data connection, such as 3G or Wi-Fi.

25.4.4 Hardware Calibration

While dust sensors exhibit the same trends, there are still significant variations between them and require calibrating against a reference sensor, at various

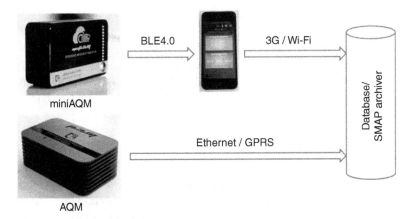

Figure 25.7 Communication between monitors and cloud.

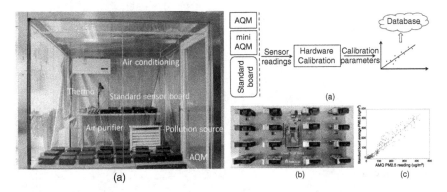

Figure 25.8 (A) Pollution source and air purifier are used to change PM2.5 concentrations, while air conditioning is used to change the temperature and humidity, we run sensors across a wide range to calibrate. (B) Hardware calibration process: (a) the hardware calibration procedure; (b) standard sensor board; (c) fitting results.

PM2.5 concentrations. In order to calibrate them, we built an 10 m³ air chamber with full internal climate control, as shown in Figure 25.8A. Using this air chamber, we can manually vary the PM2.5 concentrations across a wide range and calibrate all sensors together. Figure 25.8B shows the hardware calibration procedure. We use two-order polynomial fit to calibrate error between different PPD42NJ sensors. We use the average PM2.5 concentration of a sensor board, which contains 16 PPD42NJ sensors as the standard sensor reading. The hardware calibration parameters will be stored in a database. Before each deployment, each PPD42NJ sensor needs to be calibrated to this standard board to remove initial hardware variations.

25.5 Cloud-Side System Design and Considerations

Since our system relies heavily on the cloud for data processing, storage, calibration, and visualization, we need to carefully design our cloud-side system. To store and query data efficiently, we use sMAP [19] to define the data format and store the data. In this section, we first describe sMAP and then the details of our data exchange framework.

25.5.1 sMAP

sMAP [19] is a simple measurement and actuation profile for physical information, which enables the simple and efficient exchange of sensor data. It is a specification for transmitting physical data and describing its contents, which also provides tools for building, organizing, and querying large repositories of

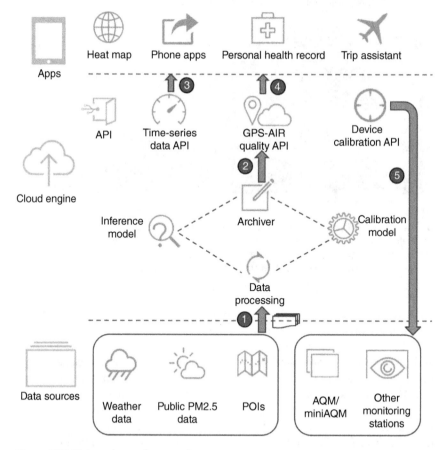

Figure 25.9 Data exchange framework.

physical data. sMAP allows instruments and other producers of physical information to directly publish their data, and it provides powerful RESTful service to get the data from sMAP archiver. The repository gives a place for instruments to send their data. It supports the following features:

- Efficient storage and retrieval of time-series data.
- Maintenance of metadata using structured key–value pairs.
- Metadata querying using ArdQuery language.

The core object in sMAP is time series, a single progression of (time, value) tuples. Each time series in sMAP is identified by a UUID and can be tagged with metadata; all grouping of time series occurs using these tags. These objects are exchanged among all components in this ecosystem. sMAP has been online for at least 3 years and has proven to be robust; all the aforementioned features make it an ideal choice for our system.

25.5.2 Data Format, Authentication, Storage and Web Services

Figure 25.9 shows the system data exchange framework, which contains three parts. (1) Data sources exchange framework: we define the data format according to sMAP specification and add authentication module. (2) System internal data exchange framework: we mainly use JSON format to exchange data among different system components. (3) API data exchange framework: to improve efficiency, we store the results in RAM and use authentication module to verify the user. We describe these in detail in the following section.

We use sMAP as our storage scheme and define the source data format as shown in Figure 25.10a. According to sMAP specification, each time series is globally identified by a universal unique identifier (UUID), which is a 128-bit name. Together with time-series readings, which contain UNIX timestamp and value, system can easily store the data. Time series are uniquely identified by UUID. However, these identifiers are unpleasant to use in practice. We add additional metadata to be attached as tags: structured key–value pairs. We add detailed and hierarchical metadata, such as instrumentID and location, to make it easier and natural to retrieve data in the future. The detailed location and InOut status also provide information to the back-end algorithm and the visualization service. To make the system more secure and stable, we add authentication module and data schema check module in the data processing framework. In Figure 25.9, all uploading data in (1) have the same data format and use specified key as authentication token.

InstrumentID uniquely identifies an instrument, such as AQM or miniAQM, and can be associated with multiple UUIDs–an instrument may have more than one data stream. Because an user may manually enter the InstrumentID to associate the device to his/her account, we designed the format for InstrumentID in a way that enables (trained) humans to read out useful information

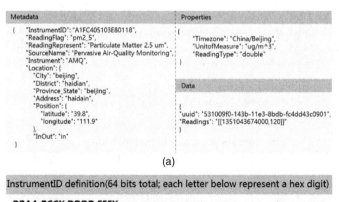

Metadata	Properties

{
 "InstrumentID": "A1FC405103E80118",
 "ReadingFlag": "pm2_5",
 "ReadingRepresent": "Particulate Matter 2.5 um",
 "SourceName": "Pervasive Air-Quality Monitoring",
 "Instrument": "AMQ",
 "Location": {
 "City": "beijing",
 "District": "haidian",
 "Province_State": "beijing",
 "Address": "haidain",
 "Position": {
 "latitude": "39.8",
 "longitude": "111.9"
 },
 "InOut": "in"
 }
	{
	 "Timezone": "China/Beijing",
	 "UnitofMeasure": "ug/m^3",
	 "ReadingType": "double"
	 }

	Data

	{
	 "uuid": "531009f0-143b-11e3-8bdb-fc4dd43c0901",
	 "Readings": "[[1351043674000,120]]"
	 }

(a)

InstrumentID definition(64 bits total; each letter below represent a hex digit)

RRAA BCCX DDDD EEFX

RR: Reserved / 8bits / 'A1' = default
AA: Manufacture ID / 8bits / 'FC' = Flex
B: Year counting from 2010 / 4bits / '4' = 2014
CC: Week of the year / 8bits / '05' = Jan. 23
X: First 4bits of CRC / 4bits
DDDD: Sequence number / 16bits / '03E8' = Start counting from 1000(DEC)
EE: Model / 8bits / '01' = AQM; '02' = miniAQM
F: Version / 4 bits / '1' = Revision 1
X: Second 4bits of CRC / 4bits

(b)

Figure 25.10 Data representations (a) and ID definitions (b).

visually, while preventing accidental input errors via CRC check, as shown in Figure 25.10b.

When the back-end server receives the data stream, it will use the cloud analytics engine to process the sensor data, produce the calibrated result, and send to sMAP archiver. To improve the response speed, we choose asynchronous celery workers to do the jobs in the back end and use tornado web server as the front end to deal with basic connection jobs. The system uses JSON format to exchange data among various system components.

In Figure 25.9, (3)–(5) represent the system API services. We have three different kinds of APIs:

- Device calibration API: each AQM, miniAQM, or other supported devices will use this API to get the calibration parameters. A key is required for authentication, and JSON is the data format.
- Time-series data API: developers can use this API to get time-series data. All data are presented in JSON format and use the authentication module.
- GPS-AIR quality API: developers can use this API to get accurate PM2.5 concentrations of the current location. To improve efficiency, we store the heatmap result in RAM once it is produced.

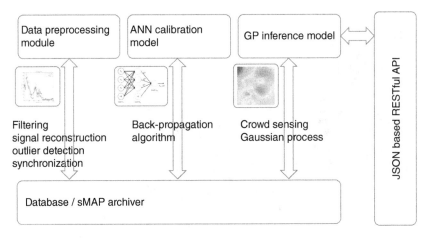

Figure 25.11 Air quality analytics engine.

25.6 Data Analytics in the Cloud

Data analytics in the cloud (i.e., air quality analytics engine) mainly consists of three components: (i) a signal reconstruction module designed to denoise and smooth the corrupted original sensor signal; (ii) an ANN-based calibration model aiming to enhance the accuracy of AQM and miniAQM in real time; and (iii) an online inference model based on Gaussian process that encompasses various source data to further improve the accuracies and to provide PM2.5 estimation for places where sensors are not available. The overall framework is shown in Figure 25.11.

25.6.1 Filtering Using Signal Reconstruction

Due to the instability of the sensor itself and the extra noise added in transmission, the original signal is extremely unstable as shown in Figure 25.12A. Therefore, eliminating the noise and reconstructing the real signal is the first step. There are several ways for smoothing out a noisy signal, such as using low-pass filters. However, due to the real-time requirements, signal reconstruction is a better choice for us. Formally, we represent the signal as x and assume that signal x is corrupted by an additive noise v:

$$x_{cor} = x + v \tag{25.1}$$

in which x_{cor} is the original corrupted signal uploaded by the sensor and we simply assume that the noise v is an unknown, small, and rapidly varying random variable. The goal here is to form an estimate the original signal x, given the corrupted signal x_{cor}. Please refer to [17] for detailed mathematical derivations.

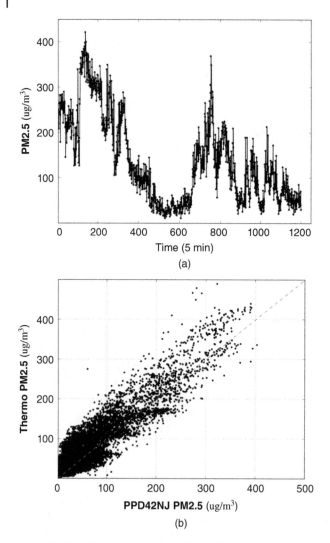

Figure 25.12 (a) The original signal of PPD42NJ during about 5 days where the signal is sampled every 5 min. (b) Thermo versus PPD42NJ. A large deviation exists between PPD42NJ and Thermo readings, which we use as ground truth.

25.6.2 Calibration Using Artificial Neural Networks

In the air quality analytics engine, the prediction is made based on the data acquired by the sensors. Thus, the precision of sensors is critical to the prediction accuracy. Unfortunately, there is usually an error in the data acquired by AQM due to the inherent randomness in particulates through the sensor's focusing window and noises in the devices. Figure 25.12b shows the concentration of PM2.5 given by AQM (PPD42NJ, after signal reconstruction) and the standard device (Thermo) at the same place, and there are obvious errors between them. Therefore, it is necessary to eliminate such errors to improve the accuracy of our engine, that is, to estimate each $h(x)$ from its observation $\hat{h}(x)$.

There are two observations from Figure 25.12b that inspire us to design the calibrator. (i) Each PPD42NJ reading may correspond to multiple Thermo readings, and thus more information is required to calibrate PPD42NJ. According to previous work [14], the concentration of PM2.5 is heavily influenced by meteorology factors, such as temperature, humidity, and pressure. This empirical knowledge inspires us to exploit these meteorology factors to help calibrate the AQM (PPD42NJ). (ii) The relationship between PPD42NJ and Thermo readings is complex and nonlinear, which makes it difficult to design the calibrator directly, and thus we resort to the learning-based method. Based on these observations, we select to simulate calibrator with a neural network, which takes the readings of PPD42NJ with colocated temperature and humidity readings as inputs and the ground truth value (given by Thermo) as output since the neural network is capable of fitting an arbitrary function.

25.6.2.1 Neural Network Model

A neural network is put together by hooking many of simple "neurons," so that the output of a neuron can be the input of another. In our experiments, we select the widely used one hidden layer network for its simplicity and generality. The architecture of the neural network is depicted in Figure 25.13.

Many types of neural networks have been investigated and extensively used in practice, among which the deep network architectures are becoming increasingly prevalent recently [20]. But in our application the widely used back-propagation (BP) network is chosen due to its simplicity and generality [21]. Denoting the readings of PPD42NJ and Thermo as $\hat{h}(x)$ and $h(x)$, respectively, the calibration function can be formulated as $h(x) = g(z(x))$, where $z(x) = [\hat{h}(x), H, T]T$, and H and T represent the humidity and temperature, respectively.

In the training process, we adopt the stochastic gradient descent algorithm [22] rather than the batch methods such as L-BFGS for the following two reasons: (i) computing the cost and gradient for the entire training set is very slow; and (ii) stochastic gradient descent allows us to incorporate new training data online as the relationship between the input and output is not constantly invariable due to the aging of device and the change of environment. In our analytics engine, we keep updating the calibrator by retraining it using online data collected at locations where both an AQM and a Thermo exist. Once the parameters of the BP network are learned, the prediction can be obtained in $O(1)$, which takes only some simple calculations. Using this model, we can effectively calibrate the AQM.

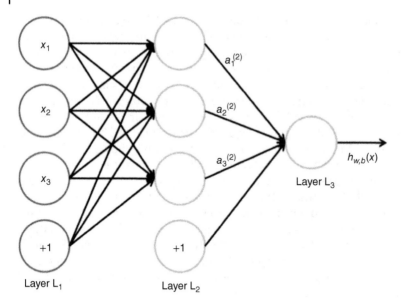

Figure 25.13 Neural network

25.6.3 Online Inference Model

The neural network calibration model attempts to improve the accuracy of the sensor by using only local point features. In the online inference model, we seek to incorporate various data sources with different confidence levels, such as the official public monitor station data, Thermo data, and Dylos data, to further improve the accuracy of AQM and miniAQM in open areas spatially. In addition, it can also provide PM2.5 estimation values for the places where the sensors are unavailable.

In our scenario, there are a variety of source data with different confidence levels, and as in a dense deployment, we can collect a relatively large amount of data for meaningful inference. The air qualities vary by locations nonlinearly and depend on multiple factors, such as meteorology, land use, and urban structures. We turn to one of the widely used Bayesian methods – Gaussian process, which can naturally bring these prior information and adaptively integrate the multiple source data into the model.

25.6.3.1 Gaussian Process

A Gaussian process is a stochastic process such that any finite subcollection of random variables has a multivariate Gaussian distribution. Formally, a collection of random variables $\{h(x) : x \in \mathcal{X}\}$ is said to be drawn from a Gaussian process with mean function $m(\cdot)$ and covariance function $k(\cdot, \cdot)$ if for any finite set of elements $x_1, x_2, \dots, x_m \in \mathcal{X}$, the associated finite set of random variables $h(x_1), h(x_2), \dots, h(x_m)$ follow the distribution

$$\begin{bmatrix} h(x_1) \\ \vdots \\ h(x_m) \end{bmatrix} \sim \mathcal{N} \left(\begin{bmatrix} m(x_1) \\ \vdots \\ m(x_m) \end{bmatrix}, \begin{bmatrix} k(x_1, x_1) & \cdots & k(x_1, x_m) \\ \vdots & \ddots & \vdots \\ k(x_m, x_1) & \cdots & k(x_m, x_m) \end{bmatrix} \right)$$

For simplicity, we denote this as

$$h(\cdot) = \mathcal{G}P(m(\cdot), k(\cdot, \cdot))$$

$m(\cdot)$ is the mean function that can be any real-valued function, and $k(\cdot, \cdot)$ must satisfy Mercer's condition. In our implementation, squared exponential function is chosen as the kernel function, which is defined as

$$k_{SE}(x, x') = \exp \left(-\sum_{i=1}^{n} \frac{\| x_i - x_i' \|^2}{2\delta_i^2} \right)$$

x_i is the ith dimension of feature vector x and n is the number of dimension. δ_i is used to control the importance of ith feature. The kernel function can be treated as a measurement of how similar two feature vectors are, and hence δ_i plays a significantly important role in characterizing the similarity of two feature vectors.

25.6.3.2 Gaussian Process Regression

Given a training set of i.i.d. (independent and identically distributed) examples, $S = \{(x^{(i)}, y^{(i)})\}_{i=1}^{m}$, from some unobserved distribution. The Gaussian process regression model can be written as

$$y^{(i)} = h(x^{(i)}) + \varepsilon^{(i)}, i = 1, \dots, m$$

where the $\varepsilon^{(i)}$ are i.i.d. noise variables(which might be generated by our sensors) with independent $\mathcal{N}(0, \delta^2)$ distributions.

The GP representation is very powerful. Given a set of test points, $T = \{(x_*^{(i)}, y_*^{(i)})\}_{i=1}^{m_*}$, drawn from the same unknown distribution as S. We could then derive equation to compute the posterior predictive distribution over the testing outputs y_* by using the properties of multivariate Gaussian distribution and Bayesian approach as

$$\begin{bmatrix} y \\ y_* \end{bmatrix} | X, X_* = \begin{bmatrix} h \\ h_* \end{bmatrix} + \begin{bmatrix} \varepsilon \\ \varepsilon_* \end{bmatrix} \sim \mathcal{N} \left(0, \begin{bmatrix} K(X, X) + \delta^2 I & K(X, X_*) \\ K(X_*, X) & K(X_*, X_*) + \delta^2 I \end{bmatrix} \right)$$

which means $y_* | y, X, X_* \sim \mathcal{N}(\mu^*, \Sigma^*)$

$$\mu^* = K(X_*, X)(K(X, X) + \delta^2 I)^{-1} y \tag{25.2}$$

Table 25.1 **Category of POIs.**

C1: Culture and education	C5: Shopping malls and supermarkets
C2: Parks	C6: Entertainment
C3: Sports	C7: Decoration and furniture markets
C4: Hotels	C8: Vehicle services (gas station, repair)

where

$$K(X,X)_{ij} = k_{SE}(x^{(i)}, x^{(j)}), K(X,X_*)_{ij} = k_{SE}(x^{(i)}, x_*^{(j)})$$

$$X = \begin{bmatrix} -(x^{(1)})^{T}- \\ -(x^{(2)})^{T}- \\ \vdots \\ -(x^{(m)})^{T}- \end{bmatrix} \in R^{m \times n}, X* = \begin{bmatrix} -(x_*^{(1)})^{T}- \\ -(x_*^{(2)})^{T}- \\ \vdots \\ -(x_*^{(m_*)})^{T}- \end{bmatrix} \in R^{m_* \times n}$$

As Gaussian process is nonparametric and hence can model essentially arbitrary functions of input points. In our spatial model, we divide the region into grids and collect air quality-related features for each grid. The features consist of GPS coordinates, location-dependent humidity, temperature, and POI. Intuitively, a grid tends to show similar air quality with its neighbors, and grids with similar meteorology and POI normally take the same air quality level. However, as these features do not have the same effect on PM2.5, to treat them equally is not ideal. Therefore, we can select different δ_i in our kernel function to distinguish these features.

Table 25.1 indicates the category of POIs that we studied in our experiment.

We can separate all the features into two different classes: one is spatial-related features such as location coordination and POI, which will not vary with time, and the other is temporal related such as humidity and temperature, which will change dynamically.

25.6.4 Evaluation of Effectiveness

The overall performance of the analytics engine is evaluated over 2 months' deployment. Then, the performance of the analytics engine is shown in an intuitive way – PM2.5 data through each calibration step in time domain, as shown in Figure 25.14. In addition, we visualized the confusion matrix after each calibration step, as shown in Figure 25.15. Finally, the numerical accuracy numbers are shown in the end.

Figure 25.14a shows the raw data of AQM compared with Thermo; it can be seen that PPD42NJ not only deviates from the Thermo significantly but is also noisy. And after the signal reconstruction, it becomes much more stable as

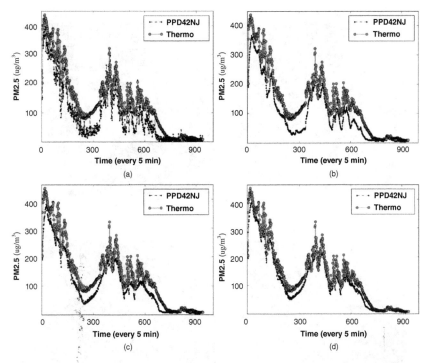

Figure 25.14 AQM PM2.5 in time domain. (a) Raw data; (b) reconstructed data; (c) calibrated by ANN; (d) calibrated by GP inference model.

shown in Figure 25.14b. These data are then calibrated by the neural network model shown in Figure 25.14c. It shows that the ANN calibration model is able to recover the real PM2.5 value to certain extent from the associated temperature and humidity. The final calibrated data given by the online inference model are shown in Figure 25.14d.

The confusion matrix of the 2 months' deployment dataset is displayed in Figure 25.15. Figure 25.15a shows that the predicted results of raw data have large variances. After these data are processed by the online-based GP inference model, the predicted levels are much better aligned with the true levels, as shown in Figure 25.15d.

The overall accuracy improvement from each step of the analytics engine is shown in Figure 25.16a. The prediction accuracy achieved by RAW (raw data), SR (signal reconstruction), ANN, and GP (inference) is 0.532, 0.603, 0.641, and 0.817, respectively. The improvement results from the previous step for SR, ANN, and GP are 13.4%, 6.3%, and 27.5% respectively. These results from the deployment dataset show that (i) raw AQM data contain significant noise and can meaningfully benefit from signal reconstruction; (ii) while ANN-based calibration is effective for historical dataset, it is less useful for real deployment;

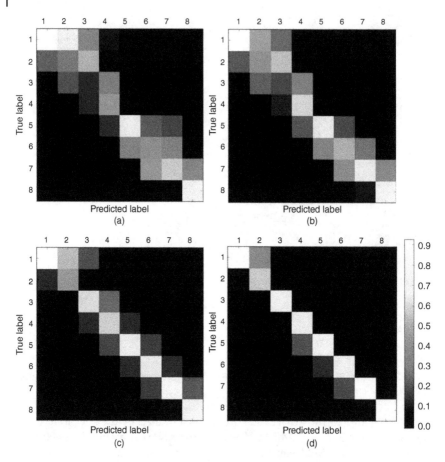

Figure 25.15 Confusion matrix of the deployment dataset. (a) Raw data; (b) reconstructed data; (c) calibrated by ANN; (d) calibrated by GP inference model.

and (iii) AQM data can benefit significantly from an online inference-based calibration model. As shown in Figure 25.16a, the combined improvement using SR, ANN, and inference is 53.6% in real deployment.

Using our Gaussian process-based inference model, AirCloud can infer air qualities at places where sensors are unavailable and further improve the accuracies of low-confidence AQM or miniAQM sensors. Figure 25.16b shows the estimation of air quality generated by the GP inference model in our deployment area using a heatmap. It demonstrates that the PM2.5 concentration deviation can be as large as 100 even in a small area of 4 km×4 km. The heatmap is valuable for a variety of applications such as pollution tracking, trip planning, and locating pollution sources.

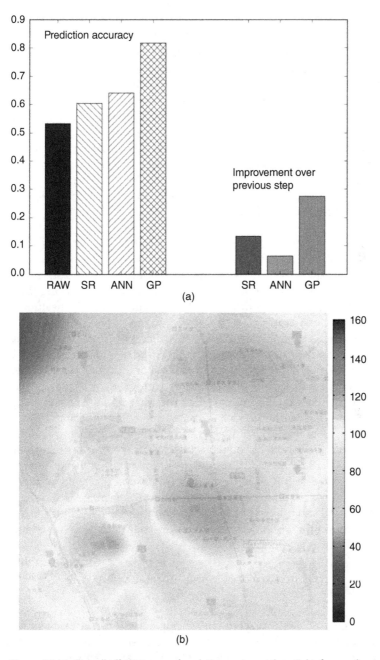

(a)

(b)

Figure 25.16 Overall effectiveness of analytics engine and spatial inference heatmap. (a) Overall prediction results of each step in the analytics engine; (b) heatmap generated using spatial inference.

25.7 Applications and APIs

In addition to the air quality analytics engine, AirCloud is an open cloud platform for storing, accessing, and sharing of air quality-related data. We create a set of APIs to enable third-party developers to innovate on top of AirCloud.

AirCloud supports two types of APIs–time-series-based access and GPS-based access, both in RESTful style and use JSON as the data format. The time-series-based APIs inherit the powerful querying capabilities from sMAP and support direct data access using an SQL-like language, or directly via device IDs or metadata lookups. The GPS-based API provides access to one of the built-in services of AirCloud, which compute an estimate of air quality given a GPS coordinate. This service essentially exports the inference functionality of the air quality analytics engine.

Using these APIs, many applications can be built on top. We present a couple of applications running on the AirCloud platform. Figure 25.17a is a visualization app to view time-series data stored on AirCloud, we can see the real-time data or historical data of each AQM or miniAQM device. Figure 25.17b is a trip planning app that plans a route between (a) and (b) based on the least PM2.5 intake instead of shortest distance, which will give users the healthiest routes.

Figure 25.18a is an iOS game called "AirPet." AirPet is a virtual pet that lives in your iPhone. She will grow up health and happy if fed at locations with good air (using the GPS API) or unhealthy and sad if fed at locations with bad air. This game encourages the player to go to places with good air. Figure 25.18b is a crowd-sourcing-based iOS game called AirFace that asks the user to guess the current PM2.5 concentration, with animated smile faces and sounds that correspond to different levels of air quality. AirFace then compares your guess with others in your area and tells you the actual PM2.5 value, again, using our GPS API. Figure 25.18c is a quadcopter that flies around the city and "warns" citizens with a loud siren sound if the air quality is bad at that GPS location. Figure 25.18d is the AQM phone app used to connect to the miniAQM device; it will show the local and calibrated PM2.5 concentrations. Figure 25.18e is the MyAir app; it shows the PM2.5 concentrations at your current location and gives visualized warnings.

Final Thoughts

Large-scale air quality monitoring is a challenging but important problem for every city of the future. Traditional approaches are often unable to meet both accuracy and cost requirements. In this chapter, using air quality monitoring as a vehicle, we introduce a number of concepts for environmental monitoring in smart cities – distributed sensing using Internet-connected monitors,

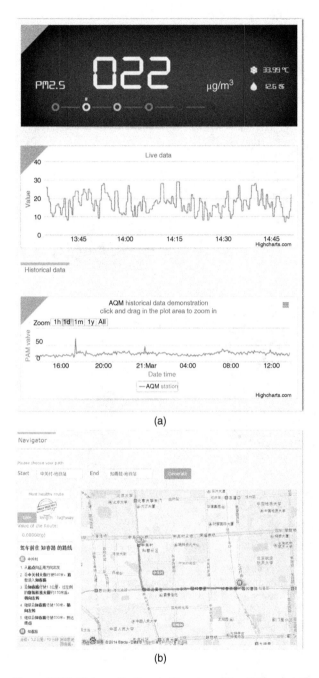

Figure 25.17 Web applications built on top of the AirCloud platform. (a) Visualization web app to view time-series data. (b) Trip planning web app – select the healthiest route.

Figure 25.18 (a) and (b) are iOS games, while (c) is an air quality-aware aerial drone, all using AirCloud APIs and by third-party developers. (d) AQM phone app used to connect to the miniAQM device; (e) MyAir app used to show PM2.5 information in detail. (a) AirPet; (b) AirFace; (c) aerial drone; (d) AQM; (e) MyAir.

cloud-offloading of computation, data analytics of large amounts of physical sensor data, storage of data both local and remote, data schema and protocols for interoperability, and designing APIs and services to enable third-party applications.

Questions

1 What are some ways to quantify air quality and how does air quality affect our health?

2 How is the concentration of particulate matter related to temperature and humidity, and why do we use a learning-based method such as ANN instead of linear functions?

3 Given a limited number of sensors deployed across a large city, describe how to use Gaussian process regression to infer or estimate air quality.

4 Why are APIs and web services important for enabling third-party application development?

5 How can techniques described in this chapter apply to other types of environmental monitoring such as water and soil monitoring?

References

1 Business Insider and Ramsey, L. (2015) *These 10 Cities Have the Worst Air Pollution in the World, and It is Up to 15 Times Dirtier Than What is Considered Healthy*, http://www.businessinsider.com/these-are-the-cities-with-the-worst-air-pollution-in-the-world-2015-9?IR=T (accessed 20 December 2016).

2 Greenpeace.org, C.O. (2015) *Air Pollution Increases Risk of Premature Death in Chinese Cities*, http://energydesk.greenpeace.org/2015/02/04/air-pollution-increases-risk-premature-death-chinese-cities/ (accessed 20 December 2016).

3 American Lung Association (2015) *Most Polluted Cities*, http://www.stateoftheair.org/2015/city-rankings/most-polluted-cities.html (accessed 20 December 2016).

4 Boldo, E., Medina, S., Le Tertre, A., Hurley, F., Mücke, H.G., Ballester, F., Aguilera, I. *et al.* (2006) Apheis: health impact assessment of long-term exposure to PM2. 5 in 23 European cities. *European Journal of Epidemiology*, **21** (6), 449–458.

5 Sørensen, M., Daneshvar, B., Hansen, M., Dragsted, L.O., Hertel, O., Knudsen, L., and Loft, S. (2003) Personal PM2. 5 Exposure and markers of oxidative stress in blood. *Environmental Health Perspectives*, **111** (2), 161–165.

6 EPA *The National Ambient Air Quality Standards for Particle Pollution*, http://www3.epa.gov/airquality/particlepollution/2012/decfsstandards.pdf (accessed 20 December 2016).

7 Wong, E. (2013) *On Scale of 0 to 500, Beijing's Air Quality Tops 'Crazy Bad' at 755*, http://www.nytimes.com/2013/01/13/science/earth/beijing-air-pollution-off-the-charts.html?_r=1 (accessed 20 December 2016).

8 WHO (2006) *Air Quality Guidelines: Global Update 2005. Particulate Matter, Ozone, Nitrogen Dioxide and Sulfur Dioxide*, World Health Organization.

9 Martin, R.V. (2008) Satellite remote sensing of surface air quality. *Atmospheric Environment*, **42** (34), 7823–7843.

10 Patashnick, H., Meyer, M., and Rogers, B. (2002) Tapered Element Oscillating Microbalance Technology. Mine Ventilation: Proceedings of the North American/Ninth US Mine Ventilation Symposium, Kingston, Ontario, Canada, pp. 625–631.

11 Dylos Corporation http://www.dylosproducts.com (accessed 20 December 2016).

12 TSI http://www.tsi.com/Optical-Particle-Sizer-3330 (accessed 20 December 2016).

13 Vardoulakis, S., Fisher, B.E., Pericleous, K., and Gonzalez-Flesca, N. (2003) Modelling air quality in street canyons: a review. *Atmospheric Environment*, **37** (2), 155–182.

14 Zheng, Y., Liu, F., and Hsieh, H.P. (2013) *U-Air: When Urban Air Quality Inference Meets Big Data*. Proceedings of the 19th ACM SIGKDD International Conference on Knowledge Discovery and Data Mining, ACM, pp. 1436–1444.

15 Hasenfratz, D., Saukh, O., Sturzenegger, S., and Thiele, L. (2012) Participatory Air Pollution Monitoring Using Smartphones. Mobile Sensing: From Smartphones and Wearables to Big Data. ACM, Beijing, China.

16 Jiang, Y., Li, K., Tian, L., Piedrahita, R., Yun, X., Mansata, O., Lv, Q., Dick, R.P., Hannigan, M., and Shang, L. (2011) *MAQS: A Personalized Mobile Sensing System for Indoor Air Quality Monitoring*. Proceedings of the 13th International Conference on Ubiquitous Computing, ACM, pp. 271–280.

17 Cheng, Y., Li, X., Li, Z., Jiang, S., Li, Y., Jia, J., and Jiang, X. (2014) *Aircloud: A Cloud-Based Air-Quality Monitoring System for Everyone*. Proceedings of the 12th ACM Conference on Embedded Network Sensor Systems, ACM, pp. 251–265.

18 ThermoFisher Scientific http://www.thermoscientific.com/ (accessed 20 December 2016).

19 Dawson-Haggerty, S., Jiang, X., Tolle, G., Ortiz, J., and Culler, D. (2010) *sMAP: A Simple Measurement and Actuation Profile for Physical Information*. Proceedings of the 8th ACM Conference on Embedded Networked Sensor Systems, ACM, pp. 197–210.

20 Bengio, Y. (2009) Learning deep architectures for AI. *Foundations and Trends® in Machine Learning*, **2** (1), 1–127.

21 Bishop, C.M. (2006) *Pattern Recognition and Machine Learning*, Springer.

22 Zhang, T. (2004) *Solving Large Scale Linear Prediction Problems Using Stochastic Gradient Descent Algorithms*. Proceedings of the 21st International Conference on Machine Learning, ACM, p. 116.

26

The Smart City Production System

Gary Graham[1], Jag Srai[2], Patrick Hennelly[2], and Roy Meriton[1]

[1] Dept – Business School, Leeds University Business School, Leeds, UK
[2] Institute for Manufacturing, Cambridge University IfM, Cambridge, UK

CHAPTER MENU
Introduction, 755 Types of Production System: Historical Evolution, 757 The Integrated Smart City Production System Framework, 761 Production System Design, 763 Chapter Summary, 767

Objectives

- To classify and categorize the different types of industrial production system.
- To develop a smart city production system framework.
- To critically assess the role of the smart city in reconfiguring supply chain design.

26.1 Introduction

Historically, cities have provided the input resources for production and allowed urban logistics connections for manufacturing to occur [1]. This fostered creative cooperation among firms and enabled the spontaneous aggregation of firms into industrial systems. Furthermore, cities encouraged new firm start-ups, lowered entry barriers, and provided quasi-public goods (i.e., infrastructure and utility services) to firms. The concept of the industrial city came to the fore in the 1920s with the rise of the manufacturing philosophy of Fordism. This led to the development of the modern economic and social system based on industrialized, standardized mass production and mass

Smart Cities: Foundations, Principles and Applications, First Edition.
Edited by Houbing Song, Ravi Srinivasan, Tamim Sookoor, and Sabina Jeschke.
© 2017 John Wiley & Sons, Inc. Published 2017 by John Wiley & Sons, Inc.

consumption. In the United States, United Kingdom, Germany, and the Soviet Union, the spread of Fordism involved the growth of core industrial regions comprising large metropolitan areas surrounded by networks of smaller industrial cities. These regions were dominated by the leading Fordist firms and their suppliers; drew in raw materials and, on a growing scale, migrants or foreign labor, from the rest of the world; and churned out mass-produced goods for global markets [2, p. 132].

As the Fordist regime developed, firms allocated activities, sourced supplies, and sought markets on an increasingly global scale. Perhaps the ultimate expression of this was the concept of the "*world car.*" Closely related to the organization of production was the nature of urban life in the Fordist era. Although national differences existed, Fordism was primarily associated with suburbanization (especially in the United States) and high-density urban renewal based on industrial construction techniques (especially in Europe) [2, p. 132].

Since the 1980s, the dramatic decline in manufacturing, deindustrialization, and shifting modes of production from West to East hit hard the city wealth of these former industrial power houses.[1] However, they have not wanted to live in the past and dwell on criticisms of their decline. Many have sought to reinvent themselves as a future or smart city [3]. With advances in postindustrial technology, the need for more sustainable supply chains (with a low carbon footprint), and economic regeneration, many cities around the world are aiming to become "smart cities." A smart city can be defined as "a city seeking to address public issues via ICT - [information and communication technology] based solutions of a multi-stakeholder, municipally based partnership" [4, p. 9].

While much of the policy focus of smart cities has been on improving local services such as transport, healthcare, and energy supply, there is a clear lack of discussion on industry (supply chains) and manufacturing. Therefore, this chapter takes a manufacturing and industrial perspective of the smart city. Furthermore we propose an integrative smart city production system framework. This is an initial first step in understanding how postindustrial technologies could be integrated into a city production system. This will enable more localized manufacturing that is both sustainable and resilient. In the next section, we document the historical evolution of production systems from Fordism to smart cities. Then we present integrative smart cities manufacturing framework developed through our conceptualizations of the interplay between the smart city and production system design.

---

1 Most notably the well-documented decline of Detroit [3].

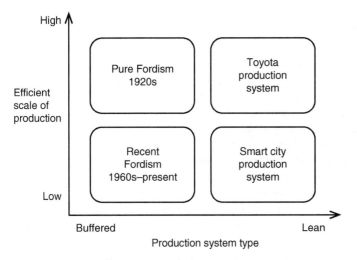

Figure 26.1 A categorization of the different types of production system.

26.2 Types of Production System: Historical Evolution

A production system can be defined "… as the integration and coordination of the different operational and logistics processes involved in network design, manufacturing, distribution and life cycle services of an industrial product or consumer good" [5]. Before Ford in the 1920s, most manufactured products such as automobiles were practically custom assembled in a job shop operation by a master craftsman. Ford brought in the notion of standardizing the sequence of activities in the production system so that scale economies and low costs per unit of output could be achieved.

In Figure 26.1 there are four different types of industrial production systems[2] positioned in the "2 × 2" matrix. The size of the operation (capacity) and growth in scale have driven production systems design since 1900 (as firms strove to drive down production costs), while throughout the postwar period, the stock was buffered against virtually everything. Inventory and stock levels were very high and production systems frequently had problems of overcapacity. We discuss the key characteristics of each type of production system in turn.

26.2.1 Pure Fordism 1920s Onward

A Fordist production system is based on optimizing scale economies often through vertical (moving upstream or downstream the supply chain) or horizontal (acquiring competitors) integration. The core resources of Fordism are

---

2 We have updated the Krafcik's [5] industrial matrix to incorporate our conceptualization of a smart city production system.

its physical assets such as capital plant and machine tools. These represented a large share of the total production factor investments.

Shearer [6] suggests that for this system of mass production to work properly, it needs to be broken down to single-task operations, with labor organized and machines configured to perform each of these operations in a repetitive standard manner. He notes that "… what made the Fordist system successful was the welter of auxiliary processes that grew up to support the production system, mechanized transport systems, extensive service facilities, a phenomenal growth in managerial and accounting staffs and a new place in the factory hierarchy for highly skilled workers who worked in the tool rooms, filing and machining the mass produced" (p. 2).

However, less successful outcomes were achieved by the plans to modernize the nationalized steel, coal, and car industries in the 1970s/1980s. In each case, it has been argued that the strategic plan was based on a dogmatic commitment to "… a Western model of mass production" (Fordism) and an obsessive belief that modernization along these lines would give "a productive opportunity to move down a long run average cost by employing large-scale capital investment techniques" [6, p. 171].

Shearer notes that the planners overestimated the likely returns to scale and ignored the problems of product range and markets. This produced overcapacity in large-scale plants and so failed to boost productivity significantly [6, pp. 185, 190]. The value streams tended to be discontinuous as a result of inaccurate forecasting, long lead times, and large volumes of parts and finished goods held as inventory.

26.2.2 Toyota Production System (TPS)

The Japanese in developing their Toyota Production System (TPS) also followed a scale economy strategy. Compared with American manufacturers, they kept their inventory levels at an absolute minimum. This enabled production costs to be saved and quality problems quickly detected. Toyota created bufferless assembly lines accompanied by continuous flow production with production networks that came to be defined by low waste and process efficiency.

Toyota initially struggled to match the high product volumes that Ford/GM managed out of their Detroit plants. Likewise, they did not have the same level of vertically integrated control over the supplier and raw material networks. It therefore did not focus its resources and production capabilities on one "dominant" manufactured product (i.e., the Model T Ford, the Chevrolet); they built their capabilities across a wide range of products and also started to build a more flexible, efficient, and time-focused production system (i.e., just-in-time (JIT)) that retained the continuous flow method designed by Ford.

It was not until the early 1990s and the release of the book *The Machine That Changed the World* by Womack *et al.* [7] that Western firms acknowledged the widening productivity gap between themselves and the Japanese. If they were to survive, this would mean a radical altering of their production philosophy, network design, manufacturing, distribution, and industrial services strategy.

26.2.3 Post-Fordism

For most of the period after WW II and before the early 1980s, there was a clear gap in the efficiency levels achieved by the production system of Toyota and leading Western manufacturers. This was not only in automobiles but also across many other sectors (including electronics, machine tools, and textiles). Recently these efficiency differences have begun to diminish as many Western producers began implementing their interpretation of the TPS.

Post-Fordism can be defined as a flexible production process based on flexible machines or systems and an appropriately flexible workforce. Its crucial hardware is microelectronic-based information and communications technologies. These are relevant to "… manual and non-manual work, to small, medium and large businesses, at corporate, divisional and workplace level, to managements and unions, and so on" [5, p. 41].

Kuhnle [8] suggests that recent Fordist interpretations compared to a JIT system operate with a lower plant scale. They configure their production to accurately supply goods with short lead times and be agile to meet actual market demand. When linked with electronics-based telecommunications systems, these "real-time" technologies can also affect "enhanced information, links and flows across space, integrating activities across departments and sites, and between individuals and organizations in different countries" [8, p. 6]. This allowed new or enhanced flexible specialization, by small firms or producer networks, even in small-batch production and, indeed, outside manufacturing, which could promote flexibility in the production of many types of services in the private, public, and so-called third sectors.

While Fordism proposed a standardized production system, its modern equivalent applies a more flexible production philosophy, with growing productivity based on "economies of scope." It optimizes technological rents and proposes the full utilization of flexible capacity, reinvestment in more flexible production equipment and techniques and/or new sets of products, and a further boost to economies of scope.

Flexible production seems to be avoiding the old Fordist production centers, and it is typically located in the suburban extensions of Fordist metropolitan areas, in relatively non-industrialized hinterland areas, and, at least in services, in central business districts [9]. These new sites of production are rearticulated into the global circuit of capital, and only its central nodes can function as locally integrated, agglomerated, and self-generating growth poles; other sites are becoming more fragmented and are being inserted at various lower points in the global hierarchy [10].

The resulting scope economies would enable even a small- or medium-sized firm to rival a larger firm organized along Fordist lines, for instance, the firm that relied on producing a standardized product and whose profits came from economies of scale. This is only effective where the goods produced have a short life or will soon become obsolete so that economies of scale are limited: otherwise a larger firm equipped with flexible machinery would outperform small or medium firms [11].

By contrast, dynamic flexibility operates on a longer time horizon and involves "... a production line which is able to evolve rapidly, in response to changes in engineering of products or of processes" [11, p. 1]. It is ideal for new products with growing demand and/or stable volumes of demand to have periodic shifts in the features offered or demanded on the potential for flexible mass production.

In both cases a declining demand would make it hard to introduce flexible manufacturing technology and also to provide the necessary investment. More generally, there are also financial limits on such investment due to the greater capital intensity of flexible machinery and plant that put it beyond the capacity of many firms to introduce profitably. This point is reinforced when the increasing velocity of the circuit of capital and the associated pressures to amortize investments quickly are taken into account. Lying even further beyond the reach of flexible specialization is the "production of large, lumpy investment goods" such as public telecommunications switching systems [12, p. 675].

These may benefit from computer integration of different stages in production, but their actual manufacture will be beyond the scope of small or medium firms even when organized flexibly in industrial districts. Nonetheless the changes occurring in the labor process, even if limited in scope and often realized only partially, do seem to involve significant departures from Fordist practice. Even if one is not justified in talking about a novel post-Fordist accumulation regime or mode of regulation, the evidence does indicate certain genuine post-Fordist trends in the labor process.

In the next section we build on the post-Fordist and TPS production system paradigms to focus on the development of the smart city production system. This is identified in the matrix and has been characterized by flexible and low scale production runs. There are low levels of inventory as the point of production is located close to consumption points and within city boundaries (local production). Through customization and lean approaches to sourcing, the point of production is located close to the consumption point, and it focuses on supplying products (in style, quality, and cost) that a consumer actually demands. This is a production system that is designed to be ambidextrous and adaptable to shifting demand patterns and also dissolvable once consumer demand has been met.

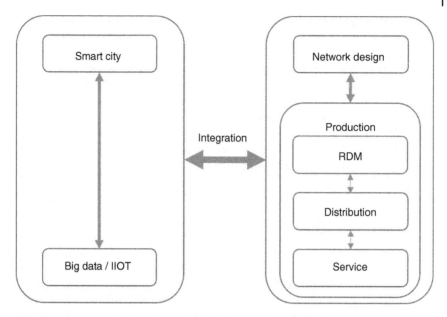

Figure 26.2 The integrated smart city production system framework.

26.3 The Integrated Smart City Production System Framework

The integrated smart city production framework is presented in Figure 26.2. This links together the smart city with production system design. Our integrated framework fits the main objective of this chapter, which is to explore the interplay of smart city technology with production organization and design. We have different categories of smart city technologies (i.e., smart cities infrastructure, big data, and the industrial Internet of Things (IIoT) interacting with production characteristics (i.e., network design capacity and redistributed manufacturing (RDM), distribution, and services).

26.3.1 Smart City Infrastructure

The "smart cities" expression has taken on some of the digital dimensions of connected systems and flexible computing infrastructures. It also incorporates elements of sustainability and inclusivity, as well as responding to the rise of new Internet technology interfaces [13]. Although the term has been popularized since the mid-1990s, in 2010 and 2011 the term really gained impetus as more places began to compete on sustainable innovation, not least the high-profile smart city projects in Singapore and Abu Dhabi [14]. According to Giffinger *et al.* [15], a smart city incorporates at least one of the following dimensions: smart economy (i.e., innovation, entrepreneurship, productivity), smart mobility (i.e., accessibility, sustainable transport systems), smart environment (i.e., pollution, sustainable resource management), smart people (i.e., level of qualification, creativity, flexibility), smart living (i.e., quality of life), and smart governance (i.e., public and social services, transparent governance).

Smart city initiatives are largely dependent on collecting and managing the right kinds of data, analyzing patterns, and optimizing systems functioning [16]. Here, the additional two-key elements are the data *per se* (i.e., big data) and the process of examining this data (i.e., big data analytics (BDA)).

26.3.2 Big Data

The concept of "big data" can be defined as large pools of unstructured data that can be captured, stored, managed, and analyzed [17]. Big data *per se* cannot be useful if it is not complemented by a process of examination and assessment ("big data analytics"). BDA is a holistic approach to managing, processing, and analyzing the "5Vs" of manufacturing data-related dimension (i.e., volume, variety, velocity, veracity, and value) in order to create actionable ideas for delivering sustained value, measuring performance, and establishing industrial competitive advantages [18]. Some practitioners have gone as far as to suggest that industrial BDA is the "fourth paradigm of science," a "new paradigm of knowledge assets," or "the next frontier for innovation, competition and productivity" [17, p. 1].

All these assertions are primarily driven by the ubiquitous adoption and use of BDA-enabled tools, technologies, and infrastructure, including social media, mobile devices, automatic identification technologies enabling the IIoT, and cloud-enabled platforms for firms' operations to achieve and sustain competitive advantage. In manufacturing, BDA is considered to be an enabler of asset and business process monitoring, supply chain visibility, enhanced manufacturing and industrial automation, and improved business transformation [18]. Smart cities provide an ideal background for exploitation of big data, while the operational interactions in production networks can generate "exhaust data."[3]

26.3.3 Industrial Internet of Things

The basic idea of the IIoT is to connect devices and things to build communication between device, sensor, and other physical objects. It is the full production and logistics potential that especially the IIoT brings to the factories that will be deployed when smart devices, smart systems, and smart automation entirely merge with the physical machines, service, fleets, and networks by the implementation of cyber–physical production systems [19]. The IIoT platform links

---

3 Unstructured information or data that is the by-product of the online activities of suppliers.

essential materials, sub-tier operational and inventory information together through its integrated systems, and processes functionality [20]. What is clearly important to its operational success is that there is a commitment and trust between manufacturer, supplier, and designer within the whole production system. That is, there is an agreement between nodes to share data, information, and knowledge.

In theory, if there is a partnership culture within the production system, then an IIoT platform offers a seamless integration of several components. Partners of the connected world would be able to build strong partnerships by implementing their services and products for efficient and quick cooperation. Although the main components of IIoT are the physical things (i.e., by linking devices, sensors, networks, and actuators), it provides multiple new opportunities to reduce inventory and improve lead times and operational efficiency [21]; it is the intangible relational components such as trust, level of social capital, and partnership culture that are probably more important if it is to achieve the "hyped" performance improvements in delivery and lead time efficiency.

26.4 Production System Design

In this section we explore the interplay of smart city technologies with four production system characteristics. These include network design, manufacturing, distribution, and service.

26.4.1 Network Design

Kim *et al.* [22] defined network design to be "the location and pattern of relationships between nodes in a production system" (p. 196). Thus, with a densely designed network, all nodes would be geographically connected to each other locally (i.e., within a defined urban area), and there would be strong relationships between them. Smart cities, big data, and IIoT appear to have a mutually reinforcing relationship on network capacity. The smart city positively affects the establishment of ties in the network. For example, logistics providers can use big data to compensate for the lost links due to retailers' evolution to direct-to-consumer service [23]. The reverse also applies.

High network density can positively affect smart city initiatives, and having more ties increases access to information [24]. For example, open data initiatives are enabled by a high-density network [4]. This relationship can also be analyzed under the theoretical lens of social embeddedness [25]. A highly embedded city production system increases network cooperation [26].

Furthermore, network effects (i.e., the value of the network increases when the number of participants increases) could also explain firm adhesion to smart city production initiatives. In particular, embeddedness is a result of repeated interactions between the parties [25]; thus more network capacity will imply more interactions and reinforce network effects, motivating firms to adhere to smart city initiatives. For example, Cooperative Intelligent Transport Systems and Services (C-ITS) are based on the principle that all cooperative parties locally exchange information between each other.

As for creating and consolidating, the smart city initiatives could be expected to lead to the creation of new relationships and companies moving to take advantage of new positions in networks. This is also seen in the earlier examples of smart cities, where companies have taken on new tasks and moved to more sustainable solutions. Companies (and other organizations) will grasp opportunities related to creating and offering solutions in line with smart city objectives: smart economy, mobility, environment, people, living, and governance.

This all would mean that sectors related to health, energy, water, waste, communications, buildings, and transport would attract new solutions, new actors, and new interactions. Therefore new actors will be expected to enter into present structures, and companies may shift sectorial belonging as radical new solutions are needed and potentially adopted from other sectors.

Parallel to new sectors, solutions, and businesses being formed, others will decrease in importance or be out of business based on how contrast fundamental principles of smart cities are. Logistics firms about transportation are expected to be less, and those focusing on long-distance transports may have difficulty to remain in the market or opt for radical changes to what business they conduct and how they perform it. Here, smart city initiatives would either mean that companies move together with their interaction partners or break new ground and hence disband from previous interaction partners to find new ones. In the wider development of smart city initiatives (but not yet seen in practice), interaction among companies (and other organizations) would expectedly increasingly be determined by geographical location.

Companies would work more with other firms geographically close to them so as to minimize environmental impact from transportations [27], with this also follows less interaction at a global level and with firms geographically distant to the company. In turn, this prescribes a change to current interaction patterns. Companies currently acting at a distance may choose to change interaction partners or colocate their businesses within city boundaries, for instance.

26.4.2 Redistributed Manufacturing (RDM)

Demand for more individuality and customer-specific product variants, coupled with localized manufacturing, requires new paradigms of production that supplant long-established methods [28]. Small, flexible, and scalable geographically distributed manufacturing units are capable of exhibiting the characteristics desired by modern operating systems – JIT delivery, nimble adjustments of production capacity and functionality with respect to customer needs, and sustainable production and supply chains. However, even in today's dispersed manufacturing, the production location often appears to be far from the point of consumption. RDM entails a deviation from conventional mass production in terms of not only scale and location but also consumer–producer relationship [29]. This implies a shift from long, linear supply chains, economies of scale and centralization tendencies, and a move toward a networked paradigm.

The user interface is also changing, with the blurring of the boundary between consumers and producers (leading to the term "prosumer" [30]), with consumers empowered to provide design input into production, enabled in large part by digitalization and the smart city, and leading to greater product personalization and customization. Concomitant with these enablers is an emerging culture of sharing community manufacturing facilities, with RDM offering the potential to transform the industrial and urban landscape. Threadless – an online apparel/print company – is an example of a company that facilitates this level of customer feedback, enabling the customer to set their preferences and partake in the design of the product.

This evolving RDM paradigm will be characterized by new business models operating in "distributed economies" [31], whose small-scale, flexible networks will have a more local dimension, utilizing local materials and other resources, thereby offering environmental benefits and leading to more sustainable forms of production, that is, energy-efficient and resource-saving manufacturing systems [29]. The network element inherent in the RDM paradigm can lead to a reduction of emissions through a reduction of transports. These developments are arising against the backdrop of future supply chain design, which are in part geared around managing scarcity of resources.

The emerging circular economy aims to make better use of resources/materials through recovery and recycling and also to minimize the energy and environmental impact of resource extraction and processing [17]. Innovation and new technology in the circular economy will also have a community impact. It could be argued that RDM in this sense represents a growing democratization and decentralization of manufacturing and to some extent the transition to a circular economy.

Manufacturing can be understood to be an activity that is not just about making things but also where multiple people, including end users, can come together and do things in a codified way, making things through quantified processes. Here lies the difference in context between old and new forms of distributed manufacturing – instead of the know-how being associated with the person doing the work, manufacture is achieved by means of modern processes and digitalization, enabling multiple people being able to do things in a codified way across many locations, most notably including the end user. In defining what RDM is today, it is particularly characterized by technological developments in engineering and computing that bring new capabilities to manufacturing in terms of automation, complexity, flexibility, and efficiency.

One of the significant enabling technologies of RDM is 3D printing, which is emblematic of a shift to on-demand, smaller-scale localized manufacturing. The redistribution of manufacturing is being enabled and driven today by this and other advanced manufacturing technologies, such as digital fabrication technologies, continuous manufacturing in previous batch-centric operations, stereolithography, laser-cutting machinery, and tools for electrical component assembly. Not only are such technologies changing how and where goods are produced, but also established organizational practices and value chains are being disrupted by the adoption of these technologies. Literature on RDM is fragmented because of its demonstrable applicability in a wide variety of sectors and in varying contexts.

26.4.3 Manufacturing Scale and Inventory

Complications arise when the capacity and scale of the operation is considered in the smart city. Compared with a mass production facility, Fordism, or TPS that can produce thousands of units of output per week, the local operation is likely to be much smaller in output given its RDM focus. Due to the personalized nature of the city production and consequently fluctuating demand, it is harder for the operation to enjoy economies of scale and the associated cost advantages, through the spreading of fixed costs [32].

Although inventory could be stored, this is not advised since each product is unique and not universally useful. However, a chase-demand plan could solve this issue of unwanted inventory, turning it into a competitive advantage. Working capital would not be tied up in stock, allowing for faster expansion in opening more 3D printing centers across the country, shortening lead times even further [32]. Conversely, mass production must hold greater amounts of stock, increasing the risk and stress on the company's cash flow position.

26.4.4 Distribution and Service

The distribution of products in the smart city production system will be composed of "swarms" of small distributors specialized in low volume and customized deliveries. This would be enabled by the widespread adoption of drone and robotic technologies [33]. Flexibility will stem from local concentrations of extremely specialized small firms that can be recombined into multiple configurations according to changing market demand and to the requirements of the lead manufacturing firms in the network. The highly fragmented organizational structure allows flexibility to meet the distribution requirements of a manufacturing system built upon small-batch runs, short lead times, fast delivery, and quick market entry and exit.

Dissolvable supply chains will be facilitated in which manufacturing configurations are temporarily constructed (i.e., the analogy would be to that of a film project or a musical event network) to capture value from rapidly emerging city "product-demand" segments (i.e., niches). Therefore lead time would replace costs of production as the key factor in determining plant location. Moreover, smart manufacturing would be at a very small boutique or batch production levels located within or close to retail outlets. Manufacturing/distribution capacity will be twinned or shared. For instance, designers are "nomadic" by nature, and they would be free to move around within or between smart cities. They would have the capacity to manufacture their designs and distribute small production runs of their products directly to consumers. Big data-driven demand and retail information on output levels and consumer preferences will drive this flexible distribution strategy, while the IIoT would ensure that the correct amounts of raw materials were delivered to the different city production points daily to enable manufacturing to take place.

In the context of industrial services, we suggest that RDM will enable replacement parts or components to be made closer to the consumer (e.g., replacement car or bicycle parts), while big data technology will be used to monitor ongoing product performance. This information will be fed directly into the design and development of new manufactured products, for instance, Volkswagen and its connected car fleet. Service relationships will center on the role that consumers and crowds can play in co-creating improved design concepts, product development, and performance throughout its life cycle. Manufacturing firms within smart cities will need to offer life cycle services from "design inception" to "after sales service" [34]. The enhanced network data capabilities through smart city and big data technology together with the IIoT can facilitate much more trusting interactions between all the different production network actors (i.e., "social capital") that are needed for enhanced service value creation to occur.

26.5 Chapter Summary

The dominant manufacturing orthodoxy in the 20th century was that of scale economies and global production systems. However the environmental risks and social costs of the global manufacturing paradigm are clearly evident. For instance, there is emerging work on the environmental challenges facing cities and regions in China [33]. These risks include its vulnerability to climate change and the damage such systems are having on the natural environment and on worker's rights and incomes. Furthermore social pressure for more sustainable and resilient cities and production systems from governments, regulatory bodies, and the general public is growing. Interestingly, Toyota may have indicated a sign of things to come with its planned development of a more localized and sustainable TPS2 production system.

According to a study from the European Commission, more than two thirds of smart city projects in Europe are in the planning or pilot testing phases [4]. Cohen and Levinthal [35] note that critical to the development of smart city production systems is the "absorptive capacity" of its manufacturing actors. How fast and successfully the local suppliers internalize and translate transferred smart city production knowledge into their own capability through learning will be largely determined by their absorptive capacity and their ability to upgrade it continuously. As RDM machinery becomes more accessible within the smart city, this improves the feasibility of boutique and batch production. If this encourages more designers and new innovative suppliers to enter production and build high quality and high value niches with customized products, then the demand for locally produced goods has the potential to grow.

Final Thoughts

This chapter explored manufacturing in the smart city. We have presented a framework that integrates distributed manufacturing with smart city technologies (such as big data and the IIOT). The unique characteristics of smart city manufacturing include (i) networks of micro-factories (small plant size), that are characterized by fewer supplier nodes, dispersed, and organized by city-based demand segmentation; (ii) dissolvable supply chains and flexible supplier capacity; and (iii) products that are highly customized (co-created).

Finally, we believe that "smart" policy interventions are needed to incentivize local designers, SMEs, and entrepreneurs to build local manufacturing capacity.

Questions

1 What are the main types of production system?

2 Which categories of product could be manufactured in a smart city production system?

3 How could a smart city production system change supply chain design?

4 Why is actor transformation so important in a smart city production system?

References

1 Toms, S. (2005) Financial control, managerial control and accountability: evidence from the British cotton industry 1700–2000. *Accounting, Organizations and Society*, **30** (7/8), 627–653.

2 Harvey, D. (1989) From managerialism to entrepreneurialism: The transformation in urban governance in late capitalism. *Geografiska Annaler Series B, Human Geography*, **71** (1), 3–17. Retrieved from http://www.jstor.org/stable/490503 (accessed: 11/01/16).

3 Katz, B. and Bradley, J. (2013) *The Metropolitan Revolution*, Brookings Institution Press, New York.

4 Manville, C., Cochrane, G., Cave, J., Millard, J., Pederson, J.K., Thaarup, R.K., Liebe, A., Wissner, M., Massink, R., and Kotterink, B. (2014) *Mapping Smart Cities in the EU*. http://www.rand.org/pubs/external_publications/EP50486.html (accessed 10 February 2016).

5 Krafcik, J.F. (1988) Triumph of the lean production system. *Sloan Management Review*, **30** (1), 41–52.

6 Shearer, D. (1996) *Industry, State, and Society in Stalin's Russia, 1926–1934*, Cornell University Press, Ithaca.

7 Womack, J., Jones, D.T., and Roos, D. (1990) *Machine that Changed the World*, Simon and Schuster, London.

8 Kuhnle, H. (2010) *Distributed Manufacturing: Paradigm, Concepts, Solutions and Examples*, Springer, London.

9 Storper, M. and Scott, A.J. (1989) *The Power of Geography: How Territory Shapes Social Life*, Unwin Hyman, Boston.

10 Amin, A. and Robins, K. (1990) The re-emergence of regional economies? The mythical geography of flexible accumulation. *Environment and Planning D: Society and Space*, **8** (1), 7–34.

11 Coriat, B. (1990) *The Revitalization of Mass Production in the Computer Age*. Paper presented to the UCLA Lake Arrowhead Conference Center, Pathways to Industrialization and Regional Development in the 1990s March 14–18.

12 Sayer, A. (1989) Postfordism in question. *International Journal of Urban and Regional Research*, **13** (4), 666–695.

13 Deakin, M. and Al Waer, H. (eds) (2012) *From Intelligent to Smart Cities*, Routledge, London.

14 Joss, S. (2010) *Eco-Cities: A Global Survey 2009. The Sustainable City VI: Urban Regeneration and Sustainability*, WIT Press, Southampton, pp. 239–250.

15 Giffinger, R., Fertner, C., Kramar, H., and Meijers, E. (2007) *City-Ranking of European Medium-Sized Cities*, Centre of Regional Science, Vienna, UT, pp. 1–12.

16 Dirks, S., Gurdgiev, C., and Keeling, M. (2010) *Smarter Cities for Smarter Growth*, https://www.zurich.ibm.com/pdf/isl/infoportal/IBV_SC3_report_GBE03348USEN.pdf (accessed 18 March 2015).

17 Manyika, J., Chui, M., Brown, B. *et al.* (2011) *Big Data: The Next Frontier for Innovation, Competition and Productivity*, McKinsey and Co, New York.

18 Fosso-Wamba, S., Akter, S., Edwards, A. *et al.* (2015) How "big data" can make big impact: Findings from a systematic review and a longitudinal case study. *International Journal of Production Economics*, **165**, 234–246.

19 Hessman, T. (2013) The dawn of the smart factory. *New Equipment Digest*, **78** (9)Online at: http://search.ebscohost.com/login.aspx?direct=true&db=bth&AN=90417201&site=ehost-live (accessed: 22 January 2016)..

20 McKinsey Report (2015) *How to Navigate Digitization of the Manufacturing Sector*. McKinsey Digital Report. https://www.mckinsey.de/files/mck_industry_40_report.pdf (accessed 01 October 2015)

21 Behmann, F. and Wu, K. (2015) *Collaborative Internet of Things (C-IoT): For Future Smart Connected Life*, Wiley, London.

22 Kim, Y., Choi, T.Y., Yan, T., and Dooley, K. (2011) Structural investigation of supply networks: A social network analysis approach. *Journal of Operations Management*, **29** (3), 194–211.

23 Burnson, P. (2013) New study released at CSCMP addresses "Big Data". *Supply Chain Management Review*, **17**, 76–78.

24 Borgatti, S.P. and Li, X. (2009) On social network analysis in a supply chain context. *Journal of Supply Chain Management*, **45** (2), 5–22.

25 Granovetter, M. (1985) Economic action and social structure: The problem of embeddedness. *American Journal of Sociology*, **91** (November), 481–510.

26 Sarkis, J., Zhu, Q., and Lai, K.H. (2011) An organizational theoretic review of green supply chain management literature. *International Journal of Production Economics*, **130** (1), 1–15.

27 Bonilla, D., Keller, H., and Schmiele, J. (2015) Climate policy and solutions for green supply chains: Europe's predicament. *Supply Chain Management: An International Journal*, **20** (3), 249–263.

28 Matt, D.T., Rauch, E., and Dallasega, P. (2015) Trends towards distributed manufacturing systems and modern forms for their design. *Procedia CIRP*, **33**, 185–190.

29 Kohtala, C. (2015) Addressing sustainability in research on distributed production: An integrated literature review. *Journal of Cleaner Production*, **106**, 654–668.

30 Benkler, Y. (2006) *The Wealth of Networks: How Social Production Transforms Markets and Freedom*, Yale University Press.

31 Johansson, A., Kisch, P., and Mirata, M. (2005) Distributed economies – a new engine for innovation. *Journal of Cleaner Production*, **13**, 971–979.

32 Slack, N., Chambers, S., and Johnston, R. (2010) *Operations Management*, Pearson education, London.

33 Kumar, M., Graham, G., Hennelly, P., and Srai, J. (2016) How will smart city production systems transform supply chain design: A product level

investigation. *International Journal of Production Research*, **54** (23), 7181–7192.

34 Smirnova, M., Naude, P., Hennenberg, S.C. *et al.* (2011) The Impact of Market Orientation on the Development of Relational Capabilities and Performance Outcomes; the Case of Russian Industrial Firms. *Industrial Marketing Management*, **40** (1), 44–53.

35 Cohen, W.M. and Levinthal, D.A. (1990) Absorptive capacity: A new perspective on learning and innovation. *Administrative Science Quarterly*, **35**, 128–152.

27

Smart Health Monitoring Using Smart Systems
Carl Chalmers

Department of Computer Science, Liverpool John Moores University, Liverpool, UK

CHAPTER MENU
Introduction, 773 Background, 775 Integration for Monitoring Applications, 786 Conclusion, 788

Objectives

- To investigate how smart gird technologies can be utilized for applications beyond generation, distribution, and consumption.
- To understand how electricity usage data can facilitate independent living, early intervention practice (EIP) for people living with self-limiting conditions.
- To investigate the use of smart meters for the behavioral analysis of individual patients with healthcare conditions.
- To become familiar with the end-to-end smart metering infrastructure and the role it plays in identifying behavioral trends.

27.1   Introduction

Each year, the number of people living with self-limiting conditions, such as dementia, Parkinson's disease, and mental health problems, is increasing [1]. This is largely due to individuals living longer and improvements in diagnosis and treatments. The number of populace living with dementia worldwide [2] is currently estimated at 35.6 million, and this number is set to double by 2030 and

Smart Cities: Foundations, Principles and Applications, First Edition.
Edited by Houbing Song, Ravi Srinivasan, Tamim Sookoor, and Sabina Jeschke.

more than triple by 2050. Additionally, one in four people currently experience some kind of mental health problem each year [3]. Supporting these sufferers places a considerable strain on organizations such as the National Health Service (NHS), local councils, frontline social services, and carers/relatives [4]. In monetary terms, dementia alone costs the NHS over £17 billion a year [5], exacerbated by the cost of depression patients, which is predicted to increase to 1.45 million in the United Kingdom, adding a further £2.96 billion cost to essential services by 2026.

This figure excludes other mental health conditions such as anxiety disorders, schizophrenic disorders, bipolar-related conditions, eating disorders, and personality disorders. Effective around-the-clock monitoring of these conditions can be a considerable challenge and often leads to patients having to reside in care homes and other accommodations. A safe independent living environment can be hard to achieve [6]. Currently, 20–30% of individuals with dementia are living alone, yet no technology exists that enables the automated monitoring. The same problems exist for patients with depression. Additionally, the number of people living alone has doubled over the last three decades, amounting to one in three people in the United Kingdom and United States. This is a growing concern as solitary living is proven to produce increased number of patients with depression [7].

In addition, the need to detect accurately sudden or worsening changes in a patient's condition is vital for early intervention. Community mental health groups, crisis and home resolution teams, assistive outreach teams, and early psychosis teams all play a key role in preventing costly inpatient admissions. If any changes are not dealt with early, the prognosis is often worse, and, as a result, costs for treatment will undoubtedly be higher [8]. An early intervention approach has been shown to reduce the severity of symptoms, improve relapse rates, and significantly decrease the use of inpatient care. Evidence suggests that a comprehensive implementation of early intervention practice (EIP) in England could save up to £40 million a year in psychosis services alone. Being able to detect deteriorating conditions in dementia patients earlier enables physicians to better diagnose and identify stage progression for the disease. This enables earlier intervention for the illness before cognitive deficits affect or worsen mental capacity, supporting the individual and their family in adapting to the illness simultaneously.

Analyzing a patient's electricity using smart meters ensures for accurate around-the-clock monitoring of patients by identifying certain and often subtle changes in behavior. This motoring can be utilized not only for safety but also for enabling the prediction of immediate, mid-, and long-term prognosis. In addition, this type of active monitoring enables the detection of certain patterns and trends to help facilitate early intervention. Monitoring patients using this method offers a vast improvement over existing assistive technologies as monitoring the physical and mental well-being of a patient becomes possible. By the end of 2020, the government aims to install smart meters in every household throughout the United Kingdom. This is also true for a large number of countries around the world such as Italy, United States, Netherlands, and Australia. The challenge, therefore, is how to interpret, analyze, and make use of the data collected by smart meters and the wider advanced metering infrastructure (AMI) for use in such monitoring applications.

Smart meters provide granular energy usage readings at 30 min intervals. However some countries such as Canada set their intervals as low as 15 min, which can be used to monitor and profile consumers. Additional devices, such as smart plugs, facilitate the identification of electrical appliances being used in the home, along with their duration of use and the amount of electricity being consumed. This information is extremely beneficial in determining abnormal user behavior, such as a device being left on for atypically long periods or devices not being used at all.

27.2 Background

27.2.1 Advanced Metering Infrastructure

The AMI provides bidirectional communication between the consumer and the rest of the smart grid. A smart grid is a complex modern electricity system [9]. It uses sensors, monitoring, communications, and automation to improve the electricity system. Smart grids fundamentally change the way in which we generate, distribute, and monitor our electricity. This enhanced communication removes the traditional need for energy usage readings to be collected manually. Instead, a robust automatic reporting system with greater granularity of readings is offered [10].

There are many advantages of deploying the AMI, some of which include reduced costs for meter readings (possibility to access meters otherwise difficult to attend due to position or security reasons), support for real-time pricing, increased fraud detection, and reduced read-to-bill time, to name a few. As part of the larger smart grid, the AMI can be broken down into three specific areas, each with their own roles and functions, as shown in Figure 27.1.

Home Area Network (HAN): The HAN is housed inside the consumer premises and is made up of a collection of different devices. Firstly, the in-home display (IHD) unit, which is the most visible and accessible part of the AMI. It provides the consumer with up-to-date information, in real time, on electricity usage, as well as the units of energy that are being consumed. Secondly, the smart meter provides real-time energy usage to both the consumer and all of the stakeholders. Smart meters are able to store 13 months of data, keeping a record of total energy consumption. In addition to smart

Key:

- Home area network (HAN)
- Wide area network (WAN)
- Utility companies and organizations

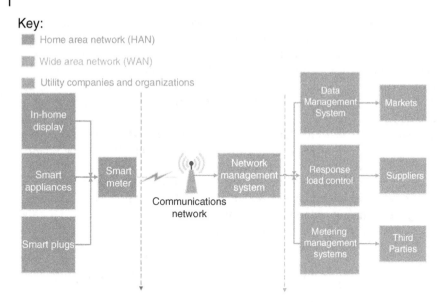

Figure 27.1 Advanced metering infrastructure.

meters, smart plugs enable the identification of the individual devices that are responsible for the reported energy usage. Additionally they provide detailed information about their duration of use.

Wide Area Network (WAN): This section of the AMI handles the communication between the HAN and the utility companies. The WAN is responsible for sending all polled meter data to the utility companies and other grid stakeholders, using a robust backhaul network, such as Carrier Ethernet, GSM, CDMA, or 3G [11]. The geographical location of the smart meter dictates what WAN technologies need to be implemented due to the constraints of certain communication technologies. The data aggregator unit (DAU) is the communication device that is used to collect the energy usage data from the home gateway or the smart meter. The acquired data is transmitted using one of the communication technologies mentioned above to the control center. The meter data management system (MDMS) is the central control center, which provides the storage and data processing capabilities for the obtained smart meter data. The MDMS also collects information regarding the status for the generation, transmission, and distribution of the wider smart grid.

Service users are a number of organizations and utility companies that have access to the data for analysis purposes. Energy suppliers communicate remotely with the smart metering equipment in order to perform a number of tasks such as taking meter readings, updating price information on the IHD, and identifying readings on a change of supplier or change of tenancy.

Energy network companies access data to evaluate the loads on their network, at the local level, and to respond appropriately to loss of supply. Consumers can allow other organizations to have access to the data from their smart meter. For example, energy switching sites could use the accurate information generated from smart meters to establish the amount of energy used by the consumer and advise on the best tariff based on their individual energy requirements.

27.2.2 Smart Meters

Smart meters are seen as one of the most important components of the AMI and smart grid [12]. They are the foundation of any future smart electricity grid and provide consumers with highly reliable, flexible, and accurate metering services. Smart meters provide real-time energy usage readings at granular intervals to both the consumers and all of the smart grid stakeholders [13]. They obtain information from the end users' load devices and measure the energy consumption. Added information such as any home-generated electricity is provided to the utility company and/or system operator for better monitoring and billing process. This is achieved by monitoring the performance and the energy usage characteristics of the load on the grid.

Smart meters are able to store 13 months of historical energy usage, which allows for the creation of detailed energy usage profiles [14]. Currently readings are taken at 30-min intervals; however as these meters become more sophisticated, they are able to measure household power consumption at ever finer timescales [15]. Smart meters are able to report energy usage as low as 1 min intervals, even though this is not currently widely deployed due to the vast amount of data it would generate [16]. This is important when trying to identify individual devices and their duration of use. Figure 27.2 highlights the additional information obtained from increasing the frequency of reading. This can have a significant impact on the ability of identifying individual devices.

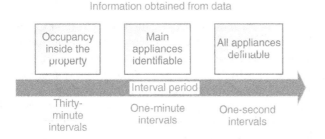

Figure 27.2 Information obtained by increasing interval reading.

Smart meters are able to perform a wide variety of roles and bring many benefits over the traditional electricity meter. Some of these roles and benefits include the following:

- Accurately record and store information for defined time periods (to a minimum of 30 min). This enables remote, accurate meter readings with no need for estimates [14]. As these meters become more sophisticated, they are able to measure household power consumption at ever finer timescales.
- Offer two-way communications to and from the meter so that, for example, suppliers can read meters and update tariffs remotely [17].
- Enable customers to collect and use consumption data by creating a HAN to which they can securely connect data access devices [18].
- Communicate with microgeneration, home appliances, and equipment within the property [19]. Smart meters will be able to control smart home appliances and communicate with other smart meters within reach. This enables devices to be switched on when grid demand is low and turned off when demand is high.
- Allow customers to collect and use consumption data by creating a HAN to which they can securely connect data access devices.
- Enable other devices to be linked to the HAN, permitting customers to improve their control of energy consumption [20].
- Support time-of-use tariffs, under which the price varies depending on the time of day at which electricity is used [21]. Energy prices are more expensive during peak times. Billing consumers by time, as well as usage, encourages them to change their consumption habits.
- Support future management of energy supply to help distribution companies manage supply and demand across their networks [22]. This is achieved automatically through previously agreed demand response actions.

Figure 27.3 shows the different metering infrastructures; it highlights the key differences between the conventional energy meter and the smart meter. The newer smart meter removes the need for manual data collection. Instead energy usage data is collected automatically, providing instant readings and automatic billing.

Smart meters collect and upload a wide variety of data from consumer usage to power generation. This presents an opportunity to accommodate independent living while monitoring for safety. As previously stated smart meters will be installed in every UK property by the year 2020. Within this frame, the smart meter becomes a potential tool for the implementation of policies relevant to the community as generator and storage of information and sensitive data. Being able to interface with these devices enables a wide variety of applications and services. A selection of some of the data parameters that are collected are shown in Table 27.1

Conventional energy meter

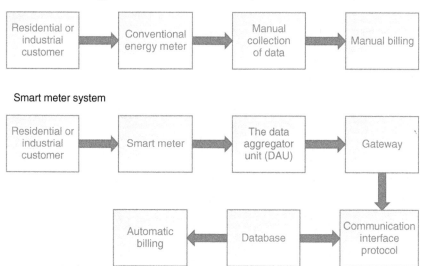

Smart meter system

Figure 27.3 Metering architectures of a conventional energy meter and a smart meter.

Table 27.1 Smart meter data parameters.

Reading	Description
Generated interval data (kW)	Half-hour interval held on meter for 13 months – average kW demand over half-hour period
Generated kilovolt–ampere–reactance (kVAr)	Reactive power measurement in half hourly interval held on meter for 13 months – average kVAr demand over half-hour period
Generation technology type	For example, solar PV, micro-CHP, wind, hydro, anaerobic digestion
Import demand (kW)	Load being drawn from grid
Export (kW)	kW being exported to grid
Total consumption today (kW h)	Import + generated – export
Cost of energy imported (£/h) and £ today	Net cost of imported energy and less value of exported energy. Pushed to the IHD via SMS for the consumer
Value of exported energy (£/h) and £ today	Calculated from meter export value and sell rate. Net cost of imported energy, less value of exported energy. Pushed to the IHD via SMS for the consumer
Total saved by generation	Calculated from (generated – export) × import (ppu)

Table 27.2 Single smart meter data sample taken from a dataset containing 78,000 individual smart meters.

CUSTOMER_KEY	Time of reading	General supply (kW h)
8150103	05:59:59	0.042
8150103	06:29:59	0.088
8150103	06:59:59	0.107
8150103	07:29:59	0.040
8150103	07:59:59	0.042
8150103	08:29:59	0.041
8150103	08:59:59	0.049
8150103	09:29:59	0.189
8150103	09:59:59	0.051
8150103	10:29:59	0.050

Table 27.2 shows a data sample from a smart meter during a single period. This sample shows the granularity of the data collected compared with traditional meters, where the readings are submitted collectively over a much larger period (e.g., monthly, quarterly, or yearly). It displays the data parameters obtained at each 30-min interval, totaling of 48 individual readings in a 24-h period (10 readings are shown). Customer key identifies the individual smart meter device within the AMI; time of reading indicates the time and date of the reading, while general supply highlights the amount of on peak electricity being used in kW h.

Figure 27.4 shows readings taken from an individual's smart meter over a single 24-h period. Each of the 48 individual readings represents the total amount of electricity being consumed in kW h at each 30-min interval. This frequency of readings makes it possible to identify certain daily activities as shown above.

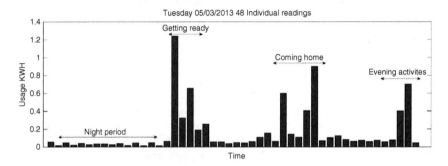

Figure 27.4 Forty-eight individual readings showing a single 24-h period.

Table 27.3 Home plug readings over a 1-h period.

Plug name	Reading time	Total (kW h)
Oven	17/06/2013 09:00	13.099
TV2	17/06/2013 09:00	2.787
Washing	17/06/2013 09:00	0.553
Aircon	17/06/2013 10:00	12.873
Computer	17/06/2013 10:00	1.423
Dishwasher	17/06/2013 10:00	2.641
Dryer	17/06/2013 10:00	0.583
Hot water system	17/06/2013 10:00	37.734
Microwave	17/06/2013 10:00	0.461
Oven	17/06/2013 10:00	13.099
TV	17/06/2013 10:00	1.744
TV2	17/06/2013 10:00	2.797
Washing	17/06/2013 10:00	0.553

Any change in energy usage will enable the identification of any change in routine and habit.

In addition to smart meters, smart plugs that monitor individual electronic devices in the home can be used to establish a more accurate profile. Data from these smart plugs enable the identification of each electrical device and which appliance is responsible for the electrical load at each 30 min reading. Smart plug reading frequencies can be reduced from minutes to seconds to provide a more detailed analysis of a person's behavior while showing the exact duration of use for each electrical device.

Smart plugs can interface directly with the smart meter using ZigBee Smart Energy; the ZigBee Alliance forms a collection of device descriptions and functions that allow energy providers to manage and monitor energy loads to optimize consumption. Over a million ZigBee electric meters are deployed by many utility companies in the United States with UK Department of Energy and Climate Change (DECC) announcing SMETS 2, which cites ZigBee Smart Energy 1.x. Table 27.3 shows an example of the data generated by home plug readings over 1 h.

27.2.3 AMI Implementation Challenges

Using smart meter data to examine the behaviors of vulnerable people enables patients to live independently while safe in the knowledge that they are being actively monitored. This however presents many technical, ethical, and privacy challenges.

Firstly, the scale and size of data collected from smart meters and the AMI presents real and complex difficulties in terms of storing, structuring, and analyzing the data. New methods for analyzing and modeling data will need to be developed with the focus on using cloud platforms and data centers. Cloud platforms, such as Azure, currently have the ability to analyze large datasets and assign virtually unlimited resources in order to process the data in a timely manner.

Secondly, due to the vast scale of the smart grid, ensuring standards, interoperability, and continuity throughout the system is a challenge. This is largely due to the integration of interchangeable components from a variety of different providers [23]. Additionally, there is an ever-increasing interdependency between control systems, such as SCADA, and other commercial networks.

Thirdly, there are many ethical and privacy concerns associated with the smart meter role out, which could potentially leave consumers vulnerable to exploitation [24]. For example, criminals could process data that is generated by the AMI to identify when households are unoccupied, helping to facilitate burglary or some other crime. Additionally, being able to identify appliances would allow burglars to identify and target households with the most electronic devices. Criminal activities are not the sole concern however. Law enforcement agencies could exploit the data for a variety of purposes. These include categorizing properties being used for the purpose of producing drugs due to the implementation of heat lamps, hydroponics, and so on or proving/disproving premises occupation. Commercial entities, using targeted advertising at a specific household, or simply being able to know when someone is home to take a sales call, are further possible applications.

Being able to identify legitimate and beneficial applications is imperative for long-term success of the AMI. People justifiably have concerns on how their data will be utilized and accessed. Any misuse of data or incidents of security breach will lead to large-scale resistance of their use. This resistance is already being seen in the United States, for example, where groups of consumers are actively refusing smart meter installations [25]. In Philadelphia, some customers have opted out of having the smart meters installed because of its two-way communication capabilities. Residents are concerned that government organizations will use the information for spying on their activities. These concerns will need to be addressed by providing transparency on how the collected data will be used. In one significant case, the First Chamber of the Dutch parliament rejected two smart metering bills in 2009 because of privacy concerns, forcing the government to add significant privacy protections to the revised bills that were passed in 2011 [26].

27.2.4 Patient Behavior and Uses

As previously discussed, smart meter data enables active in-home monitoring. By analyzing past behaviors, an improved prediction of worsening conditions is made possible. Analyzing the data in this manner facilitates early intervention and an improved outcome for the patient by ensuring that their medical and care needs are sufficient. Being able to detect and predict these changes requires a detailed understanding of the symptoms and behaviors that are expected for each condition. The following section discusses the features for each condition and the monitoring applications.

27.2.4.1 Active Monitoring for Behavioral Changes with Dementia

There are a common set of features of Alzheimer's disease and other dementias. These include agitation, anxiety, depression, apathy, delusions, sleep and appetite disturbance, elation, irritability, disinhibition, and hallucinations [27]. The severity of each symptom differs at various stages of the disease. Therefore, any system would need to be fully adaptable to these changes, as patient's progress through the different stages of the illness [28]. Particularly, as later stages of the disease are regarded as important as (if not more important than) the earlier stages, they tend to harbor unique characteristics and events, which occur. These affect the lives of the patients and their carers. Behavioral problems, such as agitation, become more pronounced in the later progressive stages of the disease.

Currently, the Mini–Mental State Examination (MMSE) [29] is used by clinicians to help diagnose dementia and assess its progression and severity. The 6-Item Cognitive Impairment Test (6CIT) is also used for similar purposes [30]. These tests would need to be performed at regular intervals to identify the stage of the disease. The process involves identifying the correct characteristics from patient datasets.

Figure 27.5 highlights the MMSE in more detail, showing the need for changes in the monitoring techniques, as the severity of the disease increases.

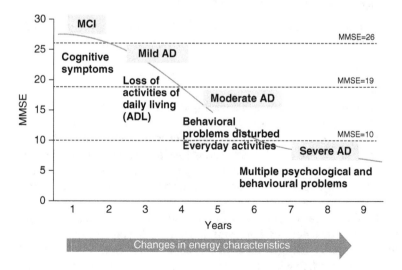

Figure 27.5 MMSE graph.

The feature vectors would need to change regularly, along with the algorithms used, in order to maintain system accuracy for each stage of the disease.

Figure 27.5 displays, although each person with dementia experiences the illness in their own individual way, their common traits in behavior that can be identified [31]. These symptoms are listed below and can be detected through their electricity usage:

- *Loss of mobility* – People with dementia gradually lose their ability to perform everyday tasks. They will usually perform tasks at a much slower rate and are more likely to fall due to a reduction in mobility.
- *Eating* – People with dementia often lose weight due to a number of factors, which range from forgetting to make meals to finding it hard to eat. Sufferers often need constant help and encouragement to ensure they consume enough food and liquids. Failing to monitor and ensure that the patient consumes regular meals can increase the likelihood of further falls and other complications.
- *Unusual behavior* – A sufferer's behavior can drastically change especially in later stages of dementia. A person might become more agitated and confused in the late afternoon and early evening, so extra attention is needed during these periods. In addition, some sufferers might experience hallucinations and delusions and may alter how they interact with their environment. Some patients might become more restless because they often need more physical activity; by contrast the patient might have long periods of physical inactivity.
- *Side effects of medication* – Drugs that are prescribed to dementia sufferers can have severe side effects, which can often increase a person's confusion. Patients can often be prescribed with doses that are too high or drugs that are no longer appropriate to the patient's needs.

Being able to detect changes in a patient's habits, routines, and features, as highlighted above, will ensure the active monitoring of their well-being.

27.2.4.2 Active Monitoring for Behavioral Changes in Depression and Other Mental Illness

Severe depression exhibits many behavioral symptoms similar to dementia, for example, memory problems and social disengagement. Additionally, depression can cause physical complications, such as chronic joint pain, limb pain, back pain, gastrointestinal problems, tiredness, sleep disturbances, psychomotor activity changes, and appetite changes [32]. These changes can be reflected in how the sufferer interacts with people, their environment, and their electric devices.

Specifically, during periods of severe depression, the sufferer might interact less with their electrical devices. For example, they might stay in bed for longer durations (insomnia or hypersomnia) or not cook meals (change in appetite) [33]. Changes in sleep behaviors and appetite are all reflected through energy usage. Such behavior could be easily identified and flagged for further investigation where appropriate. Being able to detect early any erratic of sudden behavior change caters for better intervention and can lead to an early diagnosis of psychosis. Each individual is different and, as such, presents their own set of symptoms and warning signs; however one or more warning signs are likely to be evident:

- Memory problems
- Severe distractibility
- Severe decline of social relationships
- Dropping out of activities – or out of life in general
- Social withdrawal, isolation, and reclusiveness
- Odd or bizarre behavior
- Feeling refreshed after much less sleep than normal
- Deterioration of personal hygiene
- Hyperactivity or inactivity, or alternating between the two
- Severe sleep disturbances
- Significantly decreased activity.

27.2.4.3 Prediction for EIP

For dementia sufferers, knowledge about a person's ability to undertake normal activities of daily living (ADL) is an essential part for the overall assessment [34]. This is imperative in determining the diagnosis and enabling the accurate evaluation of any changes. Being able to detect subtle changes early and predict future cognitive and noncognitive changes facilitate much earlier intervention. Often, dementia sufferers in hospital are admitted due to other poor health caused by other illnesses [35].

These illnesses are often a result of immobility in the patient; most common infections cause additional complications and can also speed up the progression of dementia [36]. Additionally, immobility leads to pressure sores, which can easily become infected, other serious infections, and blood clots, which can be fatal. With any of these complications, early intervention for both preventative care and early treatment is vital to ensure a good prognosis and safe independent living.

27.2.5 Current Assistive Technologies

In this section, an investigation into current devices (assistive technology), which enable or aid independent living for people with self-limiting conditions and their associated benefits, is presented [37]. The term assistive technology refers to any device or system that allows an individual to perform a task that they would otherwise be unable to do or increases the ease and safety with which the task can be performed. There are many benefits associated with assistive technologies from the ability to promote independence and autonomy, both for the person with dementia and those around them [38]. Additionally they help manage potential risks in and around the home.

Surprisingly [39], assistive technology covers a wide range of equipment from simple low-tech devices, such as handrails and grips, to high-tech equipment that includes power wheelchairs and robots. Limited technology exists, which enables proactive monitoring for people living with these conditions. Monitoring technologies include video monitoring, fall detectors, and health monitors [40]. Environmental (passive) motion sensing devices, such as a smart floor, have pressure sensors located in the floor that track the movement and location of the resident within the home. While most of these solutions help with the monitoring of physical impairments, none cater for mental and inactivity monitoring, which is crucial for conditions like Alzheimer's disease and depression.

27.3 Integration for Monitoring Applications

It is clear that the number of people living with self-limiting conditions is increasing exponentially. Being able to actively monitor these patients to facilitate independent living and enable EIP is a huge challenge. In its current form it is both costly and impossible to achieve a realistic outcome as current assistive technologies do not take into account the routines and habits of a patient. Utilizing smart meter data provides the ability to detect changes in:

- Sleep patterns
- Eating
- Activity
- Social interaction
- Routines
- Behavioral changes.

Detecting changes in behavior, routine, and immobility is possible by monitoring a patient's electricity usage. This, in turn, enables earlier intervention, where needed, and the possibility for independent living. Figure 27.6 highlights how the data can be utilized to achieve different monitoring applications.

The first application facilitates the need for active real-time monitoring to enable independent living. Should certain conditions or circumstances arise, such as no electricity usage or excessive usage, immediate emergency intervention from the patient's support network would be required. The second application enables behavioral changes, such as reduced or prolonged inactivity, to be identified to facilitate early intervention. This ensures a better outcome for the patient and reduced cost due to less hospital admissions and a reduction in expensive care. Lastly performing historical behavioral analysis assists healthcare professionals in identifying worsening conditions. Additionally, should a

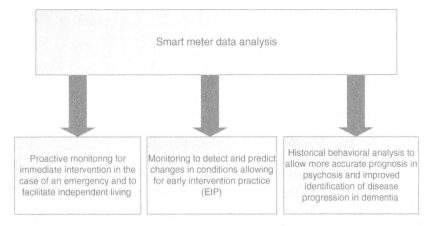

Figure 27.6 Data applications for health monitoring.

patient's medication change, this type of analysis can assist in identifying suitability or complications from possible side effects.

27.3.1   Case Study

In this section, a case study is presented for one application and condition, which highlights an individual's habits and routines using smart meter data. Being able to recognize changes and patterns in behavior aids in the ability to ensure a better outcome by ensuring a more personalized healthcare treatment/package for the patient. Normally, people who suffer mental illness exhibit certain behavioral changes during periods of heightened severity. One of these most usual changes is the alterations in sleep patterns. Typically, a sufferer will awaken much earlier than normal. Figure 27.7 shows the total energy consumption between the hours of 1:30 a.m. and 4:00 a.m. over a 1-year

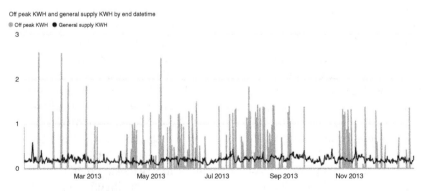

Figure 27.7 Energy usage over a 1-year period between the hours of 1:30 a.m. and 4:00 a.m.

period. Each of the larger peaks displays an increase in electricity during the early hours of the morning.

This type of result highlights changes in the person's sleep patterns, which could provide an indication to a healthcare professional whether intervention is required. This method can also be used to identify long periods of inactivity that keep on recurring. In many conditions early intervention is important where inactivity is prolonged and reoccurs frequently.

27.4   Conclusion

The smart grid addresses the constraints imposed by the current energy generation and distribution infrastructure by allowing detailed monitoring and consumer energy usage profiling. This leads to more efficient energy usage, planning, and fault tolerance. Being able to collect and analyze sufficient amounts of usage data makes it possible to identify reoccurring patterns and trends. This can be used to address many problems not just in the electricity and generation paradigm but also in health monitoring. However, its implementation produces extensive datasets, which can also be used by researchers, customers, and energy providers to build an accurate representation of user behavior. We have demonstrated how this representation has applications in the field of health, as explored here, and elsewhere.

As smart meters are largely financed by energy suppliers and the government, there would be little expense involved in adapting the implementation for social care needs. Utilizing smart meters for monitoring may provide a low cost solution to the problems discussed previously. Additional considerations need to be carefully addressed in relation to privacy and data protection. Being able to identify a consumer's daily habits and routines could place the individual at serious risk should the data be accessed by unauthorized parties. Strict policies and appropriate technologies should be investigated and deployed in order to protect the consumer.

Processing and storing all of the data that is generated by the AMI is a huge technical challenge [41]. New data and communication traffic is introduced at every level, which will generate billions of data points from thousands of system devices. Efficient data transport and analysis will need to be introduced in order to cope with the vast amount of additional data [42]. In the United Kingdom the communications and data storage providers have already been contracted by the DECC, which will be regulated by Ofgem. We have discussed how a new communication standard ZigBee Smart energy can be used to interface additional devices such as smart plugs with the smart meter to help achieve more detailed monitoring. This new communication gateway could potentially be utilized to extract the relevant data to help facilitate health monitoring applications. This method could potentially bypass the need to retrieve the data from the MDMS by installing a custom device to revive and submit the data from within the patient's home. In order to utilize the data in the best manner, a detailed understanding about a patient's condition and expected behaviors is essential to ensure accurate monitoring.

Final Thoughts

In this chapter we discussed the concept of using smart grid technologies for applications beyond their intended use. In particular we focused on one key and important element of the smart grid, the AMI. We investigated how the end-to-end smart metering system could be used to accurately profile a person's behavior by identifying how the individual interacts with their electrical devices. Finally we discussed how this data could be used in the field of healthcare, particularly for the purposes of facilitating independent living and providing a sensor-free health monitoring environment.

Questions

1   What are the four main benefits associated with the AMI implementation?

2   Name the three main components of the AMI and describe their individual functions.

3   Describe what behavioral activities can be identified for each of the following reading frequencies: 30 min, 1 min, and 1 s.

4   List the three main data applications for health monitoring.

References

1 Department of Health (2012). *Report. Long-Term Conditions Compendium of Information*: 3rd edition.
2 Chan, M. (2012) *Dementia A public health priority*, World Health Organization.
3 Mental Health Foundation, *The Fundamental Facts The Latest Facts and Figures on Mental Health*, http://www.mentalhealth.org.uk/content/assets/ PDF/publications/fundamental_facts_2007.pdf?view=Standard (accessed June 2015).
4 Knapp, M. and Prince, M. (2007) Dementia UK the Full Report.

5 McCrone, P., Dhanasiri, S., Patel, A., Knapp, M. and Lawton-Smith, S., *PAYING THE PRICE The Cost of Mental Health Care in England to 2026*, King's Fund.

6 Lehmann, S.W., Black, B.S., Shore, A. *et al.* (2010) Living alone with dementia: Lack of awareness adds to functional and cognitive vulnerabilities. *International Psychogeriatrics*, **22**, 778–784.

7 People who live alone more likely to develop depression, http://www.nursingtimes.net/nursing-practice/specialisms/public-health/people-who-live-alone-more-likely-to-develop-depression/50 43117.article (accessed June 2015).

8 Mental Health Network NHS CONFEDERATION (2011) *Early intervention in psychosis services*.

9 Anas, M., Javaid, N., Mahmood, A., Raza, S.M., Qasim, U., and Khan, Z.A. (2012) *Minimizing Electricity Theft using Smart Meters in AMI*, Seventh International Conference on P2P, Parallel, Grid, Cloud and Internet Computing.

10 Popa, M. (2011) *Data Collecting from Smart Meters in an Advanced Metering Infrastructure*, INES 2011 15th International Conference on Intelligent Engineering Systems.

11 Bennett, C. and Highfill, D. (2008) *Networking AMI Smart Meters*, IEEE Energy 2030, November 2008.

12 Molina-Markham, A., Shenoy, P., Fu, K., Cecchet, E., and Irwin, D. (2010) *Private Memoirs of a Smart Meter*, BuildSys '10 Proceedings of the 2nd ACM Workshop on Embedded Sensing Systems for Energy-Efficiency in Building, pp. 61–66.

13 Wang, L., Devabhaktuni, V., and Gudi, N. (2011) *Smart Meters for Power Grid – Challenges, Issues, Advantages and Status*, IEEE/PES Power Systems Conference and Exposition (PSCE), pp. 1–7, March 2011.

14 Benzi, F., Anglani, N., Bassi, E., and Frosini, L. (2011) Electricity smart meters interfacing the households. *IEE transactions on Industrial Electronics*, **58** (10), 4487–4494.

15 McKenna, E., Richardson, I., and Thomson, M. (2012) Smart meter data: Balancing consumer privacy concerns with legitimate applications. *Energy Policy*, **41**, 807–814.

16 Vojdani, A. (2008) Smart integration. *IEEE Power & Energy Magazine*, **71** (9), 71–79.

17 Bennett, C. and Highfill, D. (2008), *Networking AMI Smart Meters*, November 2008, IEEE Energy 2030.

18 Krishnamurti, T., Schwartz, D., Davis, A. *et al.* (2011) Preparing for smart grid technologies: A behavioral decision research approach to understanding consumer expectations about smart meters. *Energy Policy*, **41**, 790–797.

19 Sreenadh, S.S., Depuru, R., Wang, L., and Devabhaktuni, V. (2011) Smart meters for power grid: Challenges, issues, advantages and status. *Renewable and Sustainable Energy Reviews*, **15**, 2736–2742.

20 Venables, M. (2007) Smart meters make smart consumers. *Engineering & Technology*, **2** (4), 23.

21 Olmos, L., Reuster, S., Liong, S.J., and Glachant, J.M. (2011) Energy efficiency actions related to the rollout of smart meters for small consumers, application to the Austrian system. *Energy*, **36** (7), 4396–4409.

22 Anas, M., Javaid, N., Mahmood, A., Raza, S. M., Qasim, U., and Khan, Z. A. (2012) *Minimizing Theft using Smart Meters in AMI*, Seventh International Conference on P2P, Parallel, Grid, Cloud and Internet Computing.

23 MacDermott, Á., Shi, Q., Merabti, M., and Kifayat, K. (2009) *Intrusion Detection for Critical Infrastructure Protection, Neural Networks*, IJCNN 2009, pp. 14–19.

24 Lisovich, M.A., Mulligan, D.K., and Wicker, S.B. (2010) Inferring Personal Information from Demand-Response Systems. *IEEE Security & Privacy*, **8**, 11–20.

25 Is your home's energy meter spying on you? http://www.foxnews.com/us/2014/04/17/is-your-home-energy-meter-spying-on/ (accessed July 2015).

26 Brown, I. (2013) Britain's smart meter programme: A case study in privacy by design. *International Review of Law, Computers & Technology*, **2**, 172–184.

27 Mega, M.S., Cummings, J.L., Fiorello, T., and Gornbein, J. (1996) The spectrum of behavioural changes in Alzheimer's disease. *Neurology*, **46**, 130–135.

28 Byrne, E.J., Collins, D., and Burns, A. (2006) Behavioural and psychological symptoms of dementia-agitation, in *Severe Dementia* (eds A. Burns and B. Winblad), Wiley, pp. 51–61.

29 MacDowell, I., Kristjansson, B., Hill, G.B., and Hebert, R. (1997) Community screening for dementia: The Mini Mental State Exam (MMSE) and modified Mini-Mental State Exam (3MS) compared. *Journal of Clinical Epidemiology*, **50**, 377–383.

30 Abdel-Aziza, K. and Larner, A.J. (2015) Six-item cognitive impairment test (6CIT): pragmatic diagnostic accuracy study for dementia and MCI. *International Psychogeriatrics*, **27**, 991–997.

31 Alzheimer's Society (2014) Changes in behaviour fact sheet, October 2014.

32 Trivedi, M.H. (2004) The link between depression and physical symptoms. *Primary Care Companion to the Journal of Clinical Psychiatry*, **6**, 12–16.

33 Dowrick, C. (2004) *Beyond Depression a New Approach to Understanding and Management*, Oxford University Press, Oxford.

34 Early Psychosis, http://www.earlypsychosis.ca/pages/curious/warning-signs-of-psychosis (accessed October 2015).

35 Selikson, S., Damus, K., and Hameramn, D. (2015) Risk factors associated with immobility. *Journal of the American Geriatrics Society*, **8**, 707–712.

36 Alzheimer's Society (2012) The later stages of dementia, Factsheet, March 2012.

37 Royal Commission on Long Term Care (1999) *With respect to old age: long term care – rights and responsibilities*, The Stationery Office, London.

38 Alzheimer's Society (2013) *Assistive technology – devices to help with everyday living*, Alzheimer's Society.

39 Hurst, A. and Tobias, J. (2011) *Empowering Individuals with Do-It-Yourself Assistive Technology*, The proceedings of the 13th international ACM SIGACCESS conference on Computers and accessibility.

40 Daniel, K., Cason, C.L., and Ferrell, S. (2009) *Assistive Technologies for Use in the Home to Prolong Independence*, Proceedings of the 2nd International Conference on PErvasive Technologies Related to Assistive Environments.

41 Bouhafs, F., Mackay, M., and Merabti, M. (2012) Links to the Future Communication Requirements and Challenges in the Smart Grid. *IEEE Power & Energy Magazine*, **10**, 24–32.

42 Department of Energy and Climate Change (2013) Smart Metering Implementation Program.

28

Significance of Automated Driving in Japan
Sadayuki Tsugawa

Intelligent Systems Research Institute, National Institute of Advanced Industrial Science and Technology, Tsukuba, Japan

CHAPTER MENU

Introduction, 793
Definitions of Automated Driving Systems, 794
A History of Research and Development of Automated Driving Systems, 795
Expected Benefits of Automated Driving, 804
Issues of Automated Driving for Market Introduction, 805
Possible Market Introduction of Automated Driving Systems in Japan, 808
Conclusion, 815

Objectives

- To introduce the history and the state of the art of the development of automated driving systems.
- To broaden the readers' understanding for the process of introducing automated or semiautomated driving systems to the market.
- To point out the benefits of automated driving.
- To explore which kind of automated or semiautomated driving are realizable for the middle and the far future.

28.1 Introduction

Automated driving systems of road vehicles [1, 2] can be briefly defined as the introduction of automation to road transportation systems, with the objective to provide automated solutions to traffic accidents and congestion. Automation will not only facilitate the elimination of human errors and latency during

Smart Cities: Foundations, Principles and Applications, First Edition.
Edited by Houbing Song, Ravi Srinivasan, Tamim Sookoor, and Sabina Jeschke.
© 2017 John Wiley & Sons, Inc. Published 2017 by John Wiley & Sons, Inc.

human driving, thus preventing traffic accidents, but also enable precise lateral and longitudinal vehicle control. This will increase road capacity, thus mitigating congestion and, in doing so, saving energy and reducing the environmental impact. Road vehicle automation can be regarded as a new transportation means for vulnerable road users like the physically disabled and elderly who cannot drive.

The first proposal of automated driving systems was made at the New York World Fair in 1939, and research and development (R&D) in this field has a long, but stop–start, history beginning in the 1950s. In the 21st century, R&D started to aim to introduce automated systems in the near future. One such project was the implementation of automated driving systems in automated truck platoons in order to save energy, which was undertaken by public projects in the European Union, United States, and Japan. Initial competitions regarding the creation of robotic vehicles were sponsored by DARPA. As a result, Google has been developing autonomous passenger cars since 2009, and, probably inspired by Google cars, other automobile companies and suppliers have been developing autonomous cars with the aim to implement them into the market soon. However, besides legal and institutional issues, there are still technological issues like hardware and software reliability to resolve. Japan has a number of unique characteristics that cannot be observed in any other industrialized country. These characteristics include the decreasing size of the Japanese population and the increasing size of the senior citizen population. This chapter will discuss the issues specific to this field and the near-future market introduction, including a short history and the expected benefits of automated driving systems, and the significance of automated driving systems of road vehicles in Japan.

28.2 Definitions of Automated Driving Systems

Automated driving can be defined as a type of driving where all perception, decision making, and operation tasks normally carried out by a human driver while driving a vehicle are replaced by a system. A human driver is out of the control loop, and the responsibility of driving no longer lies with the driver. This is a very narrow definition and in a wider sense, automated driving also includes driver assistance systems where the responsibility still remains with the driver. In this chapter, we focus on the narrow sense of automated driving.

Table 28.1 shows the definitions of automated driving levels as defined by the Japanese government, that is, the automated driving at Levels 2–4 can be classed as automated driving; however, automation Level 2 is in essence just driver assistance and automation Level 3 is automated driving in the narrow sense along with the possibility of transitioning control over to driver assistance systems. This chapter mainly deals with Level 4.

Table 28.1 Automated driving levels and their definitions (Japanese government Cabinet Office, 2015).

Level	Definition by cabinet office	Systems	
Level 1	One of the operations of acceleration, steering, and braking is performed by the system	Driver assistance	
Level 2	Plurality of the operations of acceleration, steering, and braking are performed by the system	Semiautomated driving	Automated driving
Level 3	All of the operations of acceleration, steering, and braking are performed by the system, and upon request from the system human drivers overtake the control		
Level 4	All of the operations of acceleration, steering, and braking are performed by the system, and human drivers do not take the control at all	Fully automated driving	

28.3   A History of Research and Development of Automated Driving Systems

The world's first proposal for an automated driving system was probably "Fururama," which was exhibited during the World Fair of 1939 in New York by General Motors. The term Fururama is a compound of two words, "future" and "panorama," and the objectives were not to find solutions to road transportation issues, accidents, and congestion but instead to display the possibilities of a future society and living style in 1960. It was in the United States in the 1950s that the earliest research started looking into automated driving systems with the aim of finding solutions to road transportation issues. This began with a proposal by Dr. V. K. Zworykin, the vice president of RCA at that time. In Japan, the research on automated driving started in the 1960s, and the history of R&D in Japan will be described later in more detail [3].

28.3.1   Classification of Automated Driving Systems

Automated driving systems can be classified into autonomous systems or cooperative systems involving lateral control. The autonomous system is defined as a system that employs only onboard intelligence for lateral control and utilizes existing components on roadways like lane markers and guard rails, which are not installed for an automated driving purpose. It can also employ GPS and use precise digital road maps. On the other hand, a cooperative system is defined as a system that uses roadway intelligence specifically installed for automated driving, like inductive cables, magnetic markers, and onboard intelligence.

A cooperative system also exists among vehicles and between roadways and vehicles where communication plays an important role. Communication among vehicles will improve safety and vehicle efficiency, while communication between roadways and a vehicle will provide the driver with more traffic information.

The author divides a history of automated driving systems into four periods according to the technologies employed: the first period covers the 1950s and 1960s, the second period begins in the 1970s and ends in the mid-1980s, the third period continues until the end of the 1990s, and the fourth period starts in 2000. The main focus of the first period was R&D of cooperative systems; the second period focused on R&D of autonomous systems; and the third period on large-scale demonstrations of various systems as a part of huge national projects in intelligent transport systems (ITS). The fourth period was interested in field operation tests of systems involved with applications aiming for near-future market introduction. Table 28.2 summarizes the history of automated driving.

28.3.2 The First Period of Automated Driving Systems

The automated driving systems created in the first period were cooperative systems involving inductive cables for lateral control being embedded under a road

Table 28.2 A history of automated driving systems and technology.

Period	Technologies	Objectives	Vehicles
1939–1940 New York World Fair	"Fururama"	Future society of 1960	
First period 1950s–1960s	Cooperative, inductive cable	Safety	Single passenger car
Second period 1970s–1980s	Autonomous, machine vision	Safety	Single passenger car
Third period 1980s–1990s	Testing of various systems	Safety, efficiency	Platoon, passenger car, transit bus, truck
Fourth period 2000–	Verification of feasible technologies	Energy saving, environment impact reduction, efficiency, convenience	Platoon, transit bus, truck

(a)

(b)

Figure 28.1 The automated driving system with inductive cable: (a) the vehicle and (b) the inductive cable and onboard sensors at the front bumper.

surface. In the late 1950s and 1960s, the main body of research was conducted in the United States [4–6], the United Kingdom, West Germany [7], and Japan [8].

Figure 28.1 shows the automated vehicle developed by the Mechanical Engineering Laboratory in the early 1960s. The sensing equipment attached at the front bumper and the inductive cable embedded under the road surface for lateral control can also be seen. It drove at 100 km/h in 1967 on a test track.

The main advantage of the systems created in the first period was the active indication of the vehicle's driving course even under adverse conditions; however disadvantages included the laying of cables under road surfaces, the current feeding into the cables, and maintenance. Due to these disadvantages, the systems have only been employed in restricted areas like test tracks. Some vehicle testing systems with the embedded cable have been developed for test tracks.

A few systems with the cable were employed on public roads in the 1980s. In Halmstad (Sweden) and Fuerth (West Germany), the system was employed for transit buses in order to perform precision docking. Precision docking is automated driving only in the vicinity of bus stops in order to make a bus's approach to the platform as close as possible in order to facilitate easy embarkment and disembarkment of passengers in wheelchairs and with baby buggies. The short-distance installation of the cable system justifies the cost.

28.3.3 The Second Period of Automated Driving Systems

The main outcome of the second-period systems was the creation of vision-based autonomous vehicles in the 1970s and mid-1980s. Onboard machine vision systems enable the configuration of autonomous vehicles that do not require any special equipment along the roadways. Thus, they can be seen as an antithesis to the first-period systems.

The first vision-based autonomous vehicle was an "intelligent vehicle" developed by the Mechanical Engineering Laboratory in Japan [9]. It is depicted in Figure 28.2, and it drove about 30 km/h on a test track. The vehicle was steered to drive along the guard rails that are detected by the stereo vision system. In the 1980s, the vehicle was followed by an autonomous land vehicle (ALV) for military purposes and Navigation Laboratory (NavLab) [10] in the United States and Versuchsfahrzeug fuer autonome Mobilitaet und Rechnersehen (VaMoRs) [11], a mini bus employed in West Germany.

The second period can be regarded as a bridge to the third period, where automated driving systems were significant in ITS.

28.3.4 The Third Period of Automated Driving Systems

In the middle of the 1980s in Europe, the United States, and Japan, huge national projects in ITS were initiated in order to provide solutions to traffic accidents and congestion using complex technologies like communication, information processing, sensing, and control. As a result of these projects, in each region and the country, many new systems were developed, introduced, and deployed; such technology included traffic control, driver information, electronic toll collection, driver assistance, and logistics. Automated driving systems were one of the most significant systems focused on in the projects because they had the potential to provide substantial solutions to the required issues, although they

Figure 28.2 Intelligent vehicle.

could not be introduced to public roadways in the near future due to legal and institutional hindrances.

In Europe, as part of *programme for a European traffic with highest efficiency and unprecedented safety* (PROMETHEUS), a vision-based, autonomous vehicle named "VITA II" was developed [12]. The autonomous feature of VITA II was the vision system, which consisted of 18 cameras that covered the entire vehicle, and a main camera system consisting of 2 cameras with different focal lengths. The algorithm of the machine vision that was developed for VaMoRs and its successor, a passenger car-type "VaMP," was facilitated by the use of the Kalman filter [13]. In 1995, VaMP drove from Munich to Odense along public roadways using the automated driving system [14].

In the United States, with the passing of an act named Intermodal Surface Transportation Efficiency Act (ISTEA) in 1991, a demonstration of automated highway systems (AHSs) was held in San Diego in 1997. During this demonstration, automated vehicles drove along the HOV lane of the I-15 interstate freeway. The course for the demonstration was about 12 km long, and seven teams took part in the demonstration. The automated vehicles were not based on passenger cars but on transit buses, a truck, and a construction vehicle. The technology used for the automated driving included a cooperative system with magnetic markers, radar-reflective tape on the roadway, and an autonomous system with machine vision. Many teams demonstrated the plurality of functions in automated vehicles, like vehicles following adaptive cruise control (ACC) or a platoon. California Partners for Advanced Transit and Highways (PATH) [15] demonstrated automated platooning with eight passenger cars driving at 96 km/h with a gap of 6.3 m. Lateral control uses magnetic markers, whereas the longitudinal control uses millimeter-wave radar and vehicle-to-vehicle communication. The platooning was expected to increase road capacity and, in doing so, decrease congestion.

In Japan during 1995 and 1996, the Ministry of Construction developed AHS. The system, consisting of 11 passenger cars, was demonstrated along an expressway before it was open for public use. A cooperative system employing magnetic markers was used. In 2000, a cooperative driving system with five autonomous vehicles was developed by the Ministry of International Trade and Industry [16]. The vehicles were linked by a 5.8 GHz vehicle-to-vehicle communication system – dedicated short-range communication (DSRC). This system exchanged localization data for each vehicle on a real-time basis, which was measured by RTK-GPS. The vehicles autonomously drove on a test track with the help of a map database, though without machine vision. The automated vehicles performed platoon driving, lane changing, splitting a platoon, and merging into a platoon. The automated vehicles drove at a speed of 40–60 km/h, and the gaps were 20–34 m. The objective of the system was to test the compatibility of the safety and efficiency features of the ground transportation. Figure 28.3 shows the merging scene of the cooperative driving system.

The activities undertaken in the third period largely featured feasibility testing of various technologies on a variety of vehicles in a number of driving formations, including a platoon. Feasible technologies, like magnetic markers that are inexpensive, stand-alone intelligence on the roadside, were used for the lateral control. In the second period, a single passenger car was the objective of

Figure 28.3 A merging scene of the cooperative driving system.

automated driving, but in the third period, it was not only passenger cars but also transit buses, a truck, and a construction vehicle that were automated. In addition, platooning, or driving a number of automated vehicles with a small gap in between, was proposed and experimented on with the aim of being able to save energy and increase throughput.

28.3.5 The Fourth Period of Automated Driving Systems

The third period had ended by the year 1998 after the San Diego demonstration due to the lack of implementation and introduction expectations in the near future. However, in this century, new trends have emerged [17]: automated transit buses and tracks have been developed for implementation in the near future. The main objectives of automated transit buses are to dock precisely in order to ease embarkment and disembarkment and drive along narrow lanes provided specifically for buses in order to improve punctuality. Automated trucks are energy saving when driving as part of a platoon as aerodynamic drag is larger than rolling resistance when driving at a high speed.

In the United States, California PATH has been conducting research on automated transit buses and an automated truck platoon since the beginning of the 21st century. The automated transit buses, as shown in Figure 28.4, are for precision docking and narrow-lane driving specifically. The lateral control of the automated buses uses magnetic markers like the platoon displayed in San Diego. Currently, automated transit buses are being operated in Eugene, Oregon.

The longitudinal control of a truck platoon in California PATH [18] uses millimeter-wave radio and LIDAR, though the lateral control of the truck is not automated. Driving experiments using a platoon of three heavy trucks along a closed roadway were conducted to show the energy-saving potential. The experiments showed that the lead truck saved about 4.5%, the middle truck about 11.9%, and the tail truck about 18% when driving at a speed of 85 km/h with gaps of 6 m.

In Europe, there was a project named "Chauffeur" [19] running from 1995 to 2004, which aimed to save energy and personnel expenses through the use of an automated truck platoon. The lead truck was operated by a human driver and the following ones were automated and programed to follow the preceding one. The lateral and longitudinal control of the trucks involved detecting optical markers mounted on the back of the preceding truck with machine vision as well as through the use of vehicle-to-vehicle communication. The experiments on the trucks were conducted along a public road.

In Germany, another project concerning automated truck platooning named "KONVOI" was conducted between 2005 and 2009 [20]. The objective of KONVOI was to increase the throughput of roadways. The lateral control was based on machine vision and the longitudinal control was based on radar and LIDAR.

Figure 28.4 An automated transit bus by California PATH.

A platoon of four automated trucks drove at a speed of 80 km/h with gaps of 10 m.

In Japan since the mid-1990s, Toyota has been developing an automated platoon of transit buses. The objective is to implement a dual-mode transit system, which is named intelligent multimodal transportation system (IMTS); in a residential area, a human drives each bus, but along a dedicated lane on the road, the buses drive themselves, thus making an automated platoon. The lateral control was based on magnetic markers. The automated bus platoon operated along a dedicated lane for guests to a theme park between 2002 and 2008, as shown in Figure 28.5, and also during an international exposition held in Nagoya, Japan, in 2005.

In 2008 the Japanese Ministry of Economy, Trade, and Industry launched a new project named "Energy ITS" aiming to save energy and reduce CO_2 emissions in order to help prevent global warming [21]. As part of the project, an automated platoon of heavy trucks was developed. The lateral control used machine vision to detect lane markers, and the longitudinal control is based on the gap measured by a millimeter-wave radar and a laser scanner. The vehicle-to-vehicle communication used in the DSRC project and infrared were utilized in order to transmit data like acceleration, which may be difficult or even impossible to measure with sensors. An automated platoon of three

Figure 28.5 IMTS bus driving along a dedicated lane on a theme park.

Figure 28.6 An automated platoon of trucks within "Energy ITS."

heavy trucks and a single light truck drove at a speed of 80 km/h with 4.7 m gaps, as shown in Figure 28.6. The energy savings brought about by platooning reached about 9% in the lead truck, about 5% in the middle, and about 23% in the tail truck.

There is another important feature of automated driving in the 21st century. Besides the aforementioned public activities, a number of automated vehicle competitions were held and sponsored by DARPA including the Grand Challenge held in both 2004 and 2005 and the Urban Challenge in 2007, the focus of which was on unmanned operation of military ground vehicles. Tens of teams from various universities attended each competition. After the competitions, the team from the Technical University of Braunschweig, who attended Urban Challenge, carried out experiments on the urban streets in Braunschweig using their automated vehicle entered in the competition [22]. Google developed automated cars based on the technology developed by Carnegie Melon University and Stanford University and drove them on public roadways in California, Nevada, and Texas in the United States. Many car manufacturers and suppliers followed Google's lead in announcing the introduction of automated cars into the market. In August 2013, an automated passenger car developed by Daimler drove from Mannheim to Pforzheim along a variety of roadways, with a total distance of 103 km [23].

28.4 Expected Benefits of Automated Driving

As the proposal by Dr. Zworykin states, the introduction of automation into road transportation brings benefits. Regarding social aspects, they are safe and efficient, and in regard to drivers and passengers, they are convenient and comfortable. Automated driving systems can also be regarded as a new transportation system that can provide transportation means to vulnerable road users.

28.4.1 Safety

One of the benefits of introducing automation is the elimination of human errors. Thus, automated driving will eliminate most accidents, as human driver errors account for more than 90% of the causes of accidents. However, this requires discussion about the mean time between failures (MTBF) of human drivers, which will be described later.

Another benefit of introducing automation is that there is much less latency in perception time, the tasks of decision making, and vehicle operation than if there is a human driver. The results from analyses into traffic accidents in Germany show that if a driver started to maneuver the vehicle in order to avoid an accident by 2 s or more earlier, vehicle-on-vehicle accidents, junction accidents, frontal collisions, and rear-end collisions could be avoided [24]. Even in an accident avoidance maneuver, one second earlier could prevent 90% of junction accidents and rear-end collisions and 50% of frontal accidents.

28.4.2 Efficiency

The efficiency of automated road traffic means the elimination of congestion, resulting in energy saving and reduced CO_2 emissions. The introduction of automation will provide precise vehicle control. Precise lateral control of vehicles will require narrower lanes than those at present, leading to an increased number of lanes on a roadway, resulting in an increase in road capacity.

Precise longitudinal control of vehicles will facilitate driving with much smaller gaps between cars, which provides not only an increase in road capacity that can be expressed as a macroscopic effectiveness but also a reduction of aerodynamic drag that can be expressed as a microscopic effectiveness. Simulation studies show that when a string of three vehicles drives at a speed of 25 m/s with gaps of 2 m, the road capacity doubles, and when a string of 10 vehicles drives at a speed of 25 m/s with gaps of 6 m, the road capacity triples [25].

When a string of vehicles is driving with small gaps between each car, the coefficient of drag (CD) values of the vehicles will decrease. Wind tunnel experiments show that when a string of two vehicles of the same type are driving with a gap half the length of the vehicle, not only does the CD value of the following vehicle reduce by more than 30% but also the CD of the lead vehicle reduces by about 15% of a single vehicle [1]. Since aerodynamic drag accounts for more than half of the resistance when a vehicle is driving at a high speed, driving a string of vehicles with small gaps at a high speed will save fuel. Since human drivers in the following vehicles drive as they watch roadways and respond to sudden braking, automated driving is required.

28.5 Issues of Automated Driving for Market Introduction

Recent worldwide R&D activities on automated driving systems may suggest that the market introduction of automated vehicles might be possible in the near future, but there still remain serious issues to be resolved regarding legal and institutional challenges before the introduction of Levels 3 and 4 automated driving systems.

28.5.1 Performance of Human Drivers

Before discussing issues surrounding automated driving systems, the performance of human drivers will be presented. Current Japanese traffic accident statistics show that the fatalities per 100 million vehicle-km traveled are about 0.6, and the fatalities and the injured per 100 million vehicle-km traveled are about 113 (in 2012). In other words, assuming that a human driver drives at a speed of 30 km/h, 24 h a day, 7 days per week, it would take more than 500 years to cause a fatality and over 3 years to cause a fatality or injure a person. Thus, the MTBF by human drivers is more than 500 years for fatalities and over 3 years for fatalities and injury. However, when a human driver falls asleep or becomes distracted, the MTBF becomes seconds or minutes, hence the essence of driver assistance systems and automated driving systems for safety.

28.5.2 What we Learned from Experiments

The two major benefits of automated driving systems are safety and increased efficiency in road traffic. Regarding safety, currently there is almost no evidence

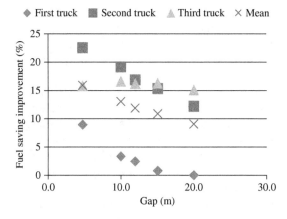

◆ First truck   ■ Second truck   ▲ Third truck   ✕ Mean

Figure 28.7 **Fuel-saving improvement by platooning.**

that automated driving systems contribute to increased safety. That is because this kind of safety validation of automated driving will require tremendously long driving test times under various conditions. Japanese statistics state that, as indicated in the previous section, there is one fatality for every 160 million vehicle-km driven. Significantly, there is only one of a few cases in Japan that shows that an automated braking system with stereo vision reduces rear-end collisions with injury by 60% (2015, Subaru public relations).

Regarding efficiency or energy saving, some experiments with platoons of heavy trucks conducted in the United States, Europe, and Japan have shown that platooning will contribute to energy saving. Figure 28.7 shows the energy-saving improvements measured in an automated platoon of three heavy trucks within the Energy ITS project [21]. The measurements correspond to the experiments with a wind tunnel, showing that when the gap is small enough, energy can be saved, even on the lead truck.

28.5.3 Reliability and MTBF Requirements of the Systems and the Devices

The reliability and MTBF of the systems and devices will be decided based on the safety level required of automated driving systems. The MTBF of human drivers regarding fatal accidents is more than 500 years on all kinds of roadways, that is, assuming they are driving at a speed of 30 km/h. If the automated driving system must provide safer driving than human drivers, the MTBF of automated driving devices must be longer than that of human drivers. Thus, for validation of the safety of automated driving systems, testing of automated driving will require a very long driving distance (160 million vehicle-km).

It is quite difficult or impossible to make bug-free software. In the years between 2009 and 2013, program bugs accounted for 3.8% of recall causes

(Ministry of Land, Infrastructure, Transport, and Tourism, 2015). The software and database for automated driving will be much more complicated than that of current driver assistance systems, and the validation will be much more difficult to obtain and take a longer amount of time.

Sensing systems, software, and automated driving databases are essential parts of automated driving systems. Automated driving will require the quantification of the operational ranges of sensors because if a sensor becomes unable to detect an object that the sensor is supposed to detect while driving automatically, automated driving has to be suspended. Transitioning from manual driving to automated driving and vice versa depends on the function of the sensing devices. For example, if a computer vision could not detect a lane marker, transition to automated driving should be prohibited.

In order to investigate the reliability of the devices and the systems, long-term field testing will be necessary. One method of carrying out these tests may involve a fully automated vehicle with suppressed functions of actuators being driven by a human driver and then a performance comparison being drawn between the driver the automated driving system.

28.5.4 Issues on Human Factors

Since Level 4 of automated driving is fully automated, there are no issues with human factors; however in Level 3 there is a chance to transition from fully automated driving to manual driving with driver assistance. While driving at Level 3, a driver – or rather a passenger at the driver's seat – has to be alert enough and ready to drive as the driver has to be able to drive the vehicle upon request from the system or in the case of an emergency. Thus, it would be difficult for a human driver to use an automated driving system at Level 3.

Level 3 has another issue regarding the transition from automated driving to manual driving – the transition may be a cause of accidents. Automation has already been introduced to modern passenger airplanes, though accidents have occurred after the transition from autopilot to manual operation. In 1985, for example, after an engine failed and manual operation began, a serious accident occurred with Boeing 747, though it did survive. In 2009 a speedometer broke and autopilot switched to manual operation on Airbus A330, which fatally crashed (cited from Wikipedia). These accidents were due to improper operational procedures by the pilots because they could not understand the situation they were in. When the transition from automated driving to manual driving occurs suddenly, automated vehicles can cause such accidents. Thus, as a result of the Level 3 automated driving systems, transition systems are more difficult for human drivers to use.

28.6   Possible Market Introduction of Automated Driving Systems in Japan

When making a scenario of the market introduction of automated driving systems, a taxonomy of factors will be helpful. Some of the axes of the taxonomy are as follows:

- Kinds of vehicles: passenger cars, transit buses, touring coaches, heavy trucks, delivery trucks, specialty vehicles (snow plows, road maintenance vehicles, etc.), small vehicles, and so on
- Kinds of roadways: expressways, highways, urban streets, suburban streets, rural roadways, special roadways dedicated to automated vehicles, campuses, plazas, and so on
- Methods of automated driving: autonomous systems, vehicle-to-infrastructure cooperative systems, and vehicle-to-vehicle cooperative systems
- Speeds of vehicles: high speed and low speed
- Kinds of weather: all weather, clear, cloudy, foggy, rainy, snowy, stormy, and so on
- Time of day: all day, daytime, nighttime, and so on.

In Japan as well as the United States and Europe, car companies often announce that they will introduce automated passenger cars in the near future; however, these cars may only be at Level 2 of automation, rather than Level 3 or 4 where fully automated vehicles are placed. The introduction of Levels 3 and 4 automated passenger cars, which can drive along urban streets during all weather conditions, is the product of the far future. Here, the introduction of automated truck platoons and automated small, low-speed vehicles for vulnerable road users will be discussed, that is, after an introduction is given about the population issues in Japan. Therefore, automation in these types of vehicles is very necessary.

28.6.1   Population Issues

Issues specific to Japan are that it has an ever-decreasing population size and increasing number of elderly citizens. Figure 28.8 shows the estimated future population of Japan and Figure 28.9 shows the estimated future percentage of elderly citizens (over 65) (Cabinet Office, 2014). The estimations state that the population of Japan will continue to decrease, and in 35 years it will be three-quarters of the size of the present population and two-thirds of the size in 50 years. The rate of the elderly citizens at present is 25%, which will increase to 30% in 10 years and 40% in 50 years.

Another issue is the gravitation of people toward cities. Figure 28.10 shows the gravitation trends of people toward the "mega cities" of the Tokyo area and Tokyo/Nagoya/Osaka areas (Cabinet Office, 2014). In the rural areas of Japan

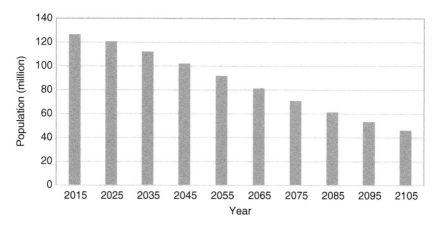

Figure 28.8 The estimated future population of Japan.

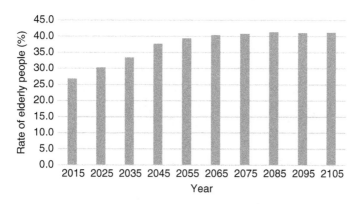

Figure 28.9 The estimated future rate of elderly people.

in particular, the population will decrease and the rate of elderly citizens will increase. The rural areas account for more than 40% of the country's cities, towns, and villages and occupy half of the area of Japan. However, the population of the rural areas accounts for only 8% of the whole population of Japan, and more significantly the number of elderly citizens in rural areas accounts for 33.3% of the population, as opposed to 23.6% nationwide (in 2012).

28.6.2 Small, Low-Speed Automated Vehicles

The increase of elderly citizens will create a larger proportion of vulnerable road users, and consequently transportation systems must be provided for them in urban, suburban, and rural areas. Many such transportation systems have already been developed and implemented in Europe in particular since the late 1990s. Figure 28.11 shows examples of how small, low-speed automated

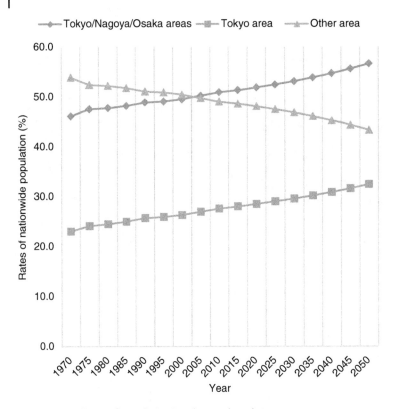

Figure 28.10 Rates of population in urban and rural areas.

vehicles operate in these systems. These vehicles are automated at Level 4 and since the speed of the vehicle is low, the safety regulations are relaxed. The fatality rate of vehicles traveling at a speed below 30 km/h, for example, is less than 10% (Institute for Traffic Accident Research and Data Analysis, 2011). Therefore, after 2020 small, low-speed automated vehicles can be introduced in campuses, plazas, residential areas, and sparsely populated communities. Small, low-speed vehicles can be justified because the mean number of drivers and passengers in a passenger car is 1.30 (Ministry of Land, Infrastructure, Transport and Tourism, 2010), 93.7% of the daily traveling distance of passenger cars is within 30 km, and 97.7% is within 50 km. Moreover, the mean traveling distance per trip is only 14.1 km (Ministry of Land, Infrastructure, Transport and Tourism, 2005).

An experiment with an automated passenger car, depicted in Figure 28.12, not a small, low-speed vehicle, was conducted in a rural area in order to validate the role of automated vehicles in a sparsely populated city. The area, population, and density of the city are 247.26 km², 14,401 (in 2015), and 58.2 km^{-2}

(a)

(b)

Figure 28.11 Examples of small, low-speed automated vehicles: ParkShuttle in the Netherlands (late 1990s) (a) and CyberCars (2011) (b).

(the density of Japan is $343\,\mathrm{km}^{-2}$), respectively, and the proportion of elderly citizens is 46.1% (in 2015). In the city, since railways are not available and transit buses are available but not convenient, a new public transportation system for vulnerable road users is necessary. This information contextualizes the experiment. New transportation systems including small, low-speed automated vehicles will also be necessary in suburban and urban communities and will be used not only by elderly citizens but also by citizens with disabilities.

One of the issues associated with such a system is the ownership and management of the vehicles. Vehicle ownership through sharing is preferable in this situation. The service area and the number of users are also problems to this

(a)

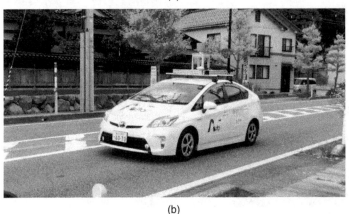

(b)

Figure 28.12 An automated vehicle for the transportation means in a sparsely populated city (a) and its test drive on a public roadway (b). Image Courtesy of Prof. Hiroki Suganuma of kanazawa University.

system. An appropriate service area and the number of users and vehicles must be decided upon. In an area too large or too small, with too many or too few users and vehicles, the system will not work.

The combination of a number of small, low-speed vehicles can comprise a large, virtual vehicle [26]. Figure 28.13 shows an example of such a large, virtual vehicle, in which four small, low-speed vehicles are laterally and/or longitudinally combined through vehicle-to-vehicle communication and small gaps. The objective is to optimize the size of the vehicle. Since the mean number of drivers and passengers in a passenger car is 1.30, as is previously stated, a small,

(a)

(b)

Figure 28.13 Examples of a virtual big vehicle composed of four small vehicles. Image Courtesy of Prof. Manabu Omae of Keio University.

one-seater vehicle would be enough in most cases. Such a vehicle system will also contribute to energy saving.

28.6.3 Automated Truck Platoons

In addition to the general issues associated with road transportation safety and efficiency, there are specific issues relevant to the trucking industry in Japan, including expense details and the increasing number of aging drivers. Personnel and fuel costs accounted for 36% and 18%, respectively, of the total expenses in the trucking industry in 2010 (Japan Trucking Industry Association, 2013), and the annual mean expense for fuel per heavy truck is about 7.5 million yen. Therefore, fuel saving through automated platooning offers a desirable outcome. Additionally, while the number of elderly heavy truck drivers is increasing, the number of younger drivers is decreasing wherein only 22.9% of all drivers were aged over 50 years in 1993 and increased to 35.1% in 2010 due to the decrease in size of the Japanese population, with a concurrent increase in population percentage only seen in elderly citizens. Thus, the introduction of truck automation could be a way of counteracting the shortage of heavy truck drivers. Truck automation can also be justified when looking at the impacts of truck accidents, which are often far more serious in terms of not only fatalities and injuries but also the suspension of traffic.

As described in Section 28.3.5, in Japan an automated truck platoon was developed within the Energy ITS project. As a possible scenario for introducing automated heavy truck platoons while taking into account the technological feasibility, a possible scenario could be that after the year 2025 platoons of automated trucks are used. This platoon's lead truck would be driven by a human driver and followed by two automated trucks, with gaps of around 4–6 m. These trucks would drive on dedicated roadways for automated vehicles and will be able to function under almost all weather conditions. The automated trucks in the platoon are completely autonomous and drive by detecting lane markers, as was the case in the Energy ITS project. Regarding the number of trucks in a platoon, three trucks per platoon will be appropriate because long platoons require large spaces in order to assemble and disassemble them, and it also makes cargo handling more difficult. After 2035, these automated truck platoons will be able to drive on expressways, and by 2035 automated, unmanned heavy trucks may drive on dedicated roadways for automated vehicles and on expressways in 2050.

28.6.4 Cooperative Adaptive Cruise Control

Cooperative adaptive cruise control (CACC) is an extension of ACC. It involves collecting data about acceleration, for example, that is difficult or impossible for onboard sensors to measure from the preceding vehicle. Therefore, the preceding vehicle transmits this data to the truck following it. The data is necessary in order to precisely control the gap between the vehicles and it is also regarded as the first step of platooning. One of the benefits of CACC is the increase in road capacity made possible by shortening the gaps between vehicles. A simulation study with passenger cars shows that the capacity is increased by about 50% when the penetration rate of CACC is 100% [25].

One of the issues with CACC is the scope of the vehicle-to-vehicle communication. As long as the installation of the communication units is not mandatory on passenger cars, it will remain difficult to implement this method. When the

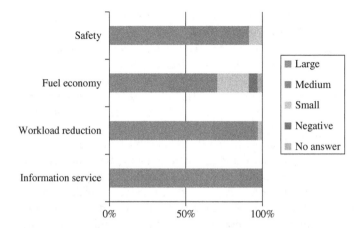

Figure 28.14 Evaluation of CACC by truck drivers in Energy ITS survey (information service means that situations ahead of a lead truck are transmitted to drivers on following trucks via vehicle-to-vehicle communications).

penetration rate of the unit is low, the chance of the communication is less and it is rare that the user receives any benefits. Vehicles with the potential to introduce CACC in the near future are heavy trucks and sightseeing buses on expressways. These vehicles drive on expressways, which form a string of vehicles and make installation easier. The benefits reaped are safety, energy saving, and workload reduction. Figure 28.14 shows an evaluation of CACC being used in heavy trucks during the Energy ITS project [21]. In the evaluation, four heavy trucks drove on a test track using CACC at 80 km/h with the gap being set at 30 m. These vehicle-to-vehicle communication systems enable emergency braking and transmit signal from a preceding vehicle to the vehicle following it in order to avoid rear-end collisions. In terms of heavy trucks and sightseeing buses, it will also ease the installation process if the decision to install the technology is made by the company owners rather than the drivers.

28.7 Conclusion

Japan has a long history in the research and development of automated driving systems of not only passenger cars but also heavy trucks, transit buses, and small low-speed vehicles. Fully automated vehicles have technological, legal, and institutional issues to overcome; therefore, the safety validation of fully automated vehicles will require tremendously long-distance driving tests, which must be undertaken during almost all kinds of weather. Therefore, fully automated passenger cars, whose main objective is safety, will not be introduced to the market for a while. Automated vehicles that could be introduced into the market in the near or middle future will be automated truck platoons and small, low-speed vehicles. These vehicles will meet the requirements necessary for use. The automated truck platoons will consist of a manned lead truck and unmanned, automated trucks following the lead truck on expressways over long distances. The objectives are to save energy and cut personnel costs. The small, low-speed vehicles will be a new transportation means for elderly citizens, the physically disabled, and other vulnerable road users. The systems will be operated in sparsely populated communities, residential areas, campuses, and plazas in urban and suburban areas.

The key factors affecting the market introduction of automated driving systems as well as other ITS-related systems are affordability and system acceptance. If the systems are not widespread enough, the benefits of the systems will not be brought about. Even if it takes a long time, the introduction of automated driving systems to various kinds of road vehicles including passenger cars, heavy trucks, and small low-speed vehicles will be essential for the safety and efficiency of road transportation systems and transportation means available to vulnerable road users.

Questions

1 How can automated driving systems be classified and what are the major aspects?

2 In which periods can the history of automated driving systems be divided and what are the objectives?

3 What are the expected benefits of the introduction of automation in road transportation systems?

4 Which factors have to be considered for market introduction of automated driving systems?

References

1 Ioannou, P. (1997) *Automated Highway Systems*, Plenum, pp. 247–264.
2 Shladover, S.E. (2012) *Recent International Activity in Cooperative Vehicle–Highway Automation Systems*. US DOT Report No. FHWA-HRT-12-033.
3 Tsugawa, S. (2011) Automated driving systems: Common ground of automobiles and robots. *International Journal of Humanoid Robots*, **8** (1), 1–12.
4 Flory, L.E. *et al.* (1962) Electric techniques in a system of highway vehicle control. *RCA Review*, **23** (3), 293–310.

5 Morrison, H.M. *et al.* (1961) Highway and driver aid developments. *SAE Transactions*, **69**, 31–53.

6 Fenton, R.E. *et al.* (1968) One approach to highway automation. *Proceedings of the IEEE*, **56** (4), 556–566.

7 Drebinger, P. *et al.* (1969) Europas Erster Fahrerloser Pkw. *Siemens-Zeitschrift*, **43** (3), 194–198.

8 Ohshima, Y. *et al.* (1965) *Control System for Automatic Automobile Driving.* Proc. IFAC Tokyo Symposium on Systems Engineering for Control System Design, pp. 347–357.

9 Tsugawa, S. *et al.* (1979) *An Automobile with Artificial Intelligence.* Proceedings of the 6th International Joint Conference on Artificial Intelligence, pp. 893–895.

10 Thorpe, C. *et al.* (1990) *Vision and Navigation – The Carnegie Mellon Navlab*, Kluwer Academic Publishers.

11 Graefe, V. (1993) *Vision for Intelligent Road Vehicles.* Proc. IEEE Intelligent Vehicles '93 Symposium, pp. 135–140.

12 Ulmer, B. (1994) *VITA II – Active Collision Avoidance in Real Traffic.* Proc. the Intelligent Vehicles '94 Symposium, pp. 1–6.

13 Dickmanns, E.D. *et al.* (1992) Recursive 3D road and relative ego-state recognition. *IEEE Transactions on Pattern Analysis and Machine Intelligence*, **14** (2), 199–213.

14 Behringer, R. *et al.* (1996) *Results on Visual Road Recognition for Road Vehicle Guidance.* Proc. IEEE Intelligent Vehicles '96 Symposium, pp. 415–420.

15 Rajamani, R. *et al.* (2000) Demonstration of integrated longitudinal and lateral control for the operation of automated vehicles in platoons. *IEEE Transactions on Control Systems Technology*, **8** (4), 695–708.

16 Kato, S. *et al.* (2002) Vehicle control algorithms for cooperative driving with automated vehicles and inter-vehicle communications. *IEEE Transactions on Intelligent Transportation Systems*, **3** (3), 155–161.

17 Bishop, R. (2005) *Intelligent Vehicle Technology and Trends*, Artech House, Boston and London.

18 Shladover, S.E. (2010) *Truck Automation Operational Concept Alternatives.* Proc. IEEE Intelligent Vehicles Symposium, pp. 1072–1077.

19 Fritz, H., Bonnet, C., Schiemenz, H., and Seeberger, D. (2004) *Electronic Tow-Bar Based Platoon Control of Heavy Duty Trucks Using Vehicle.* Vehicle Communication: Practical Results of the CHAUFFEUR2 Project, Proc. ITS World Congress.

20 Kunze, R. *et al.* (2009) in *Organization and Operation of Electronically Coupled Truck Platoons on German Motorways* (eds M. Xie *et al.*): ICIRA 2009, LNAI 5928, Springer, pp. 135–146.

21 Tsugawa, S. (2014) *Results and Issues of an Automated Truck Platoon within the Energy ITS Project.* Proc. 25th IEEE Intelligent Vehicles Symposium, pp. 642–647.

22 Wille, J.M., Saust, F., and Maurer, M. (2010) *Stadtpilot: Driving autonomously on Braunschweig's inner ring road.* Proc. Intelligent Vehicles Symposium, pp. 506–511.

23 Ziegler, J. *et al.* (2014) *Making Bertha Drive? An Autonomous Journey on a Historic Route. IEEE Intelligent Transportation Systems Magazine,* **6** (2), 8–20.

24 Metzler, H. (1988) *Computer Vision Applied to Vehicle Operation.* SAE Technical Paper 881167.

25 Shladover, S. (2012) *Highway Capacity Increases from Automated Driving.* TRB Workshop on Future of Road Vehicle Automation, Irvine, CA, July 25.

26 Tsugawa, S. *et al.* (1991). *Super Smart Vehicle System – Its Concept and Preliminary Works.* Proc. Vehicle Navigation and Information Systems Conference, 2, pp. 269–277.

29

Environmental-Assisted Vehicular Data in Smart Cities

Wei Chang[1], Huanyang Zheng[2], Jie Wu[2], Chiu C. Tan[2], and Haibin Ling[2]

[1] Department of Computer Science, Saint Joseph's University, Philadelphia, PA, USA
[2] Department of Computer and Information Sciences, Temple University, Philadelphia, PA, USA

CHAPTER MENU

Location-Related Security and Privacy Issues in Smart Cities, 820
Opportunities of Using Environmental Evidences, 822
Challenges of Creating Location Proofs, 823
Environmental Evidence-Assisted Vehicular Data Framework, 825
Conclusion, 841

Objectives

- To become familiar with the security and privacy issues of location information in smart cities
- To become familiar with the concept of location proof for trajectories in context of smart cities
- To become familiar with the conventional approaches for generating location proofs and their shortcomings
- To become familiar with the idea of environmental element-based location proof
- To become familiar with the optimal RSU deployment problem, which aims to generate secure and privacy-preserved location proofs by using a minimum number of RSUs

Smart Cities: Foundations, Principles and Applications, First Edition.
Edited by Houbing Song, Ravi Srinivasan, Tamim Sookoor, and Sabina Jeschke.
© 2017 John Wiley & Sons, Inc. Published 2017 by John Wiley & Sons, Inc.

29.1 Location-Related Security and Privacy Issues in Smart Cities

For the past two decades, the term *smart cities* has gained an increasing attraction from academia, government [1, 2], and industry [3, 4]. Within future smart cities, people are expecting the usage of data, not only from some static pre-deployed roadside sensors but also from intelligent vehicles moving within the cities day after day. A typical intelligent vehicle is equipped with multiple sensing devices, such as on-car cameras and gyroscopes, and also has wireless communication capabilities, such as Wi-Fi/LTE. All these devices record every incidence within the city. Unlike the conventional location-based services or mobile social networks, where data is related with a location spot, the recorded data in intelligent vehicles is a continuous and integrated observation along a vehicle's trajectory (or trajectory segments). In the foreseeable future, these vehicle-based data sequences will support a considerable number of new applications, ranging from criminal scene reconstruction to smart traffic management to environmental monitoring.

The intelligent vehicles will inevitably generate an enormous amount of data. Moreover, the data itself may also bring plenty of security and privacy issues. In order to control the data, a carefully designed data management system is urgently needed. Such a management system must be able to balance the trade-off among privacy, security, and data utility, which is extremely hard. For example, in the application of criminal scene reconstruction, when an incident occurs, how can the data management system efficiently and accurately find all related data, meanwhile providing the privacy of these witnesses? When a driver reports an illegal littering from a vehicle, how does the smart city's system verify that the claimer indeed was at the reported location and time, instead of a frame-up?

In order to preserve data searching privacy, the existing works adopt homomorphism-based data encryption. However, the scheme does not suit intelligent vehicle systems, not only because there is a huge amount of data and searching over cipher text is time consuming, but also because there is no solution for extracting semantic information directly from encrypted images or videos. In addition, from the existing location-based services and mobile social networks, we have already seen the motivations for an adversary to misstate their spatiotemporal claims [5–7]; the encryption-based scheme cannot handle situations in which the data itself is maliciously tampered with by an adversary. Consequently, a key requirement for the intelligent vehicle-involved smart cities involves its abilities (i) to verify the spatiotemporal claims made by a vehicle, (ii) to quickly locate the corresponding queried records from an enormous amount of data, and (iii) to simultaneously provide strong privacy protections to the data owners.

According to certain features of applications, the data owner (i.e., a vehicle and its driver) must be in one of two modes: *proactive* or *reactive*. In the proactive mode, the data owner proactively claims a set of spatiotemporal data, and the data management system should be able to verify these claims, while in the reactive mode, the system searches for the data of the vehicles that are likely to have appeared at a specific location during a specific time period.

In this chapter, we study the use of *environmental factors* to develop "evidence of presence" for the intelligent vehicle system. More specifically, we consider how to use the measured wireless signal from roadside units (RSUs) to *verify* and *index* data. The evidence of presence is a means for a vehicle to demonstrate that it was indeed at a specific location and time. For instance, given a car that claims to have witnessed a particular car collision accident at a specific location at a specific time, we would be able to verify such a claim by comparing the claimer's captured surrounding environmental factors against a known database of environment features. The malicious users, who did not pass by the specific location and time, should be unable to generate the same evidence.

Unlike the existing approaches, where the location proof is constructed by using cryptographic keys, the content of environmental evidences is not linked to the identity of any vehicle, because many applications in smart cities are only interested in the correctness of where and when data is collected. We take a novel approach that relies on measuring the wireless properties inherent in environments, for example, due to the multipath effect, to generate data index (for the reactive mode) or to demonstrate evidence of presence (for the proactive mode). The only task that each vehicle should take is to passively record the surrounding environmental information.

For example, when a vehicle takes a short video at some place and time, the surrounding features (e.g., shadows, colors, brightness, etc.) exhibited in the video frames will be different from those taken at other locations or times, due to the differences caused by environmental factors such as weather condition and random obstructions by physical objects. Similarly, the vehicle moves in a region, and the quality of its received wireless signals from the same transmitter will differ due to factors like interferences and multipath effect. Here, we focus on using wireless signal features to index or verify spatiotemporal data about sequences of observations in vehicle networks. Ideally, when the received wireless signal features at every road stretch are unique, one can easily use the features to index or verify the data of any location at any time. However, in reality, it is too expensive to achieve such a dense coverage on a road stretch. In order to minimize the deploying costs, we further study the optimal placement problem of the roadside signal transmitters and the synchronization problem among different transmitters.

The contributions of this chapter are as follows. We propose a novel approach by exploring the spatiotemporally varied environmental signals to index or verify vehicle networks' data. We also provide an approximation algorithm to provide a near-optimal placement of the signal transmitters in this system. Finally, we also study the time synchronization issue among different signal transmitters.

29.2   Opportunities of Using Environmental Evidences

Due to the existence of malicious users, every piece of spatiotemporal data should be verifiable by authorities. For instance, when a car accident occurs, the police should not only verify the evidence of presence of a witness (i.e., location claimer) but also check how well the claimer's provided information corroborates with additional evidence, such as the data records of nearby vehicles, surveillance cameras, and environmental factors.

The evidence of presence can be verified via either direct witness (DW) or indirect support (IS). The DW comes from the directly recorded location proofs from nearby attestors, the construction of which is the core of the conventional cryptographic key-based approaches. Considering a group of nearby vehicles, whenever one of them wants to create location-based data, all these vehicles need to exchange some encrypted and spatiotemporal-bounded messages to build the location proof. Although this type of scheme provides high-level security protection, it inevitably discloses the nearby vehicles' location privacy during verification, since each cryptographic key is uniquely linked to a vehicle. In addition, there are also key management issues, such as the revocation and renewal of certain keys.

Considering that not all applications in smart cities need to know the identities of data owners, in this chapter, we construct the evidence of presence for each spatiotemporal *data* rather than vehicles. We use the impacts of some unpredictable environmental factors on the recorded data as IS for the evidence of presence: the adversary, who did not physically appear at the claimed location during the claimed time, is not able to generate the data with the corresponding "environmental marks." The IS-based verification is conducted by checking the consistency of spatiotemporal data's embedded environmental factors against a known database of historical environment features. Admittedly, the IS-based evidence of presence cannot provide a security protection as high as DW does. But it can successfully hold back the attackers who easily make location claims without any physical appearance. Generally, at least six environmental facts can be used as IS:

- *Environment signals*: The control messages of the received environment signals are unpredictable, by which only the vehicles that have physically appeared at that location at that time can possess the data. Moreover, due to the multipath fading and shadowing conditions, the received signals' qualities can also be considered.

- *Road patterns*: The claimer and attestors are driving on the same road segment (straight road, right curve, or left curve) and therefore should have the same turning pattern. Based on the readings of a gyroscope, one can discriminate the cases of driving on a curve from changing lanes [8] and extract the corresponding road patterns in that period.
- *Non-overlapping trajectories*: Since each vehicle takes a space, the claimer and attestors' trajectories should not overlap with other vehicles' trajectories in a spatiotemporal domain.
- *Local co-viewing*: The claimer and attestors are on the same road segments and, therefore, should have similar local views, such as the same front cars or similar nearby scenes, in their camera videos. We also consider the imperfect recording conditions, such as bad weather and unpaved roads, and use them to check the existence of inconsistency.
- *Landmark co-viewing*: Police can also find other vehicles that are at different locations but relatively close to the reported region. From the vehicles' carrying cameras, police may be able to extract the unique random statuses of some landmarks and use them as the indirect supporters of a location claim.
- *EZpass-based PO*: On roadside, there are some randomly deployed EZpass readers (i.e., POs). The reader cannot obtain each vehicle's account number due to the inference, but we can use it to measure the number of cars that have been passed within a short period of time [9]. If the reported number of attestors differing from the reader's measured number is greater than a threshold, the claimer's statement should not be accepted.

However, the randomness of the environmental factors may not be able to provide full distinguishability among a given set of vehicle flows. For instance, if one solely uses weather conditions as an IS-based evidence of presence, the granulite must be at least on a city level. In other words, if the vehicle flows do not consist of the paths through several far away cities, the weather condition-based evidence becomes useless since every vehicle in a city is very likely to experience the same weather. In this chapter, we focus on using wireless signal features to index or verify spatiotemporal data in vehicle networks. Besides the existing cellular towers and Wi-Fi access points, we intentionally deploy several RSUs (i.e., wireless signal transmitters) on certain road stretches and let them generate spatiotemporal-bounded random signals. The signals from both the RSUs and the existing wireless network infrastructures will be used as environmental evidence of presence in our system.

29.3 Challenges of Creating Location Proofs

Location verification, also known as location proof [10], is a well-known problem in the mobile computing communities. The goal of location verification is to securely prove that a claimer has indeed appeared at a specific location at a specific time. For verifying the spatiotemporal claims, different types of schemes are designed. Using the distance-bounding protocols [11–13] is a common approach, which measures the physical times/distances for messages to transmit between a claimer and verifiers and estimates the claimer's real physical location based on these times. In this type of solution, the verifiers could be other participants, such as mobile phone users and vehicles [14], or some special infrastructures [15]. However, the accuracy of the distance-bounding approaches relies on the deploying density of the verifiers and their trustworthiness.

Cryptographic key-based approach [16–18] is another popular way to generate the location proof, where the claimer and verifiers share a set of spatiotemporal-bounded messages. Although this kind of approach avoids the deployment issues with certain measuring infrastructures, it still has trustworthiness and key management issues.

Recently, people have begun to consider using unique impacts of environmental factors on surrounding objects to create evidence for location verification. Unlike the conventional schemes, the environment-based approaches [19, 20] do not require storing of any cryptographic keys/certificates nor do they require the participants to perform any cryptographic processing, which is very time consuming. Instead, claimers only need to capture some environmental features, such as received signal strength (RSS) or the control messages in 802.11/4G LTE networks, which will be verified later against a known database of features to establish the validity of local claims. Note that all these schemes focus on built evidences for verifying the physical presence on a single location spot.

However, for the vehicular data, it is essentially a sequence of records about the surroundings, from the last data uploading location to the next one. Due to the fact that, for certain applications like criminal scene reconstruction, no one knows which piece of information is useful at the time of recording, we need to create a set of location verifications for the vehicle. Clearly, directly adopting the existing schemes for single spot is too expensive since a vehicle would frequently create plenty of location proofs. The proposed system in this chapter has been inspired by an indoor-tracking paper [21], where authors use a set of collected Wi-Fi data to associate identities with different moving objects in surveillance videos. More specifically, we verify the presence of a vehicular trajectory by providing spatiotemporal-bounded messages only on some crucial road stretches, the combination of which can uniquely distinguish a trajectory from others. From the consideration of computing complexity, our system adopts the RSS-based environmental evidence scheme [20] to generate the messages on crucial road segments, and only the roadside infrastructures possess keys, instead of vehicles. In order to make the vehicular data indexable and privacy preserving, we embed location and environment information into some time-bounded random numbers and use them as both the index and location proof of the vehicular data. Unlike the time-bounded random numbers generating approach in paper [18], our system's random numbers are bounded to certain preknown locations (i.e., the physical locations of RSUs), and our random numbers are more secure even if the initial random number generating parameters is obtained by attackers. Note that exploring roadside infrastructures is a commonly used approach in smart cities. However, the existing works [22–24] mostly focus on the improvement of data transmissions by using RSUs, while this chapter considers how to use the roadside infrastructures to securely verify/index trajectory data.

29.4 Environmental Evidence-Assisted Vehicular Data Framework

29.4.1 System Model and Attack Model

Our intelligent vehicle-based smart city system consists of three components: vehicles, roadside infrastructures, and a supporting data management system. The proposed system integrates the existing devices on a vehicle and provides a more comprehensive description of a city. We assume that each vehicle is equipped with a video camera, which keeps recording all surrounding events; an EZpass tag, which is associated with a driver's account number; and multiple sensing and communication devices, such as a gyroscope, accelerator, and wireless signal transmitter. Without loss of generality, we use D_i to represent a piece of data. Note that, in our model, a single vehicle can return zero or multiple pieces of data, and how to use the data is determined by the smart city applications, which is out of this chapter's scope.

The centralized data management system is responsible for collecting, searching, and verifying the data pieces D collected from vehicles. Each D_i is implicitly associated with certain temporal and spatial information, (T_i, L_i). In order to protect the location privacy of a data owner and provide the capability of spatiotemporally verifiable evidence of presence, we embed the time and location information into environmental wireless signals (i.e., $\mathcal{E}(\cdot)$) and use them as both data index and verification evidence. Each vehicle only needs to send the location claims to the data management system in the form $(\mathcal{E}(T_i, L_i), D_i)$ rather than directly and explicitly uploading the (T_i, L_i) to the data collector.

For constructing the spatiotemporal-embedded wireless signals, we consider two types of roadside infrastructures: the existing wireless communication infrastructures and some RSUs, which are specially installed by the smart cities. RSUs are wireless transmitters, and the only task they conduct involves continually broadcasting certain specially designed random signals to the passing vehicle.

There are two types of attackers: privacy prier (PP) and fake claimer (FC). The objective of PP is to establish a connection between a vehicle and its reported data without physically dogging the victims. In this chapter, we focus on the scenarios, where PP tries to find others' location privacy by querying the data management system with some well-designed spatiotemporal query demands.

We also assume that there are a small number of malicious vehicles controlled by FC, who is able to manipulate any value of the controlled vehicles. According to the exact applications, reporting fake spatiotemporal data can be beneficial to the adversary in different ways. For example, in the application of smart traffic management, FC can maliciously create an illusion of having several accidents on a road stretch such that the routing paths may be recalculated or the traffic lights may falsely adjust their switching frequencies or lengths. For the crime scene reconstruction, FC may use some tampered data to frame some victims or exculpate outlaws. In this chapter, we want to prevent the attackers who make fake location claims without any physical presence at the claimed location during the claimed time.

29.4.2 Roadside Unit-Based Environmental Evidence Construction

The construction of the embedded signals is based loosely on [20]. Let us assume that RSU and the central data management system use public/private keys. Here, we only consider the keys of RSU and the data collectors, instead of individual vehicles, the number of which is significantly greater than that of RSU. The physical location of RSU is preknown by the data management system, and a special control message will be randomly generated and sent from the data management system to the corresponding RSU. The control message to RSU_i contains a future time T_0, an initial value u_i, and an increment Δu_i. At run-time T, when a passing vehicle is detected at RSU_i, the RSU randomly selects a transmission power $p > 0$ and uses this signal power to broadcast certain spatiotemporal-embedded messages to the vehicle. Based on these four variables, the corresponding RSU will generate a series of time-dependent random numbers: from future time T_0, each moment T will be represented as $R_i(T) = u_i + \Delta u_i \times \sum_{T_0}^{T} p$. Instead of explicitly using the time value T, our system will take the random number $R_i(T)$ as a time indicator; only the data management system and the corresponding RSU_i can extract the spatiotemporal information from it.

The spatiotemporal message is defined as

$$M_i(T, L) \leftarrow \langle \text{RSU}_i, R_i(T), \text{Enc}_i(p), H_i(T, p) \rangle$$

where $\text{Enc}_i(\cdot)$ represents an encryption by using RSU_i's key, and $H_i(T, p)$ is a hash signature of $(\text{RSU}_i, R_i(T), \text{Enc}_i(p))$. Upon receiving $M_i(T, L)$, a vehicle first measures the RSS and then constructs evidence of presence as the following:

$$\mathcal{E}(T, L) \leftarrow \langle M_i(T, L), \text{RSS} \rangle$$

In the proactive mode, a user can make a location claim by $\langle T, L, \mathcal{E}(T, L), D \rangle$, and then, the data management system will verify whether the claimed location and time is consistent with $\mathcal{E}(T, L)$. In the reactive mode, users stochastically upload a sequence of data records, $\langle \mathcal{E}(T, L), \{D\} \rangle$, to the data management system whenever they have Wi-Fi access.

29.4.3 Environmental Evidence-Assisted Application Models

29.4.3.1 Location Claim Verification

The environmental factor-based evidence of presence provides a strong protection against FC: for an individual attacker, unless the keys of both data management system and RSU are compromised at the same time, any fake location claim can always be identified.

The verification has two phases. The first phase is a simple filtering, which simply checks the information consistency within message $\mathcal{E}(T, L)$. The system first verifies whether the claimed local L is under the signal range of the reported RSU_i. Next, it extracts the cipher text $\text{Enc}_i(p)$ from the claimed message, decrypts it by using RSU_i's key, and compares the result with the reported RSS value. If they match, it is likely that the claimer had been physically present at the claimed time and location.

However, considering that the number of possible power levels is very limited, there is an extreme situation that an adversary may correctly guess the value of p at some moments. To solve the situation, regarding a local claim that has passed the first phase of verification, the data management system conducts a second round of verification, which requires direct communication with RSU. Upon finishing the first phase of verification, the data management system further checks the data consistency between the reported random number $R_i(T)$ and RSU_i's historic records. The system requires RSU_i to send a list of its selected transmitting powers from T_0 to the claimed time T, reconstructs the spatiotemporal-bounded random number as $R'_i(T)$, and compares it with the reported value $R_i(T)$. If no inconsistency is detected during this phase, the location claim is trustworthy, unless the adversary compromises both RSU and the data management system.

29.4.3.2 Privacy-Preserved Data Collecting

In the conventional location proof for mobile users, a user intends to prove his physical appearance at a location spot in a moment. However, for many applications in smart cities, including on-car camera-based city surveillance, crime scene reconstruction, searching for abducted children, and smart traffic controls, any spot on a vehicle's trajectory may contain critical information, which is unknown at the time of recording. Therefore, along the moving path of a vehicle, a series of on-road records will be generated, and the whole data segment will be stored as a data unit on a server. In the reactive mode, the crux becomes how to use environmental evidence to index and retrieve the data about a period of walking in a privacy-preserved way, which directly affects the efficiency of the system.

For the design of a privacy-preserved data collecting/searching system, the trade-off between users' privacy and data utility is an important issue: on the one hand, when an accident occurs, the surveillance system should be able to quickly identify the accident's witness and the corresponding location proofs, which are the evidences showing that the witness was indeed at the region near the accident; on the other hand, the surveillance system must also consider the privacy of individual users, whose historical visiting sequences must be hidden. Based on this trade-off, our surveillance system separately stores the index of a vehicle's historical data from the whole data.

In this chapter, we build a multiagency privacy-preserved system, where different agencies are unable to see the content of any vehicle without the cooperation of others. Basically, there are three components: (i) a data management server, which is responsible for collecting, searching, or verifying spatiotemporal data; (ii) a special storage system for supporting smart city applications, which consists of several clouds; and (iii) a bookkeeping server, which stores the mapping between each data record and its access address in the clouds.

For the users in reactive mode, they stochastically upload their recorded surrounding data, from the last uploading time to the current time onto their own selected clouds in the form of $\langle \mathcal{E}(T, L), \{D\} \rangle$. Note that the users do not need to provide any information about themselves during uploading. For each record $\langle \mathcal{E}(T, L), \{D\} \rangle$, it must be associated with one environmental evidence $\mathcal{E}(T, L)$, and data $\{D\}$ are partitioned into different segments according to the closest $\mathcal{E}(T, L)$. Upon receiving the record, the cloud returns the corresponding access address A to the user. Clearly, only the vehicle's owner knows the full access addresses of his data. Next, the user sends A to the bookkeeping server, and the server will generate a unique random number R and send it back. Data pair (R, A) is locally saved in the bookkeeping server. For the last step of data collecting, the user sends the indexing $\langle \mathcal{E}(T, L), R \rangle$ to the data management server. Figure 29.1 shows the structure of our environmental evidence-based data collecting system. Since the data collecting process does not involve any user's identity, our system is privacy preserved.

For the reactive mode, a data management server essentially is an index server, and all indexes are sorted according to the values of RSU_i and $R_i(T)$ in $\mathcal{E}(T, L)$. Note that for each initial value pair (T_0, u_i), the values of the following time-dependent random numbers are strictly increasing, which partially reflects the temporal visiting orders of different vehicles at the same RSU.

Figure 29.1 Environmental
evidence-based data
collecting process.

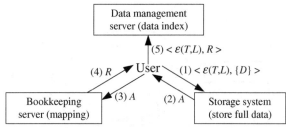

Figure 29.1 Environmental
evidence-based data
collecting process.

29.4.3.3 Environmental Index-Based Data Retrieval

The basic environmental index-based data retrieval process is as follows. When an incident occurs at $\{T,L\}$, the law enforcement will query the data management server by using $\mathcal{E}\{T,L\}$, and the server will return a set of record numbers $\{R\}$, whose environmental indexes match the query content. Based on the return results, the law enforcement will contact the bookkeeping server and find the physical storing addresses $\{A\}$. Finally, the data segments can be obtained from the storage system by using $\{A\}$. However, in reality, most incidences do not happen around RSUs. How to find out a set of potential witnesses by using the environmental indexes must be considered, which essentially relates with RSU-based localization.

The main idea of the RSU-based localization is that, at a given past time T, the location of any vehicle can be estimated based on the traveling distances toward one or several RSUs. Let $T_i(\cdot)$ and $L_i(\cdot)$ be the record time and location of RSU_i's environmental evidence, respectively, and let L and T be the vehicle's current location and time. Assume that the vehicle had received a set of evidences from $\{\mathrm{RSU}_i\}$, $1 \le i \le l$, and $T_{i-1} < T_i$. Based on the environmental evidences, the location of the vehicle at T can be pinpointed at the locations satisfying the following set of equations:

$$\| L - L_i \| = \int_{T_i}^{T} s(\tau)\, d\tau, \forall i \in [1, l]$$

where $\| L - L' \|$ is the length of the road from L to L' and $s(\cdot)$ represents the vehicle's historical speed at any moment. Similarly, for every RSU_i near the incident's location, we can compute a time window for the users, who potentially may be witnesses, as the following: $T_i \in [T - \frac{\|L-L_i\|}{S_{\min}}, T - \frac{\|L-L_i\|}{S_{\max}}]$, where T and L represent the time and location of the incident. Take Figure 29.2 as an example. Suppose that a vehicle received two consecutive environmental evidences, respectively, from RSU_c and RSU_e. Let T' be a moment between the receiving times of the tags. Based on the vehicle's speed information, we can compute the traveling distance from RSU_e to the vehicle's location at T'. In Figure 29.2, there are three possible destinations at which the vehicle may arrive from RSU_e based on the distance, and there are two other locations that

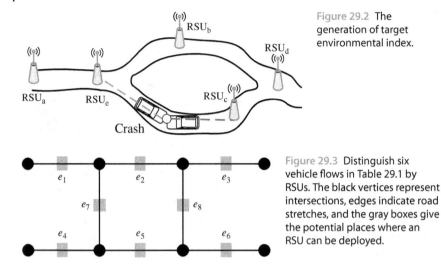

Figure 29.2 The generation of target environmental index.

Figure 29.3 Distinguish six vehicle flows in Table 29.1 by RSUs. The black vertices represent intersections, edges indicate road stretches, and the gray boxes give the potential places where an RSU can be deployed.

the vehicle could arrive at RSU_c. The intersection of these two sets gives the estimated location at time T'.

29.4.4 Optimal Placement of Roadside Units

In the previous section, our investigations assumed an ideal environment in which sufficient RSUs cover every road stretch in a city. However, in practice, such a situation is unlikely to occur. In order to minimize the deployment costs of the RSUs, a special optimally placing algorithm is needed.

However, this problem is not trivial since not every road segment needs an RSU, and although it is hard for an attacker to forge environmental evidence, the attacker can still easily hide some received evidence in order to pretend that he was somewhere else. Take Figure 29.3 as an example. Assume that there is a map consisting of eight road stretches and there are six vehicle trajectories on the map, which are given in Table 29.1. Our objective is to

Table 29.1 RSU-based tags of given flows in Figure 29.3.

f_1	$e_1 \to e_7 \to e_5 \to e_6$	$\emptyset$	e_7	e_7, e_6
f_2	$e_4 \to e_5 \to e_6$	e_4	e_4	e_4, e_6
f_3	$e_4 \to e_5 \to e_8 \to e_3$	e_4, e_3	e_4, e_8	e_4, e_8
f_4	$e_1 \to e_2 \to e_8 \to e_6$	e_2	e_8	e_8, e_6
f_5	$e_1 \to e_7 \to e_5 \to e_8 \to e_3$	e_3	e_7, e_8	e_7, e_8
f_6	$e_4 \to e_7 \to e_2 \to e_3$	e_2, e_3	e_4, e_7	e_4, e_7

install a minimal number of RSUs on some road stretches such that every trajectory can be uniquely identified according to the received environmental evidences. More specifically, considering that some RSUs may use the same transmitting powers or time-dependent random numbers, here we focus on the distinguishability exclusively based on RSU identity numbers within the environmental evidences. For the ease of description, we name the received environmental evidences' RSU identities as *tags*.

Table 29.1 gives three different methods of RSU displacement in Figure 29.3. If only honest users are considered, the optimal RSU placement set is $\{e_2, e_3, e_4\}$, and the received tag sequences of each flow are given in "tags 1" column of Table 29.1. Clearly, all of them have different tag sequences, and therefore, they are fully distinguishable. However, this displacement has a problem when the system contains malicious users: any attacker can easily pretend to be flow f_1 by using an empty tag set. As a result, when an attacker exists, all flows must be covered by some tags. Column "tags 2" shows an optimal placement by deploying RSUs on stretches e_4, e_7, and e_8, and this placement provides full distinguishability and coverage on the given flows. However, in terms of security, the requirements of full coverage and full distinguishability are not enough. For the attackers who travel along the flow f_6, they are able to be disguised as either f_1 or f_2 by intentionally dropping tags from e_4 or e_7, since the tag sequence of f_6 is a super-sequence of that of f_1 and f_2. The secure and optimal RSU placement in Figure 29.3 is to deploy RSUs on $\{e_4, e_6, e_7, e_8\}$, and the corresponding tag sequence of each flow can be found in the "tags 3" column of Table 29.1's.

Generally, the optimal placement of RSUs must guarantee three conditions. First, a minimal number of RSUs are deployed on certain road stretches. Second, the vehicles traveled along different routes must be distinguishable according to their received environmental evidences' RSU identities. Last, considering that an adversary may intentionally drop certain RSUs' messages in order to create fake location claims, a flow's tag sequence cannot be the subsequence of any other flow's received tag sequence.

29.4.4.1 Problem Formulation

Let graph $G = (V, E)$ denote a map, where node set V is a set of road intersections, and edge set $E = \{e\}$ represents all road segments on G with $E \subseteq V^2$. G contains m predefined vehicle flows $F = \{f_1, f_2, \dots, f_m\}$. Each flow is represented as a *walk*, which is a sequence of edges, $f_i \in (e_1, e_2, \dots)$. For the ease of description, we redefine the symbol "$\subseteq$" to represent a subsequence relationship, and we have $f_i \subseteq f_i$, $\emptyset \subseteq f_i$. Note that all flows in F satisfy: $f_i \not\subseteq f_j$ for $\forall i, j, i \neq j$, and in a walk, both nodes and edges can be repeated, as illustrated in Figure 29.4. For instance, in Figure 29.4, if all edges are deployed with RSUs, the tag sequence for f_2 is $(e_1, e_5, e_8, e_6, e_2, e_5, e_8, e_6, e_3, e_4)$, which is a walk.

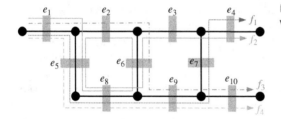

Figure 29.4 Distinguish four vehicle flows by roadside stations.

In order to securely distinguish vehicles of different vehicle flows, several RSUs are deployed on E: whenever a vehicle passes an RSU, the vehicle will receive an environmental evidence, which contains the unique ID of the RSU [25]. Let x_e denote whether road segment e contains an RSU (i.e., $x_e = 1$) or not (i.e., $x_e = 0$) and f' be the road tag sequence of f. f' is a subsequence of f, where only the elements e of f with $x_e = 1$ are kept. We say flows f_i and f_j are *securely distinguishable* if their tag sequences are not in subsequence with each other: $f'_i \not\sqsubseteq f'_j$ and $f'_j \not\sqsubseteq f'_i$.

Our objective is to securely distinguish all flows in F by deploying a minimum number of RSUs on E. The optimal RSU placement problem can be formulated as follows:

$$\min \quad \sum_{e \in E} x_e$$

$$\text{s.t.} \quad f'_i \not\sqsubseteq f'_j, \quad \forall i, j, i \neq j$$

$$x_e \in \{0, 1\}$$

In the environmental evidences, RSU tags (i.e., RSU_i) show the spatial relationship among different flows [26], and the time-bounded random numbers (i.e., $R_i(T)$) issued from different RSUs offer temporal relationships to indicate the direction of each flow. Note that, in practice, a flow may contain multiple vehicles, and each vehicle will save a series of on-road data records, from the beginning to the end of the flow, as one data unit.

29.4.4.2 Properties

Theorem 29.1 The optimal RSU placement is NP-hard.

Proof: For each pair of flows f_i and f_j, we define a distinguish set as $d_{ij} = \{e_k | e_k \in f_i, e_k \notin f_j\}$, which gives a set of possible locations on which deploying RSUs can distinguish flow f_i from f_j. Due to the requirement of non-subsequence relation, $d_{ij} \neq d_{ji}$. The whole distinguish set for all flows is $D = \{d_{ij}, \forall i, j, i \neq j\}$. We say d_{ij} is *covered* by an RSU placement set $d' = \{e | x_e = 1\}$ if $\exists e \in d'$ such that $e \in d_{ij}$.

The optimal RSU placement problem is to find an optimal set $d^* = \{e|x_e = 1\}$ such that every element $d_{ij} \in D$ is covered by d^*. Clearly, it is a variation of the classic maximum independent set problem [27], which is NP-hard.

Note that the optimal placement of RSUs only under the constraints of full coverage and full distinguishability is also an NP-hard problem [28]. But from the consideration of securities, we must consider the constraint about non-subsequence. As any two flows in F are distinct and any flow is not the subsequence of others, an optimal RSU placement always exists. In the worst case, one can simply install RSUs on every road stretch, and then all flows are securely distinguishable. In addition, we have the following bounds on the minimum and maximum numbers of RSUs.

Theorem 29.2 The minimum number of roadside stations, which can provide distinguishability to F, must be no less than $\lceil \log_2 m \rceil$, where m is the cardinality of F.

Proof: We prove it by contradiction. Suppose there is an optimal placement using $\lceil \log_2 m \rceil - 1$ stations. We give these stations an order and use a binary number with length $\lceil \log_2 m \rceil - 1$ to represent whether a flow received the corresponding tags. There are totally $2^{\lceil \log_2 m \rceil - 1}$ possible values of this number. Let $k = \lceil \log_2 m \rceil$, then we have $2^{k-1} < m \leq 2^k$. Because $2^{\lceil \log_2 m \rceil - 1} = 2^{k-1} < m$, there must exist at least one pair of flows, f_i and f_j, having received a same set of tags. In other words, f_i and f_j are indistinguishable. Contradiction occurs, and therefore, the minimum number of roadside stations should be greater than or equal to $\lceil \log_2 m \rceil$.

The result shows the limit of binary coding to distinguish m flows.

Theorem 29.3 The minimum number of roadside stations, which can provide distinguishability to F, must be no more than $\min\left(\frac{m(m-1)}{2}, |E_F|\right)$, where m is the number of flows and E_F is the edge set of F.

Proof: In the worst case, for every pair of flow, we need to build a new roadside station to distinguish them. Therefore, there are at most $\frac{m(m-1)}{2}$ stations. In addition, for the given flow set $F = \{f_1, f_2, \dots, f_m\}$, since $f_i \neq f_j, \forall f_i, f_j \in F$, the set of the optimal solution used edges must be a subset of $E_F = \{e|e \in f_i, \forall f_i \in F\}$.

The results show two worst cases: (i) one station is needed to separate every pair of flows for a total number of m flows, and (ii) each edge has one station in place.

Algorithm 1 Distinguishability-Oriented Greedy (DOG) Approximation

---

1: Construct distinguish set $\{d_{ij}\}$
2: **while** $D \neq \emptyset$ **do**
3:    Select one edge $d^* \leftarrow d^* \cup \{e_i\}$ covered most number of sets in $\{d_{ij}\}$
4:    Update $\{d_{ij}\}$ by removing the sets which have been covered

---

29.4.4.3 Approximation for the Optimal RSU Placement

Algorithm 1 is a Distinguishability-Oriented Greedy Algorithm, which always selects the edge covering the most number of elements in the remaining set. However, unlike the maximum independent set problem, in the optimal RSU placement problem, the constructed tag sequence of each flow cannot be the subsequence of any other flow's tag sequence. To approximate the optimal result, we first construct the overall distinguish set D for all flows, which is given by Algorithm 2, lines 3–5. Consider that for the flows satisfying $d_{ij} \subseteq d_{i'j'}$, when an optimal set d^* covers d_{ij}, it must also cover $d_{i'j'}$. In other words, the RSU placements satisfying d_{ij} must also provide both distinguishability and coverage for $d_{i'j'}$. Therefore, in Algorithm 2, lines 7–10, we eliminate the subsequence relations from D. For the remaining elements of D, assuming d_i, if it contains one and only one edge, then this edge e must be associated with an RSU; otherwise, the corresponding flows related with the d_i will not be securely distinguished or fully covered. We call these types of edges *requisite edges*, and lines 12–17 create the optimal RSU placement set d^* by including all requisite edges. The overall distinguish set D is updated by eliminating all elements that are covered by the constructing set d^*. Finally, from lines 20 to 26, within the resulting set D, we construct d^* by greedily selecting the edges covering the largest number of remaining elements in D. The process stops when all elements of D are covered by d^*, and d^* gives the approximated optimal locations for placing RSUs.

Let's consider an example in Table 29.1 and Figure 29.3. For the six flows, their pairwise distinguish sets are shown in Table 29.2. We eliminate any d_{ij} from D if $\exists d_{i'j'} \in D$ s.t. $d_{i'j'} \subseteq d_{ij}$. The resulting $D = \{\{1,7\}, \{4\}, \{7,5\}, \{2,8\}, \{6\}, \{8,3\}, \{5,8\}, \{7,2\}\}$. Since some d_{ij} values only contain e_4 or e_6, they are the requisite edges of the given flows. Therefore, at the third part of Algorithm 2, we create $d^* = \{4,6\}$ and update D to $\{\{1,7\}, \{7,5\}, \{2,8\}, \{8,3\}, \{5,8\}, \{7,2\}\}$. Find the remaining edges in D, which are $E_D = \{1,2,3,5,7,8\}$, and compute the edges' appearing times in D: $|Q(e_1)| = 1$, $|Q(e_2)| = 2$, $|Q(e_3)| = 1$, $|Q(e_5)| = 2$, $|Q(e_7)| = 3$, and $|Q(e_8)| = 3$. e_7 and e_8 appear the most times; we randomly select e_7, and d^* becomes $\{4,6,7\}$. Updating D, E_D, and $Q(e)$, we have $D = \{\{2,8\}, \{8,3\}, \{5,8\}\}$, $E_D = \{2,3,5,8\}$, $|Q(e_2)| = 1$, $|Q(e_3)| = 1$, $|Q(e_5)| = 1$, and $|Q(e_8)| = 3$. Since e_8 has the highest appearing frequency, we put another RSU on e_8, $d^* = \{4,6,7,8\}$. After another round of updating, D becomes an empty set and Algorithm 2 terminates. The final optimal edges for deploying RSUs are e_4, e_6, e_7, and e_8.

Algorithm 2 RSU Optimal Placement Approximation

1: $D \leftarrow \emptyset, d^* \leftarrow \emptyset$
2: /* Construct distinguish set: lines 2-5 */
3: **for** $\forall f_i, f_j \in F$ **do**
4:    $d_{ij} \leftarrow \{e | e \in f_i, e \notin f_j\}, d_{ji} \leftarrow \{e | e \in f_j, e \notin f_i\}$
5:    $D \leftarrow D \cup \{d_{ij}, d_{ji}\}$
6: /* Eliminate subsequence relation: lines 6-10 */
7: Sort $D = \{d_1, d_2, \dots, d_k\}$ s.t. $|d_i| \geq |d_j|$ if $i < j$
8: **for** $i \leftarrow 1 \dots k$ **do**
9:    **if** $\exists d_j \subseteq d_i, j > i, d_j \in D$ **then**
10:       $D \leftarrow D \backslash \{d_j\}$
11: /* Include requisite edges in d^*: lines 11-17 */
12: **for** $\forall d_i \in D$ **do**
13:    **if** $|d_i| == 1$ **then**
14:       $d^* \leftarrow d^* \cup d_i$, find e s.t. $e \in d_i$
15:       **for** $\forall d_j \in D$ **do**
16:          **if** $e \in d_j$ **then**
17:             $D \leftarrow D \backslash \{d_j\}$
18: /* Find other elements of d^* by a greedy scheme: lines 18-26 */
19: **while** $D \neq \emptyset$ **do**
20:    Create edge set $E_D \leftarrow \{e | \exists d \in D, e \in d\}$
21:    For $\forall e_k \in E_D$, compute $Q(e_k) \leftarrow \{d | \exists d \in D, e_k \in d\}$
22:    Find $e_i \in E_D$ s.t. $|Q(e_i)| \geq |Q(e_j)|$ for $\forall e_j \in E_D, i \neq j$
23:    $d^* \leftarrow d^* \cup \{e_i\}$
24:    **for** $\forall d_j \in D$ **do**
25:       **if** $e_i \in d_j$ **then**
26:          $D \leftarrow D \backslash \{d_j\}$

Theorem 29.4 By deploying RSUs on the edges found by Algorithm 2, the tag sequences of any two flows $f_i, f_j \in F$ surely satisfy $f_i' \not\subseteq f_j'$.

Proof: We prove the theorem by contradiction. Assume that there is at least one pair of flows $f_i, f_j \in F$, whose tag sequences satisfy $f_i' \subseteq f_j'$. There are totally three conditions that may cause a subsequence relation: (i) $f_i' = \emptyset$, (ii) $f_i' = f_j'$, and (iii) $f_i', f_j' \neq \emptyset, f_i' \neq f_j', f_i' \subseteq f_j'$. For the given flow set F, any two flows are unique and non sequence $f_i \not\subseteq f_j$, and, therefore, there is at least one pair of edges e_i and e_j satisfying $e_i \in f_i, e_i \notin f_j, e_j \in f_j$, and $e_j \notin f_i$. According to the definition of distinguish sets, we have $d_{ij} \cap d_{ji} = \emptyset, e_i \in d_{ij} \neq \emptyset$, and $e_j \in d_{ji} \neq \emptyset$. Since Algorithm 2 requires $d^* \cap d \neq \emptyset$ for $\forall d \in D$, the resulting set d^* must contain edges e_i^* and e_j^* such that $e_i^* \in d_{ij}$ and $e_j^* \in d_{ji}$, which also means that neither f_i' nor f_j' can be empty; it is impossible for condition (i) to occur. Since $d_{ij} \cap d_{ji} = \emptyset, f_i'$ must possess at least one tag e_i^*, which f_j' does not contain. So,

Table 29.2 The construction of d_{ij} for flows in Figure 29.3.

d_{ij}	$j = 1$	$j = 2$	$j = 3$	$j = 4$	$j = 5$	$j = 6$
$i = 1$		$\{1, 7\}$	$\{1, 7, 6\}$	$\{7, 5\}$	$\{6\}$	$\{1, 5, 6\}$
$i = 2$	$\{4\}$		$\{6\}$	$\{4, 5\}$	$\{4, 6\}$	$\{5, 6\}$
$i = 3$	$\{4, 8, 3\}$	$\{8, 3\}$		$\{4, 5, 3\}$	$\{4\}$	$\{5, 8\}$
$i = 4$	$\{2, 8\}$	$\{1, 2, 8\}$	$\{1, 2, 6\}$		$\{2, 6\}$	$\{1, 8, 6\}$
$i = 5$	$\{8, 3\}$	$\{1, 7, 8, 3\}$	$\{1, 7\}$	$\{7, 5, 3\}$		$\{1, 5, 8\}$
$i = 6$	$\{4, 2, 3\}$	$\{7, 2, 3\}$	$\{7, 2\}$	$\{4, 7, 3\}$	$\{4, 2\}$	

both conditions (ii) and (iii) cannot happen. As a result, by using Algorithm 2, any two flows must satisfy $f_i' \not\subseteq f_j'$.

29.4.4.4 Extension: Optimal RSU Placement with Package Loss

In the real world, moving objects such as trucks can block the communication line of sight between an RSU and cars [13]. Therefore, wireless signal-based environmental evidences may fail to be delivered to the passing vehicles. Missing an RSU's signal may cause location verification or the data retrieval of a vehicle flow to fail. For example, in Figure 29.2, losing the tags from RSU_b or RSU_c will cause the system to be unable to determine whether the corresponding vehicles were traveling along the upper or the lower path, especially when the paths have a similar length.

The package loss rates on different road stretches may not be the same due to their traffic densities and topographies. We denote r_i as the package loss rate for each RSU that is placed on the road stretch e_i, and b_i is used to represent the billing (i.e., cost) for deploying an RSU on e_i. Multiple RUSs can be placed on the same road stretch to mitigate package losses. For vehicles, receiving several RSUs' environmental evidences on a road stretch is functionally equivalent to obtaining a single tag in the ideal model, where package loss rate is zero. Moreover, we assume that the tag losses for different RSUs are independent of each other. Let $k_i \in \{0, 1, 2, \ldots\}$ denote the number of RSUs placed on the road stretch e_i. Then, the probability of the tag delivery on e_i can be calculated as $1 - r_i^{k_i}$, and the corresponding deploying cost is $b_i \times k_i$. Since vehicles may receive multiple environmental evidences on one road stretch, we redefine the road tag sequence as the following: f' is a subsequence of vehicle flow f, where only the elements e_i of f with $k_i > 0$ are kept.

The problem of optimal station placement with tag loss is defined as follows: using a minimal RSU deployment costs such that every different vehicle flow is theoretically distinguishable and that the average recognizing probability on-road stretches, on which RSUs are deployed, is no less than a predefined threshold. The problem of optimal station placement with tag loss can be reformulated as the following:

$$\min \quad \sum_{e_i \in E} (b_i \times k_i)$$

$$\text{s.t.} \quad f_i' \not\subseteq f_j', \qquad\qquad \forall i, j, i \neq j$$

$$1 - r_i^{k_i} > \tau, \qquad\qquad \forall k_i \neq 0$$

$$k_i \in \{0, 1, 2, \dots\}, \qquad\qquad \forall e_i \in E$$

The approximation algorithm for the optimal RSU placement problem with package loss is given in Algorithm 3. Since the requisite edges must be deployed with RSUs in order to provide a full distinguishability, the beginning parts of Algorithms 2 and 3 are the same. However, for the construction of the remaining part, Algorithm 2 always selects the edge covering the most number of distinguish sets in the remaining D, while Algorithm 3 picks the one with the least cost per set coverage. Algorithm 3 line 4 computes the total costs $B(e_k)$ for achieving the required RSU-based recognizing probability on edge e_k. In line 5, it finds out the distinguish sets $Q(e_k)$ that would be covered after the edge e_k has been selected. The final set d^* is constructed by using the edge with the lowest $B(e_i)/|Q(e_i)|$ value, where $|\cdot|$ is the cardinality of a set.

29.4.4.5 Performance Analysis

In this section, we test the performance of the RSU placement algorithms, Algorithms 2 and 3. We use Figure 29.3 as the regional map, which consists of 8 road stretches, and use the 6 flows in Table 29.1 as the given flows within the region.

We first test Algorithm 2. First, we consider the impotentness for the deploying locations of RSUs. In Figure 29.5, we gradually increase the number of RSUs, which are deployed within the given region and compare the difference of RSU tag-based distinguishability by using the deploying strategies of Algorithm 2 and a random approach, which randomly selects several road stretches to install RSUs. In order to measure the distinguishability among flows, we propose a concept, called securely distinguishable rate (SDR). Recall that, for any pair of vehicular flows f_i and f_j, if their RSU tag sequences satisfy $f_i' \not\subseteq f_j'$, then we say f_i and f_j are securely distinguishable. Similarly, SDR computes the percentage of securely distinguishable flow pairs out of all possible pairs, and

---

Algorithm 3 RSU Placement with Package Loss

1: Algorithm 2 lines 1-17.
2: **while** $D \neq \emptyset$ **do**
3:     Create edge set $E_D \leftarrow \{e | \exists d \in D, e \in d\}$
4:     For $\forall e_k \in E_D$, compute $B(e_k) \leftarrow b_i \times \lceil \frac{\log(1-\tau)}{\log r_i} \rceil$
5:     For $\forall e_k \in E_D$, compute $Q(e_k) \leftarrow \{d | \exists d \in D, e_k \in d\}$
6:     Find $e_i \in E_D$ s.t. $\frac{B(e_i)}{|Q(e_i)|} \leq \frac{B(e_j)}{|Q(e_j)|}$ for $\forall e_j \in E_D, i \neq j$
7:     $d^* \leftarrow d^* \cup \{e_i\}$
8:     **for** $\forall d_j \in D$ **do**
9:       **if** $e_i \in d_j$ **then**
10:         $D \leftarrow D \backslash \{d_j\}$

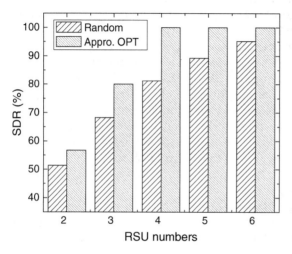

Figure 29.5 Securely distinguishable rate (SDR).

simulation results are shown in Figure 29.5. From the figure, we can see that the SDR values of both random scheme and Algorithm 2 are initially very close to each other when only few RSUs are used; however, with the growth of the RSU numbers, the SDR of Algorithm 2's deployment significantly and quickly goes up to 100%. Since different road stretches possess diverse impotentness for flows' distinguishability, the deploying locations of RSUs must be carefully selected.

In Figure 29.6, we randomly assign a tag loss rate to each road stretch and check the impact of RSU construction costs on the deployment locations. In this part of simulation, we first let all stretches have the same deploying costs, and then, we intentionally add some random extra costs on the critical edges, which are selected by Algorithm 2. The average extra costs are 25% and

Figure 29.6 Alternate
deploying rate (ADR).

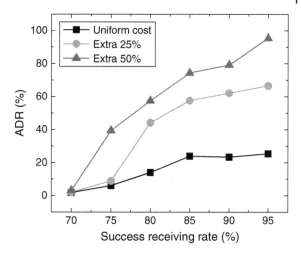

50%, respectively. For observing the change of the deployment locations, we
further propose another concept, called alternate deploying rate (ADR), which
counts the percentage of Algorithm 3's results using an alternate deployment
other than the previous Algorithm 2's result. The greater the ADR is, the
more impacts of construction costs on the RSU deployment. In Figure 29.6,
we gradually increase the minimal tag acceptance rate τ and the ADR values
under different construction costs. Figure 29.6 clearly shows that, with an
increasing τ, more and more cases drop the previous deploying result (i.e.,
Algorithm 2) and turn to use some cheaper stretches for achieving a fully
secure distinguishability.

29.4.5 Time Synchronization among Roadside Units

In reality, some RSUs, especially the ones deployed in less-traveled regions, may
not be able to access the Internet. For using the environmental evidence-based
verifiable data indexing in smart city, some special cars are used to periodically
collect the historically used RSS time sequence $\{p\}$ from these RSUs and reas-
sign the random number-related parameters $(T_0, u_i, \Delta u_i)$ to them. Although
the verification process of the evidence of presence in these regions takes more
time, the whole environmental evidence-based system works normally.

The functionality of our system is based on a crucial assumption that all
RSUs are time synchronized. For most regions, this requirement can be easily
achieved, as long as there are Internet connections or GPS signals. In practice,
RSUs are usually cheap devices without high-accuracy atomic clocks [29]. So,
there may be time drifting issues for the RSUs without any network connection,
which inevitably results in time inconsistency among vehicles and RSUs. Recall
that, for avoiding an environmental evidence being modified by attackers, the

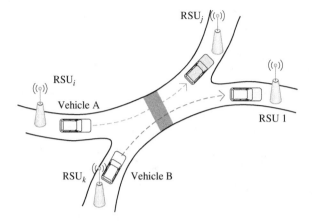

Figure 29.7 Clock synchronization problem among RSU stations. All four roadside stations are not synchronized. The system can never tell whether vehicle A or B passed the shadowed region first.

RSUs' messages are associated with a stations' own time-bounded random number. When the local clocks of RSUs are unsynchronized, not only the data management server is unable to verify the authenticity of some location claims in the proactive mode but also the inconsistency can cause data ambiguities or disorder in the reactive mode.

Take Figure 29.7 as an example. First, the numerical relationship of two RSUs' time stamps may cause disorder in the interpretation of the real visiting sequence. Suppose that the vehicle A consecutively received two environmental evidences $\mathcal{E}\{PC_a, L_a\}$ and $\mathcal{E}\{PC_c, L_c\}$, respectively, from RSU_a and RSU_c at times T_a and T_c, where PC_i represents the physical clock of RSU_i. Since the car moved from RSU_a to RSU_c, the real generation time of the evidences must satisfy $T_a < T_c$. However, due to unsynchronized time of RSU_a and RSU_c, the time information embedded in the evidences may become $PC_a > PC_c$, which could be interpreted as A driving from RSU_c to RSU_a. Second, the inconsistent local clocks make data incomparable. Also in Figure 29.7, assume that RSU_a and RSU_c are synchronized and so are RSU_b and RSU_d, but RSU_a and RSU_b are not. Vehicle A moved from RSU_a to RSU_c, and vehicle B drove from RSU_b to RSU_d. However, since the time is unsynchronized, we cannot determine whether A or B passed the shadowed region first. In real life, the passing order of a region is critical in criminal investigations. Therefore, the clocks of RSUs must be synchronized.

Here, we use special vehicles to periodically synchronize the local clocks between different RSUs. For each RSU, if no vehicle passes RSU_i at physical time t, then PC_i is differentiable at t and $dPC(t)/dt > 0$ [30]. If there is a car A that passed RSU_i at physical time t, then A contains $PC_i(t)$. A vehicle A' from RSU_j arrives RSU_i at time t, and the vehicle's local time is PC_j. The RSU_i sets PC_i to $\max(PC_i(t), PC_j + \Delta t)$, where Δt is the traveling delay of a vehicle from RSU_j to RSU_i. In the meantime, the special vehicles record the time difference

$\max(PC_i(t), PC_j + \Delta t) - PC_i(t)$ and historical RSS values, which will be used for data verification and indexing at the central data management server.

29.5 Conclusion

Vehicular data provide a new perspective for many applications in smart city. Unlike the conventional data, where each data is a discrete record, vehicular data is usually a sequence of spatiotemporal records about the surroundings. Compared to location-based services or mobile social networks, the moving trajectories of vehicles within a region are more predicable due to traffic restrictions. Regarding the aspects of security and privacy, this unique feature makes the construction of verifiable vehicular data indexes become cheaper than that of a series of location proofs in a mobile social network, which is strongly dependent on the cryptographic keys among different participants. In this chapter, we propose a new management system by using wireless signal-based environmental data to verify and index vehicular data. Considering that many applications in smart cities are only interested in the correctness of where and when data is collected, in our system, vehicles do not possess any cryptographic key; instead, they simply listen and collect the environmental evidences along their trajectories, and in our system, only the recorded environmental evidences are used to verify/index the vehicular data. In order to guarantee that the collected evidences of different vehicle flows are unique, we deploy several roadside signal transmitters to generate the environmental evidences. Considering the deploying costs and accuracy, we further study the optimal placement problem and time synchronization problem of the transmitters. We believe that the proposed environmental evidence-based vehicular system can bring many new research opportunities to smart cities.

Final Thoughts

In this chapter we discussed the concept of location proof for vehicular trajectory data in smart cities. We overviewed the existing approaches for generating the location proof for a single location spot and provided a set of surrounding environmental information, which can potentially be used to generate the unpredictable, verifiable, and indexable location proofs. We also presented a detailed framework by using the wireless signals from Road Side Units (RSUs) to generate the location proof. And finally we discussed the optimal RSU placement problem and the location–time synchronization problem among RSUs. The concept of environmental evidence-based location proof for vehicular trajectories provides a brand-new research direction in smart cities.

Questions

1 What is the definition of location proof for vehicular trajectory data?

2 How does the conventional location proof disclose user's location privacy?

3 What are the definitions of full distinguishability, full coverage, and secure distinguishability?

4 What is the definition of the optimal RSU placement problem?

5 Why does the synchronization of RSUs' local clocks matter?

References

1 Caragliu, A., Del Bo, C., and Nijkamp, P. (2011) Smart cities in Europe. *Journal of Urban Technology*, **18** (2), 65–82.

2 Lombardi, P., Giordano, S., Farouh, H., and Yousef, W. (2012) Modelling the smart city performance. *Innovation: The European Journal of Social Science Research*, **25** (2), 137–149.

3 Chourabi, H., Nam, T., Walker, S., Gil-Garcia, J.R., Mellouli, S., Nahon, K., Pardo, T., Scholl, H.J. et al., (2012) *Understanding Smart Cities: An Integrative Framework*. IEEE HICSS.

4 Nam, T. and Pardo, T.A. (2011) *Conceptualizing Smart City with Dimensions of Technology, People, and Institutions*. ACM DG.O.

5 Chang, W., Wu, J., and Tan, C.C. (2012) Wormhole defense for cooperative trajectory mapping. *International Journal of Parallel, Emergent and Distributed Systems*, **27** (5), 459–480.

6 Carbunar, B. and Potharaju, R. (2012) *You Unlocked the Mt. Everest Badge on Foursquare! Countering Location Fraud in Geosocial Networks*. MASS, IEEE.

7 He, W., Liu, X., and Ren, M. (2011) *Location Cheating: A Security Challenge to Location-Based Social Network Services*. ICDCS, IEEE.

8 Chen, D., Cho, K.T., Han, S., Jin, Z., and Shin, K.G. (2015) *Invisible Sensing of Vehicle Steering with Smartphones*. ACM MobiSys.

9 Abari, O., Vasisht, D., Katabi, D., and Chandrakasan, A. (2015) Caraoke: an E-toll transponder network for smart cities. *ACM SIGCOMM*, **45** (5), 297–310.

10 Khan, R., Zawoad, S., Haque, M.M., and Hasan, R. (2014) *OTIT: Towards Secure Provenance Modeling for Location Proofs*. ACM ASIACCS.

11 Song, J.H., Wong, V.W., and Leung, V. (2008) *Secure location verification for vehicular Ad-Hoc networks.* IEEE GLOBECOM.

12 Hubaux, J.P., Capkun, S., and Luo, J. (2004) The security and privacy of smart vehicles. *IEEE Security & Privacy Magazine,* **2** (3), 49–55.

13 Abumansoor, O. and Boukerche, A. (2012) A secure cooperative approach for nonline-of-sight location verification in VANET. *IEEE Transactions on Vehicular Technology,* **61** (1), 275–285.

14 Xiao, B., Yu, B., and Gao, C. (2006) *Detection and Localization of Sybil Nodes in VANETs.* ACM DIWANS.

15 Schäfer, M., Lenders, V., and Schmitt, J. (2015) *Secure Track Verification.* IEEE S&P.

16 Zhu, Z. and Cao, G. (2011) *Applaus: A Privacy-Preserving Location Proof Updating System for Location-Based Services.* IEEE INFOCOM.

17 Talasila, M., Curtmola, R., and Borcea, C. (2012) Link: location verification through immediate neighbors knowledge, in *Mobile and Ubiquitous Systems: Computing, Networking, and Services,* Springer.

18 Malandrino, F., Casetti, C., Chiasserini, C.F., Fiore, M., Yokoyama, R.S., and Borgiattino, C. (2013) *A-VIP: Anonymous Verification and Inference of Positions in Vehicular Networks.* IEEE INFOCOM.

19 Zheng, Y., Li, M., Lou, W., and Hou, Y.T. (2012) Sharp: private proximity test and secure handshake with cheat-proof location tags, in *Computer Security–ESORICS 2012,* Springer.

20 Zhang, Y., Tan, C.C., Xu, F., Han, H., and Li, Q. (2015) Vproof: Lightweight privacy-preserving vehicle location proofs. *IEEE Transactions on Vehicular Technology,* **64** (1), 378–385.

21 Higuchi, T., Martin, P., Chakraborty, S., and Srivastava, M. (2015) *AnonyCast: Privacy-Preserving Location Distribution for Anonymous Crowd Tracking Systems.* ACM UbiComp.

22 Ahmed, S.H., Bouk, S.H., and Kim, D. (2015) RUFS: RobUst forwarder selection in vehicular content-centric networks. *Communications Letters,* **19** (9), 1616–1619.

23 Ahmed, S.H., Bouk, S.H., and Kim, D. (2015) Target RSU selection with low scanning latency in WiMAX-enabled vehicular networks. *Mobile Networks and Applications,* **20** (2), 239–250.

24 Bouk, S.H., Ahmed, S.H., Omoniwa, B., and Kim, D. (2015) Outage minimization using bivious relaying scheme in vehicular delay tolerant networks. *Wireless Personal Communications,* **84** (4), 2679–2692.

25 Sohn, K. and Kim, D. (2008) Dynamic origin–destination flow estimation using cellular communication system. *IEEE Transactions on Vehicular Technology,* **57** (5), 2703–2713.

26 Popa, R.A., Balakrishnan, H., and Blumberg, A.J. (2009) *Vpriv: Protecting Privacy in Location-Based Vehicular Services.* USENIX Security.

27 Lokshantov, D., Vatshelle, M., and Villanger, Y. (2014) *Independent Set in p 5-Free Graphs in Polynomial Time.* ACM-SIAM SODA.

28 Zheng, H., Chang, W., and Wu, J. (2016) *Coverage and Distinguishability Requirements for Traffic Flow Monitoring Systems.* IEEE/ACM IWQoS.

29 Zhou, T., Sharif, H., Hempel, M., Mahasukhon, P., Wang, W., and Ma, T. (2011) A novel adaptive distributed cooperative relaying MAC protocol for vehicular networks. *IEEE Journal on Selected Areas in Communications,* **29** (1), 72–82.

30 Wu, J. (1998) *Distributed System Design,* CRC Press.

Index

a

ability *(A)* level, occupants   483
acoustic models
  acoustic event detection pipeline
      680
  Auditeur   678–679
  ensemble approach   681
  fault detection   674
  faults, in air handling room   673
  mel-frequency cepstral coefficients
      (MFCCs)   680
  multi-output inference model   681
  point anomaly   679
  sound library   679
  soundscape capture and analysis
      672
  SoundSense   678
ad hoc networks   276, 422
advanced metering infrastructure
      (AMI)   395, 775–777. *see also*
      smart meters
  advantages   775
  challenges   781–782
  home area network   775–776
  service users   776–777
  wide area network   776
AIA 2014 Progress Report   646
AirCloud
  air quality analytics engine   732
  APIs–time-series and GPS-based
      access   748

data representation and storage
    731–732
data sources   731
RESTful APIs   732
web applications   749
AirPet   748, 750
air quality analytics engine   732
  artificial neural networks
      740–742
  effectiveness   744–747
  filtering using signal reconstruction
      739, 740
  online inference model   742–744
air quality-aware aerial drone   750
air quality, in Delhi   726
Air Quality Index (AQI) for PM2.5
    727
algorithmic clustering
  information   465
  robustness   466–467
  scaling   465–466
alternate deploying rate (ADR)   839
ambient assisted living (AAL)
    627–629
Analysing Transition Planning and
    Systemic Energy Planning Tools
    57
Apache Hadoop Software   46–48
API data exchange framework   737
application-aware networks   40

Smart Cities: Foundations, Principles and Applications, First Edition.
Edited by Houbing Song, Ravi Srinivasan, Tamim Sookoor, and Sabina Jeschke.
© 2017 John Wiley & Sons, Inc. Published 2017 by John Wiley & Sons, Inc.

AQM and miniAQM
  data communication   734, 735
  hardware calibration   734–735
  mechanical design   734
  sensor selection   733–734
ArcPy methods   220
area control error (ACE)   541
artificial neural networks (ANN)   246,
    446, 740–742
ASHRAE Great Energy Predictor
    Shootout Challenge   265
attendance policy   138
automated driving systems
  advantage   798
  classification   795–796
  definition   793–795
  efficiency   804–805
  fourth period systems   801–804
  history   796
  with inductive cable   797
  issues   805–807
  in Japan, market introduction of
    cooperative adaptive cruise
      control   814–815
    population issues   808–809
    small, low-speed automated
      vehicles   809–813
    truck platoons   813–814
  research and development   794
  safety   804
  second-period systems   798
  third-period systems   798–801
automated meter reading (AMR)
    technology   397
automated network identification
    (ANI)   437, 451
automated truck platoons   813–814
automated vehicles   423. *see also*
    autonomous car
automatic generation control (AGC)
    541

autonomous car
  cost saving   423
  MaaS   423
  negative impacts   424
available transfer capacity (ATC)
    analysis   557

b

balanced sustainability assessment
    507, 508
base stations (BSs)   103, 196, 201
battery threshold   139
big data analytics (BDA)   762
  ambulances   30
  Apache Hadoop Software
    Framework   46–48
  application-aware networks   40
  application coordination   45
  business acceleration and
    augmentation mechanisms   29
  business innovations   30
  city-specific data systems   25
  Civitas   42–43
  customer satisfaction analysis   32
  data analysis   41
  data collection   44
  data integration   41
  data warehousing (DW)   31
  electric power data   44
  EMC Greenplum Data Computing
    Appliance (DCA)   31
  enterprises   31
  epidemic analysis   32–33
  Hadoop   39
  Hadoop-based analytical products
    31
  in healthcare   33–34
  Hitachi Smart City platform   43
  IBMPureData System   31
  implications   27–28
  information and communication
    infrastructures   28–29
  infrastructure   41

in-memory and in-database analytics 39

machine data
  characteristics 34–35
  digital publishing 36
  E-Commerce 36
  environmental monitoring sensors 35
  firefighting 35
  growth of 35
  human-to-machine (H2M) interactions 35
  machine-to-machine (M2M) 35
  sensor data 35
  smart sensors 36
  software application/hardware device records 36
  Software as a Service (SaaS) 36
  Splunk reference architecture 36, 37

market research and analyst groups 31

market sentiment analysis 32

multiplicity and heterogeneity 25

NoSQL databases 41, 48–50

open data, for next-generation cities 38–39

real-time analytics 31

robust and resilient productivity-enhancing methods and models 29

social networks and multifaceted devices 29

stakeholders and end users 39

value 26

variety 26

velocity 26

volume 26

bi-level optimization 563

building automation systems (BAS) 480

building data 253

building information modeling (BIM) 231, 440, 682

business intelligence 201–202

business package 417

C

CACC. *see* cooperative adaptive cruise control (CACC)

car2Go 428–429

carpooling services. *see* ride-sharing services

C40 Cities Climate LeadershipGroup (C40) 645

circular economy 765

citizen-centric smart services 392

city performance management (CPM) 26

Civitas 42–43

climate resilience
  carbon inventory
    C40 Cities Climate LeadershipGroup (C40) 645
    emissions reductions 644
    energy conservation measures (ECMs) 646
    fuel consumption 645
    space conditioning 646
  Chicago multifamily building 658–660
  climate change
    greenhouse gas (GHG) emissions 643
    IPCC 642
    Paris climate agreement 643
    resilience plans 643
    resilient infrastructure 643
    zero-net-energy buildings 644
  climate scenarios and impacts 651–653
  Fort Collins multifamily building 660–662
  future cities and net-zero building 662–664

climate resilience (*contd.*)
  future climate predictions
    AIA 2014 Progress Report   646
    building simulation modeling
      648
    energy models   647
    energy simulation   647
    simulation software   647
  modeling methodology
    ASHRAE Research Project 1051
      and Guideline   14, 650
    DOE2 models   649
    HVAC system   649
    Nelder–Mead simplex
      optimization algorithm   650
    SSC models   649
  Southern Mississippi Space Center
    652–658
cloud-based air quality monitoring
    system
  air quality analytics engine   732
  data representation and storage
    731–732
  data sources   731
  RESTful APIs   732
cloud computing   129, 277, 347, 395,
    670, 677
cloud-side system design and
    considerations
  API data exchange framework   737
  data sources exchange framework
    737
  InstrumentID   737–738
  sMAP   736–737
  system internal data exchange
    framework   737
  universal unique identifier   737
clustering
  clustering occupants' MOA levels
    486–487
  criteria   466–467
  measures used for   517
  nonspatial clustering

allocation, scatterplot from matrix
    451
  clustering principle, EEM   448
  clusters, denogram   449
  HAC   448
  reduction in gCO_2 450
  spatial clustering
    ANI   451
    geographically close buildings
      452
    nonuniform effect dimensions
      452
    one-step   452–454
    two-step   454–456
CO_2 emissions   411
cognitive radio (CR)   105
cognitive radio networks (CRNs)
  coexistence   109–110
  PU/SU dichotomy   105
  rendezvous problem   109
  spectrum management and Handoff
    108–109
  spectrum sensing
    automated metering
      infrastructure (AMI)   108
    cooperative spectrum sensing
      107
    detection methods   106–107
    vehicular *ad hoc* networks
      (VANETs)   108
common control channel (CCC)   109
Common European Sustainable
    Building Assessment (CESBA)
    55
compressed sensing   107
computer modeling   216
conditional probability distribution
    514
connected cars
  business intelligence   201–202
  Green V2X Communications
    198–200

multi-hop communication
195–198
vehicular communications
infrastructure reliability
200–201
connected vehicle 424–425
consumerism-fostered economies
385
context identification over time,
iterative process 519
context inference 142
contextual and temporal balance
factors, neighborhood context 513
macrolevel analysis 514
microlevel analysis 514–517
neighborhoods comparison, in
Gainesville 517–518
control layer attacks
communication network 549
cyber attacks, classes 550
monitoring layer 551–552
control–monitoring (CM) attack 548
conventional building stock
management 440, 441
conventional cryptographic key-based
approaches 822
cooperative adaptive cruise control
(CACC) 814–815
cooperative intelligent transport
systems and services (C-ITS)
764
cooperative land cover (CLC) map
219
cooperative spectrum sensing (CSS)
107
crenel filtering 111
crowd-sensing applications
Here-*n*-Now framework 144–145
McSense 146
supporting framework 148–149
XMPP pub/sub model 145
cultural services 215

curtailment service provider (CSP)
244
customer satisfaction analysis 32
cyber-physical systems (CPS) 392
big data sets 4
economic and social challenges
fundamental caveats 14
global competition 11
international goods and
information flows 13
national economic conditions 13
national framework 11
regional assets and location
factors 11
smart education 15
smart governance approaches 13
smart industries and economies
14
smart metabolism 11
smart mobility systems 13
societal attitudes 14
spatial proximity 11
e-government services and public
information 16
home automation 606
ICT and networking systems 2
Internet of Things 4
planning and implementation 17
remote control 607
smart city process fields
CPS elements 7
economic context 12
human capital 9
smart construction 4, 5, 8
smart consumption 4, 6, 9
smart economy 4, 6, 9
smart education 4, 6, 9
smart energy 4, 5
smart governance 4, 6, 9
smart ICT infrastructure 4, 5, 8
smart industries 4, 6, 8, 9
smart metabolism 4, 5, 8
smart mobility 4, 5, 8

cyber-physical systems (CPS) (*contd.*)
  smart security   4, 5
  system-enhancing powers   7
  smart energy and smart energy   607
  smart home
    ambient assisted living (AAL)
      627–629
    context awareness   609
    distributivity   611
    heterogeneous smart devices
      611
    human–machine interaction   613
    "Industry 4.0," concept of
      624–627
    interoperability   611, 615–618
    IoT   610
    location detection system   609
    multi-agent systems (MAS)
      619–624
    object-oriented design and
      service-oriented architectures
      612
    remote management   609
    resources scarcity   611
    scalability   611
    scenarios and use cases   613–615
    security   611
    security issues   609
    and smart cities   629–631
    user-centered design   609
    wireless networks   610
  smart home notion   607
  social acceptability   16
  software and hardware components
    3
cyber security life-cycle, smart cities
  399
  cyber security laws and regulations
    392
  cyber security management
    audit and compliance checking
      403

cyber security metrics generation
  403
cyber security requirements
  identification   400, 402
detailed security policy
  formulation   401
disaster recovery   402
incident management   401–402
proportionate defense   398
risk management   400–402
scope and cyber security policy
  formulation   398–400
security measures implementation
  401
service continuity management
  402
ICT   391
smart city services   393–394
  smart transportation   393
  smart water management systems
    394
  vehicular technologies   393
smart services security issues
  396–397
  AMR technology   397
  potential attack   396
  TPMS   396
smart services technologies
  AMI   395
  V2V technology   394
  wireless technologies   394
cyber security management
  audit and compliance checking   403
  cyber security metrics generation
    403
  cyber security requirements
    identification   400, 402
  detailed security policy formulation
    401
  disaster recovery   402
  incident management   401–402
  proportionate defense   398
  risk management   400–402

scope and cyber security policy
formulation 398–400
security measures implementation
401
service continuity management
402

d

DA. *see* distribution automation(DA)
data acquisition
automation tasks 184
GPS 185
traffic flow data 184
traffic model 185
data aggregator unit (DAU) 776
data analysis 41
data analytics in cloud. *see* air quality
analytics engine
data collection 44
data communications
data acquisition 183–185
government authorities and agencies
182
intelligent transport system
182–183
traffic surveillance 185–186
data integration 41
data mediation 142
data mining methods
application case 442–443
approach 438–440
clustering
nonspatial 448–451
spatial 451–456
datamining conditions, building
stock 439
data sources and preprocessing
construction of feature space
446–447
data sources and modeling
443–446
EEM 438
Fuzzy reasoning

energy retrofit strategies 456
nonspatial 456–458
spatial 458–459
large building stocks 467
methods 441–442
postprocessing
datamining postprocessing
461–462
data visualization 459–461
policy development and retrofit
programs 463–464
strategy development 463
transformation strategy
development 463
smart cities and datamining 438
spatial clustering 451–456
data network
CCN 277
DONA 278–279
ICT services 276–277
NDN 277–278
network of information (NetInf)
279–281
data-oriented network architecture
(DONA) 278–279
data sources and preprocessing
construction of feature space
CO_2 emissions 446
feature space 447
metamodelling 446–447
modelling
bottom-up modeling 445–446
protocols and surveys 444–445
public databases and datasets
444
sensor measurements 445–446
data sources exchange framework
737
defender–attacker–defender model
563
defense methods against threats
control layer 564–565
monitoring layer

defense methods against threats
  (*contd.*)
  assessing system vulnerability
    565
  detecting anomalies   565–566
  identifying meters   566–567
  restoring the contaminated
    measurements   566
  physical grid   562–564
demand and supply yield market   380
demand response (DR)
  ANN and regression models   249
  baseline prediction   247, 250–251
  closed-loop optimal control strategy
    248
  control complexity and scalability
    245–246
  control synthesis   247–248
  curtailment service provider   244
  data-driven models
    data description   252–253
    DR baseline   253
    DR evaluation   253–254
  DR-Advisor   247
    baseline tab   263
    evaluation   266–268
    model identification tab   262
    model validation, with Penn data
      264–265
    Penn campus building   264
    report generation   263
    strategy evaluation   263
    strategy synthesis   263
    synthesis   268–71
    upload data   261
    workflow   262
  experimental validation and
    evaluation   248
  machine learning algorithms   249
  modeling complexity and
    heterogeneity   245
  model interpretability and control
    246

PJM ISO   244
  residential electricity consumers
    249
  rule-based DR, limitation of   245
  strategy evaluation   247, 251
  strategy synthesis   251
  synthesis, regression trees
    data and variables   260
    fast computation times   260
    missing data   260
    model-based control   254–256
    outliers   261
    synthesis optimization   256–259
dementia
  activities of daily living   785
  early intervention practice   774
  eating   784
  loss of mobility   784
  medication, side effects of   784
  prevalence   773–774
  unusual behavior   784
deployment
  abnormal event detection   187–188
  cross-platform data analytics   187
  micromobility data communications
    assistive disability access vehicle
      191
    automated booking system   193
    bicycle monitoring system   194
    Bonferroni method   194
    data-driven methodology   190
    data-driven techniques   190
    demand estimation modeling
      192
    efficient detection and prognostic
      methods   194
    efficient spatiotemporal
      surveillance algorithms   194
    electrically assisted
      bicycle-sharing service system
      192
    electric bicycle service location
      193

features   190
fusion prognostic operation   192
personalized flexibility vehicle
    191
PoF   190
prognostic prediction   192
sharing scheme   190
spatial and temporal surveillance
    194
surveillance approaches   194
network failure   188–189
return on investment (ROI)   187
technological changes   187
V2X communication
    Green V2X Communications
        198–200
    multi-hop communication
        195–198
    personal mobility vehicles
        194–195
depression   314, 784–785
developed strategies, average cost
    efficiency   464
device calibration API   738
digital publishing   36
Directive 2009/28/EC of the European
    Union (EU)   576
direct witness (DW)   822
distributed energy resource (DER)
    562
distributed generation (DG)   561
distributed networking protocol 3.0
    (DNP3)   395
distribution automation(DA)
    543–544
distribution management systems
    (DMS)   538, 540
DoE Commercial Reference Building
    (DoE CRB)   264
Douglas-fir simulator (DFSIM)   216
3D printing   766
drainage   218–219
dust sensor   733

Dylos   729, 730
dynamic-SIM platform   521
dynamic-sustainability information
    modeling (Dynamic-SIM)
    platform   212
dynamic tolling   419

e
early intervention practice (EIP)   774
E-Commerce   36
economic modeling   385
economic optimization   376
economic scarcity   378
ecosystem goods and services (EGS)
    214
ecosystem services   214–215
effective energy policy tools   484
e-governance   520
electric grids
    falling mechanisms   556–557
    in smart cities
        power systems operation
            540–544
        power systems structure
            537–540
    threats to
        cyber layers, power systems   547
        physical grid   544–546
electricity markets   244
EMC Greenplum Data Computing
    Appliance (DCA)   31
e-mobility   411
energy conservation measures (ECMs)
    646
energy demand   74
energy detection (ED)   106
energy efficiency measures (EEM)
    438
energy industry   244
energy management systems (EMS)
    vendors   174
energy policy making   493
energy policy tools   474

energy production   74
energy storage
  energy-efficient radiation heating
    adaptive comfort standard (ACS)
      593
    applications for   591
    coefficient of performance (COP)
      592
    intelligent control system   593
    panel heating system   588
    plasmadust technology   590
    predictive mean vote (PMV)   592
    process plasmadust®   589
    residents' heating requirements
      587
  residential sector refrigeration and
      freezing appliances   595–597
  smart homes
    agent-based energy grid.   585
    en:key system   594
    in Germany   586
    heating control systems   593
    learning algorithm   595
    microcontrollers   584
    photovoltaic plants and small
      wind turbines   586
    ProStumers   584
    self-learning approach   593
    supervisory control and data
      acquisition (SCADA) system
      583
    technical approach   595
  stable grid operation   586–587
  thermal storage heating systems
    climate and energy objectives
      581
    communication   582, 583
    energy generation and energy
      consumption   582
    in Germany   582
    old coal-fired plants   581
    standard and modified charging
      583

ventilation, with heat recovery
  axial/radial fans   597
  conventional heat exchanger   597
  friction ventilator   598–599
environmental evidence-assisted
    vehicular data framework
  environmental index-based data
      retrieval process   829–830
  intelligent vehicle-based system and
      attack model   825–826
  location claim verification   827
  privacy-preserved data collecting
      827–829
  roadside unit-based environmental
      evidence construction
      826–827
  roadside units
    placement of   830–839
    time synchronization   839–841
Environmental Protection Agency
    (EPA)   159
epidemic analysis   32–33
European Railway Traffic Management
    System   420
externalities
  internalization strategy   381, 382
  market failure   381
  negative   380

f
fake claimer (FC)   826
fault detection, isolation, and
    restoration (FDIR)   543
Florida Land Cover Classification
    System   220
Fordism   756
  core resources   757–758
  manufacturing philosophy   755
fourth period automated driving
    systems   801–804
frequency envelope modulation (FEM)
  crenel filtering   111
  experimental results   115–116

finite impulse response (FIR) filter
111
network configuration time 110
network self-configuration
112–113
network topologies 110
physical layer performance
113–115
frequency instability 554–555
friction ventilator 598–599
Fuerzas rmadas Revolucionarias de
Colombia (FARC) 554
Fuzzy reasoning
energy retrofit strategies 456
information 465
nonspatial
defuzzification 458
ramp-shaped membership
functions 457
retrofitting strategies, in Zernez
457
robustness 466–467
scaling 465–466
spatial 458–459

g
GDP. *see* gross domestic product (GDP)
Geographic Hash Tables (GHT) 291
geographic information systems (GIS)
438, 440
Global Infrastructure Basel (GIB) 644
Global Positioning System (GPS) 185,
426
GPS-AIR quality API 738
Greatest Generation 413
greenhouse gas (GIIG) emissions 54,
506, 643
Green V2X Communications
198–200
grid-integrated vehicle (GIV) 199
grid resilience enhancement 559–560
grid side converter (GSC) control 171

gross domestic product (GDP) 374
5Gwireless technologies 103–104

h
Hadoop 39
heat and mass balance modeling 75
heating/ventilation/air conditioning
(HVAC) systems
acoustic models
acoustic event detection pipeline
680
Auditeur 678–679
ensemble approach 681
fault detection 674
faults, in air handling room 673
mel-frequency cepstral
coefficients (MFCCs) 680
multi-output inference model
681
point anomaly 679
sound library 679
soundscape capture and analysis
672
SoundSense 678
decision support system 671,
682–685
failure detection 671–672
failures and acoustics 672–673
longer-term deployment 687–689
low-cost sensing platform 671
privacy and security 678
quality of data and trust 678
RoomZoner 675
sensing modalities 675
spectral analysis 685–687
heterogeneous networks (HetNets)
103
hierarchical agglomerative clustering
(HAC) 448, 459
hierarchical clustering 448, 449
high-level probabilistic models 546
high-reward and low-reward group
479

high-voltage (HV) transformers   539
Hitachi Smart City platform   43
home area network (HAN)   775–776
homomorphism-based data encryption
   820
horizontal coexistence   110
Housing and Development Board
   (HDB)   444
human context sensing
   applications   311
   emotive context   314–315
   emotive sensing detects   312
   functional context
      assess activities of daily living
         (ADL)   315
      cognitive decline   316
      person's gait   318
      self-care activities   316
   functional sensing   312
   location context   316–317
   physiological context
      biological signs   313
      medical investigations   314
      smart health applications   313
      social influence factors   313
      social media   313
      vital signs   313
   physiological sensing   312
   presence and location sensing
      detects   312
   sensing technologies
      environment   328–331
      smartphones   324–328
      video and audio   317–320
      wearable sensing technology and
         devices   320–324
human-in-the-loop approach   681
"human-to-human" (H2H)
   communication   104
Hydro-Quebec power system   174,
   545

i

IBMPureData System   31
ICT-based governance   520
IEC   61850 395
independent systemoperators (ISO)
   244
indirect support (IS)-based evidence
      verification
   environment signals   822
   EZpass-based PO   823
   local and landmark co-viewing   823
   non-overlapping trajectories   823
   road patterns   823
industrial city, concept of   755
industrial internet of things (IIoT)
   762–763
industrial production systems
   definition   757
   post-Fordism   759–760
   pure Fordism   757–758
   Toyota Production System
      758–759
"Industry 4.0," concept of   624–627
informational asymmetry   379
information and communication
      technology (ICT)   391, 412
   digitization   420, 421
   ecosystem of application   421
in-memory and in-database analytics
   39
Institut Bauen und Umwelt (IBU)   440
"institutional" sustainability   504
InstrumentID   737–738
integrated smart city production
      system framework
   big data   762
   industrial internet of things
      762–763
   smart city infrastructure   761–762
intelligent electronic devices (IEDs)
   393, 538
intelligent transportation systems (ITS)
   186, 393

intelligent vehicles 820
Intergovernmental Panel on Climate
    Change (IPCC) 54, 642
Internet of Things (IoT) 104–105
Internet Protocol Version 6 (IPV6)
    104
interoperability 422–423
in-vehicle wireless sensor networks
    396
6-Item Cognitive Impairment Test
    (6CIT) 783

j

Japan, automated driving systems in
    cooperative adaptive cruise control
        814–815
    population issues 808–809
    small, low-speed automated vehicles
        809–813
    truck platoons 813–814

k

k-means clustering 519
knowledge-based interventions
    feedback methodology 478
    informative messages 477
    peer comparison 478
    reducing energy use 477
knowledge translation theory 476
Kyoto Protocol 576

l

large scale air-quality monitoring
    AirCloud 731–732
    classical dispersion models 730
    cloud-connected air quality
        monitors 733–735
    cloud-side system design and
        considerations
        API data exchange framework
            737
        data sources exchange framework
            737

InstrumentID 737–738
        sMAP 736–737
        system internal data exchange
            framework 737
        universal unique identifier 737
    Dylos 729, 730
    satellite remote sensing 729
    tapered element oscillatory
        microbalance 729, 730
large scale energy savings, building
    stocks
    energy policy tools 474
    objectives 475–476
LiDAR datasets 216
light-emitting diode (LED) 562, 698
load shedding 541
load tap changers (LTCs) 543
local area network (LAN) 547
location claim verification 827
location verification
    cryptographic key-based approach
        824
    distance-bounding protocols 824
    goal 823–824
    RSS-based environmental evidence
        scheme 824–825
long term evolution (LTE) 103

m

machine data
    characteristics 34–35
    digital publishing 36
    E-Commerce 36
    environmental monitoring sensors
        35
    firefighting 35
    growth of 35
    human-to-machine (H2M)
        interactions 35
    machine-to-machine (M2M) 35
    sensor data 35
    smart sensors 36

machine data (*contd.*)
  software application/hardware
    device records   36
  Software as a Service (SaaS)   36
  Splunk reference architecture   36,
    37
machine-to-machine communication
  (M2M)   35, 104
MAC protocols   108
MapReduce   47
market-driven economy   380
market sentiment analysis   32
master data management (MDM)   26
McSense data back end   147–148
McSense mobile app   147
McSense task control console   148
mel-frequency cepstral coefficients
  (MFCCs)   680
mental illness   784–785
Miami Central Station (MCS)   420
Miami Intermodal Center   420
Miami International Airport (MIA)
  420
millennials   413–415
Mini–Mental State Examination
  (MMSE)   783
MOA approach   475
MOA model   482
mobile *ad hoc* network (MANETs)
  105
  applications and services   290, 291
  CCN Data structures   291
  CCN scheme   292
  CHANET architecture   292
  GHT   291
  GPSR protocol   292
  LFBL   290
  SCALE   292
mobile cellular networks (MCN)   103
mobile crowd-sensing (MCS)
  big data   129
  classification   128
  commercial organizations   126

definition   126
environmental applications   134
geo-social model   135
government bodies   126
Here-*n*-Now framework   144–145
incentive mechanisms   140–141
infrastructure applications
  131–133
localized analytics   141–142
McSense   138, 146–148
personal sensing and community
  sensing   127
role of   137
security and privacy   142–144
social applications   134–135
task life cycle   137, 138
three-tier architecture   129
tier 1 devices   131
tier 2 devices   130
tier 3 devices   130
user profiling and trustworthiness
  139–140
workflow   130
mobility as a service (Maas)   409
  autonomous car   422–423
  case studies   428–432
  concept of   415–418
  connected vehicle   424–425
  four packages   417
  global urbanization process   418
  ICT   420–422
  interoperability   422–423
  millennials   413–415
  service package   417
  sharing mobility   425–427
  transportation infrastructures
    418–420
  transportation sector   409, 412, 416
model-based control with regression
  trees (mbCRT) algorithm   247
modeling and vulnerability assessment
  methods   557–558
monetary incentives   140, 478

monitoring–control (MC) attack   548
Monte Carlo simulation   163
motivation *(M)* level, occupants   482,
    484
multi-agent systems (MAS)
    agent-based control paradigm   619
    artificial intelligence (AI)   619
    smart building applications
        620–621
    smart building environment
        621–624
multidisciplinary approach   482
multilevel perspective modeling (MLP)
    56–57
multiple access schemes
    optical code division multiple access
        (OCDMA)   711–713
    orthogonal frequency-division
        multiple access (OFDMA)   713
    space division multiple access
        (SDMA)   711
    time division multiple access
        (TDMA)   710
multiple levels sustainability
    assessment   507
multiple mobility patterns   410

n

named data networking (NDN)
    vs. CNN   284
    content caching   301–302
    content discovery   300–301
    data structures   283, 299
    dynamic network topology   301
    evaluation methods   303
    handling subscriber mobility   283
    Interest message   283
    Internet protocol   282
    in IoT   285–287
    in MANETs   290–293
    message forwarding   299–300
    mobile ad hoc network (MANET)
        283

naming content/data   298–299
naming scheme   282
NDNcontent/data   297–298
potentials and applications   285
security and privacy   302–303
smart grid   287–288
in VANETs   293–296
WSN   288–290
National Institute of Standards and
    Technologies (NIST)   129
National Land Cover Database (NLCD)
    216
National Research Council (NRC)
    544
natural threats   400
negative externalities   380
neighborhood level   522
neighborhood sustainability
    assessment systems (NSAs)
    211
Nelder–Mead simplex optimization
    algorithm   650
network of information (NetInf)
    name resolution and data routing
        280
    SAIL project   279
    subscriber mobility   281
network self-configuration   112–113
neural network model   741, 742
New York Independent System
    Operator (NYISO)   545
Nordic Power System   174
normalized difference vegetation index
    (NDVI)   217
North American Regional Climate
    Change Assessment Program
    (NARCCAP)   651
NoSQL databases   48–50
    auto-sharding   49
    integrated caching   50
    query support   50

o

occupancy-focused energy efficiency interventions   474–476, 482, 490
  building energy use interventions 480–481
  case study   490–493
  conceptual framework, delivery
    clustering occupants' MOA levels 486–487
    identifying multilevel building energy use   487–490
    occupancy characteristics impact, building energy use   483–486
  intervention strategies   493
  knowledge-based interventions 477–478
  penalty interventions   479
  persuasion interventions   478–479
  technology interventions   480
occupants characteristics, building energy use   481–483
  ability *(A)* level   483
  impacts
    energy policy tools framework 484
    measuring occupancy ability and opportunity level, energy conservation   485
    metrics, MOA energy   483
  MOA model   482
  motivation *(M)* level   482
  multidisciplinary approach   482
  opportunity *(O)* level   483
occupants' energy use characteristics 492
on-and off-peak costs   479
on-demand rides   425
one-step spatial clustering
  single feature space   452
  for spatial aspects   453
  uniform weighting   453, 454
online inference model

Gaussian process
  description   742–743
  regression model   743–744
online peer-to-peer economic activities 425
on–off keying (OOK)
  m-ary pulse amplitude modulation (M-PAM)   709
  pulse position modulation (PPM) 710
on-peak electricity   479
Opal-RT real-time simulator   164
open spectrum
  fixed frequency allocation scheme 102
  "fixed" spectrum regime   101
ozone depletion   54

p

Paris climate agreement   643
particulate matter (PM2.5)   726–727. *see also* large scale air-quality monitoring
  WHO air quality guidelines and interim targets   728
penalty interventions
  dynamic building control, energy costs   479
  unfavorable behavior   479
personal mobility vehicles   194–195
persuasion interventions
  favorable behavior   478
  monetary incentives   478, 479
  rewards to favorable behavior   478
phasor approach   159–161
photovoltaic (PV)   544
physical grid, threats
  from malicious attacks   545–546
  models for   546
  weather hazards   544–545
PM10 filtration   215, 217
post-Fordism   759–760

postprocessing, interpretation and
strategy identification
datamining postprocessing
cluster postprocessing 461
heat maps 462
thermal microgrids, cost efficiency
461
data visualization
clustering information 461
mixed identification procedure
460
policy development and retrofit
programs 463–464
strategy development 463
transformation strategy development
463
power grids
circuit faults 553–554
frequency instability 554–555
monitoring layer of 548
voltage instability 555
power systems
cyber attacks, classification
547–548
operation of
control 541
DA 543–544
distribution automation
543–544
protection 542–543
scheduling 542
resilience, core technologies
distribution level, restoration
561–562
emergency control and protection
560–561
structure of
distribution 539–540
station and substation 539
transmission 538–539
vertical structure 537–538
Preventive Care for Accountable Care
(ACO) 34

primary and supporting assets 400
primary energy, electric power
(electricity) conversion 538
Principles of Political Economy, 377
privacy-preserved data collecting
827–829
privacy prier (PP) 826
proactive mode, data owner 821
pro forma analysis 235
programmable logic controller (PLC)
537
Publish Subscribe Internet Technology
(PURSUIT) 281–282
pulse oximetry 321
PV. *see* photovoltaic (PV)

r
random policy 138
Raster Calculator tool 220
R&D policies 57
reactive mode, data owner 821
real-time electricity market 244
recency policy 138
redistributed manufacturing (RDM)
764–766
remedial action scheme (RAS) 561
rendezvous network (RENE) 281
renewable sources, electricity 411
rental car center (RCS) 420
ResiSTAT 351
resolution handlers (RHs) 278
resource value, market outcomes and
price
common goods 382–383
externalities
internalization strategy 381, 382
market failure 381
negative 380
market distortions 379–380
market prices 383–384
response surface method (RSM) 446
retrofit strategies 451
RideScout 431–432

ride-sharing services   426
roadside units
  optimal placement of   835–836
    alternate deploying rate   839
    conditions   831
    distinguishability-oriented greedy
      approximation   834
    with package loss   836–837
    performance analysis   837–839
    problem formulation   831–832
    properties   832–833
    RSU displacement   830, 831
    securely distinguishable rate
      837–838
  roadside unit-based environmental
    evidence construction
    826–827
  time synchronization   839–841
road vehicle automation. *see* automated
    driving systems
rotor side converter (RSC) Control
    171

S
Salesforce.com   36
satellite remote sensing   729
scatterplot matrix of, *n* variables   460
schedule data   252–253
scope and cyber security policy
    formulation   400
second-period automated driving
    systems   798
securely distinguishable rate (SDR)
    837–838
self-driving cars. *see* autonomous car
sensing algorithms   107
sensing technologies
  environment
    child's eating habits   328
    functional sensing   329–330
    HealthChair   331
    location sensing   330
    physiological sensing   329

SmartFloor   330
    WiTrack system   330
  smartphones   324–328
    ambulatory activities   326
    applications   324–325
    explicit method   325
    facial recognition   326
    GPS sensor   327
    implicit method   325
    in-built microphone   325
    modulated ultrasonic waves   325
    onboard accelerometer   326
    peripheral devices   325
    person's social interactions   324
    privacy and security   328
    social group emotions   326
    spirometry   325
    Wi-Fi radio   327
  video and audio
    cameras   317
    emotive human context   319
    functional human context   319
    heart rate detection   319
    infrared-capable camera   318
    location information, on humans
      319
    physiological sensing   318, 320
    placement and processing   318
    targeted microphones, audio
      information   320
  wearable sensing technology and
    devices
    ambulation detection   322–323
    ambulatory aspects, functional
      health   322
    blood glucose measurement   321
    contact-based sensing methods
      320
    electrodermal activity (EDA)
      sensor and accelerometer
      sensor   321
    emotional states   321
    GPS sensor   323

heart rate monitoring 321
MoodWings flutters 321
non-ambulation activities 323
person's mood 321
physical health quantification 324
physiological context sensing 321
pulse oximetry 321
RFID and magnetic beaconing methods 324
sleep tracker 321
wearer biometric data 321
wearer's functional activities 322
wearer's stress level 322
wireless-based technologies 324
service users 776–777
shared parking 427
sharing economy 416, 417
sharing mobility
car-sharing services 425
GPS 426
shared parking 427
sharing economy 425
sigmoid function 452
signal-to-noise ratio (SNR) 107, 163
small-signal analysis (SSA) 557
sMAP 731, 736–737
Smart Audio SEnsing-based Maintenance (SASEM) system 670
smart cities 527
BDA (*see* Big Data Analytics (BDA))
city services
smart transportation 393
smart water management systems 394
vehicular technologies 393
in CPSs (*see* cyber-physical systems (CPS))
cyber security life cycle 399
definition 756
dimensions 510, 511

infrastructure 761–762
integrated smart city production system framework 761–763
large scale air-quality monitoring (*see* large scale air-quality monitoring)
location-related security and privacy issues
cryptographic keys 821
homomorphism-based data encryption 820
intelligent vehicles 820
production system characteristics
distribution of products 766
manufacturing scale and inventory 766
network design 763–764
redistributed manufacturing 764–766
service relationships 767
smart ecology
building sector
building construction professionals 210
Dynamic-SIM platform 212
ecosystems services 211
multidisciplinary approach 212
NSAs 211
ratings systems 211
water and energy consumption 211
carbon sequestration
Alachua County carbon sequestration. 223, 224
ArcPy methods 220
Capstone land cover and parcel dataset 221
finer-scale analysis 219
FNAI CO-OP land cover 220
looping process 222
NumPy raster calculation 222
CO_2 sequestration 215–217
drainage 218–219

smart ecology (*contd.*)
   Alachua parcels   226–227
   CN determination   224
   NDVI method   228
   raster and polygon soil data   226
   soil type and characteristics   225
   zonal statistics tool   225
  dynamic-SIM platform
   BIM   231
   building's site location   233
   EnergyPlus energy simulation
    software   231
   land development impacts   232
   lynchpin neighborhoods   233
  ecosystem goods and services (EGS)
   214
  ecosystem services   214–215
  environmental issues   210
  micromanage ecosystem services
   213
  PM10 filtration   217, 228–231
  soil chemistry remediation   214
  state-of-the-art assessment   213
  urban ecosystems   213
smart economic development
  conscious consumption   384–388
  contextual significance   377
  economics in cultural context
   375–376
  economic theory and historical
   context   376–377
  education for sustainability
   374–375
  GDP   374
  needs and limitations   386
  next steps theory   387–388
  resource value, market outcomes and
   price
   common goods   382–383
   externalities   379–382
   market distortions   379–380
   market prices   383–384

sustainability foundation of smart
  cities   384–388
smart energy
  CHP, using ORCs
   efficiency   578
   German household   579
   heating system   579
   ORCAN Energy AG   580
   power demand   581
  wind turbines
   aerodynamic power performance
   577
   in Germany   577
   H-Darrieus turbine   578
   power generation and durability
   577
   renewable energy   576
   sound prediction by surface
    integration (SPySI)   578
   vertical axis wind turbines
    (VAWTs)   577
smart governance. *see also* ICT-based
  governance
  circumspect implementation   344
  citizen engagement   346
  climate change   344
  municipal water and sanitation
   systems   345
  planning and design, role of
   347–348
  and smart cities   347
  Somerville, Massachusetts
   deindustrialization and
    depopulation   349
   professionalize City Hall
    349–352
   smart cities and planning
    363–365
   Somerville by Design (SBD)
    process   352–363
   SomerVision   352
  technological developments   344
smart grids   55, 393, 775

smart health monitoring
  active real-time monitoring 786
  advanced metering infrastructure
    775–777
    advantages 775
    challenges 781–782
    home area network 775–776
    service users 776–777
    wide area network 776
  assistive technology 785–786
  case study 787–788
  data applications 786, 787
  dementia
    activities of daily living 785
    behavioral changes with
      783–784
    early intervention practice 774
    eating 784
    loss of mobility 784
    medication, side effects of 784
    prevalence 773–774
    unusual behavior 784
  patient behavior and uses 782–785
  smart meters 777–781
smart homes
  Cyber-physical systems (CPS)
    ambient assisted living (AAL)
      627–629
    context awareness 609
    distributivity 611
    heterogeneous smart devices
      611
    human–machine interaction 613
    "Industry 4.0," concept of
      624–627
    interoperability 611, 615–618
    IoT 610
    location detection system 609
    multi-agent systems (MAS)
      619–624
    object-oriented design and
      service-oriented architectures
      612

    remote management 609
    resources scarcity 611
    scalability 611
    scenarios and use cases 613–615
    security 611
    security issues 609
    user-centered design 609
    wireless networks 610
  energy storage
    agent-based energy grid. 585
    en:key system 594
    in Germany 586
    heating control systems 593
    learning algorithm 595
    microcontrollers 584
    photovoltaic plants and small
      wind turbines 586
    ProStumers 584
    self-learning approach 593
    supervisory control and data
      acquisition (SCADA) system
      583
    technical approach 595
  and smart cities 629–631
smart lighting
  dimming and brightening 700
  indoor and outdoor environments
    698
  indoor and outdoor positioning
    700
  interior point algorithm 699
  light-emitting diode (LED) 698
  Newton–Raphson method 699
  outdoor smart lighting systems 699
  power saving measurement 699
  in vehicles 700
  visible light communication
    cellular structure 713–716
    indoor and outdoor
      communications 701
    indoor channel model 702–706
    intersymbol interference (ISI)
      716–717

smart lighting (*contd.*)

    MIMO technology 706–708

    multiple access schemes 710–713

    nonlinearity 717–718

    on–off keying (OOK) 708–710

    radio frequency (RF) communications 701

    transmitter and receiver model 702

smart meters

    data parameters 778, 779

    data sample during single period 780

    description 777

    home plug readings 781

    infrastructures 778, 779

    roles and benefits 778

smart services security issues

    AMR technology 397

    potential attack 396

    TPMS 396

smart services technologies

    AMI 395

    V2V technology 394

    wireless technologies 394

smart water management systems 394

social learning 57–58

Society of Automotive Engineers (SAE) International 423

software as a service (SaaS) 36

SomerSTAT 351

Somerville, Massachusetts

    deindustrialization and depopulation 349

    smart cities and planning 363–365

    SomerSTAT 351–352

    Somerville by Design (SBD) process

        Curtatone's approach 350–352

        neighborhood planning 354–361

        SomerSTAT 353

        technology 361–363

SomerVision 352

SoundSense 678

South African National Energy Development Institute (SANEDI) 683

spatial clustering

    ANI 451

    geographically close buildings 452

    nonuniform effect dimensions 452

    one-step 452–454

    two-step 454–456

special protection system (SPS) 561

spirometry 325

Splunk 36–38

SSA. *see* small-signal analysis (SSA)

standard behavior pattern 475

static compensator (STATCOM) 561

static security assessment (SA) 557

static VAR compensator (SVC) 561

supervisory control and data acquisition (SCADA) systems 394, 397, 538

supply and demand relationship 383–384

support vector regression (SVR) 446

sustainability

    achieving balanced

        contextual and temporal balance 512–518

        integrational balance 518–520

        procedural balance 512

    assessment 507–508

    definition 54

    information modeling platforms 520–521

    in smart cities 508–511

    sustainable development, three pillars of sustainability 505

sustainability-oriented policy 411

sustainability transitions

    built environment

        advanced workflows 80

automation systems and sensors 77–78

building design and operation practices 76

building energy balance analysis 74–75

forward/inverse modeling and visualization techniques 76

innovative technologies and practices 60

lean production 63

life-cycle phases 68–70

model integration 80

models and tools 59

model types 63

numerical techniques 83

4P model of lean production 59

process/scales 62

reference building 70–74

research advances, modeling and computing 81–82

stakeholders, role of 61

virtual prototyping, for design optimization 67–68

workflows, state-of-the-art of models 78

data and modeling techniques

advanced modeling and computing tools 65

automation *vs.* autonomation 64

forward and inverse building energy models 66

information exchange and decision-making processes 64

KPIs and visualization techniques 64

performance indicators 64

energy modeling 56

low-carbon and sustainable society 55

multidisciplinary system thinking 58–59

multilevel perspective modeling 56–57

technological and social learning 57–58

sustainable community 531

synchronous generators (SGs) 162

system internal data exchange framework 737

t

tapered element oscillatory microbalance 729, 730

technology interventions 480

telematics 202

test system

Monte Carlo simulation 163

Opal-RT real-time simulator 164

oscilloscope traces 164, 165

proposed phasor technique 163

SNR effect 164

synchronous generators (SGs) 162

The95 Express 419

thermostats 490

third-period automated driving systems 798–801

Thyristor-Controlled Series Compensation (TCSC) 163

time-series data API 738

TinyEARS 678

tire-pressure monitoring systems (TPMS) 396

Toyota Production System (TPS) 758–759

TrafficInfo 130

traffic surveillance 185–186

transformation strategy development 463

transient instability 555

transient security assessment (TSA) 557

transmission system 538

transportation infrastructures
  economic investment   418
  mobility solution   418
two-step spatial clustering
  economic thermal microgrids   456
  energy networks identification   455
  first clustering procedure   454
Typical Meteorological Year 2 (TMY2)
  647
Typical Meteorological Year 3 (TMY3)
  647

u

Uber   430–431
UbiGo   427–428
United Nations Educational, Scientific
  and Cultural Organization
  (UNESCO)   375
United Nations Population Fund   504
universal unique identifier (UUID)
  737
US Residential Energy Consumption
  Survey   445

v

vehicle to infrastructure (V2I)   393
vehicle to vehicle (V2V)   393
vehicular *ad hoc* networks (VANETs)
  bloom-filter-based advertisements
  295
  CarSpeak   294
  characteristics   293
  climate data communications
  296–297
  content-centric approaches   293
  content segmentation–reassembly
  295
  hierarchical and hash-based content
  naming scheme   296
  hierarchical Bloom-filter routing
  (HBFR)   294
  hierarchical naming scheme   294
  last encounter content routing (LER)
  295

reliable content delivery scheme
  295
RobUst Interest Forwarder Selection
  (RUFS) scheme   295
traffic violation ticketing (TVT)
  application   296
WAVE   293
vehicular technologies   393
vertical axis wind turbines (VAWTs)
  577
vertical coexistence   109
virtual participant credit (VPC)   140
virtual private network (VPN)   547
visible light communication (VLC)
  cellular structure
    centralized power allocation
    algorithm   714
    distributed power allocation
    algorithm   714–716
  indoor and outdoor communications
    701
  indoor channel model   702–706
  intersymbol interference (ISI)
    716–717
  MIMO technology   706–708
  multiple access schemes
    optical code division multiple
    access (OCDMA)   711–713
    orthogonal frequency-division
    multiple access (OFDMA)   713
    space division multiple access
    (SDMA)   711
    time division multiple access
    (TDMA)   710
  nonlinearity   717–718
  on–off keying (OOK)
    m-ary pulse amplitude
    modulation (M-PAM)   709
    pulse position modulation (PPM)
    710
  radio frequency (RF)
    communications   701
  transmitter and receiver model   702

visualization techniques   76
voltage security assessment (VSA)
      557
VPN, virtual private network (VPN)
V2X communication
   Green V2X Communications
      198–200
   vehicular communications
      infrastructure reliability
      200–201

W

wearable sensing technology and
      devices
   ambulation detection   322–323
   ambulatory aspects, functional
      health   322
   blood glucose measurement   321
   electrodermal activity (EDA) sensor
      and accelerometer sensor   321
   emotional states   321
   GPS sensor   323
   heart rate monitoring   321
   MoodWings flutters   321
   non-ambulation activities   323
   person's mood   321
   physical health quantification   324
   physiological context sensing   321
   RFID and magnetic beaconing
      methods   324
   sleep tracker   321
   wearer biometric data   321
   wearer's functional activities   322
   wearer's stress level   322
   wireless-based technologies   324
weather data   252
Western System Coordinating Council
      (WSCC)   556
wide-area damping control
   latency compensation   157–158

adaptive phasor POD (APPOD)
      framework   166
   GPS receiver   166
   Opal RT simulator   169
   TCSC   168
   with wind farms   159
      control input   172
      coordinated control design   172
      DFIG-based wind farm modeling
         169–171
      grid side converter (GSC) control
         171
      linear analysis   172
      modal controllability, of DFIG
         rotor currents   173
      modal observability, of local and
         remote power flow signals   173
      rotor side converter (RSC) Control
         171
wide-area monitoring
   communication approach   157
   damping ratio   162
   mode-shape vector   162
   narrow band filter (NBF)   157
   phasor approach   157
   transfer function (TF) approach
      157
wide area network (WAN)   776
wireless access in vehicular
      environments (WAVE)   293
wireless sensor networks (WSNs)
      105, 288–290, 445, 675
workers ratio   139
world car concept   756

X

XMPP pub/sub model   144–145

Z

ZigBee Smart Energy   781, 788

CPSIA information can be obtained
at www.ICGtesting.com
Printed in the USA
BVHW040315130619
550896BV00011B/135/P